THE IDEAS OF PHYSICS

THIRD EDITION

HBJ

HARCOURT BRACE JOVANOVICH, PUBLISHERS

San Diego New York Chicago Austin
London Sydney Toronto

Douglas C. Giancoli

THE IDEAS OF PHYSICS

THIRD EDITION

Front Cover: Temple of Hera (fifth century B.C.), Agrigento,
Photo by D. Giancoli.

ISBN: 0-15-540562-4

Library of Congress Catalog Card Number: 85-60813
Printed in the United States of America

Line art by Pat Rogondino of Rogondino & Associates.

Photograph Credits: 1–1, Courtesy of the Art Institute of Chicago; 2–1, American Museum of Natural History; 2–2, Yerkes Observatory; 2–5(a) and (b), 2–7, Istituto e Museo di Storia della Scienza; 2–8, Yerkes Observatory; 2–12, AIP, Niels Bohr Library; 3–1, 3–5, D. Giancoli; 4–1, National Portrait Gallery, London; 5–10, PSSC, Physics; 6–7, © 1962 by the Regents of the University of California, reprinted by permission of the University of California Press; 8–1, Russ Kinne, Photo Researchers, Inc.; 9–12, HBJ photo; 10–12, 10–15, 10–18, D. Giancoli; 10–20, National Gallery of Art, Washington, D.C.; 10–21, D. Giancoli; 12–21, D. Giancoli; 13–1, 13–2, Istituto e Museo di Storia della Scienza; 16–9, The Bettman Archive; 16–10, D. Giancoli; 16–20, USAF/Cambridge Research Laboratories; 16–22, 16–23, D. Giancoli; 17–13, Photo Researchers, Inc.; 17–17, Kim Stewart-Miller; 19–1, 19–2, The Bettman Archive; 20–5, HBJ photo; 21–16, Robert Smull; 22–7, Joanne Pinkowitz; 22–16, 22–18, D. Giancoli; 23–18(b), 23–19(a) and (b), Istituto e Museo di Storia della Scienza; 24–1, HBJ photos; 24–12, Courtesy of Bausch & Lomb; 24–18, D. Giancoli; 24–21, 24–23, © Fritz Golo, Life Magazine; 24–24, D. Giancoli; 24–26, The Metropolitan Museum of Art, Bequest of Mrs. H. O. Havemeyer, 1929, the H. O. Havemeyer Collection; 25–1, American Institute of Physics; 26–11(a) and (b), Phototake; 29–5, American Institute of Physics; 29–11, Courtesy of the Brookhaven National Laboratories; 30–5, Courtesy of Fermilab; 30–7, Courtesy of R. Turgeon; 30–8, Courtesy of J. Van't Hof; 30–10, Kim Stewart-Miller; 31–2(a) and (b), Courtesy of Fermilab; 31–7, Lawrence Radiation Laboratory; 32–2(a) and (b), NASA.

PREFACE

This Third Edition of *The Ideas of Physics* is intended to give nonscience students "scientific literacy" in physics without engulfing them in a great deal of mathematics. My experience suggests that students do not really achieve even a limited grasp of physics simply by studying the concepts or ideas. They need to actually use the physics, to work with it, even if only on a very limited basis. Retained from earlier editions are the experiments and projects within the text intended to involve the students with direct experiences. In addition, for this new edition I have added material that uses a bare minimum of mathematics, mainly arithmetic and the simplest algebra: Students are shown very simple derivations, and, more frequently, are shown how to use physics in simple calculations. (Some of this material is marked "optional," or appears in footnotes, so that it can be bypassed easily by instructors who prefer the "conceptual approach" or who simply don't feel their students have the time or interest to cover it.)

The intent of this book, as before, is to lead students into physics as gently and easily as possible. And the basic objective remains: to present physics in an interesting, understandable, and enjoyable manner to readers who have little or no background in science. Great care has been taken to avoid overwhelming the student with specialized language or mathematics: jargon is avoided, and necessary terms are defined carefully; and math, though its use is slightly increased, is still used only to a very limited extent, and its use is carefully explained through the use of simple worked-out examples in the text. This slight increase in level, and the accompanying increase in detail, make the book more practical and useful for students: they will now be able to look up a topic and get enough detail to really understand it, and, if they wish, to use it in everyday life.

It is clear to nearly everyone now, at this time in history, that it is crucial for students to be scientifically literate. We live in a highly technological age. Home computers and other technical devices are commonplace. The choices open to people in their pursuit of a profession (or just a job) are far broader if they have some understanding of science. And the incredible complexity of science and technology, in war and peace, from weapons (and their avoidance) to medicine, requires intelligence in the voting booth. We can no longer teach platitudes as physics.

This new edition differs from its two predecessors in several other ways. First of all, the book has been divided into more, and shorter, "bite-sized" chapters. Each chapter is thus more quickly and easily absorbed and digested by students. This reorganization has allowed an expanded treatment of many crucial topics, including a much fuller treatment of rotational motion (Chapter 9) and of energy and momentum (in separate Chapters, 7 and 8), and a whole new chapter (11) on the structure of matter. This last subject is taken up again later, using the quantum theory, in the chapters on modern physics (which are now all together, beginning with Special Relativity in Chapter 25), and includes atomic theory, molecules, semiconductors and their use in electronics, lasers, nuclear physics, and the most recent developments in elementary particle physics.

There are also a number of entirely new topics. The most important are contained in the two final chapters, Chapter 32 on astrophysics and general relativity, and Chapter 33 on modern cosmology, which includes the very latest results using the "standard model." Also new are a variety of applications to other fields, such as medicine, which includes ultrasound imaging, CAT scanning, and nuclear medicine and dosimetry.

This new edition also includes eight pages of full color photographs. My hope is that these will help draw the student into the subject and will show the richness of physics within our general culture. These color photographs illustrate a number of beautiful and important instruments in the history of science, ancient and contemporary cosmology, different aspects of color, and the relationship between physics and other fields such as art and architecture.

The book is entirely metric in this third edition, although British units are mentioned (that is, defined). The figures have all been redrawn completely to provide vastly improved clarity. In the figures, the second color (blue) is used to represent real, tangible objects; black is used to represent analysis or theoretical ideas (such as vectors and electric fields); this convention should help students who sometimes have trouble distinguishing theories from facts.

All areas of the book have been updated, some of which have been mentioned above. Even the intuitive approach—starting each topic with everyday observations that students have or can make, and only then proceeding to the great generalizations and laws of physics—has been perfected further. In fact, almost no topic has passed into this third edition untouched.

Each chapter ends with a brief summary of the material of that chapter. This is followed by a set of questions requiring a little thought and verbal answers, and then by a set of exercises requiring mathematical calculations (usually very simple, although slightly more difficult ones are marked by an asterisk) which can be assigned at the option of the instructor. Answers to selected questions (more than half), and to the odd-numbered exercises, are found in Appendix C.

As in previous editions, I have attempted to present the basic concepts of physics in their historical and philosophical context, rather than simply stating the latest viewpoint. I believe that students can better understand the motion of celestial bodies or our present view of the atom, for example, if they have traced the development of ideas that have led to our present view. It is through this approach, too, that a student can get the flavor of what physics is about and come to realize that it is indeed a human endeavor. Also retained are ample discussions on the practical side of physics: "how things work." The many practical examples include optical instruments (cameras, telescope, microscope, eye), engines and motors, radio and TV, electronics, loudspeakers, other electrical devices, nuclear power and other power production, environmental pollution, arches and domes, and the effects of radiation, among others.

The Ideas of Physics can be used readily in a one-term course. The book contains much material, but at the same time it allows instructors considerable flexibility in choice of topics. The wide range of coverage also allows students to read topics of interest even though they are not covered in class; so the book can be a useful resource for them. Chapters are written so that some sections, and even whole chapters, can be omitted without losing continuity. For a very brief course, I suggest covering, for classical physics, most of Chapters 3 to 7, 11, 13, 16, 18 to 20, and the early parts of Chapters 8, 12, 14, and 21 to 24. Any course in physics should also include some modern physics, including special relativity (Chapter 25) and quantum theory of the atom (at least parts of Chapters 26 and 27 and the first two sections of Chapter 28). Also recommended would be at least a little coverage of radioactivity and nuclear fission. If there is time, students will respond to some of the latest topics such as quarks (Chapter 31) and the Big Bang (Chapter 33). Beyond this list, the instructor (and/or the students in their spare time) can choose other topics of importance and interest. Physics is a rich and fascinating field, and I hope that students will come to see it as such.

This revision of *The Ideas of Physics* owes a lot to the many users and reviewers of its previous editions. I wish to thank them all, in particular Royal Albridge, Vanderbilt University; Paul H. Barrett, University of California, Santa Barbara; Raymond Chang, Honolulu Community College; Robert K. Cole, University of Southern California; Robert Dawson, Western Kentucky University; Thattil J. Devassy, Central Connecticut State College; Gerald N. Estberg, University of San Diego; Merle Fisher, Ricks College; A.T. Fromhold, Jr., Auburn University; Junichiro Fukai, Auburn University; J. Ronald Galli, Weber State College; Edward G. Grimsal, University of South West Louisiana; Sigmund P. Harris, Los Angeles Pierce College; Richard G. Hills, Weber State College; John A. Howell, Franklin and Marshall College; Patrick F. Kenealy, Wayne State University; Robert E. Kribel, Auburn University; Daryl L. Letham, Weber State College; Robert Martin, Tarrant County Junior College; Alan K. Miller, Pasadena City College; Nathan Ockman, City College, City University of New York; James Putnam, Southern Vermont College; Douglas O. Richstone, University of Michigan; Cecil G. Shugart, Memphis State University; Larry R. Sill, Northern Illinois University; Gary Swanson, Auburn University; Ed Winter, San Antonio College; Clarence Wolff, Western Kentucky University; John K. Wood, Utah State University; Roger D. Woods, San Bernardino Valley College; Stephen G. Wukovitz, Bloomsburg University.

I owe special thanks to Professors Richard Marrus, John Heilbron, Howard Shugart, Tito Arecchi, and Paolo Galluzzi for valuable discussions and for generous hospitality at the University of California, Berkeley, Physics Department, and at the Museum and Institute for the History of Science in Florence. I also wish to thank the staffs of these institutions, in particular Ms. Franca Principe for her beautiful photographs of instruments in the superb collection of the History of Science Museum in Florence. Finally, I gratefully acknowledge the fine and helpful work of the staff at Harcourt Brace Jovanovich, in particular Ellen Aleksic, Avery Hallowell, Cheryl Solheid, and Robert C. Miller.

D. C. Giancoli

CONTENTS

23 LENSES AND OPTICAL INSTRUMENTS *363*

24 THE WAVE NATURE OF LIGHT *382*

25 SPECIAL THEORY OF RELATIVITY *403*

26 EARLY QUANTUM THEORY *430*

31 ELEMENTARY PARTICLES *521*

32 ASTROPHYSICS AND GENERAL RELATIVITY *541*

33 COSMOLOGY *561*

Appendixes

Index

THE IDEAS OF PHYSICS

THIRD EDITION

PHYSICS: A HUMAN ENDEAVOR

CHAPTER 1

"Science is as integral a part of the culture of our age as the arts are."
Jacob Bronowski
Science and Human Values

In the words of the famous humanist-scientist Jacob Bronowski, "for any man to abdicate an interest in science is to walk with open eyes toward slavery."[†] These foreboding words are especially telling if we think of such fictional accounts of the future as *Brave New World*, *Nineteen Eighty-Four*, and *Star Wars*. These words also have meaning in the world as it is with the threat of nuclear war, the dangers of pollution, and the advances of medical and biophysical research in the fields of human behavior and disease. Politicians, and others, must consider problems that relate to science. They must make decisions regarding space exploration, pollution, the generation of electric power, the uses of nuclear energy, and genetic engineering. Yet the majority of people in decision-making positions know little or nothing about science.

Can it be a desirable state of affairs that so few people have an acquaintance with science and its practice?

A person on the street may be willing to pass judgment on a modern painting or a recent novel; yet it is unlikely that the same person will render a judgment on a recent scientific theory. That science should be accessible to only a few does not benefit the public nor does it serve the purposes of science itself. It might well be a better world if the nature of science, its uses and its limitations, were understood and appreciated by people in all walks of life. For government officials, businesspersons, and simply voters casting ballots, a knowledge of science can help effect more rational decisions about the many problems facing today's world

[†] Jacob Bronowski, *Science and Human Values* (New York: Harper and Row, 1972).

The aim of this book is to lessen the isolation of science and to bring the ideas of science and how it is practiced to nonscientists. The paragraphs above set forth an incentive for studying science, including physics. But there are other sound reasons for studying physics, and these are more personal. After reading this book you may feel more confident about performing minor repairs on a toaster or bicycle. And perhaps the most important reason for studying physics is an aesthetic one—physics is an intellectually stimulating activity; it is an exciting field of human endeavor. Like artists and poets, physicists and other scientists seek to comprehend and interpret the world around them. In their own fields they each attempt to increase our degree of consciousness. Great scientists—like great artists, poets, or musicians—make us aware as we have not been aware before. They extend certain aspects of our experience and deepen our appreciation of life.

1. SCIENCE: A CREATIVE ACTIVITY

The principal aim of all sciences, including physics, is generally considered to be the ordering of the complex appearances detected by our senses—that is, an ordering of what we often refer to as the "world around us." Many people think of science as a mechanical process of collecting facts and devising theories. This is not the case. Science is a creative activity that in many respects resembles other creative activities of the human mind.

Let's take some examples to see why this is true. One important aspect of science is *observation* of events. But observation requires imagination, for scientists can never include everything in a description of what they observe.[†] Hence, scientists must make judgments about what is relevant in their observations. As an example, let us consider how two great minds, Aristotle (384–322 B.C.) and Galileo (A.D. 1564–1642), interpreted motion along a horizontal surface. Aristotle noted that objects given an initial push along the ground (or on a tabletop) always slow down and stop. Consequently, Aristotle believed that the natural state of a body is at rest. Galileo, in his reexamination of horizontal motion in the early 1600s, chose rather to study the idealized case of motion free from resistance. In fact, Galileo imagined that if friction could be eliminated, an object given an initial push along a horizontal surface would continue to move indefinitely without stopping. He concluded that for an object to be in motion was just as natural as to be at rest. By seeing something new in the same "facts," Galileo is often given credit for founding our modern view of

[†] This is something like the well-known circumstance where five witnesses to an automobile accident give five different stories of what happened. No one person can grasp all the details, and what each *does* observe depends partly on what that person *expects* to observe; i.e., it depends on one's background.

motion (further details in Chapters 3 and 4). This seeing of something new was surely inspired thinking.

Theories are never derived from observations—they are *created* to explain observations. They are inspirations that come from the minds of human beings. For example, the idea that matter is composed of atoms (the atomic theory) was certainly not arrived at because someone observed atoms. Rather, the idea sprang from a creative mind. The theory of relativity, the electromagnetic theory of light, and Newton's law of universal gravitation were likewise the result of inspiration.

As creative achievements, the great theories of science may be compared with great works of art or literature. But how does science differ from these other creative activities? One important difference is that science requires *testing*† of its ideas or theories to see if predictions are borne out by experiment. Indeed, careful experimentation is a crucial part of physics.

Although the testing of theories can be considered to distinguish science from other creative fields, it should not be assumed that a theory is "proved" by testing. First of all, no measuring instrument is perfect, so precise confirmation cannot be possible. Furthermore, it is not possible to test a theory in every possible circumstance. Hence, a theory can never be absolutely verified. In fact theories are generally not perfect—a theory rarely agrees with experiment exactly, within experimental error, in every case in which it is tested. Indeed, the history of science tells us that theories come and go, and that long-held theories are replaced by new ones. The process of one theory replacing another is an important subject* in the philosophy of science today; we can discuss it here only briefly.

A new theory is accepted by scientists in some cases because its predictions are quantitatively in much better agreement with experiment than is the older theory. But in many cases, a new theory is accepted only if it explains a greater *range* of phenomena than does the older one. Copernicus's sun-centered theory of the universe, for example, was no more accurate than Ptolemy's earth-centered theory for predicting the motion of heavenly bodies. But Copernicus's theory had consequences that Ptolemy's did not: for example, it made possible a determination of the order and distance of the planets and predicted the moon-like phases in the appearances of Venus. A simpler (or no more complex) and richer theory—one which unifies and explains a greater variety of phenomena—is more useful and beautiful to a scientist. And this aspect, as well as quantitative agreement, plays a major role in the acceptance of a theory.

---

† Some philosophers therefore emphasize that testing of a theory can be used only to falsify it, not to confirm it—and/or to put a limit on its range of validity.

* This question is discussed at length in the stimulating book by Thomas Kuhn, *The Structure of Scientific Revolutions* (Chicago: University of Chicago Press, 1970).

An important aspect of any theory is how well it can quantitatively predict phenomena; from this point of view, a new theory may often seem to be only a minor advance over the old one. For example, Einstein's theory of relativity gives predictions that differ very little from the older theories of Galileo and Newton in nearly all everyday situations; its predictions are better mainly in the extreme case of speeds close to the speed of light. From this point of view, the theory of relativity might be considered as mere "fine-tuning" of the older theory. But quantitative prediction is not the only important outcome of a theory. Our view of the world is affected as well. As a result of Einstein's theory of relativity, for example, our concepts of space and time have been completely changed; and we have come to see mass and energy as a single entity (via the famous equation $E = mc^2$). Indeed, our view of the world underwent a major change when the relativity theory came to be accepted.

2. SCIENCE AND THE HUMANITIES

To many people, the sciences and the arts and humanities are two very different fields. Nonetheless, a good argument can be made that they are much more alike than different. As we discussed earlier, scientists do not work from a simple collection of facts; they observe the world and interpret it. Just as artists and poets attempt to represent reality in their work, so too do scientists. Artists and poets look at the world and seek relationships and order. But they translate their ideas to canvas, or to marble, or into poetic images. Scientists try to find relationships between different objects and events. To express the order they find, they create hypotheses and theories. Thus the great scientific theories are easily compared to great art and great literature.

To carry the similarity a step further, note for example that artists do not represent exactly everything they see. They emphasize those aspects of reality they feel are most important or relevant. Neither do scientists include everything in the physical world in their theories. They too judge what is relevant.

Of course differences exist between the humanities and sciences. Although scientists are not mere recorders of facts, they are constrained to conform to observations of nature. Artists, however, are not so constrained. Novelty is often legitimate in art but not in science. Most scientists believe that their goal is to seek a single, accurate account of nature. Yet two artists can view the same subject and arrive at two different creations—and each may be recognized as valid. The sciences and the arts differ in the critique or testing of the finished painting, poem, or scientific theory. Artists do not necessarily seek approval of their work by other artists; but scientists do, and must, seek confirmation of their ideas from other scientists.

Despite these differences, scientists and artists think and work in a basically similar way. Each deals with the world and each creates a vision of reality, be it a painting, a poem, or a scientific theory. But art and science emphasize different aspects of reality. Scientists recreate one side of this world, poets and artists another. Neither has a monopoly on the correct view. But each vision reveals different aspects of reality.

The apparently different ways in which scientists and artists think can sometimes overlap, as pointed out by C. H. Waddington† in his book on the relationship between painting and the natural sciences in the twentieth century. He considers it no accident that Picasso founded Cubism in 1905, the same year Einstein proposed his theory of relativity. And Cubism, he points out, has its roots in Cézanne's struggle to find a way to paint the *structure* of objects rather than their appearance (Figure 1-1, Color Plate I). Structure has also concerned physicists in the twentieth century, as we shall see beginning in Chapter 26. Around 1910 artists of the Futurist school attempted to represent speed and movement in their paintings, particularly as they relate to technology. Later painters absorbed from the new physics of the quantum theory the idea that matter is less solid and more transparent than previously thought, and that motion cannot really be frozen into a timeless second.

FIGURE 1-1
See Color Plate I.

In our study of physics in the remaining chapters of this book, we will try to illustrate the humanness of science. We will, where possible, follow the historical development of ideas. We will see that the scientists of the past whose theories are now considered outmoded were not fools. Their vision of reality was different from ours, and perhaps more limited, but they were not necessarily wrong. After all, science, like the arts, develops by building on the ideas of the past. And we will see that scientific development is not the dogmatic process it is popularly thought to be.

3. MODELS, THEORIES, AND LAWS

When scientists are trying to understand a particular set of phenomena, they often make use of a **model**. A model, in the scientists' sense, is a kind of analogy or mental image of the phenomena in terms of something we are familiar with. One example is the wave model of light. We cannot see waves of light as we can water waves; but it is valuable to think of light as if it were made up of waves because experiments on light indicate that it behaves in many respects as water waves do.

The purpose of a model is to give us a mental or visual picture—something to hold onto—when we cannot see what actually is happening. Models often give us a deeper understanding: The analogy to a known

† C. H. Waddington, *Behind Appearance, A Study of the Relations Between Painting and the Natural Sciences in this Century* (Cambridge, Mass.: MIT Press, 1970).

system (for instance, water waves as analogy for light waves) can suggest new experiments to perform and can provide ideas about what other related phenomena might occur.

No model is ever perfect, and scientists are constantly trying to refine their models or to think up new ones when old ones seem inadequate. The atomic model of matter has gone through many refinements. At one time or another, atoms were imagined to be tiny spheres with hooks on them (to explain chemical bonding) or as tiny billiard balls continually bouncing against each other. More recently, the "planetary model" of the atom visualized the atom as a nucleus with electrons revolving around it—just as the planets revolve about the sun.

You may wonder what the difference is between a theory and a model. Sometimes the words are used interchangeably. Usually, however, a model is fairly simple and provides a structural similarity to the phenomena being studied, whereas a **theory** is broader, more detailed, and attempts to solve a set of problems usually with mathematical precision. Often, as a model is developed and corresponds more closely to experiment over a wide range of phenomena, it may come to be referred to as a theory. The atomic theory is an example, as is the wave theory of light.

Models can be very helpful, and they often lead to important theories. But it is important not to confuse a model or a theory with the real system or the phenomena themselves.

Scientists give the title **law** to certain concise but general statements about how nature behaves (that energy is conserved, for example); sometimes the statement takes the form of a relationship or equation between quantities (such as Einstein's famous equation, $E = mc^2$).

To be called a law, a statement must be found experimentally valid over a wide range of observed phenomena; in a sense, the law brings a unity to many observations. For less general statements, or ones that cover a more restricted range of phenomena, the term **principle** is often used (such as Archimedes' principle). Where to draw the line between laws and principles is, of course, arbitrary, and there is not always complete consistency.

Scientific laws are different from political laws in that the latter are *prescriptive*: They tell us how we must behave. Scientific laws are *descriptive*: They do not say how nature *must* behave, but rather describe how nature *does* behave. As with theories, laws cannot be tested in the infinite variety of cases possible. So we cannot be sure that any law is absolutely true. We use the term law when its validity has been tested over a wide range of cases and when any limitations and the range of validity are clearly understood. Even then, as new information comes in, certain laws may have to be modified or discarded.

Scientists normally do their work as if the accepted laws and theories were true; but they are obliged to keep an open mind in case new information should alter the validity of any given law or theory.

4. THE ROLE OF MATHEMATICS

Because this book emphasizes the *ideas* of physics, little mathematics is used.[†] However, we do need the simple mathematical concept of proportion.

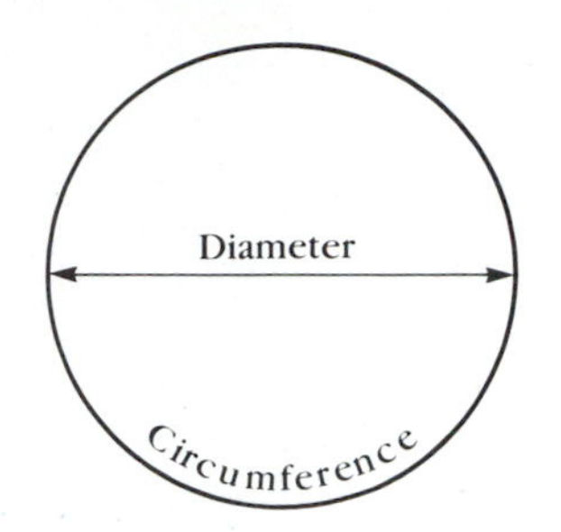

FIGURE 1-2
A circle

Direct Proportion

When an increase or a decrease in one quantity* results in the same relative change in a second quantity, we say that the second quantity is proportional to the first. For example, when water runs from a faucet into a bathtub at a constant rate, the tub becomes fuller the longer the water is on. We say the amount of water in the tub is proportional to the length of time the faucet was on. Using the symbol $\propto$ to mean *is proportional to*, we can write this relationship as

$$\text{amount of water in tub} \propto \text{length of time faucet is on.}$$

If the faucet is left on for two minutes, twice as much water will be in the tub than if the faucet is left on for one minute (assuming, of course, that the rate of flow from the faucet doesn't change). If the faucet is on for only half as long, there will be half as much water in the tub. This kind of proportionality is usually referred to as **direct proportion** to distinguish it from inverse proportion, which we will discuss shortly. In direct proportion, the two quantities increase or decrease together in the same ratio.

The circumference and the diameter of a circle are also related by a direct proportion (Figure 1-2). If one circle has twice as large a diameter as that of a second circle, the first will also have twice the circumference. If the diameter of one circle is half as large as that of another, the circumference will be half as large—and so on. We can write this proportionality as

$$\text{circumference} \propto \text{diameter.}$$

Constants of Proportionality

The ancients found by careful measurement that the ratio of the circumference of any circle to its diameter equals approximately 3.1416, and they called this special number π (pi). The proportionality between the circumference and diameter of a circle can then be written as an equality—

$$\text{circumference} = 3.1416 \times \text{diameter,}$$

or

$$\text{circumference} = \pi \times \text{diameter.}$$

[†] A brief appendix at the end of the book describes the "powers of ten notation," which we occasionally use.

* By a *quantity* we mean an observable property in nature that can be measured or specified numerically. For example, length, time, weight, temperature, and energy are all quantities.

You may have seen this relation written as

$$C = \pi D,$$

where C stands for circumference and D for diameter. This number, $\pi = 3.14\ldots$ (keeping only three digits), is called a **constant**, because it does not change. The diameter and circumference can be referred to as **variables**, because they are different for different circles.

We can always make a direct proportionality into an equality by inserting a constant called a **proportionality constant**. The constant of proportionality for the circumference and diameter of a circle is called π and has the approximate value 3.1416. In the example of water running into a tub, we might call the proportionality constant K. Then we can write the proportionality as an equality—

$$\text{amount of water in tub} = K \times \text{time faucet is on.}$$

To find the numerical value of the constant of proportionality K would require a simple measurement (namely, the volume or weight of water that flows from the faucet each second). Of course K will have a different value depending on the flow rate of the water; but it will be a constant at any particular rate. This example differs slightly from the situation with the circle because the proportionally constant π is the same for every circle. We therefore say that π is a *universal* constant.

Inverse Proportionality

Sometimes a relationship between two quantities is such that if one quantity is increased the other is decreased. Such a relationship is known as **inverse proportion**. For example, if you have a fixed distance to drive, the time the trip takes is inversely proportional to how fast you drive. Suppose you are driving 100 kilometers between two cities. If you average 60 kilometers per hour (abbreviated km/hr), the trip will take half as long as it would if you averaged 30 kilometers per hour. When the speed is doubled, the time is halved. If the speed is tripled, it takes only a third as much time. We can write an inverse proportionality in the following way—

$$\text{time for a trip} \propto \frac{1}{\text{speed}}.$$

We get *inverse* proportionality by putting the speed in the denominator of a fraction. Notice that when the denominator of a fraction is increased, the fraction as a whole is decreased. For example, one over two ($\frac{1}{2}$) is less than one over one ($\frac{1}{1}$); $\frac{1}{3}$ is less than $\frac{1}{2}$; and so on. Thus, if the speed is doubled in the above example, $\frac{1}{\text{speed}}$ is only half as large. This relationship corresponds to the fact that the time required for the trip is only half as much.

An inverse proportionality, like a direct proportionality, can be changed into an equality by introducing a proportionality constant.

SUMMARY

The aim of science is to find order in the physical world that surrounds us. The work of science involves observation of the physical world, the formulation of theories or hypotheses, and the careful testing of those theories. But scientific progress is achieved not in a routine, mindless fashion but rather with creativity and perception, as in the arts.

A scientific *model* is a kind of analogy in which a familiar phenomenon is used to explain more complicated phenomena, such as the wave model for light. A scientific *theory* involves a group of ideas used to explain a wide range of phenomena, often with mathematical precision. A scientific *law* is a general and concise statement about how nature behaves; such a statement that is of less generality, or covers a smaller range of phenomena, is often called a *principle*.

Although the emphasis of this book is the *ideas* of physics, some mathematical tools are useful in grasping those ideas. Two quantities are *proportional* if a change in one implies or produces a change in the other. The two are *directly proportional* if an increase in one leads to a corresponding increase in the other. They are *inversely proportional* if an increase in one leads to a corresponding decrease in the other. A *proportionality constant* is a number (or constant) used to express a proportionality as an equality.

QUESTIONS

1. In what ways might the world, and science too, benefit if the general public were more familiar with science and its practice? Discuss.
2. It is sometimes said that science is the new religion, complete with high priests and mysteries known only to the select few—the trained scientists. Do you agree? Discuss.
3. Discuss generally the limitations of science.
4. It has been said that society's ills are the result of science. Scientists often respond that their work is of a pure and intellectual nature and practical applications, i.e. technology, create the problems. Discuss. (It might be useful to distinguish first between science and technology.)
5. Carefully draw several circles of different sizes using a compass. Measure both the circumference and the diameter of each. Is the ratio of $\frac{\text{circumference}}{\text{diameter}}$ the same for each?

6. What type of proportion is involved (a) when the increase in one quantity is always accompanied by a corresponding decrease in a second quantity and (b) when the increase in one quantity always results in an increase in a second quantity at the same rate?
7. How is the total cost of a bag of oranges related to the total weight of oranges purchased?
8. Suppose you are driving your car at a steady 90 kilometers per hour. What kind of proportion is there between how long you drive and the total distance you drive? What do you think is the constant of proportionality in this relationship?
9. What kind of proportion (approximately) relates the amount of water remaining in a once-full bathtub to the amount of time the plug has been out?
10. Approximately what kind of relationship might exist between the steepness of a hill and the speed with which a cyclist can ride up the hill?
11. Give some simple examples of direct and inverse proportions (or at least approximate ones) that you have observed in everyday life.
12. In the equation $A = Kr^2$ relating the area of a circle to its radius, what is the constant of proportionality?

CELESTIAL MOTION: OUR VIEW OF THE HEAVENS

CHAPTER

2

Motion of automobiles, animals, people, or the sun is an obvious phenomenon in nature. Motion has fascinated people at least since the time of the ancient civilizations in Asia Minor. But not until the late Renaissance was the problem of motion attacked with vigor and important concepts clarified. Although many contributed to this understanding, Galileo Galilei (1564–1642) and Sir Isaac Newton (1642–1727) stand above the others.

Motion was perhaps the first aspect of nature to be studied thoroughly and understood. Other phenomena such as electricity, magnetism, heat, light, and the nature of matter remained mysteries for a longer time.

Since the time of the ancient Greeks, two distinct types of motion have seemed central to an understanding of the universe: the motion of celestial bodies and the motion of objects on earth. These two kinds of motion were considered as separate matters until the work of Galileo and Newton.

A study of the development of ideas about celestial and terrestrial motion from the ancient Greeks through the great ideas of Galileo is fascinating and forms the subject matter of the next few chapters.

1. THE EARLY OBSERVATIONS OF HEAVENLY BODIES

Have you ever escaped the lights of the city to gaze late at night at the multitude of stars above you? It is a moving experience that we rarely enjoy today, but it must have deeply impressed the ancients. They saw this sight every cloudless night, for they had neither street lights nor smog to obscure it. The starry heavens have been a source of inspiration for poets and scientists alike for thousands of years. The ancient Greeks and

Babylonians were fascinated by the stars and carefully traced their motions.

If you look at the stars on a moonless night, you may notice that they seem to be arranged on a huge inverted bowl. Indeed, the ancient Greeks pictured the stars as fixed on the inside of a giant rotating sphere. Early observers, as they examined this celestial sphere night after night, allowed their imaginations to roam freely. In certain groupings of the stars they saw the outlines of wild animals, great warriors, and beautiful women. These groupings are the *constellations*—the big and little dippers, Virgo, Orion the hunter, Cassiopeia (Figure 2-1).

Most stars appear to be fixed with respect to each other. The shape of the big dipper, for example, never seems to change. Scientific evidence today, however, indicates that the stars do indeed move with respect to each other. In fact their speeds seem to be great, but they are so far away that movement is imperceptible to the naked eye. Thus the Babylonians and Greeks believed that the stars were fixed in position on the inverted

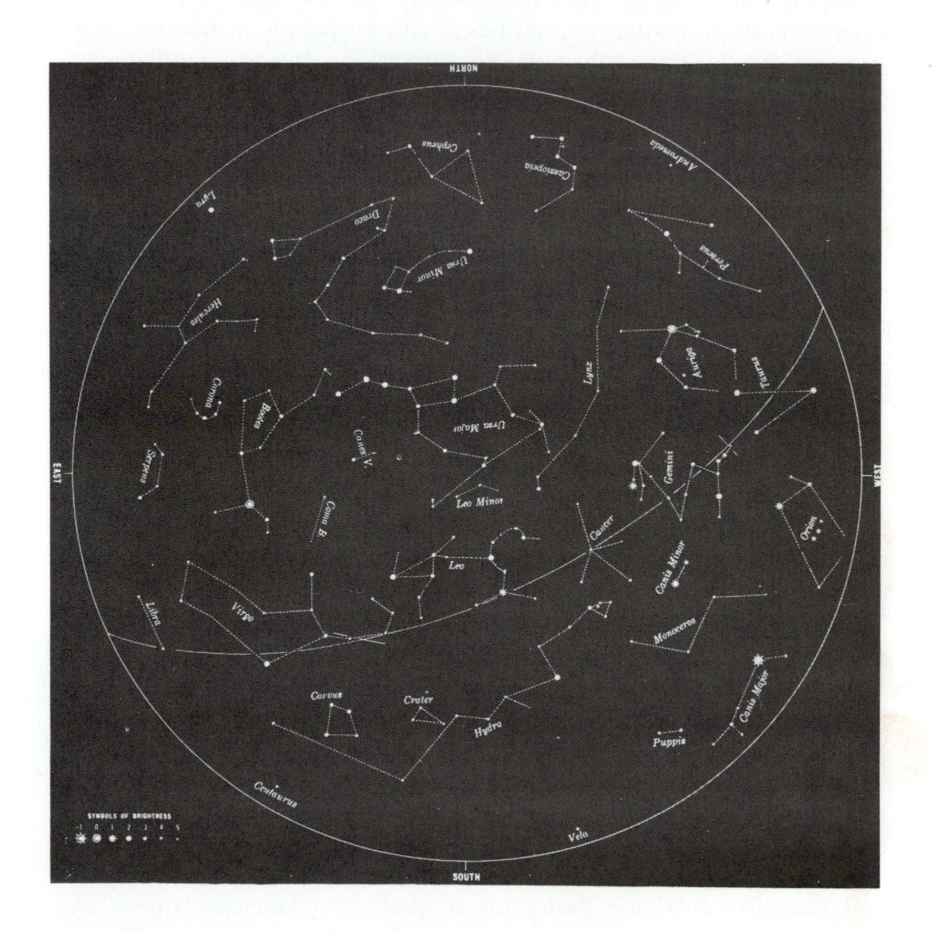

FIGURE 2-1
A chart of the stars.

bowl of the heavens. But they did observe a very small number of starlike objects that move with respect to the "fixed" stars. They discovered five such nomadic objects and called them **planets** (from the Greek *planetes*, "wanderer"). Although their motion was very slow, the planets clearly changed position over a period of several days or months. The five planets were named for the Roman gods Mercury, Venus, Mars, Jupiter, and Saturn. The sun, the moon, and the planets still serve as the names of the days of the week: *Sun*day, *Mo(o)*nday, Tuesday (from *Tys*, Germanic equivalent of *Mars*), Wednesday (from *Wotan*, meaning *Mercury*), Thursday (from *Thor*, equivalent of *Jove* or *Jupiter*), Friday (*Fria*, the Germanic *Venus*), and *Satur(n)*day.

FIGURE 2-2
Time exposure showing movement of stars.

The fixed stars, as well as the sun and moon, appear to rotate from east to west. You can detect this very slow westerly motion by observing the sky at various times during the night. Successive observations will reveal that the stars have indeed moved to the west; on a time exposure photograph of the heavens, Figure 2-2, each star makes a curved line because of this motion. A complete revolution of the *celestial sphere* takes 24 hours, after which the same stars are again directly overhead. The moon also seems to move from east to west, but more slowly; a complete revolution takes about 25 hours. Because the moon moves westwardly more slowly than the stars, its position with respect to the stars seems to move *eastward*. On successive nights the moon's position will be observed to be slightly to the east with respect to the fixed stars, as shown in Figure 2-3(a), (b) and (c). About a month later the moon will have returned to approximately the same position among the stars, Figure 2-3(d). Careful observations of which stars appear on the horizon just after the sun sets reveal that the sun, too, changes position with respect to the stars. The sun returns to the same position among the stars after one year. In fact, a year could be defined as the time it takes for the sun to pass through the celestial sphere and return to the same position with respect to the fixed stars.

The motion of the planets among the stars is more irregular than that of the sun and moon; but it, too, is generally in an easterly direction with

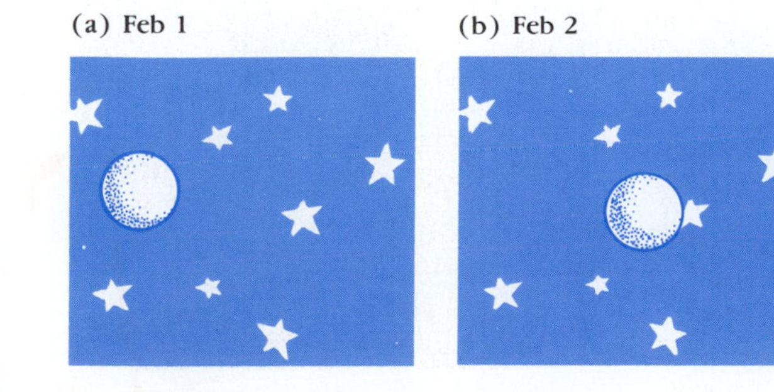

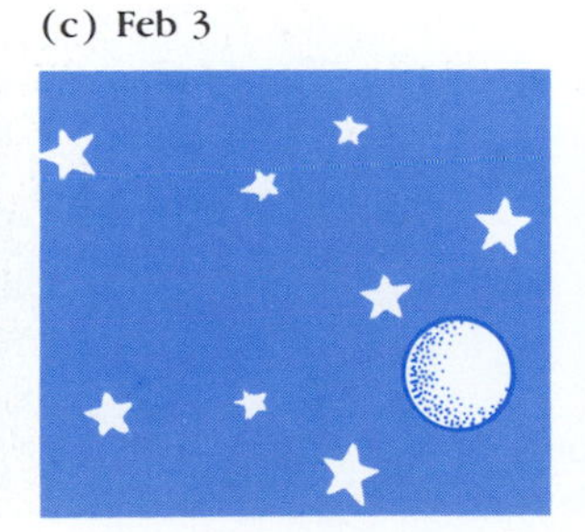

FIGURE 2-3
The moon appears to move slowly among the stars (a), (b), and (c). After a month (d), it has returned to the same place among the stars.

respect to the stars, as shown in Figure 2-4. The occasional westerly motions, the loops in Figure 2-4, are referred to as *retrogressions*.

You should not accept these statements as true without examining the night sky yourself. Even without instruments you can make some of the observations described. There are *experiments* suggested throughout this book that are easily performed. Some may seem trite or trivial, yet it is surprising how much is learned from performing a simple experiment. Some experiments will take only a minute or two to perform and should be done immediately upon reading them. Others will take longer and are more in the nature of a project, such as this one.

EXPERIMENT-PROJECT

Look at the nighttime sky for five or ten minutes. Note if there is any obvious motion of the stars and moon. It might help to lie on the ground and carefully observe a star above a power line or telephone wire; notice if the star moves, even slowly, past the wire.

Draw a diagram of a fairly large region of the sky—say near the eastern horizon—indicating the position of the brighter stars and any constellations you can see, as well as the moon if it is visible. Be sure to include some reference points such as a fence, a mountain, or a telephone pole.

About three hours later observe the sky again from the same location and draw a *second* diagram showing the same stars (and constellations) and the moon, and note any changes in position.

Finally, on the next night, at precisely the same hour as either of the first two observations, again observe the nighttime sky and draw a *third* diagram of the moon and stars. Compare the three diagrams and note any change in the positions of moon and stars as a whole or with respect to each other. From your observations, briefly describe the motion of the moon and stars.

This section has given a brief description of the motion of the sun, moon, planets, and stars as they appear to an observer on the earth. But it seems that we humans are not content with mere observations. We want to know what is "behind" them. We seek some kind of order or unifying principle for the diverse observations we make. We are thus led to create hypotheses to explain the observations.

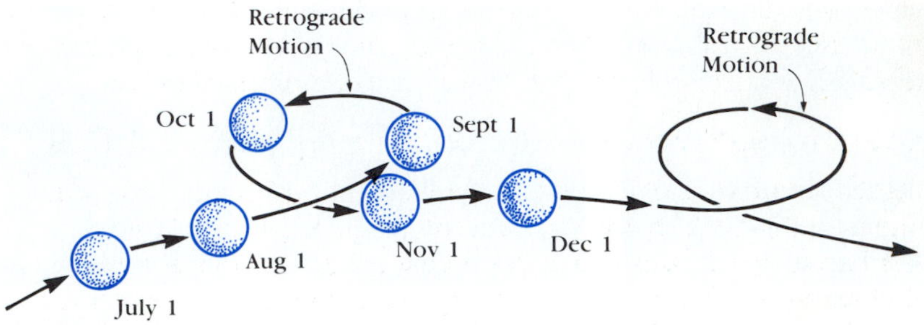

Retrograde Motion
Retrograde Motion
Oct 1
Sept 1
July 1
Aug 1
Nov 1
Dec 1

FIGURE 2-4
The path of the planet Mars among the stars during the course of several months, showing its position on different dates.

2. GEOCENTRIC SYSTEMS OF THE UNIVERSE

The Greeks Conceived of Great Celestial Spheres

FIGURE 2-5(a)
See Color Plate I.

Scientific theories about celestial motion can be traced back at least to the time of the ancient Greeks, and more specifically perhaps to the sixth and seventh century B.C. philosophers; the most famous were Thales and Pythagoras. By the fourth century B.C. a detailed theory of the heavenly bodies had been developed. According to this theory the heavens were composed of eight concentric transparent spheres. The sun, the moon, and the five planets were each on one of these spheres and all the fixed stars were attached to the eighth sphere [Figure 2-5(a) and (b), Color Plates I and II]. Each sphere rotated on a different axis and at a different rate, but the center of each was at the earth. This earth-centered, or **geocentric**, view of the universe was to persist with minor modifications for well over two thousand years. Indeed, references to the "seven celestial bodies and their orbs" (or "spheres") are found in Shakespeare and other great literature over the centuries.

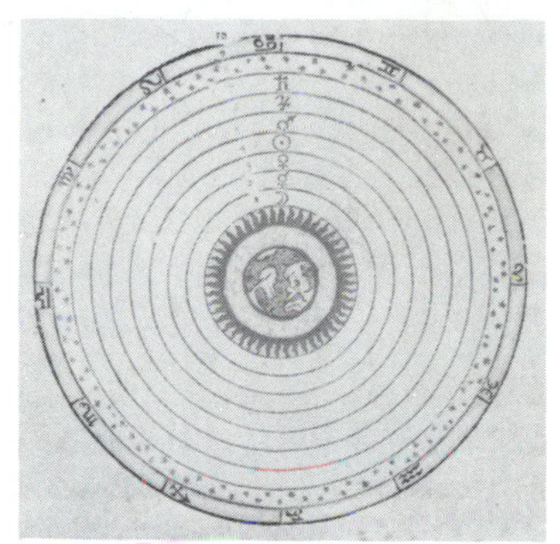

FIGURE 2-5(b)
See Color Plate II.

The two greatest philosophers of Greek civilization, Plato and Aristotle—and especially certain of their students—contributed further detail to this concept of the universe. They were very aware that this simple model did not explain adequately the detailed motion of the heavenly bodies, particularly the retrograde motion of the planets. At the same time, Plato and his students raised to the level of dogma the idea that the heavenly bodies move in perfect circles. This view was only natural because their philosophy was largely concerned with form and perfection. For example, it was inconceivable to them that the orbits of the planets should follow oval paths. To preserve the idea of perfect motion in a circle and to explain the retrograde motion of the planets, Aristotle and his contemporaries hypothesized a complicated set of circular motions. Each planet, the sun, and the moon were attached not to one sphere but to several, each of which rotated about a different axis. The theory required some 50 spheres in all. This combination of motions accounted for many of the observed features of heavenly motion, but by no means solved all the problems.

Another ancient idea (often attributed to Aristotle and also based on the doctrine of perfection) was the notion that only the "sublunary" (Latin: "below the moon") region of space—the space between the earth and the moon, including the earth—was subject to change and imperfection. The moon and the universe beyond were considered unchangeable and eternal. This idea persisted well into the second millennium A.D.

Ptolemy Used Circles—The Ultimate Geocentric System

The geocentric view of the universe reached its high point in the work of Ptolemy (100–170). By the second century A.D. the cultural center of the West had shifted to Alexandria; there, Ptolemy made extensive astronomical observations and worked out his system of the universe. Ptolemy, too,

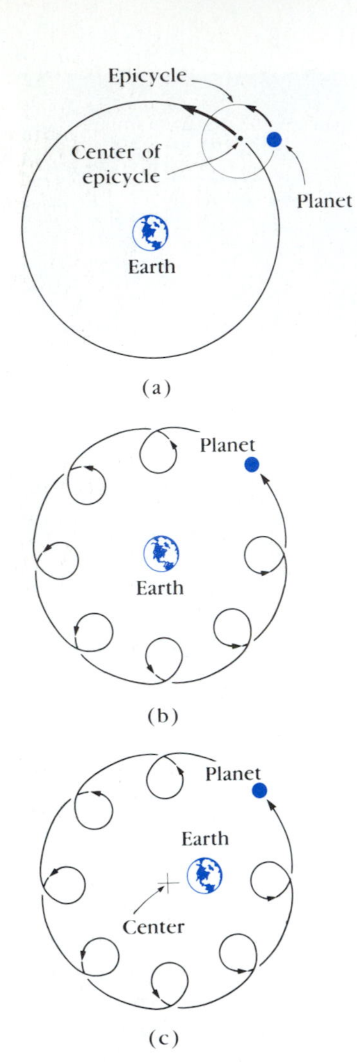

FIGURE 2-6
(a) Retrograde motion of planets explained with the concept of an "epicycle"—a small circle turning around a center which itself moves on a larger circle. (b) The net motion of a planet is thus a series of loops. (c) Ptolemy proposed that the earth may be displaced a certain distance from the center.

accepted the notion that the planets must move around the earth in circular orbits. But he did away with the need for spheres to which the planets, the sun, and the moon were attached. Instead he proposed that the planets move in a combination of simple circles, one superimposed on the other. As shown in Figure 2-6(a), a planet revolves in a small circle, called an *epicycle*, about a point—which itself describes a larger circle. The combined motion is a series of loops, Figure 2-6(b). This model could account for the retrograde motion of the planets, at least in general outline. To bring his model into better agreement with observation, Ptolemy tried other schemes, such as placing the center of each larger circle off to one side of the earth, Figure 2-6(c).

Although Ptolemy's system was rather complicated, it was a successful theory. It was based on accurate observations, and from it he was able to predict past and future positions of the planets that agreed remarkably well with observations. Although the agreement with experiment was not perfect, it was sufficiently close that almost no one doubted that Ptolemy's theory was basically correct. Indeed, no essential changes were made until Copernicus conceived of a sun-centered, or **heliocentric**, view of the universe 1300 years later.

But we should not leave our discussion of the ancients with the idea that every thinker accepted the geocentric view of the world. A number of ancient philosophers, notably Aristarchus (*c.* 310–230 B.C.), hypothesized a sun-centered system in which the earth was just another planet orbiting the sun between Venus and Mars. It was a remarkably modern view, but it did not fit in with the prevailing philosophic views of that time. Most thinkers felt it unreasonable to consider the earth anything but the center of the universe. Aristarchus's heliocentric theory was thus not widely accepted and consequently had little impact on the future development of science.

It is sometimes said, or implied, that the Greeks lacked observational power or were somehow blind or ignorant because they believed in an earth-centered universe. Such a judgment lacks historical perspective and overlooks the natural development of ideas. Our own contemporary vision of the universe did not suddenly occur—it developed, one step at a time, one idea on top of another, with the theories of the ancients as a base. We must also realize that there was no experimental evidence to indicate the superiority of a heliocentric system, such as Aristarchus's, over the geocentric system. Nor was Aristarchus's theory any less complicated than the prevailing system. The ancients were not stupid; they simply lived in a different age.

3. THE RENAISSANCE: HELIOCENTRIC SYSTEMS

Copernicus's Heliocentric System

Nicolaus Copernicus (1473–1543), a Pole, started the revolution which changed man's conception of the universe. In his book, *De Revolutionibus*

Orbium Caelestium ("On the Revolution of the Heavenly Spheres"), published just before his death, Copernicus revealed his controversial system to the world.

According to Copernicus the earth is just one of the planets, each of which revolves about the sun (Figure 2-7, Color Plate III). It takes the earth exactly one year to make a complete revolution around the sun and return to the same position. Furthermore, the earth rotates on its own axis once a day and it is this motion which gives rise to the apparent daily rotation of the sun, the moon, and the stars. To account for the monthly motion of the moon among the stars, Copernicus postulated that the moon itself revolves around the earth.

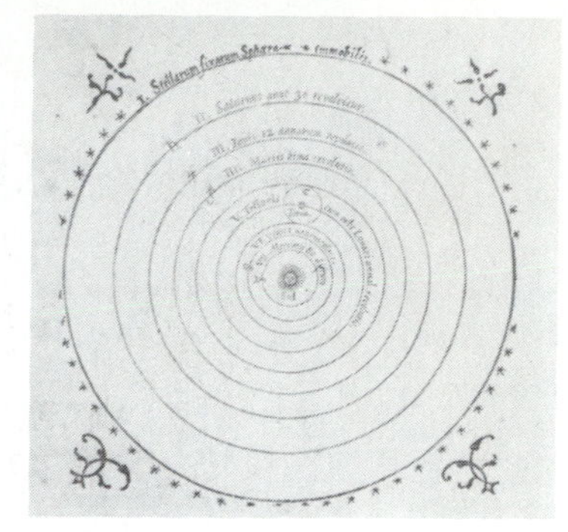

FIGURE 2-7
See Color Plate III.

Copernicus was not the first to conceive of a heliocentric system, however; there were others before him, as far back as the ancient Greeks, as we have seen; and the idea had again begun to flower in Italy during the fifteenth century. The Italy of the late Renaissance was alive with new thought. It was the time of Leonardo, Michelangelo, and Columbus. And it was there that Copernicus went to study and where he learned of the idea that perhaps the earth moves. Although the heliocentric idea had been discussed by thinkers for a century, it was Copernicus who built it into a comprehensive system.

Many other ideas of Aristotle were beginning to be questioned at this time. But Copernicus overturned only one: the concept of the earth-centered universe. The fact that Copernicus placed his heliocentric idea amidst medieval Aristotelian ideas does not diminish his greatness. But it does illustrate that the evolution of ideas, like biological evolution, occurs one step at a time. Only later would the other Aristotelian ideas be replaced, one by one, particularly through the work of Kepler, Galileo, and Newton. Copernicus, however, deserves the credit for initiating this revolution. His system remains a landmark in the history of science.

Because he was still predominantly an Aristotelian, Copernicus preserved the requirement that celestial bodies must move in perfect circles. But a single circle was not adequate to describe each planet's orbit around the sun. Consequently, Copernicus's system, like Ptolemy's, utilized some 50 circles to describe the motion of the planets. Copernicus's system was as complicated as Ptolemy's.

But, did Copernicus's system predict the motion of the planets better than Ptolemy's? By the sixteenth century, serious deviations from Ptolemy's predictions had been observed. Unfortunately, over a period of time Copernicus's model by itself did little better. With the increased accuracy of measurement then available, Copernicus's orbit of Mars particularly differed from that observed.

At the time, the Copernican model did not have a great deal to recommend it over the Ptolemic, although it did give predictions that Ptolemy's did not, such as that Venus ought to show moon-like phases (later observed by Galileo). Copernicus's model, in its simplest form, also gives a simpler and more natural explanation for the retrograde motion of the planets. The planets only *appeared* to be going backward. This would occur when the earth and the planet in question were passing close to one another, and their relative speeds would give the impression of backward motion. (A similar impression occurs when traveling in a car or train as you pass a row of trees—the trees seem to move backward.) The Copernican system thus answered one of the most difficult questions of astronomy—why the planets should appear to turn around and go backward. Nonetheless, opposition to the Copernican system was widespread. Objections, particularly in northern Europe, centered on the fact that it violated contemporary philosophic thought and on the "obvious absurdity" that the earth moves about a fixed sun.

Kepler Used Tycho's Data to Establish an Accurate Heliocentric System

Amidst the controversy, a great astronomer, Tycho Brahe (1546–1601), was born in Denmark. He was one of the first scientists to realize the power of precise observation. Tycho constructed a large device called a quadrant (Figure 2-8) with which to view the sky. Although the telescope had not yet been invented, Tycho's instruments were so precise that he could measure the angular position of a planet or star with an accuracy of 1/100 of a degree. For over 20 years he painstakingly plotted the paths of the planets from his observatory near Elsinore Castle, not far from Copenhagen.

But it was an assistant of Tycho's—Johannes Kepler. (1571–1630)—who at last brought order out of chaos. Kepler, a bit of a mystic, was deeply interested in the motion of the heavenly bodies and was a strong supporter of Copernicus. He sought to meet Tycho because he believed that Tycho was the one man whose observations might hold the key to confirming the Copernican hypothesis.

FIGURE 2-8
Tycho Brahe and his quadrant.

Kepler spent years examining Tycho's data, and from it he was able to improve upon the Copernican system. Kepler published a number of works dealing with his investigations, among which were *A New Astronomy* and *Harmony of the World* published in 1609 and 1618, respectively. Tucked away among these writings were three findings now famous as **Kepler's Laws**.

Kepler's first law deals with the shape of the planetary orbits. Like other thinkers of his day, Kepler did not doubt that the only proper curve for a planetary orbit was a circle. He tried to fit the orbits of the planets, as measured by Tycho, to a combination of circles and epicycles. He focused his attention on Mars, the planet that deviated most from a perfect circle.

Because all planetary observations are made from the earth and the Copernican theory regards the planets as moving about the sun, a painstaking translation of the observational data was necessary. Kepler had to take each of Tycho's measurements of Mars and determine its position with respect to the sun. For each measurement he had to do a new

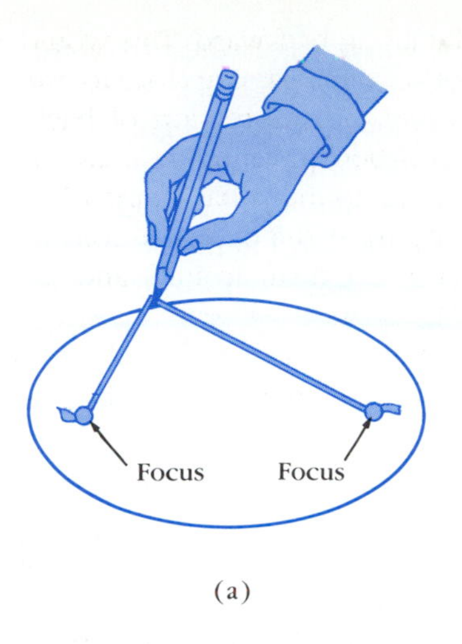

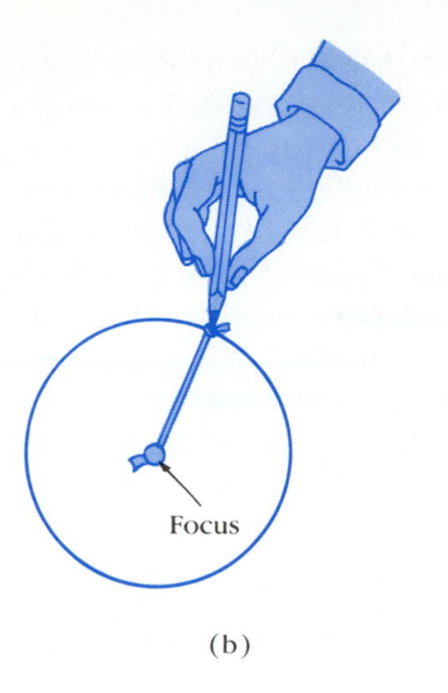

FIGURE 2-9
(a) The sum of the distance from a point on an ellipse to one focus plus the distance to the other focus is the same for each point on the ellipse. An ellipse can be made by choosing any two points as foci and tacking two ends of a piece of a loose string at these two points; a pencil held taut against the string and moved under this constraint will outline the shape of an ellipse. (b) A circle is a special kind of ellipse in which the two foci are at the same point.

calculation until he had a complete picture of the planet's orbit around the sun.

Once he had done this, Kepler tried to combine various-sized circles to reproduce on paper the actual orbit of Mars. He tried and tried to fit the data to a combination of circles, but gave up in despair. At this point Kepler reluctantly made a break with the ancient dogma of perfect circles. He decided to see if the actual shape of Mars' orbit might fit some other simple curve. And sure enough, Mars' orbit fit the shape of an ellipse—a particular kind of oval shape well known to mathematicians (Figure 2-9). The other planets, he found, fitted elliptical orbits as well. This became the tradition-breaking hypothesis now known as Kepler's first law: *the paths of planets around the sun are ellipses.*

The ancient rule of the circle had come to an end. Kepler's elliptical orbits were far simpler than the complicated epicycles hypothesized by Copernicus and Ptolemy. The heliocentric theory now possessed a considerable advantage over the geocentric theory.

If a theory is to be viable, one must be able to make accurate predictions based upon it. Kepler's second and third laws dealt with the quantitative aspects of the planetary orbits. In order to predict the past or future motion of a planet in its orbit, it is necessary to find a relationship between the position of the planet and its speed. Kepler sought to determine if the speed was constant on the elliptic orbits, and if not, whether a simple relationship existed between speed and position. He did indeed find a simple relationship and stated it quantitatively as follow: *each planet moves so that an imaginary line drawn from the sun to the planet sweeps out equal areas in equal times.* This, the second of Kepler's laws, is illustrated in Figure 2-10. Simply stated, the law says that when a planet is close to the sun it moves faster than when it is far from the sun. For example, when the earth is at its closest point to the sun,

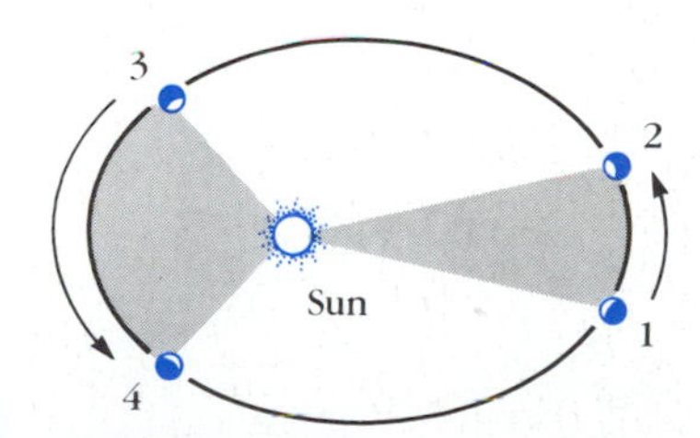

FIGURE 2-10
Illustration of Kepler's second law. The two shaded regions have equal areas. It takes the planet the same time to go from point 1 to point 2 as it does to go from point 3 to point 4.

FIGURE 2-11
Earth moves fastest in its orbit when it is closest to the sun and slowest when farthest from the sun.

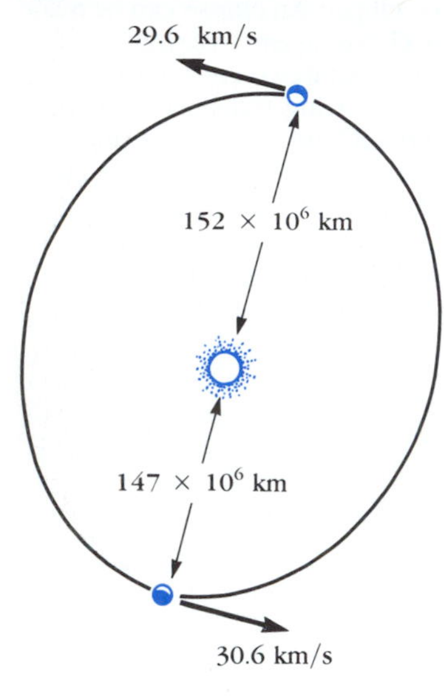

147,000,000 km away, its speed is 30.6 km/s; when it is farthest from the sun, 152,000,000 km away, its speed is 29.6 km/s (Figure 2-11).

In his search for order in the universe, Kepler brought to light a third law that quantitatively specifies how the period of the different planets increases with their increased distance from the sun. (The period is the time required for a complete revolution about the sun.) Kepler stated his third law very precisely in the following way: *the square of the ratio of the periods of any two planets is equal to the cube of the ratio of their average distances from the sun.*

Kepler's view of the solar system, as embodied in his three laws, is essentially the view we hold today. Kepler's view is both a simple and an elegant one. The laws that carry his name are significant, too, because they link astronomy to physics rather than to theology.

However, Kepler's laws lacked some sort of "reason" or unifying principle. It was soon to be supplied by Newton, as we shall see in Chapter 6. Although Newton was to put the finishing touches on the work begun by Copernicus, Kepler had greatly simplified the heliocentric view of the world. Yet the controversy between the heliocentric and geocentric systems was not over. We should point out, however, that today the controversy is not really an argument about which system is right and which is wrong, although it was probably considered so in the sixteenth and seventeenth centuries.† Today we realize that the choice is one of ease in visualization, or of elegance and simplicity. Modern relativity, which we shall discuss in Chapter 25, has shown us that it is as legitimate to say that the sun goes around the earth as it is to say that the earth goes around the sun. It all depends on your point of view. To describe the orbits of the sun and planets as they go around the *earth* is very complicated. But to describe the motion of the earth and the other planets going around the sun is very simple.

A Simple Numerical Example (optional)

Kepler's laws were mathematical laws, and here let us take an example of how they can be used.

Kepler's second law can be written in algebraic form as

$$\left(\frac{T_1}{T_2}\right)^2 = \left(\frac{d_1}{d_2}\right)^3.$$

T_1 and T_2 refer to the periods (time for one complete revolution about the sun) for two planets, say number 1 and number 2; d_1 and d_2 represent the average distances of planets 1 and 2 from the sun.

Let us consider the case of Saturn, which is approximately 10 times farther from the sun than the earth on average. Let us calculate approxi-

† Even today it is commonly taught that the heliocentric system is right and the geocentric wrong.

mately how long it takes Saturn to make a complete revolution about the sun. Let 1 be Saturn and 2 be earth. Then $(\frac{d_1}{d_2}) = 10$ and so $(\frac{d_1}{d_2})^3 = 1000$. Then, from Kepler's second law we have:

$$\left(\frac{T_1}{T_2}\right)^2 = 1000;$$

therefore,

$$\left(\frac{T_1}{T_2}\right) = \sqrt{1000} = 30.$$

We have rounded off, since our answer is only approximate ($\frac{d_1}{d_2}$ is more nearly equal to 9.2). Since the period of the earth is $T_2 = 1$ year, we have $T_1 = 30 \times T_2 = 30$ years.

Galileo Supports Copernicus

The story of our view of the universe and how it developed would not be complete without mention of Galileo Galilei (Figure 2-12) and his confrontation with the Church. Galileo was the son of a nobleman of low rank who was, however, highly cultured and a composer of merit (his works are still played occasionally). Galileo grew up with considerable contempt for authority and was considered a radical by his colleagues. He particularly scorned the old-fashioned Aristotelian professors who were still common in the universities.

Although Galileo did not actually invent the telescope, he developed it into a useful instrument and was the first to point it at the heavens, where he made a number of revolutionary observations. He saw, among other things, that there are many more stars in the sky than are visible to the naked eye and that the Milky Way is made up of a huge number of closely spaced stars. He observed that four moons orbit Jupiter and that the planet Venus exhibits phases just as the moon does. He found that the moon, contrary to prevailing opinion, is not a perfect sphere but is irregular and contains mountains and valleys like the earth. And finally he observed that the sun has spots. These last three observations led to the downfall of the Aristotelian doctrine that the heavenly bodies are perfect spheres and only the sublunary sphere, including the earth, is imperfect. Galileo had shown that the earth and the heavenly bodies are not of a fundamentally different nature. The overthrow of this ancient Aristotelian dogma was a significant achievement.

Galileo felt that his observations of the phases of Venus and the moons of Jupiter—which, he noted, resemble a miniature solar system—were strong evidence for the heliocentric system.

Galileo was a firm supporter of the Copernican system, and it was here that he got into trouble. At first, his teaching of the Copernican theory only aroused the anger of those of his fellow professors who maintained

FIGURE 2-12
Galileo Galilei (1564–1642).

old-fashioned ideas. The Church did not strongly oppose him on the Copernican system as long as he treated it as a hypothesis and not as established fact. In fact, Galileo was encouraged in his studies by a number of cardinals, one of whom later became pope. But in his zeal to discredit the old-fashioned thought of some of his contemporaries, he inevitably stepped on the toes of the clergy. A strong case has been made[†] that Galileo's confrontation with the Church was one of personality conflict rather than one of content.

In 1632, Galileo published his great defense of the Copernican system, *Dialogue Concerning the Two Great World Systems*. The dialogue was between three characters, one of whom was portrayed as a simple-minded defender of Ptolemy and Aristotle, and was an obvious caricature of the pope himself. Galileo, by then nearly 70, was immediately summoned before the Inquisition and decided to back down. We can be thankful, for soon after the trial he began a new book, *Dialogues Concerning Two New Sciences*, which summarized his detailed findings about the motion of objects on the earth. This book contained Galileo's greatest contributions to science and was important in our present appraisal of Galileo as the "father of modern science." We discuss these contributions in Chapters 3 and 4.

It is interesting to note that besides his great contributions to science, Galileo also initiated a new style of writing. In his books he used a straightforward style that was much more readable than the long-winded writings of his contemporaries and predecessors.

In this chapter we have covered the history of ideas on the motions of heavenly bodies, culminating in the heliocentric view established by Copernicus, Kepler and Galileo. This view of the universe (refined by Isaac Newton) remained almost unchanged until the twentieth century when significant new ideas came forth. We discuss them in the final two chapters of this book. In the next few chapters, we look at the development of ideas about motion on an earthly scale.

SUMMARY

The ancients viewed motion as being of two distinct types: celestial (motion of heavenly bodies) and terrestrial (motion of bodies on the earth). Observations of heavenly bodies showed that they appear to move from east to west. The sun, the moon, and the planets, which move at different rates from the stars, appear to move relative to them. Early views of the universe, culminating in that of Ptolemy about A.D. 150, were *geocentric*—i.e., earth-centered. Later, Copernicus proposed a *heliocen-*

[†] See Arthur Koestler, *The Sleepwalkers* (New York: Macmillan, 1968).

tric, or sun-centered, view of the universe in which the earth and the other planets were viewed as moving in circular orbits about the sun. Using the observations of Tycho Brahe, Kepler refined the heliocentric view by hypothesizing that the planets move in elliptical orbits and working out their relative speeds. Galileo, who made many great contributions to scientific thought, was the first to use the telescope to observe heavenly bodies, which further supported the heliocentric viewpoint.

QUESTIONS

1. What does it mean when we say that the moon moves eastward with respect to the stars?
2. On the basis of your own observations of the stars and the moon (see the experiment in this chapter), do you think the geocentric view of the universe developed by the ancient Greeks was reasonable? Why or why not?
3. Have you observed anything that would lead you to believe that the earth travels around the sun?
4. On the basis of Kepler's second law, determine whether the earth is moving around the sun faster in the summer or in the winter. Note: The earth is farther from the sun in summer than in winter.
5. Write the days of the week in one of the Latin languages (Spanish, Italian, French). Do you recognize the names of the planets in them? Explain.

EXERCISES

1. On the average, Mars is about 1.5 times farther from the sun than the earth is. Approximately how long is the Martian year?
2. Jupiter's period is about 12 years. On the average, approximately how far is it from the sun?
3. The asteroid Icarus, though only a few hundred meters across, orbits the sun like the other planets. Its period is about 410 days. What is its average distance from the sun?
4. Venus is an average distance of 1.08×10^8 km from the sun. Estimate the length of the Venusian year using the fact that the earth is 1.50×10^8 km from the sun on the average.
5. Use Kepler's second law and the period of the moon (27.4 days) to determine the period of an artificial satellite orbiting near the earth's surface. The earth's radius is 6380 km, and the moon is 384,000 km from earth.

DESCRIPTION OF MOTION

CHAPTER

3

In the previous chapter we dealt with the development of ideas about the motion of heavenly bodies; in this chapter we begin our study of the motion of ordinary objects on earth. We will see that the description of the motion of objects on earth can also be applied to heavenly bodies, and that the work of Galileo and Newton overturned the ancient doctrine that the motion of heavenly bodies was somehow different from the motion of objects on earth.

The study of the motion of objects, and the related concepts of force and energy, forms the field called *mechanics*. Mechanics is customarily divided into two parts: *kinematics*, the description of how objects move, and *dynamics*, why objects move as they do. In this chapter we deal primarily with the description of motion, kinematics, which involves the concepts of speed, velocity, and acceleration.

1. SPEED

Speed Relates Distance and Time

When an object is moving from one place to another, we say it is in motion; when it stops moving, we say it is at rest. The *distance* that an object moves is an important thing to know when describing its motion. Equally important is the *time* it takes to move that distance. Distance and time are related to another important property of motion, **speed**. Speed refers to how fast an object is moving; it is the *rate* at which distance is covered. The average speed of an object is defined as the distance the object travels divided by the time it takes to travel this distance

$$\text{speed} = \frac{\text{distance}}{\text{time}}.$$

If a car is driven a distance of 90 kilometers (about 55 miles) in one hour, its average speed is then 90 kilometers per hour.[†] However, saying that a car is moving at a speed of 90 kilometers per hour (90 km/hr) does not imply that the car will necessarily travel for an entire hour. For example, a car that travels 45 kilometers in a half hour also travels at an average speed of 90 km/hr because speed = (45 km)/($\frac{1}{2}$ hr) = 90 km/hr.

The definition of speed as the distance traveled divided by the time required is the definition of **average speed**. **Instantaneous speed**, on the other hand, is the speed an object has at any instant of time (this is what the speedometer of a car indicates). Perhaps a clearer way to define instantaneous speed is that it is the average speed taken over a very short time interval—an interval so short that the speed can be considered constant for that short time. The instantaneous speed and the average speed may be quite different. If you were to drive your car 80 kilometers in one hour, the average speed would indeed be 80 km/hr. But during that time the instantaneous speed, as registered by the speedometer, may have varied from 70 to 90 km/hr, or even more, depending on the traffic.

If a car moves at the same speed for a period of time with no speeding up and no slowing down, we say it has a *constant speed*. In this case the average speed and the instantaneous speed are the same, and either is referred to simply as the speed.

If a car moving with an average speed of 80 km/hr maintains this speed for one hour, it will travel a distance of 80 kilometers. If the speed of 80 km/hr is maintained for a half hour, how far will the car have gone? The answer, of course, is 40 kilometers. Thus, if we know the average speed of an object and how long the object traveled at that speed, we can figure the distance it traveled:

$$\text{distance} = \text{average speed} \times \text{time}.$$

This relation is the inverse of the one defining speed. In the example of a car traveling at 80 km/hr for a half hour, the relation tells us that distance = 80 km/hr × $\frac{1}{2}$ hr = 40 km. This result is easily figured in your head without using this relation. But we use relations to keep concepts straight in our minds and to help avoid mistakes in more complicated situations.

Relationships Between Quantities Can Be Abbreviated

Often it is convenient to write relationships—such as the ones described above that relate speed, distance, and time—in an abbreviated form. For example, instead of writing out "distance," we simply use its first letter d. For time we use t and for speed we use v (v is short for velocity, which is often taken as a synonym for speed; a distinction can be made between these two words, as we shall see shortly). Thus the definition of average

[†] Notice that "per" means "divided by".

speed can be written in abbreviated form as

$$\bar{v} = \frac{d}{t}.$$

(The bar over the v means "average.") This equation can be used to calculate the speed just as the word-equation was earlier. Suppose, for example, that we want to calculate the average speed of a car that travels 300 km in 4 hours. Then the distance $d = 300$ km, the time $t = 4$ hr, and $\bar{v} = \frac{d}{t} = (300\text{ km})/(4\text{ hr}) = 75\text{ km/hr}$. Thus the letters in this equation can be considered symbols for the quantities speed, distance, and time. This equation can be solved (algebraically) for the distance d by multiplying both sides by the same quantity t: $\bar{v} \times t = \frac{d}{t} \times t$. The time t appears in both the numerator and the denominator on the right-hand side and thus cancels out. We get

$$d = \bar{v}t,$$

which is the relationship we had before: distance = average speed × time. As an example, let's calculate how far a jet plane goes in 3.5 hours if its average speed is 800 km/hr. The distance $d = \bar{v}t = (800\text{ km/hr})(3.5\text{ hr}) = 2800\text{ km}$.

If you are not comfortable using symbols, use the words instead. The physics is in the concepts, not in how we write them.

In the example we just worked (the jet plane), notice that we must always be consistent in the units we use. That is, if the time is given in hours, then before any calculation is done, the speed, too, must have hours in it: kilometers per hour or miles per hour—not, say, meters per second. Also, note how the units of "hour" cancel out. Keeping track of the units in this way helps to avoid mistakes.

And now, before going much further in our study of physics, we must discuss the subject of units and measurement.

2. UNITS AND MEASUREMENT

Standards and Units

The measurement of distance, time intervals, weight, and other quantities is an important aspect of science. The measurement of any quantity is made relative to a particular standard or unit, and this unit must be specified along with the numerical value of the quantity. For example, we can measure length in such units as inches, feet, or miles, or in the metric system in centimeters, meters, or kilometers. To specify that the length of a particular object is 18.6 is meaningless; the unit *must* be given. Clearly, 18.6 meters is quite different from 18.6 inches or 18.6 millimeters.

Until about 200 years ago the units of measurement were not standardized, and that made precise scientific communication difficult. Different people used different units: cubits, leagues, hands, and even the length of the foot varied from place to place. (Henry VIII and Louis XIV apparently didn't wear the same size shoe.)

The first real international standard was the establishment of the standard *meter* (abbreviated m) by the French Academy of Sciences in the 1790s. In a spirit of rationality, the standard meter was originally chosen to be one ten-millionth of the distance from the earth's equator to either pole.† A platinum rod to represent this length was made, and is now kept near Paris.* Copies of the standard meter were sent around the world. A copy "for the people" remains in the Place Vendôme in Paris (Figure 3-1), placed there originally so Paris shopkeepers would have access to a standard of measurement.

The British units of length are defined today in terms of the meter: 1 inch = 2.54 centimeters. Because there are 12 inches in a foot, there are 12×2.54 centimeters = 30.48 centimeters in one foot (ft). The relationships between other metric and British units are given in Table 3-1.

In the metric system, larger and smaller units are defined in multiples of 10, 100, and so on, from the standard unit. Thus, a centimeter

† Modern measurements of the earth's circumference reveal that the intended length was off by about one-fiftieth of 1 percent.

* In 1983, the meter was redefined more precisely in terms of the speed of light.

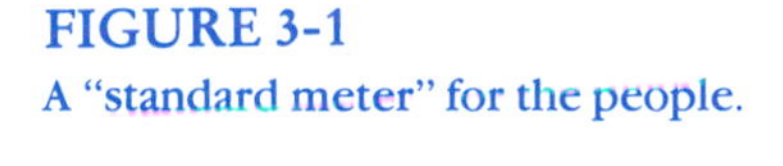

FIGURE 3-1
A "standard meter" for the people.

TABLE 3-1 Units of Length and Volume

LENGTH

Metric		British
1 cm = 10 mm	=	0.394 in.
1 m = 100 cm = 1000 mm	=	39.37 in = 3.28 ft
1 km = 1000 m	=	3,280 ft = 0.621 mi
British		**Metric**
1 in.	=	2.54 cm = 25.4 mm
1 ft = 12 in.	=	30.48 cm = 304.8 mm
1 mi = 5,280 ft	=	1.61 km = 1,610 m

VOLUME

1 liter = 1000 cm^3 = 1.06 quarts
1 quart = 0.946 liter

(abbreviated cm) is 1/100 of a meter (m); a millimeter (mm) is 1/1000 of a meter; and a kilometer (km) is 1000 meters. The prefixes **milli-** (meaning 1/1000), **centi-** (1/100), and **kilo-** (1000) can be attached to other kinds of metric units. For example, a kilogram is one thousand grams. Table 3-2 shows other prefixes that are used, but these three are the most common.

The relation between various units in the British system makes calculations difficult (for example 5,280 feet = 1 mile). For this reason, and to make communication and commerce simpler, scientists, and nearly all countries of the world, have adopted the metric system.

TABLE 3-2 Metric (SI) prefixes (multipliers)

PREFIX	ABBREVIATION	VALUE[‡]
Tera	T	10^{12} = 1,000,000,000,000
Giga	G	10^{9} = 1,000,000,000
Mega	M	10^{6} = 1,000,000
Kilo	k	10^{3} = 1,000
Hecto	h	10^{2} = 100
Deka	da	10^{1} = 10
Deci	d	10^{-1} = 1/10
Centi	c	10^{-2} = 1/100
Milli	m	10^{-3} = 1/1000
Micro	μ	10^{-6} = 1/1,000,000
Nano	n	10^{-9} = 1/1,000,000,000
Pico	p	10^{-12} = 1/1,000,000,000,000
Femto	f	10^{-15} = 1/1,000,000,000,000,000

‡ A discussion of the powers of ten notation is provided in Appendix A.

Sometimes it is necessary to convert from one system to the other. Table 3-1 will help here. For example, suppose you measure something as 20 centimeters long and you want to know how many inches it is. From Table 3-1 we see that 1 centimeter is about 0.4 inches. Therefore 20 centimeters equal $20 \times 0.4 = 8$ inches.

To give you some practice with the metric system, try the following simple exercise.

EXPERIMENT-PROJECT

Use a metric ruler to measure your height in meters or in centimeters. (Do *not* measure in feet and inches and then convert to metric.) Then, take the result of your measurement and use Table 3-1 to convert to British units (feet and inches). Is this how tall you thought you were?

We have so far mentioned only units of length. We will discuss other units when appropriate. For now we add that the units of time—hours (hr), minutes (min), and seconds (s)—are the same in the metric and British systems.

Although learning metric units may seem a chore, it may be encouraging that many everday units are already metric. For example, the units volts and amperes used in electricity are metric and do not require conversion.

Units of Measurement for Speed

The units most commonly used for speed in everyday life are kilometers per hour and miles per hour. Other common units for speed are meters per second (m/s) and feet per second (ft/s), both of which are particularly useful when distances are not great.

Sometimes speed is given in one set of units when another set of units would be more convenient. For example, when figuring out braking distances in automobiles, it is often more useful to know the speed of the automobile in meters per second than in kilometers per hour. Table 3-3 gives comparative speeds in different sets of units. As an example of how these conversions are made, note that a speed of 1 km/hr is equivalent to 1000 m in 1 hr, or 3600 seconds (since there are 60 seconds in a minute and 60 minutes in an hour, and $60 \times 60 = 3600$). Thus 1 km/hr equals 1000 m/3600 s equals 0.28 m/s. A speed of 1 km/hr is therefore equivalent

TABLE 3-3 Equivalent Speeds in Different Units

km/hr		m/s		mi/hr		ft/s
20 km/hr	=	5.6 m/s	=	12 mi/hr	=	18 ft/s
40 km/hr	=	11 m/s	=	25 mi/hr	=	37 ft/s
60 km/hr	=	17 m/s	=	37 mi/hr	=	55 ft/s
80 km/hr	=	22 m/s	=	50 mi/hr	=	73 ft/s
100 km/hr	=	28 m/s	=	62 mi/hr	=	92 ft/s

to 0.28 m/s. Similarly, a speed of 10 km/hr is equivalent to a speed of 2.8 m/s, and so on. In fact, in Table 3-3 you will see that all speeds in meters per second are 0.28 times the speed in kilometers per hour. The number 0.28 is a *conversion factor*, used to convert kilometers per hour to meters per second. Conversion factors are also useful in changing miles per hour to kilometers per hour or feet per second.

EXPERIMENT-PROJECT

Measure the average speed of an ant or another insect or animal. Use an ordinary ruler to measure the distance and a watch with a second hand to measure the time. It will be easier if you find an ant that moves in a straight line.

OR

Check the accuracy of the speedometer of your car. You can do this simply by maintaining a constant speed and measuring the time it takes to go an accurately measured distance. Many highways and superhighways have accurate mileage markers or speedometer check areas in which signs are posted exactly one mile or one kilometer apart. You must use a watch with a second hand so that you can measure the time accurately. To find the speed in miles or kilometers per hour, the time in seconds must be changed to the proper fraction of an hour. Two people are necessary for this experiment: one to drive the car at constant speed, the other to measure the time on the watch. This experiment can be done just as well on a motorcycle or motorbike.

OR

Calculate your maximum average speed while riding a bicycle by measuring the time it takes you to ride a known distance at top speed.

3. VELOCITY

Velocity Has Direction as Well as Magnitude

When specifying the motion of an object, both the speed and the direction of motion are important. For example, if a friend leaves Los Angeles by plane traveling at 600 mi/hr, you must also know where your friend is going: i.e., in what direction. Is she going to San Francisco, the Grand Canyon, Mexico City? North, east, south, or west? Similarly, when you drive a car you may drive north, west, southeast, or whatever. Thus it is important to specify *direction* as well as magnitude. (Magnitude means the numerical value, such as 50 km/hr.)

We can use the term **velocity** to signify both the speed and direction of a moving object. Speed then refers to the magnitude of the velocity; the direction of the velocity is the direction in which the object is moving. When drawing a diagram showing the motion of an object, we often use an arrow to indicate the direction of motion. For example, in Figure 3-2 a car moving along a curvy road is shown at various positions along the road. At each position an arrow is drawn to show the direction of the car's velocity at that instant.

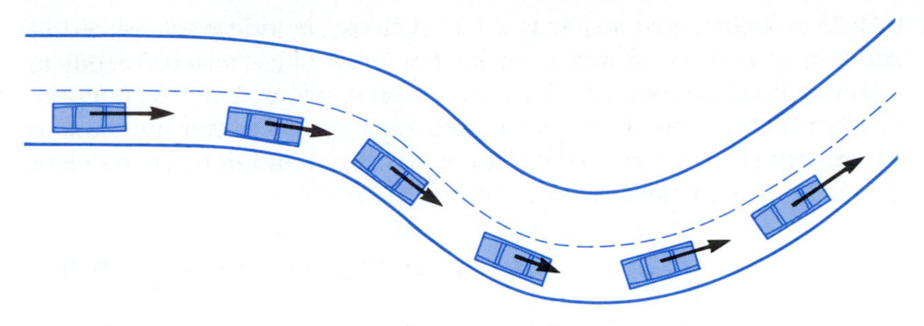

FIGURE 3-2
Arrows represent the velocity of the automobile at various times as it goes around a curve. Notice that the direction of the velocity changes as the car moves through the turn. The magnitude of the velocity (length of the arrow) also changes.

Notice that constant speed is not the same as constant velocity. For example, an automobile moving at a constant speed of 60 km/hr as it rounds a curve does not have a constant velocity, because its direction is changing at each instant. An object has a constant velocity only if its speed *and* its direction do not change.

Although the words speed and velocity are often used interchangeably, it is helpful to remember the distinction between them.

Vectors

Quantities such as velocity that have both magnitude and direction are referred to as vector quantities, or **vectors**. Force, as we shall see, is also a vector. Not all quantities are vectors. Time, temperature, and energy have no direction associated with them; they are specified merely by a number and are referred to as **scalars**.

Arrows are often used to represent vectors, as in Figure 3-2 for the velocity. The arrow is always drawn to point in the direction of the vector it represents. The length of the arrow is proportional to the magnitude of the vector. Thus, in Figure 3-2 the arrows representing velocity are longer where the speed is greater.

4. ACCELERATION

Acceleration is the word used to refer to a rate of change in velocity. Whenever a moving object is changing speed, we say it is accelerating. For example, when an automobile starts from rest and speeds up to, say, 80 km/hr, it is accelerating. If one car can accelerate from rest to 80 km/hr in less time than another, it is said to undergo greater acceleration. Acceleration, then, is defined as the change in velocity divided by the time required to make that change. More specifically, this defines the *average* acceleration:

$$\text{average acceleration} = \frac{\text{change in velocity}}{\text{time}}.$$

Like velocity, acceleration is a rate. Velocity is the rate at which the *position* of an object changes. Acceleration is the rate at which the *velocity* changes. Be careful not to confuse acceleration with velocity. Stepping on the "accelerator" pedal of your car, for example, may or may not cause an acceleration because maintaining a constant speed (which is zero acceleration) requires you to keep your foot on the pedal.

Let us consider an example. Suppose a car accelerates from 0 to 80 km/hr in 10 s (Figure 3-3). Its change in velocity is simply the final velocity minus the initial velocity: 80 km/hr − 0 km/hr = 80 km/hr. Therefore, we can calculate that the average acceleration $= \frac{80\text{ km/hr} - 0\text{ km/hr}}{10\text{ s}} = \frac{80\text{ km/hr}}{10\text{ s}} = \frac{8\text{ km/hr}}{\text{s}}$. The average acceleration is 8 kilometers per hour per second. This means that on the average the velocity changes by 8 km/hr during each second. During the first second the car's velocity increased from zero to 8 km/hr; during the next second its velocity increased by another 8 km/hr to 16 km/hr; and so on. This of course will be precisely true only if the acceleration is *uniform*. The *instantaneous* acceleration may be different from the average of (8 km/hr)/(s) we just calculated. But if the acceleration *is* uniform, the instantaneous acceleration during the 10 seconds will be the same as the average acceleration.

Notice that there are two "pers" and two time units associated with acceleration, because acceleration is the "rate of a rate." If we change kilometers per hour to meters per second (from Table 3-3, 80 km/hr = 22 m/s), then the acceleration becomes: acceleration = (22 m/s − 0 m/s)/(10 s) = 2.2 m/s/s: that is, 2.2 meters per second per second. This result is more simply written 2.2 m/s^2 which is read as "2.2 meters per second squared." This calculation tells us that the velocity changes by 2.2 m/s during each second for a total change of 22 m/s over the 10 seconds.

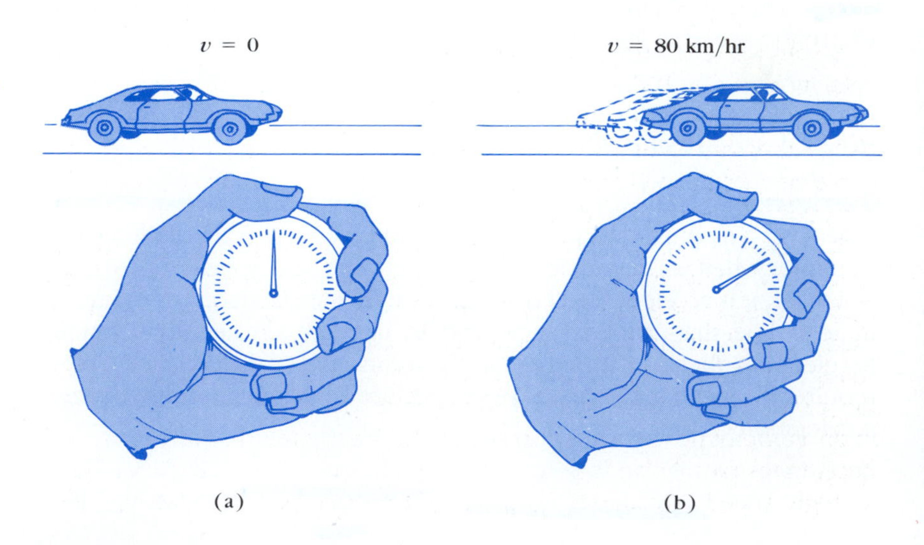

FIGURE 3-3
The automobile accelerates from rest up to 80 km/hr in 10 s.

The definition of average acceleration can also be written as

$$\text{average acceleration} = \frac{\text{final velocity} - \text{initial velocity}}{\text{time}}.$$

If the initial velocity is zero, as it often is, then this relation reduces to: acceleration $= \frac{\text{final velocity}}{\text{time}}$ In symbols, $a = v_f/t$ where a stands for average acceleration, t for time, and v_f for the instantaneous final velocity at the end of the time t. This relation can be turned around to yield: final velocity = acceleration × time, or in symbols,

$$v_f = at.$$

For example, if a bicyclist accelerates at 1 m/s^2 starting from rest, after 5 seconds the speed will be $(1\ m/s^2) \times (5\ s) = 5\ m/s$.

When an object slows down (for example, when the brakes are applied on an automobile), its velocity decreases. This is sometimes called "deceleration" and is merely another example of acceleration. In this case, the final velocity is less than the initial velocity, so the acceleration is negative.

EXPERIMENT-PROJECT

With the help of a friend who has a watch with a second hand, measure the time it takes for your car to accelerate from rest to 50 km/hr (or 30 mi/hr), and then calculate the average acceleration.

OR

Calculate your acceleration on a bicycle in a similar way as you accelerate from rest to top speed (which you measured in an earlier experiment).

So far we have considered only cases in which the *magnitude* of the velocity changes. If the speed remains constant and the *direction* of the velocity changes, this constitutes an acceleration as well. For example, a child riding on a merry-go-round and a person riding in a car rounding a curve at high speed are both aware of an acceleration because the direction of the velocity is changing.

It is an interesting physiological phenomenon that the semicircular canals that lie behind our ears are quite sensitive to accelerations. When these canals are affected, the brain sends messages to the stomach, which then feels tickled or nauseous when excess acceleration occurs. The same strange feeling occurs on a rapidly rotating merry-go-round or barrel-of-fun, or when an elevator starts or slows (accelerates) too rapidly. Similarly, in an automobile we lurch forward or backward when it accelerates or decelerates rapidly, and we lurch to the side when going around a curve at high speed. We lurch because of the acceleration. Clearly, then, acceleration results when either the magnitude or the direction of the velocity, or both, changes. We will discuss acceleration due to change in the direction of velocity in more detail later.

5. FALLING BODIES

Among the earliest studies of terrestrial motion were those made by the ancient Greeks, in particular, Aristotle. Although these early studies revealed insight, many of the conclusions were based on what was thought to be common sense or at least "obvious" from everyday experience.

It was Galileo, the same Galileo who made the first telescopic observations of the skies, who clearly established the modern scientific method. Galileo advocated the idea that no theory or model of nature was meaningful unless it predicted results in accord with experiments. Galileo felt that one must examine natural phenomena with great care. In fact, many phenomena are so complex that, Galileo reasoned, the simplest situations must be examined first. He sought the simplest examples of natural phenomena and set up simple experimental arrangements to examine them. For analyzing observational results, Galileo introduced an important new tool—that of *idealizing* a situation, such as imagining motion in the absence of friction. This theoretical method of idealizing became a cornerstone of science, and its use has allowed otherwise complex phenomena to be analyzed successfully.

To the early investigators there existed two main problems relating to the motion of bodies on earth: the motion of bodies along the ground or on a tabletop—that is, *horizontal motion* (which we discuss in the next chapter); and the motion of falling bodies, *vertical motion* (which we discuss now).

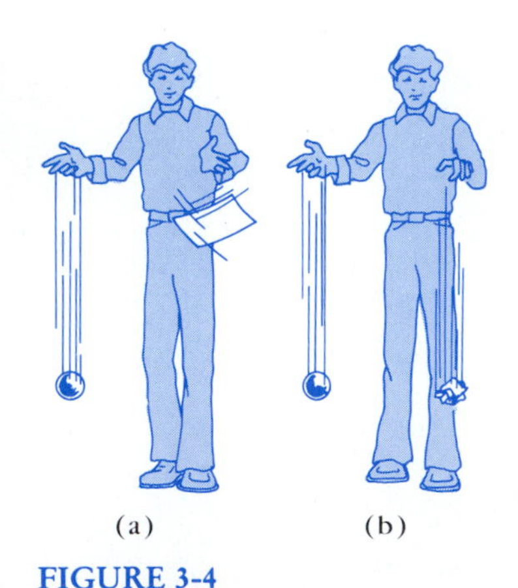

FIGURE 3-4
(a) A heavy object and a piece of paper are dropped simultaneously. (b) The same heavy object and the same piece of paper—this time crumpled into a wad—once again dropped simultaneously.

Galileo's Analysis of Falling Bodies

We are all familiar with the fact that if we hold an object above the ground and let it go, it falls downward. We say that it falls because of gravity. The gravitational force, or gravitational pull, of the earth pulls the object downward.

This well-known fact leads to many questions that we can try to answer. What is the nature of the motion of falling bodies? How fast do bodies fall? Do they undergo acceleration? Do some bodies fall faster than others?

It is commonly thought that heavy objects fall faster than light objects. Aristotle held as a basic principle that bodies fall with speeds proportional to their weight. Let us test this principle.

EXPERIMENT

Take in one hand a reasonably heavy object such as a baseball, a stone, or an eraser; in your other hand take a flat piece of paper, holding it horizontally. Hold the two objects at equal heights above the floor and let them go at the same time [Figure 3-4(a)]. Which reaches the floor first? This experiment seems to be a clear confirmation of Aristotle's principle that lighter objects fall more slowly than heavier objects.

Now modify the experiment. Take the same heavy object and the same piece of paper but this time crumple the sheet of paper into a small wad. Hold the heavy object and the wad of paper at equal heights and let go [Figure 3-4(b)]. What do you observe in this case? The result is quite different this time!

FIGURE 3-5
The Tower of Pisa.

What can you conclude from this experiment? Apparently the speed of a falling body does not depend on its weight. The paper, whose weight is the same whether it is flat or crumpled into a wad, falls faster when it is wadded up; that is, it falls faster when it has a smaller cross-sectional area.

Galileo broke[†] with the (then) traditional view that heavier objects always fall faster than lighter objects. Using his new and creative technique of abstracting to simple situations, Galileo postulated that in the absence of air, or if air resistance is negligible, *all objects fall at the same rate.*

It is air resistance, a kind of friction, that slows down the flat piece of paper. If the flat piece of paper had been dropped where there was no air, in a vacuum, it would have dropped as fast as the heavy object (Figure 3-6). Apparently, air resistance is only significant when a body has a large cross section compared to its weight.

Galileo was able to overturn another of Aristotle's conclusions—that a falling body acquires its speed immediately after it is dropped and maintains that speed throughout its fall. Galileo showed that a falling body is constantly increasing in speed and that the increase occurs at a constant rate. In other words, bodies fall (in the absence of air resistance) with a *constant acceleration.* Galileo's discoveries on the motion of falling bodies can be summarized in the following sentence:

In the absence of air resistance, all bodies fall with the same constant acceleration.

This conclusion was clearly at variance with previous beliefs. Even today people who have not studied physics might be more likely to accept Aristotle's view. Doubters can be shown the experiment of Figure 3-4. The fact that objects increase in velocity as they fall was made graphic by Galileo when he pointed out that a block allowed to drop from a height of ten feet will drive a stake much farther into the ground than the same block dropped from a height of one inch. Clearly the block is going faster when dropped from the greater height.

FIGURE 3-6
A feather and a rock drop at the same rate in a vacuum (b).

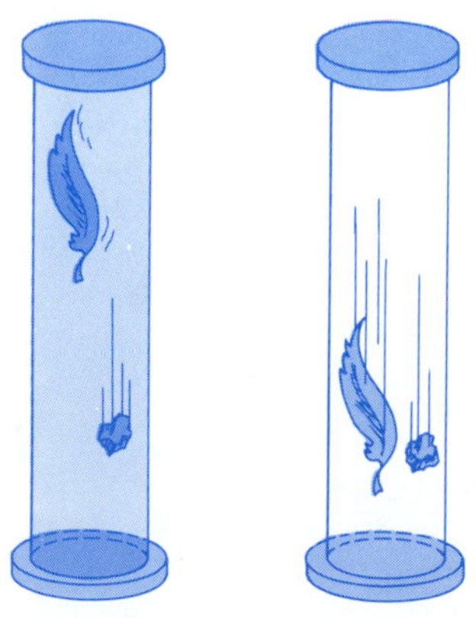

[†] Although Galileo performed many experiments, it is interesting that historians are not convinced that he ever used the Leaning Tower of Pisa (Figure 3-5), as is so often reported, to measure the times of fall for objects of different weights.

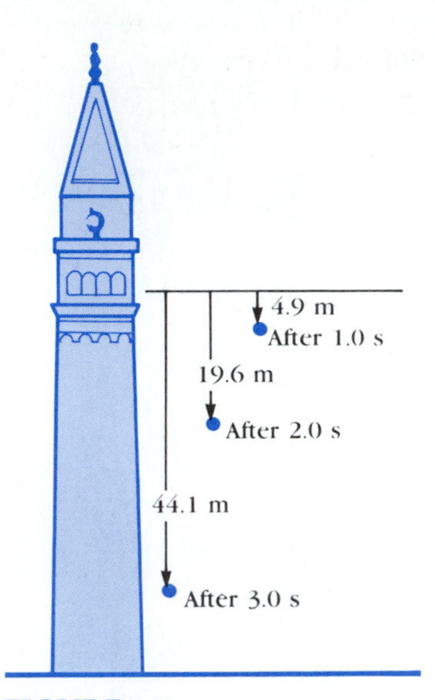

FIGURE 3-7
When an object is dropped from the top of a tower, it falls with progressively greater speed and covers greater distance with every successive second.

It was stated earlier that "in the absence of air resistance, all bodies fall with the same constant acceleration." What about air resistance? For most practical cases, it has a negligible effect. Only in the special case of a very light object with a large cross section (such as a feather or a piece of paper) or in the case of an object that falls a great distance (such as a skydiver) does air resistance retard an object significantly. Other than those exceptional cases, objects fall with the same acceleration. This acceleration is known as the **acceleration due to gravity** (on earth) and is given the symbol g. The acceleration due to gravity has been measured to be

$$g = 9.8\ \text{m/s}^2.$$

In British units, $g = 32\ \text{ft/s}^2$.

Some Simple Examples (optional)

For those who wish to get a feel for using the results of this chapter in a quantitative way, we now consider some simple numerical examples.

Suppose that an object is dropped from the top of a high tower (Figure 3-7). Let us calculate the speed of the object after it has been falling for one second and after it has been falling for two seconds. We have learned that when an object starts from rest, the velocity after a given time equals the acceleration times the time—in symbols, $v_f = at$. In the present case, the acceleration is simply the acceleration due to gravity: 9.8 m/s². Then after one second the speed will be $v_f = at = 9.8\ \text{m/s}^2 \times 1.0\ \text{s} = 9.8\ \text{m/s}$. After two seconds the speed will be $v_f = 9.8\ \text{m/s}^2 \times 2.0\ \text{s} = 19.6\ \text{m/s}$. Thus the object is going 9.8 m/s after one second and 19.6 m/s after two seconds.

Let us next calculate the *distance* the object will have fallen after one second and after two seconds. We have learned that distance equals average velocity times time, or $d = \bar{v}t$. First, we must determine the average velocity. Because the velocity increases at a constant rate, the average velocity is simply the average of the initial and final velocities. To find the average value of two numbers we add the two numbers and divide by 2. Thus the average velocity *during* the first second is $\frac{9.8\ \text{m/s} + 0\ \text{m/s}}{2} = 4.9\ \text{m/s}$. Therefore, after one second the object will have fallen a distance ($d = \bar{v}t$) of $4.9\ \text{m/s} \times 1.0\ \text{s} = 4.9\ \text{m}$. To find out how far it will have fallen after two seconds, we must calculate its average velocity during these two seconds, which is $\frac{19.6\ \text{m/s} + 0\ \text{m/s}}{2} = 9.8\ \text{m/s}$. Therefore, the object will have fallen a distance of $9.8\ \text{m/s} \times 2.0\ \text{s} = 19.6\ \text{m}$; this is 14.7 additional meters since the end of the first second. After three seconds an object will have fallen 44.1 m, and so on (see Figure 3-7 and Table 3-4). Notice that the distance covered by a falling object *during* each second increases in time. This occurs because the object's speed is continually increasing.

TABLE 3-4 Calculation of Distance Traveled by Dropped Object

ELAPSED TIME	ACCELERATION (CONSTANT)	SPEED AFTER THIS TIME (use $v_f = at$)	AVERAGE SPEED SINCE BEGINNING $\bar{v} = \frac{0 + v_f}{2}$	TOTAL DISTANCE TRAVELED ($d = \bar{v}t$)
0	9.8 m/s^2	0	$\bar{v} = 0$	$d = 0$
1 s	9.8 m/s^2	$v_f = 9.8 \times 1 = 9.8$ m/s	$\bar{v} = \frac{0 + 9.8}{2} = 4.9$ m/s	$d = 4.9 \times 1.0 = 4.9$ m
2 s	9.8 m/s^2	$v_f = 9.8 \times 2 = 19.6$ m/s	$\bar{v} = \frac{0 + 19.6}{2} = 9.8$ m/s	$d = 9.8 \times 2.0 = 19.6$ m
3 s	9.8 m/s^2	$v_f = 9.8 \times 3 = 29.4$ m/s	$\bar{v} = \frac{0 + 29.4}{2} = 14.7$ m/s	$d = 14.7 \times 3.0 = 44.1$ m

A quick way to determine how far an object has moved after a given time while accelerating is to use a relationship that relates distance, acceleration, and time. This relation† was first determined by Galileo:

$$\text{distance} = \frac{1}{2} \times \text{acceleration} \times \text{time squared.}$$

This is valid whenever the body starts from rest. In symbols this can be written

$$d = \frac{1}{2}at^2.$$

For falling objects, the acceleration is that due to gravity, 9.8 m/s^2. Thus $d = \frac{1}{2} \times 9.8 \times t^2$, or $d = 4.9t^2$; this gives the distance in meters, and the time must be given in seconds. Try putting $t = 1.0$ s, $t = 2.0$ s, and $t = 3.0$ s into this formula. You should obtain distances of 4.9 m, 19.6 m, and 44.1 m, just as before.

Terminal Velocity

In our discussion of falling bodies, we may have been unfair to Aristotle. He claimed that bodies fall with a constant speed in any particular medium and that this speed depends on the mass of the object. We saw that this is not generally the case for ordinary objects falling in air. The media

† This formula is easily derived, and we use symbols to make it briefer. From the definitions of velocity and acceleration we know that $d = \bar{v}t$ and $v_f = at$, where v_f represents the final velocity *after* any particular time, and $\bar{v}$ is the average velocity during that time period. The average velocity, $\bar{v}$, is simply the average of the initial and final velocities, but since the initial velocity is zero (the object starts from rest), the average velocity is $\bar{v} = v_f/2$. Thus $d = \bar{v}t = (v_f/2)\ t$. If we substitute $v_f = at$ into this relation we get

$$d = \frac{v_f}{2}t = \frac{at}{2}t = \frac{1}{2}at^2.$$

Aristotle used, however, included water and other liquids. We know now that when a body falls through a fluid it increases in speed. However, especially in liquids, a maximum velocity can be reached. This velocity, called **terminal velocity**, is attained when the force of gravity is just balanced by the resistive (or frictional) force† of the medium. At this point the net force becomes zero, and therefore the velocity will be constant after that. The terminal velocity depends on the size and weight of the object and on how viscous the fluid is—that is, on how much of a resistive force it exerts. In viscous liquids even a heavy object will reach terminal velocity fairly quickly. A feather reaches its terminal velocity in air quite rapidly. But most objects falling in air or other gases reach their terminal velocity only after a tremendous distance of fall. A skydiver without a parachute, for example, reaches a terminal velocity of about 200 km/hr (120 mi/hr) after falling about 300 m (1000 ft). For most practical cases an object hits the ground long before the terminal velocity is reached.

Was Aristotle Wrong?

Except for his assumption that terminal velocity is reached immediately, Aristotle's observations were not really erroneous. But he didn't probe as deeply as Galileo did. And because of Galileo we understand motion far better.

The difference between Galileo's and Aristotle's views of motion is not really one of right and wrong. The real difference lies in the fact that Aristotle's view was almost a final statement; one could go no farther. But the view established by Galileo could be extended to explain many more phenomena. It was a far more useful view, and it made nature more a whole. Before Galileo, vertical motion seemed to be governed by different laws from those governing horizontal motion. Because of Galileo we see that the two are linked. Indeed, three-dimensional space is symmetric, and the vertical dimension appears different only because of the gravitational force.

As we shall see in Chapters 4 through 6, another great scientist, Sir Isaac Newton, extended Galileo's results. Adding his own ideas to Galileo's, Newton synthesized a beautiful and all-encompassing theory of motion and its causes.

SUMMARY

To describe the motion of an object we use the concepts of speed, velocity, and acceleration. The *average speed* is defined as the distance traveled divided by the elapsed time. The *instantaneous speed* (the "speed at a

† The resistive force that fluids exert on an object falling through them is quite complicated and is a field of study in itself. We should note here, however, that this force increases as the velocity of the object increases.

given instant") is the average speed taken over a very short time interval. The term *velocity* refers to both the speed of an object and the direction in which it is moving. *Acceleration* is the rate at which velocity changes. The *average* acceleration over some time interval is equal to the change in velocity divided by the elapsed time. Quantities that have both magnitude and direction, such as velocity and acceleration, are called *vectors*. Quantities that have magnitude only, such as time and temperature, are called *scalars*.

Objects that fall vertically near the surface of the earth fall with increasing speed, with the same constant acceleration of gravity: $g = 9.8\ m/s^2$, independent of their weight. This, along with its consequence that the distance fallen is proportional to the square of the elapsed time, was first determined by Galileo. These results are valid only in the absence of air resistance or when air resistance is small enough to be ignored. If the air resistance is large or the objects fall a great distance, they may eventually reach a constant *terminal velocity*.

Two *systems of measurement* are ordinarily used today. The *British system* uses such units as inches, feet, and pounds. In the *metric system*, more widely used throughout the world and in science, the different-sized units are related to each other by simple powers of ten.

QUESTIONS

1. Discuss the advantages and disadvantages of the metric system compared to those of the British system of units.
2. Measure the dimensions (length, width, height, diameter—whatever is appropriate) of at least three objects in a room. State what the objects were and the measurements you obtained. (This exercise is intended to give you a feel for the metric system.)
3. Does a greater speed necessarily imply a greater acceleration? Give examples.
4. Can you assume that a car is not accelerating if its speedometer shows a steady speed of 40 km/hr? Explain.
5. Can the velocity of an object be zero at the same instant its acceleration is not zero? Give an example.
6. Does the odometer of a car measure a scalar or a vector quantity? What about the speedometer?
7. One car travels due east at 40 km/hr and a second car travels north at 40 km/hr. Are their velocities equal? Explain.
8. A car travels around a curve at a speed of 30 km/hr. If it rounds the same curve at 40 km/hr, is its acceleration greater? Why or why not?
9. Will acceleration be the same when a car rounds a *sharp* curve at 50 km/hr as it is when it rounds a *gentle* curve at the same speed? Why or why not?

10. A ball is thrown straight up in the air. What is its velocity when it reaches its highest point? What is its acceleration at this point?
11. Seasickness and airsickness are often referred to as motion sickness. Is it motion per se that makes people sick, or is it a particular aspect of motion? What might be a more appropriate term?
12. Compare the acceleration of a car that speeds up from 60 to 70 km/hr in 5 seconds with a cyclist who accelerates from rest to 10 km/hr also in 5 seconds.

EXERCISES

[Exercises marked with an asterisk (*) are slightly more difficult.]

1. How many millimeters are there in one meter?
2. How many millimeters are there in one inch?
3. It is 8.5 miles from my house to school. How many kilometers is it?
4. A car is 5.2 m long. How long is it in feet?
5. What is the average speed of a sprinter who runs the 100-m dash in 9.8 seconds? What would be the time for 1500 m at this pace?
6. At an average speed of 15 km/hr, how far will a bicyclist travel in $4\frac{1}{2}$ hours?
7. What must be the average speed of a sprinter if the 200-m dash is to be run in 18.0 seconds?
8. How far will an automobile travel in 15 minutes if its average speed is 90 km/hr?
9. The distance from New York to Chicago is 1400 km. How long will this trip take if you average 90 km/hr?
10. The distance from San Francisco to Los Angeles is 610 km. How fast must you drive to make the trip in $6\frac{1}{2}$ hours?
11. What is the speed of the earth around the sun in kilometers per hour and meters per second? Use the average of the values given in Figure 2-11.
12. At an average speed of 11.8 km/hr, how far will a bicyclist travel in 175 min?
13. If you are driving 90 km/hr and you look to the side for 2.0 s, how far do you travel during this inattentive period?
14. 55 mph is how many (a) km/hr, (b) m/s, (c) ft/s?
15. Determine the conversion factor between (a) km/hr and mi/hr, (b) m/s and ft/s, (c) mi/hr and m/s.
16. What is the acceleration of a car that travels on a straight road from rest to 80 km/hr in 8 seconds?

17. What is the acceleration of a bicycle that starts from rest and reaches 6.0 m/s after 5.0 seconds?
18. What is the acceleration of a car that increases its speed from 60 km/hr to 90 km/hr in 6 seconds?
19. A horse accelerates at a rate of 3.0 m/s^2. What will be its speed after 2.5 seconds if it starts from rest? How far will it have traveled?
20. A stone is dropped from a cliff and is seen to hit the ground 4.5 s later. How high was the cliff?
21. How long does it take a stone to fall 100 m starting from rest?
22. A stone is dropped from the roof of a building 80 m high. A second stone is dropped exactly 1.0 s later. How far above the ground is the second stone when the first one hits the ground?
23. For how long does an object dropped from a high tower have to fall before it reaches a speed of 90 km/hr (25 m/s)?

*24. Calculate the carrying capacity (number of cars passing a given point per hour) on a freeway with three lanes (in one direction) using the following assumptions: The average speed is 90 km/hr, the average length of a car is 6.0 m, and the average distance between cars should be 65 m.

*25. An Indian shoots an arrow at a cattle rustler 50 m away. The Indian hears the thief's cry 1.5 sec later. If the speed of sound is 330 m/s, what was the speed of the arrow?

*26. An advertisement claims that a sportscar can accelerate from rest to 100 km/hr in 7.6 s. What is its acceleration in m/s^2?

*27. A sportscar is advertised to be able to stop from a speed of 100 km/hr within 45 m. What is its acceleration in m/s^2? How many *g*'s is this ($g = 9.8$ m/s^2)?

*28. A person who is properly constrained by a shoulder harness has a good chance of surviving a car collision if the deceleration does not exceed 30 *g*'s. Assuming uniform deceleration at this rate, calculate the distance over which the front end of the car must be designed to collapse if a crash occurs at 100 km/hr.

---

* The asterisk (*) denotes a more difficult exercise.

FORCE: NEWTON'S LAWS OF MOTION

CHAPTER

4

Sir Isaac Newton (Figure 4-1) is generally recognized as having been one of the great thinkers of the Western world. Born on Christmas Day, 1642, he entered Trinity College at Cambridge in 1661 and there became interested in the problem of motion. The college was closed in 1665, however, because of an outbreak of the Plague. Newton returned to the farm on which he had been raised and there, between the ages of 23 and 24, he worked out the ideas that would bring him lasting fame. He formulated the laws now known as Newton's three laws of motion, which deal with motion and force in general. Unlike Galileo, who thrived on controversy, Newton was shy and retiring. He was aware of the inevitable criticism that falls on new and creative ideas and sought to avoid any kind of controversy. In fact, some of his most important work was published only at the vigorous urging of his friends. In spite of his fears, Newton's work was accepted comparatively rapidly.

In Chapter 3 we discussed the quantities that are used to describe the motion of an object: speed, velocity, and acceleration. In the present chapter we discuss *dynamics*, the subject of *why* objects move as they do. This involves Newton's three great laws of motion and the concept of *force*.

1. FORCE

What do we mean by "force"? Usually we mean a push or a pull. For example, a person pushing a grocery cart exerts a force on it. When a child pulls a wagon or an automobile pulls a trailer, a force is being exerted. When a hammer hits a nail, the hammer exerts a force on the nail. When the wind blows against a leaf, it is air pushing on the leaf and exerting a force on it. Another example is the *force of gravity* that pulls objects

FIGURE 4-1
Isaac Newton (1642–1727).

toward the ground. We will later encounter other examples of forces, such as those due to electricity and magnetism.

One way you can tell a force is acting is that when a force acts on an object at rest, it may cause that object to start to move. Another way to tell that a force is acting is that a force will change the shape of an object, at least a little. When you exert a force on a balloon, it compresses—it changes shape. The same thing happens if you push on a mattress or a piece of soft wood. Careful measurement shows that even hard materials such as steel are compressed when pushed on. We can tell that a force is acting on an object if the object is set into motion or its shape is distorted.

Force, like velocity, is a vector quantity. Force has both magnitude and direction. The magnitude is the strength of the force; the direction is the direction in which the force is exerted. The direction of a force is as important as its magnitude. For example, if you exert a sufficient force sideways on this book, it moves sideways. But if you exert the force upward, the book moves up (Figure 4-2). When we draw a force vector, we conventionally put the tail of the vector—not the arrowhead—on the object that feels the force.

One way to quantitatively measure the magnitude (or strength) of a force is to make use of a spring scale (Figure 4-3). Normally such a spring scale is used to find the weight of an object; by weight we mean the force

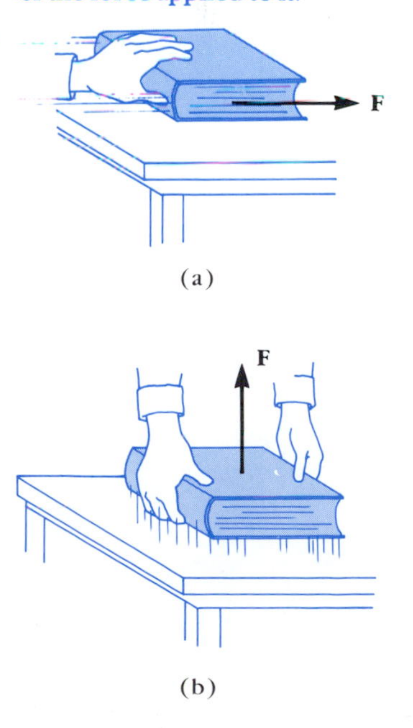

FIGURE 4-2
The book moves in the direction of the force applied to it.

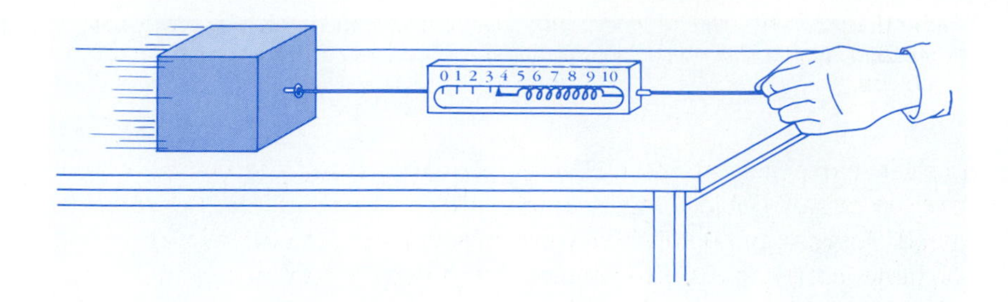

FIGURE 4-3
A spring scale used to measure a force.

of gravity acting on the body. The spring scale, once calibrated†, can be used to measure other kinds of forces as well—such as the pulling force shown in Figure 4-3.

Although the definition of force as a push or pull is adequate for the moment, a more precise definition is discussed later.

2. NEWTON'S FIRST LAW: THE LAW OF INERTIA

What is the exact connection between force and motion? Aristotle believed that a force was required to keep an object moving along a horizontal plane. He would argue that to make a book move across the table, you would have to exert a force on it continuously. To Aristotle, the natural state of a body was at rest, and a force was believed necessary to keep a body in motion. Furthermore, Aristotle argued, the greater the force on the body, the greater its speed.

Some 2000 years later, Galileo, skeptical about these Aristotelian views just as he was of those on falling bodies, came to a radically different conclusion. Galileo maintained that it is just as natural for an object to be in horizontal motion with a constant speed as it is to be at rest. To understand Galileo's idea, let us do a simple experiment involving motion along a horizontal plane (where the effects of gravity won't enter).

EXPERIMENT

Take an object, such as a book, and push it across the floor or top of a table. Clearly you must exert a force to make it move. When you stop pushing, you will probably notice that the object either stops immediately or slows down and comes to rest after a short distance. Now try pushing other objects across

† A spring scale is calibrated by hanging from it a series of identical objects of equal weight or mass, say 1 pound or 1 kilogram. The positions of the pointer when 0, 1, 2, 3, . . . units of weight or mass are hung from it are marked. Although the amount the spring stretches is very nearly proportional to the amount of weight hung from it (as long as it isn't stretched too far), we don't have to make use of this fact in this calibration method. We only assume that the pointer rests at the same position when the same weight (force) acts.

the floor or table. Use objects with rough surfaces and objects with smooth surfaces. Notice the differences in the amount of force needed to move the objects.

In such an experiment, you no doubt found that it took a certain amount of force to push an object with a rough surface along a tabletop at constant speed. To push an equally heavy object with a very smooth surface across the table at the same speed requires less force. Furthermore, if a layer of oil or other lubricant is placed between the surface of the object and the table, then almost no force is required to move the object. In each of these steps, the force required is less and less. The next step is to extrapolate from these data to a situation in which the object does not rub against the table at all—or where there is a perfect lubricant between them—and theorize that once started, the object would move across the table at constant speed with *no* force being applied. A steel ball bearing rolling on a hard horizontal surface approaches this situation closely.

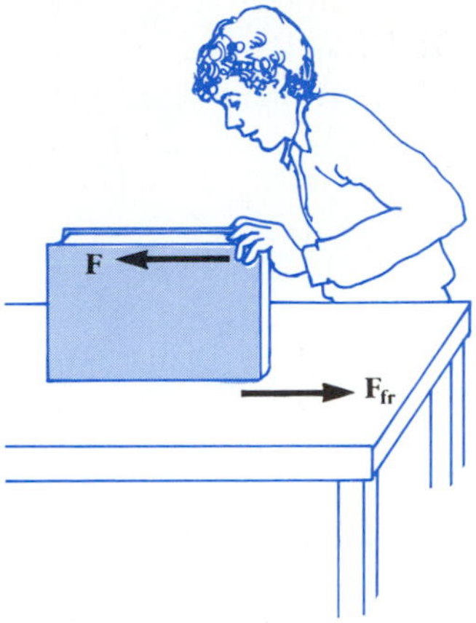

FIGURE 4-4
F is the force applied by the persons; F_{fr} is the force of friction.

It was Galileo's genius to imagine an idealized world—in this case, one where there is no friction—and to see that it could produce a more useful view of the real world. It was this idealization† that led him to his remarkable conclusion that if no force is applied to a moving object, it will continue to move with constant speed in a straight line. An object slows down only if a force is exerted on it. Galileo thus interpreted friction as a force akin to ordinary pushes and pulls.

To push an object across a table at constant speed requires a force from your hand only to balance the force of friction. The pushing force is equal in magnitude to the friction force but they are in opposite directions, so the *net* force on the object is zero (Figure 4-4). This is consistent with Galileo's viewpoint, for the object moves with constant speed when no net force is exerted on it.

The difference between Aristotle's view and Galileo's is not simply one of right or wrong. Aristotle's view was not really wrong, for our everyday experience indicates that moving objects do tend to come to stop if not continually pushed. The real difference lies in the fact that Aristotle's view about the "natural state" of a body was essentially a final statement—no further development was possible. Galileo's analysis, on the other hand, could be extended to explain a great many more phenomena. By making the creative leap of imagining the experimentally unattainable situation of no friction, and by interpreting friction as a force, Galileo was able to reach his conclusion that an object will continue moving with constant velocity if no force acts to change this motion.

† This was a great leap of the imagination on Galileo's part. After all, a perfect lubricant does not exist, so friction between two surfaces cannot be entirely eliminated. For Galileo to imagine what would happen if friction were eliminated required considerable intuition and imagination.

Upon this foundation, Isaac Newton built his great theory of motion. Newton's analysis is summarized in his famous "three laws of motion." In his great work, the *Principia* (published in 1687, it contains nearly all his work on motion), Newton readily acknowledged his debt to Galileo. In fact, Newton's **first law of motion** is very close to Galileo's conclusions:

> **Every body continues in its state of rest or of uniform speed in a straight line unless it is compelled to change that state by forces acting on it.**

The tendency of a body to maintain its state of rest or of uniform motion in a straight line is called **inertia**. As a result, Newton's first law is often called the **law of inertia**.

3. INERTIA, MASS, WEIGHT

Inertia and Mass

The harder it is to change the state of motion of a body, the more inertia the body is said to have. An object such as a refrigerator or a heavy truck has a large amount of inertia, whereas a pencil or a safety pin has very little inertia. An object with a large inertia, such as an automobile, is hard to start moving, even if there were no friction. It is also hard to stop it moving once it is in motion. (Have you ever tried to stop a coasting automobile singlehandedly?) Furthermore, if an object with a large inertia is moving with constant speed in a straight line, it is difficult to shift the object from its straight-line path by pushing sideways on it (Figure 4-5). Thus, *inertia is the property or tendency of a body to resist any change in its state of motion, be it starting, stopping, or changing it from a straight-line path.*

FIGURE 4-5
An object with a large inertia is (a) hard to start moving, (b) hard to stop moving once it is in motion, and (c) hard to deviate from a straight-line path by pushing sideways on it.

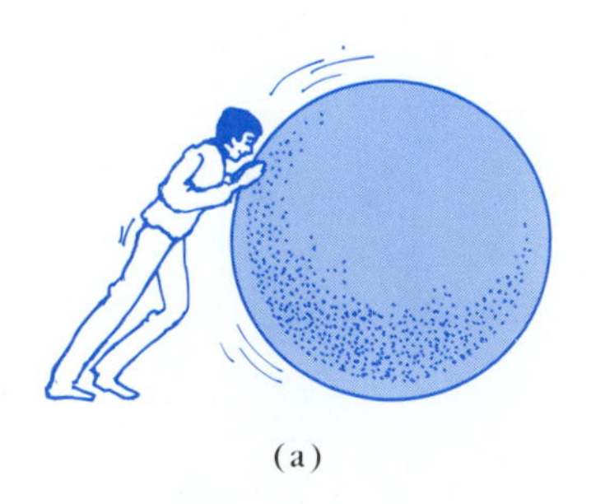

(a)

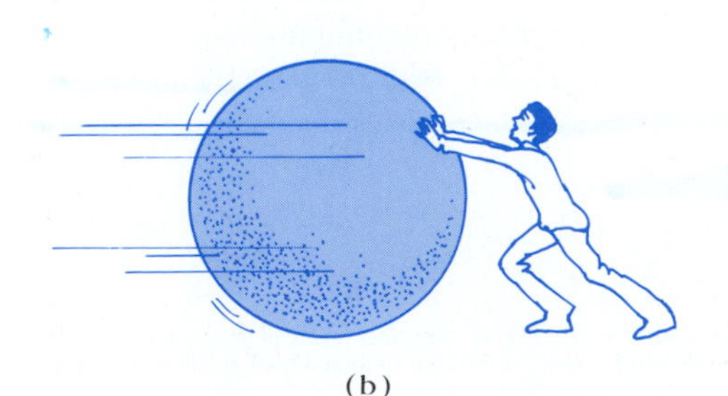

(b)

(c)

The term **mass** is used to refer to the amount of inertia an object has. The more inertia a body has, the greater is its mass. The term mass can also be said to refer to the intuitive concept of "quantity of matter" in an object. Mass is a very important property of any body, and it plays a prominent role in Newton's theory of motion.†

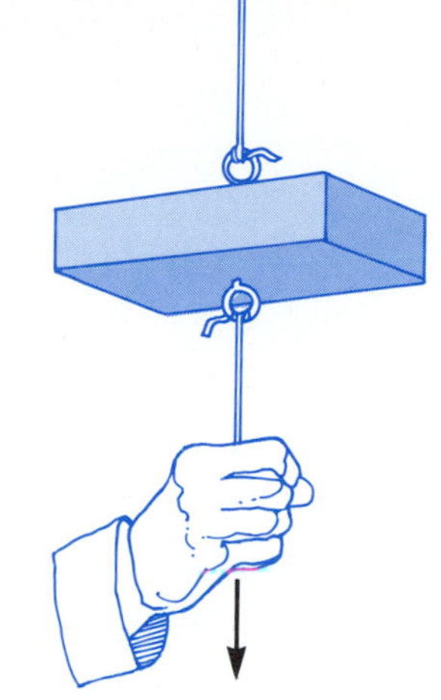

FIGURE 4-6
Checking the law of inertia.

EXPERIMENT

Place a sheet of paper under a book on the edge of a table. Pull the paper (a) gradually (b) all the way out with a sudden jerk. Are your observations consistent with the law of inertia?

OR

Suspend a block of wood or a rock by a piece of thread as shown in Figure 4-6. Connect a second piece of thread to the bottom of the object. If you pull this thread with a slowly increasing force, will the thread break above or below the block? If instead you exert a sharp powerful tug on the thread, where will the thread break? After guessing the outcomes, do both experiments. Explain your observations. Were your predictions fulfilled?

Weight versus Mass

A body with a large mass is hard to start, hard to stop, and resistant to any change in its motion; it is also hard to lift. The property of a body that makes it easy or hard to lift is called its **weight**. To be more specific, weight is defined as *the gravitational force acting on an object*. The greater the pull of gravity on a body, the harder it is to lift. Weight and mass are often confused; yet they are different concepts. They are related in that the weight of an object is directly proportional to its mass. The greater the mass of a body the greater is its weight and the harder it is to lift. But a careful distinction must be made between weight and mass. Weight is a force—the force of gravity acting on a body—whereas mass is the amount of inertia (or quantity of matter) in the object.

To see more clearly the difference between mass and weight, consider what happens when an object is taken to the moon. It will have lost none of its matter, so its mass will be unchanged. In the absence of friction, it will be just as hard to start or stop moving as on the earth, since its inertia (mass) remains the same. But it will be much easier to lift, because the

† Another important property of an object is its **volume**, which is the amount of space it takes up. We tend to associate mass and volume because objects with a large volume usually have a large mass. But this is not always true. A balloon, for example, may have a greater volume than an apple, but the apple has the greater mass.

force of gravity on the moon is only one-sixth as great as on the earth. Thus the object will weigh only one-sixth as much on the moon as it did on earth, even though its mass is the same.

4. NEWTON'S SECOND LAW: $F = ma$

Using Galileo's ideas as a basis, Newton found it relatively easy to state his first law. By building on the foundation laid by his predecessor, Newton was able to proceed more deeply into the nature of motion. Galileo had sought a correct *description* of motion and succeeding in this, was satisfied. But Newton, starting with Galileo's findings behind him, was able to ask a deeper question: what is the *cause* of a change in motion? What is it that makes a body at rest start moving or, once moving, change its speed or direction? To find an answer to this question Newton considered the role of force.

Newton's first law tells us that if no force is exerted on an object, it will continue to stay at rest or continue to move with constant velocity in a straight line. If the velocity is zero at one instant, it stays zero; if it is 14 m/s in a northeasterly direction, it stays that way—as long as no forces act on the object. But what happens when a force is exerted on the body? Newton saw that the velocity will change. A force exerted on an object may make it speed up or, if the force opposes the motion, slow down. In either case, the velocity changes. If the force acts sideways on a moving object, the *direction* of its velocity can be changed. Thus, when a force is exerted on an object, the object experiences an *acceleration*.

Our own experience tells us that forces give rise to acceleration and even suggests that a simple relationship exists between force and acceleration. For example, if you push a child's wagon very gently in a straight line for a few seconds it will accelerate from rest to some velocity, maybe 5 km/hr. If you push twice as hard, the acceleration is twice as great—it only takes half the time to accelerate the wagon from zero to 5 km/hr. Newton observed that the acceleration of a body invariably is in direct proportion to the net applied force. If the net force is doubled, the acceleration is doubled; if the net force is tripled, the acceleration is tripled, and so on. We use the term **net force** in case there is more than one force acting. For example, friction is usually present, so the net force on the wagon would be the force applied by the person minus the friction force.

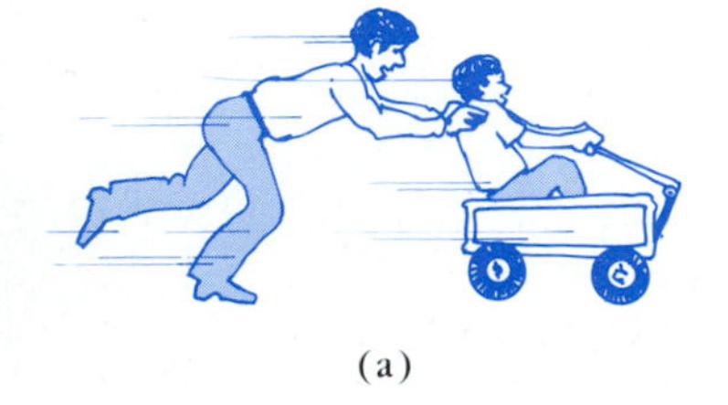

(a)

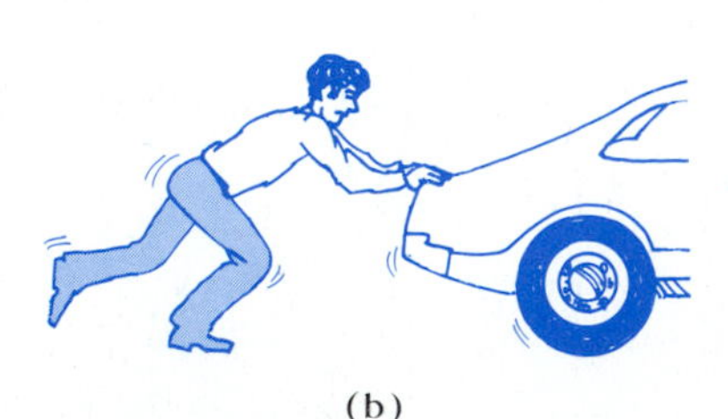

(b)

FIGURE 4-7
The same force applied to a child's wagon gives rise to a greater acceleration than when applied to an automobile.

The acceleration of a body also depends on its mass. Suppose you push a stalled automobile and a child's wagon with the same net force (Figure 4-7). The automobile will undergo a much smaller acceleration than the wagon. The car's velocity will change much more slowly than the wagon's. The more mass a body has, the more slowly it accelerates. Newton argued that the acceleration of a body is always inversely proportional to its mass. With these observations, Newton was able to state his **second law of motion**:

> **The acceleration of a body is directly proportional to the net force acting on it and inversely proportional to its mass. The direction of the acceleration is in the direction of the applied force.**

Newton's second law can be written in symbols as

$$a \propto \frac{F}{m},$$

where a stands for acceleration, F for net force, and m for mass; the symbol $\propto$ means "is proportional to." Notice in this proportionality that if the net force F is doubled, the acceleration will be doubled. But because the mass is in the denominator, if the mass is doubled the acceleration will be halved—this is what is meant by inverse proportion. If the appropriate units are chosen for force and mass, the above proportionality† can be written as an equality: $a = \frac{F}{m}$. By multiplying both sides of this equation by the mass m, the familiar form of Newton's second law is obtained:

$$F = ma.$$

In words, *the mass of an object multiplied by its acceleration is equal to the net applied force*. This is an alternative and exactly equivalent way of stating Newton's second law.

Newton's second law basically relates the description of motion—velocity and acceleration—to the cause of motion, force.

Earlier in this chapter we discussed force in an intuitive way. Now we can give a more precise definition to force: *force is an action capable of accelerating an object*.

Units of Force and Mass

Mass, in the metric system, is measured in grams or kilograms (a kilogram is 1000 grams). It is of course necessary to define exactly what is meant by one gram or one kilogram. We use as a standard a particular block of platinum and iridium alloy kept at the International Bureau of Weights and Measures near Paris. By definition its mass is one kilogram.

Force, in the metric system, is measured in *newtons* (abbreviated N). In accordance with Newton's second law, $F = ma$, one newton is the force required to impart an acceleration of one meter per second per second to a one-kilogram mass. That is, $1\ \text{N} = 1\ \text{kg} \cdot \text{m/s}$.

† Ordinarily when a relationship is found, such as the proportionality $a \propto \frac{F}{m}$, a constant of proportionality is needed when the equality is written; for example, $a = k\frac{F}{m}$, where k is a constant. However, in this situation it is possible to choose the unit of force so that $k = 1$.

In the English system, the common unit of force is the pound. We are most familiar with using the pound to specify weight, but it is used for other forces as well. The unit of mass is called the "slug," but it is rarely used. One slug is defined as that quantity of mass such that a force of one pound will impart to it an acceleration of one foot per second per second. Thus 1 pound = 1 slug·ft/s^2.

It is very important that only one set of units be used in a given calculation. If the force is given in newtons and the mass in grams, then before attempting to solve for the acceleration the mass must be changed to kilograms. For example, if the force is given as 2.0 N and the mass is 500 g, we change the latter to 0.50 kg and the acceleration will then automatically come out in m/s^2 when Newton's second law is used:

$$a = \frac{F}{m} = \frac{2.0\ \text{N}}{0.50\ \text{kg}} = 4.0\ \text{m/s}^2.$$

5. NEWTON'S THIRD LAW: ACTION AND REACTION

Newton's second law of motion describes quantitatively how forces affect motion. But where, we may ask, do forces come from? Observations suggest that a force applied to any object is always applied *by another object*. A horse pulls a wagon, a person pushes a grocery cart, a hammer strikes a nail, a magnet attracts an iron nail. In each of these examples one object exerts the force and the other body feels it; for example, the hammer exerts the force and the nail feels it.

But Newton realized things cannot be so one-sided. True, the hammer exerts a force on the nail. But the nail evidently exerts a force back on the hammer as well, for the hammer's speed is rapidly reduced to zero upon contact. Only a strong force could cause such a rapid deceleration. Thus, said Newton, the two bodies must be treated on an equal basis. The hammer exerts a force on the nail and the nail exerts a force on the hammer. This led Newton to his third law of motion:

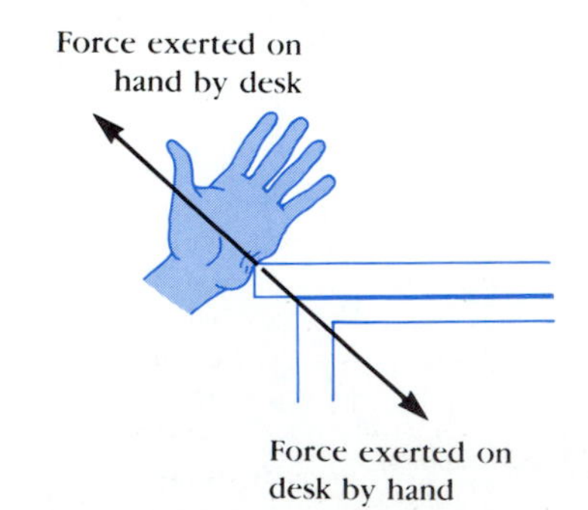

FIGURE 4-8
When your hand pushes on a table, the table pushes back. Notice how indented the hand looks.

> **Whenever one object exerts a force on a second object, the second exerts an equal and opposite force on the first.**

This law is sometimes paraphrased as "to every action there is an equal and opposite reaction." To avoid confusion, it is very important to remember that the "action" force and the "reaction" force are acting on *different* objects.

As evidence for the validity of this third law, consider your hand when you push firmly against a grocery cart or against the edge of a table (Figure 4-8). You know that your hand is pushing against the table—but is the table pushing back against your hand? If you look at your hand, you will see (Figure 4-8) that its shape is somewhat distorted. This is clear evidence that a force is being exerted on it. You can *see* the edge of the table pressing into your hand. You can even *feel* the table exerting a force on your hand: it hurts. The harder you push against the table the harder the table pushes back on your hand.

Consider next the ice skater in Figure 4-9. Since there is very little friction between her skates and the ice, she will move freely if a force is exerted on her. She pushes against the wall, and then *she* starts moving backwards. Clearly there had to be a force exerted on her to make her move. The force she exerts on the wall cannot make *her* move, for that force pushes on the wall and can only affect the wall. Something had to exert a force on her to make her move, and that force could only have been exerted by the wall. The force with which the wall pushes on her is equal and opposite to the force she exerts on the wall. Again Newton's third law.

FIGURE 4-9
The ice skater pushes on a wall and moves backward because the wall exerts a force on her.

When a person throws a package out of a boat, the boat moves in the opposite direction (Figure 4-10). The person exerts a force on the package; the package exerts an equal and opposite force back on the person; and this force propels the person and the boat backward slightly. Rockets work on the same principle. A common misconception is that rockets accelerate because the gases rushing out the back of the engine push against the ground or the atmosphere. Actually, a rocket accelerates because it exerts a strong force on the gases, expelling them. The gases exert an equal and opposite force *on the rocket* (Newton's third law). And it is this force *on the rocket* that propels the rocket forward. Thus, the motion of a rocket is analogous to the motion of the boat in Figure 4-10. A space vehicle can maneuver in empty space simply by firing its rockets in the direction opposite to that in which it wants to accelerate.

FIGURE 4-10
Throwing a package onto shore from a boat that was previously at rest causes the boat to move outward from shore (Newton's third law).

Life would be very different if for some reason Newton's third law did not hold. For example, let us analyze how a person walks. Suppose you are standing still; your velocity is zero. In order to start walking, you push your foot against the ground. You exert a force on the ground. By Newton's third law, the ground exerts an equal and opposite force on you (Figure 4-11). This latter force, on *you*, causes you to move forward.

The fact that it is the force the ground exerts on you that makes you move forward is set forth by Newton's second law. In order to accelerate an object—in this case you—a force must be exerted on that object. The force exerted on you that makes you start moving is the one exerted by the ground.

FIGURE 4-11
We can walk forward because the ground pushes forward on our feet when we push backward against the ground.

Similarly, when an automobile engine causes the wheels to turn, the tires exert a force against the ground. The ground, in turn, exerts an equal and opposite force on the car; and this latter force causes the car to move forward. In these last two examples we also see the importance of friction. Without friction, we would be unable to exert a horizontal force on the ground and so the ground could not exert an equal and opposite force on us to make us move.

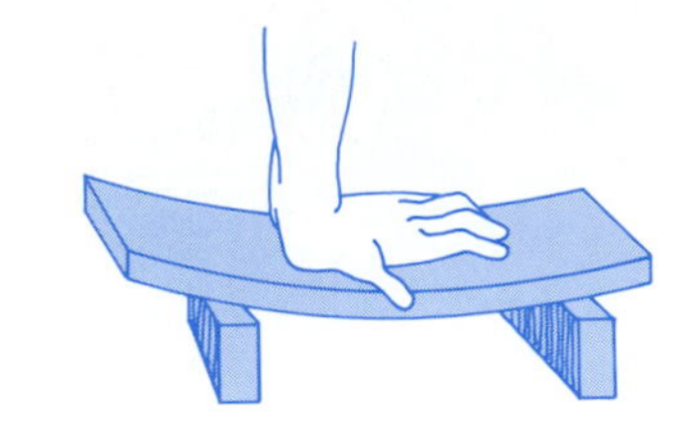
FIGURE 4-12
A hand bends a board.

How Can Inanimate Objects Exert Forces?

It is difficult for most people to accept the fact that an inanimate object such as a wall can exert a force. We tend to associate force with active bodies. Humans, animals, and engines can exert forces; so can an object in motion, such as a flying rock or a falling wall. But can a wall or table *at rest* exert a force? The dented hand in Figure 4-8 is clear evidence that it can. And so is the skater moving away from the wall in Figure 4-9. But how does the wall or table do it?

An object such as a wall can exert a force because it is elastic, although not very elastic. Pushing on it is like pushing on a stiff piece of stretched rubber. No one can deny that a stretched piece of rubber can exert a force. A slingshot, or merely the shooting of a wad of paper with a rubber band, testifies to this. But all materials are elastic, at least to some extent—rubber is only more so. Because of its elasticity, a solid material can exert a force on an object. When you pull a wad of paper against a rubber band, the band exerts a force on the paper and your hand. (If you release the paper, it is accelerated and flies across the room.) Similarly, the force your hand feels when it pushes on a wall is the elastic force due to the "stretched" wall. The stretching is often too small to be noticeable. But you may notice the stretch when you push on a thin board (Figure 4-12), the side of a refrigerator, a soft piece of wood, or when you bend a paper clip slightly.

Action and Reaction Forces Act on Different Bodies

As discussed, Newton's third law is sometimes misapplied. A famous example is that of the horse and cart (Figure 4-13). This particular horse is apparently well read and thinks to itself, "When I exert a forward force on the cart, the cart exerts an equal and opposite force backward. So, how can

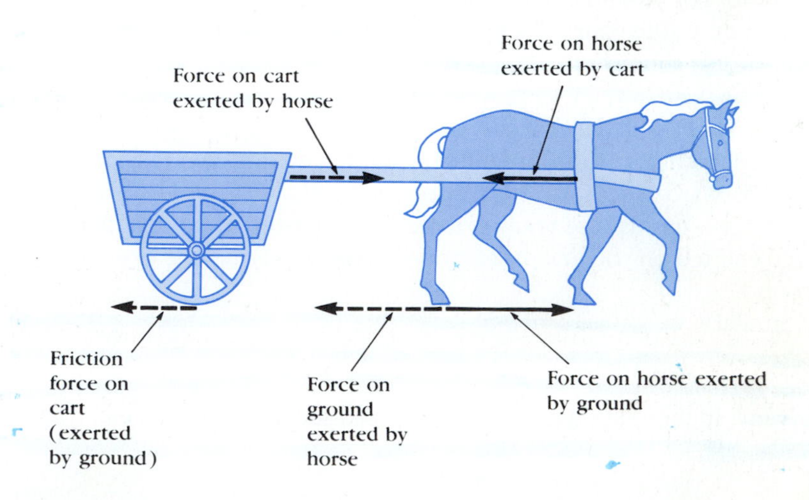

FIGURE 4-13
The horse and cart argument. The forces acting *on the horse* are shown as solid arrows; the forces acting *on the cart* are shown as dashed arrows; only one of the forces acting on the ground is shown (also dashed).

I move? No matter how hard I pull, the backward reaction force always equals my forward force and the net force must be zero. I'll never be able to pull this cart." Although it is true that the action and reaction forces are equal, the horse has forgotten that they are exerted on different bodies! The forward action force is exerted by the horse *on the cart*, whereas the backward reaction force is exerted by the cart *on the horse*. To determine whether the *horse* moves or not we must consider only the forces *on the horse* and then apply $F = ma$, where F is the net force on the horse, a is the acceleration of the horse, and m is the mass of the horse. The forces on the horse that will affect its forward motion are the force of the ground pushing forward on it (the reaction to its pushing backward on the ground) and the cart pulling backward (Figure 4-13). When the ground pushes on the horse harder than the cart pulls back, the horse moves forward. The cart, on the other hand, moves forward because the pulling force exerted by the horse is greater than the frictional force pulling backward.

This example illustrates that whenever you want to determine how an object will move, you must consider *only* the forces that act *on* that one object. The horse was wrong because it was confused about which forces acted on which object. It is necessary, therefore, to specify *on* what object the force is acting and *by* what object the force is being exerted. For example, we should say "the force exerted *by* the horse *on* the cart," or "the force exerted *by* the earth *on* the horse." The prepositions *on* and *by* are very important in this situation.

SUMMARY

Force, which is a vector quantity, is a push or a pull and may distort the shape of objects as well as change their motion. Galileo reasoned that a body moving horizontally would continue moving at constant velocity if no friction or other forces acted on it.

Sir Isaac Newton formulated three basic laws that describe the principles of motion. *Newton's first law of motion* states that if the net force on an object is zero, an object at rest will remain at rest and an object in motion will continue in motion at constant velocity. This tendency to resist a change in motion is called *inertia*. *Mass* is a quantitative measure of the inertia of a body. *Weight*, on the other hand, is defined as the gravitational force (or pull) on an object.

Newton's second law of motion states that the net force acting on a body will cause it to accelerate: the acceleration of a body is directly proportional to the net force acting on it and inversely proportional to the mass ($F = ma$).

Newton's third law of motion states that whenever one body exerts a force on another body, the second body exerts an equal and opposite force on the first.

QUESTIONS

1. Can an object keep moving without being pushed?
2. Is it possible to test directly Galileo's assertion that a body in motion will continue in motion with unchanging speed as long as no external force is acting on it?
3. How does your answer to the above question affect the validity of Galileo's assertion?
4. Why does a child in a wagon fall backward when another child suddenly pulls the wagon forward?
5. (a) Is it proper to say that "bodies at rest tend to stay at rest and bodies in motion tend to stay in motion as long as no force is acting" *because of* inertia? Or should we say that we use the word inertia to *describe* this property of bodies? (b) Comment on the statement "We often think that by giving a name to some concept we then understand it."
6. A rear-end collision between a soft-drink truck and a car occurs. A lawsuit develops over who is at fault. The truck driver claims the car backed into him. The auto driver claims the truck hit him from behind. The only evidence is that quite a number of soda bottles fell forward into the truck driver's seat. From the evidence, can you tell who was at fault?
7. What is the principle behind a magician's ability to pull a tablecloth from beneath china and glassware without breaking them?
8. Explain why you are flung sideways when your car travels around a sharp curve.
9. Whiplash is a common result of an auto accident when the victim's car is struck from the rear. Explain why the victim's head seems to be thrown backward in this kind of accident. Is it really?
10. Why does a person wearing a cast on an arm or leg feel more tired than usual at the end of the day?
11. Analyze the motion of your leg during one stride while walking, using Newton's first and second laws.
12. Why do automobiles use more gasoline for city driving than for highway driving?
13. Why do you exert more force on the pedals of a bicycle when you first start out than when you have reached a constant speed?
14. When a tennis ball is dropped to the floor, it bounces back up. Was a force required to make it come back up? If so, what exerted the force, and how?
15. A horse pulls on a cart with a force of 200 N. With what force does the cart pull back?
16. When you are running and you want to come to a quick stop, you must

decelerate rapidly. Analyze, in the light of Newton's second and third laws, the origin of the required force.

17. When you lift a bag of groceries, you exert an upward force on the bag. Newton's third law says that there is a "reaction" force to your upward force on the bag. What object exerts this reaction force and what object feels it? In which direction does this force act?
18. Explain, using Newton's third law, why when you walk on a log floating in water, the log moves backward as you move forward. Draw a diagram of the log and a person walking on it, and show the motion of each.
19. Why does it hurt your toe when you kick a rock?
20. According to Newton's third law, each team in a tug-of-war pulls with equal force on the other team. What, then, determines which team will win?

EXERCISES

1. What is the net force being exerted on a 1400-kg car accelerating at 3.5 m/s^2?
2. How much force is needed to accelerate a bicycle of mass 70 kg (including rider) at 1.5 m/s^2?
3. What is the acceleration of a 6000-kg rocket if the net force on it is 30,000 N?
4. How much force must a rope be able to withstand if it is not to break when used to accelerate a 1200-kg car at 1.0 m/s^2?
5. A force of 200 N is applied to a full grocery cart whose mass is 40 kg. There is a friction force of 100 N acting in the opposite direction. What is the acceleration of the cart?
6. What net force is needed to bring a 1000-kg car to rest from 25 m/s (90 km/hr) in 5.0 s?
7. What is the acceleration of a freely falling 70 kg skydiver if the force of air resistance is 200 N?
8. A person has a good chance of surviving an automobile accident if the deceleration does not exceed 30 times g, where g is the acceleration of gravity. What would be the force in this case on a 50-kg person?
9. A net force of 26.4 N accelerates an object to 10.8 m/s^2. What is the mass of the object?
10. What is your mass in kilograms, and your weight in newtons?
11. How much force is needed to accelerate a 4.0-gram object at 10,000 g (say in a centrifuge)?

*12. An elevator (mass 4250 kg) is to be designed so that the maximum acceleration is 0.0500 g. What are the maximum and minimum forces the motor should exert on the supporting cable?

*13. A 40-kg child wants to escape from a third story window to avoid punishment. Unfortunately, a makeshift rope of sheets can support a mass of only 30 kg. How can the child use this "rope" to escape?

*14. A 0.10-gram spider is descending on a strand which supports it with a force of 5.6×10^{-4} N. What is the acceleration of the spider? Ignore air resistance.

*15. A baseball pitcher throws a fastball with a speed of 40 m/s. In the throwing motion, the 145-gram ball is accelerated over a distance of about 3.5 m. What is the force on the ball?

*16. A 0.14-kg baseball traveling 30 m/s strikes the catcher's mitt which, in bringing the ball to rest, recoils backwards 10 cm. What is the average force applied by the ball on the glove?

*17. What is the average force exerted by a shotputter on a 7.0 kg shot if the shot is moved through a distance of 2.7 m and is released with a speed of 14 m/s?

USING NEWTON'S LAWS

CHAPTER

5

Newton's three laws of motion were a milestone in the history of science; indeed, they laid the foundation for the study of motion. In this chapter, and also the next, we will explore the use of Newton's laws to understand various types of motion.

1. WEIGHT: THE FORCE OF GRAVITY

Galileo claimed that the acceleration of any freely falling object at the surface of the earth, in the absence of air resistance, is the same regardless of the mass of the object. Applying Newton's second law ($F = ma$) to the gravitational force, for a we can use g, the acceleration due to gravity. Then the force of gravity on an object, which is its **weight**, can be written as the mass of the body times the acceleration due to gravity; in symbols,

$$\text{weight} = mg.$$

Thus the weight of a body, which is defined as the gravitational force acting on it, is directly proportional to its mass.

Since g is $9.8\ \text{m/s}^2$ in the metric system, Newton's second law tells us that a mass of 1 kilogram will be pulled earthward by a force $F = mg = 1\ \text{kg} \times 9.8\ \text{m/s}^2 = 9.8$ newtons. Thus one kilogram weighs 9.8 newtons. Similarly, the weight of a 2-kilogram mass will be 19.6 newtons, and so on. (In the English system, Newton's second law tells us that the gravitational force on a mass of 1 slug will be $F = mg = 1\ \text{slug} \times 32\ \text{ft/s}^2 = 32\ \text{lb}$; a mass of 1 slug thus has a weight of 32 pounds.)[†]

[†] Although we will not have much occasion to use British units, we note that for practical purposes a 1-kg mass has a weight of about 2.2 lb on the surface of the earth. (On the moon. 1 kg weighs about 0.4 lb.)

The force of gravity acts on an object when it is falling. When an object is at rest on the earth, the gravitational force does not disappear, as we know if we weigh it on a spring scale. The same force, mg, continues to act. Why then doesn't the object move? From Newton's second law, the net force on an object at rest is zero; there must be another force on the object to balance the gravitational force. For an object resting on a table, the table exerts this upward force—Figure 5-1(a). The table is compressed slightly beneath the object, and due to the table's elasticity it pushes up on the object as shown. (The force exerted by the table is often called a *contact force*, since it occurs when two objects are in contact; the force of your hand pushing against a table is also a contact force.) We say the object is in *equilibrium* ("equal forces" in Latin).

But note carefully in Figure 5-1(a) that the two forces shown are *not* the action-reaction pairs of Newton's third law. Both forces are acting on the same body, and they do not have to be equal. (If they were *not* equal, the object would begin moving, according to Newton's second law $F = ma$; only when these two forces are equal, so the net force is zero, can the object remain at rest.) Figure 5-1(b) shows one of the reaction forces[†], the force the statue exerts on the table. By Newton's third law, it *must* be equal and opposite to the force the table exerts on the statue.

[†] The reaction to the gravitational force on the statue is a force exerted on the earth by the statue, which is also a gravitational force, and can be considered to act at the earth's center, as we shall discuss in Chapter 6.

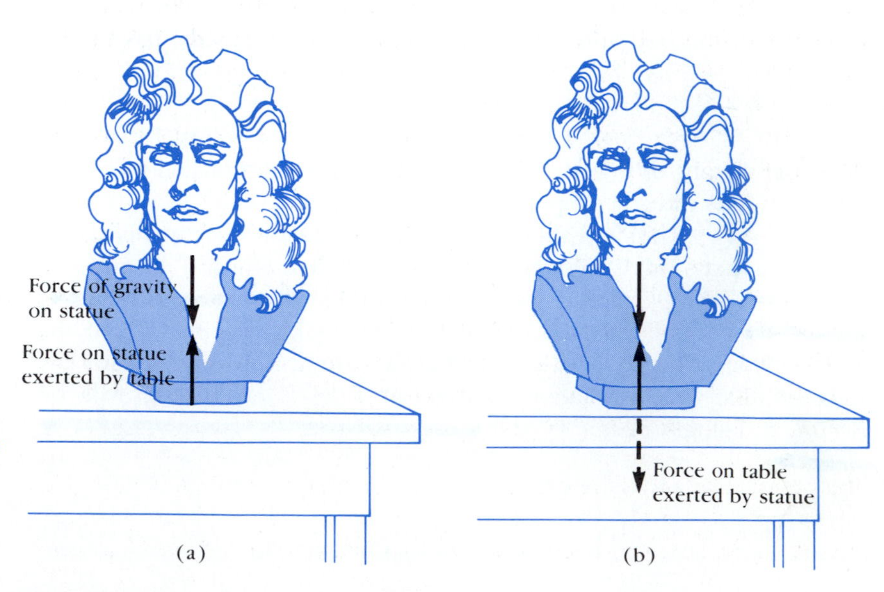

FIGURE 5-1
The net force on an object at rest is zero according to Newton's second law. Therefore, the downward force of gravity on the statue must be balanced by the upward force exerted by the table (a). The reaction force to the latter—which according to Newton's third law must be equal in magnitude—is shown in (b).

2. NET FORCE; ADDITION OF VECTORS

Net Force

Newton's second law states that the acceleration of a body is directly proportional to the net force acting on the body and inversely proportional to the mass of the body. What is meant by the term **net force**? If only one force is acting on the object, that force is the net force. If two or more forces are acting, *the net force is the sum of all those forces, taking into account both the magnitude and the direction of each*. Because force, like any vector, has both magnitude and direction, adding forces is not as simple as adding ordinary numbers. We must now investigate how to add vectors. We are interested in force vectors at the moment, but the rules we will learn also apply to other types of vectors, such as velocity.

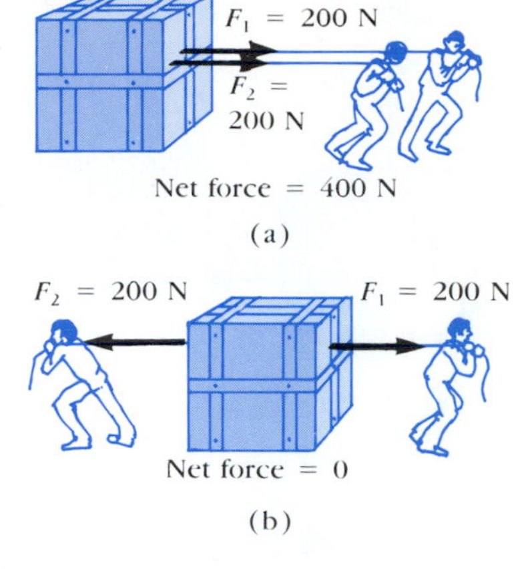

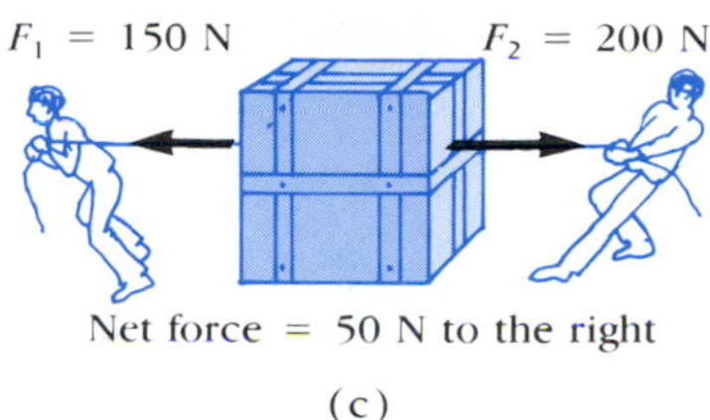

F_1 = 150 N F_2 = 200 N

Net force = 50 N to the right

(c)

FIGURE 5-2
Two men pulling on a heavy load.

Addition of Vectors

As mentioned in Chapter 3, we can draw an arrow to represent a vector such as a force. The arrow is drawn so that (1) the direction of the arrow is the direction of the vector and (2) the length of the arrow is proportional to the magnitude of the vector. For example, a force of 10 newtons might be represented by an arrow 2 cm long; then a 30-N force would be represented by an arrow 6 cm long.

To see how two vectors can be added, we first consider an object being pulled by two workers—Figure 5-2(a). They each exert a force of 200 N in the same direction; the total force on the object in this case is 400 N. That is, the two 200-N forces have the same effect on the object as a single 400-N force. Thus the net force, which is the sum of the two individual forces, is 400 N. If the two men pull in opposite directions, each with a force of 200 N—Figure 5-2(b)—the object won't move at all; the net force is zero. In Figure 5-2(c), the two men are pulling in opposite directions, but the one on the left is pulling with a force of only 150 N. The net force in this situation is 50 N to the right.

These examples suggest that two vectors can be added by the following set of operations, which are illustrated in Figure 5-3.

1. Draw the two force vectors to scale; call one of them $\mathbf{F_1}$, the other $\mathbf{F_2}$. (It doesn't matter which is called $\mathbf{F_1}$ and which $\mathbf{F_2}$.)
2. Carefully move $\mathbf{F_2}$ so that its tail is at the tip of $\mathbf{F_1}$. Be sure you do not change the length or direction of $\mathbf{F_2}$ or $\mathbf{F_1}$.
3. The sum of the two force vectors is the arrow drawn from the tail of $\mathbf{F_1}$ to the tip of $\mathbf{F_2}$. The sum vector is labeled $\mathbf{F_1} + \mathbf{F_2}$. (We use boldface type to remind us that we are adding vectors.)

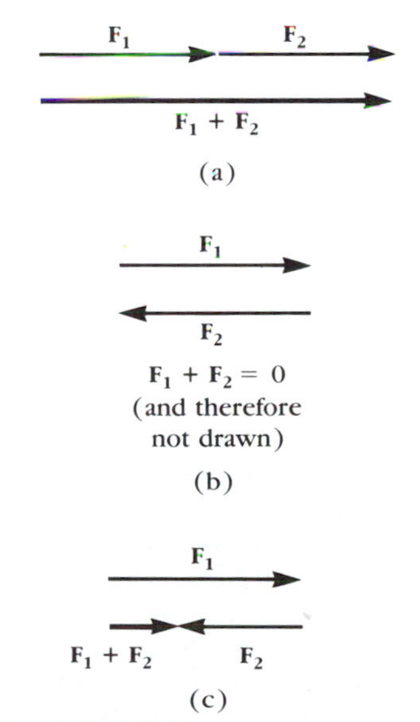

FIGURE 5-3
Tail-to-tip method of adding vectors.

This set of rules gives the expected result for each example in Figure 5-2. These rules also apply when the two forces do not act along the same line. For example, suppose two 200 N forces act at right angles (90°) to one

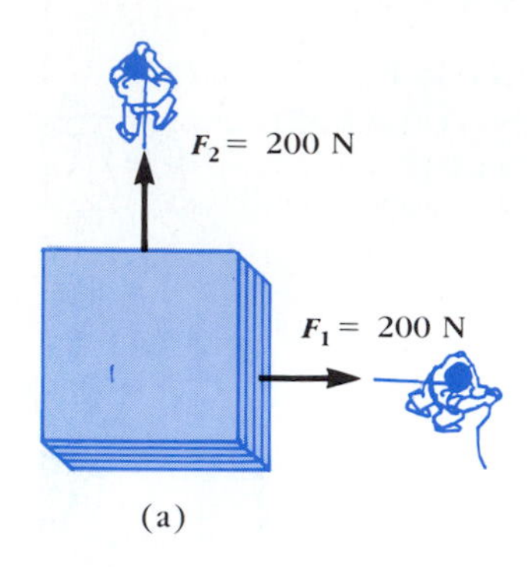

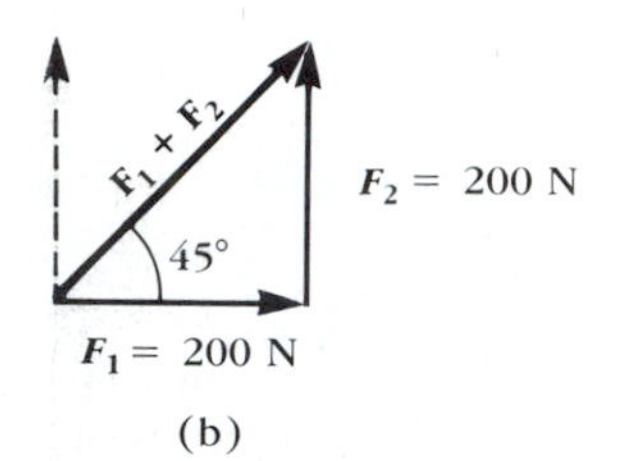

FIGURE 5-4
Adding two nonparallel forces together to get the net, or total, force using the tail-to-tip method.

another as in Figure 5-4(a). In this case, it is clear that the object will move along a line at 45°. Therefore the sum of the two forces must be along the line at 45°. When our rules for adding two vectors are used, Figure 5-4(b) results. The sum of the two forces $\mathbf{F_1} + \mathbf{F_2}$ is indeed at 45° according to our rules. The magnitude of the total force, $\mathbf{F_1} + \mathbf{F_2}$, can be measured with a ruler directly on Figure 5-4(b). If the scale is one centimeter = 50 N, $\mathbf{F_1}$ and $\mathbf{F_2}$ are each 4.0 cm long. The sum $\mathbf{F_1} + \mathbf{F_2}$ measures about 5.6 cm, which corresponds to a total force of about 280 N.† This result can be checked experimentally. Indeed, experiment shows that a single force of 280 N at an angle of 45° has exactly the same effect as two 200 N forces at right angles to one another. In fact, experiment confirms the validity of our vector addition rules in general.

The fact that two 200-N forces should add up to a total force of 280 N may at first seem strange. But remember that forces are vectors and their direction is important. Two men who pull with a force of 200 N in the same direction do exert a total force of 400 N, as in Figure 5-2(a). If they pull at an angle to one another as in Figure 5-4, some of their cooperative effect is lost and we would expect the total, or net force, to be less than 400 N. If the angle between the two forces is 180°—that is, if they pull in opposite directions—the net force is zero, as in Figure 5-2(c). Thus, when the two forces are at an angle of 90° to one another, we expect the force to be less than 400 N but more than zero. Only when the two forces are in the same direction will they add as in ordinary arithmetic.

Figure 5-5(a) shows a more general case of two unequal vectors acting at a particular angle to one another, in this case 60°. These two forces can be added in the usual way to obtain the net force: $\mathbf{F_1} + \mathbf{F_2}$ as in Figure 5-5(b). When adding two vectors, it doesn't really matter which one is moved, $\mathbf{F_1}$ or $\mathbf{F_2}$, as long as the tail of the moved vector is placed at the tip of the other. Figure 5-5(c) shows that the same result is obtained if the vector $\mathbf{F_1}$ is moved. Use a ruler to measure the length of $\mathbf{F_1} + \mathbf{F_2}$; you should find its magnitude is about 140 newtons.

This technique for adding vectors is called the **tail-to-tip** method. Another way to determine the sum of two vectors is the **parallelogram**

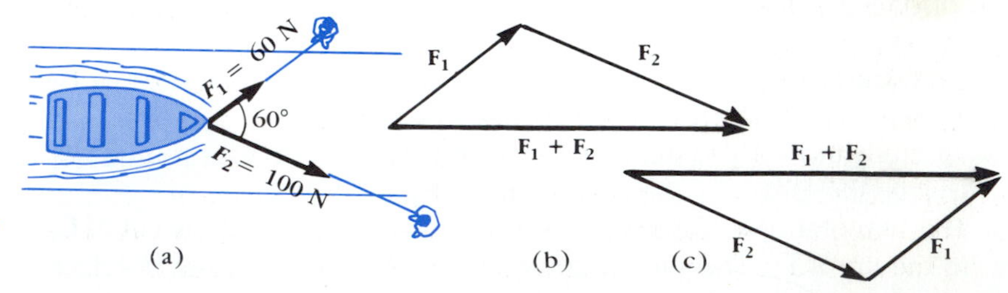

FIGURE 5-5
(a) Two unequal forces act on a boat. (b) and (c) Adding the two forces together to get the net force using the tail-to-tip method. Notice that the same result is obtained no matter which force is drawn first.

† The same result can be obtained using arithmetic and Pythagoras' theorem that the sum of the squares of the sides of a right triangle equals the square of the hypotenuse: $c^2 = a^2 + b^2$. Since $\mathbf{F_1}$, $\mathbf{F_2}$, and $\mathbf{F_1} + \mathbf{F_2}$ in Figure 5-4(b) make up a right triangle with $\mathbf{F_1} + \mathbf{F_2}$ being the hypotenuse, the length or magnitude of $\mathbf{F_1} + \mathbf{F_2}$ equals the square root of $200^2 + 200^2 = 40{,}000 + 40{,}000 = 80{,}000$ and the square root of 80,000 is 283, which rounds off to 280 N.

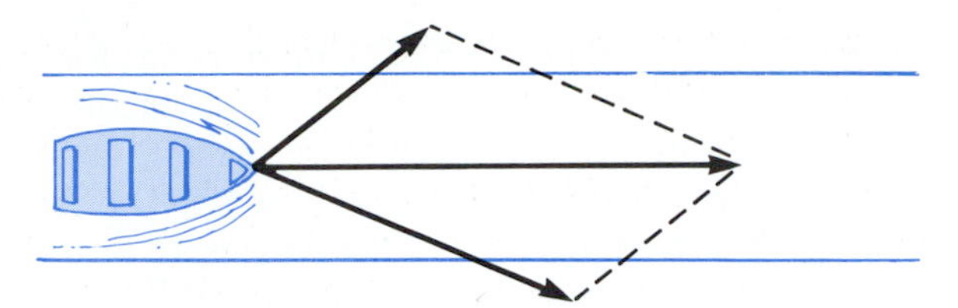

FIGURE 5-6
The same net force as in Figure 5-5 is obtained here using the "parallelogram" method of adding vectors.

method. A parallelogram is a four-sided plane figure in which each pair of opposite sides is parallel and equal; for example, ▱ . In the parallelogram method, the two vectors are left just as they are (Figure 5-6). Then a parallelogram is drawn with the two vectors as two of its sides. The sum vector is then the diagonal as drawn from the common origin to the opposite corner of the parallelogram, as shown. Because the opposite sides of a parallelogram are of equal length and parallel (i.e., at the same angle), the construction of Figure 5-6 is equivalent to the tail-to-tip method of Figure 5-5 and the same net force is obtained.

When three or more vectors must be added, the tail-to-tip method is the easiest to use. Figure 5-7(a) shows three forces acting on a skier. Figure 5-7(b) shows how to obtain the sum, which is the net force on the skier.

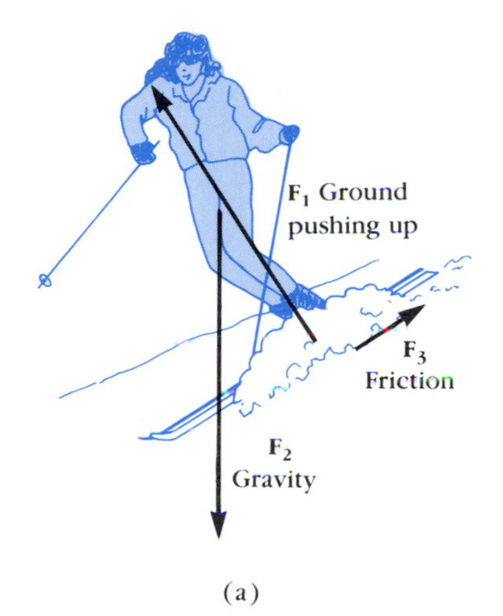

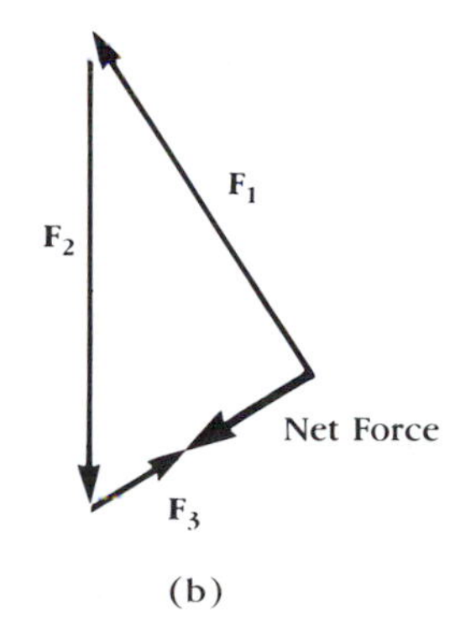

FIGURE 5-7
(a) Forces acting on a skier.
(b) The net force on the skier is found using the tail-to-tip method of adding vectors.

Resolution of Vectors

We have seen that two (or more) vectors can be added to give a single vector, the sum, whose effect is exactly the same as the original two vectors. It is often useful to do the opposite—that is, to consider any single vector as the sum of two other vectors. These latter vectors are called the **components** of the original vector. Most often the components of a vector are chosen so that they are perpendicular to each other, with one component horizontal and the other vertical. For example, consider Figure 5-8(a) where a man pushes his lawn mower with a force **F** directed along the handle. We want to find the components of **F** in the horizontal and vertical directions. The process of determining these components is called the **resolution** of the vector, or *resolving the vector into its components*. Keep in mind that the two components must add up to the original vector. Here is how to resolve a vector. Draw a rectangle around the given vector, with that vector as a diagonal—Figure 5-8(b). The components are then the two sides of the rectangle, with the origin at the same point as the original vector. It should be clear that the horizontal and vertical components, labeled $\mathbf{F_h}$ and $\mathbf{F_v}$ respectively, add up to give the original vector **F** by the parallelogram method of adding vectors.

This example also illustrates the usefulness of resolving vectors. The force exerted on the handle of the lawn mower had horizontal and vertical components, as indicated by $\mathbf{F_h}$ and $\mathbf{F_v}$. That is, the man pushes on the lawn mower in the horizontal and vertical directions at the same time. However, only the horizontal component of the force, $\mathbf{F_h}$, propels the lawn mower forward. The vertical component, $\mathbf{F_v}$, merely pushes the lawn mower against the ground and in a sense is wasted. If the man exerted the same force at a smaller angle, Figure 5-8(c), more of his force would go into horizontal motion; thus he could exert a smaller total force **F** along the handle and still obtain the same $\mathbf{F_h}$ and therefore the same horizontal motion. However, there are practical disadvantages to the smaller angle. His back would probably get tired from bending over; furthermore, the vertical component of force ensures good contact of the lawn mower with the ground and thus a more even cutting job. In this case a proper balance between the horizontal and vertical components of the force is advantageous.

Newton's second law ($F = ma$), as this example illustrates, applies to each component separately. For example, when dealing with objects moving horizontally on the earth, we can apply $F = ma$ to the horizontal component. If there is no vertical motion, so the acceleration $a = 0$, then the net force in the vertical direction is zero: the downward force of gravity in this case is just balanced by an upward force exerted by the ground or a table, as discussed previously.

(a)

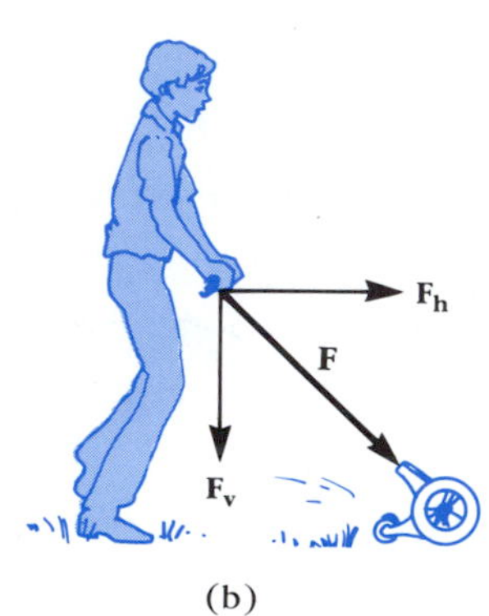

(b)

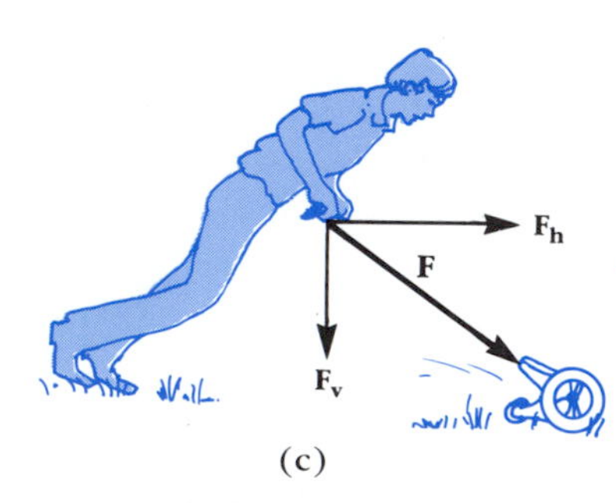

(c)

FIGURE 5-8

(a) A person pushes a lawn mower with a force F directed along the handle. (b) The force is resolved into vertical and horizontal components. (c) The horizontal component can be increased by reducing the angle without increasing the total force on the mower.

3. PROJECTILE MOTION

An object moves in a straight line whenever the net force is either zero or is along the direction of motion. But when the net force on an object acts *sideways* to the direction of motion, the object begins moving in a curved path. If, for example, gravity were not acting when you threw a baseball, the baseball would continue outward in a straight line indefinitely. But of course the force of gravity does act on the baseball and brings it back to earth. A thrown baseball, a kicked football, a speeding bullet are all examples of *projectiles* acted on by the vertical force of gravity. We will consider now the motion of projectiles whose path does not take them far from the earth's surface.

Consider first a ball rolling off the edge of a table (Figure 5-9). It was Galileo who first realized that this kind of motion is most simply dealt with by considering the horizontal and vertical motions separately and then combining them to get the complete motion. The velocity vector always points in the direction the object is moving at any given point and hence it is always tangent to the path of the object. Following Galileo, we consider the vertical and horizontal components of the velocity, v_v and v_h, separately at each point. The velocity vector and its components are shown in Figure 5-9 at several points along the path; these points are separated by equal time intervals. The force of gravity is acting vertically downward at every point and therefore gives rise to a vertically downward acceleration. Because acceleration means a changing velocity, the vertical component of velocity is continually increasing. Thus the arrows representing v_v in Figure 5-9 are longer for each successive point on the path. On the other hand, if we ignore air resistance, there are no forces in the horizontal direction and v_h does not change at all; it has the same magnitude at each point shown in Figure 5-9.

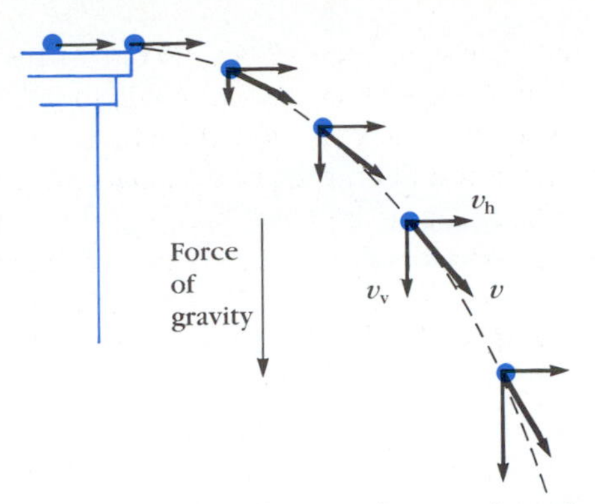

FIGURE 5-9
The velocity vector and its horizontal and vertical components are shown at several equally spaced time intervals.

One result of this analysis is the prediction that if a second ball is dropped straight down from the edge of the table just as the rolling ball passes the edge, the two balls will reach the ground at the same time (Figure 5-10). Galileo first predicted this because he realized that the

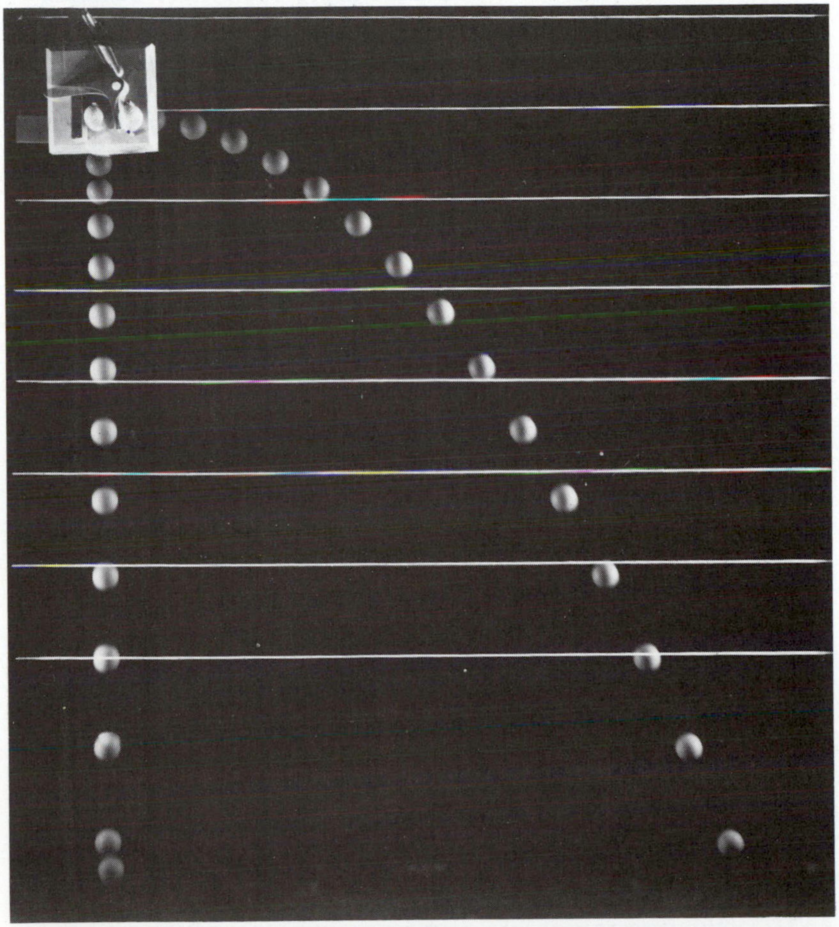

FIGURE 5-10
A ball is projected horizontally at the same time that a second ball is dropped straight down. Both reach the ground at the same time.

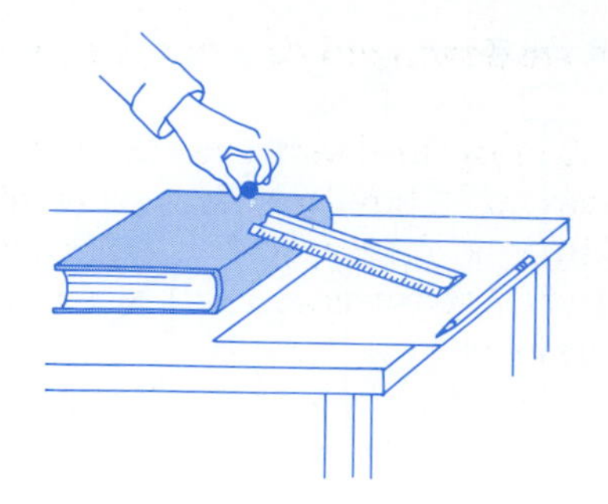

FIGURE 5-11
Experiment on projectile motion.

vertical motion of the ball in Figure 5-9 is exactly the same as that of a vertically falling object. You can check this yourself:

EXPERIMENT-PROJECT

Place a small pen or pencil on the edge of a table or top of a bookshelf (the higher the better), so that it is almost falling off (Figure 5-11). Now roll a marble (the bigger the better) swiftly so that it just barely bumps the end of the pen—which then falls vertically. Listen for the sounds they make when they hit the ground. To assure a simultaneous start, you may find it helpful to roll the marble down a ruler to guide it as shown and to balance the pen on the edge of a sheet of paper.

OR

Stand on top of your house or another building and throw a tennis ball out horizontally at the same time you drop one vertically.

When an object is projected upward at an angle, such as when a football is kicked, the analysis is essentially the same. Figure 5-12 shows this situation. The horizontal component of velocity remains constant as before. The vertical component of velocity is initially quite large. But because the force of gravity acts downward, this component decreases in time until the football reaches a maximum height, at which point the vertical component of velocity is zero. The vertical component then

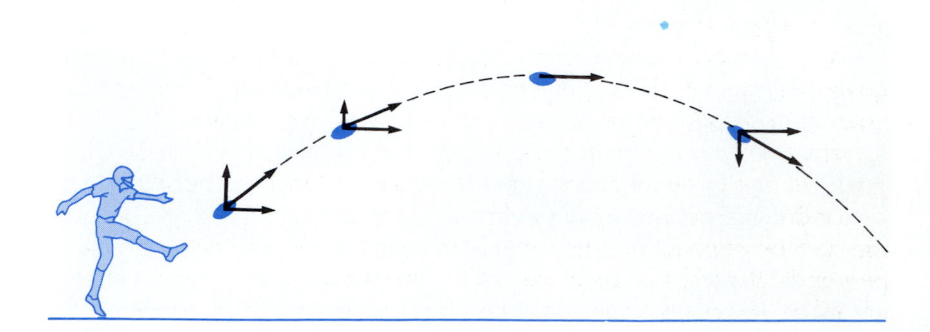

FIGURE 5-12
Path taken by a kicked football, showing the velocity and its components at several points along the path.

begins to increase in the downward direction, becoming larger and larger as time goes on.

Air resistance is often so small that it can be ignored, as we have done above. It becomes important, however, if the projectile is not too massive compared to its volume, or if the projectile is to go a long distance—for example, an artillery shell. Air resistance acts as a retarding force and thus decreases both the horizontal and vertical components of velocity. The path of a projectile is modified by air resistance in the manner shown in Figure 5-13.

FIGURE 5-13
Air resistance reduces the distance a projectile will travel.

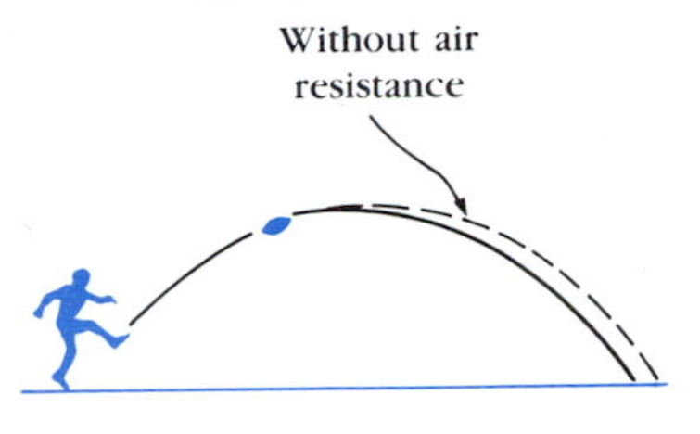

A Calculation (optional)

To see how a numerical exercise is done, consider the following. Suppose a baseball is thrown horizontally with a speed of 25 m/s from the top of a cliff 44 m high. How far from the base of the cliff will the ball land? This may at first seem like a difficult problem, but it is actually quite easy because we can analyze the horizontal and vertical motions separately. In the vertical direction, the object is accelerating at $g = 9.8\ \text{m/s}^2$. (The downward direction is here taken to be positive.) Since the vertical motion of our projectile is just the same as a falling object, we can use $d = \frac{1}{2}gt^2$ with $g = 9.8\ \text{m/s}^2$ (see Section 5 of Chapter 3). We solve for t and find

$$t = \sqrt{\frac{2d}{g}} = \sqrt{\frac{(2)(44\text{m})}{9.8\ \text{m/s}^2}} = \sqrt{9.0\ \text{s}^2} = 3.0\ \text{s}.$$

Now that we know the time the ball is in flight, we can calculate how far it goes horizontally. Its horizontal velocity is 25 m/s initially, and it stays at this speed until it hits the ground 3.0 s later. Thus the horizontal distance traveled is $d = vt = (25\ \text{m/s})(3.0\ \text{s}) = 75\ \text{m}$. So the ball hits the ground 75 m from the base of the cliff (air resistance ignored).

SUMMARY

The force of gravity acting on a body (its *weight*) near the surface of the earth can be written as the product of its mass times the acceleration of gravity: weight = mg.

The *net force* is the *vector sum* of all the forces acting on an object. Vector quantities can be added graphically by such methods as the parallelogram method and the tail-to-tip method. In the latter method, the tail of the arrow representing each successive vector is placed at the tip of the previous one (without altering its orientation or length). The sum is the vector drawn from the tail of the first vector to the tip of last one. A vector can also be *resolved* into its *components*, which are two vectors (usually perpendicular to each other) that add to give the original vector. When the net force on an object is zero, the object is said to be in *equilibrium*.

Projectile motion—that of an object moving through the air—can be analyzed as two separate motions in the horizontal and vertical directions. If air resistance can be ignored, the horizontal motion is at constant velocity whereas the vertical motion is uniformly accelerated and is the same as for a body falling vertically under the action of gravity.

QUESTIONS

1. How much would you weigh on the moon? Compare to your weight on earth.
2. Determine your mass in kilograms. What would be your mass on the moon?
3. When you are standing on the ground, how large is the upward force exerted on you by the ground? Why doesn't this force make you fly up in the air?
4. A rock weighs 50 N. What is the net force on it when you hold it? What is the force on it just after it is dropped? What is the force on it just before it hits the floor?
5. If we know the magnitude and direction of the net force acting on a body of given mass, what does Newton's second law allow us to calculate?
6. Can two vectors, neither of which is zero, add up to a vector that is zero? Show how on a diagram.
7. Can *two* vectors of unequal magnitude add up to a vector of zero magnitude? Can *three* unequal vectors?
8. Can two vectors of equal length add up to a sum vector whose length is the same as each of the original vectors? Show on a diagram.
9. Can the magnitude of a vector ever (a) equal or (b) be less than one of its components?
10. At what point in its path is the speed of a projectile the least?
11. You are hiding in a tree from a sniper who is quite far away. You see the sniper aim directly at you and fire. Should you remain where you are or drop from the tree as the rifle is fired?
12. When a football is kicked, the vertical component of its velocity is greatest at what point?
13. What factors are important for an athlete doing the broad jump? What about the high jump?
14. An object that is thrown vertically upward will return to its original position with the same speed as it had initially, if air resistance is negligible. If air resistance is appreciable, will this result be altered, and if so, how?

EXERCISES

1. What is the weight of a 70-kg astronaut (a) on earth, (b) on the moon where $g = 1.7\ \mathrm{m/s^2}$, (c) on Venus where $g = 8.7\ \mathrm{m/s^2}$, (d) in outer space?
2. A force of 400 N is exerted on an object in one direction and a force of 300 N is exerted on the same object in a perpendicular direction. What is the net force on the object? Use a diagram to show the direction of the net force.
3. You push on a small car with a force of 600 N at a 45° angle to the horizontal. How much force actually goes into moving the car along a level road?
4. A child is pulling a sled in the snow. Draw a diagram showing the force the child exerts on the sled, and the horizontal and vertical components of this force.
5. A 200-N force acts in a southeasterly direction. A second 200-N force must be exerted in what direction so that the net force points due south? Draw a diagram.
6. Figure 5-14 shows the forces acting on an automobile climbing a hill. F_3 is the force of friction. (a) What are the other forces due to? (b) Use the tail-to-tip method of adding vectors to find the net force on the car.
7. A man pushes on a lawn mower at an angle of 45° to the ground with a force of 50 N. What is the component of this force in the horizontal direction?
8. A 60-kg sprinter exerts a force of 800 N at an angle of 45° on a starting block. (a) What is the horizontal component of this force? (b) What is the acceleration of the sprinter?
9. A diver running 3.6 m/s dives out horizontally from the edge of a cliff and reaches the water below 2.0 s later. How high was the cliff and how far from its base did the diver hit the water?
10. A tiger leaps horizontally from a 15-m high rock with a speed of 7.0 m/s. How far from the base of the rock will it land?
11. The pilot of an airplane traveling at about 200 km/hr wishes to drop supplies to flood victims stranded on a small island. Draw a diagram showing the airplane and the island; show the approximate point where the plane must release the package of supplies and the subsequent path of the package through the air.

*12. (a) A skier is accelerating down a 30.0° hill at $3.60\ \mathrm{m/s^2}$. What is the vertical component of her acceleration? (b) Assuming she starts from rest and accelerates uniformly, how long will it take her to reach the bottom of the hill if the elevation change is 150 m?

*13. A ball thrown horizontally at 18 m/s from the roof of a building lands 24 m from the base of the building. How high is the building?

FIGURE 5-14

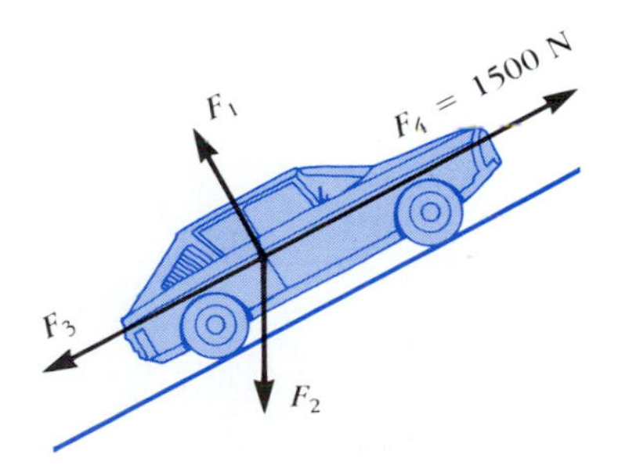

*14. A hunter aims directly at a target (on the same level) 220 m away. If the bullet leaves the gun at a speed of 550 m/s, by how much will it miss the target?

*15. A person in an airplane traveling 50 m/s wants to drop supplies to flood victims on an island 200 m below. How many seconds before the plane is directly overhead should the supplies be dropped?

*16. A child wishes to determine the speed a slingshot imparts to a rock. How can this be done using only a meter stick?

CIRCULAR MOTION; GRAVITATION AND NEWTON'S SYNTHESIS

CHAPTER

6

An object moves in a straight line if the net force on it acts in the direction of motion, or is zero. If the net force acts at an angle to the direction of motion at any moment, then the object moves in a curve. We have already discussed one example of this in projectile motion (Chapter 5). Another important situation is that of an object moving in a circle, such as a ball at the end of a string revolving around one's head or the nearly circular motion of the moon about the earth. These are examples of circular motion.

In this chapter we study the circular motion of an object, and how Newton's laws of motion apply to it. We will also discuss how Newton conceived of another great law by applying the concepts of circular motion to the motion of the moon and the planets. This new law was his *law of universal gravitation*, which was the capstone of Newton's analysis of the physical world. Indeed, Newtonian mechanics—with its three laws of motion and the law of universal gravitation—was accepted for centuries (until the early 1900s) as the way the universe works.

1. CIRCULAR MOTION

The simplest circular motion occurs when an object moves in a circle with constant speed. (This is called *uniform circular motion*.) Even though the velocity is not changing in magnitude it is continuously changing in direction—Figure 6-1(a). Therefore, the object is accelerating and there must be a force acting on it. If there were no force acting on the object, it would go in a straight line. But for the object to move in a circle, a force must pull it out of a straight-line path. This force is directed toward the center of the circle—Figure 6-1(b). At each point along the

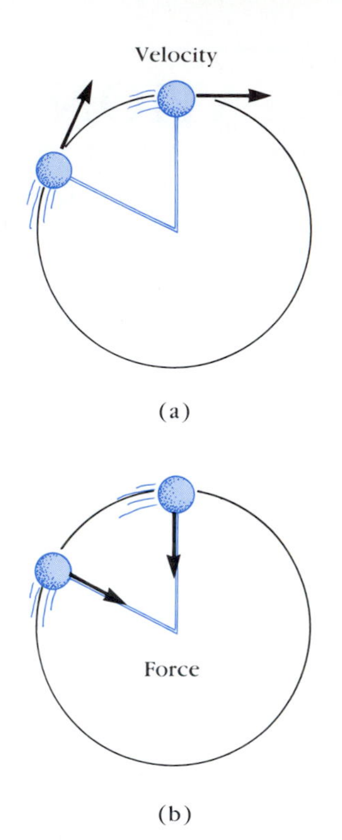

FIGURE 6-1
A ball on the end of a string is swung in a circle. In (a) the velocity vectors are shown at two different places; in (b) the force vectors that keep the ball moving in the circle are shown.

path, this force (always pulling toward the circle's center) pulls the object away from its natural tendency to go in a straight line.

The magnitude of the force required to keep an object moving in a circle depends on how fast the velocity vector is changing; that is, on how large the acceleration is. By using geometry, it is possible to show that the acceleration of an object moving in a circle is given by the formula $a = \frac{v^2}{r}$, where v is the speed of the object and r is the radius of its circular path. Although we will not prove this formula, we can indicate that it does make sense. First of all, the greater the speed, v, the more rapidly the velocity vector is changing direction and therefore the greater the acceleration. (Why the acceleration is proportional to the speed *squared* is not quite so obvious.) Similarly, the larger the radius r for a given speed v, the less rapidly the velocity vector changes and the less the acceleration; this accounts for the inverse proportionality to r. To cause an object moving in a circle to accelerate requires a force $F = ma = \frac{mv^2}{r}$. Because the acceleration points toward the center of the circle, it is sometimes called a "center seeking" or **centripetal** acceleration. This is very different from the projectile situation (Chapter 5) in which the acceleration, and the force that causes it, always points in the same downward direction.

There is a common misconception that some kind of mysterious *"centrifugal force"* ("center-fleeing" force) is pulling outward on the object rotating in a circle. To see why this is a misconception, let us consider a simple example. Figure 6-2 shows a person swinging a ball on the end of a string in a horizontal circle. If you have ever done this, you know that you feel a force pulling outward on your hand. The misconception arises when this pull is interpreted as an outward "centrifugal" force pulling on the ball, that is transmitted along the string to the hand. But this is not what is happening at all. To keep the ball moving in a circle, the person continually pulls *inwardly* on the ball. Because of Newton's third law, the ball exerts an equal and opposite force on the hand. This is the force your

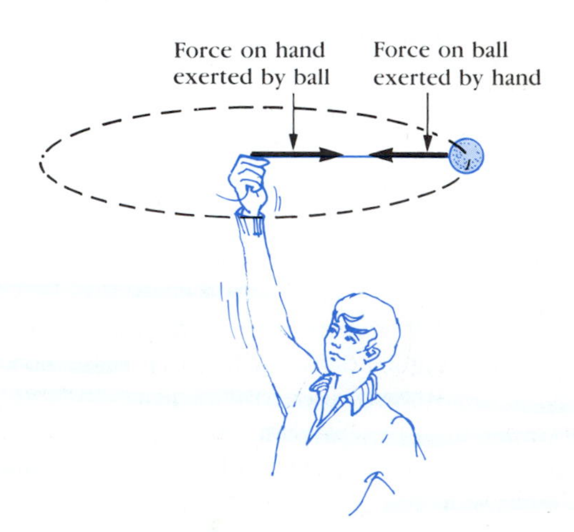

FIGURE 6-2
Ball swung in a circle, showing force acting on the ball and the "reaction" force acting on the hand.

hand feels—the reaction to your pulling in on the ball. The only force on the ball (except for gravity) is the one you exert inwardly on it. There are no others. In particular, there is no outward "centrifugal force" acting on the ball.

EXPERIMENT

Tie an object such as a rock or a tin can to the end of a string and swing it horizontally around your head, as in Figure 6-2. As the object is swinging, suddenly let go of the string. Note very carefully the position of the object at the instant you let go and observe in which direction it moves.

If a centrifugal force were pulling outward on the object, you would expect it to fly outward when you released it, as shown in Figure 6-3(a). But if our discussion is correct, the object will fly off tangentially in a straight line, because there is no longer a force pulling it inward—Figure 6-3(b). Which figure, 6-3(a) or 6-3(b), represents what you observed in your experiment?

Another example of centripetal acceleration occurs when a fast-moving automobile rounds a curve. In such a situation you may feel that you are thrust outward. This is not some mysterious centrifugal force pulling on you. What is happening is that you tend to move in a straight line whereas the car curves "in front of you," so to speak. To make you go in the curved path, the back of the seat or the door of the car exerts a force on you (Figure 6-4). The car itself must have an inward force on it if it is to move in a curve. On a flat road this force is supplied by friction between the tires and the pavement. If the friction force is not great enough, as under icy conditions, sufficient force cannot be applied and the car will skid out of a circular path into a more nearly straight path. A lower velocity means less force is required and therefore there is less likelihood of skidding.

FIGURE 6-3
If centrifugal force existed, the ball would fly off as in (a) when released. In fact, it flies off as in (b).

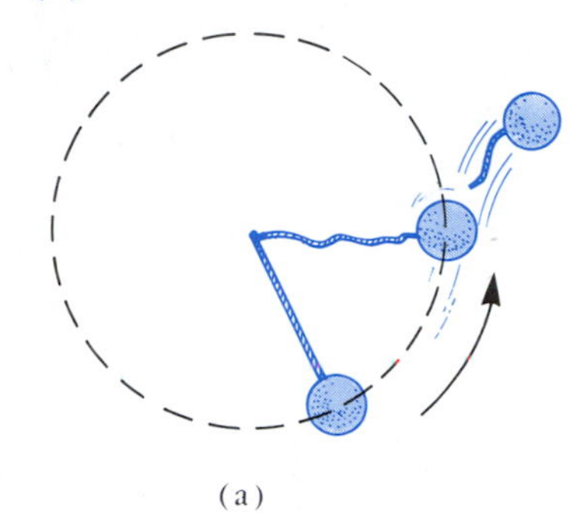
(a)

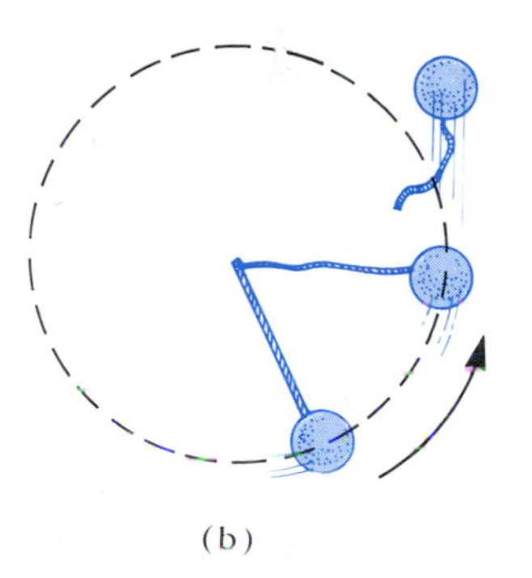
(b)

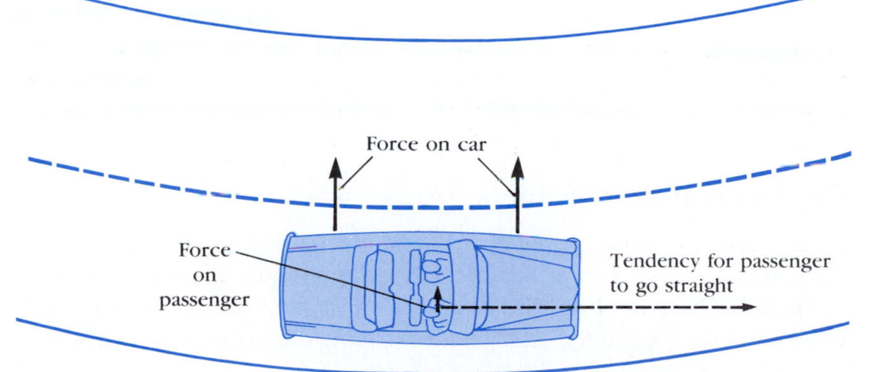

FIGURE 6-4
The road exerts an inward force on a car to make it move in a circle; and the car exerts an inward force on the passenger.

FIGURE 6-5
The force of gravity. The earth pulls on the apple; the apple pulls on the earth.

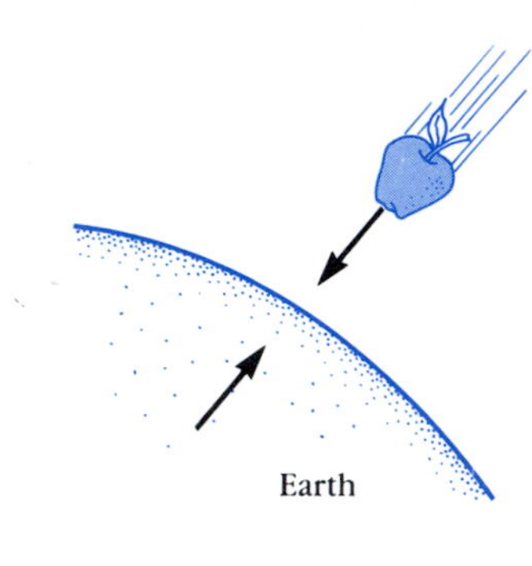

2. NEWTON'S LAW OF UNIVERSAL GRAVITATION

It may not have been circular motion per se that started Newton thinking about the nature of gravity, but it did play a role. Among other things, Newton was concerned with understanding the motion of the planets. Although some thinkers still held to the ancient idea that the orbits of the planets needed no explanation since they merely represented a *natural* motion, many thinkers in Newton's time recognized that some sort of force was needed to keep the planets in their nearly circular orbits. Newton felt that the origin of this force was the sun. And on the basis of Kepler's laws (Section 3 of Chapter 2) Newton was able to show that the strength of this force must be inversely proportional to the square of the distance between a planet and the sun. Furthermore, Newton was also wondering about the nature of the force that must act to keep the moon in its nearly circular orbit around the earth.

Newton was also thinking about the apparently unrelated problem of gravity. Since falling bodies accelerate, Newton had concluded that they must have a force exerted on them—a force we call the force of gravity. But what, we may ask, *exerts* this force of gravity? For, as we have seen, whenever a body has a force exerted *on* it, that force is exerted *by* some other body. Every object on the surface of the earth feels this force of gravity, and no matter where the object is, the force is directed toward the center of the earth. Newton concluded that it must be the earth itself that exerts the gravitational force on objects at its surface (Figure 6-5).

According to an early account, Newton was sitting in his garden and noticed an apple drop from a tree. He is said to have been struck with a sudden inspiration: If the effect of gravity acts at the tops of trees, and even at the tops of mountains, then perhaps it acts all the way to the moon! Whether this story is true or not, it does capture something of Newton's reasoning and inspiration. Indeed, we can perceive here the creative act: the bringing together of two separate ideas and the finding of unity in them. Newton saw the moon as falling toward the earth just like any other body. This statement may seem strange, but remember that an object tends to go in a straight line if no forces act on it. The moon would move in a straight line if no forces were acting on it. Instead, it keeps *falling out* of a straight line path because of the gravitational attraction of the earth (Figure 6-6). Newton thus came to the remarkable conclusion that *the gravitational force of the earth holds the moon in its nearly circular orbit.*

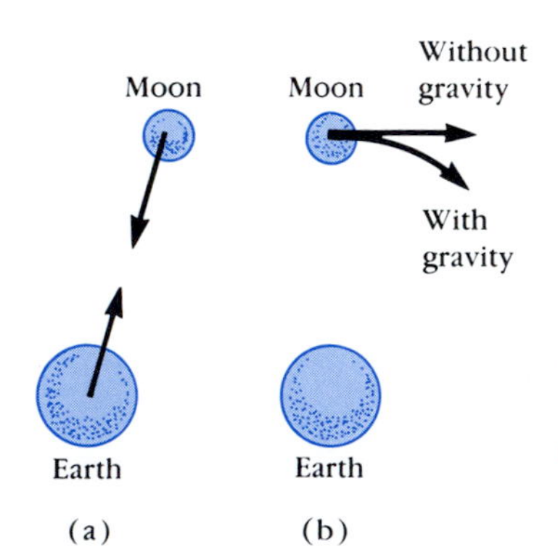

FIGURE 6-6
(a) The earth pulls on the moon; the moon pulls on the earth. (b) The moon is pulled out of a straight line path by the force of gravity.

The Moon as a Projectile

To substantiate his argument that the moon is held in its orbit by the gravitational pull of the earth, Newton used the drawing shown in Figure 6-7. It shows that an object projected with a large horizontal velocity will travel part way around the earth before striking the ground (lines VD and

FIGURE 6-7
Drawing from Newton's *Principia* showing the path of a projectile projected horizontally at various high speeds.

VE in the figure); the larger the velocity of the projectile, the farther it travels. If its velocity is great enough, the projectile will continue all the way around the earth without ever hitting the ground. It has then taken a circular or elliptical path (Figure 6-7). Thus, Newton was the first to recognize the possibility of artificial satellites circling the earth, but the high velocity necessary to attain an orbit was not achieved for almost three centuries (see section following). However, what Newton was trying to make plausible was not the possibility of artificial satellites but the idea that the moon is merely a high-speed projectile that circles the earth under the influence of the gravitational force.

Force of Gravity Decreases with Distance (Squared)

At the surface of the earth, the force of gravity accelerates objects at the rate of 9.8 m/s^2. But what is the acceleration of the moon in its orbit? Using the formula for the acceleration of an object moving in a circle, $a = \frac{v^2}{r}$, Newton calculated the acceleration of the moon. He found that it was 3600 times smaller than the acceleration of objects at the surface of the earth (0.0027 m/s^2 versus 9.8 m/s^2). Newton explained this much smaller acceleration by proposing that the gravitational pull of the earth is weaker on more distant objects.

Since the radius of the earth is 6400 km (4000 miles) and the moon is 384,000 km away, the moon is 60 times farther from the earth's center than are objects at the surface of the earth—Figure 6-8(a). Now $60 \times 60 = 60^2 = 3600$—again that number 3600. This suggested to Newton that the gravitational attraction of the earth for any object must

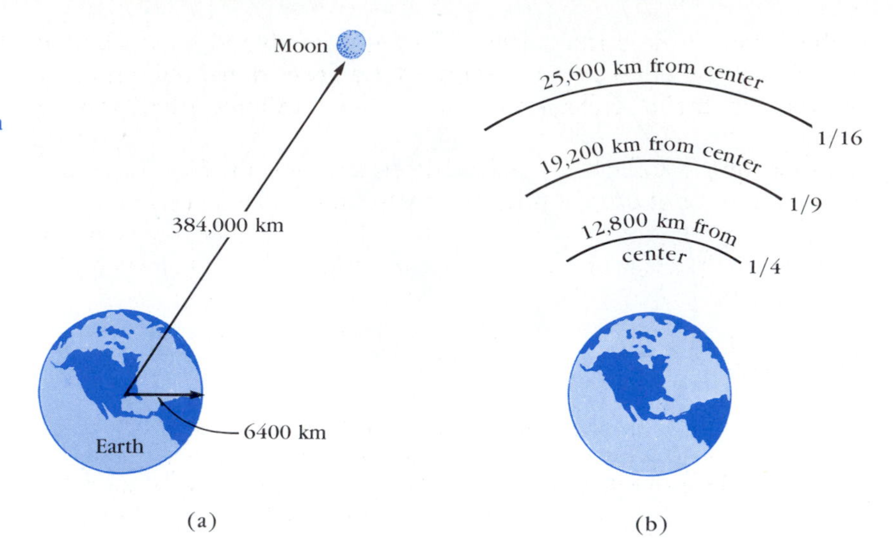

FIGURE 6-8
(a) Objects at the earth's surface are 6400 km from the center of the earth; the moon is 384,000 km away. (b) Force of gravity due to the earth decreases inversely as the square of the distance. For example 6400 km from earth's surface (12,800 km from its center), the force of gravity is $\frac{1}{4}$ what it is at earth's surface.

decrease inversely with the square of its distance from the earth's center:

$$\text{Force of gravity} \propto \frac{1}{(\text{distance})^2}.$$

This is illustrated in Figure 6-8(b). The moon, which is 60 earth radii away, will feel a force only $\frac{1}{60^2} = \frac{1}{3600}$ times as strong as it would if it were at the surface of the earth. This theoretical result is what is found experimentally.

The Law of Universal Gravitation

In the same way that all objects on the surface of the earth have the same acceleration due to gravity of 9.8 m/s^2, so all objects 384,000 km from the earth will have the same acceleration due to the earth's gravity, 0.0027 m/s^2. The actual force on the object, as we saw earlier, depends also on its mass. It is, in fact, proportional to the mass.

When the earth exerts its gravitational force on an object, such as the moon, then according to Newton's third law the other object must exert an equal and opposite force on the earth—Figure 6-6(a). Because of this symmetry, Newton argued that the force of gravity must be proportional to *both* the masses involved. Thus, the force of gravity is proportional to the product of the earth's mass and the mass of the object, and inversely proportional to the square of the distance between their centers:

$$F \propto \frac{\text{mass of earth} \times \text{mass of object}}{(\text{distance})^2}.$$

The distance used here is the distance from the center of the earth to the center of the object. This is not to say that the force of gravity originates at the center of the earth. Rather, as Newton was able to show with the use of calculus, all parts of the earth exert a gravitational pull on other objects—but the net effect is the same as if the total force acted from the center.

As mentioned earlier, Newton had determined that the force that holds a planet in its orbit decreases by the inverse square of the distance of the planet from the sun. This led him to believe that the same kind of gravitational force was acting between the sun and each planet. It also reinforced the idea that the gravitational force decreases with the square of the distance. With all these arguments in hand, Newton proposed his famous **Law of Universal Gravitation**:

> **Every body in the universe is attracted to every other body with a force that is proportional to the product of their masses and inversely proportional to the square of the distance between them. This force acts along the line joining the two bodies.**

In symbols, this law can be written as

$$F \propto \frac{mm'}{d^2},$$

where m and m' are the masses of the two bodies and d is the distance between them. To make this proportionality into an equality, we need only introduce a constant of proportionality, which we call G. Thus,

$$F = G\frac{mm'}{d^2}.$$

The constant G is a universal constant; that is, it has the same value no matter what bodies are involved. According to Newton, then, the force of gravity is not unique to the earth; it exists between all objects.

Be careful not to confuse G, which is a universal constant, with g, which is the acceleration due to gravity at the surface of the earth.

The law of universal gravitation might seem at first a rather outlandish claim. *Every* body attracts every other body? That means a pencil attracts a paper clip; and a rock attracts other rocks and all other objects as well. But we are not aware of any such attraction in our daily lives. The only attraction that we notice is the attraction of the earth for other objects. This is because the constant G is very small. So the gravitational force between two objects will be very small unless the mass of one of them is extremely large, as the earth's is.

The force between two ordinary-sized objects was first measured experimentally in 1798 by Henry Cavendish (1731–1810), who thus confirmed Newton's hypothesis (Figure 6-9). At the same time, Cavendish was able to measure the value of the constant G. The accepted value today is $G = 6.7 \times 10^{-11}\ \mathrm{N \cdot m^2/kg^2}$. Once he had measured G, Cavendish was able to determine the mass of the earth because the force F exerted on any

FIGURE 6-9
A schematic diagram of Cavendish's apparatus. A light horizontal rod, to which two spheres are attached at the ends, is suspended at its center by a thin fiber. When another sphere (labeled A) is brought close to one of the suspended spheres, the attraction causes the latter to move which twists the fiber slightly. Cavendish had earlier determined what force was required to twist the fiber a given amount. He was thus able to measure the magnitude of the gravitational force between two objects whose mass is known.

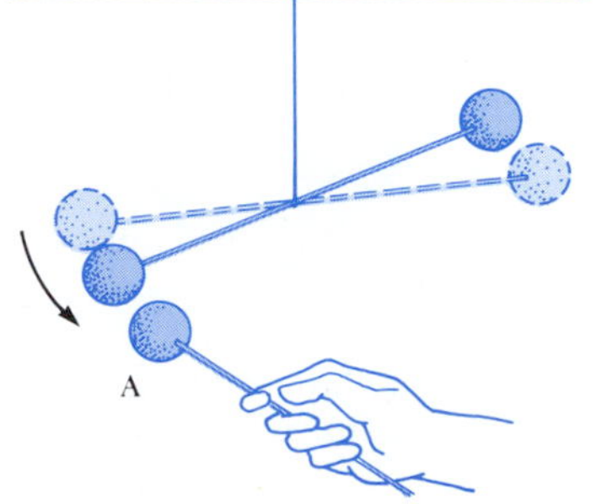

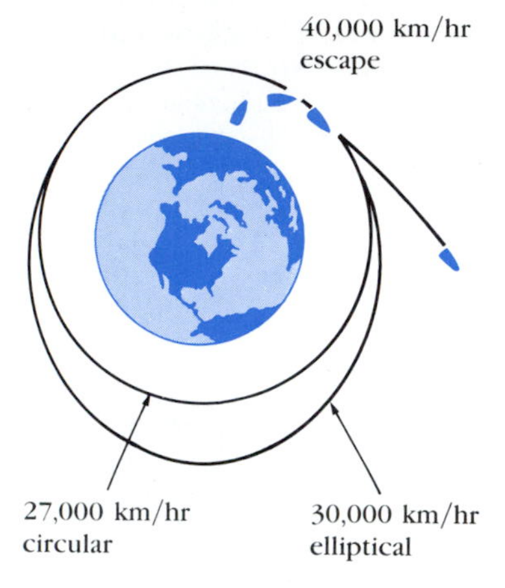

FIGURE 6-10
Artificial satellites.

body of mass m at the surface of the earth, and d (radius of the earth), were known. Only m', the mass of the earth, was not known, and it could be calculated by the above formula. The result is 6.0×10^{24} kg. Thus, Cavendish was said to be the first to "weigh" the earth.

The law of universal gravitation should not be confused with Newton's second law of motion, $F = ma$. The former describes a particular force, gravity, and how its strength varies with the distance and the masses involved. Newton's second law, on the other hand, does not describe a particular kind of force. Instead, it relates the force applied to a body—be it a pushing force, a magnetic force, or a gravitational force—to the mass of the body and its acceleration. It is a more general law; it relates any force whatever to the motion of the body on which it acts.

Gravity Is a Force Acting at a Distance

One aspect of this gravitational theory bothered even Newton. Other common forces acted by contact—a push or a pull. Or at least there was something material to transmit the force; for example, a person can tie a rope to a rock and then pull on the rock by pulling on the rope. The rope transmits the force from the person to the rock. But with gravity, the force can be exerted over a distance without any matter to transmit it. A falling apple feels the pull of the earth even though it isn't touching the earth. The moon, 384,000 km from the earth, feels the gravitational attraction of the earth. The fact that the gravitational force should be a "force acting at a distance," is difficult to understand or accept. Partly because of this difficulty, Newton did not publish his results immediately.

3. SATELLITES AND "WEIGHTLESSNESS"

Earth Satellites

Newton was the first to indicate the possibility of an earth satellite (Figure 6-7). To launch a satellite, a rocket carrying the satellite is fired off vertically. When the rocket reaches a height of a few kilometers, where the frictional drag of the earth's atmosphere is very small, it gives a horizontal thrust to the satellite (Figure 6-10). If its velocity is large enough, the satellite will go into a circular orbit. The velocity required for a circular orbit not far above the earth's atmosphere is about 27,000 km/hr. If its velocity is less than this, the satellite returns to the earth like an ordinary projectile. If its velocity is larger, the satellite will go into an elliptical orbit around the earth—just as the planets revolve in elliptical orbits around the sun. However, if its velocity is greater than 40,000 km/hr, the satellite will completely escape the earth, never to return. The gravitational force of the earth will not be strong enough to hold it back. This minimum velocity that an object must have to escape is known as the *escape velocity*.

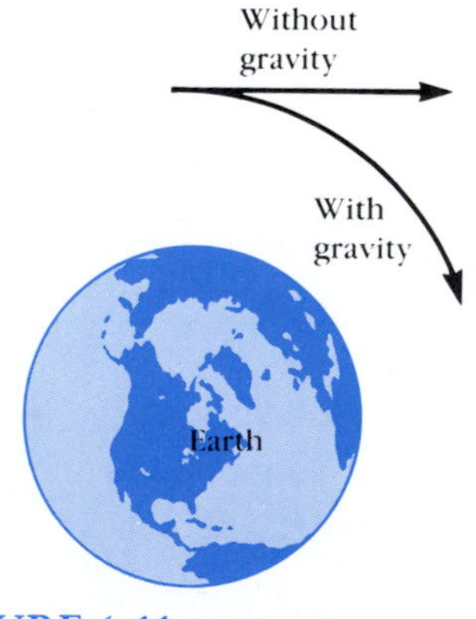

FIGURE 6-11
A moving satellite "falls" out of a straight line path, toward the earth.

It is sometimes asked: "What keeps a satellite up?" The answer is its speed and the fact that the earth is round. If a satellite stopped moving, it would of course fall down to earth. At the high speed a satellite has, it would fly off into space if the gravitational force of the earth did not act on it (Figure 6-11), which is identical to Figure 6-6(b) for the moon. And if the earth were flat (and sufficiently long) the satellite would be pulled down until it hit the earth. But the earth is round. So the satellite—though "falling" under the action of gravity—"falls" around the earth as shown, if its speed is sufficiently high.

FIGURE 6-12

The objects in an elevator at rest (a) have two forces on them: gravity and the force exerted by the floor (or whatever they are attached to) which balances it. But when the elevator undergoes free-fall (b), only gravity acts on them.

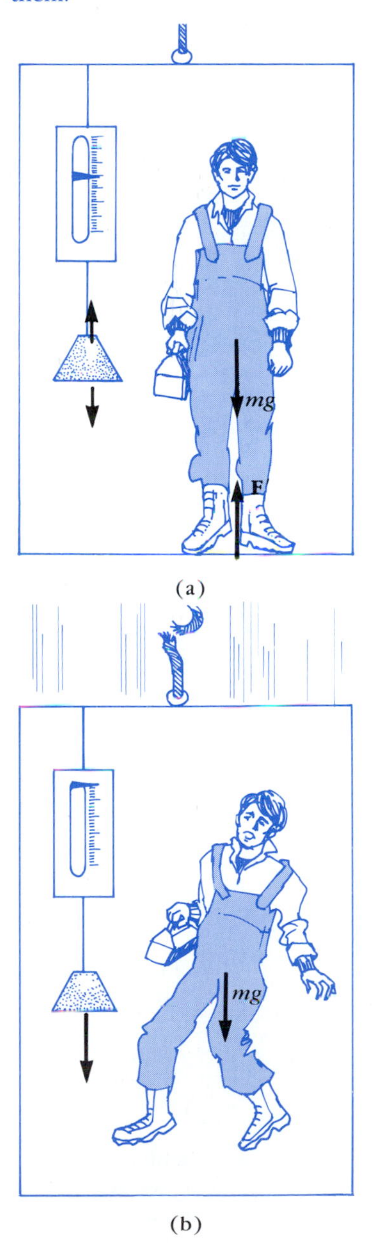

"Weightlessness"

People and other objects in a satellite circling the earth are said to experience apparent weightlessness. Before tackling the case of a satellite, however, let us first look at the simpler case of a falling elevator. In Figure 6-12(a) we see an elevator at rest. If the elevator cable were to break and the elevator were to fall freely, we would have the situation shown in Figure 6-12(b). The elevator, and all objects in it, are accelerating downward with an acceleration equal to g (9.8 m/s^2). The force on each object would then be only the force of gravity, mg, where m is the mass of the object. In other words, the floor of the elevator would not push up on the man as it does when the elevator is at rest—Figure 6-12(a); see also Figure 5-1. We can show this from Newton's second law. Suppose the elevator floor did exert a force upward on the man, call it F'. Then by Newton's second law, $ma = mg - F'$, where mg is the force of gravity on the man and $mg - F'$ is then the net force on the man. If the elevator is at rest, $a = 0$ so $0 = mg - F'$ and $F' = mg$ just as in our discussion concerning Figure 5-1. But when the elevator is in free fall, $a = g$. So Newton's second law in this case, $ma = mg - F'$, becomes $mg = mg - F'$, so $F' = 0$. Thus, no force acts upward on the man; nor does the man (by Newton's third law) exert a downward force on the floor. In fact, the man has difficulty standing on the floor at all. Sure, gravity is acting to pull the man downward. But gravity is pulling the floor out from under him at just the same rate! The man and all other objects in the elevator are experiencing "apparent weightlessness."

If the person in the elevator let go of a pencil, say, it would not fall to the floor. True, the pencil would fall with acceleration g. But so would the floor of the elevator and the person. The pencil would hover right in front of the person. This is called *apparent weightlessness* because, in fact, gravity is still acting on the objects. The objects seem weightless only because the elevator is accelerating at g.

The "weightlessness" experienced by people in a satellite orbiting close to the earth is the same apparent weightlessness experienced in a freely falling elevator. It may seem strange, at first, to think of a satellite as freely falling. But a satellite is indeed falling toward the earth, as is shown in Figure 6-11; the force of gravity causes it to "fall" out of its natural straight-line path. The acceleration of the satellite must be the acceleration due to gravity, since the only force acting on it is gravity. Thus, although the force of gravity acts on objects within the satellite, the objects experience an apparent weightlessness because they, and the satellite, are accelerating as in free-fall.

When a spacecraft is out in space far from the earth and other attracting bodies such as the moon, the force of gravity due to these bodies will be quite small because of the large distances involved. Persons in such a spacecraft will experience real weightlessness, although the effects will be the same as for the apparent weightlessness described earlier.

The effects of weightlessness (whether real or apparent makes no difference) on human beings are interesting. In ordinary circumstances, for example, a person can become quite tired from holding out his arms horizontally. But for a person experiencing weightlessness, no effort is needed. The arms will just "float" there because there is no sensation of weight. This effect has many applications in athletics. During a jump or a dive, while on a trampoline, and even between strides while running, a person is experiencing apparent weightlessness or free-fall—although only for a short time. During these brief periods, limbs can be moved much more easily because inertia only, not gravity, must be overcome. The loss of control because of lack of contact with the ground is compensated for by the increased mobility. Prolonged weightlessness in space, however, can have deleterious effects on health. Red blood cells diminish, blood collects in the thorax, bones lose calcium and become brittle, and muscles lose their tone. These effects are at present being carefully studied.

4. EARTH'S TIDES AND GRAVITY

The reason for the earth's tides, and why high and low tides occur at regular intervals, was a subject of interest to thinkers for thousands of years. But it was only in the seventeenth century that a clear explanation was given—by Newton. Newton argued that the two high tides (Figure 6-13) each day are due to the gravitational attraction of the moon. This force has only a slight effect on the solid land masses (and it is detectable), but the great masses of water that make up the oceans are free to move. Because the gravitational force decreases with the square of the distance, the moon pulls harder on that part of the earth closest to it.

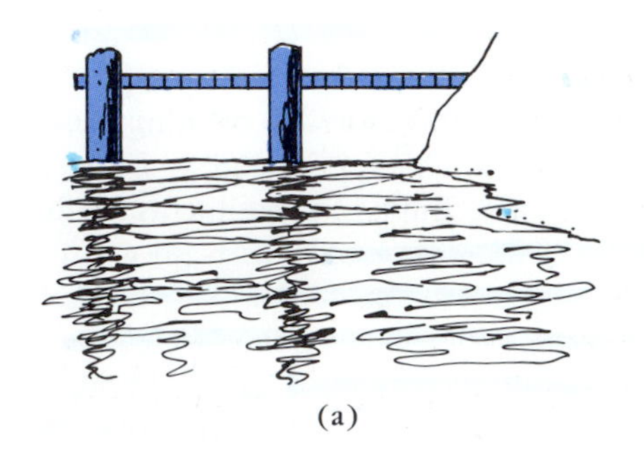

(a)

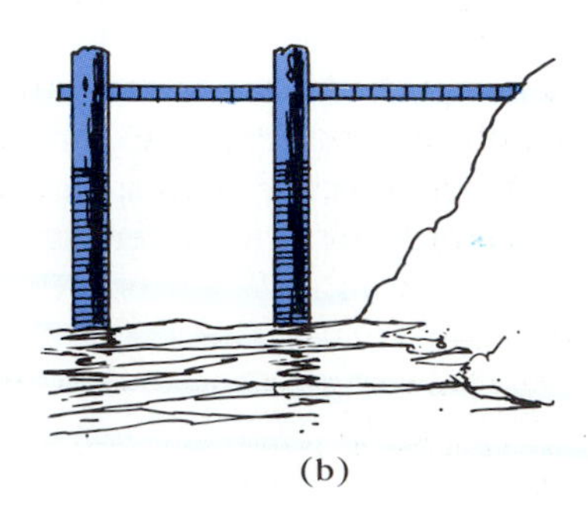

(b)

FIGURE 6-13
A pier or dock reaching out into the ocean showing water level (a) at high tide and (b) at low tide.

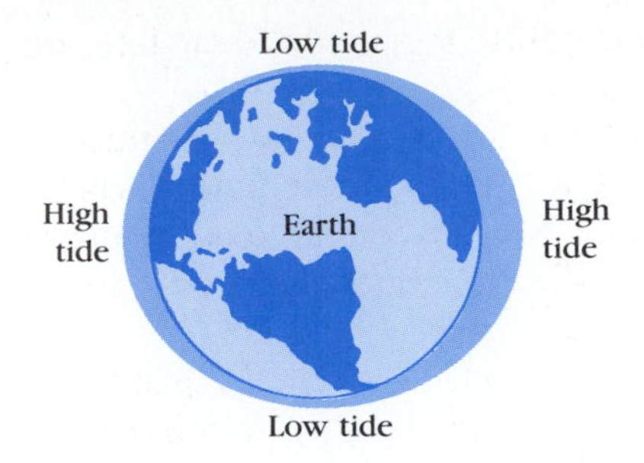

FIGURE 6-14
The tides on the earth are due to the movement of water attracted to the moon by gravity.

The water in the oceans is pulled toward the moon, resulting in a high tide on the side of the earth facing the moon as shown in Figure 6-14. Simultaneously a high tide occurs on the opposite side of the earth. Why this second high tide? Because the pull of the moon on the far side of the earth is less (since it is farther from the moon) than the pull on the central (nearer) parts of the earth. We might visualize this as the earth being pulled away from the water on the far side, just as the water on the side of the earth near the moon is pulled away from the earth a bit. The ocean water thus tends to collect on the sides closest and farthest from the moon, and is taken from the regions between, where there will be low tides (Figure 6-14).

Because the moon moves relative to the earth so that it returns to the same position overhead after about 25 hours, there are two high and two low tides at any point every 25 hours† (approximately two of each per day). Because the high tides stay more or less in line with the moon, it is as if the solid earth moved beneath the tidal bulges. High tides do not necessarily correspond precisely to the time at which the moon is directly overhead because other forces are also acting; among the complicating factors are the rotation of the earth, friction between the solid earth and the moving water, and the gravitational pull of the sun. These factors also affect the height of the tides. For example, the highest and lowest tides occur when the sun and moon are lined up so both are pulling in the same direction (called "spring tides"); when the sun and moon are at right angles, the tides are smallest ("neap tides")—see Figure 6-15.

FIGURE 6-15
The tides are greatest when the sun, moon, and earth are in line (a) and (b), and least when they make a right angle (c).

5. THE PLANETS AND GRAVITY

Kepler had discovered experimentally (see Section 3 of Chapter 2) that the planets trace elliptical orbits around the sun. Newton showed theoretically that a force law that depends on the inverse square of the distance necessarily implies that the orbits are ellipses. This he used as evidence for

† According to an amusing historical anecdote (probably apocryphal), Napoleon (or was it Philip of Spain?) was supposedly told by his admirals that he had nothing to fear from the British navy because it was run by superstitious people who believed the tides had something to do with the motion of the moon!

FIGURE 6-16
Gravitational force due to one planet can cause *perturbation* of another planet's orbit.

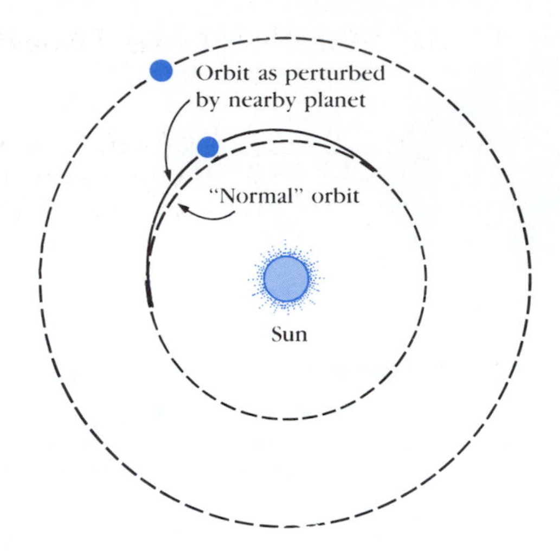

his law of universal gravitation. However, because each body in the universe attracts every other body, each planet must exert a force on the other planets as well. Since the mass of the sun is much greater than that of any of the planets, the force on one planet due to any other planet will be very small by comparison. But because of this small force, each planet should show some departure from a perfectly elliptical orbit, especially when a second planet is fairly close to it (Figure 6-16). Such deviations, or *perturbations*, as they are called, from perfect ellipses are indeed observed. In fact, Newton's recognition of perturbations in the orbit of Saturn was a hint that helped him formulate the law of universal gravitation that all bodies can attract gravitationally. Observation of other perturbations later led to the discovery of Neptune and Pluto. Deviations in the orbit of Uranus, for example, could not be accounted for by perturbations due to the other known planets: careful calculation in the nineteenth century indicated that these deviations could be accounted for if there were another planet farther out in the solar system. The position of this planet was predicted from the deviations in the orbit of Uranus, and telescopes focused on that region of the sky quickly found it: the new planet was called Neptune. Similar but much smaller perturbations of Neptune's orbit led to the discovery of Pluto in 1930.

In addition to the nine known planets, a large number of tiny objects known as *asteroids* orbit the sun between Mars and Jupiter. A number of comets, which are also relatively small objects and often have very elongated orbits, also orbit the sun. The most famous is Halley's comet, which passes the earth every 75 years (Figure 6-17).

6. NEWTON'S SYNTHESIS

The development by Newton of the law of universal gravitation and the three laws of motion was a major achievement in Western thought. For

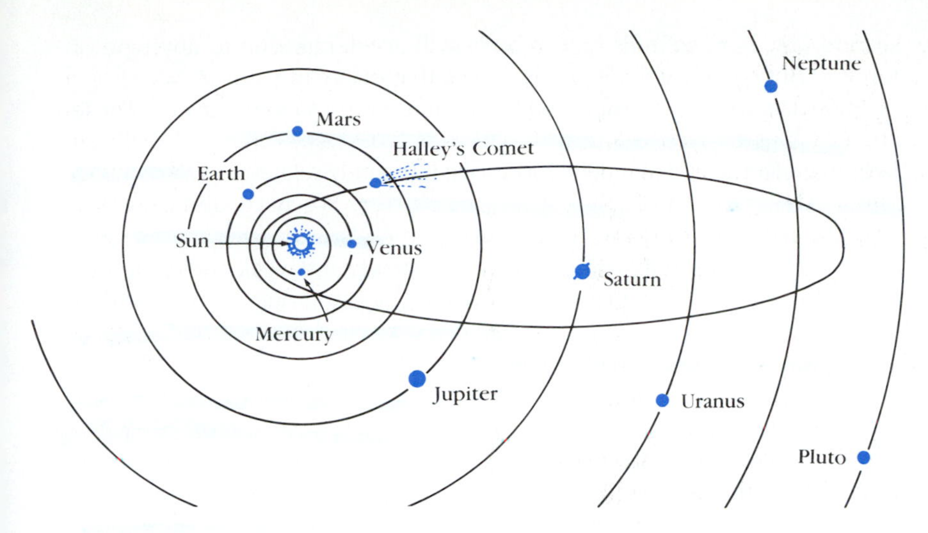

FIGURE 6-17
The solar system, including the orbit of Halley's comet.

with these laws, Newton was able to describe the motion of objects on earth and in the heavens. The motions of heavenly bodies and bodies on earth were seen to follow the same laws (something not previously recognized generally, although Galileo and Descartes had argued in its favor). For this reason (and also because Newton integrated the results of earlier workers into his system), we sometimes speak of Newton's "synthesis."

Newton's work was so encompassing that it constituted a theory of the universe, and influenced philosophy and other fields. The laws formulated by Newton are referred to as *causal laws*. By *causality* we mean the idea that one occurrence can cause another. We have repeatedly observed, for example, that when a rock strikes a window, the window almost immediately breaks. We infer that the rock *caused* the window to break. This idea of "cause and effect" took on more forceful meaning with Newton's laws. For the motion— or rather the acceleration—of any object was seen to be *caused* by the net force acting on it. As a result the universe came to be pictured by many scientists and philosophers as a big machine whose parts moved in a predictable and predetermined way—according to natural laws. However, this *deterministic* view of the universe had to be modified by scientists in the twentieth century, as we shall see in Chapters 26 through 28.

7. TYPES OF FORCES IN NATURE

We have already discussed the fact that Newton's law of universal gravitation describes how a particular type of force—gravity—depends on the distance between, and the masses of, the objects involved. Newton's second law, $F = ma$, tells how a body will accelerate due to any type of force F. But what are the types of forces that occur in nature?

Physicists today recognize only four different forces in nature.[†] These are: (1) the gravitational force, (2) the electromagnetic force (we shall see later that electric and magnetic forces are intimately related), (3) the strong nuclear force, and (4) the weak nuclear force. We have discussed the gravitational force in detail. The nature of the electromagnetic force will be discussed in detail in later chapters. The strong and weak nuclear forces operate at the level of the atomic nucleus, and although they manifest themselves in such phenomena as radioactivity and nuclear energy, they are much less obvious in our daily lives.

But what about such forces as ordinary pushes and pulls, and friction? What category do they fit in? We call these types of force *contact forces* because they occur when one object is in contact with another object which exerts the force. According to modern quantum theory, these forces are due to the electromagnetic force. For example, the force your fingers exert on a pencil is the result of electrical repulsion between the outer electrons of the atoms of your fingers and those of the pencil.

SUMMARY

An object moving at constant speed in a circle has an acceleration toward the center of the circle, or a **centripetal acceleration**. A force on the object acting toward the center of the circle produces this acceleration, thereby keeping the object moving in its circular path.

Newton's **law of universal gravitation** states that every body in the universe attracts every other body with a force proportional to the product of their masses and inversely proportional to the square of the distance between them. The direction of this force is along the line joining the two objects.

It is this gravitational force that keeps the moon and the artificial satellites in their orbits revolving around the earth and the planets in their orbits around the sun. It is also responsible for the earth's tides.

QUESTIONS

1. If a bucket of water is swung in a vertical circle at a high enough speed, the water won't spill at the top of the circle when the bucket is upside down. Explain.
2. Why are curves on a highway banked?

[†] Physicists are now working on theories that would unify these four forces; that is, to consider some or all of these forces as different manifestations of the same basic force. This is discussed in Chapter 31.

3. Why does a car tend to skid on an icy curve?
4. Explain why a car is less likely to skid when traveling around an icy level curve if it does so at a low speed.
5. Sometimes it is said that water is removed from clothes in a spin dryer by centrifugal force. Why is this incorrect?
6. Will the acceleration of a car be the same if it travels around a sharp curve at 60 km/hr as when it travels around a gentle curve at the same speed? Explain.
7. Suppose a car moves at constant speed along a mountain road. At what places does it exert the greatest and least forces on the road: (a) at the top of a hill (b) at a dip between two hills (c) on a level stretch near the bottom of a hill?
8. Describe all the forces acting on a child riding a horse on a merry-go-round.
9. Does an apple exert a gravitational force on the earth? If so, how large a force? Consider an apple (a) attached to a tree (b) falling.
10. Describe how careful measurements of the variation in g in the vicinity of an ore deposit might be used to estimate the amount of ore present.
11. Distinguish between the gravitational force exerted on the moon by the earth and the gravitational force exerted by the moon on objects on the moon's surface.
12. Is the acceleration due to gravity the same everywhere?
13. Does an egg exert a gravitational force on the earth? How large a force?
14. Would your weight on top of Mt. Everest be more or less than your weight at sea level?
15. Would you expect the escape velocity from the moon to be greater than, less than, or equal to the escape velocity from the earth? Why?
16. If you were in a satellite orbiting the earth, how might you cope with walking, drinking, or putting a pair of scissors on a table?
17. An antenna loosens and becomes detached from a satellite in a circular orbit around the earth. Describe the antenna's motion subsequently. If it will land on earth, describe where; if not, describe how it could be made to land on earth.
18. The sun is directly below us at midnight, in line with the earth's center. Are we then heavier at midnight, due to the sun's gravitational force on us, than we are at noon? Explain.
19. Astronauts who spend long periods in outer space could be adversely affected by weightlessness. One way to simulate gravity is to shape the spaceship like a bicyle wheel that rotates about an axis with the astronauts walking on the inside of the "tire." Explain how this

simulates gravity. Consider (a) how objects fall, (b) the force we feel on our feet, and (c) any other aspects of gravity you can think of.

20. People sometimes ask: "What keeps a satellite up in its orbit around the earth?" How would you respond?
21. Explain why a running person experiences "free-fall" or "apparent weightlessness" between steps.
22. If the moon rises shortly after the sun sets, do you think the tides will be large or small? What will the tides be like if the moon is directly overhead when the sun sets?
23. The sun's gravitational pull on the earth is much larger than the moon's. Yet the moon is mainly responsible for the tides. Explain.

EXERCISES

1. A racing car travels around a curve of radius 30 meters at a speed of 30 m/s. What is its centripetal acceleration? Express the answer in "g's"; that is, how many times larger than the acceleration due to gravity, $g = 9.8\ \text{m/s}^2$, is this?
2. What force is required in the above exercise if the car's mass is 1400 kg?
3. If a stone is spun with a speed of 2.5 m/s in a circle of radius 0.5 m, what is its centripetal acceleration? If the stone's mass is 2 kg, what force is being exerted on it and in what direction?
4. A jet plane traveling 300 m/s pulls out of a dive in an arc of radius 2500 m. What was the plane's acceleration in "g's" (i.e., what multiple of the acceleration of gravity, g)?
5. A child moves with a speed of 1.80 m/s when 12 m from the center of a merry-go-round. Calculate (*a*) the centripetal acceleration of the child and (*b*) the net force exerted on the child (mass = 25 kg).
6. The moon is 384,000 km from the earth and takes about 27 days to make a complete revolution about the earth. Calculate the centripetal acceleration of the moon.
7. Calculate the force on the moon due to the earth (*a*) using the result of the previous exercise and $F = ma$; (*b*) using the law of universal gravitation. The mass of the moon is about 7.4×10^{22} kg.
8. What is the centripetal acceleration of the earth in its orbit around the sun? The earth moves in a circle of radius approximately 1.5×10^{11} m and takes 365 days to make a complete revolution. With what force must the sun pull on the earth?
9. Calculate the force of gravity between two 200-kg refrigerators that are 1.0 m apart (center to center).

10. How strong is the force of gravity on you right now in newtons? How strong would it be if you were 6400 km above the earth's surface?

*11. At what distance above the earth's surface will the force of gravity be only half what it is at the earth's surface?

12. If the moon changed its orbit around the earth so that it was only half as distant as it is now, by what factor would the gravitational force on the moon due to the earth be changed?

13. Suppose both the mass of the earth and the mass of the moon were double their present values, but that the distance between them remained the same. By what factor would the force of gravity between them change? Would the speed of the moon around the earth have to increase, decrease, or remain the same?

14. Calculate the acceleration due to gravity on the moon. The moon's radius is about 1.7×10^6 m and its mass is 7.4×10^{22} kg.

15. A hypothetical planet has a radius 1.6 times that of the earth, but has the same mass. What is the acceleration due to gravity near its surface?

16. A hypothetical planet has a mass of 2.2 times that of earth, but the same radius. What is g near its surface?

*17. Tarzan plans to cross a gorge by swinging in an arc from a hanging vine. If his arms are capable of exerting a force of 1500 N on the rope, what is the maximum speed he can tolerate at the lowest point of his swing? His mass is 85 kg; the vine is 4.0 m long.

*18. Is it possible to whirl a bucket of water fast enough in a vertical circle so the water won't fall out? If so, what is the minimum speed?

*19. What minimum speed must a roller coaster be traveling when upside down at the top of a circle if the passengers are not to fall out? Assume a radius of curvature of 8.0 m.

*20. Use Newton's second law, $F = ma$, and the law of universal gravitation to show that

$$G\frac{mm_e}{r^2} = m\frac{v^2}{r}$$

if an artificial satellite of mass m is to travel in a circular orbit of radius r with speed v. Next, solve this equation for v, and determine what speed is required for a satellite to orbit 200 km above the earth's surface (6600 km from the earth's center).

WORK AND ENERGY

CHAPTER

7

The concept of energy is basic to every one of the sciences. And yet it is difficult to define exactly what energy is. We usually think of energy as distinct from matter. Yet matter can *contain* energy. For example, we say that a child who runs around a great deal is "very energetic." Or we say that it "takes lots of energy" to pedal a bicycle uphill. Energy is thus associated with motion or activity. But we also say that food contains energy; this is the energy our bodies use when we do work. And we say that gasoline contains energy; when the engine of a car burns gasoline, this energy is released and makes the car go. Thus energy can also be associated with objects at rest. In this chapter we will develop the concept of energy and discover how useful a concept it is. But first we examine the related concept of work.

1. WORK

Work Is Force Acting Through a Distance

In everyday language the word **work** has various meanings. In physics, however, work is defined in a very specific way: work is what is accomplished by the action of a force. Whenever a force is applied to an object, the work done on the object by that force is defined as *the product of the magnitude of the applied force times the distance through which the force acts*:

$$\text{Work} = \text{force} \times \text{distance}.$$

The man pushing the refrigerator in Figure 7-1(a) does an amount of work equal to the magnitude of the force he applies times the distance he moves the refrigerator. If he pushes the refrigerator 10 m with a force of 50 N, he does $50\text{ N} \times 10\text{ m} = 500\text{ N}\cdot\text{m}$ of work.

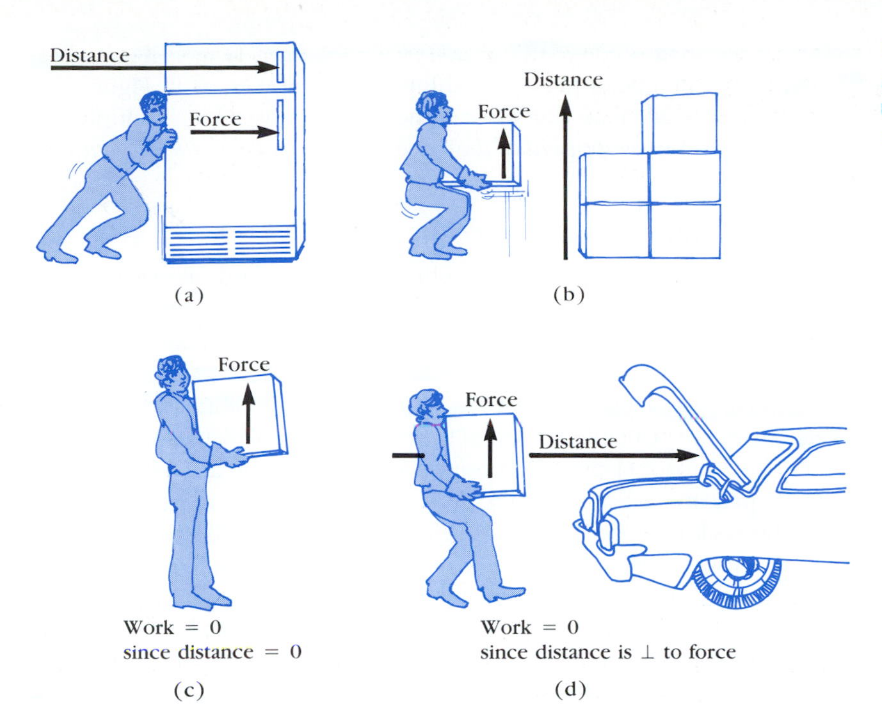

FIGURE 7-1
Work is done on the object in (a) and (b). No work is done on the object in (c) and (d).

As can be seen in this example, in the metric system work is measured in newton · meters. A special name is given to this unit: the joule (J). One joule equals one newton · meter (1 J = 1 N · m). In the British system of units, work is measured in foot · pounds.

Work is done on an automobile to accelerate it. Work is done to push a grocery cart or pedal a bicycle. A person who lifts a package vertically does work on the package—Figure 7-1(b). However, a person may exert a force and yet do no work, in the specific sense of the word, if the force does not act through a distance. For example, a man pushing against a brick wall does no work on the wall if the wall does not move. The product of force and distance is zero because the distance is zero. Similarly, when you hold an object, as in Figure 7-1(c), you do no work on it even though you may get tired holding it. Only a force that gives rise to motion is doing work.

Another example of a force that does no work is shown in Figure 7-1(d). A person carrying a package is walking at constant speed. Thus no horizontal force is needed to move the package (Newton's first law). But the person does exert a force on it that is vertically upward, perpendicular to the direction of motion. This upward force has nothing to do with the horizontal motion of the package. Thus we say that the upward force is doing no work. Whenever the force is perpendicular to the motion, no work is done. That is what we meant when we defined work as the product of the applied force times the *distance through which the force acts*.

The force does not necessarily have to be parallel to the direction of motion in order for work to be done by that force. Work will be done even if the force is at an angle to the direction of motion as long as the angle is not 90°. For example, a person pushing a lawn mower, as in Figure 5-8, does work. To calculate the work done in such cases, we multiply the distance times the *component of the force that is parallel to the direction of motion*.

Simple Machines: The Lever

When two children of the same weight sit on either end of a seesaw, the board balances if it is supported in the middle. But if one child is heavier than the other, the support must be moved closer to the heavier child to keep the board balanced. If one child is twice as heavy as the other, then the lighter child must be twice as far from the balance point as the heavier child. This same principle is used in the simple lever, which can be used to lift very heavy loads (Figure 7-2). Suppose we want to lift a 100-kg rock. An ordinary person would have difficulty exerting the upward force necessary to lift the rock directly. The force required is equal to the weight of the rock, which is $mg = 100\ \text{kg} \times 9.8\ \text{m/s}^2$ or about 1000 N. (We rounded off g to $10\ \text{m/s}^2$.) But by using a lever with a long "lever arm," as shown in Figure 7-2(b), that same person can lift the 1000-N rock by exerting a much smaller force. For example, a force of only 200 N is needed to lift the 1000-N rock if the distance of the applied force from the support is five times the distance of the rock from the support.

Simple machines like the lever are said to offer a "mechanical advantage." Other simple machines are the wheel and axle, pulley systems, and the inclined plane. The mechanical advantage of a simple machine is defined as the output force divided by the input force. With the lever in Figure 7-2, the output force is the force that lifts the rock, namely 1000 N; and the input force is the force applied by the person, 200 N. The mechanical advantage is $\frac{1000}{200} = 5$.

Do Simple Machines Save Us Work?

By using a lever we can multiply a force (as in the above example where the applied force is multiplied by a factor of 5). It may look as if we are getting something for nothing. But are we? Look at Figure 7-2(c). The man pushes down with a force of 200 N over a distance of 1.5 m while the 1000-N rock goes up only 0.3 m. We don't really get something for nothing.

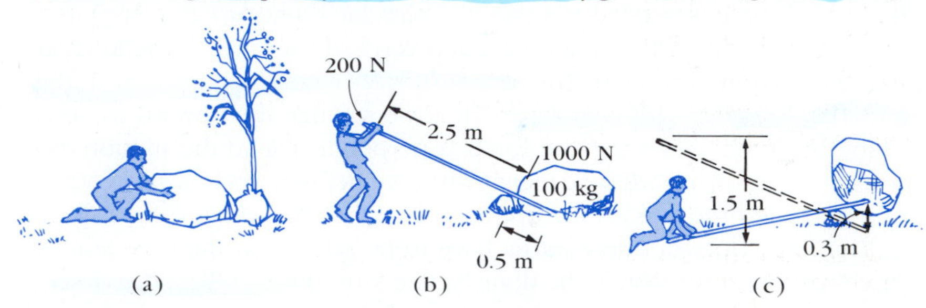

FIGURE 7-2
A heavy rock (a) can be lifted using a lever (b), which in this case offers a mechanical advantage of 5. The distance the rock is raised (c) is only $\frac{1}{5}$ of the distance through which the man pushes.

What is gained in force is lost in the distance the object can be moved. This does not lessen the usefulness of the lever—indeed, the rock couldn't be lifted at all without it.

Although the force has been multiplied, the work done has not been increased. The work done by the person, called the work input, is the force he exerts multiplied by the distance through which he exerts the force: 200 N × 1.5 m = 300 N · m. The work done on the rock, the work output, is also 300 N · m (= 1000 N × 0.3 m). So, for this simple machine, we can say

$$\text{work input} = \text{work output}.$$

We can make the same statement about other simple machines as well.†

Why is it that this quantity we call work should be the same for the input as it is for the output? Is there some larger rule or law operating here? Yes. As we shall see presently, what we have here is an example of the "conservation of energy."

2. ENERGY

Energy can be defined as the *ability to do work*. Energy takes many forms. We begin with the simplest: kinetic energy and potential energy.

Kinetic Energy: Energy of Motion

When a moving object strikes another object, it can do work on that object. A moving hammer, for example, can do work on a nail (Figure 7-3). When the hammer strikes the nail, it exerts a force on the nail and moves it through a distance. Similarly, when a flying cannonball strikes a brick wall, it exerts a force on the wall and knocks it down. Thus, an object in motion has the ability to do work and, by definition, it therefore has energy. This kind of energy, energy of motion, is called **kinetic energy** ("kinetic" comes from the Greek word for "motion"). Mathematically, the kinetic energy of an object is defined as:

$$\text{Kinetic energy} = \text{KE} = \frac{1}{2}mv^2,$$

where m is the object's mass and v is its velocity. Notice that because the velocity term is squared, when the velocity of an object is doubled, its kinetic energy is quadrupled ($2^2 = 4$); and, as we will see below, the object is able to do four times as much work.

It was a Dutch scientist, Christian Huygens (1629–1695), who first suggested that this particular combination of mass and velocity, which we

FIGURE 7-3
When the moving hammer strikes the nail it exerts a force on the nail, moving it through a distance. The kinetic energy of the hammer goes into work.

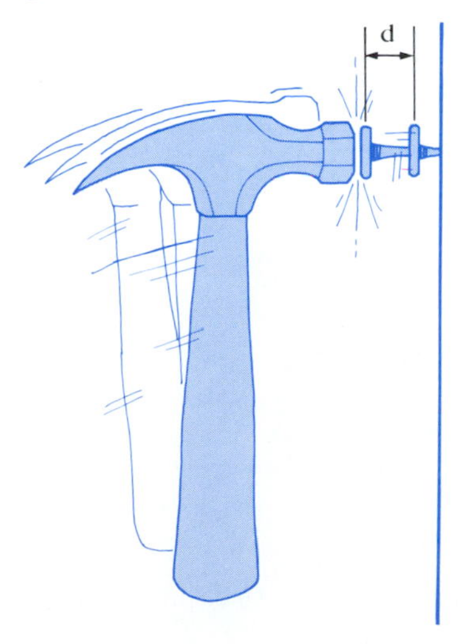

† In our example, we have assumed that there is negligible friction. If friction is significant, the work output will be less than the work input.

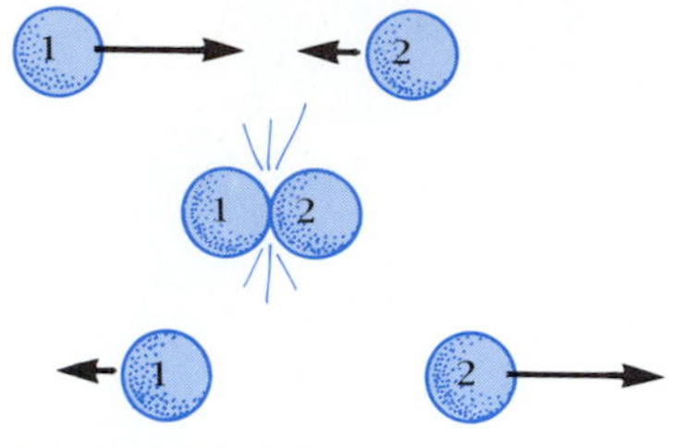

FIGURE 7-4
When two hard elastic balls collide, the total kinetic energy of the two balls does not change. What one ball loses in kinetic energy, the other ball gains.

call kinetic energy, was somehow special. He found that during the collision of two hard elastic balls, such as billiard balls, the total kinetic energy (he called it "vis viva," or "living force") remains constant. That is, the sum of the kinetic energies of the two balls—the kinetic energy of ball 1 plus the kinetic energy of ball 2—before the collision is equal to the sum of their kinetic energies after the collision, even though the kinetic energy of each of the balls may change as a result of the collision (Figure 7-4). When a physical quantity (such as total kinetic energy in this example) remains unchanged during a process, that quantity is said to be **conserved**. That something should be conserved, even under the special circumstances of two colliding billiard balls, is remarkable. That's why Huygens thought "vis viva," or kinetic energy, must somehow be a special quantity in nature.

The Connection Between Work and Kinetic Energy

If a net force F acts on an object over a distance d, work is done on the object and it is accelerated from rest to some velocity v. It can be shown† that the kinetic energy acquired by the body as a result of the net force F acting on it is exactly equal to the net work that is done on it: $\frac{1}{2}mv^2 = Fd$. (By the *net* work we mean the work done by the *sum* of all forces acting on the body, rather than, say, the work done by only one of the forces.) If the body is already moving when the force is applied, the work done goes into changing its kinetic energy. Thus,

Net work done on a body = change in kinetic energy of the body.

FIGURE 7-5
(a) When you throw a baseball you exert a force on it through a distance; that is, you do work on it. (b) As a result, the ball acquires kinetic energy in an amount equal to the work done on it.

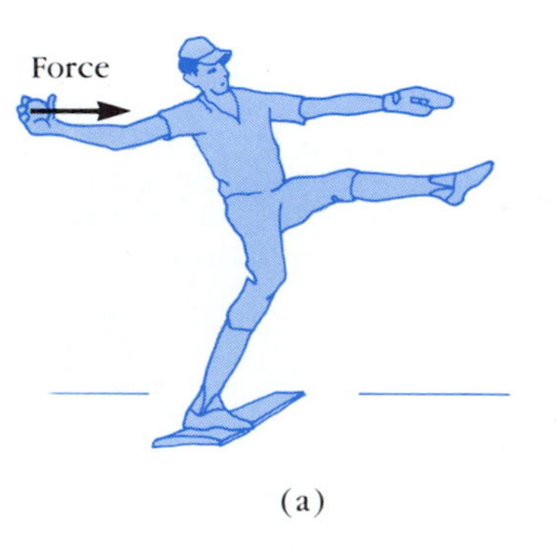

(a)

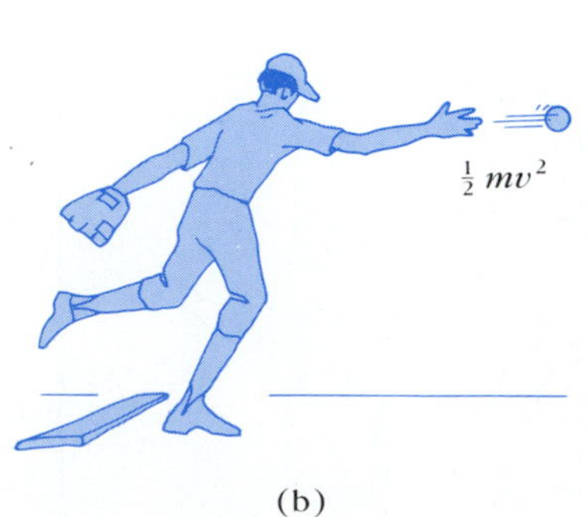

(b)

As an example, consider what happens when you throw a baseball—Figure 7-5(a). Your hand exerts a force on the baseball, and this force acts through a distance. Because of the work you do on the ball, the ball leaves your hand with a certain velocity and with a kinetic energy, $\frac{1}{2}mv^2$; this is just equal to the work done on it by your hand—Figure 7-5(b). The reverse process occurs when the ball strikes a fielder's glove; the ball exerts a force on the glove, and the glove moves backward a short distance. The ball, in the process of stopping, does work on the glove.

Thus the connection between work and kinetic energy operates in two directions. First, if work is done on an object, the kinetic energy of the object increases. Second, if an object has kinetic energy, it can do work on something else.

A similar example is the motion of the hammer shown in Figure 7-3. The carpenter does work on the hammer, which thereby gains in kinetic

† The net work done on the object is $W = Fd$. The net force F on the body gives rise to an acceleration because of Newton's second law, $F = ma$; the acceleration is given by $a = \frac{v}{t}$, where v is the final velocity after the force has acted and the initial velocity is assumed to be zero. The distance d is given by $d = \bar{v}t = \frac{v}{2}t$. Substituting these into $W = Fd$ we find $W = Fd = mad = m\frac{v}{t}d = m\frac{v}{t}\frac{v}{2}t = \frac{1}{2}mv^2$. That is, the kinetic energy of the body equals the work done on it.

energy. The hammer, because it has kinetic energy, can then do work on the nail.

The connection between work and kinetic energy has a direct bearing on automobile safety. A car traveling 80 km/hr has four times the kinetic energy of a car traveling 40 km/hr. So it will take four times as much work to stop a car going 80 km/hr than to stop one going 40 km/hr. Since, for a given car, the force that can be applied by the brakes is essentially the same at all speeds, the car traveling 80 km/hr will go *four* times as far as the car going 40 km/hr before it comes to a stop.

Because of the direct connection between work and energy, energy is measured in the same units as work: joules in the metric system and foot • pounds in the English system. However, we must make a distinction between work and energy. Energy is something a body *has*. Work is something a body *does* (to some other body). And a body can *do* work only if it *has* energy. Note, too, that both work and energy are scalars (they do not have a direction).

Potential Energy: Energy of Position

An object can have energy not only by virtue of its motion (kinetic energy) but also by virtue of its position or shape. This form of energy is called **potential energy** (PE).

A wound-up watch spring, for example, has potential energy because of its shape. As the spring unwinds, it releases this energy and does work in moving the watch hands around. The spring acquired its potential energy because work was done on it by the person who wound the watch. Just as in the case of kinetic energy, an object can acquire potential energy when work is done on it; and an object that has potential energy can do work on something else.

An ordinary coil spring has potential energy when it is compressed. A ball placed against a compressed spring, and then released, acquires kinetic energy, as shown in Figure 7-6. The compressed spring has potential energy; when the spring is released, it does work on the ball, which acquires kinetic energy.

FIGURE 7-6
A compressed spring has potential energy; it can do work on an object, giving it kinetic energy.

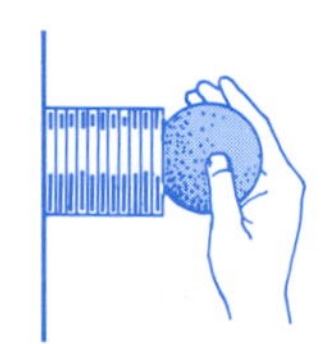

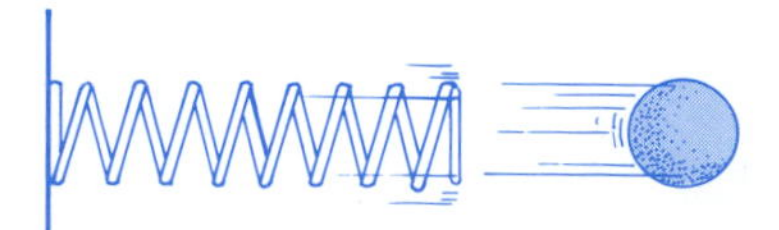

An object held high in the air has potential energy because of its position. The object is said to have **gravitational potential energy**, because it has potential energy by virtue of the gravitational attraction of the earth. When the object is released, it falls with increasing speed toward the ground. The object's potential energy is changed into kinetic energy, and the object can do work. A pile driver has an enormous amount of potential energy when it is lifted into the air; and when it is released, it can do a great deal of work.

EXPERIMENT

Find three empty soft-drink cans (the ones that are rather weak and "spongy"). Lift a large stone and a small stone from the floor and place them on a table. Put a can on the floor below each stone. Then push each stone off the table so that it falls on the can below. Which does more work? Now place the third can on the floor and drop one of the stones on it from twice the height as before. Does it do more work this time? On what two factors does the potential energy of a stone depend? Remember that energy—in this case, potential energy—is defined as the ability to do work.

When you lift an object a height h from the floor, you are exerting a force equal to its weight. So the work you do is the product of the object's weight times the height h. Similarly, if you let the object fall, it can do an amount of work equal to the product of its weight times the height it has fallen. Thus we can say that the gravitational potential energy of an object is equal to the product of its weight times its height above the ground. Since its weight is the product of its mass times the acceleration of gravity, mg, we can say:

$$\text{Gravitational potential energy} = \text{weight} \times \text{height} = mgh.$$

A 1500-kg pile driver lifted 20 m into the air has $(1500\ \text{kg})(9.8\ \text{m/s}^2)(20\ \text{m}) \approx 300{,}000$ joules of potential energy. A 5-kg stone 20 meters above the ground has $(5\ \text{kg})(9.8\ \text{m/s}^2)(20\ \text{m}) \approx 1000\ \text{J}$ of potential energy.

Gravitational potential energy plays a role in all sorts of practical situations. One example is the action of a pile driver. Another is the potential energy of water at the top of a waterfall or in a reservoir behind a dam; when the water falls, it can do work. If the water falls into turbines at the bottom of a dam, the work done can be transformed into electric energy, as we shall see in Chapter 21.

In each of these examples an object has the *capacity* to do work even though it may not actually be doing work and isn't even in motion. That is why energy of position is called "potential" energy. A coiled-up watch spring isn't doing any work when it is coiled up, nor is the compressed spring shown in Figure 7-6. But each has the *potential* to do work. Its potential energy *could* be transformed into kinetic energy. Similarly, a rock or a pile driver high above the ground may just be resting there, but each has the potential for doing a lot of work.

These examples show that energy can be *stored*, for later use, in the form of potential energy.

Other Forms of Energy

There are other forms of energy besides the kinetic and potential energy of ordinary-sized objects. These include thermal energy, electrical energy, nuclear energy, and the chemical energy stored in food and in fuels such as gasoline, oil, and natural gas. With the rise of the atomic theory of matter in the last century, scientists have come to the view that these other forms of energy are simply kinetic energy or potential energy at the molecular level. For example, according to the atomic theory, thermal energy is merely the kinetic energy of molecules. When an object is heated, the molecules that make up the object move faster and the object feels hot.

The energy stored in food and gasoline, on the other hand, is potential energy—energy stored by virtue of the relative positions of the atoms within a molecule. In order for this potential energy to be used to do work, it must be released, usually through a chemical reaction. This is much like a compressed spring which can do work when it is released. In our own bodies, enzymes allow the release of energy stored in food molecules. In an automobile, the violent spark of the spark plug allows the mixture of gas and air to release its energy and do work against the piston to propel the car forward.

Electric, magnetic, and nuclear energy are other examples of kinetic and potential energy. In later chapters we will have more to say about those forms of energy.

Energy Can Be Transformed

In the last few pages we discussed several examples of the change, or transformation, of one type of energy into another. A rock held high above the ground has potential energy. As it falls it loses potential energy because its height above the ground decreases; at the same time it gains kinetic energy because its velocity increases. Its potential energy is being transformed into kinetic energy.

Often the transformation of energy involves a transfer of energy from one body to another. The potential energy of the compressed spring in Figure 7-6 is transformed into the kinetic energy of the ball. Water plunging to the bottom of a dam turns the blades of a turbine; the kinetic energy of the water is transformed into the kinetic energy of the rotating turbine blades. The potential energy stored in a bent bow can be transformed into the kinetic energy of an arrow.

In each of these examples, the transfer of energy is accompanied by the performance of work. The spring does work on the ball; water does work on the turbine blade; and the bow does work on the arrow. This observation gives us a new insight into the nature of work: *Work is done whenever energy is transferred from one object to another.* This is true when a person throws a baseball (Figure 7-5) or pushes a heavy object across the floor—Figure 7-1(a). In both cases, chemical energy within the person, obtained from food, is transformed into kinetic energy of the object that is pushed. The work done on the object is a manifestation of the transfer of energy from the person to the object.

3. CONSERVATION OF ENERGY

Energy Is Conserved

Whenever energy is transformed from one form to another, or transferred from one body to another, we find that no energy is lost in the process. As an example, let us look at what happens when a rock is dropped from a height h above the ground (Figure 7-7). Before it is dropped, it has a potential energy equal to its weight times height; as it falls, it loses potential energy because the height is decreasing. Meanwhile, its kinetic energy is increasing, and just before it strikes the ground it has only kinetic energy. The kinetic energy it has at the bottom is exactly equal to the amount of potential energy it had originally at the top. The potential energy was completely transformed into kinetic energy. No energy was lost.

FIGURE 7-7
As an object falls it loses potential energy and gains kinetic energy. This energy can be used to do work—for example, to drive a stake into the ground.

When the object has fallen halfway to the ground, it has both potential and kinetic energy. Since its height is only half what it was originally, it has only half as much potential energy; the other half has been transformed into kinetic energy. At this point it has equal amounts of kinetic and potential energy. The sum of its kinetic energy plus its potential energy at this halfway point is equal to the amount of potential energy it had originally. In fact, the sum of its kinetic and potential energy, which we call its *total energy*, remains the same for all points along its path. Thus, although the kinetic energy and the potential energy each change throughout the motion, the sum of the two remains constant.†

This example illustrates one of the most important principles of physics, the law of the **conservation of energy**:

> **Energy is never created or destroyed. Energy can be transformed from one kind into another, but the total amount of energy remains constant.**

Like all laws of physics, this one is based on experimental data. It has been found to hold true in every experimental situation in which it has been tested.

Other Examples of the Conservation of Energy

When the billiard balls in Figure 7-4 collide, the sum of the kinetic energies of the two balls, as found by Huygens, remains essentially constant. Since there is no potential energy involved, the law of conservation of energy applies to their kinetic energies alone. If, as a result of the collision, one ball loses kinetic energy and slows down, the other ball will have its kinetic energy increased by exactly that amount. Any energy lost by one ball is gained by the other.

Another example is illustrated in Figure 7-8, which shows a car on the top of a hill, on the left. It could be a real car, with its engine off, or it could be a toy car. In either case we will assume that friction is so small that it can be ignored. The car begins to coast downhill, increasing in speed and in kinetic energy (KE) as it loses potential energy (PE). It reaches its maximum

† If you are wondering what happens to the energy when the object strikes the ground, you will find the answer in just a few pages.

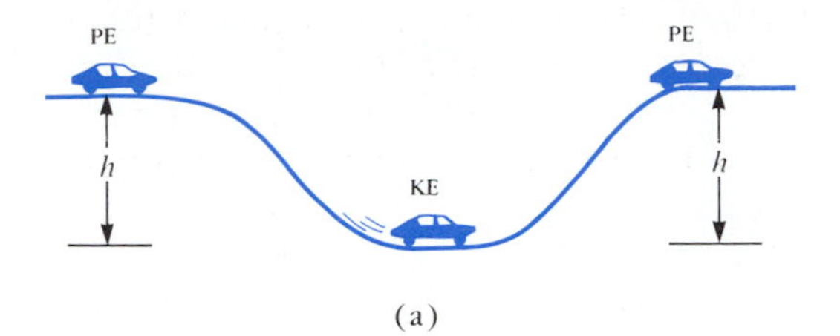

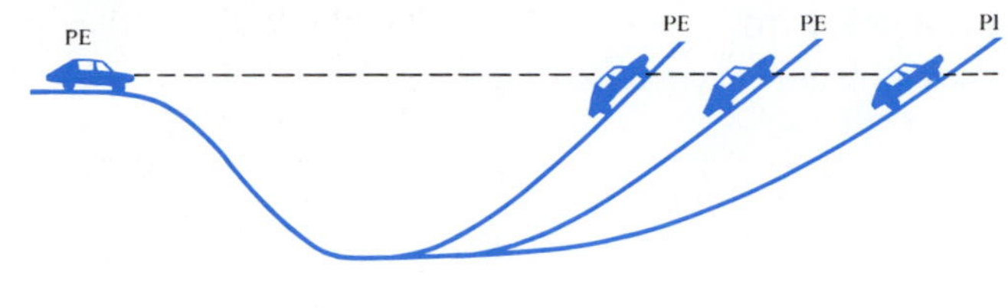

FIGURE 7-8

(a) A car loses potential energy and gains kinetic energy as it goes down the hill; as it ascends the hill on the right, it loses kinetic energy and gains in potential energy, arriving at the top with the same potential energy it had earlier. (b) Starting from rest on the hill at the left, the car will reach the same *height* on the hill at the right no matter what the angle of that hill. (Friction ignored).

KE at the bottom of the valley. As the car climbs up the hill on the other side, its KE is gradually changed back to PE. When the car comes to rest, its KE is zero and all of its energy is again PE. Since energy is always conserved, the car will have the same amount of PE as it had initially on the hill. And, because PE is proportional to height, it will come to rest at a height above the ground equal to its original height. If the second hill has the same height as the first, the car will just reach the top as it stops. If the second hill is lower, not all of the car's KE will be transformed into PE and the car will continue over the top and down the other side. If the second hill is higher, the car will reach a vertical height equal to its original height on the first hill (and then roll back down). The length and angle of the second hill do not affect how high the car will go vertically, as shown in Figure 7-8(b). This is because the PE depends only on the vertical height h.

Another example, and one that more closely approximates a friction-free system, is that of a simple pendulum, which is merely an object hung from a support (Figure 7-9). If the pendulum is pulled to one side and then released, it swings to the other side and then back, reaching an equal height on each side (as long as friction is negligible). At the lowest point of its swing it has its maximum KE, at the greatest height of its swing on either side it has no KE and a maximum PE. Its KE at the bottom is equal to its PE at the top. At intermediate points it has both KE and PE, the sum of which remains constant throughout the swing.

There are many interesting examples of the conservation of energy in athletics, one of which is the pole vault illustrated in Figure 7-10. In terms of energy, the sequence of events is as follows: KE of the running athlete is transformed into elastic PE of the bending pole, and as he leaves the ground, into gravitational PE; then, as he reaches the top and the pole straightens out again, it has all been transformed to gravitational PE. (We ignored his low speed as he travels over the bar.) The pole does not supply any energy, but it acts as a very convenient device to *store* energy and thus aid in the transformation of KE into gravitational PE, which is the net result. The energy required to pass over the bar depends on how high the center of mass[†] (c.m.) of the vaulter must be raised. By bending their

FIGURE 7-9

The simple pendulum illustrates conservation of energy.

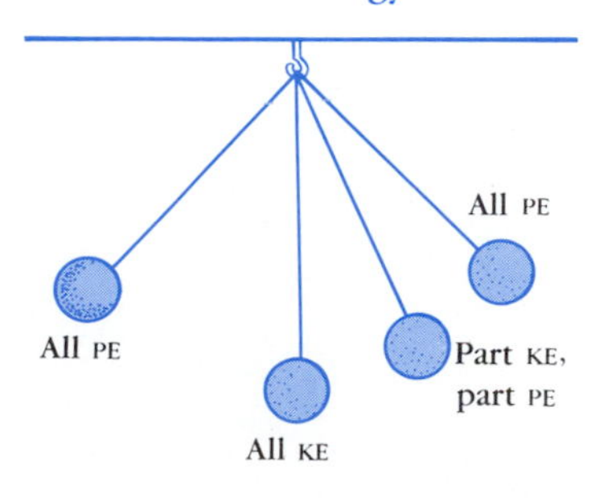

[†] The center of mass (c.m.) of a body is that point where the entire mass of the body can be considered to be concentrated for the purpose of describing its motion.

FIGURE 7-10
Transformation of energy during a pole vault.

bodies, pole vaulters can keep their c.m. so low that it actually passes slightly beneath the bar, thus enabling them to cross over a higher bar than would otherwise be possible.

Friction Gives Rise to Thermal Energy

Up to now we have neglected friction; but it is always present in real situations, although sometimes it is small enough to ignore.

Since friction affects the swing of a pendulum, the pendulum bob does not reach the same height on each swing and will eventually come to rest. The car moving between the two hills in Figure 7-8(a) will not reach quite the same height on the second hill because of friction.

Do these observations throw doubt on the concept of the conservation of energy? After all, aren't we saying that the sum of the kinetic and potential energies of an object decreases instead of staying constant?

It was questions of this kind that prevented a comprehensive law of conservation of energy from emerging until the mid-nineteenth century. Not until then was heat, which is an inevitable product of friction, recognized as being related to energy. That friction produces heat is readily observed when you rub your hands together rapidly. Whenever friction is present, heat is produced. Quantitative studies by nineteenth-century scientists, particularly the Englishman James Prescott Joule (1818–1889), demonstrated that energy is indeed conserved in all natural processes when heat is considered a form of energy. This form of energy is properly called thermal energy.

An object moving on a level surface slows down because some of its kinetic energy becomes thermal energy. In Figure 7-8, as the car travels between the two hills some of its kinetic energy is transformed into thermal energy because of friction. Both the ground and the car's tires heat up (the heating up of the tires is easily noticeable when a car is driven a considerable distance). Thus, at the bottom of the hill the car's energy is slightly less than it was at the top of the hill, because some of its energy was transformed into thermal energy. And the car won't quite make it to the top of the second hill, since all along its path part of its energy is slowly being transformed to thermal energy. When the car comes to rest on the second hill, its potential energy will therefore be less than its initial potential energy. But the total energy, KE plus PE of the car, plus thermal energy (some of which is absorbed by the car, some by the road), remains constant.

And so it is with the pendulum. Friction at the point of suspension and friction on the bob due to air resistance give rise to a transformation of kinetic energy into thermal energy. In the case of the colliding billiard balls (Figure 7-4), if the balls are not very hard and highly elastic, a significant amount of energy will be transformed into heat and so kinetic energy alone will not be conserved.

Another example of the transformation of kinetic energy into thermal energy occurs when the rock in Figure 7-7 strikes the ground. As the rock falls, its original potential energy changes into kinetic energy. But when the rock hits the ground, its kinetic energy is transformed almost instantaneously into thermal energy. Both the rock and the area of the ground where the rock hits become warmer.

EXPERIMENT

To observe the transformation of kinetic energy into thermal energy, strike a nail or a piece of metal several times in quick succession with a hammer and then touch the metal gently. What do you observe?

Other Forms of Energy: Energy Is Conserved in All Processes

When other forms of energy are involved, such as chemical or electrical energy, we find that the law of conservation of energy is still valid. The kinetic energy of rushing water can be transformed into electrical energy (Figure 7-11). Electrical energy can be transformed into light energy. The chemical energy of gasoline can be transformed into the kinetic energy of an automobile. In all these processes the total amount of energy is conserved; none is lost. However, in any energy transformation some thermal energy is always produced. For example, most of the kinetic energy of the rushing water is transformed into electrical energy, but some goes into thermal energy. The sum of the electric energy and the thermal energy produced in the process is just equal to the loss of kinetic energy of the water. The total amount of energy remains constant.

The lever problem we discussed at the beginning of this chapter, shown in Figure 7-2, is another example of the conservation of energy. The man does work to push the lever down. To do this work, he has to expend

FIGURE 7-11
Kinetic energy of water falling over a dam can be transformed into electrical energy by a generator (see Chapter 21).

energy, which comes ultimately from the chemical energy of the food he has eaten. The chemical energy that is used up is just equal to the work he does on the lever plus that transformed to thermal energy; and the work done on the lever goes into increasing the potential energy of the rock as it is raised. Chemical energy is transformed into the potential energy of the rock, plus a small amount of thermal energy. No energy is lost in the process.

Why Is the Energy Concept Important?

We speak of energy as something that cannot be created or destroyed, and as something that can be transferred from one object to another. But this does not mean that energy is a kind of material. Rather, energy is an abstract concept invented by human minds. Kinetic energy, for instance, is defined as a product of the factors mass and velocity squared. Gravitational potential energy involves the weight and the height of an object above the ground. Some physicists might say that energy is a mathematical function. This, of course, doesn't lessen its importance. That we can define a quantity that can take various forms (such as $\frac{1}{2}mv^2$ or mgh) but that always remains constant is very remarkable—and very useful.

What makes energy so valuable is the fact that it is *conserved*. What is so valuable about a quantity that is conserved? To a scientist who is looking for order in the universe, the existence of a quantity that does not change, no matter what happens, is gratifying. The universe seems in a state of constant flux, one thing interacting with another; but within this complex of interactions there is a quantity that remains constant—energy. It is a unifying principle and helps to bring order to the natural world.

Furthermore, the fact that energy is conserved makes it easier to solve practical problems. For example, you can figure out the velocity of the car in Figure 7-8 at any point by using the conservation of energy principle. Suppose, for instance, that the car starts from rest at the top of the hill. If the height of the hill is, say 30 m, then the car loses potential energy equal to mgh where h is the height of the hill. This is all transformed to kinetic energy, so $\frac{1}{2}mv^2 = mgh$. We solve for v and get $v = \sqrt{2gh}$. And if we set $h = 30$ m, then $v = \sqrt{2(9.8 \text{ m/s}^2)(30 \text{ m})} = 24$ m/s.

It can be shown that the conservation of energy principle is essentially equivalent to Newton's laws; therefore it gives us an alternative, and sometimes easier, method of analyzing nature and of solving practical problems.

The conservation of energy is not the only conservation law scientists have conceived. A number of others have come to light, and in later chapters we will discuss some of them. Today many physicists feel that the conservation laws are the most basic laws in nature. Therefore, it may be that the law of conservation of energy is more basic than Newton's laws themselves.

4. POWER

Power is defined as the rate at which work is done, or as the rate at which energy is transformed:

$$P = \frac{\text{work}}{\text{time}} = \frac{\text{energy transformed}}{\text{time}}.$$

For example, power refers to how much work a horse or an engine can do per second; or to how much electrical energy per second an electric heater transforms into heat energy; or to the rate at which an electric light bulb transforms electric energy into light energy.

In the metric system, power is measured as joules per second. This unit is given a special name, the *watt*: 1 watt = 1 joule/second (1 W = 1 J/s). The watt is used most often in electricity, where the electric power transformed by an appliance or a light bulb is given in watts. For mechanical systems such as engines and automobiles, power is sometimes specified in English units. The usual English unit is foot · pounds per second; for most practical purposes a larger unit, the horsepower, is normally used. The horsepower unit was devised by James Watt (1736–1819), who also developed the steam engine. Watt needed some way to compare the work of his engines with the work a horse could do, since most work at the time was done by horses. He made various measurements of the ability of horses to do work and came to the conclusion that a typical horse could work all day at the rate of about 360 ft · lb/s. But, being a careful and conscientious man, he did not want to overrate his steam engines. So he multiplied his measured value for the power output of an average horse by about $1\frac{1}{2}$ and thus defined the horsepower (hp) unit as 1 hp = 550 ft · lb/s. (The capabilities of Watt's early steam engines ranged from about 4 to 100 horsepower.) The metric horsepower is defined as 750 watts and is almost identical to the British horsepower (which equals 746 watts).

To clarify the difference between power and energy, let us consider a simple example. A person can walk quite a distance, or climb many flights of stairs, before becoming tired. But eventually so much energy is used up that the person will have to stop. A person is limited in physical exertion by the amount of energy stored in the muscles that can be used for doing work. But this is not the only limitation. A person may be able to climb 20 flights of stairs slowly; but if the person runs up the stairs very rapidly, exhaustion may occur after only one flight. Thus a human being is limited not only by the total amount of energy required for a task, but also by the *rate* at which the energy is used, or power.

The maximum rate at which energy can be expended puts a limit on the acceleration of an automobile. That is why automobile engines are rated in horsepower. The brightness of a lamp depends on the rate at which energy is transformed. Thus, light bulbs are rated by power—60 watts, 100 watts, and so on.

EXPERIMENT-PROJECT

Measure your maximum power output by timing how many seconds it takes you to run up a flight of stairs. Your power output will be equal to the increase in your potential energy (your weight times the *vertical* height of the staircase) divided by the time it took you to reach the top. Give your power output in horsepower.

For example, if a 50 kg woman can run up a 4 m high flight of stairs in 5 s, her power output is her change in potential energy [$mgh = (50\ \text{kg})(9.8\ \text{m/s}^2)(4\ \text{m}) = 2000\ \text{J}$] divided by the time [5s]: Power = $2000\ \text{J}/5\text{s} = 400$ watts, or slightly more than $\frac{1}{2}$ hp.

SUMMARY

The *work* done by a force is defined as the product of the magnitude of the force and the distance through which the force acts. *Simple machines*, such as a lever, do not save us work but they offer a *mechanical advantage*, which is the ratio of the output force to the input force.

Energy can be defined as the ability to do work. *Kinetic energy* (KE) is the energy of motion. For a body of mass m and velocity v, its KE is $\frac{1}{2}mv^2$. *Potential energy* (PE) is the energy an object has because of its shape or position. One type of PE is gravitational PE. A body of mass m held a height h above the ground can be said to have a PE equal to mgh. Other forms of energy include thermal, electrical, and nuclear energy.

Although energy can be transformed from one kind to another, or transferred from one object to another, the total amount of energy remains constant; this is the *law of conservation of energy*.

Power is defined as the rate at which work is done or at which energy is transformed.

QUESTIONS

1. Work depends on what two factors?
2. The pyramids of Egypt were built with the help of inclined planes, or ramps. Explain why an inclined plane can be considered a simple machine. Does it have a mechanical advantage?
3. In which case is more work done: when a 50-kg bag of groceries is lifted 50 cm or when a 50-kg crate is pushed 2 m across the floor with a force of 50 N?

4. How does the meaning of the term "energy" as used in this chapter differ from everyday use of the term? In what ways is it the same?
5. In the old West, a rifle with a long barrel was considered extremely valuable. Explain why, using the fact that the gases from the exploding shell act on the bullet over a longer distance than they do in a rifle with a short barrel.
6. A moving billiard ball strikes a second ball and imparts some of its kinetic energy to the latter. Did it do work on the second ball?
7. Why do the stopping distances of a car when the brakes are applied increase with the square of the car's speed?
8. Which has more kinetic energy, a 1000-kg car traveling 80 km/hr or a 2000-kg car traveling 40 km/hr?
9. A compressed spring has potential energy. How do you know this? Why don't we call it "gravitational" potential energy?
10. Does a stretched rubber band have potential energy? How do you know?
11. An overloaded car sometimes has difficulty climbing a steep hill. Explain.
12. Describe all the energy transformations that take place when you throw a ball into the air, when the ball reaches its maximum height and descends, and finally, when you catch it.
13. Describe all the energy transformations that occur when a sled accelerates as it slides down a steep hill, reaches soft snow halfway down, and slowly comes to rest before it reaches the bottom.
14. Describe the transformations of energy that occur when a child hops around on a pogo stick.
15. What is the purpose of a spring that must be wound up in a pendulum clock?
16. A skier at the top of a hill is ready to descend. Does she have kinetic or potential energy? What happens to this energy as she descends the hill?
17. What does "conservation" mean?
18. A roller coaster car is usually pulled by a chain drive to the top of the first rise. It then coasts on its own (see Figure 7-12). Where does it reach maximum kinetic energy? Where does it reach maximum potential energy? Assuming little or no friction, will the roller coaster make it over all the rises shown? What will happen if a large amount of friction is present?
19. When a car is braked, why do the brakes heat up?
20. A toy rocket is fired into the air. What kind of energy does it have just before it hits the ground on its return? What happens to this energy when it crashes into the ground?

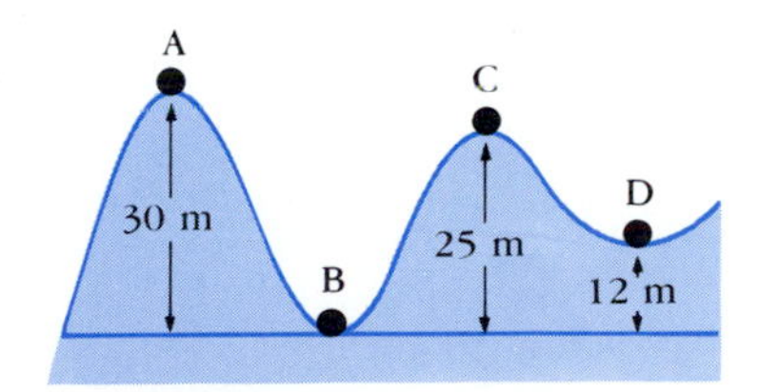

FIGURE 7-12
A roller-coaster track.

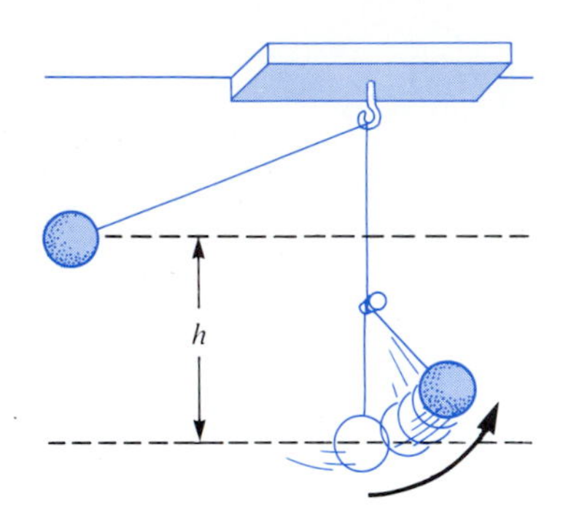

FIGURE 7-13
A pendulum cord strikes the peg.

21. A stone that is thrown with a particular speed from the top of a cliff will enter the water below with the same speed whether it is thrown horizontally or at an angle. Why?
22. A pendulum bob is swung so that it reaches a maximum height h above its lowest point. Suddenly a peg is placed as shown in Figure 7-13. How high will the pendulum rise on the right side now? Show this on a diagram.
23. When a "superball" is dropped, can it rebound to a greater height than its original height?
24. When an experienced hiker comes to a log across his path, he prefers to step over it rather than surmounting it and jumping down on the other side. Why?
25. Why is it easier to climb a mountain via a zigzag trail rather than to climb straight up?
26. Can an electric heater be rated in horsepower as well as in watts? Can an automobile be rated in watts as well as in horsepower?

EXERCISES

1. A horse exerts a 900-N force to pull a 600-kg load 2.5 km. How much work did the horse do?
2. A 50-kg woman climbs a flight of stairs 6.0-m high. How much work is required?
3. A 550-kg crate rests on the floor. How much work is required to move it at constant speed (a) 2.0 m along the floor against a friction force of 150 N and (b) 2.0 m vertically?
4. How far must a 200-kg pile driver fall if it is to be capable of doing 13,000 J of work?
5. Six bricks, each 6.0 cm thick with mass 1.5 kg lie flat on a table. How much work is required to stack them one on top of another?
6. Calculate the kinetic energy of a 55-kg person running 9.0 m/s.
7. When a bicycle's speed is doubled, by what factor does its kinetic energy change?
8. If the mass of a moving object is doubled without changing its speed, by what factor does its kinetic energy change?

9. How much farther will a car travel after the brakes are applied when it is going 90 km/hr than when it is going 30 km/hr? (Hint: Use ratios.)
10. A pile driver is raised to a height of 7.0 m. A pile driver of only half the mass must be raised to what height to have the same amount of potential energy?
11. How much work is required to stop a 1000-kg car traveling at 100 km/hr?
12. A 100-kg rocket has 40,000,000 joules of kinetic energy. How fast is it moving?
13. A 10-kg monkey swings from one branch to another 220 cm higher. What is its change in potential energy?
14. A woman skies without friction down a 30° hill 10 m high. What is her speed at the bottom?
15. A car rolls from rest down a hill with no friction. At the bottom of the hill its speed is 10 m/s. How high is the hill?
16. A 65-kg hiker starts at an elevation of 1500 m and climbs to the top of a 2600-m peak. (*a*) What is the hiker's change in potential energy? (*b*) What is the minimum work required of the hiker? (*c*) Can the actual work done be more than this? Explain.
17. Two railroad cars, each of mass 4200 kg and traveling 80 km/hr, collide head-on and come to rest. How much thermal energy is produced in this collision?
18. A roller coaster is shown in Figure 7-12. Assuming no friction, calculate the speed (of its car) at points B, C, D, assuming it starts from rest at point A.
19. Suppose a car starts coasting from the top of the hill that is 50 m high. How fast will it be going at the bottom of the hill if there is no friction?
20. Calculate the speed of the car in Exercise 19 when it is halfway down the hill.
21. A 65-kg man runs up a 3.5 m high flight of stairs in 5.0 s. What is his power in horsepower?
22. How long will it take a 2-hp motor to lift a 300-kg piano 12 m to a fifth story window?
23. One horse does 1200 joules of work in 2 hours. A second horse does 1000 joules in $1\frac{1}{2}$ hours. For which horse is the power output greater?
24. To move a car along a level road at 100 km/hr requires a force of about 400 N. Approximately how much horsepower is required? Do you think high-powered engines capable of 200 or more horsepower are justified?
25. A car generates 10 hp when traveling at a steady 20 m/s. What must be the average force on the car due to air resistance and friction?

*26. Suppose the car in Exercise 19 actually reaches the bottom of the hill with a speed of only 20 m/s because of friction. How many joules of thermal energy were produced? Assume the mass of the car is 1500 kg.

*27. A 2-kg rock is thrown with a speed of 10 m/s from the top of a cliff 30 m high. With what speed does it hit the ground below? (Hint: The rock has both KE and PE initially.)

*28. Calculate the kinetic energy and the velocity required for a 70-kg pole vaulter to pass over a 5.0-m-high bar. Assume the vaulter's center of mass is initially 0.90 m off the ground and reaches its maximum height at the level of the bar itself.

MOMENTUM AND ITS CONSERVATION

CHAPTER

8

The law of conservation of energy, which we discussed in the previous chapter, is one of several great conservation laws in physics. Among the other quantities found to be conserved are linear momentum, angular momentum, and electric charge. We will eventually discuss all of these because the conservation laws are among the most important laws in all of science. In this chapter we discuss linear momentum (often simply called "momentum") and its conservation. The law of conservation of momentum is particularly useful when dealing with two or more bodies that interact with each other, such as in collisions. Our focus up to now has been mainly on the motion of a single object. But in this chapter we will deal with systems of two or more objects. By a *system* we simply mean any group of objects we choose to deal with.

1. MOMENTUM AND ITS RELATION TO FORCE

The **linear momentum** of a body is defined as the product of its mass (m) and its velocity (v):

$$\text{momentum} = \text{mass} \times \text{velocity}.$$

In symbols we write $p = mv$, where p stands for momentum. The momentum of an object, as we shall see shortly, is often a more useful quantity than its mass or its velocity alone. Note that because velocity is a vector, so is momentum; the direction of the momentum of an object is the same as the direction of its velocity.

A 3000-kg truck traveling north along a highway at 15 m/s has momentum of magnitude $p = mv = (3000\text{ kg})\ (15\text{ m/s}) = 45{,}000\text{ kg·m/s}$; the direction of the momentum is north, the direction of the velocity.

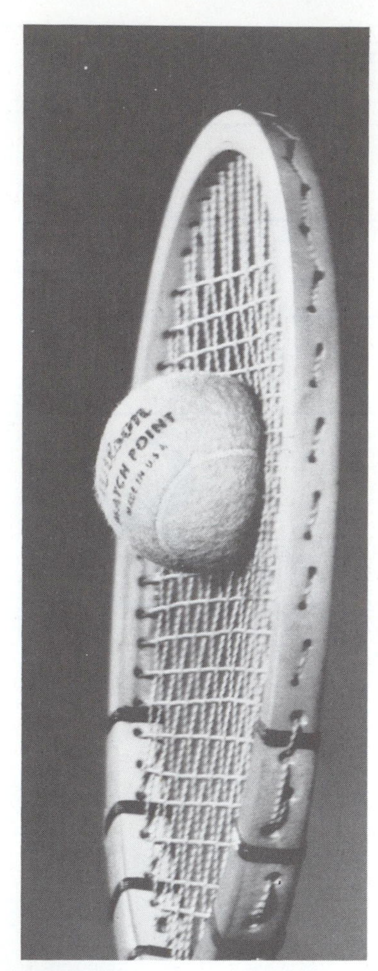

FIGURE 8-1
Tennis racket striking a ball. Note deformation of both ball and racket due to the large force each exerts on the other. (Photograph by Russ Kinne, Photo Researchers, Inc.)

According to our definition, a moving object will have a large momentum if either its mass or its velocity, or both, is large; this is in accordance with our everyday use of the term momentum. A fast-moving car has more momentum than a slow-moving car; and a heavy truck has more momentum than an automobile moving at the same speed. The more momentum an object has, the greater will be its effect if it strikes a second object. For example, in a collision, a fast-moving car will cause a more serious accident than a slow-moving car, and a heavy fast-moving truck will cause even more damage. Similarly, a football player is more likely to be stunned when he is tackled by a huge opponent running at top speed than by a lighter or slower-moving tackler. It is the product of mass times velocity—*momentum*—that is important.

A force is required to change the momentum of an object whether it is to increase the momentum, decrease it (such as to bring a moving object to rest), or to change its direction. Newton in fact originally stated his second law in terms of momentum (although he called the product mv the "quantity of motion"). Newton's statement of the second law, translated into modern language, is as follows: "the rate of change of momentum of a body is proportional to the net force applied to it." We can write this as

$$\text{Force} = \frac{\text{change in momentum}}{\text{time}}.$$

This is equivalent to our earlier statement of Newton's second law, $F = ma$, if the mass of the object doesn't change. To show that this is true, we note that because momentum is mass times velocity ($p = mv$), change in momentum equals the mass times the change in velocity†. Thus we can write, starting from the relation above and using symbols:

$$F = \frac{\text{change in } p}{t} = m \times \frac{\text{change of } v}{t}.$$

Since change of velocity per time is acceleration (Chapter 3), we have $F = ma$, which is our earlier statement of Newton's second law.

2. IMPULSE

We have just seen that Newton's second law can be written in terms of momentum as

$$F = \frac{\text{change in momentum}}{t},$$

where F is the net force acting on the body during the time t. If we multiply

† If $p = mv$ is the momentum of an object at one moment, and $p' = mv'$ is the momentum of the object a short time later, then the change in momentum is $p' - p = mv' - mv = m(v' - v)$, which is mass times the change in velocity.

both sides of this relation by t, we get another useful relation:

$$\text{change in momentum} = F \times t.$$

In words, the change in momentum of an object is equal to the net force acting on the body multiplied by the time during which this force acts. The same change in momentum can be accomplished by a large force acting over a short time, or a small force acting over a long time. The quantity $F \times t$ is called the **impulse**. Thus we can say that the change in momentum of a body is equal to the impulse it receives.

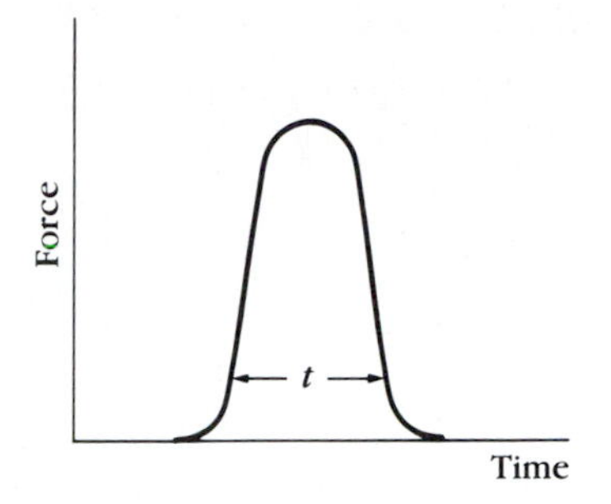

FIGURE 8-2
Force as a function of time during a collision; t is the time during which the force acts (the impulse time).

The concept of impulse is most useful when the force acts over a very short time, such as when there is a collision between two billiard balls, a hammer strikes a nail, or a tennis racket hits a tennis ball (Figure 8-1). The force each object exerts on the other in such a collision normally acts for a very short time only. But this brief force can be very large, and thus can cause considerable deformation of the objects while it acts; note the deformation of the tennis ball and racket in Figure 8-1. When the collision occurs, the force normally jumps from zero at the moment of contact to a very large value within a short time, and then abruptly drops to zero again, as shown in Figure 8-2.

As an example, suppose a tennis ball (mass = 60 gram = 0.060 kg) traveling at 40 m/s is struck by a racket and flies off in the opposite direction with the same speed (Figure 8-3). Assuming a reasonable contact time between racket and ball of 0.030 s, let us determine the average force on the ball during the collision. We use the preceding relation between change in momentum and impulse,

$$\text{change in momentum} = F \times t.$$

We can solve for F because we are given t and we can readily determine

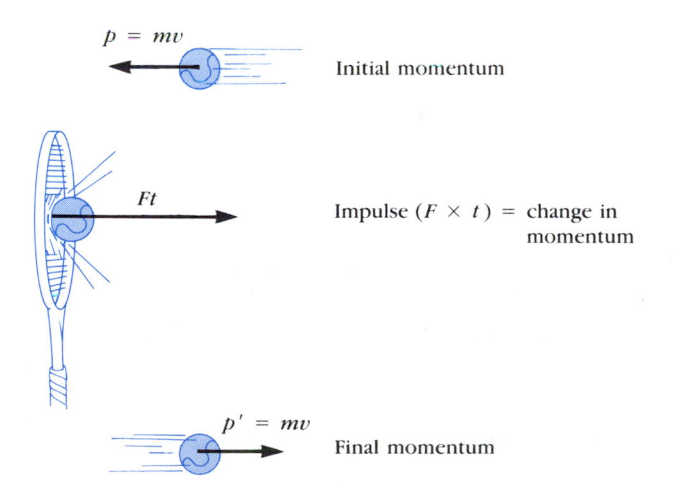

FIGURE 8-3
Tennis ball with momentum $p = mv$ strikes a racket that gives it an impulse (Ft) directed to the right, which changes the ball's momentum so that it then has momentum $p' = mv$ in the opposite direction.

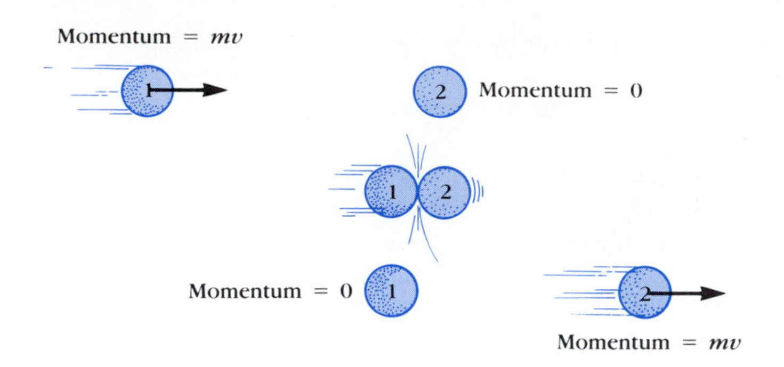

FIGURE 8-4
Collision of two equal-mass hard elastic balls.

the change in momentum of the ball[†]. In Figure 8-3, it takes an impulse of mv, pointing to the right, to bring the ball to rest; and it takes another impulse of mv to the right to speed it back up to v going to the right. So the total impulse required is $2mv$. Thus, since $Ft = 2mv$, the average force F over the time $t(= 0.030\ \text{s})$ is $F = 2mv/t = 2(0.060\ \text{kg})(40\ \text{m/s})/(0.030\ \text{s}) = 160\ \text{N}$. This is the average force exerted by the racket on the tennis ball during the impulse. (It is a force large enough to lift a 16-kg mass.)

3. CONSERVATION OF MOMENTUM

The real importance of the concept of momentum lies in the fact that, under certain circumstances, it is a conserved quantity. In the mid-seventeenth century, shortly before Newton's time, it had been observed that the sum of the momenta of two colliding objects *remains constant*. Although the momentum of each of the two bodies may change as a result of the collision, the sum of the two momenta is the same after the collision as it was before. Figure 8-4 illustrates the conservation of momentum for the head-on collision of two hard elastic balls of equal mass, one of which is initially at rest. In this particular case, the cue ball comes to rest as a result of the collision and ball 2 moves off with the same speed that the cue ball had initially. The momentum of the cue ball is completely transferred to ball 2. Momentum is conserved in the collision. If ball 2 had been

---

[†] Another way to see this is as follows: Although the magnitude of the ball's momentum (mv) doesn't change in this example, the direction does. To take direction into account, we consider a vector pointing horizontally to the right as positive and one pointing to the left as negative. Then the initial momentum in Figure 8-3 is $p = -mv$, whereas the momentum after being struck by the racket is $p' = +mv$. So the change in momentum of the ball is

$$\begin{aligned} p' - p &= mv - (-mv) \\ &= 2mv. \end{aligned}$$

And this is just equal to the impulse Ft:

$$Ft = 2mv.$$

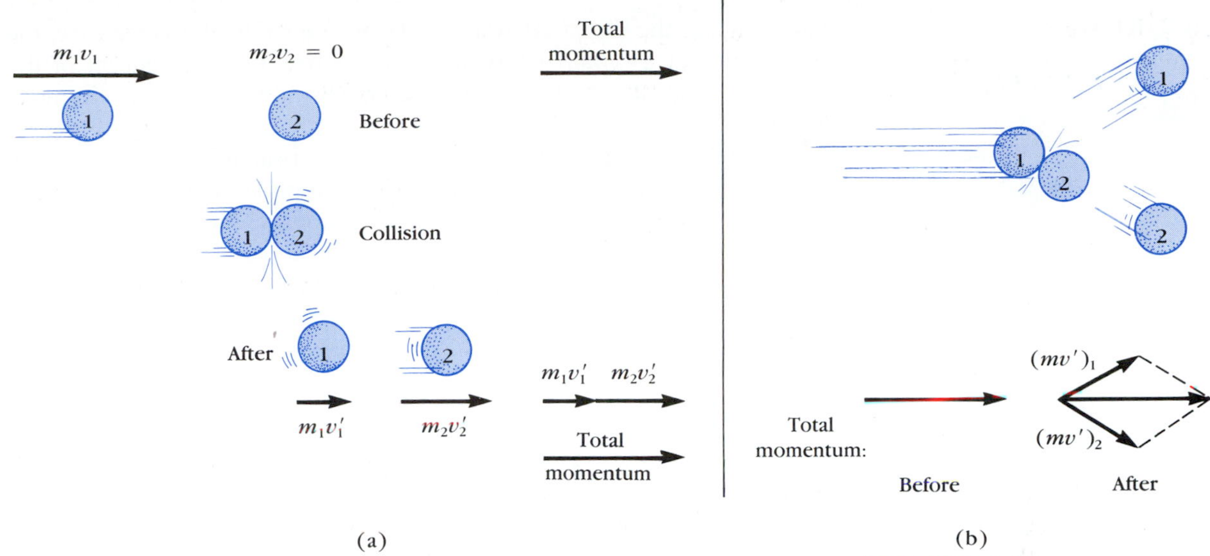

FIGURE 8-5

Momentum conservation in collisions of two balls: (a) head-on collision of balls of different masses, (b) glancing collision.

lighter than the cue ball, it would have moved off at a higher speed to compensate for its smaller mass.

If both balls are found to be moving after the collision, which is generally the case if they have different masses (Figure 8-5(a)), or if the two balls make a glancing collision and go off at an angle (Figure 8-5(b)), then we must use vector addition. In either example of Figure 8-5, the total momentum before the collision is simply the momentum of ball 1. After the collision both balls have momentum, and the momentum of ball 1 is added to that of ball 2, as a vector, to give the total momentum; this total momentum after the collision is equal to the total momentum before the collision. Both the *magnitude* and the *direction* of the *total* momentum remain constant. (If both objects are moving initially, then the total initial momentum is the vector sum of each of their momenta.)

EXPERIMENT

You can observe collisions of this sort by using billiard balls, or by skidding ice cubes across a wet table. Try to estimate the speed for a number of collisions and see if momentum is conserved. See if you can produce a collision in which momentum is *not* conserved. What happens when one of the balls or ice cubes rebounds at an angle—does the other do so as well? If so, does it rebound to the same side as the first one or to the opposite side? Explain.

Momentum is conserved even when two colliding objects stick together, as when two pieces of putty collide or when a football player

FIGURE 8-6
Momentum is conserved when the player leaps in the air to catch a pass.

(a)

(b)

leaps high in the air to catch a pass (Figure 8-6). In the latter case, the momentum of the ball before the catch equals the momentum of the ballplayer and the ball together afterward, and so they both move backward—Figure 8-6(b).

Although the conservation of momentum principle was discovered experimentally, it is closely connected to Newton's laws of motion and can in fact be derived from them. We show it now for the one-dimensional case illustrated in Figure 8-7, and we look at the general case when two objects are moving both before and after the collision. According to Newton's second law as discussed earlier in this chapter, the impulse Ft applied to, say, ball one is equal to its change in momentum:

$$Ft = m_1v_1' - m_1v_1,$$

where v_1 represents the velocity initially and v_1' the velocity after the collision. During the collision, the force F on ball 1 is exerted by ball 2; and according to Newton's third law, ball 1 exerts an equal and opposite force $(-F)$ on ball 2. Thus the change in momentum of ball 2 equals $-Ft$, which is exactly equal and opposite to the change in momentum of ball 1: $m_1v_1' - m_1v_1 = -(m_2v_2' - m_2v_2)$. Thus, any momentum lost by one ball is gained by the other. Hence, the total momentum remains constant. This equation is readily rearranged to give

$$m_1v_1' + m_2v_2' = m_1v_1 + m_2v_2,$$

which says that the total momentum of the two balls before the collision is equal to the total momentum of the two balls after the collision. This argument can be extended to include any number of interacting bodies.

The above argument is valid if the only forces involved are those between the two bodies. But if external forces are present—that is, forces produced by other objects—then the sum of the momenta of the two bodies will not be conserved; it will change as a result of those external forces. However, if we include the other objects in our "system," then the total momentum of all the interacting bodies will be conserved.† The general statement of the **law of conservation of momentum** is:

> **The total momentum of a group of objects is the same after they interact as it was before, so long as no external forces act.**

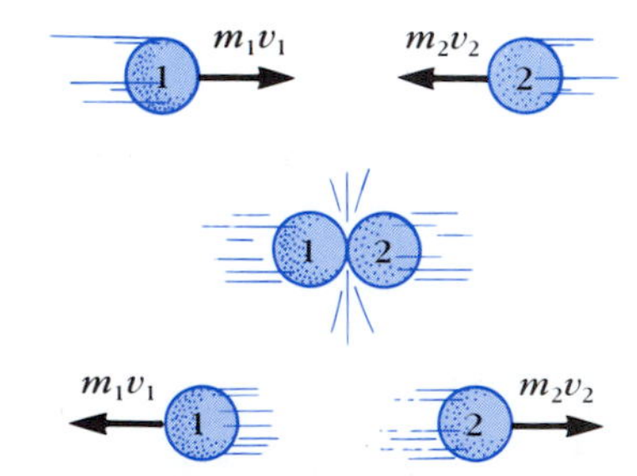

FIGURE 8-7
Momentum is conserved in collision of two balls.

Let's take a simple example. Suppose a railroad car traveling 10 m/s strikes an identical car at rest. If they couple as a result of the collision, they

† For example, if we take as our system a falling rock, it does not conserve momentum since an external force, the force of gravity exerted by the earth, is acting on it. However, if we include the earth in the system, the total momentum of rock plus earth is conserved. (This of course means that the earth comes up to meet the ball; naturally you don't notice this motion of the earth since its mass is so great that the required velocity is very tiny.)

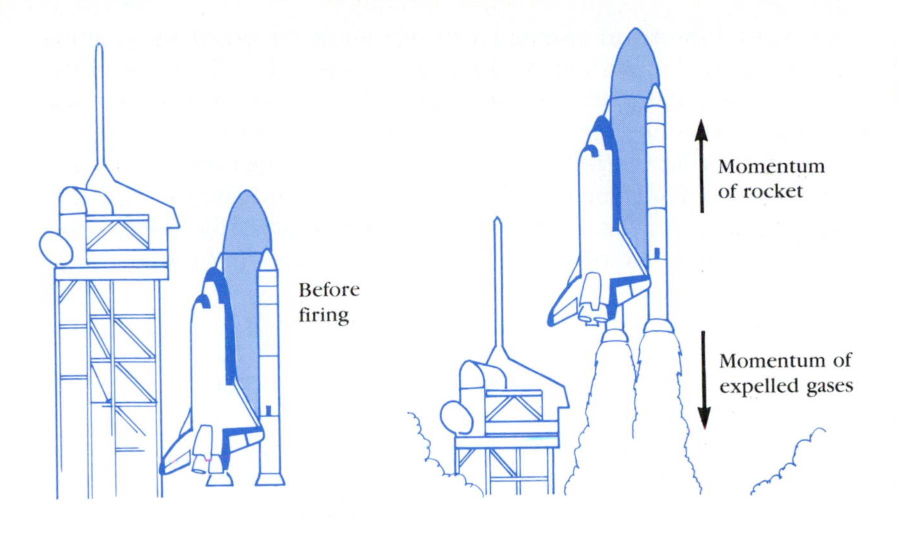

FIGURE 8-8
Total momentum is zero both before and after firing a rocket.

FIGURE 8-9
(a) Total momentum = 0. (b) The bullet moves forward and the gun recoils backward. Total momentum is still zero.

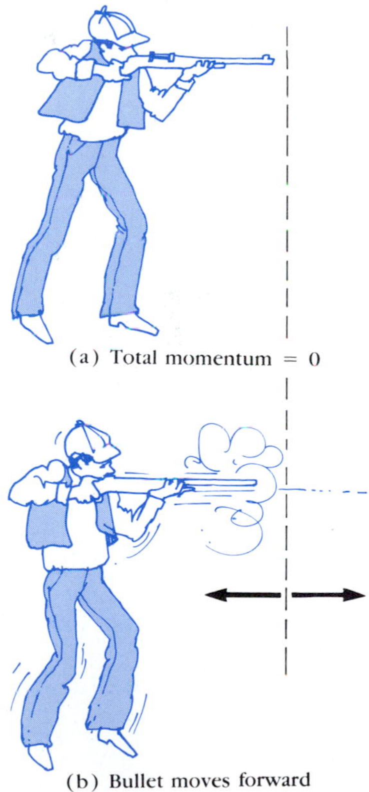

will move off with equal speeds. Since momentum is conserved, the momentum before the collision equals the momentum after:

$$(mv)_{\text{before}} = (mv)_{\text{after}}.$$

Before the collision, only one car is moving: with a speed $v = 10$ m/s. After the collision, twice as much mass is moving; therefore, the velocity must be half as much: 5m/s.

The law of conservation of momentum is particularly useful when dealing with fairly simple systems such as collisions and certain types of explosions. For example, rocket propulsion, which we saw in Chapter 4 can be understood on the basis of action and reaction, can also be explained on the basis of the conservation of momentum (Figure 8-8). Before a rocket is fired, the total momentum of rocket plus fuel is zero. As the fuel burns, the total momentum remains unchanged: the backward momentum of the expelled gases is just balanced by the forward momentum of the rocket itself. Thus, as we saw earlier, a rocket is not propelled because the expelled gases push on the ground or the atmosphere. Rather it is the conservation of momentum, or alternatively Newton's third law, that explains rocket propulsion.

The recoil of a gun is similarly an example of the conservation of momentum—Figure 8-9. Before firing, the gun and the bullet have zero momentum; they are both at rest. When the trigger is pulled, a minor explosion takes place, propelling the bullet down the barrel. The large forward momentum of the bullet must be compensated by the backward momentum of the gun, its *recoil*. The total momentum, the vector sum of the bullet's forward momentum and the gun's backward momentum, remains zero. The gun's velocity (recoil) is much less than that of the bullet's, since it has a much greater mass. For example, suppose a 4.0-kg rifle fires a 50-gram (0.050-kg) bullet at a speed of 280 m/s. The recoil speed of the rifle can be calculated using conservation of momentum. Because the total momentum before firing is zero (we assume that the rifle is at rest), it must be zero afterward; so the momentum of the bullet equals the momentum of the rifle (but they are in opposite directions):

$$(4.0\ \text{kg})\,v = (0.050\ \text{kg})(280\ \text{m/s}),$$

where v is the recoil speed of the rifle. Solving for v, we find $v = (0.050\ \text{kg})(280\ \text{m/s})/(4.0\ \text{kg}) = 3.5\ \text{m/s}$.

The laws of conservation of energy and conservation of momentum give us an alternative to Newton's laws in analyzing motion. But there is a difference between these two approaches. When we use the conservation laws, we can ignore many of the details of the process. We are mainly concerned about the positions and velocities of the objects before and after the interaction. We do not explicitly consider the forces that act during the interaction; in fact, in situations such as collisions, it is difficult to measure the strength of the forces between the bodies. This is one reason why the conservation laws are so useful. We can use them to analyze a situation even though we do not know the forces involved in detail.

4. WHICH IS MORE IMPORTANT— ENERGY OR MOMENTUM?

In the seventeenth century, scientists discovered that the quantities mv and $\frac{1}{2}mv^2$ are conserved in collisions of two hard elastic balls. Soon afterward, a disagreement arose over whether momentum, mv, or kinetic energy, $\frac{1}{2}mv^2$, was the true measure of the "quantity of motion" in a body. This discussion could have no rational resolution, of course, other than that both momentum and energy are important quantities. They are two quite different quantities. Not only does one have the velocity squared and the other simply the velocity; but momentum is a vector, whereas kinetic energy is a scalar. Both kinetic energy and momentum are useful quantities associated with the motion of a body. Both are conserved. One is not "better" than the other. They illuminate different aspects of motion. Though energy is perhaps used more often, momentum can be enlightening as well. For example, in the firing of a gun, the total energy is conserved as well as the momentum. But from energy conservation alone one would have no idea that recoil would occur; only momentum conservation can reveal that.

5. COLLISIONS: ELASTIC AND INELASTIC

During most collisions, we usually don't know how the collision force varies as a function of time. Nonetheless, given the initial motion, we can still determine some of the details of the motion after a collision by making use of the conservation laws for momentum and energy. We saw earlier in this chapter that in the collision of two objects such as billiard balls the total momentum is conserved. If the two objects are very hard and elastic and no heat is produced in the collision, then kinetic energy is conserved as well. By this we mean that the sum of the kinetic energies of the two objects is the same after the collision as before. Of course for the brief moment during which the two objects are in contact, some (or all) of the energy is stored momentarily in the form of elastic potential energy; but if we compare the total kinetic energy before the collision with the total after the collision, they are found to be the same. Such a collision is called an *elastic collision*. If we use the subscripts 1 and 2 to represent the two objects, we can write the equation for conservation of total kinetic energy as

$$\frac{1}{2}m_1v_1^2 + \frac{1}{2}m_2v_2^2 = \frac{1}{2}m_1v_1'^2 + \frac{1}{2}m_2v_2'^2 \quad \text{[elastic collision]}$$

where primed quantities (′) mean after the collision and unprimed mean before the collision.

Although at the atomic level the collisions of atoms and molecules are often elastic, in the "macroscopic" world of ordinary objects an elastic collision is an ideal that is never quite reached since at least a little thermal energy (and perhaps sound and other forms of energy) is always produced during a collision. This collision of two hard elastic balls, such as billiard balls, however, is very close to perfectly elastic and we often treat it as such. Even when the KE is not conserved, the *total* energy is, of course, conserved.

Collisions in which kinetic energy is not conserved are said to be *inelastic*. The kinetic energy that is lost is changed into other forms of energy, usually thermal energy, so that the total energy is conserved. In this case we can write that $\text{KE}_1 + \text{KE}_2 = \text{KE}_1' + \text{KE}_2' +$ thermal and other forms of energy. In this case, the total final KE is less than the total initial KE. The inverse can also happen when potential energy (such as chemical or nuclear) is released, and then the total final KE can be greater than the initial KE. If two objects stick together as a result of a collision, the collision is said to be *completely inelastic*. Two colliding balls of putty that stick together, or two railroad cars that couple when they collide (such as we discussed in Section 3 of this chapter) are examples. The kinetic energy is not necessarily all transformed to other forms of energy in an inelastic collision. In our earlier example of colliding train cars, for instance, we saw that when a traveling railroad car collided with a stationary one, the coupled cars traveled off with some KE. In that example (see Section 3) we found, using conservation of momentum, that if a railroad car traveling 10 m/s strikes an identical car at rest and if they couple as a result of the collision, they move off together with a speed of 5 m/s. If the mass of each car is, say 10,000 kg, then the kinetic energy before collision was $\frac{1}{2}mv^2 = \frac{1}{2}(10{,}000\text{ kg})(10\text{ m/s})^2 = 500{,}000\text{ J}$. After the collision, the total KE is $\frac{1}{2}(10{,}000\text{ kg} + 10{,}000\text{ kg})(5\text{ m/s})^2 = 250{,}000\text{ J}$, which is only half the original KE. The other half must have been transformed into other forms of energy, such as thermal energy and perhaps sound.

Though KE is not conserved in inelastic collisions, the total energy *is* conserved, and the total vector momentum is also conserved.

SUMMARY

The *momentum*, p, of a body is defined as the product of the body's mass and its velocity: $p = mv$. In terms of momentum, Newton's second law can be stated as, "the rate of change of momentum of a body is equal to the net force on it." This can be written:

$$F = \frac{\text{change in momentum}}{t}$$

where F is the net force on the body acting during the time t.

Another way of writing this is

$$\text{change in momentum} = Ft$$

and the product $F \times t$ is called the *impulse*.

The *law of conservation of momentum* states that the total momentum of a group of objects remains constant if no external forces act on the group.

The law of conservation of momentum is very useful in dealing with *collisions*. In a collision, two (or more) bodies interact with each other for a very short time and the force between them during this time is very large. Momentum is conserved in any collision. The total energy is also conserved, but this may not be useful unless the only type of energy transformation involves kinetic energy; in this case, kinetic energy is conserved and the collision is called an *elastic collision*. If kinetic energy is not conserved, the collision is called *inelastic*.

QUESTIONS

1. We claim that momentum is conserved. Yet most moving objects eventually slow down and stop. Explain.

2. Name three everyday experiences that illustrate the conservation of momentum principle.
3. A heavy object and a light object have the same momentum. Which has the greater speed? Which has the greater kinetic energy?
4. A light body and a heavy body have the same kinetic energy. Which has the greater momentum?
5. Is it possible for an object to have momentum without having energy? Can it have energy but no momentum? Explain.
6. Is it possible for a body to receive a larger impulse from a small force than from a large force?
7. According to our discussion in Section 2, the shorter the impact time of an impulse, the greater the force must be for the same momentum change and hence the greater the deformation of the object on which the force acts. Explain on this basis the value of "air bags" which are intended to inflate during an automobile collision and reduce the possibility of fracture or death.
8. Describe how a fish moves forward by swishing its tail back and forth.
9. In a collision between two cars, which would you expect to be more damaging to the occupants: if the cars collide and remain together or if the two rebound backward? Explain.
10. When you jump down from a tree, what happens to your momentum as you hit the ground?
11. When a ball is dropped to the ground, it is said that the earth comes up to meet the ball. Explain on the basis of the conservation of momentum.
12. It is said that in ancient times a rich man with a bag of gold coins was frozen to death stranded on the surface of a frozen lake. Because the ice was frictionless, he could not push himself to shore. What could he have done to save himself had he not been so miserly?
13. How can a rocket change direction when it is far out in space and is essentially in a vacuum?
14. It is not very useful to consider energy conservation in collisions like the one when two train cars collide and couple. Why?
15. Explain, on the basis of the conservation of momentum, why it is that when a person throws a package out of a boat, the boat moves off in the opposite direction (see Figure 4-10).
16. Suppose you find yourself teetering on the edge of a cliff or the roof of a tall building (Figure 8-10) with a heavy physics book in your hand. As you are about to fall, you suddenly remember how physics can save you. What law do you remember and how do you apply it?
17. When you release a balloon that has just been inflated, why does it fly across the room?

FIGURE 8-10
You are falling off the edge. What do you do?

EXERCISES

1. Calculate the momentum of a 120-gram robin flying with a speed of 10 m/s.
2. A 90-kg fullback is traveling 5.0 m/s and is stopped by a tackler in 1.0 s. Calculate (a) the original momentum of the fullback, (b) the impulse imparted to the tackler, and (c) the average force exerted on the tackler.
3. A car traveling 30 km/hr strikes an identical second car from the rear and the two cars lock bumpers. Assuming that the brake of the second car was off and that momentum is conserved in the collision, what is the velocity of the two attached cars after the collision?
4. If you throw a 4-kg rock from your resting boat with a speed of 10 m/s, what will be the resulting speed (and direction) of your boat? (Total mass, including you, is 110 kg).
5. A 20,000-kg loaded railroad car is traveling alone on a level, frictionless track with a speed of 20 m/sec. An additional weight of 5,000 kg falls down onto the car. What will be the speed of the car now?
6. A 120-kg tackler traveling 3 m/s tackles a 75-kg halfback running 6 m/s in the opposite direction. What is their common speed immediately after the collision?
7. A 75-kg football player running 4.0 m/s leaps into the air to catch a pass. The 2.0 kg ball is traveling 25 m/s. How fast will the player be moving immediately after catching the pass?
8. An atomic nucleus at rest decays radioactively into an alpha particle and a smaller nucleus. What will be the speed of this recoiling nucleus if the speed of the alpha particle is 6.2×10^5 m/s? Assume that the nucleus has a mass 57 times greater than that of the alpha particle.

*9. A 0.145-kg baseball pitched at 35 m/s is hit on a horizontal line drive straight back at the pitcher at 50 m/s. If the contact time between the bat and ball is 5×10^{-4}s, calculate the force (assumed to be constant) between the ball and bat. Would this force be great enough to lift an average-size person?

*10. Water strikes the turbine blades of a generator so that its rebounding velocity is zero. If the flow rate is 60.0 kg/s and the water speed is 16.5 m/s, what is the average force on the blades?

*11. Calculate the force exerted on a rocket given that the propelling gases are expelled at a rate of 1000 kg/s with a speed of 60,000 m/s.

*12. A 150-kg (including space suit) astronaut acquires a speed of 2.3 m/s by pushing off with his legs from a 2200-kg space capsule. (a) What is the change in speed of the space capsule? (b) If the push lasts 0.20 s, what was the average force exerted by each on the other? (As the reference frame, use the position of the capsule before the push.)

*13. A meteor whose mass was about 10^8 kg struck the earth ($m = 6.0 \times 10^{24}$ kg) with a speed of approximately 15 Km/s and came to rest in the earth. (a) What was the earth's recoil speed? (b) What fraction of the meteor's kinetic energy was transformed to KE of the earth? (c) By how much did the earth's KE change as a result of this collision?

*14. A ball of mass 0.440 kg moving with a speed of 8.10 m/s collides head-on with a 0.220-kg ball at rest. If the collision is perfectly elastic, what will be the speeds and directions of the two balls after the collision?

ROTATIONAL MOTION AND ANGULAR MOMENTUM

CHAPTER

Up to now we have been discussing what is called *linear*, or *translational motion*: that is, the motion of bodies as a whole without regard to any rotation they might undergo. We now consider *rotational motion*: the motion of a body rotating about an axis. Rotating wheels, a swinging door, and the earth itself exhibit rotational motion.

1. ANGULAR QUANTITIES

To describe rotational motion we make use of angular quantities such as angular distance, angular velocity, and angular acceleration. These are defined in analogy to the corresponding quantities for ordinary linear motion (Chaper 3).

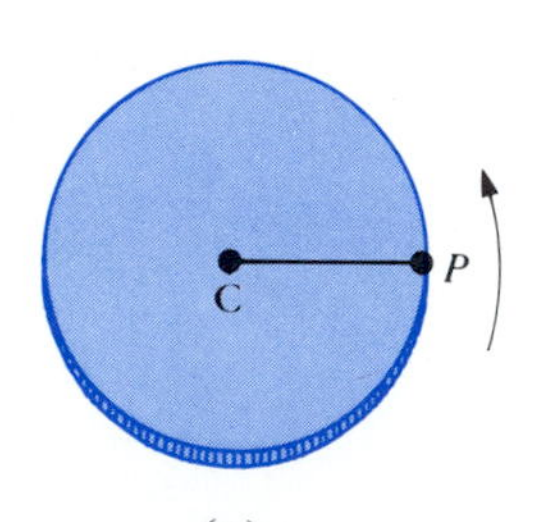

(a)

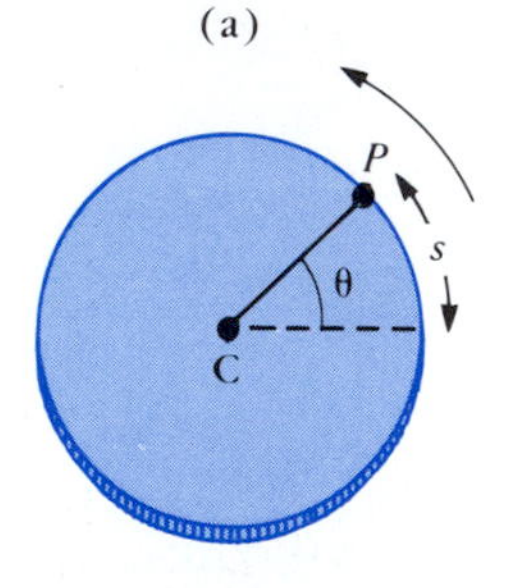

(b)

FIGURE 9-1
(a) A wheel is rotating about an axis through its center, C. (b) A short time later the wheel has rotated through an angle θ. A point P on the edge of the wheel has moved through a linear distance s.

Angular Position

Let us consider a wheel rotating about an axle (which we call the *axis of rotation*) passing through its center. A line drawn perpendicular from the axis to any point on the wheel moves through the same angle θ in a given time (Figure 9-1). This angle θ is the *angular distance* (or "change in angular position") through which the wheel has rotated during a given time. We can measure θ in degrees. But for scientific purposes, angles are often measured in **radians** (abbreviated rad), which we now explain.

When the wheel rotates through an angle θ, a point on the edge of the wheel (point P in Figure 9-1) moves a linear distance s, as shown. If r represents the radius of the wheel (distance of point P from the axis of rotation), then the angle θ, in radians, is defined as

$$\theta = \frac{s}{r}. \qquad [\theta \text{ in radians}]$$

Thus, one radian is the angle subtended by an arc (s) whose length is equal to the radius (r). How many degrees are there in one radian? To answer this, we remember the following: in a complete circle there are 360°; and this of course must correspond to an arc length, s, equal to the circumference of the circle: $s = 2\pi r$; thus $\theta = s/r = 2\pi r/r = 2\pi$ rad in a complete circle, so

$$360° = 2\pi \text{ radians.}$$

One radian is then $360°/2\pi = 360°/6.28 = 57.3°$.

Let us take an example that illustrates the usefulness of radian measure. A particular bird's eye can just distinguish objects that subtend an angle no smaller than about 3×10^{-4} radians (how many degrees is this?).† Let us calculate just how small an object the bird can barely distinguish when flying at a height of 100 m (Figure 9-2). Since $\theta = s/r$, we solve for s and get

$$s = r\theta.$$

Strictly speaking, s is the arc length; but for small angles the straightline distance is approximately the same, Figure 9-2(b). Since $r = 100$ m and $\theta = 3 \times 10^{-4}$ rad we find that $s = (100 \text{ m})(3 \times 10^{-4} \text{ rad}) = 3 \times 10^{-2} \text{ m} = 3$ cm. Thus from a height of 100 m, the bird can only make out objects that are at least 3 cm long. (Note that had the angle been given in degrees, we would first have had to change it to radians to make this calculation.)

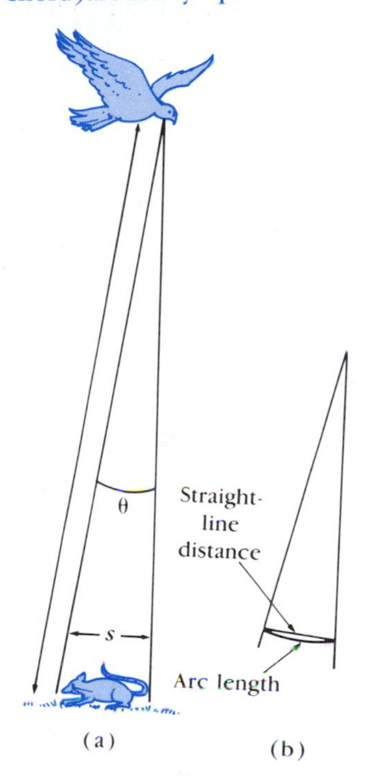

FIGURE 9-2
(a) Can the bird see the mouse? (b) For small angles, arc length s, and the straight-line distance (the chord) are nearly equal.

Angular Velocity and Acceleration

Angular velocity is defined in analogy with ordinary linear velocity. Instead of distance traveled, we use the angular distance θ, Figure 9-1. Thus the **average angular velocity** (denoted by ω, the Greek lowercase letter omega) is defined as

$$\bar{\omega} = \frac{\theta}{t},$$

where θ is the angle through which the body (say, the wheel of Figure 9-1) has rotated in the time t. Angular velocity is generally measured in radians per second.

Each particle or point of a rotating body has, at any moment, a linear velocity v that points in a direction tangent to its circular path. This is shown in Figure 9-3 for a point P on the rim of a rotating wheel. We can relate this linear velocity v of a point on the body to the angular velocity ω of the body as a whole. The linear velocity is v = distance/time. When the body rotates through an angle θ, a point P moves a distance C as shown in

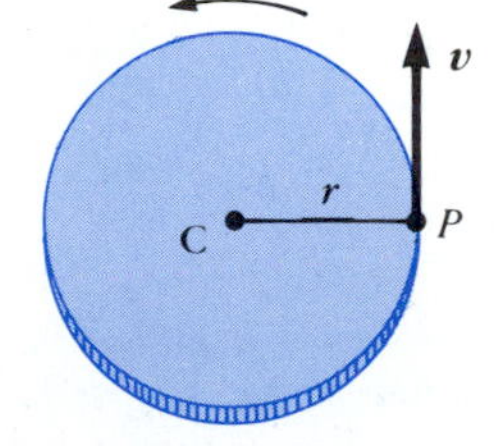

FIGURE 9-3
Wheel rotating counterclockwise. At the instant shown, point P has linear velocity v in the direction shown.

† Since 1 radian = 57.3°, then (3×10^{-4} radian) (57.3° per radian) = 0.02° or $\frac{1}{50}$ of 1 degree.

Figure 9-1(b). If P is a distance r from the axis of rotation, then $s = r\theta$ as we saw in the last subsection. Thus, the linear velocity of the point P (Figure 9-1 or 9-3) is

$$v = \frac{\text{distance}}{\text{time}} = \frac{r\theta}{t}.$$

Since θ/t is equal to the angular velocity ω, we have

$$v = r\omega.$$

In words, the linear velocity (v) of any point on a rotating body has magnitude equal to the product of the distance (r) of that point from the rotation axis times the angular velocity (ω) of the body as a whole.

Consider, for example, two horses on a merry-go-round, one on the outer edge and a second only half the distance from the rotation axis. Whatever the angular velocity of the merry-go-round, the linear velocity of a rider on the outer horse will be twice that for a rider on the inner horse. If the angular velocity of the merry-go-round is doubled, both linear velocities will be doubled, but their ratio remains the same.

Angular acceleration, in analogy to ordinary linear acceleration, is defined as the change in angular velocity divided by the time required to make this change. The **average angular acceleration** (denoted by α, the Greek lowercase letter alpha) is defined as

$$\bar{\alpha} = \frac{\omega - \omega_0}{t},$$

where ω_0 is the angular velocity initially and ω is the angular velocity after the time t has passed. When ω is measured in rad/s and t is in seconds, then α is expressed in rad/s^2.

Just as the linear velocity v of a particle on the body is related to the angular velocity ω of the body as a whole ($v = r\omega$), so the linear acceleration a of any point is related to the angular acceleration of the body by

$$a = r\alpha.$$

The **frequency** of rotation, f, is defined as the number of complete revolutions (rev) made per second. Since one revolution (of, say, a wheel) corresponds to an angle of 2π radians, and thus 1 rev/s = 2π rad/s, then in general, the frequency f is related to angular velocity ω by

$$\omega = 2\pi f.$$

For example, suppose we want to determine the linear speed of a point on the edge of a 33-rpm (revolutions per minute) phonograph record whose diameter is 30 cm. First we find the angular velocity in radians per second: the frequency $f = 33$ rpm $= 33$ rev/60 s $= 0.55$ rev/s. Then $\omega = 2\pi f = 3.5$ rad/s. The radius r is half the diameter (30 cm), so $r = 0.15$ m; then the speed v at the edge is $v = r\omega = (0.15 \text{ m})(3.5 \text{ rad/s}) = 0.52$ m/s.

2. TORQUE

The dynamics of translational motion using Newton's laws can be extended to rotating bodies. Instead of ordinary velocity and acceleration, we use the corresponding angular quantities. And in place of force we use the concept of **torque**. Torque is closely related to force, but before we define it, try the following experiment.

EXPERIMENT

Open a door slightly. Now exert a force on the door near its end. Be sure to push at right angles to the door—Figure 9-4(a). Next, exert the same force on the middle of the door as shown in Figure 9-4(b). Third, exert a force of the same magnitude at a 45° angle near the end of the door as shown in (c). Finally, exert a force on the end of the door (d). In each case, note how quickly the door opens.

You undoubtedly found that the door rotated more quickly in (a) than in (b) and (c), and that in (d) the door didn't rotate at all. Indeed, careful observations show that the angular acceleration of the door is not only proportional to the magnitude of the force. It is also proportional to the *lever arm*, which is designated ℓ in Figure 9-4. The **lever arm** is defined as

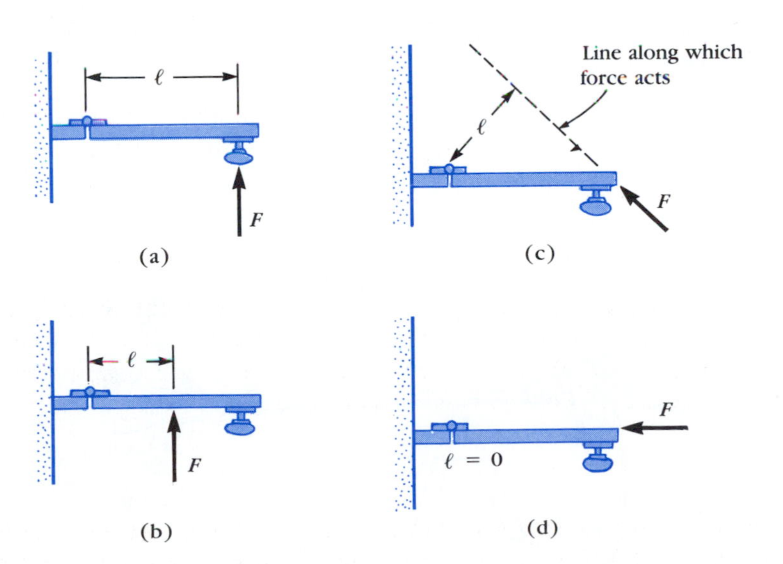

FIGURE 9-4
Applying a torque to a door.

the perpendicular distance from the axis of rotation to the line along which the force acts. Note that in Figure 9-4(d), $\ell = 0$; so no matter how hard you push it, the door will not rotate. The product of the magnitude of the force and the lever arm is defined as the **torque**:

$$\text{torque} = \text{lever arm} \times \text{force}.$$

In symbols, $\tau = \ell \times F$ where τ stands for the torque. If in Figure 9-4(a) you push on an 80-cm-wide door with a force of 20 N, the torque would be $0.80 \text{ m} \times 20 \text{ N} = 16 \text{ N} \cdot \text{m}$. In Figure 9-4(d), the torque is always zero, because $\ell = 0$.

As a simple example of torque, let us consider a child's seesaw. It is well known that if one child on the seesaw is heavier than the other, the pivot support must be closer to the heavy child. Exactly where the pivot should be located can be determined by knowing that the torques produced by the weight of each child must balance one another. In Figure 9-5, child A tends to rotate the seesaw counterclockwise and child B tends to rotate it clockwise. There will be a balance when these two opposite torques have the same magnitude (so the sum of them, or the net torque, is zero). Thus, if $\ell_A = 1.0 \text{ m}$, the torque produced by child A is $300 \text{ N} \times 1.0 \text{ m} = 300 \text{ N} \cdot \text{m}$. To balance this, $\ell_B \times 200 \text{ N}$ must equal $300 \text{ N} \cdot \text{m}$. Thus, $\ell_B = 300 \text{ N} \cdot \text{m}/200 \text{ N} = 1.5 \text{ m}$. Since child A weighs $1\frac{1}{2}$ times more than child B, then to achieve balance child B must be $1\frac{1}{2}$ times farther from the pivot.

From our discussion and the experiment you tried, it should be clear that *torque* (and not simply force) gives rise to rotational motion. That is, torque plays the same role in rotational motion that force does in ordinary translational motion. Indeed, the rotational equivalent of Newton's second law states that *the angular acceleration is proportional to the applied torque*. Now we must examine what plays the role of mass (or inertia) for a rotating body.

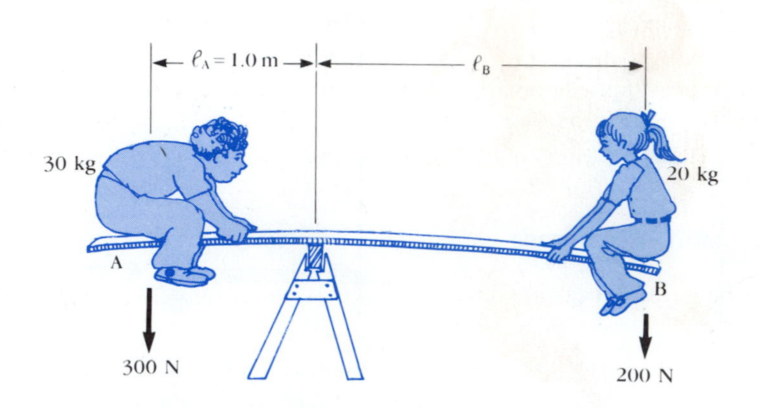

FIGURE 9-5

A seesaw. Note that the force on each child is the force of gravity, mg, where $g = 9.8 \text{ m/s}^2 \approx 10 \text{ m/s}^2$. Thus the weights of the children are $(30 \text{ kg})(10 \text{ m/s}^2) = 300 \text{ N}$ and $(20 \text{ kg})(10 \text{ m/s}^2) = 200 \text{ N}$. See Section 1 of Chapter 5.

3. ROTATIONAL INERTIA: MOMENT OF INERTIA

You have probably noticed that rotating objects, such as a spinning bicycle wheel or top or the earth itself, tend to continue spinning. This tendency is the rotational equivalent of inertia, and we call it **rotational inertia.** Indeed, the rotational equivalent of Newton's first law states that *a rotating body continues to rotate uniformly as long as no net torque acts to change this motion*.

An object of large mass normally has a large amount of rotational inertia. However, the mass is not the only thing that determines how much rotational inertia an object has. How this mass is distributed is also important. If more of the mass is concentrated a large distance from the axis of rotation, the rotational inertia will be greater. For example, a large-diameter cylinder will have much greater rotational inertia than a small-diameter cylinder of the same mass (Figure 9-6). The former will be much harder to start rotating and harder to stop once it is rotating. In other words, a greater torque will be needed to give the same angular acceleration to a larger-diameter wheel than to a smaller-diameter one; or, for the same net torque, the angular acceleration will be less if the rotational inertia is greater. Thus we can say that *the angular acceleration of a body is directly proportional to the net torque applied to it and inversely proportional to the rotational inertia*. This is the rotational equivalent of Newton's second law for translational motion, $F = ma$ (see Chapter 4) and can be written

$$\tau = I\alpha.$$

In rotational motion, the torque τ plays the role that force plays in translational motion (that is, the "cause"); and angular acceleration plays the role that acceleration a plays for linear motion. The inertia factor (mass m in linear motion) is the rotational inertia, I, which is normally called the **moment of inertia**.

The mass m of a body is easily determined by use of any type of balance or spring scale. But how do we determine the moment of inertia, I, of a body that can rotate?

First, let us consider a small mass m rotating in a circle of radius r at the end of a string whose mass we can ignore (Figure 9-7). The torque that gives rise to its angular acceleration is $\tau = rF$. If we make use of Newton's second law for linear quantities, $F = ma$, and the relation between the angular acceleration and the tangential linear acceleration (see Section 1), $a = r\alpha$, we have

$$F = ma = mr\alpha.$$

When we multiply both sides by r, we find that the torque $\tau = rF$ is given by by

$$\tau = mr^2\alpha.$$

FIGURE 9-6
Two cylinders of equal mass. The one with the larger diameter has the greater rotational inertia.

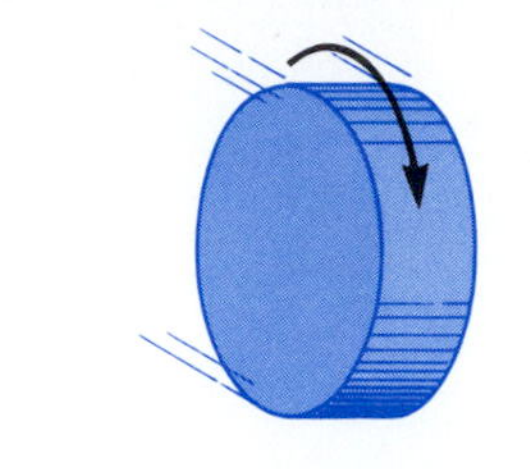

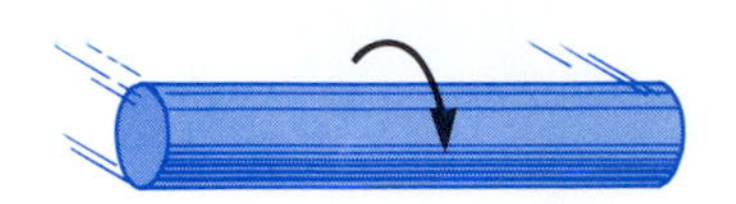

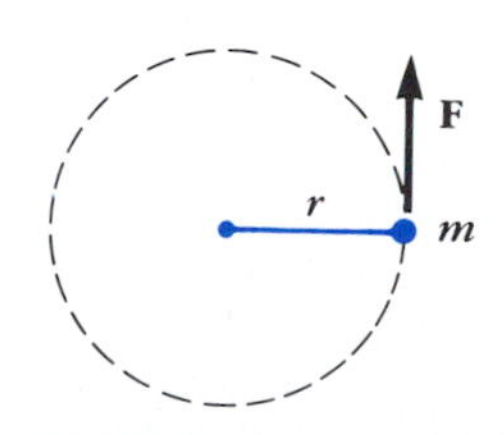

FIGURE 9-7
Rotational inertia of a single particle.

Here we have a direct relation between the angular acceleration and the applied torque τ. The quantity mr^2 is the moment of inertia I of the particle: $I = mr^2$.

An ordinary object such as a wheel rotating on its axis, a door turning on its hinges, or an arm rotating about the shoulder joint can be thought of as consisting of many particles located at various distances from the axis of rotation. We can apply the relation $I = mr^2$ to each particle and sum over all the particles to get the total moment of inertia of the whole body. This can be quite a task, often requiring the use of calculus. In Figure 9-8, we give formulas for the moments of inertia of a number of simple and symmetric bodies. These formulas are given in terms of the total mass and dimensions of the body (*outer* radius, length, and so on). Note that the moment of inertia for a given body may be different if the axis of rotation is different—see Figure 9-8(e).

The important thing to remember about rotational inertia is that it depends on two factors: the mass, and how far the mass is from the axis of rotation. The more mass there is concentrated far from the rotation axis,

	Object	Location of axis	Moment of inertia
(a)	Particle revolving in circle of radius r	Center of circle	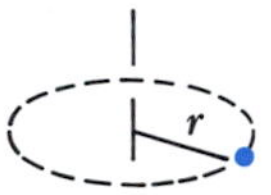 mr^2
(b)	Thin ring of radius r	Through center	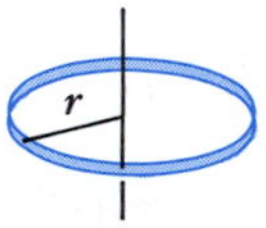 mr^2
(c)	Cylinder of radius r	Through center	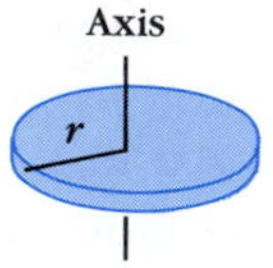 $\frac{1}{2}mr^2$
(d)	Sphere of radius r	Through center	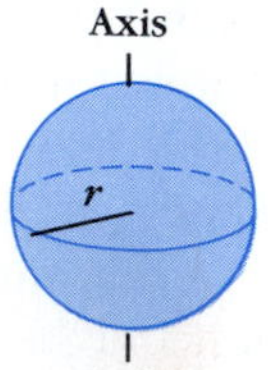 $\frac{2}{5}mr^2$
(e)	Long rod of length ℓ	Through center	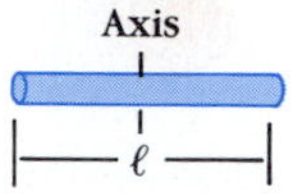 $\frac{1}{12}ml^2$
		Through end	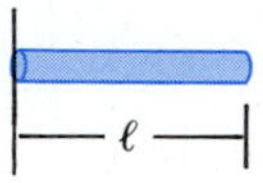 $\frac{1}{3}ml^2$

FIGURE 9-8
Moments of Inertia.

the greater the moment of inertia. A small mass far from the axis of rotation can have a greater inertial effect than a large mass concentrated near the axis of rotation (see Figure 9-6).

4. ROTATIONAL KINETIC ENERGY

The quantity $\frac{1}{2}mv^2$ is the kinetic energy of a body undergoing translational motion. A body rotating about an axis is said to have **rotational kinetic energy**, and by analogy we would expect this to be given by the formula $\frac{1}{2}I\omega^2$ where I is the moment of inertia of the body and ω is its angular velocity. This is indeed true; so we can write

$$\text{rotational KE} = \frac{1}{2}I\omega^2.$$

As an example, suppose a 33-rpm phonograph record has a mass of 100 grams (0.10 kg). The moment of inertia I of a cylindrical disk is, according to Figure 9-8(c), $I = \frac{1}{2}MR^2$. For a 12-inch diameter record, the outer radius is $R = 6$ inches = 15 cm = 0.15 m, so $I = \frac{1}{2}MR^2 = (\frac{1}{2})(0.10\text{ kg})(0.15\text{ m})^2 = 0.0011\text{ kg m}^2$. When rotating at 33 rpm, its angular velocity is $\omega = 3.5$ rad/s, as we saw at the end of Section 1 of this chapter. Thus its kinetic energy is $\frac{1}{2}I\omega^2 = (\frac{1}{2})(0.0011\text{ kg}\cdot\text{m}^2)(3.5\text{ rad/s})^2 = 0.0067\text{ J}$, which is not a lot of energy.

5. ANGULAR MOMENTUM AND ITS CONSERVATION

Throughout this chapter we have seen that by using the appropriate angular variables, the description of rotational motion is analogous to that for ordinary linear motion. In the last section we saw, for example, that rotational kinetic energy can be written as $\frac{1}{2}I\omega^2$, which is analogous to the translational KE $= \frac{1}{2}mv^2$. In like manner, the linear momentum, $p = mv$, has a rotational analog. It is called **angular momentum**, and for a body rotating about a fixed axis it is defined as

$$L = I\omega$$

where I is the moment of inertia and ω is the angular velocity.

For a single particle of mass m rotating in a circle of radius r (as in Figure 9-7), the moment of inertia I is mr^2 (see Section 3 and Figure 9-8). Thus its angular momentum will be $L = mr^2\omega$. From this we can see that since the angular momentum of each piece of an object equals $mr^2\omega$, the angular momentum is increased (1) if the mass is increased (2) if the rotational speed is increased or (3) if more of the mass is concentrated farther from the axis of rotation. Like ordinary momentum, angular

FIGURE 9-9
Conservation of angular momentum. (a) Part of the mass (arms) is far from the axis, and the speed of rotation is slow. (b) All the mass is close to the axis of rotation, and the speed is faster.

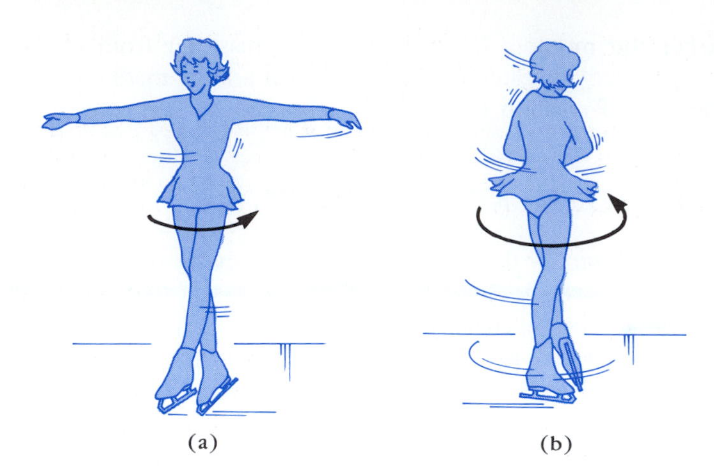

momentum obeys a conservation law, the **law of conservation of angular momentum**:

> **the angular momentum of an object remains constant so long as no external torques act to change it.**

This law is nicely illustrated by a skater doing a spin on ice (Figure 9-9). When her arms are outstretched, she rotates at a relatively slow rate; when she pulls in her arms close to her body, she begins rotating faster. She has reduced the distance (r) of her arms from the axis of rotation; and since angular momentum is conserved, she automatically speeds up to compensate. She doesn't push harder in order to go faster; her speed of rotation increases all by itself when she pulls in her arms. Conservation of angular momentum is evident here. When r is large, ω is small; when r is small, ω is large.

EXPERIMENT

Locate a chair that rotates, such as an office chair. Sit in the chair with your arms and legs extended. Now have someone spin you around. While you are rotating, pull in your arms and legs. What happens to your speed? Try the

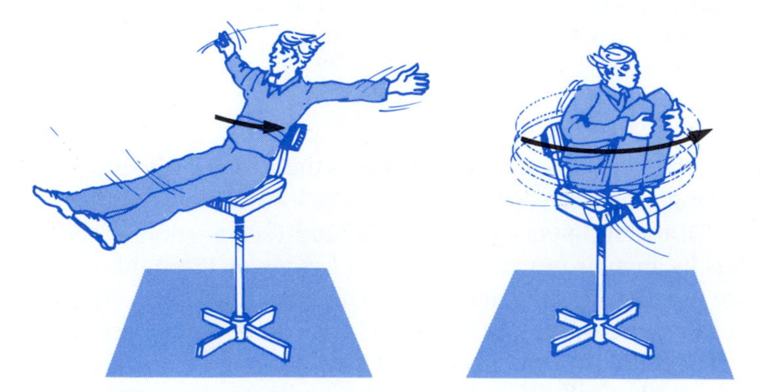

FIGURE 9-10
A person rotating on a swivel chair. How does the speed of rotation change?

experiment again, this time holding a couple of books or bricks in your hands. Explain your observations. See Figure 9-10.

OR

Tie a string around an eraser or some other small object. Hold the other end of the string between your thumb and fingers and start the object rotating. Then extend your forefinger so the string starts wrapping around it. Why does the object's speed increase as the string winds around your finger? See Figure 9-11.

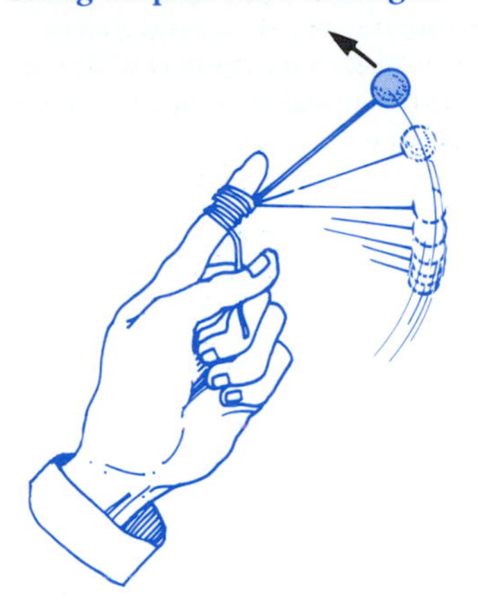

FIGURE 9-11
The ball's speed increases as the string wraps around the finger.

The law of conservation of angular momentum follows from the rotational equivalent of Newton's second law, $\tau = I\alpha$. For, by the definition of angular acceleration (Section 1), $\alpha = (\omega - \omega_0)/t$; then $\tau = I\alpha = I(\omega - \omega_0)/t = (L - L_0)/t$ where L_0 is the initial angular momentum of an object or system of objects and L is the angular momentum after a time t. From this we see that a net torque acting for a period of time will change the angular momentum of a body. If the net torque on the object is zero, then $(L - L_0) = 0$ and the angular momentum doesn't change—which is the law of conservation of angular momentum.

Angular momentum, like linear momentum, is a vector quantity. It can be represented by a vector that points along the axis of rotation. A child's top illustrates the conservation of angular momentum as a vector. If set on its tip, a top immediately falls over. But when it is spinning it remains upright for some time. If the spinning top were to fall, the axis of rotation would have changed drastically. Angular momentum is a vector, and although its magnitude might not be affected during the fall, the *direction* would change a great deal. Only a strong torque would be capable of causing such a large change in the angular momentum. Because such strong external torques are not ordinarily present, the angular momentum stays essentially constant and the top stays upright. However, the forces of friction and gravity do act, so the angular momentum changes slowly: the top gradually slows down and eventually falls over.

The situation is similar to a bicycle or a motorcycle. If you sit on a bike at rest, it falls over; but if the bike is moving, the angular momentum of the spinning wheels resists any tendency to change and helps to keep the bike upright.[†]

The gyroscope, which is used by mariners to maintain their course, works on the principle of conservation of angular momentum (Figure 9-12). The rapidly spinning wheel is mounted on a complicated set of bearings so that when the mount is moved, as when a ship pitches in heavy seas, no forces act to change the direction of the angular momentum. Thus the gyroscope wheel remains pointing in the same direction in space no matter where the ship is situated. If it is pointing toward the north star

FIGURE 9-12
A gyroscope.

[†] The spinning top and the stability of a moving bike can also be viewed from the point of view of rotational inertia. The rotating top or wheels, because of their rotational inertia, resist any change in their rotational velocity, either in magnitude or direction.

FIGURE 9-13
The motion of a body, in this case a bowling pin moving as a projectile, can be considered as the translational motion of its c.g. plus rotational motion about its c.g.

initially, it continues to point in that direction no matter how far the ship travels. Thus a gyroscope helps a captain tell in which direction the ship is moving.

6. CENTER OF GRAVITY AND CENTER OF MASS

If you throw an object through the air so that it rotates—say, a football going end-over-end or the bowling pin shown in Figure 9-13—its motion may appear wobbly, but one point in the object follows the smooth parabola curve of a projectile. This point is called the **center of gravity** or **center of mass** of the body.† This point is the average position of all the particles in the body. To determine the translational motion of a body, we can assume that *the force of gravity on the body acts at the center of gravity*. Thus, the motion of a body such as the bowling pin in Figure 9-13 can be considered the translational motion of its center of gravity (c.g.) plus rotational motion about its center of gravity. This is also illustrated by the diver in Figure 9-14 who in (a) exhibits only translation motion (all parts of the body follow the same path) but in (b) is both translating and rotating. In both cases the c.g. follows the same parabolic path characteristic of projectile motion (Section 3, Chapter 5).

The center of gravity of regularly shaped objects that are uniform in composition, such as cylinders, spheres, and rectangular solids, is at their geometric center. To determine the c.g. of an irregular body, we make use of the following fact. When an object suspended from a point is at rest, its center of gravity lies directly below (or at) the point of suspension. If this

† Technically these terms differ, but for most practical purposes they are the same.

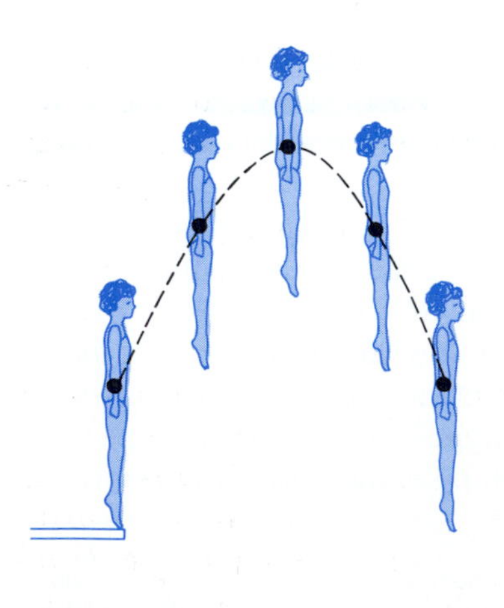

(a)

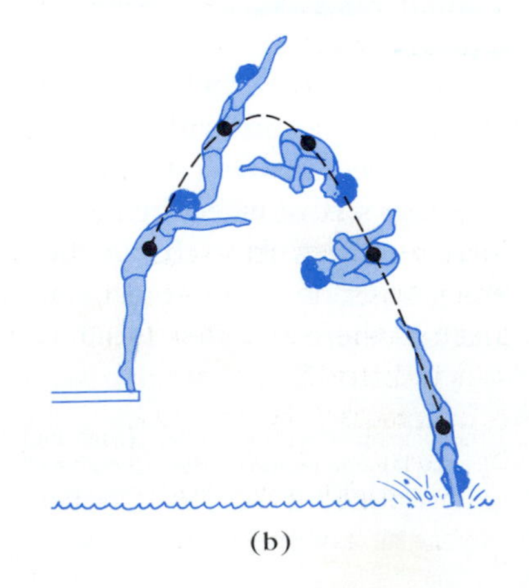

(b)

FIGURE 9-14
Motion of the diver is pure translation in (a), but is translation plus rotation in (b).

were not true, a torque would act to rotate the object as shown in Figure 9-15(a). Only when the c.g. is below the suspension point does the lever arm, and therefore the torque, equal zero—Figure 9-15(b). The center of gravity of an irregularly shaped body can thus be found by suspending it from two different points and drawing in the vertical line for each case as shown in Figure 9-16. The intersection of these lines is then the center of gravity.

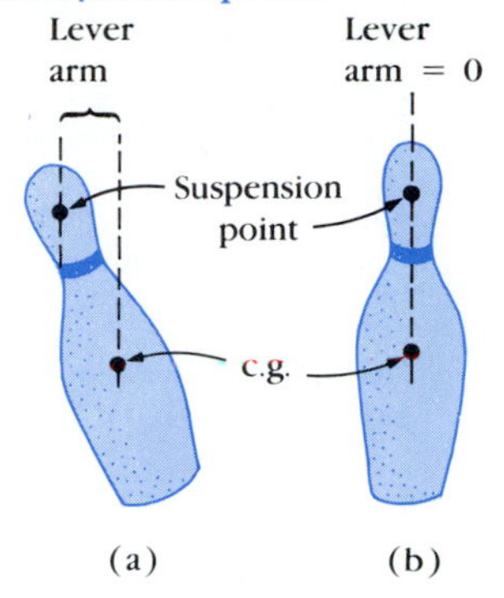

FIGURE 9-15

(a) A torque acts on a body if its c.g. is not directly below the point of support. The body can remain at rest (b) only if the c.g. is below the suspension point.

EXPERIMENT-PROJECT

Cut an arbitrary shape, such as that shown in Figure 9-15 or 9-16, from a piece of cardboard and then determine its center of gravity.

The center of gravity of an object may not lie within the body itself. This is clearly true for a doughnut—its c.g. lies at the center of the hole.

Knowing the c.g. of the body when it is in various positions is of great use in studying body mechanics. One simple example from athletics is shown in Figure 9-17. If high jumpers can attain the position shown, their c.g. would actually pass below the bar—which means that for a particular takeoff speed, they could clear a higher bar. This is indeed what they try to do.

SUMMARY

When a body rotates about a fixed axis, any lines drawn perpendicularly from the rotation axis to various points in the body will sweep out the same angle θ in any given time interval. Angles are conveniently measured in *radians* where one radian is the angle subtended by an arc whose length is equal to the radius.

Angular velocity, ω, is defined as the rate of change of angular position: $\omega = \theta/t$. All parts of a body rotating about a fixed axis have the same angular velocity at any instant. *Angular acceleration*, α, is defined as the rate of change of angular velocity.

Angular velocity is related to *frequency*, f, by $\omega = 2\pi f$. The linear velocity, v, of a point fixed at a distance r from the axis of rotation is $v = r\omega$.

The dynamics of rotation is analogous to the dynamics of linear motion: force is replaced by *torque*, τ, which is defined as the product of force times lever arm (perpendicular distance from the line of action of the force to the axis of rotation); mass is replaced by *moment of inertia*, I, which depends not only on the mass of the body but also on how the mass is distributed about the axis of rotation; and linear acceleration is replaced by angular acceleration. The **angular acceleration** of a body is directly proportional to the net torque and inversely proportional to its rotational inertia, or $\tau = I\alpha$.

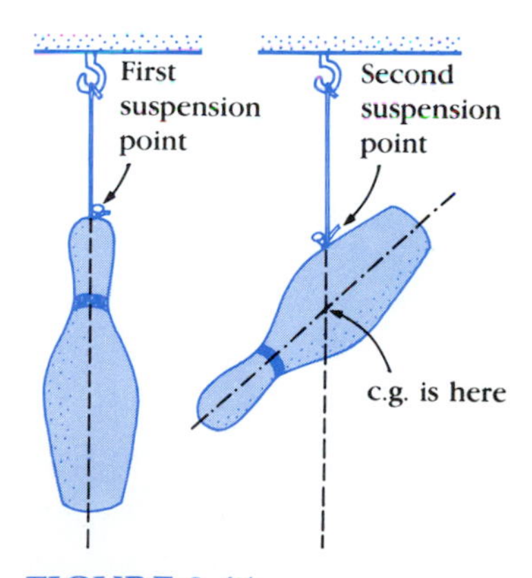

FIGURE 9-16

Determining an object's center of gravity by suspending it from two different points and drawing vertical lines.

FIGURE 9-17
The center of gravity of a high jumper or pole vaulter can actually pass beneath the bar.

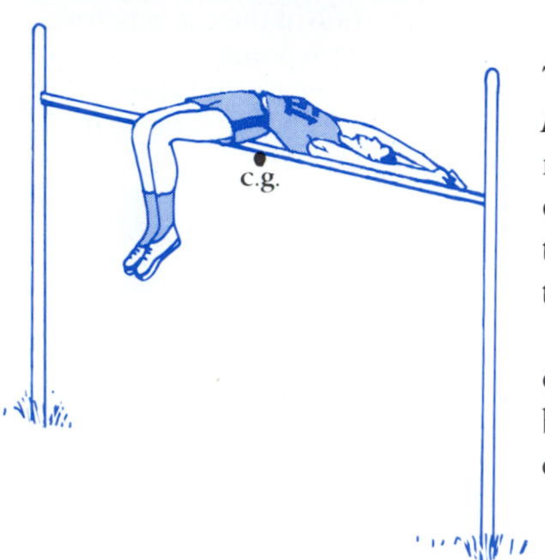

The *rotational kinetic energy* of a body rotating about a fixed axis with angular velocity ω is

$$\text{KE} = \tfrac{1}{2}I\omega^2.$$

The *angular momentum* of a body about a fixed rotation axis is given by $L = I\omega$. The *angular momentum* of a body depends on the body's mass, its rotational velocity, and how far from the axis of rotation the mass is distributed. The *law of conservation of angular momentum* states that the total angular momentum of an object remains constant if no external torques act.

The *center of gravity* of a body is that point at which the force of gravity on the entire body can be considered to act. The motion of any object can be considered the sum of two motions: translational motion of the center of gravity and rotation about the center of gravity.

QUESTIONS

1. You are standing a known distance from the Statue of Liberty. Describe how you can determine its height using only a meter stick and without moving from your location.
2. A bicycle odometer (which measures distance traveled) is attached near the wheel hub and is designed for 27-inch wheels. What happens if you use it on a bicycle with 24-inch wheels?
3. What is the advantage of using a screwdriver with a "fat" handle rather than a "skinny" one? Describe any advantages to using a screwdriver with a long blade.
4. Expert bicyclists use very lightweight "sew-up" (tubular) tires. They claim that reducing the mass of the tires is far more significant than an equal reduction in mass elsewhere on the bicycle. Explain why this is true.
5. Does the rotational inertia of a body depend on (a) its mass, (b) its angular velocity, (c) how the mass is distributed?
6. The following bodies all have the same mass. Which has the greatest rotational inertia? (a) a sphere of radius R, (b) a cylinder of radius R, (c) a thin hoop of radius R, (d) a cube whose side is length R?
7. A long round log can be rotated about various axes—for example, about the "symmetry" axis through its center, about an axis perpendicular to this (like spinning a baton), or about its end. In which situation would the rotational inertia be the least? The most?
8. A heavy metal cylinder known as a flywheel is attached to the crankshaft of a car and turns with it. What purpose does it serve?

9. Why do high-quality record turntables have a heavy platter?
10. Why does a tightrope walker carry a long horizontal pole?
11. Animals that depend on being able to run fast have slender lower legs with flesh and muscle concentrated higher up near the shoulder. Why is this distribution of mass advantageous?
12. A cylindrical hoop and a solid cylinder of the same radius and mass roll down an incline. Which do you think will reach the bottom first?
13. Suppose you are standing on the edge of a freely rotating merry-go-round. What would happen to the speed of rotation of the merry-go-round if you started walking along the edge in the opposite direction?
14. When a quarterback leaps into the air to throw a pass, he rotates the upper part of his body as he throws the ball. If you look carefully, you will notice that the lower part of his body rotates in the opposite direction. Explain.
15. If there were a great migration of people toward the equator, how would this affect the length of the day?
16. We claim that momentum and angular momentum are conserved. Yet most moving or rotating bodies eventually slow down and stop. Explain.
17. When a motorcyclist leaves the ground on a jump, if the throttle is left on (so the rear wheel spins) why does the front of the cycle rise?
18. On the basis of the law of conservation of angular momentum, discuss why a helicopter must have more than one rotor (or propeller). Discuss one or more ways the second propeller can operate in order to keep the body stable.
19. Why do you tend to lean backward when carrying a heavy load in your arms?
20. Explain why a uniformly rectangular brick can be placed so that slightly less than half its length can be suspended over the edge of a table, but no more.
21. Why is it more difficult to do a sit-up with your hands behind your head than when they are outstretched in front of you? A diagram may help you to answer this.
22. The center of gravity of a meter stick is at its midpoint. Why is this not true of your leg?
23. Show on a diagram how your center of gravity shifts when you change from a standing to a bending position.
24. When two children are balanced on a seesaw, why does one gain the advantage by leaning backward?
25. How would you describe the motion of a football kicked end-over-end?

FIGURE 9-18
Diver rotates faster when arms and legs are tucked in than when outstretched. Conservation of angular momentum.

26. Using the ideas of center of gravity and torque, explain why a ball rolls down a hill.
27. The diver shown in Figure 9-18 gains angular momentum by pushing against the board as he begins the dive. He rotates rapidly after tucking in his body; yet before entering the water he stretches out his body to reduce his speed of rotation. Explain.

EXERCISES

1. What are the following angles in radians: (a) 30° (b) 90° and (c) 390°?
2. The Eiffel Tower is 300 m tall. When you are standing at a certain place in Paris, it subtends an angle of 5°. How far are you, then, from the Eiffel Tower?
3. A laser beam is directed at the moon, 380,000 km from earth. The beam diverges at an angle 1.8×10^{-5} rad. How large a spot will it make on the moon?
4. A bicycle with 68-cm-diameter tires travels 2.0 km. How many revolutions do the wheels make?
5. A 20-cm-diameter grinding wheel rotates at 2000 rpm. Calculate its angular velocity in rad/s.
6. In Section 1 of this chapter we calculated the speed of a point on the edge of a 33-rpm phonograph record. Determine the centripetal acceleration, a, of such a point as the record rotates, using the fact that centripetal acceleration (see Chapter 6) is $a = v^2/r$. In what direction is this acceleration? What does this tell you about the force on a phonograph needle?
7. Determine the linear speed of a point 10 cm from the center of a rotating 33-rpm phonograph record.
8. What is the linear speed of a point 6.0 cm from the center of a 45-rpm record?
9. A 33-rpm phonograph record reaches its rated speed 2.8 s after it is turned on. What was the angular acceleration?
10. Estimate the angle subtended by the moon using a ruler and your finger or other object to barely blot out the moon. Describe your measurement and the result obtained, and then use the result to estimate the diameter of the moon. The moon is about 380,000 km from the earth.
11. Calculate the angular velocity of the earth (a) in its orbit around the sun and (b) about its axis.
12. How much torque does a 60-kg person exert when standing on the end of a 5-m-long diving board?
13. A bolt on an automobile engine is supposed to be tightened to a

torque of 80 N · m. Your wrench is 20 cm long. What force must you apply to be sure the bolt is tight enough?

14. Suppose a bolt must be tightened to 110 N · m. You have a 20-cm-long wrench and you know you can exert a maximum force of 180 N. So you put a pipe over the end of the wrench to make it longer. How far must the pipe extend the length of the wrench?
15. What is the maximum torque a 50-kg person can exert on the pedals of a bicycle if the person puts all the weight on one pedal. The pedals rotate in a circle of radius 0.18 m.
16. Two children balance on a seesaw when one is 2.0 m and the other is 2.8 m from the pivot point. If the first has a 22-kg mass what is the mass of the second child?
17. Two children, whose masses are 15 kg and 30 kg, want to "seesaw" with a board 5.0 m long. If the children sit at opposite ends of the board, where must the pivot point be placed?
18. Where should the pivot point be placed under a 3.0-m-long board if it is to be balanced when a 20-kg child and a 40-kg child are on either end?
19. What is the torque applied by the biceps on the lower arm in Figure 9-19(a) and (b). The axis of rotation is through the elbow joint and the muscle is inserted 5.0 cm from the elbow joint.
20. Calculate the moment of inertia of a 12.0-kg sphere of radius 0.205 m when the axis of rotation is through its center.
21. Calculate the moment of inertia of a 66.7-cm-diameter bicycle wheel. The rim and tire have a combined mass of 1.13 kg. The mass of the hub can be ignored. Why?
22. A centrifuge rotor has a moment of inertia of 4.00×10^{-2} kg · m^2. How much energy is required to bring it from rest to 10,000 rpm?
23. What is the angular momentum of a 2.1-kg uniform cylindrical grinding wheel rotating at 1500 rpm? Its radius is 27.4 cm.
24. A person stands, hands at the side, on a platform that is rotating at a rate of 1.20 rev/s. If the person now raises both arms, the speed of rotation decreases to 0.8 rev/s. (a) Why does this occur? (b) By what factor has the person's moment of inertia changed?
25. A diver such as that shown in Figure 9-18 can reduce his moment of inertia by a factor of about 3.5 when changing from the straight position to the tuck position. If he makes two rotations in 1.5 s when in the tuck position, what is his angular speed (rev/s) when in the straight position?
26. Two spheres stand at rest at the top of an incline. One has twice the radius and twice the mass of the other. Which reaches the bottom of the incline first? Which has the greater speed there? Which has the greater total kinetic energy at the bottom?

FIGURE 9-19

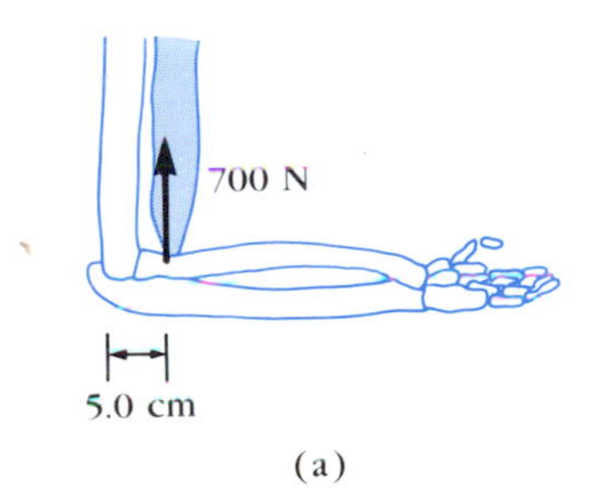

(a)

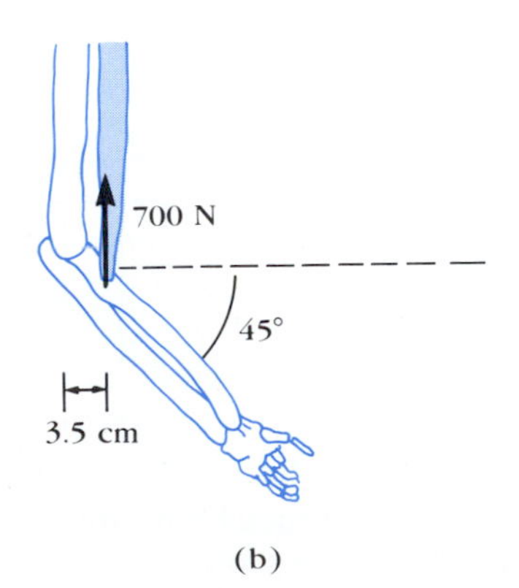

(b)

BODIES IN EQUILIBRIUM

CHAPTER

10

Until now we have been mainly concerned with bodies in motion. In this chapter we consider a body at rest. Because the velocity and acceleration are both zero in this case, Newton's second law tells us the net force on the object must also be zero. But that does not mean there are no forces acting at all. In fact, it is virtually impossible to find an object on which there are no forces acting. In everyday situations, gravity acts on every body. And if a body is at rest, some other force must be acting on it to balance the force of gravity. For example, Figure 10-1 shows a commonplace situation we have already met, an object at rest on a table. Its weight, the force of gravity, acts downward on it; and the table exerts a force upward on it which, since the net force is zero, must be equal to the force of gravity acting downward. Do not confuse this with the equal and opposite forces of Newton's third law which act on different bodies; here both forces act on the same body (see Figure 5-1). Such a body is said to be in **equilibrium** (Latin for "equal forces") under the action of these two forces.

The study of the forces acting on a body at rest is useful in many situtations ranging from the forces on muscles in the human body to the internal forces within a building or other structure. After we know one or more of the forces acting on a body, we can often determine what other forces are acting by using the fact that the sum of all the forces is zero.

One reason the subject of bodies at rest is so important is that materials change shape under the action of forces. If the forces are great enough, the object may break or fracture. Thus, architects and engineers use the ideas of equilibrium, as do physical therapists and others in the medical field.

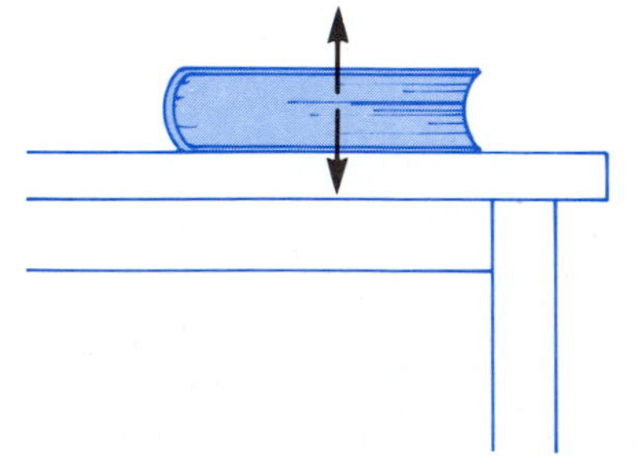

FIGURE 10-1
The book is in equilibrium; the net force on it is zero.

1. CONDITIONS FOR EQUILIBRIUM

For a body to be in equilibrium, it is clear that *the sum of all the forces acting on it must be zero*. This is called the "first condition for equilibrium"; but it is not sufficient. Consider, for example, the log in Figure 10-2. There are equal forces acting in opposite directions, so the net force on the log is zero. But there is a net *torque* on the log and so it will rotate. Thus, for a body to be at rest the *sum of all the torques acting on it must be zero also*. This is the second (and final) condition for equilibrium.

We have already seen an application of this principle in Figure 9-5. The seesaw was in equilibrium when the two torques were balanced—that is, they were equal in magnitude but acted in opposite directions.

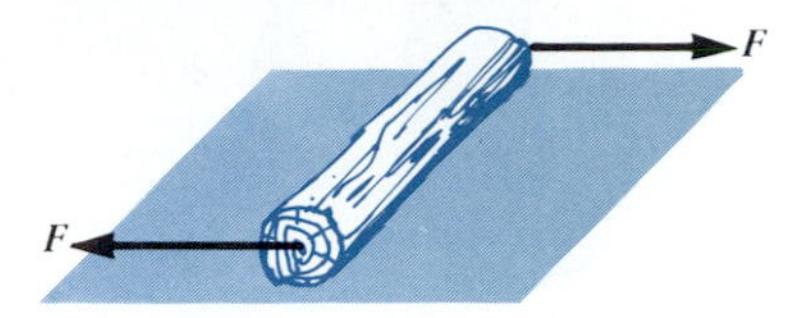

FIGURE 10-2
The net force is zero, but the log rotates because there is a net torque.

2. FORCES IN MUSCLES AND JOINTS

The principle of equilibrium, or **statics** (meaning "at rest") as it is sometimes called, can be used to find the strength of forces in the human body. Let us calculate the force exerted by the biceps muscle in your upper arm when you hold a 10-kg object (say, a rock) in your hand. Figure 10-3 shows the bones and biceps muscle in an average person's arm. The arm stays in the position shown because the biceps muscle contracts and exerts an upward force F_M on the lower arm. This force acts at the point at which the muscle is attached to the forearm; this "insertion" point is normally about 5 cm from the elbow joint, which acts as a pivot point. Two other forces act on the arm (we ignore the weight of the arm itself, as it is relatively small): (1) the force of the 10-kg rock pushing down on the hand [equal to its weight $mg = (10 \text{ kg})(9.8 \text{ m/s}^2) \approx 100 \text{ N}$]; this force acts about 35 cm from the elbow as shown; (2) the force F_J that the upper arm exerts on the lower arm at the joint. The latter gives rise to no torque, since its lever arm is zero. The torque due to the 100-N rock tends to rotate the arm downward (or clockwise about the elbow) and the torque due to F_M tends to rotate the arm upward (or counterclockwise). Because the arm is in equilibrium, these two opposite torques must be equal in magnitude:

$$(100 \text{ N})(35 \text{ cm}) = (F_M)(5.0 \text{ cm}).$$

So F_M equals $(100 \text{ N})(35 \text{ cm})/(5.0 \text{ cm}) = 700 \text{ N}$. The force exerted by the muscle is much larger (in fact, seven times larger) than the weight of the object lifted, because the muscle is connected to the lower arm at a point close to the elbow. A person whose muscle is inserted slightly farther out, say 6 cm instead of 5 cm, would have a considerable advantage in lifting and throwing. Indeed, successful athletes often have muscle insertions farther from the joint than those of the average person.

The net force on the arm in Figure 10-3 must be zero. Since $F_M = 700 \text{ N}$ acts upward and the 100-N weight acts downward, there clearly must be another force acting downward to balance them. This force must be the one exerted by the upper arm at the elbow joint F_J. Because the sum of the downward forces, $F_J + 100 \text{ N}$, must equal the upward force $F_M = 700 \text{ N}$, F_J must be 600 N. Thus, the force exerted at the joint F_J and that exerted by the muscle F_M are much greater than the weight of the object lifted.

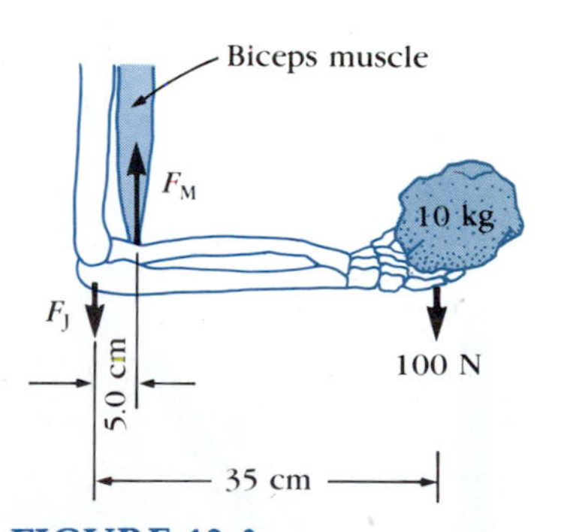

FIGURE 10-3
Forces exerted on the lower arm when a 10-kg rock is held in the hand.

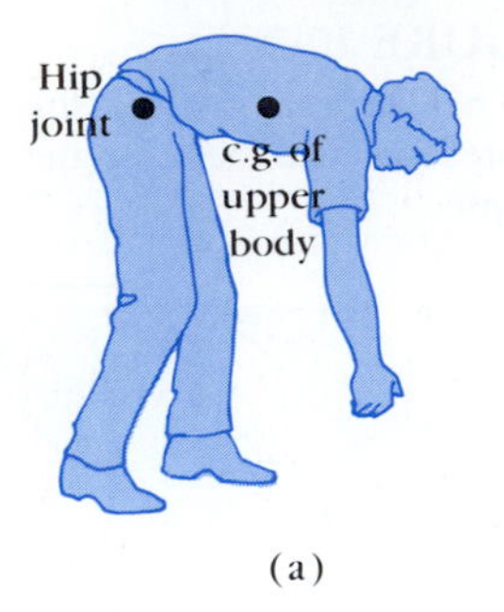

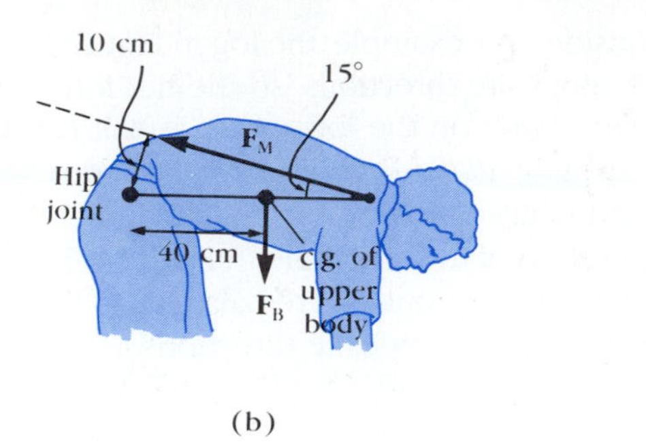

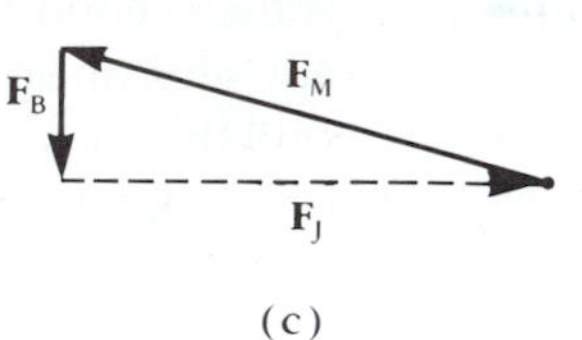

FIGURE 10-4
Forces exerted on the back when bending over.

Indeed, throughout the body the muscles and joints are subjected to forces that are often very great.

As another example, let us examine the forces involved when a person bends over to pick up something, as shown in Figure 10-4(a). The upper body is supported by back muscles that act at an angle of about 15° to the body axis. The force exerted by these muscles is represented by F_M in Figure 10-4(b). This force acts about the hip joint with a lever arm of approximately 10 cm in a person 180 cm tall; its torque balances the torque due to the weight F_B of the upper body acting at its center of gravity with a lever arm of approximately 40 cm. The upper body contains about two-thirds of the total body weight. In a 75-kg person, then, the weight of the upper body F_B is about $mg = (50 \text{ kg})(9.8 \text{ m/s}^2) \approx 500 \text{ N}$. Thus

$$(500 \text{ N})(40 \text{ cm}) = (F_M)(10 \text{ cm})$$

so $F_M = (500 \text{ N})(40 \text{ cm})/(10 \text{ cm}) = 2000 \text{ N}$. This force is equivalent to the weight of a 200-kg mass. On a vector diagram, Figure 10-4(c), we can draw the vectors F_B and F_M to scale; and the vector that completes the triangle (so the sum of all forces is zero) is the force F_J that must be exerted at the base of the spine. Using a ruler, we measure F_J to be about 1900 N. This incredibly large force is exerted on the lower part of the spine, which is composed of small bones called vertebrae that are separated from one another by flexible fluid-filled disks. These disks are thus compressed under tremendous force when a person bends over. It is no wonder that backaches are common and that occasionally a disk will rupture. You can protect your back by squatting with your knees bent to pick up an object and then letting your legs lift you, instead of bending over.

FIGURE 10-5
Objects in (a) stable equilibrium (b) unstable equilibrium.

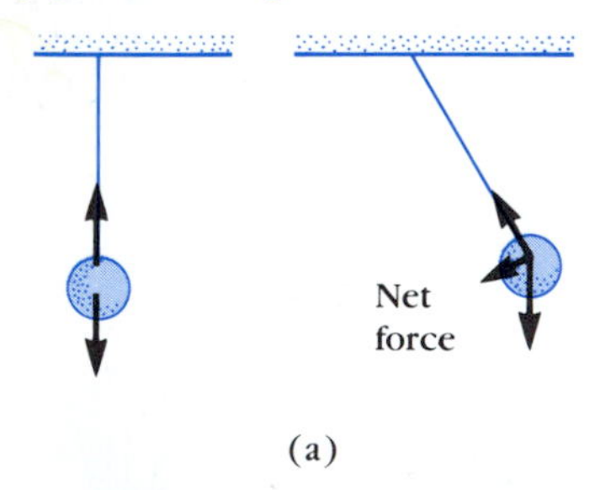

3. STABILITY AND BALANCE

If a body in equilibrium at rest is displaced slightly, three things can happen: (1) it may return to its original position, in which case we say it is in **stable equilibrium**; (2) it may move even farther from its equilibrium position of its own accord, in which case we say it is in **unstable equilibrium**; or (3) it may remain in the new position, in which case we say it is in **neutral equilibrium**.

To make this clear, let us look at examples of each possibility. Remember that the force of gravity on an object can be considered to act at its center of gravity. A ball suspended from a string as shown in Figure 10-5 (a) is in stable equilibrium, for if it is displaced to one side as shown it returns to its original position. A pencil standing on its tip is in unstable equilibrium, Figure 10-5(b), for if it is displaced from the exact balance point even slightly there will be a torque that acts to make it fall. Finally, a ball on a flat table is in neutral equilibrium, for if it is displaced to one side it remains in its new position.

In most situations, such as those that occur in the human body or in buildings, we are interested mainly in maintaining stable equilibrium, or **balance**. In general, when a body's center of gravity is below its point of support, it will be in stable equilibrium. The ball on the string in Figure 10-5(a) is an example. If an object's c.g. is above the point of support, the situation is more complicated.

Consider a milk carton standing flat as shown in Figure 10-6(a). If it is tipped slightly, Figure 10-6(b), it will return to its original position due to the torque that acts on it. But if it is tipped too far, Figure 10-6(c), it falls over. The critical point is reached when the c.g. of the object is beyond the base of support. This is because the upward force on the base of the object can act only within the area of contact; and if the force of gravity acting at the c.g. is beyond the base, there is a net torque that causes the object to fall rather than to return to its original position. Thus, *a body whose c.g. is above its base of support will be stable.*

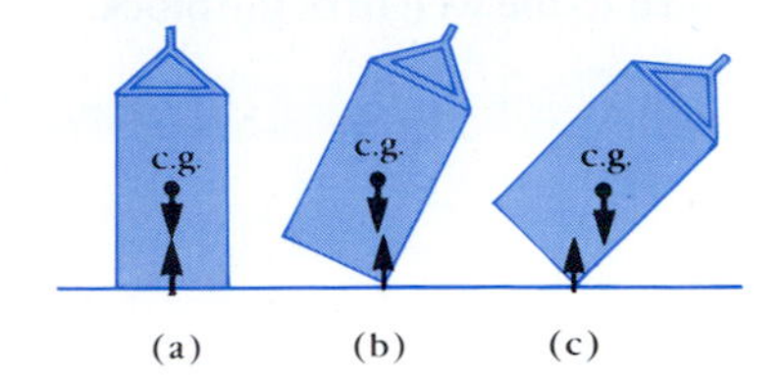

FIGURE 10-6
The cartons in (a) and (b) are in stable equilibrium. The carton in (c) is in unstable equilibrium.

When the c.g. is above the base of support, we see that stability is relative. A milk carton lying on its side is much more stable than one standing on its end, for it will take a greater effort to tip it over. The pencil in Figure 10-5(b) is an extreme case, because its base of support (the pencil tip) is so very tiny. In general, the larger the base and the lower the center of gravity, the greater is the stability of the object.

In this sense, humans are much less stable than four-legged animals, which have a broader base of support because of their four legs and a lower center of gravity. Humans have only their two feet as a base of support. A person who is walking or performing other sorts of movement must continually shift his or her body so that its c.g. is over its base of support. We are able to do this without conscious thought, although our brains had to learn how. What babies are learning when they "learn" to walk is how to keep their c.g. over their feet.

EXPERIMENT

Stand with your heels and back against a wall. Now try to bend forward. Why can't you do it without falling over? What do you normally do when you bend over that keeps you from falling?

AND

Stand facing the edge of an open door with your nose touching the door and your feet on either side of it as shown in Figure 10-7. Now try to rise up on your toes. Why can't you do it?

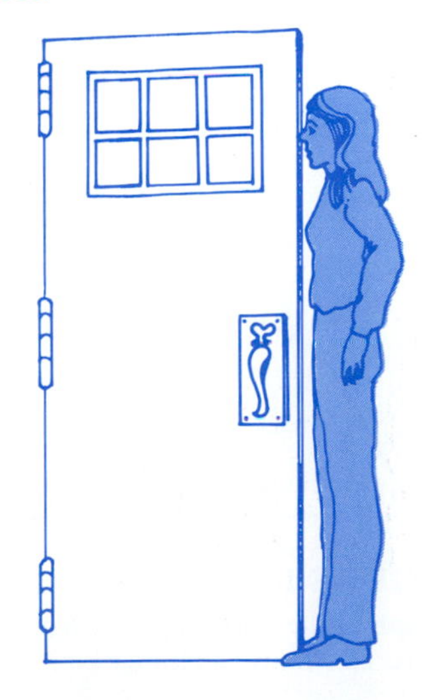

FIGURE 10-7
With your nose touching the end of the door, put your feet astride the door and try to rise up on your toes.

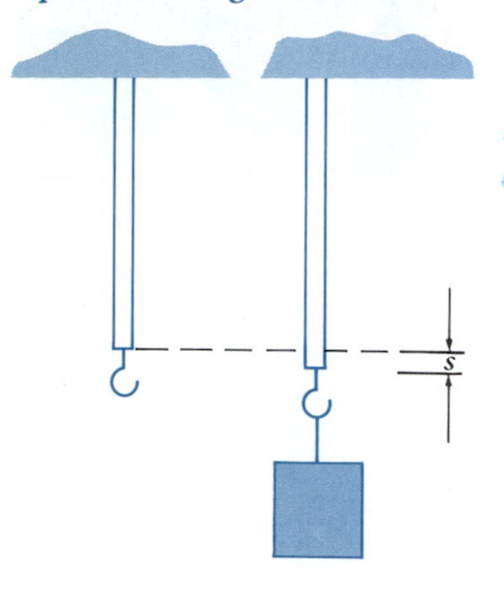

FIGURE 10-8
A wire is stretched an amount *s* proportional to the magnitude of the force *F*, which in this case is equal to the weight of the block.

4. ELASTICITY AND FRACTURE

An ordinary object in equilibrium has forces exerted on it. These forces can change the shape of the body and if strong enough can cause it to break, or fracture.

Elasticity and Hooke's Law

Robert Hooke (1635–1703) found that the amount of deformation of a body is proportional to the amount of force being applied. For example, suppose a long wire is suspended as shown in Figure 10-8. When a weight is hung on it, the wire stretches. The amount of elongation or stretch s is proportional to the force F exerted on it:

$$s \propto F.$$

This relationship is usually written as an equation,

$$F = ks,$$

and is known as *Hooke's law*. Thus, if a force of 2000 newtons stretches a wire 1 mm, a force of 4000 newtons will stretch it 2 mm, and so on.

Hooke's law is found to hold for most materials but only up to a point. If the force is too great, the object stretches excessively and eventually breaks. Figure 10-9 shows a graph of force versus elongation s for a typical material. Up to a point called the *elastic limit*, the object follows Hooke's law quite well; and when the force is released, the object returns to its original shape. This ability is called *elasticity*. If the force exceeds the elastic limit, the object becomes permanently deformed. If the force is increased much beyond the elastic limit, the object breaks. The force at the breaking point is called the *ultimate strength* of the material.

Tension, Compression, and Shear

Let us consider the wire in Figure 10-8 again. Not only is there a force pulling down on the wire, there is also a force pulling up at the top (it's in equilibrium, remember). This force is shown in Figure 10-10(a). The same force also exists within the material itself. Consider the lower part of the wire shown in Figure 10-10(b). This lower section is also in equilibrium (it is at rest); so there must be an upward force on it to balance the downward force at its lower end. What exerts this upward force? It must be the upper part of the wire. Thus we see that external forces applied to an object give rise to internal forces, or *stress*,† within the material itself.

The wire in Figure 10-10 is said to be under *tension* or *tensile stress*. Two other common types of stress are *compression* and *shear*. All three

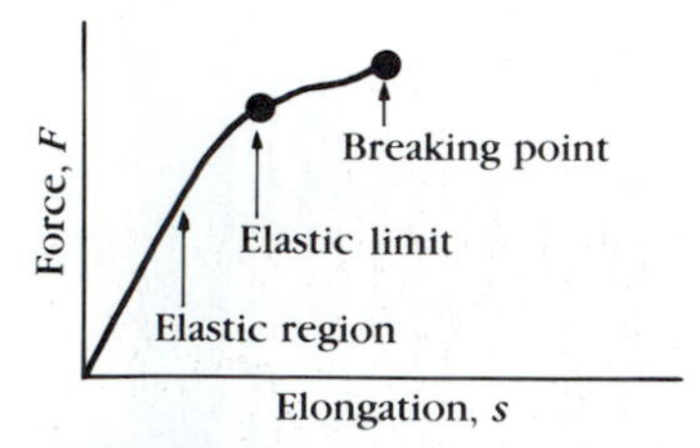

FIGURE 10-9
Graph of the force applied to a material versus its elongation.

† Stress usually refers to the force per unit area of cross section.

types are illustrated in Figure 10-11. When an object is under tension, Figure 10-11(a), it is being stretched. But when the two forces act *toward* each other, as in Figure 10-11(b), they act to shorten or compress the object. This stress is called *compression*, and Hooke's law works in this case, too. Anything that supports a weight, such as the columns of a Greek temple (Figure 10-12, Color Plate IV), is subjected to compression. *Shear*, on the other hand, occurs when equal and opposite forces act along different lines, as in Figure 10-11(c); shear tends to change the shape of the object from, say, a rectangle to a parallelogram. If you lay this book on a table and push across its top with your hand, the book will undergo shear.

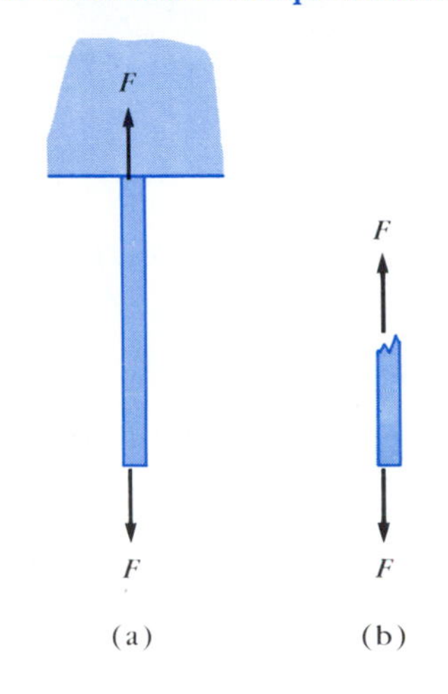

FIGURE 10-10
(a) When a force is exerted downward on the wire, the support at the top pulls upward. (b) The force exists within the wire itself because any section of the wire must be in equilibrium.

Fracture

If the stress in an object is too great, the object will break. This can happen as a result of tension, compression, or shear, as shown in Figure 10-13. The point at which fracture occurs depends on the cross-sectional area of the material—the greater the area, the greater the force it can withstand. Hence, the *ultimate strength* of a material is specified as the force per unit area (N/m^2) at which that material will normally break under a given kind of stress.

Table 10-1 gives the ultimate strengths of a number of materials under tension, compression, and shear stress. It can be seen that some materials are weak under tension and shear stress compared to their strength under compression. Because materials vary somewhat in composition, the numbers in the table are only a guide. When designing a building, architects generally use a safety factor of 5 to 10. That is, they assume a material can withstand only $\frac{1}{5}$ to $\frac{1}{10}$ of its ultimate strength.

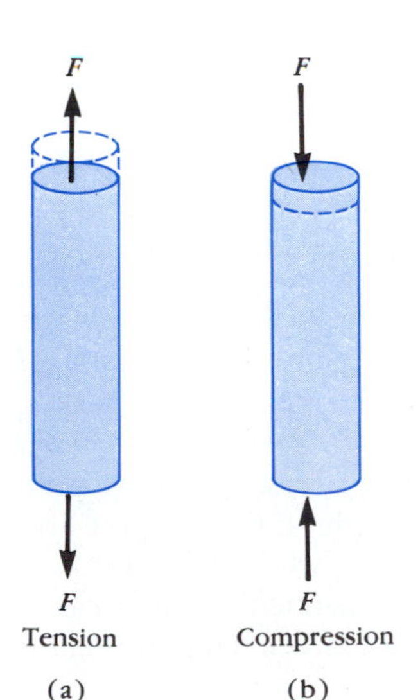

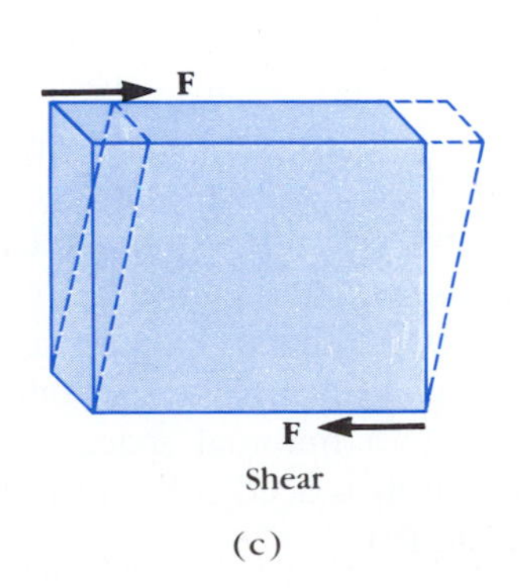

FIGURE 10-11
The three types of stress.

FIGURE 10-12
See Color Plate IV.

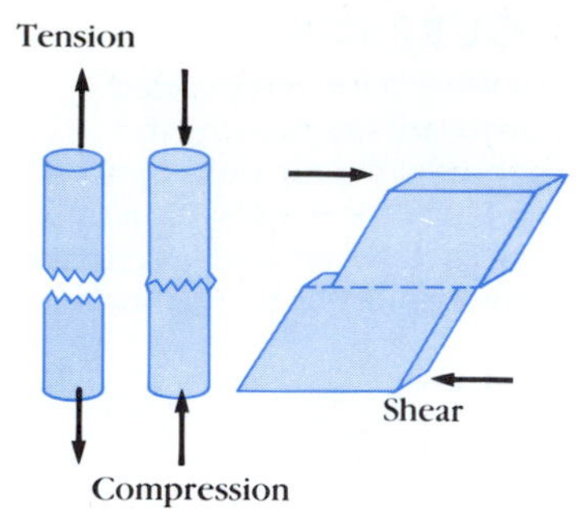

FIGURE 10-13
Fracture as a result of the three types of stress.

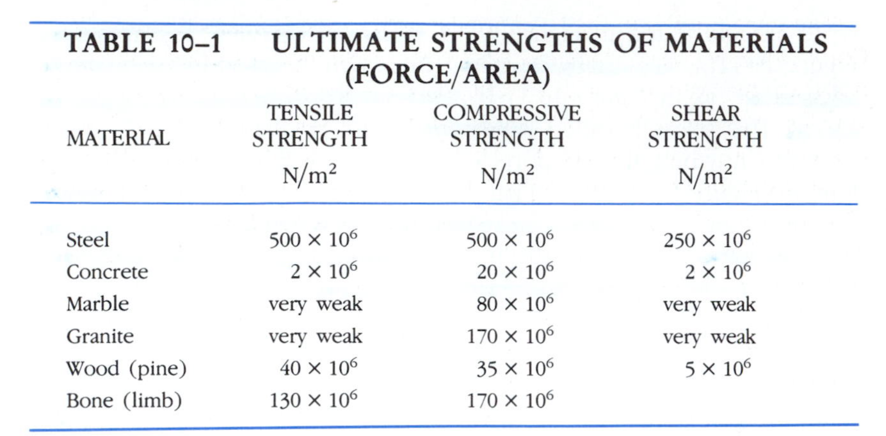

TABLE 10–1 ULTIMATE STRENGTHS OF MATERIALS (FORCE/AREA)

MATERIAL	TENSILE STRENGTH N/m^2	COMPRESSIVE STRENGTH N/m^2	SHEAR STRENGTH N/m^2
Steel	500×10^6	500×10^6	250×10^6
Concrete	2×10^6	20×10^6	2×10^6
Marble	very weak	80×10^6	very weak
Granite	very weak	170×10^6	very weak
Wood (pine)	40×10^6	35×10^6	5×10^6
Bone (limb)	130×10^6	170×10^6	

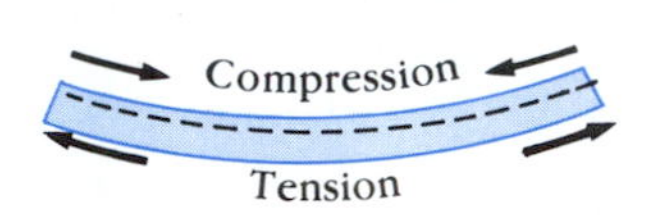

FIGURE 10-14
A beam sags under its own weight and thus undergoes compression (the upper half is shortened), tension (the lower half is elongated), and also shear.

As an example, suppose a marble column in a Greek temple (Figure 10-12) supports 600,000 kg. The total weight is then $mg = (600{,}000\ \text{kg})(9.8\ \text{m/s}^2) \approx 6 \times 10^6\ \text{N}$. Marble has an ultimate compressive strength of $80 \times 10^6\ \text{N/m}^2$ according to Table 10-1. If we use a safety factor of 10, we say the column can support $8 \times 10^6\ \text{N/m}^2$. Since it must support 6×10^6 N, the cross-sectional area needs to be at least $(6 \times 10^6\ \text{N})/(8 \times 10^6\ \text{N/m}^2) = 0.75\ \text{m}^2$. Since $A = \pi D^2/4$ for a circle, the minimum diameter D is $D = \sqrt{4(0.75\ \text{m}^2)/(3.14)} = 1.0$ m.

5. APPLICATIONS TO ARCHITECTURE: ARCHES AND DOMES

The arts and humanities often overlap the sciences, and this is particularly obvious in architecture. Many of the architectural features we admire have practical purposes as well as decorative effects. Consider, for example, the methods used to span a space from the simple beam to arches and domes.

In the so-called post-and-beam, or post-and-lintel, construction, two upright posts support a horizontal beam. It is still used a great deal today; but until the introduction of steel in the nineteenth century, the span of a beam was very limited because the strongest building materials were stone and brick. The size of a span was therefore limited by the size of available stones. Of equal importance is the fact that all these materials, though strong under compression, are exceptionally weak under tension and shear. And as shown in Figure 10-14, all these stresses occur within a beam because it sags (slightly) under its own weight. If the weight is too great, the beam will fracture on its bottom surface where the tension is greatest. That only a minimal space could be spanned using stone is evident in the closely spaced columns of the great Greek temples (Figure 10-12, Color Plate IV).

FIGURE 10-15
See Color Plate IV.

The introduction of the semicircular arch by the Romans (Figure 10-15, Color Plate IV), aside from its aesthetic appeal, was a tremendous technological innovation. It had been preceded by the so-called "triangle arch" and the "corbeled arch," but these were only slight improvements over the post-and-beam (Figure 10-16). The advantage of the well designed "true" or semicircular arch is that its wedgeshaped stones undergo stress that is mainly compressive even when supporting a large load such as the wall and roof of a cathedral. Because the stones are forced to squeeze against each other, they are, therefore, mainly under compression (Figure 10-17). Note, however, that the arch transfers horizontal as well as vertical forces to the supports. A round arch consisting of many well-shaped stones could span a very wide space. Yet, considerable buttressing on the sides was needed to support the horizontal components of the forces. The buttressing was frequently provided by heavy walls leading off laterally, and in many of the great basilicas these were conveniently used to enclose side chapels.

The pointed arch was first used in Europe about A.D. 1100 and was an essential part of all the great Gothic cathedrals. It was apparently introduced for constructional purposes, rather than for aesthetics, because it was used where heavy loads had to be supported—such as beneath the tower of a cathedral and as the central arch across the nave; the lesser arches in these early churches remained round. The builders realized that because of the steepness of the pointed arch, the forces due to the weight above would be brought down more nearly vertically. Thus, less buttressing was required. The pointed arch took the load off the walls, allowing more openness and light and the glorious use of stained glass windows. The limited amount of buttressing still required was supplied on the exterior by graceful flying buttresses (Figure 10-18, Color Plate IV), which were themselves half arches.

The technical innovation of the pointed arch was achieved not through calculation but through experience and intuition. The beginnings of the science of statics and building appeared during the Renaissance and were advanced by Galileo; it was not until somewhat later that detailed calculations came into use. To make an accurate analysis of a stone arch is quite difficult in practice. But if we make some simplifying assumptions, we can show why the horizontal component of the force at the base is less for a pointed arch than for a round one. Figure 10-19 shows a round arch and a pointed arch, each with the same span, 8.0 m. The height of the round arch is thus 4.0 m, whereas that of the pointed arch is larger and has been chosen to be 8.0 m. Each arch supports the same 40-ton weight above, which for simplicity is assumed to act as shown on the two halves of each arch. Let us focus our attention only on the right half of each arch. Since there is a downward force of 20 tons due to the weight above, the supports must exert an upward force at the base of each arch of $F_V = 20$ tons. Notice that these two forces do not act along the same line, so there seems to be a net torque. Thus the arch would rotate (and crumble) if there were

FIGURE 10-16
Various means to span a space.

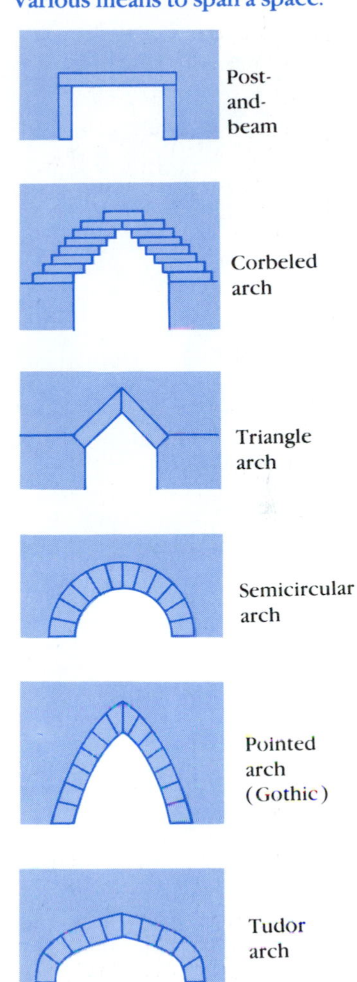

FIGURE 10-17
Stones in a round (or "true") arch undergo stress that is primarily compressive.

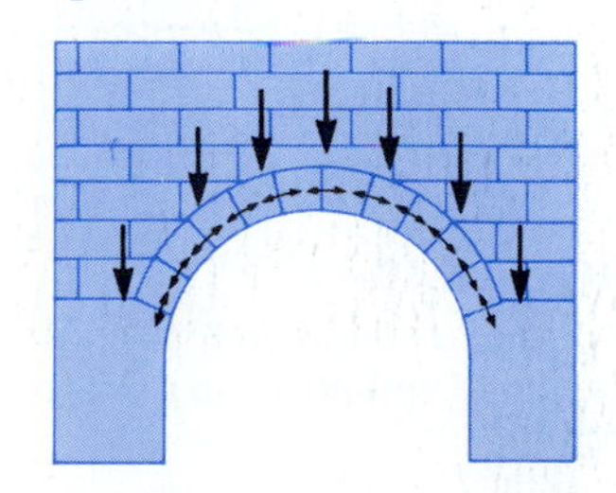

FIGURE 10-18
See Color Plate IV.

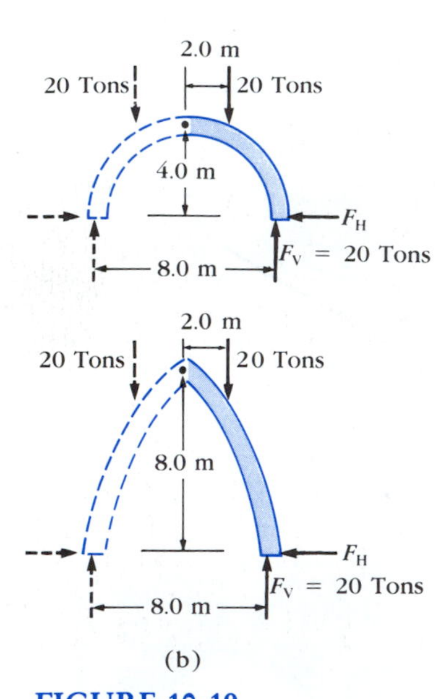

FIGURE 10-19
Forces in a round arch (a), compared with those in a pointed arch (b).

not an additional horizontal force F_H at the base of the arch. If we calculate all the torques about the apex (top) of the arch as if this were an axis, the sum of all torques must be zero. The counterclockwise torque due to F_V is greater than the clockwise torque due to the 20-ton weight, because the lever arm is greater (4.0 m versus 2.0 m). Hence F_H must point inward to give an additional clockwise torque to achieve balance. It is clear from the diagram that the lever arm for F_H in (b) is 8.0 m, whereas that in (a) is only 4.0 m. To provide the same torque, F_H need be only half as great in (b) as in (a). Thus we see that the horizontal buttressing can be less for a pointed arch because it is higher. Indeed, the steeper the arch, the less the horizontal component of the force needs to be; and hence the more nearly vertical is the force exerted at the base of the arch.

The further development of the arch was one of decline; for the subsequent flattened arches, such as the Tudor arch (Figure 10-16), were structurally weaker than the simple pointed arch. However, with the coming of advanced methods of calculation in the nineteenth and twentieth centuries, it became possible to calculate the best shape of arch for a given load condition. For example, if the load is uniform across the span, it can be shown that the stresses within the arch will be purely compressive if the arch has a parabolic shape.

Whereas an arch spans a two-dimensional space, a dome—which is basically an arch rotated about a vertical axis—spans a three-dimensional space. The Romans built the first large domes; their shape was hemispherical and some, such as that of the Pantheon in Rome, still stand (Figure 10-20). By the time of the Renaissance, the technique for constructing large domes seems to have been lost. Indeed, the dome of the Pantheon was a source of wonder to Renaissance architects. (Even today we do not know precisely how it was built.) The problem came to the fore in fifteenth-century Florence with the designing of a new cathedral that was to have a dome 43 m in diameter to rival that of the Pantheon. Now a dome, like an arch, is not stable until all the stones are in place. It had been the custom to support a dome during construction with a wooden framework. But no trees big enough or strong enough could be found to support the 43-m space required for the cathedral in Florence. In 1418, after the cathedral was finished except for the dome, a competition for the design of the dome was held and was won by Brunelleschi. One problem he had to deal with was that the dome was to rest on a drum that had been completed with no external abutments—and there was no place to put any. Hence the dome must exert the minimum of horizontal force. Brunelleschi solved this by designing a pointed dome—Figures 10-21 (a) and (b), and Color Plate IV—because a pointed dome, like a pointed arch, exerts a smaller side thrust against its base.

The other major problem was how to support the dome during construction. Instead of using a wooden framework, Brunelleschi built the dome in horizontal layers. Each layer was bonded to the previous layer, which held it in place until the last stone of the circle was placed. Each closed ring was then strong enough to support the next layer. It was an amazing feat. Not until the present century were domes larger than Brunelleschi's built, and these use newly developed building materials such as steel and prestressed concrete.

FIGURE 10-20
The Pantheon in Rome.

SUMMARY

If a body is to remain at rest, the sum of all the forces on it must be zero and the sum of all the torques on it must be zero. It is then said to be in *equilibrium*. These two conditions for equilibrium can be used to analyze the forces and torques acting on a body.

The nature of the equilibrium of a body at rest can be determined by displacing the body slightly. The body is said to be in *stable*, *unstable*, or *neutral* equilibrium depending on whether it subsequently returns to its original position, continues to move, or remains in the new position. A body whose center of gravity (c.g.) is higher than its base of support will be stable only if its c.g. is not beyond the base of its support.

All objects are *elastic* to some extent—i.e., their shape changes when a force is exerted on them, and when the force is removed they tend to return to their original shape. *Hooke's law* states that the amount of deformation is proportional to the applied force. If the force is sufficiently great, the *elastic limit* is exceeded and the object will not return to its original shape when the force is removed. If the *ultimate strength* is exceeded, the object will break. A body is under *tension*, *compression*, or *shear* stress depending on whether the forces on it tend to increase its length, decrease its length, or distort it crosswise.

The subject of statics is especially useful for calculating forces within muscles and bones, and in structures such as buildings and bridges. Many architectural innovations—for example, the round and pointed arches—resulted from a consideration of static forces and the need to avoid fracture.

FIGURE 10-21
(a) The skyline of Florence, showing Brunelleschi's famous dome.
(b) See Color Plate IV.

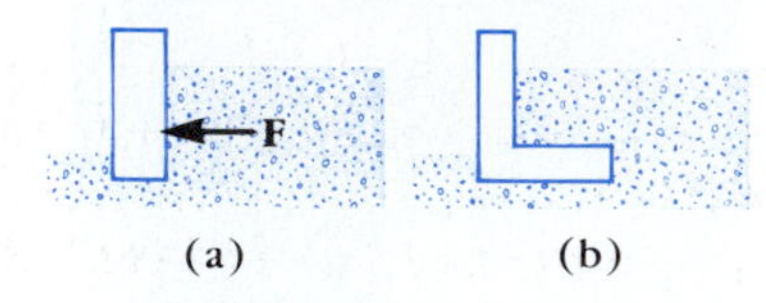

FIGURE 10-22

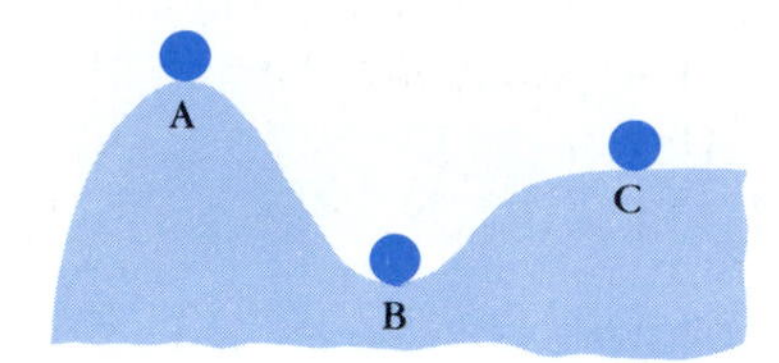

FIGURE 10-23

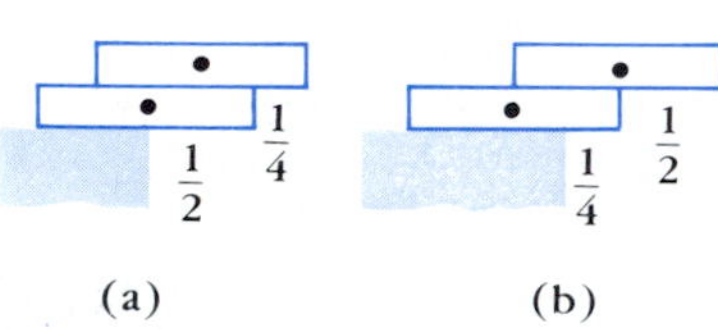

FIGURE 10-24
The dots indicate the c.g. of each brick. The fractions $\frac{1}{4}$ and $\frac{1}{2}$ indicate what portion of the brick is hanging beyond its support.

FIGURE 10-25

QUESTIONS

1. Explain why touching the toes while seated on the floor with outstretched legs produces less stress on the lower spinal column than when touching the toes from a standing position. Use a diagram.
2. An earth retaining wall is shown in Figure 10-22(a). The earth, particularly when wet, can exert a significant force F on the wall. (a) What force produces the torque to keep the wall upright? (b) Explain why the retaining wall in Figure 10.22(b) would be much less likely to overturn.
3. Give an example of a situation in which the net torque on a body is zero but the net force is not. Would such a body be in equilibrium?
4. If the following forces are exerted on a body, can the body be in equilibrium: 100 N, 200 N, 50 N? Explain.
5. Name the type of equilibrium for each position of the ball in Figure 10-23.
6. Why is it not possible to sit upright in a chair and rise to one's feet without first leaning forward?
7. Explain why a rectangular brick can be placed so that slightly less than half its length can be suspended over the edge of a table, but no more.
8. Which of the configurations of brick (a) or (b) of Figure 10-24 is the more likely to be stable? Why?
9. What physical purpose does a walking stick serve when you are hiking in rough country? Be specific.
10. Do you think the daredevil rider in Figure 10-25 is really in danger? Explain. What if the people below were removed?
11. In what state of equilibrium is a square box when it is (a) on its edge, (b) on its face?
12. Examine how a pair of scissors or shears cuts through a piece of cloth or cardboard. Is the name "shears" reasonable?
13. Why do you lean backward when carrying a heavy bag of groceries?
14. Why is it harder to overturn a body with a low c.g. and a broad base than it is to overturn a body with a high c.g. and a narrow base?
15. If you double the weight on a column of a Greek temple, how much more does it shorten?
16. Why do you bend forward when carrying a backpack?
17. Which is more likely to fracture under compression, a marble column or a wood column of the same diameter? What about under tension? (*Hint*: See Table 10-1.)

18. Why must there be a horizontal force exerted at the base of an arch?
19. A round arch and a pointed arch have the same height. Which spans the greater space? Which requires greater horizontal buttressing? Explain.
20. A round arch and a pointed arch require the same horizontal force at their bases. Which spans the greater space? Explain.

FIGURE 10-26

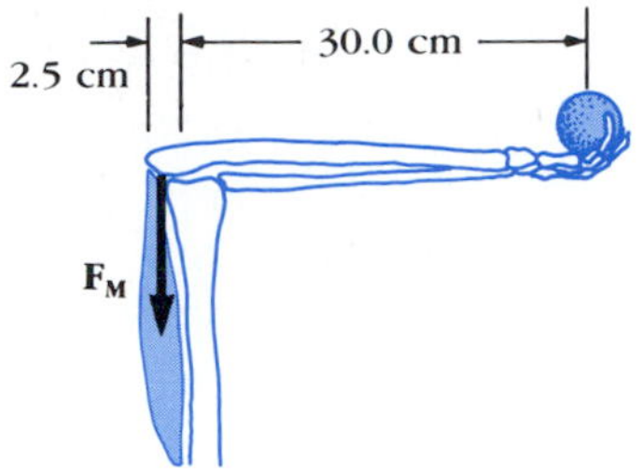

EXERCISES

1. If the biceps muscle in Figure 10-3 were attached 6.5 cm from the elbow instead of 5.0 cm, what force would be required of the biceps muscle?
2. Approximately what force, F_M, must the muscle in the upper arm exert on the lower arm to hold a 7.3 kg shotput (Figure 10-26)? Assume the lower arm has a mass of 2.8 kg and its c.g. is 12 cm from the pivot point.
3. A 50-story building is being planned. It is to be 200 m high with a base 40 m by 70 m. Its total mass will be about 1.5×10^7 kg and its weight therefore about 1.5×10^8 N. Will this building tip over in a 200-km/hr wind that exerts a force of 1450 N/m^2 over the 70-m-wide face? Assume the building's c.g. is at its geometric center and that the force of the wind (F_a) acts at the center of the building face, as shown in Figure 10-27. (*Hint*: Calculate the torque about the lower-left corner of the building—the pivot point if the building were actually to begin toppling over.)
4. The Leaning Tower of Pisa (Figure 3-5) is about 55 m tall and 7.0 m in radius. Its c.g. is 4.5 m to one side of a point directly above the center of its base. Is the tower stable? If so, how much farther can it lean before it becomes unstable? Assume the tower is uniform in composition. (Draw a diagram.)
5. Bricks are stacked as shown in Figure 10-24. (a) Show that if a four-brick stack is to be stable the bricks must extend no more than (starting at the top) one-half, one-quarter, one-sixth, one-eighth of their length beyond the one below. (b) Is the top brick completely beyond the base?
6. A piano string stretches 2 mm when a force of 400 N is applied. How much will it stretch if the force is increased to 1000 N?
7. A column is shortened by 2 cm when a 1300-kg mass rests on it. If a 3000-kg mass were placed on it, by how much would it shorten?
8. A marble column is 60 cm in diameter. In kilograms, how much mass can it support?

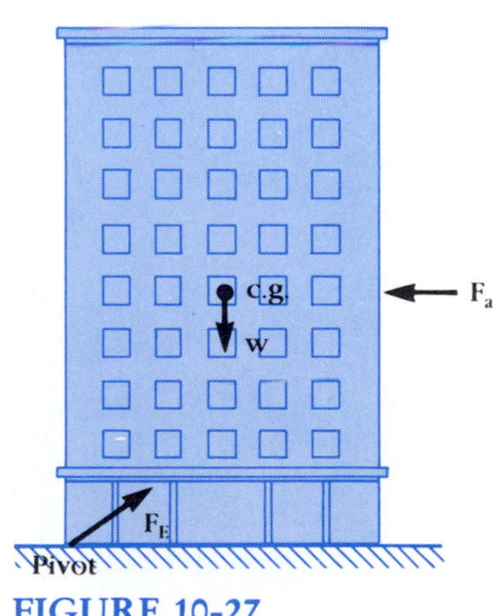

FIGURE 10-27

FIGURE 10-28

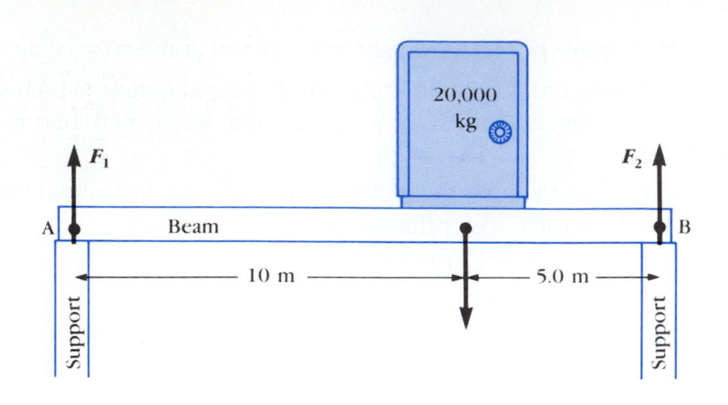

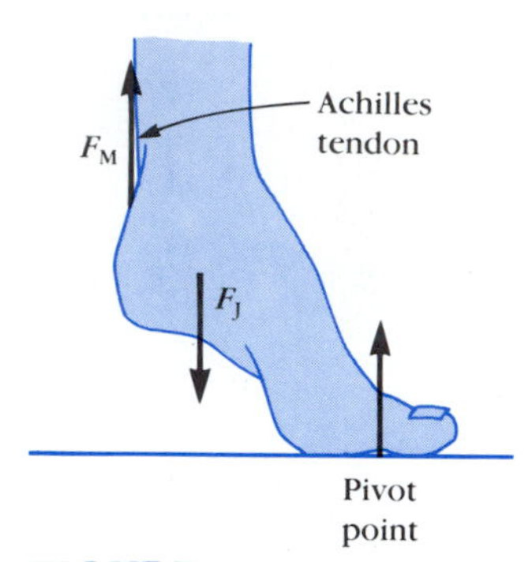

FIGURE 10-29

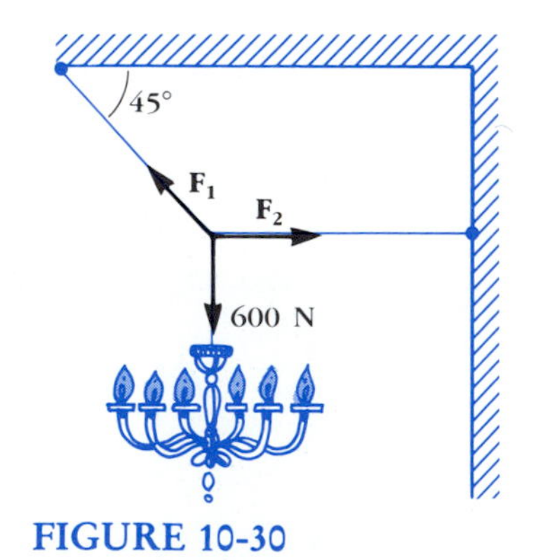

FIGURE 10-30

9. One marble column has twice the diameter of a second one. How much more force can the first one withstand than the second one without breaking?

10. Show that the horizontal force required at the base of the arches in Figure 10-19 is $F_H = 10$ tons for the round arch (a) and $F_H = 5$ tons for the pointed arch (b).

∗11. An architect has designed a building in which the floor must support a 20,000-kg load (perhaps a bank vault or heavy machinery) on the beam shown in Figure 10-28. Assume you are the architect and determine the forces, F_1 and F_2, that the two supports must withstand. (*Hint*: Use both conditions for equilibrium, giving two equations in two unknowns; calculate torque about either point A or point B.)

∗12. A man puts all his weight on one foot and stands on his toes as shown in Figure 10-29. The weight of the man is W; the ankle joint where F_J acts is $\frac{2}{3}$ of the way from the toes to the point where the force F_M is exerted upward on the Achilles tendon. Calculate both F_M and F_J in terms of W.

∗13. Find the tension in the two ropes, F_1 and F_2, in Figure 10-30. (*Hint*: Use components and the theorem of Pythagoras or a graph.)

∗14. What minimum cross section must the supports in Figure 10-28 have (see Problem 11) if they are made of concrete? Use a safety factor of 5.

∗15. How high must a pointed arch be if it is to span a space 6.0 m wide and require a horizontal force at its base that is one-third that required by a round arch?

STRUCTURE OF MATTER

CHAPTER

11

What would happen if you cut a piece of aluminum in half, then cut the halves in half, and continued to subdivide the substance into tinier and tinier pieces? Would you ever reach a "tiniest" piece of aluminum? Or can you subdivide a substance into smaller and smaller pieces indefinitely? Is matter made up of tiny particles, or building blocks, or is matter continuous? We have been asking these questions about the ultimate nature of matter since the time of the ancient Greeks.

In earlier chapters we studied the motion of material bodies. In this chapter we will study the structure of matter itself and its three principal states: solid, liquid, and gas.

1. THE ATOMIC THEORY

The earliest recorded hypothesis that matter is made up of tiny particles is credited to the Greek philosopher Democritus, who lived in the fifth century B.C. According to Democritus, there is a limit to how far matter can be subdivided. The tiniest pieces of matter beyond which no further subdivision is possible were called **atoms**, from the Greek word for "indivisible." However, most ancient philosophers, including Aristotle, believed that matter was continuous and that it consisted of four basic elements: earth, air, water, and fire. This view predominated until the Renaissance. Although Democritus and his atomic theory had a number of followers, his role in history is not unlike that of Aristarchus and the heliocentric theory (see Chapter 2).

For many centuries, acceptance of either the continuous theory or the atomic theory of matter was based more on intuition than on experimental evidence. However, toward the end of the Renaissance, philosophers began to notice that the atomic theory made certain aspects of the behavior of matter more intelligible. Galileo and Newton were avowed atomists. But the atomic theory did not really have much support until experiments were made in chemistry in the eighteenth and early nineteenth centuries.

Elements, Compounds, and Mixtures

Chemists have long been concerned with how different materials could be combined to form other materials and with whether or not a given substance could be broken down into simpler substances. They found that certain materials such as iron, gold, and silver could not be broken down by any chemical means. They called these materials **elements**. Other substances were found to be made up of combinations of elements, and these were called **compounds**. For example, water is a compound; it can be broken down into the elements hydrogen and oxygen. Ordinary table salt, or sodium chloride, is a compound; it is made from sodium and chlorine.

Compounds must be distinguished from mixtures. A compound is a wholly new substance, completely different from the substances from which it is made. In a mixture, on the other hand, the separate materials are easily recognizable. For example, the two gases hydrogen and oxygen can be mixed together to form a mixture that is still gaseous. If, however, a spark is applied to this mixture, an explosion occurs and a new substance, water, is formed. Water is a compound of hydrogen and oxygen, but it is a liquid that has very different properties from hydrogen and oxygen. The distinction between a mixture of the gases hydrogen and oxygen and the compound water is quite obvious. Many of the solid materials we deal with in everyday life are mixtures of elements and compounds: concrete, glass, pottery, wood, food, and all living matter.

The Law of Definite Proportions Supports the Atomic Theory

A crucial piece of evidence in favor of the atomic theory was the *law of definite proportions*, a summation of experimental results collected during the half-century prior to 1800. Briefly stated, this law says that

> **when two or more elements combine to form a compound, they always do so in the same proportions by weight.**

For example, salt is always formed from 23 parts by weight of sodium and 35 parts by weight of chlorine; water is formed from one part hydrogen and 8 parts oxygen; hydrochloric acid is formed from one part hydrogen and 35 parts chlorine; and so on.

The Englishman John Dalton (1766–1844) quickly recognized that the law of definite proportions lent significant support to the atomic theory. A continuous theory of matter could not readily account for the definite proportions of each element required to make a compound. But Dalton showed that the atomic theory explains it very simply. Dalton reasoned that the weight proportions of each element required to make a compound correspond to the relative weights of the combining atoms. For example, to explain why 23 parts by weight of sodium always combine with 35 parts by weight of chlorine to make the compound sodium chloride, Dalton argued that each atom of sodium combines with a single atom of chlorine, and that one atom of sodium weighs 23/35 times as much as one atom of chlorine. The combination of one chlorine atom with one sodium atom was called a **molecule** by Dalton. A molecule is thus a fixed combination of atoms.

By measuring the relative amounts of each element required for the formation of a great variety of compounds, a table of the relative weights of atoms was soon established. Hydrogen, which was found to be the lightest of all the atoms, was arbitrarily given the relative weight of 1; on this scale sodium was about 23, chlorine 35, iron 56, and so on. From the various compounds that oxygen formed, the relative weight of oxygen was found to be 16. This was hard to reconcile with the fact that in water the weight ratio of oxygen to hydrogen was 8:1. This difficulty was explained, however, by assuming that two hydrogen atoms combined with one oxygen atom to form the water molecule. A good many other molecules were also found to contain two or more atoms of the same kind (Figure 11-1).

FIGURE 11-1
A simple model of some atoms and molecules. Chemical symbols (Table 11-1) are used for each atom

Gold

Iron

Water (H_2O)

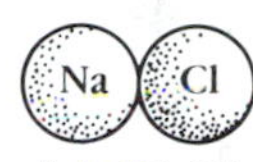

Salt (NaCl)

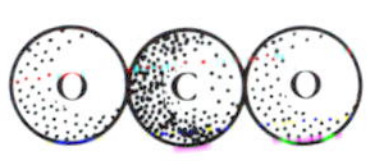

Carbon dioxide (CO_2)

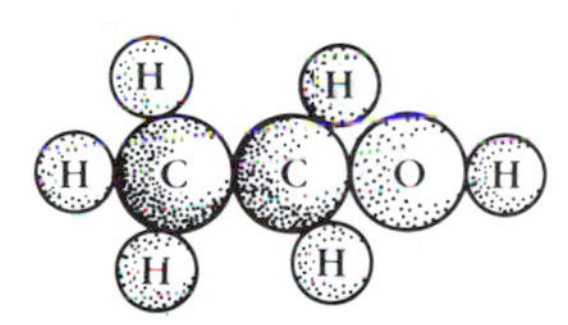

Ethyl alcohol (C_2H_5OH)

The atomic theory neatly describes the difference between elements compounds, and mixtures. Substances that contain only one kind of atom, such as iron or aluminum, are **elements**. Substances that are made up of one kind of molecule, such as water, carbon dioxide, salt, or ethyl alcohol, are **compounds**. Substances that are a combination of different kinds of atoms and molecules, such as air, milk, concrete, wood, and other organic substances, are **mixtures**. Today, there are 109 known elements. Most of the substances we know in our everyday lives are compounds or mixtures.

Table 11-1 is a list of the known elements and their chemical abbreviations or symbols. The list also contains the **atomic mass**[†] (or "relative weight") for each. The atomic masses are given relative to ordinary carbon, which is assigned the value of exactly 12.0 atomic mass units.

Brownian Movement

So much evidence was collected in support of the atomic theory of matter following Dalton's findings that by the early twentieth century most scientists accepted it as true.* It is one of the most important and influential theories in all of science.

† The term *atomic weight* is popularly used for this quantity, but technically speaking we are comparing masses.

* The last major holdout was Ernst Mach (1838–1916), an Austrian physicist and philosopher. His positivist philosophy led him to the conclusion that since atoms cannot be directly sensed, it was meaningless to believe in them.

TABLE 11-1 The Elements‡

ELEMENT	SYMBOL	ATOMIC NUMBER	ATOMIC MASS
Actinium	Ac	89	(227)
Aluminum	Al	13	27.0
Americium	Am	95	(243)
Antimony	Sb	51	121.8
Argon	Ar	18	39.9
Arsenic	As	33	74.9
Astatine	At	85	(210)
Barium	Ba	56	137.3
Berkelium	Bk	97	(247)
Beryllium	Be	4	9.0
Bismuth	Bi	83	209.0
Boron	B	5	10.8
Bromine	Br	35	79.9
Cadmium	Cd	48	112.4
Calcium	Ca	20	40.1
Californium	Cf	98	(251)
Carbon	C	6	12.0
Cerium	Ce	58	140.1
Cesium	Cs	55	132.9
Chlorine	Cl	17	35.5
Chromium	Cr	24	52.0
Cobalt	Co	27	58.9
Copper	Cu	29	63.5
Curium	Cm	96	(247)
Dysprosium	Dy	66	162.5
Einsteinium	Es	99	(253)
Erbium	Er	68	167.3
Europium	Eu	63	152.0
Fermium	Fm	100	(257)
Fluorine	F	9	19.0
Francium	Fr	87	(223)
Gadolinium	Gd	64	157.2
Gallium	Ga	31	69.7
Germanium	Ge	32	72.6
Gold	Au	79	197.0
Hafnium	Hf	72	178.5
Hahnium	Ha	105	(260)
Helium	He	2	4.0
Holmium	Ho	67	164.9
Hydrogen	H	1	1.0
Indium	In	49	114.8
Iodine	I	53	126.9
Iridium	Ir	77	192.2
Iron	Fe	26	55.8
Krypton	Kr	36	83.8
Lanthanum	La	57	139.9
Lawrencium	Lr	103	(260)
Lead	Pb	82	207.2
Lithium	Li	3	6.9
Lutetium	Lu	71	175.0
Magnesium	Mg	12	24.3
Manganese	Mn	25	54.9
Mendelevium	Md	101	(258)
Mercury	Hg	80	200.6
Molybdenum	Mo	42	95.9
Neodymium	Nd	60	144.2
Neon	Ne	10	20.2
Neptunium	Np	93	237.0
Nickel	Ni	28	58.7
Niobium	Nb	41	92.9
Nitrogen	N	7	14.0
Nobelium	No	102	(259)
Osmium	Os	76	190.2
Oxygen	O	8	16.0
Palladium	Pd	46	106.4
Phosphorus	P	15	31.0
Platinum	Pt	78	195.1
Plutonium	Pu	94	(244)
Polonium	Po	84	(209)
Potassium	K	19	39.1
Praseodymium	Pr	59	140.9
Promethium	Pm	61	(145)
Protactinium	Pa	91	231.0
Radium	Ra	88	226.0
Radon	Rn	86	(222)
Rhenium	Re	75	186.2
Rhodium	Rh	45	102.9
Rubidium	Rb	37	85.5
Ruthenium	Ru	44	101.1
Rutherfordium	Rf	104	(261)
Samarium	Sm	62	150.4
Scandium	Sc	21	45.0
Selenium	Se	34	79.0
Silicon	Si	14	28.1
Silver	Ag	47	107.9
Sodium	Na	11	23.0
Strontium	Sr	38	87.6
Sulfur	S	16	32.1
Tantalum	Ta	73	180.9
Technetium	Tc	43	98.9
Tellurium	Te	52	127.6
Terbium	Tb	65	158.9
Thallium	Tl	81	204.4
Thorium	Th	90	232.0
Thulium	Tm	69	168.9
Tin	Sn	50	118.7
Titanium	Ti	22	47.9
Tungsten	W	74	183.8
Uranium	U	92	238.0
Vanadium	V	23	50.9
Xenon	Xe	54	131.3
Ytterbium	Yb	70	173.0
Yttrium	Y	39	88.9
Zinc	Zn	30	65.4
Zirconium	Zr	40	91.2

‡ Four newly discovered elements (numbers 106, 107, 108, 109) are as yet unnamed.

Perhaps the most direct evidence in support of the atomic theory of matter was a phenomenon discovered accidentally by the biologist Robert Brown in 1827. Brown noted that tiny pollen grains suspended in water under his microscope followed a tortuous path even though the water appeared to be perfectly still, see Figure 11-2(a). This phenomenon is now called *Brownian movement*. Tiny pieces of dust or oil droplets suspended in water, and smoke particles suspended in still air, also undergo random movement.

The atomic theory explains Brownian movement if the further reasonable hypothesis is made that the atoms or molecules in any substance are continually in motion—Figure 11-2(b). Thus the tiny pollen grains of Figure 11-2(a) are buffeted about by a barrage of rapidly moving molecules of water. A piece of wood floating in still water is too large and has too much inertia to be affected by the minuscule molecular bombardment; but a tiny pollen grain is small enough so that if several molecules chance to strike it simultaneously on the same side, it will rebound. The random bombardment of water molecules from all directions thus gives rise to the observed erratic motion of Figure 11-2(a). The pollen grain is jostled about on a vast molecular sea.

In 1905 Einstein explained Brownian movement from this atomic point of view, and his quantitative studies made it possible to estimate atomic sizes and weights. It was found that atoms were extremely small, on the order of 0.00000001 cm (1×10^{-8} cm, or 1×10^{-10} m) in diameter. In the wake of Einstein's analysis, few scientists persevered in opposition to the atomic theory.

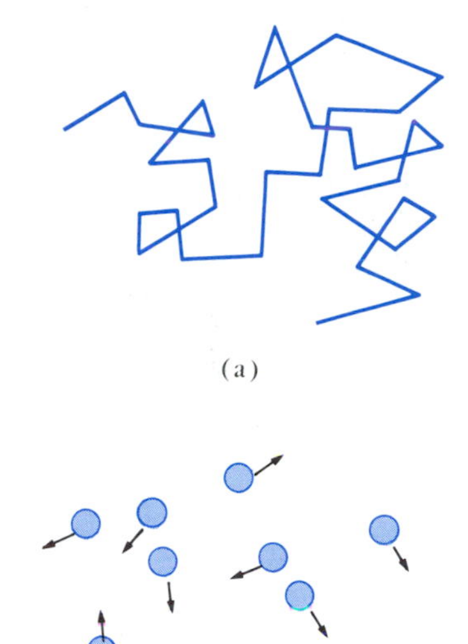

FIGURE 11-2
(a) Brownian Movement: the straight lines connect successive positions of a single pollen grain suspended in water observed under a microscope at ten-second intervals. (b) Brownian movement is explained by continual random motion of molecules.

2. STRUCTURE OF ATOMS; THE PERIODIC TABLE

A Brief Look Inside the Atom

The historical development of our understanding of the atom is a fascinating subject, and we will investigate it in detail in Chapters 27 and 28. For now we will summarize the contemporary view of the atom very briefly.

Atoms are not indivisible, as was once thought. Physicists early in the twentieth century found that an atom consists of a very tiny but heavy **nucleus**; tiny **electrons** revolve around the nucleus, much like the planets orbit the sun (Figure 11-3). However, it is not the gravitational force that keeps the electrons in their orbits; it is instead the "electrical force," a force we will discuss in Chapter 18. The electrons are "negatively charged" and the nucleus is "positively charged." A positive charge always attracts a negative charge, and it is this electrical attraction that holds the electrons in orbit around the nucleus.

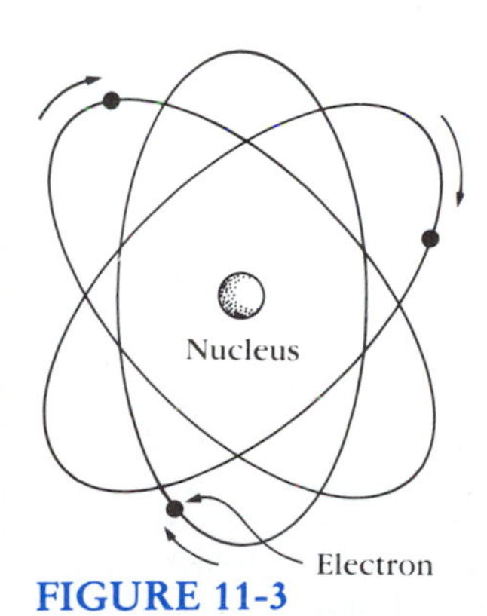

FIGURE 11-3
Model of an atom showing electrons revolving around the nucleus.

Why is iron hard and lead soft? Why is mercury a liquid at room temperature and helium a gas? Why does oxygen and not helium react readily with most metals? These and other properties of an element are determined by the number of electrons in an atom of the element. The number of electrons in an atom (called the **atomic number**) then distinguishes one kind of atom from another. Thus hydrogen has an atomic number of one because it has only one electron orbiting its nucleus; helium has two, carbon six, oxygen eight, and uranium ninety-two electrons.

The Periodic Table of the Elements

Today there are 109 known elements, corresponding to atomic numbers from one through 109. But as early as 1871 sufficient elements had been studied that the Russian chemist Dmitri Mendeleev (1834–1907) was able to find some order among them. By arranging the elements in a table according to increasing atomic weight, Mendeleev found a regular repetition (or periodicity) of chemical and physical properties. The modern version of Mendeleev's **periodic table of the elements** is shown in Table 11-2. Each vertical column contains elements with similar properties. For example, the first column contains the very reactive metals lithium, sodium, potassium, rubidium, cesium, and francium; whereas the last column on the right contains the "inert" gases—helium, neon, argon, krypton, zenon, and radon—that chemically react only very slightly if at all.

Why the periodicity contained in the periodic table of the elements should exist was not solved until the quantum theory was conceived in the early twentieth century (see Chapter 28).

In Table 11-2, the periodic table, each square contains the atomic number and then the symbol of the element; below the symbol is the relative atomic weight of that element.

FIGURE 11-4
(a) A solid. (b) A liquid. (c) A gas.

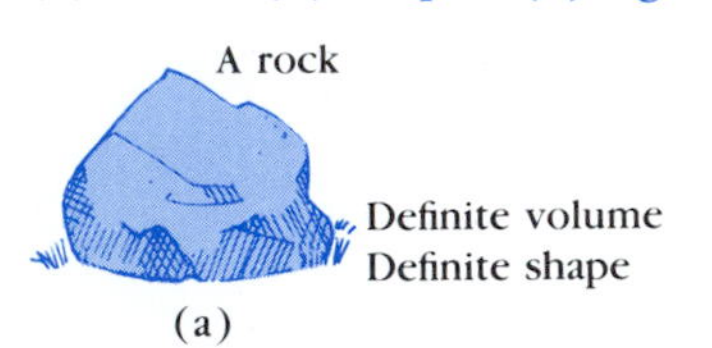

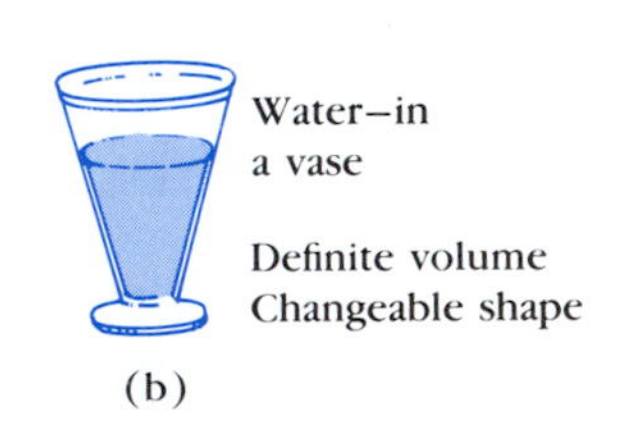

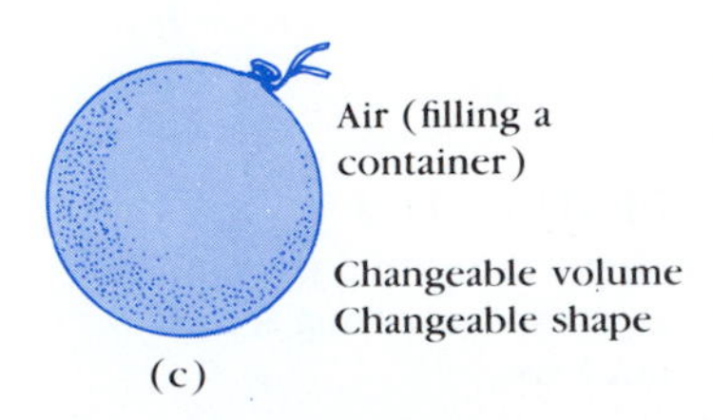

3. THE THREE STATES OF MATTER: SOLID, LIQUID, GAS

That matter is made up of atoms and molecules is regarded by many people as an established fact. Actually, it is only a theory. But it is a theory that explains a great variety of observation; and it is accepted by nearly all scientists today as an accurate picture. But what of matter itself? What happens when many atoms or molecules exist together as matter? Let us now look at matter in a general way, keeping the atomic theory in mind.

By experience, everyone knows the difference between the three states of matter—solid, liquid, and gas—at least in most cases. No one would argue that at room temperature iron is a solid, water is a liquid, and oxygen a gas. But what precisely distinguishes the three states of matter?

We can characterize and distinguish them in the following way, Figure 11-4. **Solids** maintain a fixed shape and a fixed size. Even if a force is applied to a solid, it does not readily change its shape or volume. A **liquid** does not maintain a fixed shape but takes on the shape of its container.

TABLE 11-2 The Periodic Table

1 H 1.0																	2 He 4.0
3 Li 6.9	4 Be 9.0											5 B 10.8	6 C 12.0	7 N 14.0	8 O 16.0	9 F 19.0	10 Ne 20.2
11 Na 23.0	12 Mg 24.3											13 Al 27.0	14 Si 28.1	15 P 31.0	16 S 32.1	17 Cl 35.5	18 Ar 39.9
19 K 39.1	20 Ca 40.1	21 Sc 45.0	22 Ti 47.9	23 V 50.9	24 Cr 52.0	25 Mn 54.9	26 Fe 55.8	27 Co 58.9	28 Ni 58.7	29 Cu 63.5	30 Zn 65.4	31 Ga 69.7	32 Ge 72.6	33 As 74.9	34 Se 79.0	35 Br 79.9	36 Kr 83.8
37 Rb 85.5	38 Sr 87.6	39 Y 88.9	40 Zr 91.2	41 Nb 92.9	42 Mo 95.9	43 Tc 98.9	44 Ru 101.1	45 Rh 102.9	46 Pd 106.4	47 Ag 107.9	48 Cd 112.4	49 In 114.8	50 Sn 118.7	51 Sb 121.8	52 Te 127.6	53 I 126.9	54 Xe 131.3
55 Cs 132.9	56 Ba 137.3	57–71 see below	72 Hf 178.5	73 Ta 180.9	74 W 183.8	75 Re 186.2	76 Os 190.2	77 Ir 192.2	78 Pt 195.1	79 Au 197.0	80 Hg 200.6	81 Tl 204.4	82 Pb 207.2	83 Bi 209.0	84 Po (209)	85 At (210)	86 Rn (222)
87 Fr (223)	88 Ra 226.0	89–103 see below	104 Rf (261)	105 Ha (260)	106 (263)	107 (262)	108 (265)	109 (266)									
		57 La 139.9	58 Ce 140.1	59 Pr 140.9	60 Nd 144.2	61 Pm (145)	62 Sm 150.4	63 Eu 152.0	64 Gd 157.2	65 Tb 158.9	66 Dy 162.5	67 Ho 164.9	68 Er 167.3	69 Tm 168.9	70 Yb 173.0	71 Lu 175.0	
		89 Ac (227)	90 Th 232.0	91 Pa 231.0	92 U 238.0	93 Np 237.0	94 Pu (244)	95 Am (243)	96 Cm (247)	97 Bk (247)	98 Cf (251)	99 Es (253)	100 Fm (257)	101 Md (258)	102 No (259)	103 Lr (260)	

However, a liquid does maintain a fixed volume and is thus practically incompressible,[†] like solids. When water or milk is poured from one pitcher into another, the liquid maintains its volume; as it runs to the bottom of the new pitcher, it takes on the shape of that pitcher. A **gas** maintains neither a fixed shape nor a fixed volume. A gas expands to fill whatever container it is in and is easily compressible. When air (which is a gas) is pumped into an automobile tire, it does not all run to the bottom of the tire as a liquid would; it fills the whole volume within the tire.

Because liquids and gases do not maintain a fixed shape, they both have the ability to flow. Because of this property, liquids and gases are sometimes referred to collectively as **fluids**.

FIGURE 11-5
Arrangement of atoms in (a), a crystalline solid, (b) a liquid, and (c) a gas.

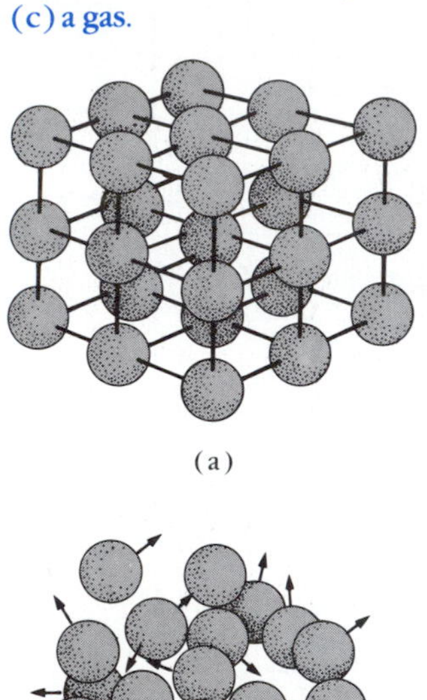

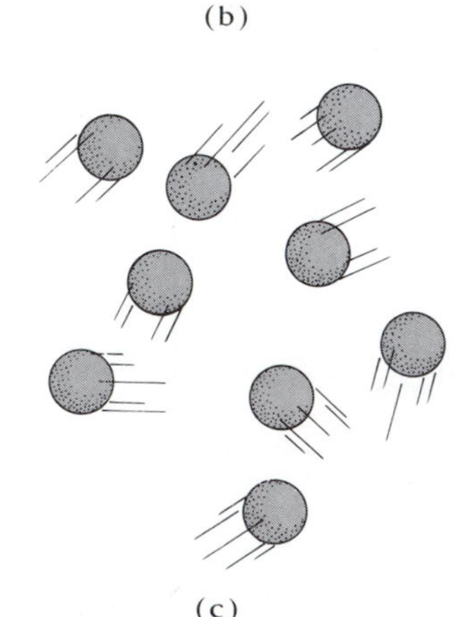

The above distinctions between solids, liquids, and gases are based on **macroscopic**, or large-scale, properties. Distinctions between them at the atomic or molecular level are also useful. We refer to the latter as **microscopic** properties, although submicroscopic might be a better word.

Microscopically, the differences between solids, liquids, and gases can be attributed to the strength of the forces existing between the atoms or molecules. That molecules exert forces on one another is apparent; otherwise an object—such as this piece of paper—would fall apart into tiny pieces. As we shall see later, the forces between molecules are electrical in nature.

In all three states—solid, liquid, and gas—the molecules (or atoms, in the case of an element) are in rapid motion. In a **solid**, the forces between the molecules are so strong that each cannot move very far, and therefore they vibrate about nearly fixed positions. Most solids occur as crystals, although the crystals are often very tiny and hard to see, as in metals; such crystalline solids have an orderly arrangement of molecules, as shown in Figure 11-5(a).*

In a **liquid**, the forces between the molecules are not strong enough to hold them in fixed positions, and so they are free to roll over one another. However, the forces are sufficiently strong so that the molecules cannot get very far apart. In a **gas**, the molecules are moving so rapidly, or the forces are so weak—or both—that the molecules don't even stay close together. They move every which way, fill any container they occupy, and occasionally collide with one another. On the average, the velocities of gas molecules are sufficiently high so that when two of them collide the forces of attraction are too weak to keep them close together and the two molecules fly off in new directions—Figure 11-5(c). A gas thus expands to fill its container simply because the molecules are moving rapidly in random directions. The speed of air molecules averages about 1500 km/hr

[†] Both external forces and temperature affect the volume of liquids and solids but the effect is much smaller than for gases

* The molecules in a few solids—such as glass, rubber, and sulfur—have no crystal structure. These *amorphous* solids have some properties characteristic of liquids.

at room temperature. Because the molecules of a gas are usually far apart on the average, gases are easy to compress compared to liquids and solids.

The division of matter into three states is not always simple. How, for example, should butter be classified? We should also mention that a fourth state of matter can be distinguished. This state is called the *plasma*, and it occurs only at very high temperatures, as we will discuss in Chapter 30. Some scientists believe that so-called colloids (suspensions of tiny particles in a liquid) should also be considered a separate state of matter. However, for our present purposes we will be interested mainly in the three ordinary states of matter.

4. DENSITY AND SPECIFIC GRAVITY

It is often said that lead is "heavier" than wood. Yet we can have a large piece of wood whose mass and weight are greater than that of a small piece of lead. What we should say is that lead is denser than wood. *Density* is an important property of materials. The **density** of a substance is defined as its mass per unit volume:

$$\text{density} = \frac{\text{mass}}{\text{volume}}.$$

We know that if we have twice the size or volume of a particular material, the mass will be twice as great as well. Objects made up of the same material have the same density. For example, several pieces of aluminum may each have a different mass and a different volume, but each will have the same ratio of mass to volume: the density is the same for each. The densities of some common materials are shown in Table 11-3. The standard metric unit for density is kg/m^3, but it is also common to use g/cm^3 (grams per cubic centimeter). Note that $1\ g/cm^3 = 0.001\ kg/(0.01\ m)^3 = 1000\ kg/m^3$. Densities in both units are given in Table 11-3.

The **specific gravity** of a substance is defined as the ratio of the density of that substance to the density of water. Specific gravity (abbreviated SG) is a pure number, without dimensions or units. Since the density of water is $1000\ kg/m^3$ or $1\ g/cm^3$, the specific gravity of any substance will be numerically equal to its density specified in g/cm^3, or 10^{-3} times its density specified in kg/m^3. For example (see Table 11-3), the density of aluminum is $2700\ kg/m^3 = 2.7\ g/cm^3$, so its SG is 2.7; similarly the SG of alcohol is 0.79.

5. GASES; THE EARTH'S ATMOSPHERE

In everyday life, gases are less obvious to us than are liquids and solids. Perhaps it is because many common gases are colorless or nearly so. Yet

TABLE 11-3 Densities and Specific Gravities of Various Substances (at 0° C and atmospheric pressure)

SUBSTANCE	DENSITY		SPECIFIC GRAVITY
	kg/m^3	g/cm^3	
Solids			
Aluminum	2700	2.7	2.7
Concrete	2300	2.3	2.3
Gold	19,300	19.3	19.3
Granite	2700	2.7	2.7
Ice	920	0.92	0.92
Iron and Steel	7800	7.8	7.8
Lead	11,300	11.3	11.3
Platinum	21,500	21.5	21.5
Silver	10,500	10.5	10.5
Wood, balsa (approx.)	200	0.2	0.2
Wood, pine (approx.)	400	0.4	0.4
Wood, oak (approx.)	800	0.8	0.8
Liquids			
Alcohol (ethyl)	790	0.79	0.79
Gasoline	680	0.68	0.68
Mercury	13,600	13.6	13.6
Oil	900	0.9	0.9
Water, pure	1000	1.0	1.0
Water, sea	1030	1.03	1.03
Gases			
Air	1.3	0.0013	0.0013
Carbon dioxide	2	0.002	0.002
Helium	0.18	0.00018	0.00018
Hydrogen	0.09	0.00009	0.00009
Nitrogen	1.3	0.0013	0.0013
Oxygen	1.4	0.0014	0.0014

gases are all around us. The air we breath, the earth's atmosphere, is a mixture of gases. The odors of food, flowers, and perfumes are due to gaseous substances. Many gases, for example chlorine, helium, hydrogen, argon, and carbon dioxide, have important commercial uses.

Properties of Gases

The molecules in a gas are in constant motion as we saw in Figure 11-5(c). Because the forces between them are relatively weak, the molecules have more than enough kinetic energy to overcome these attractive forces. The molecules normally spend little time near one another, although of course they do collide with each other and with the walls of the container. A gas expands to fill its container simply because the molecules are moving rapidly in random directions. At one moment a molecule may be near the bottom of its container, a moment later near the top. The speed of gas molecules at ordinary temperatures may average hundreds or even thousands of miles per hour.

Because the molecules of a gas are usually far apart on the average, a material in the gaseous state is mostly empty space; this explains why a gas is easily compressible.

The relatively large distance between molecules, which is reflected in the low density of gases, is also the reason that most gases are transparent and nearly colorless. The color of solids and liquids is due to light interacting with their molecules; but a gas has few molecules for light to interact with. Nonetheless, if a large amount of gas is present, it will appear colored. Air, for example, may seem colorless when you look through a thin layer of it. But if you peer up at the sky you can look through a large amount of air, and it looks blue (we will go into this in more detail in Chapter 24).

Earth's Atmosphere

The atmosphere of the earth is a mixture of gases. Near the ground level, air is made up of about 78 percent nitrogen, 21 percent oxygen, slightly less than 1 percent argon, and very tiny amounts of carbon dioxide, neon, and other gases. Water vapor is also present, but the amount varies widely from almost none to as high as 4 percent.

There is some local variability in the amounts of each gas, particularly those present in small amounts. This is quite noticeable in areas of heavy smog. The pollutants that make up smog include not only gases such as carbon monoxide, nitrogen oxides, hydrocarbons, and sulphur dioxide, but large globs of molecules—they are almost droplets—known as aerosols, and also particulate matter. The latter are mainly responsible for poor visibility under smoggy conditions.

The molecules that make up our atmosphere are moving at speeds that average about 1500 km/hr (see Chapter 13). Now a gas normally expands to fill its container. But the earth's atmosphere has no outside walls; why then doesn't all the air escape into outer space? It stays on earth because gravity holds it here. In Chapter 6 we saw that an object must attain a speed of about 40,000 km/hr to escape from the earth's gravity. Very few molecules in our atmosphere reach that speed and so they don't escape.

We are fortunate that the force of gravity is as strong as it is on the earth. The force of gravity on the moon, for example, is too weak to hold an atmosphere. The moon may once have had a gaseous atmosphere but during the past few billion years the gas molecules have all escaped into space.

Because of the pull of gravity, the molecules of the atmosphere tend to accumulate near the surface of the earth. Thus the air is densest at sea level and becomes thinner and thinner at higher elevations. At an elevation of 5500 m (18,000 feet), the air is only half as dense as at sea level. At the summit of Mt. Everest, an elevation of 8848 m (29,028 feet), the air is only one third as dense. This is why airplanes are pressurized and why Himalayan mountain climbers often use tanks of oxygen to help them breathe at high elevations. At 8800 m, for example, a lungful of natural air contains only a third as much oxygen as at sea level. This variation in density of the earth's atmosphere can be compared to liquids, like water, which maintain a nearly constant density regardless of height or depth. Of course, there is no distinct upper limit to our atmosphere. It just fades out gradually into space.

SUMMARY

The atomic theory of matter postulates that all matter is made up of tiny entities called *atoms* which are typically 10^{-10} m (10^{-8} cm) in diameter. Some substances are made up of only one type of atom, and these are called *elements*. Atoms can combine to form *molecules*, and substances made up of a single type of molecule are called *compounds*. A substance made up of more than one type of molecule is called a *mixture*. *Atomic masses* are specified on a scale where ordinary carbon is arbitrarily given the value 12.0 atomic mass units. *Brownian movement* provided the most direct early evidence in favor of atomic theory.

An atom is considered to consist of a tiny but heavy *nucleus* about which *electrons* orbit. The *periodic table of the elements* is a means of presenting the elements in order of increasing *atomic number* (equal to the number of electrons in the atom) so that elements with similar properties fall in the same column.

There are three common states of matter—solid, liquid, and gas. *Solids* have a fixed shape and volume, *liquids* have a fixed volume but take on the shape of their container, and *gases* do not have a fixed shape or volume. Because liquids and gases have the ability to flow, they are both referred to as *fluids*. Microscopically the differences between solids, liquids, and gases can be attributed to the strength of the forces existing between the atoms or molecules.

The *density* of a substance is defined as the mass per unit volume. The *specific gravity* is the ratio of the density of the substance to the density of water.

The earth's atmosphere is a mixture of gases, principally nitrogen and oxygen, with smaller amounts of other materials. And although gases normally expand to fill their containers, the earth's atmosphere is held to the earth by the force of gravity.

QUESTIONS

1. Explain how the law of definite proportions supports the atomic theory.
2. How does Brownian movement support the atomic theory?
3. What is the principal factor that determines whether the atoms or molecules of a particular substance will form a solid, liquid, or gas at ordinary temperatures?
4. Which of the following are fluids at room temperature: gold, mercury, air, glass, alcohol, carbon dioxide?
5. A kilogram of cotton compared to a kilogram of iron (a) weighs more (b) weighs less (c) has greater density (d) has less density (e) has the same density?
6. If one material has a greater density than another, does this mean that molecules of the first are necessarily heavier? Explain.
7. It is sometimes said that heavy objects sink. Is this an accurate statement? Explain.
8. Which has the greater mass, one cubic meter of ice or one cubic meter of water?
9. Which has the greater volume, 10 kg of iron or 10 kg of water?

EXERCISES

1. There are about 3.2×10^{25} molecules in a quart of water, which is a volume of about 1000 cm^3 (more precisely, 946 cm^3). Estimate the size of one water molecule.
2. An atom of copper (Cu) is how many times heavier than an atom of sulphur (S)?
3. What is the specific gravity of a material whose density is 600 kg/m^3?
4. A 750-kg solid wood barge has a volume of 1.6 cubic meters. What is its density?
5. Which has the greater mass, one cubic centimeter of platinum or one cubic centimeter of gold? How many times greater?
6. Which has the greater volume, one gram of gold or one gram of silver? By what factor is the volume greater?
7. What is the volume of 100 kg of aluminum?
8. A metallic substance has a volume of 1.3 m^3 and a mass of 13,700 kg. What metal do you think it is?
9. What is the mass of an aluminum cube that is 20 cm on a side?
10. How many kilograms of air are there in a bedroom 3 m $\times$ 4 m $\times$ 3 m?

11. What is the approximate mass of air in a living room 6.5 m × 4.4 m × 2.7 m?
12. The approximate volume of the granite monolith, known as El Capitan, in Yosemite National Park is about 10^8 m^3. What is its approximate mass?
13. If 5.0 L of antifreeze solution (SG = 0.80) is added to 4.0 L of water to make a 9.0 L mixture, what is the SG of the mixture?

FLUIDS

CHAPTER 12

Materials in either the liquid state or the gaseous state have the ability to flow. Water flows in pipes, and down rivers and streams. Blood flows in our veins and arteries. Air can flow in heating and ventilation ducts, or as drafts and great winds. Even when at rest, fluids exhibit interesting properties. For example, if you put a block of wood on the surface of a solid table, it simply sits there on top; but if you put the wood block on the surface of water, it sinks into the water, at least partially.

In this chapter we study the properties of fluids at rest and in motion. We will examine some interesting effects of fluid flow, such as how a sailboat moves against the wind and how an airplane flies through the air.

We begin with a discussion of the concept of pressure which, although it can be applied to solids, is of even greater use when dealing with fluids.

1. PRESSURE

Pressure Is Force per Unit Area

Because liquids do not have a definite shape, when dealing with them it is more convenient to use the concept of pressure rather than force. **Pressure** is defined as the force exerted per unit area:

$$\text{pressure} = \frac{\text{force}}{\text{area}}.$$

in N (newtons)
: m² (sq meters)

In symbols we write:

$$P = \frac{F}{A}.$$

In metric (SI) units, force is measured in newtons (N) and area in square meters (m²). Therefore, the SI unit for pressure is N/m^2. This unit has the official name, pascal (Pa): $1\ N/m^2 = 1\ Pa$. Automobile tire gauges are often calibrated in kilopascals ($1\ kPa = 1000\ Pa = 1000\ N/m^2$). In the British system, the common unit is the pound per square inch (lb/in^2 or psi).

As an example, consider a 60-kg person whose two feet cover an area of $500\ cm^2$. The feet exert a force on the ground equal to $F = mg = (60\ kg)(9.8\ m/s^2) \approx 600\ N$. The pressure under the feet, because the area $= 500\ cm^2 = 0.050\ m^2$, is $P = F/A = 600\ N/0.050\ m^2 = 12 \times 10^3\ N/m^2$ on the ground. If the person stands on one foot, the force is the same but the area will be half; so the pressure will be twice as much: $24 \times 10^3\ N/m^2$. If a force is exerted over a small area, the pressure will be greater than if the same force is exerted over a large area.

Pressure in Fluids

The concept of pressure is particularly useful in dealing with fluids. We will first deal with fluids at rest. To help understand pressure in a fluid imagine two containers, one whose bottom area is twice that of the other. Each is filled with a liquid, say water, to the same height (Figure 12-1). The larger container holds twice the weight of liquid as the smaller one and therefore the liquid exerts twice the force on the bottom. But the pressure on the bottom due to the weight of the liquid is the same for both since the container with twice the area has twice the force exerted on it. This is a general result: *the pressure at equal depths in a liquid is the same*.

As the following experiment shows, however, at different depths the pressure is different.

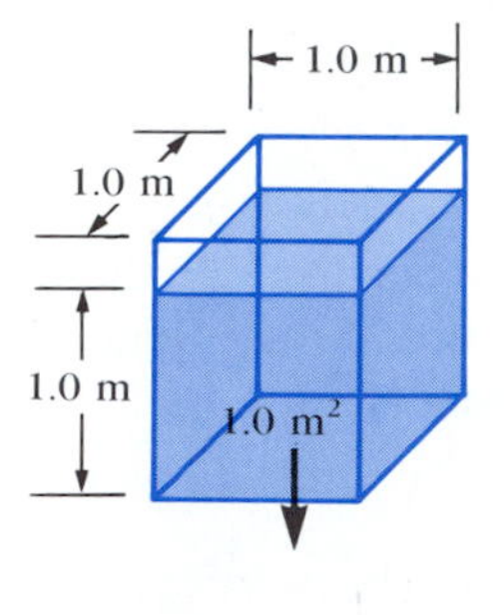

$$F = mg = (1.0 \times 10^3\ kg)(g) = 1.0 \times 10^4\ N$$

$$P = \frac{F}{A} = \frac{1.0 \times 10^4\ N}{1.0\ m^2} = 1.0 \times 10^4\ N/m^2$$

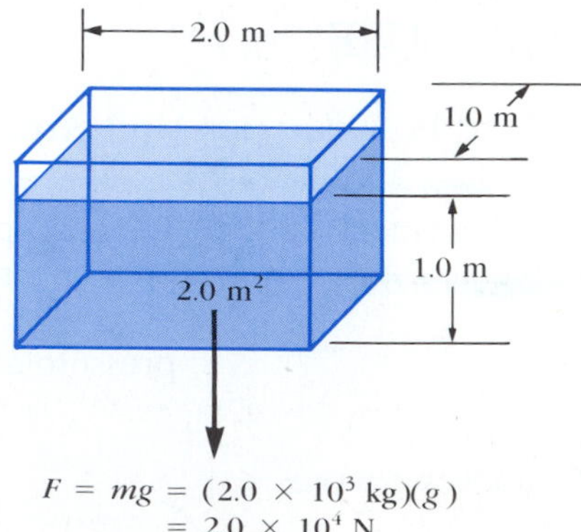

$$F = mg = (2.0 \times 10^3\ kg)(g) = 2.0 \times 10^4\ N$$

$$P = \frac{F}{A} = \frac{2.0 \times 10^4\ N}{2.0\ m^2} = 1.0 \times 10^4\ N/m^2$$

FIGURE 12-1
Pressure in a liquid depends only on depth.

EXPERIMENT

Punch three small holes in the side of an empty milk carton: one near the bottom, one in the middle, and one near the top. Cover the holes (with your fingers or tape) and fill the carton to the brim. Put the carton in the sink and uncover the holes. What do you observe? Explain.

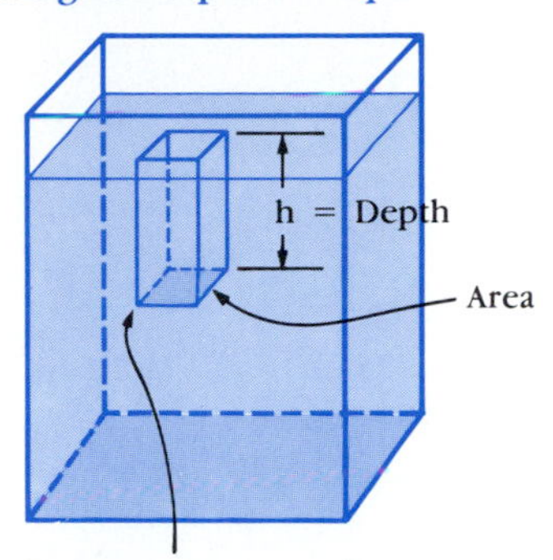

FIGURE 12-2
Calculating the pressure at a given depth in a liquid.

Divers are well aware that pressure increases as one goes deeper in water. Because water pressure at any depth is due to the weight of water above that point, the pressure is directly proportional to the depth. In fact, it is easy to show[†] that the pressure of a liquid at any depth is equal to the product of the density of the liquid times the acceleration of gravity, g, times the depth in the liquid:

$$\text{Pressure} = \text{density} \times \text{acceleration of gravity} \times \text{depth in liquid}.$$

In symbols, this can be written $P = \rho g h$ where ρ (the Greek "rho") is used for density and h is the depth in the fluid. This is illustrated in Figure 12-3. Notice that the pressure is exerted not only against the bottom of the vessel but against its sides as well. In fact, the pressure at any depth is the same in all directions; and it always acts at right angles to the surface of any submerged object, including the walls of the vessel.

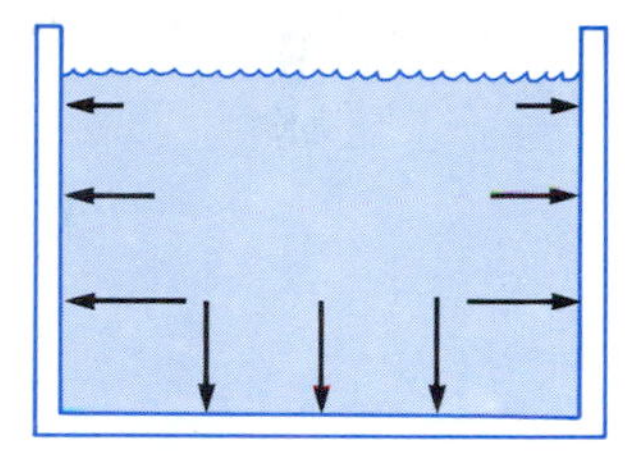

FIGURE 12-3
Pressure increases with depth and is exerted on the sides of container as well as on the bottom.

That the liquid exerts a pressure in all directions is obvious to swimmers and divers who feel the water pressure on all parts of their bodies. Figure 12-4 illustrates why the pressure must be the same in all directions at a given depth. Consider a tiny cube of the liquid that is so small that we can ignore the gravitational force on it. The pressure on one side of it must equal the pressure on the opposite side of it, as shown. If the pressures weren't equal, the tiny piece of liquid would move (i.e., the liquid would flow) until the pressure did become equal. Therefore, when a liquid is still, the pressure must have the same magnitude in all directions at the same depth.

FIGURE 12-4
Pressure is the same in every direction in a fluid at a given depth; if it weren't, the fluid would start in motion.

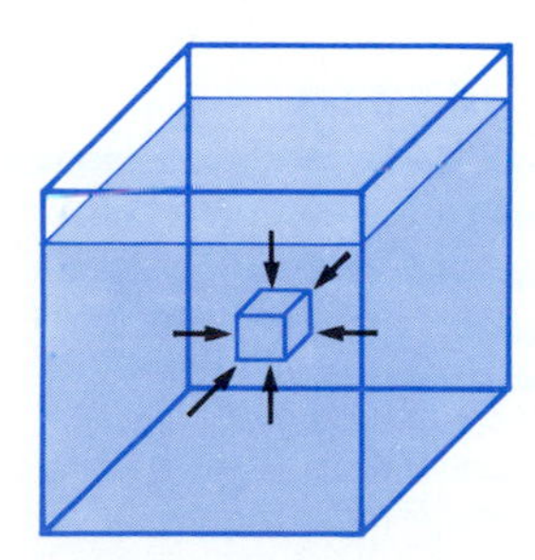

2. BUOYANCY AND ARCHIMEDES' PRINCIPLE

As the following experiment shows, objects seem to weigh less when submerged in water.

[†] The pressure at any depth is just the weight of liquid above that point, divided by the area. Thus (see Figure 12-2):

$$P = \text{pressure} = \frac{F}{A} = \frac{mg}{A} = \frac{\rho V g}{A}$$

since density $=$ mass/volume $= m/V$. From Figure 12-2 we see that the volume of liquid above the area A is $V = Ah$, where h is the depth in the liquid. Therefore $P = \rho A h g/A = \rho g h$.

EXPERIMENT

Use a pool or nearby lake or pond—or fill your bathtub with water. Take a heavy rock, one that is rather hard for you to lift, and submerge it in the water. Be sure it is completely covered with water. Now slowly lift it out of the water. Compare its weight when out of the water with its apparent weight when submerged.

A rock that you would have difficulty lifting off the ground can be readily lifted from the bottom of a lake or stream. But when the rock reaches the surface, it suddenly seems much heavier. Some objects, like wood, actually float in water. These are two examples of the effect called **buoyancy**.

Why does a rock seem to weigh less in water? Figure 12-5(a) shows the forces on a rock; these are the gravitational force, which is its normal weight, and the pressure of the water pushing on the sides, top, and bottom of it. The water pressure on the left and right sides of the rock cancel each other and do nothing to make it seem heavier or lighter. But the water pressure on the top and bottom of the rock do make a difference. Since the water pressure increases with depth, the water pressure on the bottom of the rock is greater than the water pressure on the top. Thus, the water exerts a net upward force on the rock. This upward force is the net effect of the water pressure on all parts of the rock and is called the **buoyant force**; it is shown in Figure 12-5(b). The total net force on the rock is the gravitational force downward minus the buoyant force upward. Thus the net downward force on the submerged rock is less than if it were not submerged, which is why it takes less force to lift it.

How large is this buoyant force? The answer to this question was discovered by the Greek philosopher and mathematician Archimedes (287–212 B.C.), and is known as **Archimedes' principle**:

> **the buoyant force on a body immersed in a fluid is equal to the weight of the fluid displaced by that object.**

To see what Archimedes' principle means, we first examine what is meant by the phrase "the weight of the fluid displaced by that object." If

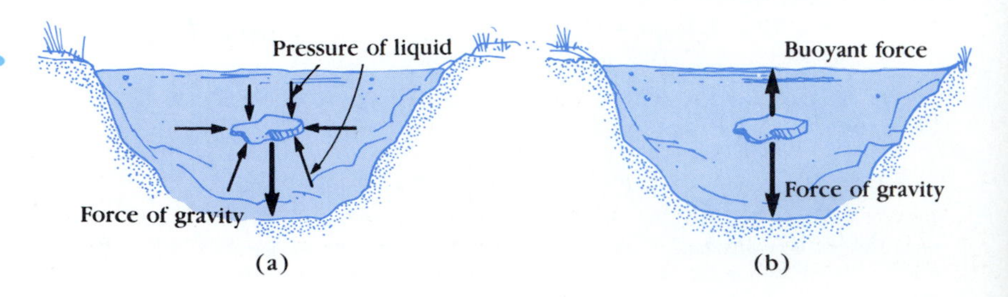

FIGURE 12-5
(a) Water pressure increases with depth, so the water pressure pushing up on the bottom of a rock is greater than the water pressure pushing down on the top. (b) Net effect of this difference of pressure is a net *buoyant force.*

you drop a stone into a glass of water, the water level rises; the water above the original level represents the water displaced by the stone. The volume of water displaced is equal to the volume of the stone†, and the "weight of the fluid displaced" is the weight of this amount of water.

Let's take a specific example and use Archimedes' principle to find the buoyant force B on the rock of Figure 12-5. If the rock weighs, say, 25 N and has a volume of $0.001\ m^3$ ($1000\ cm^3$), then the submerged rock displaces $0.001\ m^3$ of water. The mass of $0.001\ m^3$ of water is 1 kg (see Table 11-3 of Chapter 11), so the "weight of the fluid displaced" by the rock will be about 10 N ($mg = 1\ kg \times 9.8\ m/s^2 = 9.8\ N \approx 10\ N$). According to Archimedes' principle, then, the buoyant force on the rock is 10 N. The net force on the rock is thus $25\ N - 10\ N = 15\ N$ downward (see Figure 12-6). To lift the rock up through the water you would have to exert a force of 15 N; but once it is out of the water you would have to exert a force of 25 N.

What happens to an object whose density is *less* than that of the liquid in which it is placed, for example a log in water? It floats, of course. But let's see why. Suppose the piece of wood has a density that is half that of water. This means that if the volume of the wood is $0.2\ m^3$, its weight would be only half the weight of $0.2\ m^3$ of water. If you were to submerge this piece of wood in water and then let go of it, it would immediately rise to the surface and float. This happens because when the wood is submerged, as in Figure 12-7(a), the buoyant force upward on the submerged log is equal to the weight of the $0.2\ m^3$ of water—that is, a force twice as great as the weight of the log. In Figure 12-7(a), if the weight of the log is w, the

FIGURE 12-6
Because of the buoyant force, it is easier to lift a rock when it is under water than when it is out of the water.

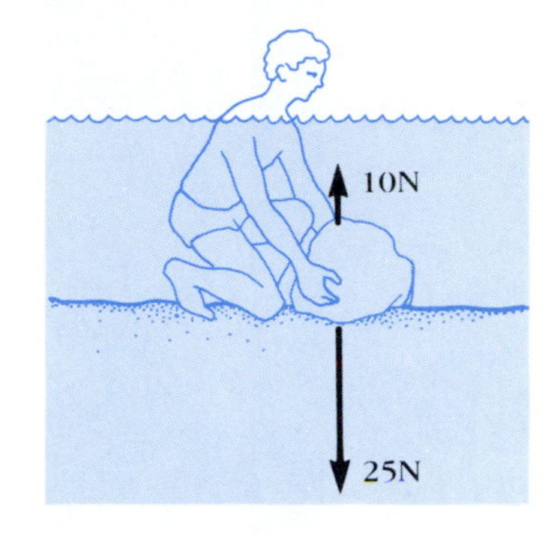

(a) Net force = 15N

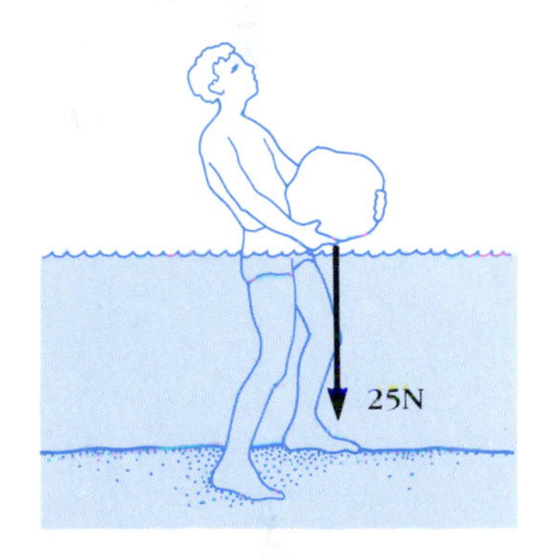

(b) Net force = 25N

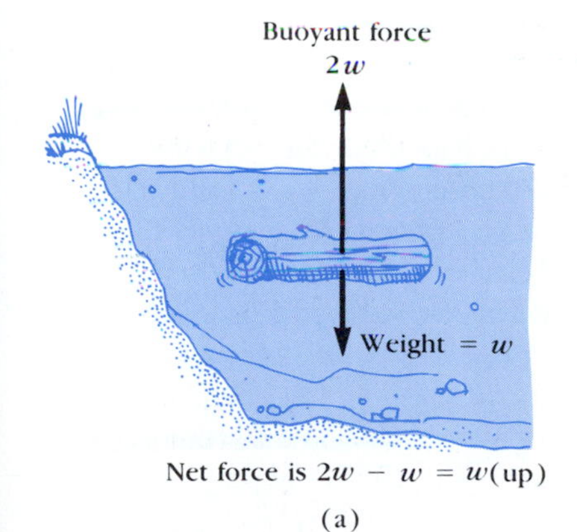

Net force is $2w - w = w$(up)

(a)

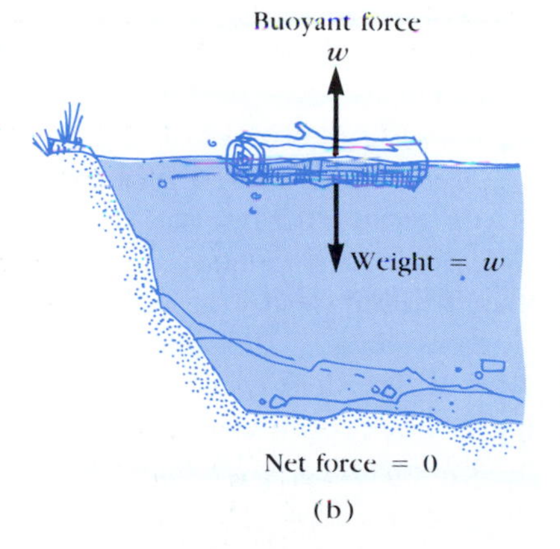

Net force = 0

(b)

FIGURE 12-7
(a) A log rises to the surface because the buoyant force is greater than the force of gravity. (b) The log reaches equilibrium when partially submerged; the weight of water displaced equals the log's total weight.

† Incidentally, this is a good way to measure the volume of an irregular solid object, like a stone. If the glass or beaker has a scale on the side, the change in volume can be read right off the scale and equals the volume of the stone.

FIGURE 12-8
A hydrometer.

buoyant force will be $2w$. There is, therefore, a net upward force on the log and it is forced to move upward to the surface. The log comes to rest when the net force on it is zero. This will happen when the buoyant force upward equals the gravity force downward—that is, when the buoyant force has been reduced to w, half its original value of $2w$. At this point, the log displaces only half as much water, and so is half submerged—Figure 12-7(b).

In general, a body floats when its density is less than that of the fluid in which it is immersed, and it sinks if its density is greater.

As an example, consider a *hydrometer*, which is a simple instrument used to measure specific gravity. A particular hydrometer (Figure 12-8) consists of a glass tube, weighted at the bottom, which is 25.0 cm long, 2.0 cm^2 in cross-sectional area, and has a mass of 45 grams. How far from the end should the 1.000 (= SG of water) mark be placed? The volume of the hydrometer is 25.0 cm × 2.0 cm^2 (= length × cross-sectional area) = 50 cm^3. Therefore, its density is 45 g/50 cm^3 = 0.90 g/cm^3 and its specific gravity is 0.90. The hydrometer will float in equilibrium when it displaces 45 g of water (Archimedes' principle), which corresponds to a volume of 45 cm^3 ($\rho = 1.0$ g/cm^3 for water). Thus 45 cm^3 of the hydrometer will be submerged, which is 0.90 (or 90 percent) of its total volume of 50 cm^3. So 0.90 × 25.0 cm = 22.5 cm of its length will be submerged when it is placed in water. And the 1.000 mark, corresponding to the SG of water, should be placed 22.5 cm from the bottom end.

This example illustrates the fact that for a floating object, the fraction of the object submerged is given by the ratio of the object's density to that of the fluid. The density of ice is 0.92 g/cm^3; hence ice floats on pure water with 92 percent submerged. An iceberg (frozen pure water) will float on seawater ($\rho = 1.03$ g/cm^3) with 0.92/1.03 = 0.89 submerged; that is, 11 percent will be visible above the surface of the sea.

How do steel ships float? After all, steel is denser than water. Steel ships are not solid steel but a steel shell; they have lots of empty space filled with air—so the density of a ship as a whole is less than that of water.

Denser fluids exert a greater buoyant force than less dense fluids. This is true because the buoyant force of an object is equal to the weight of fluid it displaces. A ship, like an iceberg, will therefore float higher in seawater than it will in fresh water. An ice cube that floats in water will sink in pure alcohol (density 790 kg/m^3) because the ice is denser than the pure alcohol. The density of an alcohol-water mixture lies in between, according to the proportion of each that is present. If someone offers you a drink in which the ice cubes have sunk, beware.

EXPERIMENT

Put an egg in a glass of water. The egg will probably rest on the bottom. Now slowly add salt to the water while stirring. Can you get the egg to float? Why does this happen?

3. PASCAL'S PRINCIPLE

The French philosopher and scientist Blaise Pascal (1623–1662) discovered another important principle at work in fluids. It is known as **Pascal's principle**:

> **Pressure applied to an enclosed fluid is transmitted throughout the fluid and acts in all directions.**

In Figure 12–9, a piston is pushing down on a cylinder of water with a force of 150 N. If the cross-sectional area of the cylinder is 0.1 m^2, then the piston exerts a pressure of $150\ N/0.1\ m^2 = 1500\ Pa = 1.5\ kPa$. At a depth of 10 cm (0.1 m), the water pressure is normally $\rho gh = 1000\ kg/m^3 \times 9.8\ m/s^2 \times 0.10\ m \approx 1000\ Pa = 1.0\ kPa$. But because of the external pressure, Pascal's principle tells us that the total pressure 10 cm below the surface is $1.5\ kPa + 1.0\ kPa = 2.5\ kPa$. Pascal's principle makes sense since we expect the pressure at any point in a fluid to be due to the weight of everything above it, fluid plus piston.

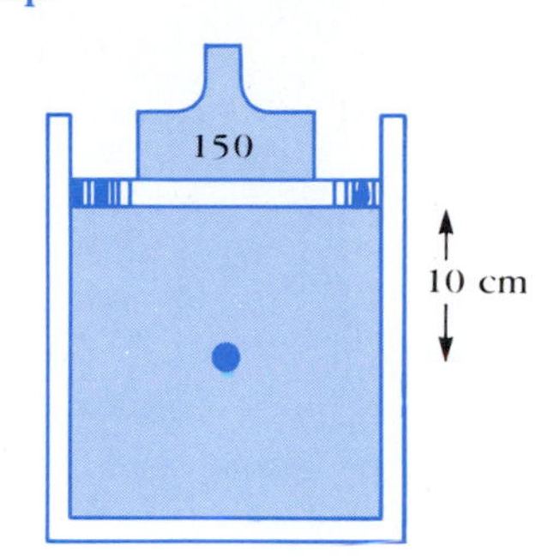

FIGURE 12-9
Pressure at a depth of 10 cm is the sum of the ordinary water pressure plus the pressure due to the external weight applied on top.

Pascal's principle applies to many practical situations, including the hydraulic brakes in an automobile and the hydraulic lift (Figure 12-10). In

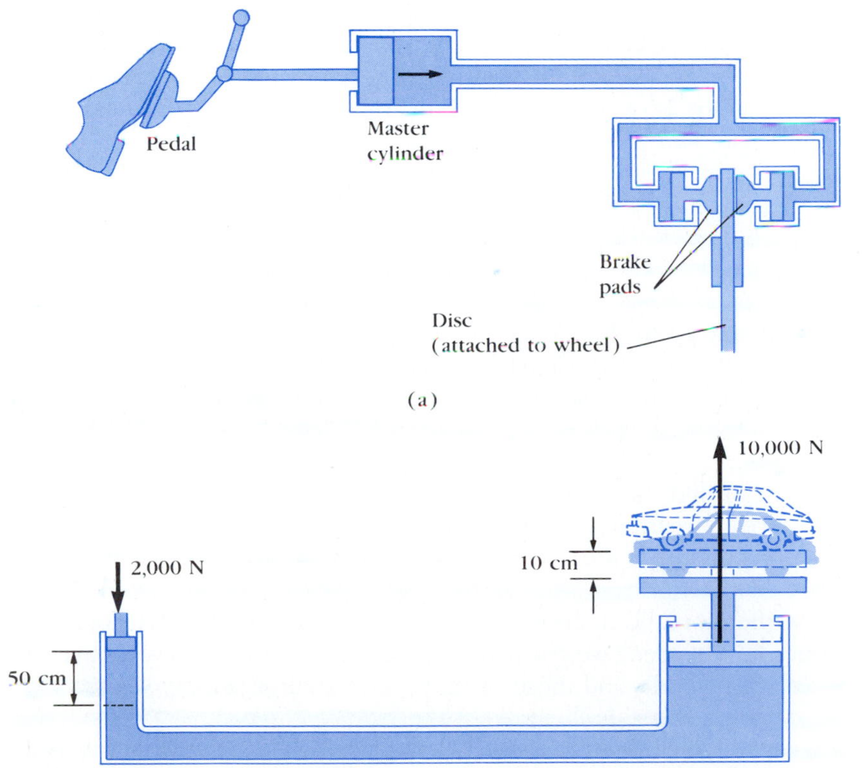

FIGURE 12-10
Applications of Pascal's principle:
(a) hydraulic brakes in a car,
(b) hydraulic lift.

the latter case, a small force can be used to exert a large force by making the area of one piston (the output) larger than that of the other (the input). Although the force can be effectively multiplied in this way, energy is still conserved—as we saw in Chapter 7—because the input force must act through a greater distance than the output.

4. ATMOSPHERIC PRESSURE; THE BAROMETER

Many of the macroscopic concepts and principles that apply to liquids apply equally well to gases. The concept of pressure, for example, is valid and is very important when dealing with gases. Archimedes' and Pascal's principles are valid as well. The rising of helium-filled balloons and the "floating" of dirigibles or blimps are examples of Archimedes' principle applied to gases. Some hydraulic lifts [Figure 12-10(b)] which operate via Pascal's principle utilize air instead of a liquid.

We tend to think of air and other gases as being weightless, but this is not so. A cubic meter of air at sea level has a mass greater than 1 kg. The air in an average living room can have a mass of 50 to 100 kg.

Just as water pressure is caused by the weight of water, the weight of all the air above the earth causes an atmospheric pressure of about 10^5 N/m^2, or 100 kPa (14.7 lb/in^2), at sea level. The average atmosphere pressure at sea level is 1.0×10^5 N/m^2, and this value is known as "normal atmospheric pressure." Sometimes pressures are specific in terms of normal atmospheric pressure; this unit is called the *atmosphere* (atm): 1 atm = 1.0×10^5 N/m^2 = 100 kPa.

At a high elevation above sea level, air pressure is less since there is less air pushing down from above. For example, at an elevation of 5500 m (18,000 ft) the air pressure, like the density, is only about half what it is at sea level. At any given altitude atmospheric pressure varies slightly according to climatic conditions. Climate, in turn, is affected by differences in air pressure at different places on the earth's surface. Wind, for example, is a result of air rushing from a region of high pressure to a region of low pressure.

We can find the total force exerted on an object due to atmospheric pressure by rearranging the definition of pressure as force/area. Thus, force = pressure × area ($F = PA$). Ordinarily the atmosphere exerts a force of 10^5 N on every square meter of surface in contact with it. There must, therefore, be an enormous total force on our bodies. Why don't we notice it? Why aren't we crushed by this huge force? The answer is that the cells of our body, and those of other terrestrial organisms, maintain an equal pressure within them; thus the pressure inside the cells equals the atmospheric pressure outside. When placed in a vacuum, many cells actually explode because of this internal pressure. Biological organisms have evolved in this high-pressure atmosphere; in order to survive they

FIGURE 12-11
Put the lid on the can when the small amount of water inside is boiling vigorously, and immediately remove the can from the burner.

had to be able to balance internally the outside pressure of the atmosphere.

We are not always aware of the pressure exerted by the air, but the following experiments attest to its existence.

EXPERIMENT

Fill a drinking straw with water and place your fingers over both ends. Now hold the straw vertically and remove your finger from the bottom end, but keep your finger tightly over the top end. Why doesn't the water run out of the straw? What happens when you release your finger from the top?

AND

Put about two inches of water in the bottom of an empty gallon can that has a tight-fitting lid (Figure 12-11). With the top of the can removed, heat the water to boiling. When steam emerges from the opening, put the cap back on and immediately remove the can from the burner. The steam has forced most of the air out of the can. As the can cools, the water condenses to the liquid state and leaves a partial vacuum. What happens to the can?

It is important to be aware that, when determining the pressure in a tire or other container of gas, tire gauges and most other pressure gauges register the pressure over and above atmospheric pressure. This is called *gauge pressure*. Thus, to get the absolute pressure P, one must add the atmospheric pressure to the gauge pressure. For example, if a tire gauge registers 220 kPa, the actual pressure within the tire is 220 kPa + 100 kPa = 320 kPa. This is equivalent to about 3.2 atm (2.2 atm gauge pressure).

Suction Is Not a Force

In the above experiment, is the can crushed by the air pressure or by the suction "created" by the vacuum? If we say that suction is pulling the walls of the can inward, we're implying that the vacuum is exerting an inward force. But a vacuum is the *absence* of matter; a vacuum is *nothing*. How can nothing exert a force? Thus the can is not crushed by suction; the air pressure on the outside forces it to cave in. Normally the can is open to the air and the pressure on the inside is the same as on the outside. But when some of the air is removed from the inside, the pressure is lower on the inside than on the outside, and therefore a net force pushes inwardly on the can.

A suction cup works on the same principle. When it is pushed onto a flat surface such as a wall, most of the air is pushed out from under it—Figure 12-12(a). When you try to pull the suction cup away from the wall, the space between the wall and the cup increases slightly; because there is very little air present, the pressure is reduced—Figure 12-12(b). Since the air pressure is greater outside than inside, there is a net force pushing the suction cup back to the wall. To remove it, you must exert a force that can

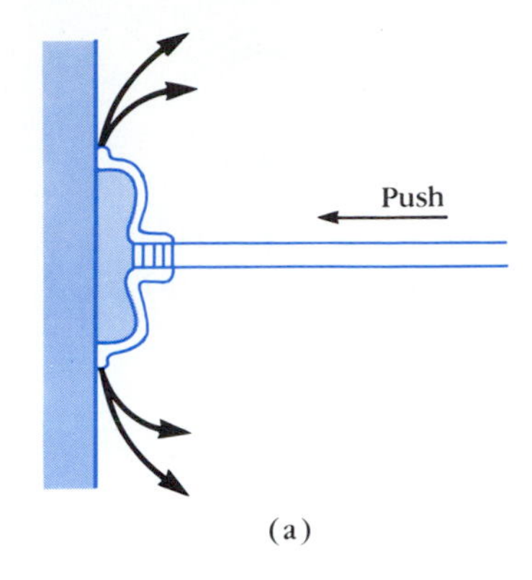

(a)

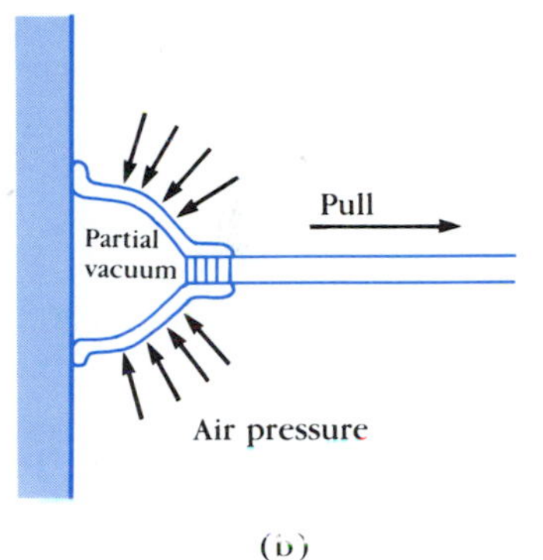

(b)

FIGURE 12-12
(a) Pushing air out from under a suction cup. (b) Removal of suction cup is difficult because the air pressure outside is greater than pressure underneath.

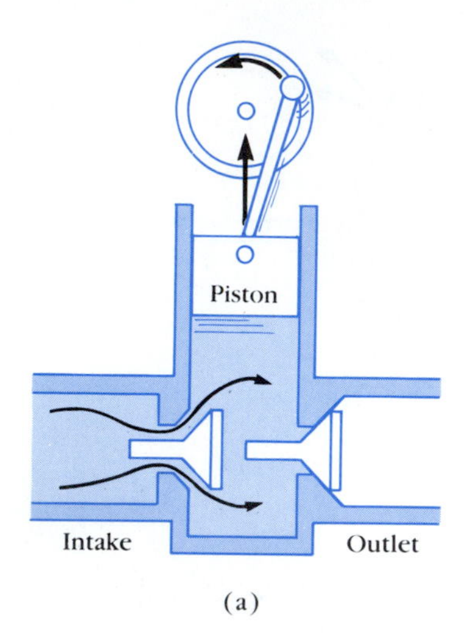

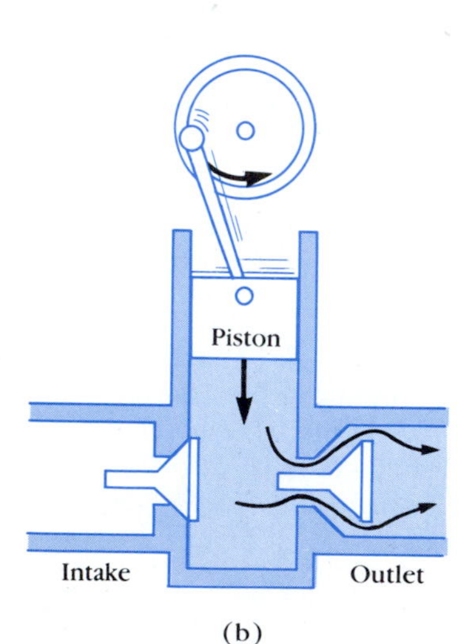

FIGURE 12-13

Diagram of one type of pump. When the piston moves up, the intake valve opens, and air (or fluid to be pumped) rushes in to fill the empty space. When the piston moves down, the fluid forces the intake valve to close, forces the outlet valve to open, and the fluid is pushed out.

overcome the force of the air pressure or bend the suction cup so that leaking develops.

The action of a vacuum pump or of a vacuum cleaner, and the drinking of a soft drink through a straw, are also examples of suction; but again suction does not act as a force. A vacuum pump works by making available empty spaces into which air can flow—remember, a gas tends to fill its container. A vacuum pump repeatedly increases the size of the space into which gas molecules can flow because of their high velocity, and then the pump pushes these molecules through another opening (Figure 12-13).† When you suck on a straw you enlarge your lungs slightly and allow more volume for the air to fill. This of course means that the air in your lungs and mouth is less dense and exerts less pressure than the air outside. The ordinary atmospheric pressure on the surface of the drink is thus greater than the air pressure at the top of the straw, which is why the drink is forced up the straw. The liquid is not drawn up but rather *pushed* up the straw.

Our lungs work because we expand them with muscles inside our body. When we breathe in, we expand our rib cage and lower our diaphragm. This action is illustrated by the following experiment.

EXPERIMENT

Find a bottle whose bottom has been cut off or use the glass "chimney" from a kerosene lamp. Cover the broad opening with rubber sheeting (from a balloon, for example) and secure it with a rubber band. Attach a balloon to the end of a glass tube (a drinking straw may work) and insert it through a cork into the smaller opening (or use modeling clay), as shown in Figure 12-14. Now pull down on the rubber sheet at the bottom (which represents your diaphragm). This is easier to do if, before stretching it over the opening, you wrap a rubber band around a small section so there is something to pull on. What does this do to the pressure in the bottle? What happens to the balloon (which represents one of your lungs)? Why?

Barometers Measure Air Pressure

Instruments for measuring air pressure are called barometers. The oldest form of barometer is the mercury barometer invented by Evangelista Torricelli (1608–1647); see Figure 12-15(a). A tube is filled with mercury and is then inverted into a bowl of mercury; some of the mercury runs out into the bowl. At standard atmospheric pressure, 100 kPa (14.7 lb/in^2), a column of mercury 760 mm (or 30 inches) high remains in the tube. Since there is no gas pressure on the top of the tube (there is a vacuum), the pressure at the base of the tube of mercury due to its own weight must be equal to atmospheric pressure. It is atmospheric pressure that keeps the mercury up in the tube. A column of air that reaches all the way to the top

† The "pump" in a vacuum cleaner is usually nothing more than a high-speed fan. Can you explain, using the principles described in this chapter, how a vacuum cleaner works?

of the atmosphere, and that has the same diameter as the tube of mercury, weighs the same as the 760-mm column of mercury. When the air pressure rises, the mercury will rise in the tube; when the air pressure drops, the mercury drops.

Because the mercury barometer was devised many years ago, pressures are still often given in millimeters of mercury (mm Hg)† rather than in kilopascals. Thus "760 mm of mercury" corresponds to 100 kPa, and so on.

Mercury is a very dense liquid; it is 13.6 times denser than water. If a barometer using water were constructed, standard atmospheric pressure would maintain a column of water 10 meters (34 feet) high; that is, 13.6 times higher than a mercury column. The column of water would be 10 meters high only if there were a perfect vacuum at the top of the tube, as there is for the column of mercury in Figure 12-15(a). If you were to try to "suck" water through a straw longer than 10 meters, no matter how strong your lungs or how perfect a vacuum pump you used, the water would not rise higher than 10 meters. It was a source of wonder and frustration a few centuries ago that no matter how good a vacuum pump was, it could not lift water more than about 10 m. It was, for example, a great practical difficulty to pump water out of deep mine shafts, which required multiple stages for depths greater than 10 m. This problem, which concerned Galileo, was first understood by Torricelli. The point is that a pump does not suck water up a tube under vacuum—it is air pressure that pushes it up, just as in the mercury barometer.

A modern barometer that is handier than the mercury barometer is the aneroid barometer shown in Figure 12-15(b). A partially evacuated metal

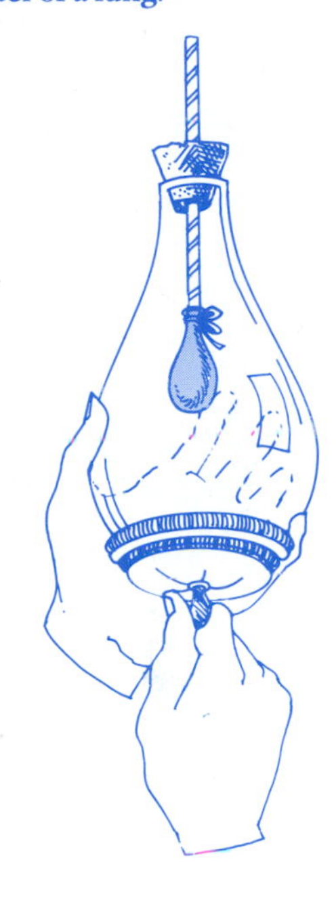

FIGURE 12-14
Model of a lung.

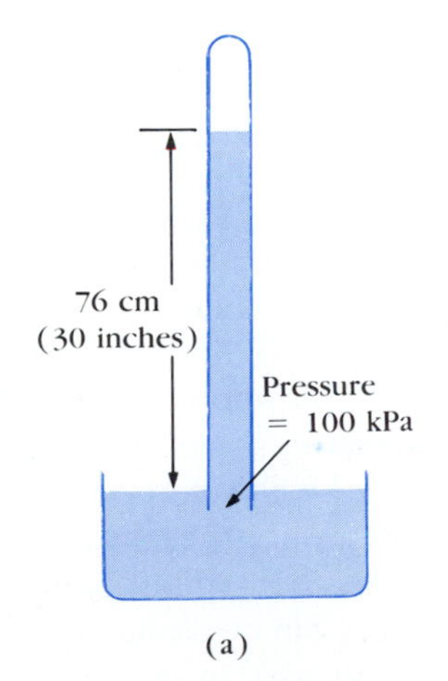

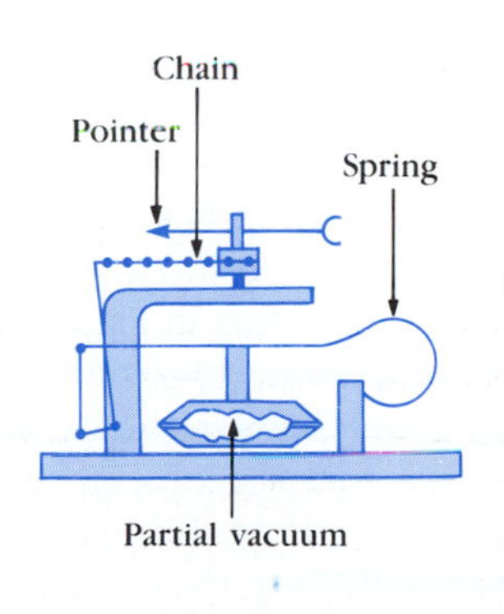

FIGURE 12-15
(a) A mercury barometer; (b) diagram and photo of an aneroid barometer.

† The unit "mm Hg" is also called the torricelli, or torr, in honor of Torricelli; 760 mm Hg = 760 torr = 1.0 atm = 1.0×10^5 N/m^2 = 100 kPa.

box whose volume is very sensitive to changes in external pressure is linked to a dial. The dials of many aneroid barometers are calibrated in inches or centimeters of mercury.

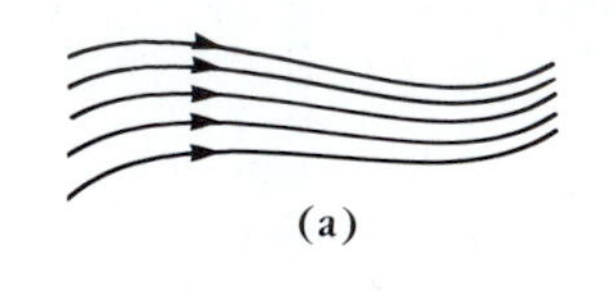

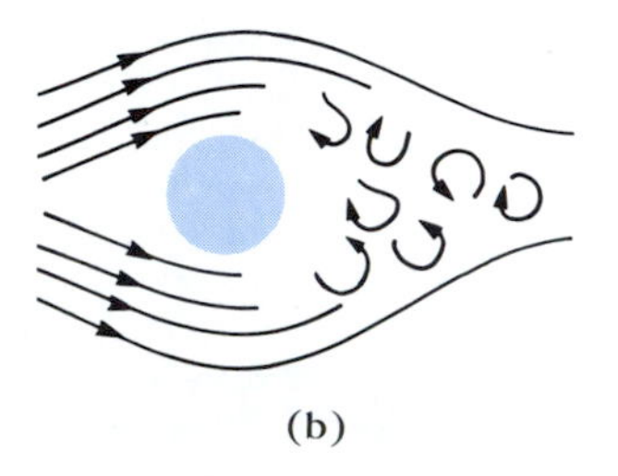

FIGURE 12-16
(a) Streamline flow; (b) turbulent flow.

5. FLUID FLOW

So far we have dealt mainly with fluids at rest. Now we examine fluids that are in motion—that is, fluids that are flowing. We usually distinguish two types of flow: streamline and turbulent. If the flow is smooth so that neighboring layers of fluid slide by each other smoothly, the flow is **streamline**, Figure 12-16(a); each particle of the fluid follows a smooth path, called a "streamline." If a fluid flows very fast, or if there are obstacles in its path, the flow becomes **turbulent**; the streamlines are distorted into small, irregular, whirling *eddies*, Figure 12-16(b). We will assume, flow is streamline unless stated otherwise.

In general, fluids flow whenever there is a difference in pressure from one area to another. We saw examples of this phenomenon in the last section (water being pushed up a straw, air rushing into your lungs). Now you might think that once water was flowing in a horizontal tube or pipe it would continue to flow so long as no outside force was applied, as Newton's first law suggests. However, there is always some friction present between the fluid and the walls of the tube and also between layers of the fluid. This internal friction is called **viscosity**. Different materials have different amounts of viscosity. Syrup is more viscous than water, for example. And gases are in general much less viscous than liquids. Because of viscosity, there must be a difference in pressure between the two ends of a horizontal tube if a fluid is to flow at a constant speed. Thus pumps are needed to force water or oil through level pipes as well as when the pipes go uphill.

Pumps

The purpose of a *force pump* or a *circulating pump* is to exert pressure on a fluid to make it flow. We can use pumps to lift a fluid (such as water) from a well or to force a viscous fluid (such as oil) through a pipe. The diagram of a pump in Figure 12-13 can in principle apply to a force pump (one that increases pressure to force a fluid to flow) as well as to a vacuum pump (one that creates a vacuum, as discussed with reference to Figure 12-13); if used as a force pump, fluid comes in through the intake valve and then is forced out through the outlet valve. Another type of pump is the *centrifugal pump* shown in Figure 12-17. It, too, can be designed as either a vacuum pump or a force pump; it is commonly used in vacuum cleaners and as a water pump in automobiles. In the latter case, it is called a *circulation pump*, since it keeps water circulating in a closed path through the engine to cool it.

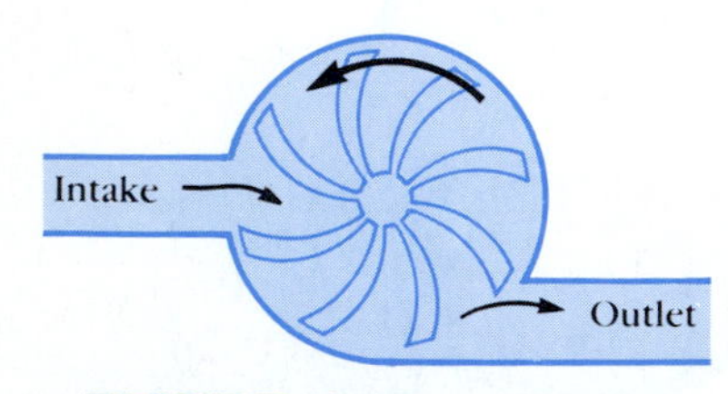

FIGURE 12-17
A centrifugal pump. The rotating blades force fluid through the outlet pipe. This type of pump is used in vacuum cleaners and to circulate the water in automobile engines.

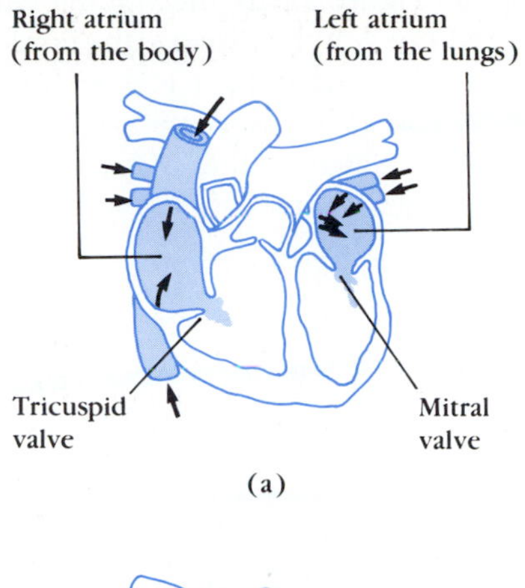

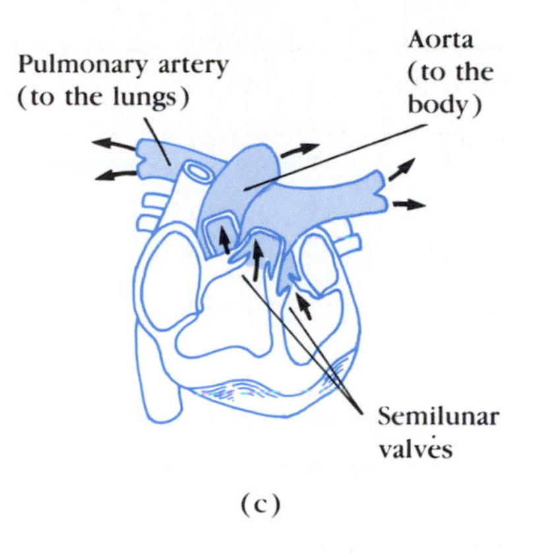

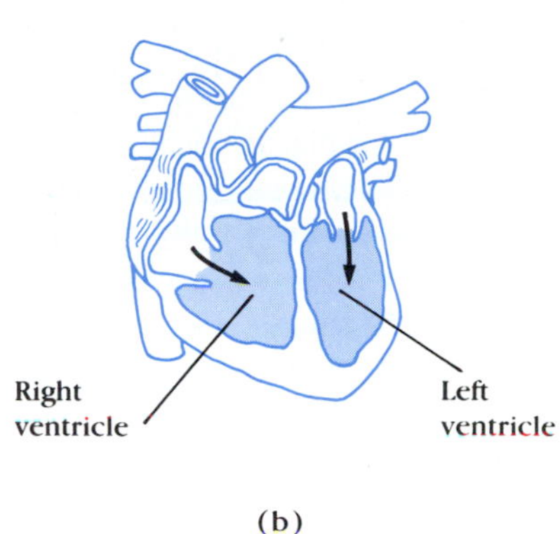

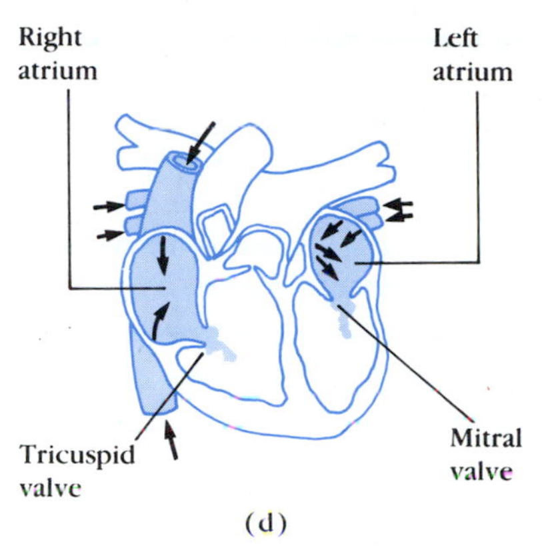

FIGURE 12-18
(a) In the diastole phase, the heart relaxes between beats. Blood moves into the heart; both atria are filled rapidly. (b) When the atria contract, the systole or pumping phase begins. The contraction pushes the blood through the mitral and tricuspid valves into the ventricles. (c) The contraction of the ventricles forces the blood through the semilunar valves into the pulmonary artery (which leads to the lungs) and to the aorta (the body's largest artery), which leads to the arteries serving all the body. (d) When the heart relaxes, the semilunar valves close; blood fills the atria, beginning the cycle again.

The human heart (and the hearts of other animals) is basically a circulation pump. The pumping action is caused by heart muscles that move the walls of the heart so the several cavities of the heart change in volume. When a cavity contracts, the increased pressure forces the blood out. When a cavity expands, blood flows into it because of the reduced pressure. Figure 12-18 shows the operation of the heart in detail. The heart pumps blood around two different paths as shown in Figure 12-19. First the blood is pumped by the contracting right ventricle to the lungs, where it picks up oxygen. The oxygen-laden blood returns to the heart and is then pumped to the rest of the body through the arteries to the capillaries, where the oxygen is transferred to tissues (and where the blood picks up waste products). Finally, the blood returns to the heart through the veins to begin another pumping cycle.

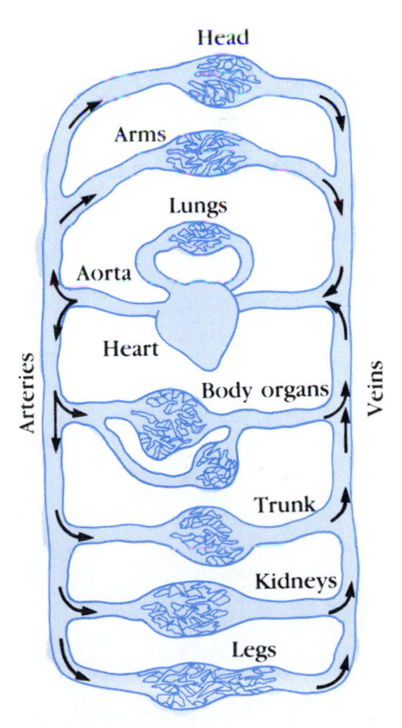

FIGURE 12-19
Human circulatory system.

Continuous Flow

The **flow rate** of a fluid is defined as how much mass (or volume) of the fluid passes a given point per second. For a fluid flowing smoothly in a continuous tube whose diameter varies, as shown in Figure 12-20, the flow

FIGURE 12-20
Fluid flow through a pipe of varying diameter.

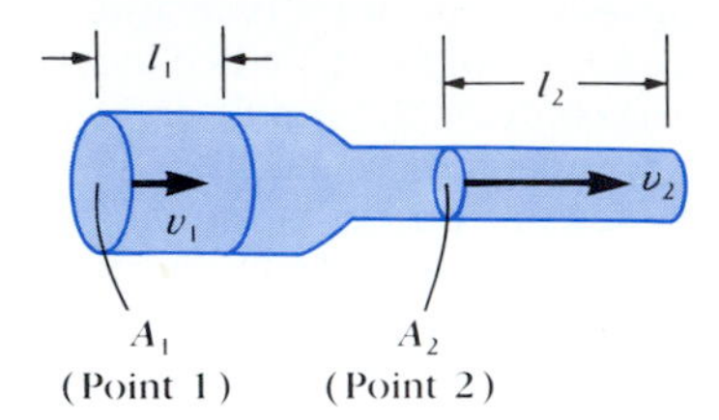

rate must be the same all along the tube. In Figure 12-20, the volume of fluid passing point 1 in a time t is just $A_1\ell_1$ where ℓ_1 is the distance the fluid moves during this time. The velocity of fluid passing point 1 is $v_1 = \ell_1/t$, so $\ell_1 = v_1 t$. Then the flow rate (volume/time) is just $A_1\ell_1/t = A_1 v_1 t/t = A_1 v_1$. Similarly, at point 2 the flow rate is $A_2 v_2$. Since no fluid flows in or out the sides, these two flow rates must be equal. Thus,

$$A_1 v_1 = A_2 v_2.$$

The speed of flow is inversely proportional to the area; the smaller the area, the faster the flow. To get through a narrower portion of the tube, the fluid must speed up. This is what happens along the course of a river. Where a river is wide, the water meanders slowly; but through a narrow gorge it rushes at high speed.

There is an interesting parallel here to the flow of blood through our bodies. As shown in Figure 12-19, blood flows from the heart through the aorta and into the arteries, which branch into tiny capillaries that bathe our tissues with blood and oxygen. Although the capillaries are tiny, there are billions of them, and their total cross-sectional area is far greater than that of the aorta. Thus, although the speed of blood in the aorta is about 30 cm/s, in the capillaries it is only about 0.05 cm/s. At this slow rate, there is plenty of time for oxygen (and other materials) to be exchanged between the blood and the tissues.

6. BERNOULLI'S PRINCIPLE

Have you ever wondered why smoke goes up a chimney or why a sailboat sails against the wind? These are just two of many interesting, and sometimes surprising, examples of a principle worked out by Daniel Bernoulli (1700–1782). Bernoulli's principle, in a simplified form, states that *where the velocity of a fluid is high, the pressure in the fluid is low; and where the velocity is low, the pressure is high*. This may at first seem surprising to you, but the following experiments will show you that it is true.

EXPERIMENT

Hold a sheet of paper horizontally with one edge just below your mouth, as shown in Figure 12-21(a). If you were to blow strongly across the top of the paper, would you be surprised if the paper moved *up*? Try it and see.

AND

Hold two pieces of paper vertically as shown in Figure 12-21(b). If you blow strongly between them, do they move apart or come closer together? Why?

In each case you confirmed Bernoulli's principle that the pressure is low where the velocity is high; the higher pressure of the still air forced

FIGURE 12-21
Experiments showing Bernoulli's principle.

the paper into the region where the air was moving swiftly and thus had lower pressure. You might have expected that where the speed is highest, the pressure would be highest. But remember that you are not blowing *at* the paper but rather *across* it.

To see why Bernoulli's principle is consistent with Newton's laws, consider the flow of fluid in a tube, as shown in Figure 12-22. Where the tube narrows, the speed is higher, as we saw earlier. Since the fluid must accelerate to this higher speed, we conclude that the pressure in the wide part of the tube, P_1, must be greater than the pressure in the narrow part of the tube, P_2.

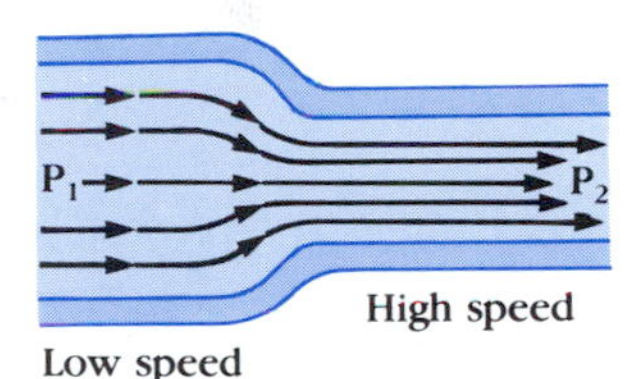

FIGURE 12-22
Flow of fluid in a tube. Where the speed is higher, the pressure must be lower.

Bernoulli's Principle can be used to explain many everyday phenomena, some of which are illustrated in Figure 12-23. The pressure in the air blown at high speed across the top of the vertical tube of a perfume atomizer, Figure 12-23(a) is less than the normal air pressure acting on the surface of the liquid in the bowl; thus perfume is pushed up the tube because of the reduced pressure at the top. A Ping-Pong ball can be made to float above a blowing jet of air (some vacuum cleaners can blow air), Figure 12-23(b); if the ball begins to leave the jet of air, the higher pressure outside the jet pushes the ball back in.

Airplane wings, and other "airfoils," such as that shown in Figure 12-23(c), are designed to deflect the air so that the streamlines are closer together above the wing than below it. We saw in Figure 12-22 that where the streamlines are forced closer together, the speed is faster. So it is with the wing: the air flows faster over the top of the wing than underneath. Thus the pressure is greater below the wing, and this greater pressure helps push the wing upward. (This, however, produces only part of the *lift*; the wing is also tilted, which means the onrushing air is deflected downward off the bottom of the wing and thus exerts an upward force on

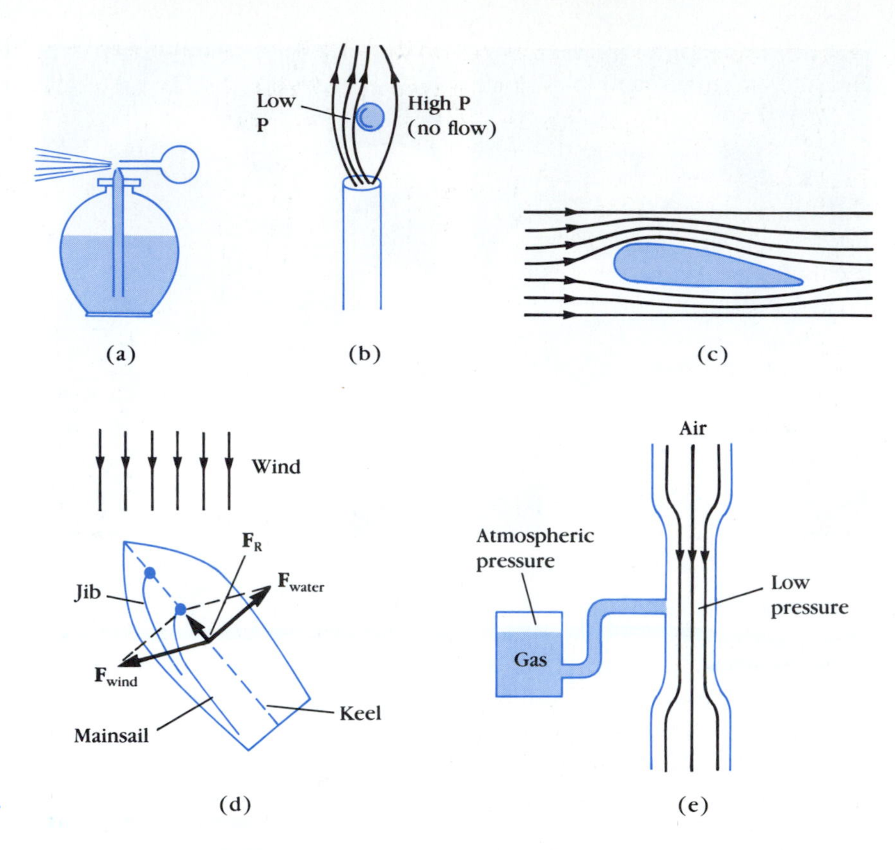

FIGURE 12-23
Examples of Bernoulli's principle.

the wing as it rebounds. Turbulence also plays a role; for example, if the wing is tilted upward too much, turbulence occurs behind the wing and causes a loss of lift.)

A sailboat can move against the wind, Figure 12-23(d), and the Bernoulli effect aids in this considerably if the sails are arranged so the air velocity increases in the narrow constriction between the two sails. (The normal pressure behind the mainsail is larger than the reduced pressure in front of it and this pushes the boat forward.) When going against the wind, the mainsail is set at an angle approximately midway between the wind direction and the boat's axis (keel line) as shown. The force of the wind on the sail (momentum change of wind bouncing off the sail), plus the Bernoulli effect, acts nearly perpendicular to the sail ($\mathbf{F}_{\text{wind}}$). This would tend to make the boat move sideways but the keel beneath prevents this—for the water exerts a force ($\mathbf{F}_{\text{water}}$) on the keel nearly perpendicular to it. The resultant of these two forces ($\mathbf{F}_{\mathbf{R}}$) is almost directly forward as shown.

A *Venturi tube* is essentially a pipe with a narrow constriction (the throat). One example of a Venturi tube is the barrel of a carburetor in a car—Figure 12-23(e). The flowing air speeds up as it passes this constriction so the pressure there is lower. Because of the reduced pressure, gasoline under atmospheric pressure in the carburetor reservoir is forced into the air stream and mixes with the air before entering the cylinders.

Why does smoke go up a chimney? It's partly because hot air rises (i.e., density). But Bernoulli's principle also plays a role. Because wind blows across the top of a chimney, the pressure is less there than inside the house. Hence, air and smoke are pushed up the chimney. Even on an apparently still night there is usually enough ambient air flow at the top of a chimney to allow upward flow of smoke.

If gophers, prairie dogs, rabbits, and other animals that live underground are to avoid suffocation, the air must circulate in their burrows. The burrows are always made to have at least two entrances. The speed of air flow across different holes will usually be slightly different. This results in a slight pressure difference which forces a flow of air through the burrow à la Bernoulli. The flow of air is enhanced if one hole is higher than the other (and this is often done by animals), since wind speed tends to increase with height.

SUMMARY

Pressure is defined as the force per unit area. In a liquid the pressure is exerted in all directions and is proportional to the density and the depth in the liquid. *Archimedes' principle* states that the buoyant force on a body immersed in a fluid is equal to the weight of the fluid displaced by the body. Whether a body sinks or floats in a liquid depends on whether the density of the body is greater or less than the density of the liquid. *Pascal's principle* says that the pressure applied to an enclosed fluid is transmitted throughout the fluid. The pressure due to the earth's atmosphere is measured using a *barometer*. Normal atmospheric pressure is 1.0×10^5 N/m^2. *Gauge pressure* is equal to the true (absolute) pressure minus normal atmospheric pressure.

Fluid flow can be classified as either *streamline* (smooth) or *turbulent*. *Viscosity* refers to the resistance of fluids to flow and results from internal friction in the fluid. Pumps, including the heart, cause fluids to flow by creating differences in pressure. In a continuous tube or channel, fluids flow fastest in those places where the cross-sectional area is smallest. *Bernoulli's principle* states that where a fluid flows fastest, the pressure is lowest.

QUESTIONS

1. Consider what happens when you push both a pin and a stick against your skin with the same force.
2. Decide what determines whether your skin suffers a cut—the net force applied to it or the pressure.

FIGURE 12-24
Water stands at the same level in each of these interconnected containers.

3. Where is the pressure greater, at the bottom of a 1-cm diameter glass tube 2 m high filled with water or at the bottom of a lake 1 m deep?
4. Legend has it that a Dutch boy held back the whole North Sea by plugging a hole in a dike with his finger. Is this possible and reasonable?
5. It is often said that "water seeks its own level" (see Figure 12-24). Explain.
6. It is sometimes said that heavy objects sink. Is this an accurate statement? Explain.
7. Why is it easier to float in salt water than in fresh water?
8. How do you suppose a submarine is able to sink and later to float on the surface of the sea?
9. Does the buoyant force on an object depend on its weight? Does it depend on its volume?
10. An overloaded ship barely stays afloat in the Atlantic Ocean and sinks when it moves up the Hudson River. Why?
11. It is harder to pull the plug out of the drain of a full bathtub than an empty one. Why? Is this a contradiction of Archimedes' principle?
12. When you are wading on a rocky beach, why do the rocks hurt your feet less when you are in deeper water?
13. Why does a helium-filled balloon rise only to a particular height in the atmosphere and go no higher?
14. The hydrometer shown in Figure 12-8 floats in a solution of grape juice, whose specific gravity is greater than that of water. Is the fluid level above or below the 1.000 mark? As the grape juice ferments, producing alcohol (see Table 11-3 in Chapter 11), does the hydrometer float higher or lower in the liquid?
15. An ice cube floats in a glass of water filled to the brim. As the ice melts, does the water overflow?
16. When you drink through a straw, is the liquid being pushed up or sucked up? Explain.
17. What keeps the mercury from running out of a barometer, Figure 12-15(a)?
18. If a single piece of newspaper is spread across a meter stick lying on a flat table, it will take quite an effort to quickly lift the stick. (Try it!) Explain.
19. Would a vacuum cleaner pick up dust from a carpet on the moon where there is no atmosphere? Explain.
20. Why can't you drink on the moon by sucking through a straw?
21. Compare the flow of air in a heating duct to the flow of air in a room.

22. Children are told to avoid standing too close to a rapidly moving train because they might get sucked under it. Is this possible? Explain.
23. If two ships travelling in parallel paths approach too closely, they may find themselves crashing into one another. Explain.
24. Why does the canvas top of a convertible bulge out when a car is traveling at high speeds?
25. Sometimes when you are taking a shower and you turn the water on hard, the shower curtain seems to be pulled in toward you. Explain.
26. Red corpuscles tend to flow in the *center* of blood vessels. Why?
27. A tornado or a hurricane does not rip off the roofs of houses by blowing against them. Explain how they are blown off; use Bernoulli's principle.

EXERCISES

1. What is the pressure exerted on the ground by a 300-newton orange crate, 0.4 m wide by 0.7 m long?
2. What is the pressure due to water 3.0 m beneath the surface of a lake?
3. Tire pressure gauges are calibrated to read the excess pressure in the tire over and above atmospheric pressure. What is the total pressure inside a tire when the gauge reads 190 kPa?
4. What is the water pressure at a faucet if the water comes from a storage tank whose water level is 20 m above the faucet?
5. Estimate the total area of your body and calculate the total force exerted on you by the air.
6. The pressure in each of the four tires of a 1500-kg car is 3.0×10^5 N/m^2. How much area of the tire is in contact with the ground?
7. What is the total force and the pressure due to the water on the bottom of an 8.0 by 15.0 m swimming pool whose uniform depth is 2.0 m? What will be the pressure against the *side* of the pool near the bottom?
8. What is the difference in blood pressure between the top of the head and the bottom of the feet of a 1.60-m-tall person standing erect? The density of whole blood is 1.05×10^3 kg/m^3.
9. (a) Calculate the total force of the atmosphere acting on the top of a table which measures 2.0 m $\times$ 1.2 m. (b) What is the total force acting upward on the underside of the table?
10. What is the approximate difference in air pressure between the top and the bottom of the World Trade Center towers in New York City which are 410 m tall and are located at sea level?
11. The arm of a record player exerts a force of (1.0 g) $\times$ g on a record. If the diameter of the stylus is 0.0013 cm (= 0.5 m = 0.5×10^{-3} in),

FIGURE 12-25

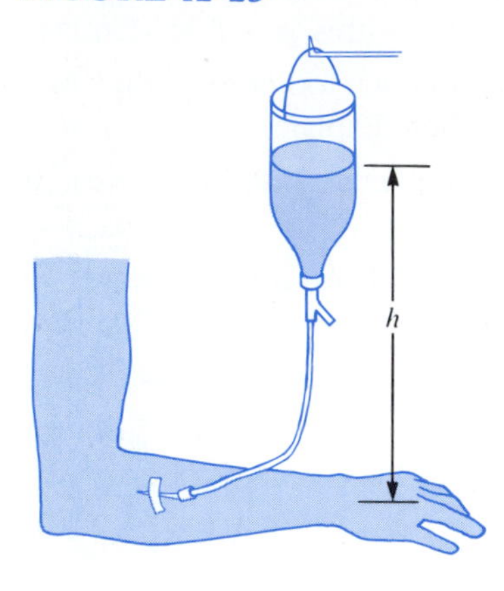

calculate the pressure on the record groove in N/m^2 and in atmospheres.

12. Intravenous infusions are often made under gravity as shown in Figure 12-25. Assuming the fluid has a density of $1.00\ g/cm^3$, at what height h should the bottle be placed so the liquid pressure is 60 mm Hg?
13. Determine the minimum gauge pressure needed in the water pipe leading into a building if water is to come out of a faucet on the twelfth floor 30 m above.
14. Will iron float on mercury? If so, what fraction of the iron will lie above the surface?
15. Draw a diagram showing all the forces (and their relative strengths) acting on an ice cube (a) as it floats in water (b) when it is held submerged beneath the water's surface.
16. What is the density of a piece of wood that is 50 percent submerged when it floats in gasoline?
17. The hydrometer discussed with reference to Figure 12-8 sinks to a depth of 22.3 cm when placed in a fermenting vat. What is the density of the brewing liquid?
18. A 65-kg person has an apparent mass of 45 kg when the legs only are submerged. What is the mass of each leg? Assume the body has a density equal to that of the water.
19. A moon rock weighs 22.5 N in air but has an apparent weight of 16.8 N when submerged in water. (a) What is the buoyant force on the rock? (b) What must its specific gravity be?
20. What volume of helium (see Table 11-3, Chapter 11) is required to lift a 500-kg load?
21. The pistons of a hydraulic pump have diameters of 5.0 cm and 20 cm. What is the mechanical advantage?
22. The maximum gauge pressure in a hydraulic lift is 18 atm. What is the largest size vehicle (kg) it can lift if the diameter of the output line is 20 cm?
23. A 4.0-N force is applied to the plunger of a hypodermic needle. If the diameter of the plunger is 1.0 cm and that of the needle is 0.25 mm, with what force does the fluid leave the needle?
24. Water travels with a speed of 2.5 cm/s in a pipe 3.0 cm diameter. How fast will it travel where the pipe narrows to a 2.0 cm diameter?
25. How fast must the air flow in a heating duct 60 cm in radius if it is to replenish the air in a room 8 m × 10 m × 3 m every 15 minutes?
26. The radius of the aorta, the vessel into which the blood flows from the heart, is about 1.0 cm, and the blood travels through it with a speed of about 30 cm/s. Calculate the average speed of the blood in the

capillaries using the fact that the total cross-sectional area of the billions of capillaries is about 200 cm^2.

27. Using the data of Exercise 26, calculate the average speed of blood flow in the major arteries of the body which have a total cross-sectional area of about 2.0 cm^2.
28. If wind blows at 25 m/s over your house, what is the net force on the roof if its area is 250 m^2?

TEMPERATURE AND KINETIC THEORY OF GASES

CHAPTER

13

The gaseous state of matter is the state that is easiest to understand from the microscopic, or atomic, point of view. In this chapter we will look at some of the properties of gases, and we will see how they can be explained from the atomic point of view using the "kinetic theory." *Kinetic theory* refers to the idea that matter is made up of atoms and molecules in motion (remember, "kinetic" is Greek for "moving"), and is one of the great scientific theories.

We begin this chapter with a discussion of *temperature*—the hotness or coldness of an object—which plays a basic role in our present study.

FIGURE 13-1
See Color Plate V.

1. TEMPERATURE

Temperature has meaning for us mainly through our sense of touch. When we touch something that feels hot—such as a hot stove or the hot air of a summer day—we say it has a high temperature. A cold object on the other hand has a low temperature.

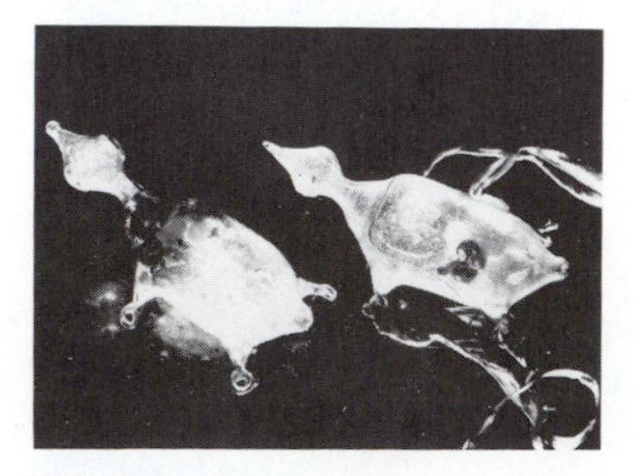

FIGURE 13-2
See Color Plate V.

Properties of Matter Change with Temperature

Nearly all materials expand when they are heated. An iron rod is longer when it is hot than when it is cold. Concrete sidewalks expand and contract with changes in temperature, which is why compressible spacers are placed at regular intervals. We can remove the metal lid from a tightly closed jar by warming it under hot water because the lid expands more than the glass. Glass itself expands, and if one part of a glass container is heated or cooled more rapidly than adjacent parts, the glass may break. The same thing can happen to an automobile engine block; water is added to an overheated automobile engine slowly and with the engine running to avoid "cracking" the block.

Another property of matter that changes with temperature is color. At very high temperatures solids become red. An example is iron; you have probably seen the burner of an electric stove turn red when it is hot. At still higher temperatures, iron and other solids turn orange, and then white. The white light from an incandescent light bulb is emitted by an intensely hot tungsten wire.

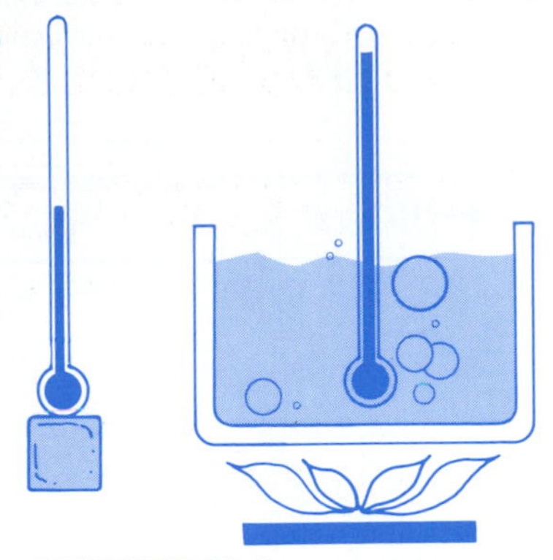

FIGURE 13-3
Liquid-in-glass thermometer.

Temperature also affects the "state" of a material—whether it is solid, liquid, or gas. For example, water at low temperatures is a solid (ice); at higher temperatures it is a liquid, and at still higher temperatures it is a gas (steam). Other materials pass through these same three states as temperature increases. However, the temperature at which the transition from one state to another occurs is different for each material.

We are all familiar with how biological organisms, particularly humans, respond to temperature. When we touch something that has either a very high or a very low temperature, it hurts. This response of our nervous system was a very important development in biological evolution, because biological tissue is readily damaged by temperature extremes. Biological organisms can tolerate a relatively limited range of temperatures. Even bacteria are sensitive to temperatures and grow well only within limited range. In fact, the pasteurization of milk products is a heating process that kills certain harmful bacteria. Viruses are much less sensitive to temperature.

Thermometers

In order to measure temperature quantitatively, we must use some property of matter that changes with temperature. A device that does this is called a **thermometer**. Most common thermometers rely on the expansion of a material with an increase in temperature. The first thermometer was invented by Galileo. (See Figure 13-1, Color Plate V.) An early clinical thermometer (around 1660) is illustrated in Figure 13-2 (Color Plate V).

Common thermometers today consists of a hollow glass tube filled with mercury or alcohol colored with a red dye (Figure 13-3). The liquid expands more than the glass when the temperature is increased, so the liquid level rises in the tube. Although metals also expand with temperature, the change in length of, say, a metal rod is generally too small to measure accurately for ordinary changes in temperature. However, a useful thermometer can be made by bonding together two dissimilar metals whose rate of expansion is different—Figure 13-4(a). When the temperature is increased, the different amounts of expansion cause the bimetallic strip to bend. Often the bimetallic strip is in the form of a coil, one end fixed and the other attached to a pointer. This kind of thermometer, Figure 13-4(b), is used for ordinary air thermometers, for oven thermometers, and in automobiles for the automatic choke. Bimetallic

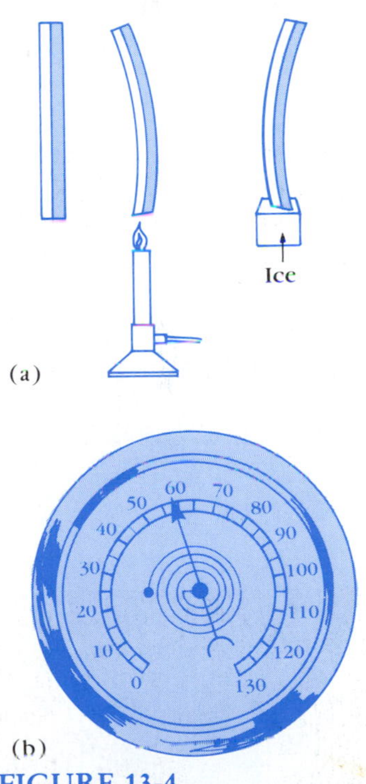

FIGURE 13-4
(a) Bimetallic strip: One metal expands or contracts more than the other when heated or cooled. (b) Common thermometer made from coiled strip.

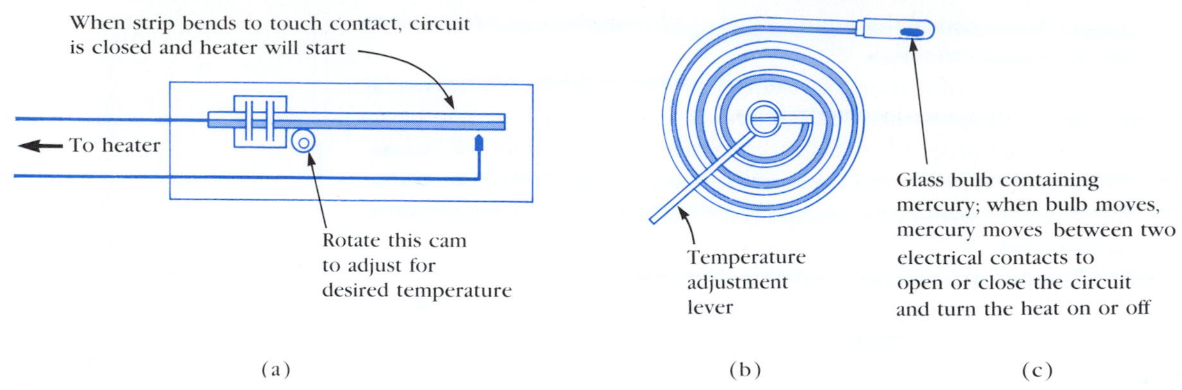
When strip bends to touch contact, circuit is closed and heater will start
To heater
Rotate this cam to adjust for desired temperature
Temperature adjustment lever
Glass bulb containing mercury; when bulb moves, mercury moves between two electrical contacts to open or close the circuit and turn the heat on or off
(a)
(b)
(c)

FIGURE 13-5
Bimetallic strip used in a thermostat. (a) At a certain temperature the strip bends sufficiently to touch the contact that closes the electric circuit and starts the heater. (b) The strip is in the form of a coil that unwinds, or winds tighter, as the temperature fluctuates: the bulb containing mercury acts as an electric switch to turn the heater on and off.

strips are also used in thermostats for home heating systems (see Figure 13-5).

EXPERIMENT-PROJECT

Construct a simple thermometer by pushing a glass tube (or a sturdy drinking straw) through a cork with a hole in it. Put the tube and cork into a small bottle filled with water (Figure 13-6). If you can't find a cork, use plasticene or clay to seal the opening. The water level will be more visible if you put a few drops of food coloring in the water. Note the water level in the tube and then place the thermometer in a pan of boiling water. Note the water level now. Finally, put the thermometer on ice or in a refrigerator and see what happens.

Temperature Scales

In order to measure temperature quantitatively, some sort of numerical scale must be defined. The most common scale today is the **Celsius** scale, sometimes called the **centigrade** scale. In the United States, the **Fahrenheit** scale is also common. Another scale that is important in scientific work is the **absolute**, or **Kelvin**, scale; it will be discussed later in the chapter.

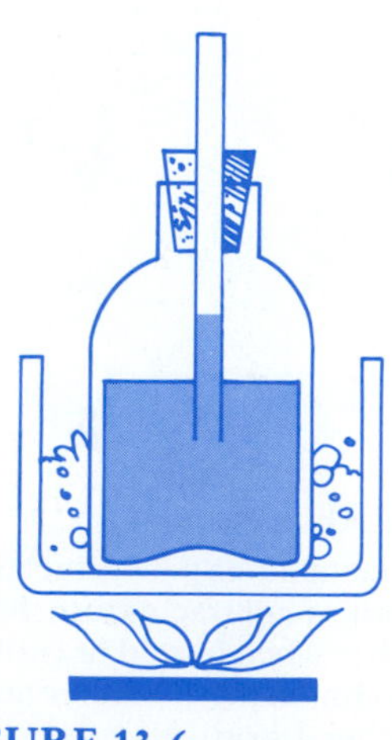

FIGURE 13-6
A homemade thermometer.

In order to define a temperature scale, two readily reproducible temperatures are assigned arbitrary values. For both the Celsius and Fahrenheit scales these two points are the freezing point and the boiling point of water, both taken at atmospheric pressure. On the Celsius scale, the freezing point of water is chosen to be 0°C ("zero degrees Celsius") and the boiling point 100°C. On the Fahrenheit scale, the freezing point is defined as 32°F and the boiling point 212°F.

A practical thermometer can be calibrated by first placing it in a container of water mixed with ice. A mark is made where the level of mercury, or the pointer, comes to rest (Figure 13-7). The thermometer is then placed in a container of boiling water and a second mark is made. The first mark represents 0°C and the second 100°C, assuming the

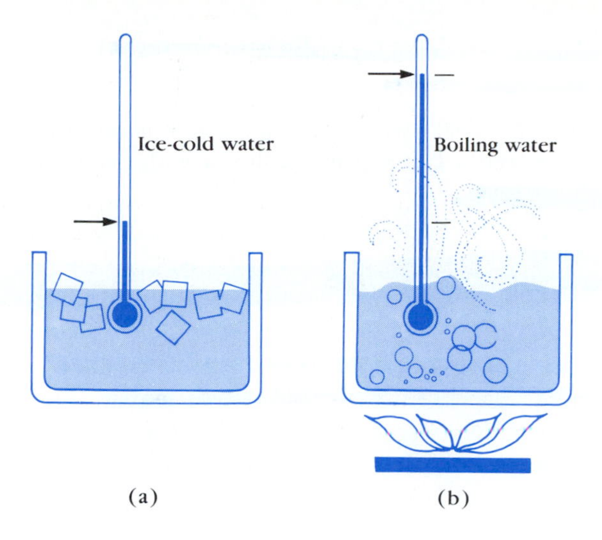

FIGURE 13-7
Calibrating a thermometer.

calibration is done at 1 atm pressure. The distance between the two marks is then divided into 100 equal intervals separated by small marks representing each degree on the Celsius scale between 0 and 100 (hence the name "centigrade scale" meaning "100 steps"). For a Fahrenheit scale, the two points are labeled 32°F and 212°F, and the distance between them is divided into 180 equal intervals. For temperatures below the freezing point of water and above the boiling point of water, the scales can be extended using the same equally spaced intervals.

Every temperature on the Celsius scale corresponds to a particular temperature on the Fahrenheit scale (Figure 13-8). It is easy to convert from one to the other if you remember that 0°C corresponds to 32°F and that a range of 100°C on the Celsius scale corresponds to a range of 180°F on the Fahrenheit scale. Thus, one Fahrenheit degree (1 F°) corresponds to $100/180 = 5/9$ of a Celsius degree; that is, 1 F° = 5/9 C°. As an example, let us determine what the normal human body temperature of 98.6°F is on the Celsius scale. First we note that 98.6°F is $98.6 - 32.0 = 66.6$ F° above the freezing point of water. Since each F° is equal to 5/9 C°, this corresponds to $66.6 \times 5/9 = 37.0$ C° above the freezing point; as the freezing point is 0°C, the temperature is 37.0°C.

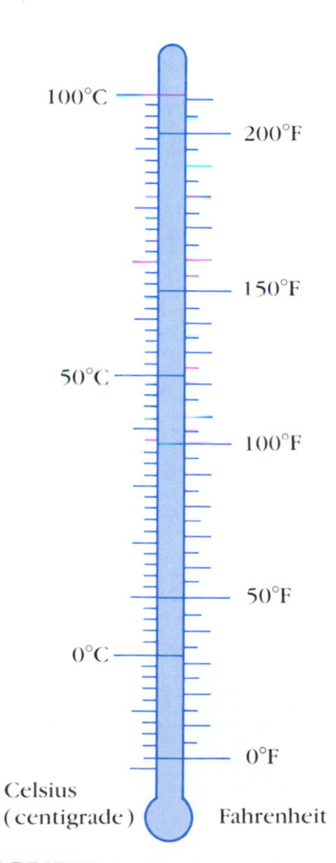

FIGURE 13-8
Celsius and Fahrenheit scales compared.

2. THE GAS LAWS

As we have just seen, temperature affects the properties of liquids and solids. But the effect of temperature on *gases* is even more striking. We will now investigate the behavior of gases in detail, and we will see that early investigations eventually led to a deeper understanding of temperature and heat.

Volume of a Gas Is Inversely Proportional to Pressure—Robert Boyle

Robert Boyle (1627–1691), who conducted some of the early experiments on gases, discovered a law that is consistent with everyday experience. **Boyle's law** states that

> **When the temperature of a gas is not changed, the volume of the gas is inversely proportional to the pressure applied to it.**

For example, if the pressure on a confined gas is doubled (Figure 13-9), the volume of the gas decreases to half of what it was; if the pressure is tripled, the volume decreases to a third of what it was; and so on. By using abbreviations or symbols, Boyle's law can be written

$$V \propto \frac{1}{P}, \qquad \text{[at constant temperature]}$$

where V refers to volume and P to pressure. Boyle's law can also be stated as follows: at constant temperature, if either the volume of a gas or the pressure applied to the gas is varied, the other varies as well, such that the product of the two remains constant. That is,

$$PV = \text{constant}. \qquad \text{[at constant temperature]}$$

Volume Is Directly Proportional to the Absolute Temperature—Jacques Charles

Temperature also affects the volume of a gas, as the following example shows.

EXPERIMENT-PROJECT

Take an "empty" bottle (which of course is filled with air) and attach a balloon or a small plastic bag to its neck (Figure 13-10). Heat the air in the

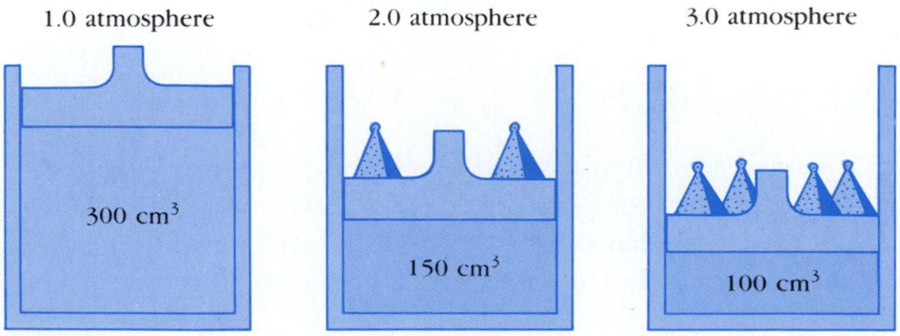

1.0 atmosphere — 300 cm³

2.0 atmosphere — 150 cm³

3.0 atmosphere — 100 cm³

FIGURE 13-9
Volume is inversely proportional to pressure.

bottle by placing the bottle in boiling water or over a flame. What happens to the balloon or the plastic bag?

OR

Blow up a balloon, tie its end firmly, and measure its diameter. Put it in a freezer for 15 or 30 minutes and then measure its diameter again. What do you find?

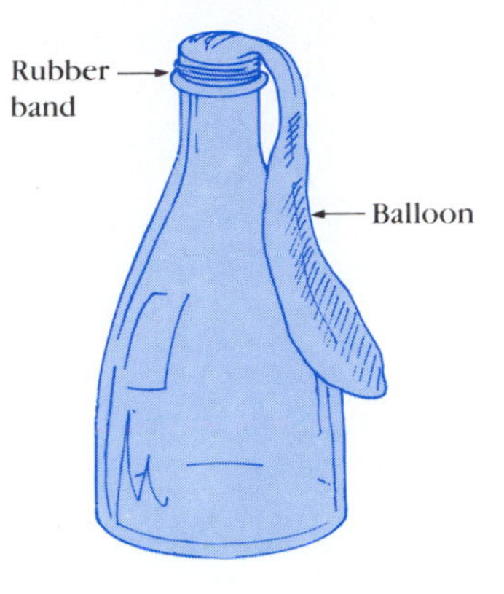

FIGURE 13-10
"Empty" bottle with a balloon over its neck, ready to be heated.

We all know that a gas expands when it is heated. Yet the quantitative relationship between the volume of a gas and its temperature was not discovered until more than a century after Boyle perceived the precise relation between volume and pressure. The French scientist Jacques Charles (1747–1823) found that when the pressure on a gas is kept constant, the volume increases with temperature at a constant rate, Figure 13-11(a). However, since gases liquefy at low temperatures (air at −196°C), the graph of Figure 13-11 cannot be carried to temperatures below the liquefaction point.

Nonetheless, when the straight line of the graph is projected to lower and lower temperatures, as shown by the dotted line of Figure 13-11(a), it

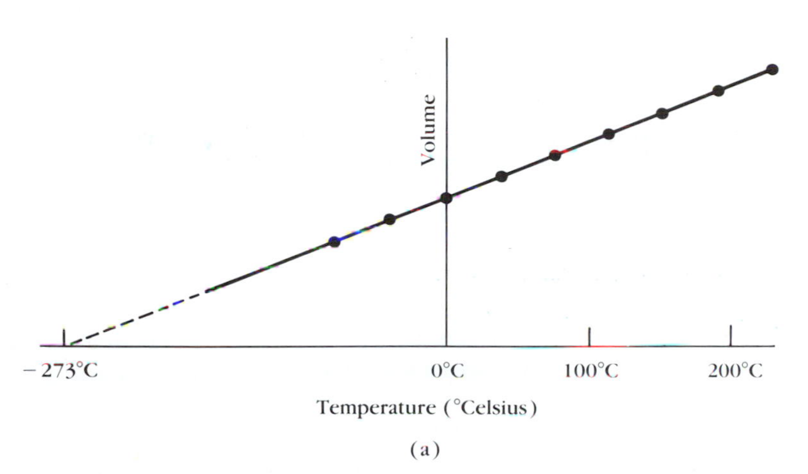

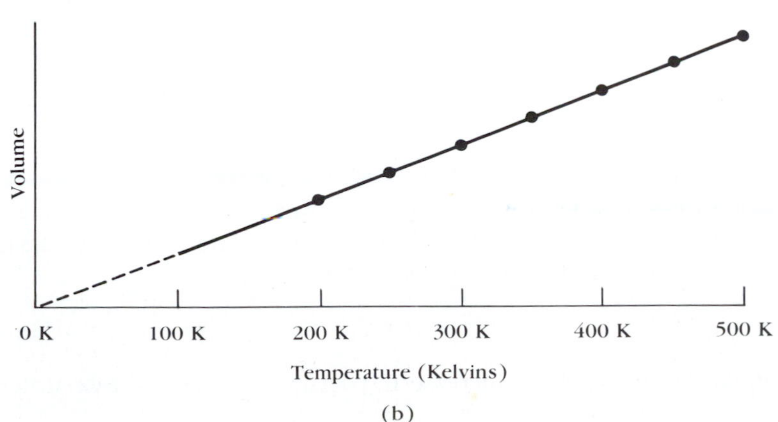

FIGURE 13-11
(a) Volume of an enclosed gas as a function of temperature (at constant pressure). The dark dots represent the measured values of the volume at various temperatures. (b) Volume of an enclosed gas at constant pressure is directly proportional to the absolute temperature.

crosses the axis at a temperature of −273°C. This kind of graph can be drawn for all gases, and the graph always projects to zero volume at −273°C. This seems to imply that if a gas could be cooled to −273°C it would have no volume! It also implies that at temperatures lower than −273°C the volume would be negative, which of course makes no sense. This led to the suggestion that perhaps −273°C is the lowest temperature attainable. A good many other experiments indicate that this is indeed the case. Therefore, −273°C is called the **absolute zero** of temperature. (This is −460°F on the Fahrenheit scale.) Although scientists have been able to reach within 0.00001°C of absolute zero, absolute zero itself apparently cannot be reached precisely.

Absolute zero is the basis of a third temperature scale called the **absolute**, or **Kelvin**, **scale**. The size of a degree on this scale is the same as on the Celsius scale. But absolute zero, −273°C, is chosen as zero, or 0 K, on the absolute scale (note that no degree sign is used with the K, which stands for "kelvins"). The freezing point of water is 273 K and the boiling point of water is 373 K. When a temperature is given in degrees Celsius, we can translate it to the absolute scale by adding 273: K = °C + 273.

If we graph the volume of a gas against the absolute temperature, Figure 13-11(b), we see that the straight line passes through the origin of the graph. In other words,

> **If the pressure on a gas is not allowed to change, the volume of the gas is directly proportional to its absolute temperature.**

This is **Charles's law**. If we let T represent the absolute temperature, Charles's law can be written

$$V \propto T. \qquad \text{[at constant pressure]}$$

In other words, if the absolute temperature of a gas were doubled, its volume would also double so long as the pressure remained constant.

At Constant Volume, Pressure Increases with Temperature

What happens when we keep the volume of a gas constant and change the temperature? This is the situation when a gas is heated in a closed (inelastic) container. We find that

> **The pressure is directly proportional to the absolute temperature.**

If the absolute temperature is doubled, the pressure of the gas inside the container is doubled. This was discovered by Joseph Gay-Lussac (1778–1850), a contemporary of Charles, and is known as Gay-Lussac's law. In symbols, Gay-Lussac's law is written

$$P \propto T. \qquad \text{[at constant volume]}$$

A familiar example of the increase in pressure with an increase in temperature when volume remains constant is what happens when a closed jar is thrown into a fire or placed on a hot stove burner: the air pressure that builds inside the jar causes it to break.

The Ideal Gas Law

Boyle's law, Charles's law, and Gay-Lussac's law are relations among the three important variables of a gas: the pressure, the volume, and the temperature. Collectively, these laws are known as the **Gas Laws**. Each law relates two of these variables while the third variable is kept constant. This technique of keeping one or more variables constant in order to observe the effect of changing one of the other variables is a useful tool for scientists. However, since it is not always practical to hold one of the variables constant, we can combine the three separate laws into a single law that relates all three variables. This single law can be stated in symbols as

$$PV \propto T.$$

When any two of the quantities—pressure, volume, and temperature—are changed, this relation tells us how the third will change. Thus it is more general than Boyle's, Charles's, and Gay-Lussac's laws separately. Note, however, that at constant temperature the above relation reduces to Boyle's law, $PV =$ constant; at constant pressure it reduces to Charles's law, $V \propto T$; and at constant volume it reduces to Gay-Lussac's law, $P \propto T$.

In addition to volume, pressure, and temperature, a gas has a fourth and equally important variable, which up to now we have assumed to be constant. This is the total mass of the gas, or the total number of molecules in the gas. If we keep the temperature and pressure of a gas constant but change the volume of the gas by adding more molecules, we find that the volume increases directly with the number of molecules, N. That is $V \propto N$ when temperature and pressure do not change. If you have ever blown up a balloon, you have discovered this for yourself (Figure 13-12). If you double the amount of air in the balloon, the volume doubles.

Similarly, if we keep the volume rather than the pressure constant, we find that the pressure is directly proportional to the amount of gas present: $P \propto N$.

This fourth variable, the number of molecules, can be combined with the relation linking pressure, volume, and temperature, and can be stated in symbols as

$$PV \propto NT.$$

Finally, putting in a constant of proportionality, k (see Chapter 1), we

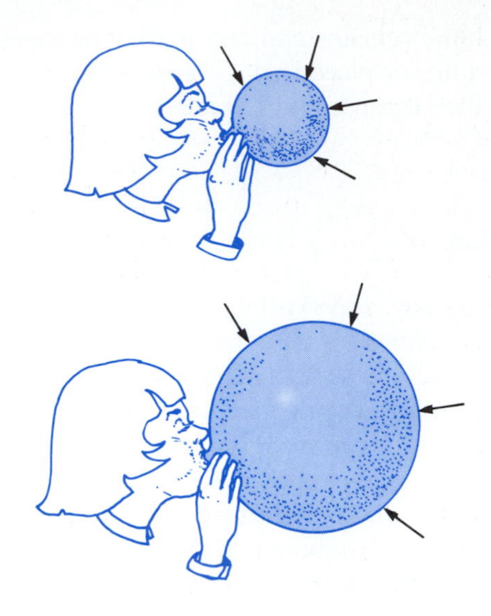

FIGURE 13-12
Volume is proportional to the amount of gas at constant pressure and temperature.

obtain

$$PV = NkT.$$

This is known as the **ideal gas law**. It is called "ideal" because real gases do not follow it precisely, particularly at very high pressures, or when the gas is near the liquefaction point. (This is also true of Boyle's, Charles's, and Gay-Lussac's laws.) However, under most ordinary circumstances real gases follow the ideal gas law quite closely.

The constant k in the ideal gas law is found experimentally to be the same for *all* gases, whether they be hydrogen, oxygen, argon, or whatever. This amazing fact was discovered by Amedeo Avogadro (1776–1856) and is known as **Avogadro's hypothesis**. Avogadro stated it in the following simple way:

> **equal volumes of gases at the same pressure and temperature contain equal numbers of molecules.**

That Avogadro's hypothesis should be valid is not at all obvious. That it *is* valid is a remarkable expression of simplicity in nature. Avogadro's hypothesis was to play an important role in the acceptance of the atomic theory.†

---

† Avogadro had no way of measuring how many molecules there are in a given volume of gas. In fact, accurate measurements were not possible until the twentieth century. These measurements indicate that there are about 2.7×10^{22} molecules in one liter of an ideal gas, or 2.7×10^{25} in a cubic meter, at STP (Standard Temperature and Pressure, meaning 0°C and 1 atmosphere).

3. KINETIC THEORY

Kinetic Theory of Gases

Kinetic theory refers to the idea that matter is made up of molecules and that these molecules are in constant motion. The theory is based on the assumption that, on the average, the molecules of a gas are rather far apart (the average distance between molecules is perhaps ten or more times greater than the diameter of a molecule) and that they frequently collide with one another and with the walls of the container. Because of these collisions, a particular molecule is assumed to be moving randomly, sometimes in one direction and sometimes in another, sometimes at high speed and sometimes at low speed, or somewhere in between.

As we mentioned earlier, this view of a gas explains why gases are so easily compressible. It is also easy to see why gases expand to fill any container and why they leak through tiny openings—it is because the molecules are constantly moving, and at rather high speeds.

By the mid-nineteenth century many scientists had come to accept the kinetic (or atomic) theory of matter and were beginning to use it to explain the behavior of gases, particularly the ideal gas law.

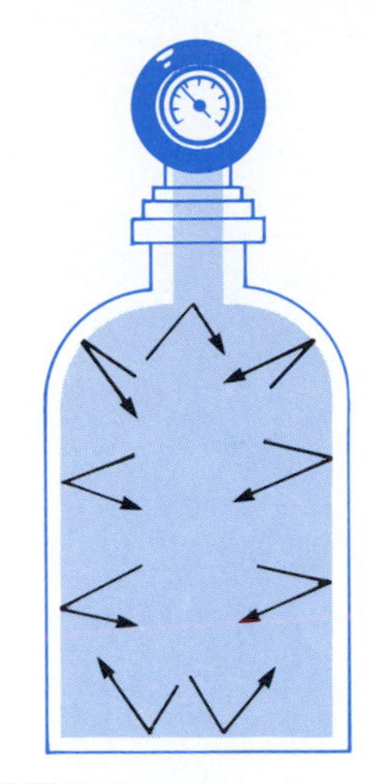

FIGURE 13-13
Pressure on the walls of a container is due to the bombardment of the rapidly moving gas molecules.

Temperature Is a Measure of the Kinetic Energy of Molecules

A crucial development in the kinetic theory of gases was an idea put forward by Rudolf Clausius (1822–1888) in 1847. Clausius showed that the entire behavior of a gas, including the behavior specified by the gas laws, could be explained by the hypothesis that

> **The temperature of a gas is proportional to the average kinetic energy of its molecules.**

This can be written in symbols as $T \propto \overline{\text{KE}} = \overline{\frac{1}{2}mv^2}$, where the bar ($\overline{\quad}$) means average.[†]

How does this relation explain the ideal gas law? Let's take a simple illustration. The pressure on the walls of a vessel containing a gas results from the constant bombardment of the gas molecules against the walls (Figure 13-13). According to the above relationship, when the temperature of a gas is increased the molecules have more kinetic energy. They move faster and strike the container walls with greater force; this means that the pressure is higher. Thus, an increase in temperature corresponds to an increase in pressure, which is just what we observe experimentally (Gay-Lussac's law).

[†] This is not to imply that *all* the molecules of a gas have this same amount of kinetic energy. Rather, it is the *average* kinetic energy of all the molecules that is proportional to the absolute temperature. At any given instant, roughly half the molecules have a kinetic energy greater than the average and the other half have a kinetic energy less than the average.

By using kinetic theory and the relation between temperature and average molecular kinetic energy we can also explain the observed behavior of gases on which Charles's and Boyle's laws are based. If we keep the pressure on a gas constant—by allowing the volume of its container to change (as with a balloon)—an increase in temperature means that the molecules will strike the walls of the container harder, forcing the container to enlarge. This is just what Charles found experimentally. Finally, at constant temperature, if a gas is compressed so that its volume is smaller, the pressure increases (Boyle's law). According to kinetic theory, the molecules have been pushed closer together, and more of them will therefore strike the walls of the vessel in any time interval; thus the force on the walls, and therefore the pressure, will be increased.

Kinetic Theory of Solids and Liquids

The kinetic theory, and the idea that temperature is a measure of the average molecular kinetic energy, can accurately explain the observed behavior of gases. Applying it to solids and liquids is a more difficult task, because the forces between molecules are much more important. For example, temperature is not in precise direct proportion to the average kinetic energy of molecules in the solid and liquid phases. Nonetheless, it is still a measure of the motion of the molecules. Thus, in a liquid an increase in temperature means that the molecules roll over one another more rapidly; in a solid it means the atoms or molecules are vibrating faster about their fixed position in the crystal.

Kinetic theory can account for the expansion of liquids and solids as temperature increases: at higher temperatures the molecules are moving faster and, on the average, they will be a little farther apart. Thus the object being heated will take up more space. Kinetic theory is also useful in explaining the processes of melting and boiling—the change of state. We will return to this in Chapter 14.

SUMMARY

The *temperature* of a body is a measure of how hot or cold it is. A *thermometer* commonly measures temperature on either the *Celsius* (centigrade) scale or the *Fahrenheit* scale. The *absolute*, or *Kelvin*, temperature scale begins at absolute zero (0 K = − 273°C), but it is believed that this temperature cannot actually be reached.

The behavior of gases is described by the "gas laws." *Boyle's law* states that at constant temperature the volume (V) of a fixed amount of gas is inversely proportional to the pressure (P) applied to it. *Charles's law* states that at constant pressure the volume of a fixed amount of gas is directly proportional to the absolute temperature (T). *Gay-Lussac's law* states that at constant volume the pressure in a fixed amount of gas is directly proportional to the absolute temperature. *Avogadro's hypothesis* states that equal volumes of gases at the same pressure and volume contain equal numbers (N) of molecules. All these laws can be combined into one relationship known as the *ideal gas law*, which is given in symbols as $PV = NkT$, where k is a universal constant—i.e., the same constant for all gases. It is an "ideal" law because real gases follow it closely only when the pressure is not too high and when the gas is not too dense.

According to the *kinetic theory* of gases, which is based on the idea that a gas is made up of molecules that are moving rapidly and at random, the average kinetic energy of the molecules is proportional to the Kelvin temperature T. At any moment there exists a wide distribution of molecular speeds within a gas.

QUESTIONS

1. Name several properties of materials that could be exploited to make a thermometer.
2. When a mercury thermometer is warmed, the mercury expands. What happens when the mercury expands? Do more molecules of mercury appear or does the distance between molecules increase?
3. Why is it sometimes easier to remove the lid from a tightly closed jar after warming it under hot running water?
4. A closed bottle placed in a fire will suddenly explode. Why?
5. On the average, on a hot day do you think air molecules are closer together, farther apart, or the same distance apart as on a cold day? Assume that atmospheric pressure is the same on both days.
6. A circular ring is heated from 20°C to 80°C. Will the hole in the ring become larger or smaller?
7. A precise steel tape measure has been calibrated at 20°C. At 40°C will it read high or low?
8. Explain the operation of the two types of thermostats shown in Figure 13-5.
9. How can Avogadro's law be true if molecules of different gases have different sizes?
10. Explain in words how Charles's law follows from the relation between average kinetic energy of molecules and absolute temperature.
11. If the average kinetic energy of molecules in a gas were to double, how would this affect the temperature?
12. If the temperature of a gas is raised from 20°C to 100°C, by what factor does the kinetic energy of the molecules increase?

13. If a copper penny has a higher temperature than a steel paper clip, does this necessarily mean that the average speed of molecules in the penny is greater than that of molecules in the paper clip? Explain.

EXERCISES

1. Room temperature is often defined as 68°F. What temperature is this on the Celsius scale? On the absolute scale?
2. On a hot day it is 90°F outside. What is this in degrees Celsius?
3. The temperature of the filament in a light bulb is about 1800°C; what is this on the Fahrenheit scale?
4. What is 30°C on the Fahrenheit scale? On the Kelvin scale?
5. The original Celsius temperature scale [due to Anders Celsius (1701–1744)] defined the freezing point of water as 100° and the boiling point as 0°. What temperature on this scale corresponds to 25°C?
6. In an alcohol-in-glass thermometer, the alcohol column has length 12.45 cm at 0.0°C and length 21.30 cm at 100.0°C. What is the temperature if the column has length (a) 15.10 cm and (b) 22.95 cm?
7. What are the following temperatures on the Kelvin scale: (a) 37°C (b) 80°F (c) −196°C?
8. Typical temperatures in the interior of the earth and sun are about 4×10^3°C and 1.5×10^7°C, respectively. (a) What are these temperatures in kelvins? (b) What percent error is made in each case if a person forgets to change °C to K?
9. If a gas is compressed to one-fifth its original volume but the temperature is kept constant, what happens to the pressure?
10. A gas is maintained at constant temperature. If the pressure on the gas is doubled, the volume is changed by what factor?
11. If the pressure on a gas is tripled, what happens to the volume if the temperature is kept constant?
12. The temperature of a gas in a glass container is increased from 300 K to 600 K. How does the pressure change?
13. If the pressure on a gas is kept constant, by what factor will its volume change if the temperature is increased from 27°C to 177°C?
14. Oxygen is compressed to a volume of 2.0 m^3 under a pressure of 10 atmospheres at 20°C. What volume will it occupy if it is allowed to expand to atmospheric pressure? Assume the temperature doesn't change.
15. Repeat the above exercise assuming the temperature drops to −10°C.
16. To what must you raise the temperature of a gas if you want its volume

to expand from 2.5 m^3 at 20°C to a volume of 4.0 m^3? Assume the pressure remains constant.

17. If 5 m^3 of a gas initially at atmospheric pressure and 0°C is placed under a pressure of 4 atm, the temperature of the gas rises to 25°C. What is the volume?

18. The pressure in a helium gas cylinder is initially 30 atmospheres. After many balloons have been blown up, the pressure has decreased to 6 atm. What fraction of the original gas remains in the cylinder?

*19. A house has a volume of 600 m^3. (a) What is the total mass of air inside the house at 0°C? (b) If the temperature rises to 25°C, what mass of air enters or leaves the house?

*20. An automobile tire is filled to a gauge pressure (see Section 4 of Chapter 12) of 210 kPa at 10°C. After driving at high speed for an hour, the temperature inside the tire rises to 40°C. What is the pressure within the tire now? (*Hint*: Be sure to use absolute pressure and Kelvin temperature.)

*21. A gas is at 20°C. To what temperature must it be raised to double the speed of its molecules?

*22. Oxygen at 0°C is cooled to −137°C. By what factor does the average speed of the molecules change?

*23. At what temperature will the Fahrenheit and Centigrade scales yield the same numerical value?

HEAT

CHAPTER 14

When a pot of cold water is placed on the hot burner of a stove, the temperature of the water increases. We say that heat flows from the hot burner to the cold water. Whenever two objects at different temperatures are put in contact, heat flows from the hotter one to the colder one. The flow of heat is in the direction that will tend to equalize the temperature. If the two objects are kept in contact long enough for their temperatures to become equal, the two bodies are said to be in equilibrium, and there is no further heat flow between them. For example, when the mercury in a fever thermometer is still rising, heat is flowing from the person's mouth to the thermometer; when the mercury stops rising, the thermometer is in equilibrium with the person's mouth, and they are at the same temperature.

We all have an intuitive notion of what *heat* means. Yet it is not easy to define. Heat is often confused with temperature, but the two are different concepts. Just what do we mean by heat?

1. HEAT AS ENERGY TRANSFER

The Caloric Theory

It is common to speak of the flow of heat—heat flows from a stove burner to a pot of coffee, from the sun to the earth, from a person's mouth into a fever thermometer. It flows from an object at higher temperature to one of lower temperature. Indeed, early theories of heat, before the nineteenth century, pictured heat as a fluid substance called *caloric*. According to the caloric theory, any object contained a certain amount of caloric; if more caloric flowed into the object, its temperature increased; and if caloric flowed out, the object's temperature decreased. When matter was broken apart, such as during burning, a great deal of caloric was believed to be released. However, the caloric fluid was never directly detected, and in the nineteenth century phenomena came to light that it could not explain satisfactorily, as we shall see shortly.

Units of Heat

Although the caloric theory has long since been discarded, remnants of the theory still remain such as in the expression "flow of heat," as if heat were a fluid. The common unit for heat that is still used is named after caloric. It is called the *calorie* (cal) and is defined as *the amount of heat necessary to raise the temperature of 1 gram of water by 1 Celsius degree, from 14.5°C to 15.5°C*. This particular temperature range is specified since the heat required is very slightly different at different temperatures. The *kilocalorie* (kcal) is 1000 calories. Thus *1 kcal is the heat needed to raise 1 kg of water by 1 C°*, from 14.5°C to 15.5°C. Sometimes a kilocalorie is called a Calorie (with a capital C), and it is by this unit that the energy value of food is specified. As an example, suppose we want to know how much heat is required to heat 2.0 liters of water from 20°C to 100°C. To solve this we note that a liter of water has a mass of 1.0 kg. Thus it takes 2.0 kcal to heat 2.0 liters of water by 1.0 C°. To heat the 2.0 liters by 80 C° requires (80)(2.0 kcal) = 160 kcal.

Heat As Transfer of Energy

One of the main problems with the caloric theory was its inability to account for all the heat generated by friction. You can, for example, rub your hands or two pieces of metal together for a long time and generate heat indefinitely. The American-born Benjamin Thompson (1753–1814), who later became Count Rumford of Bavaria, was acutely aware of this problem when he supervised the boring-out of cannon barrels. Water was placed in the bore of the cannon to keep it cool during the cutting process, and as the water boiled away it was replenished. The "caloric" that caused the boiling of the water was assumed to be due to the breaking apart of the metal. But Rumford noticed that even when the cutting tools were so dull that they did not cut the metal, heat was still generated and the water boiled away. Thus caloric was being released even though subdivision of matter was not occurring. Furthermore, Rumford realized, this process could go on indefinitely and produce a limitless amount of heat. This was not consistent with the idea that heat is a substance and that only a finite amount of it could be contained within an object. Rumford therefore rejected the caloric theory and proposed instead that heat is a kind of motion. He claimed that in some circumstances, at least, heat is produced by doing mechanical work (for example, rubbing two objects together). This idea was pursued by others in the early 1800s, particularly by an English brewer, James Prescott Joule (1818–1889).

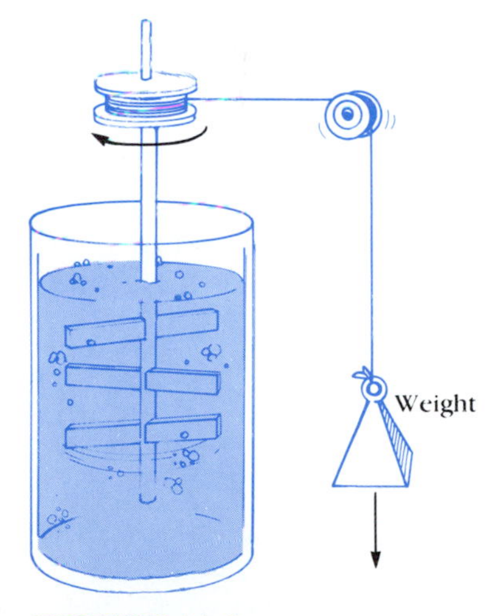

FIGURE 14-1
Joule's paddle wheel experiment. A falling weight turns the paddle wheel that heats the water because of "friction" between the paddles and water.

Joule performed a number of experiments that were crucial in establishing our present view that heat, like work, represents a transfer of energy. One of Joule's experiments is illustrated in Figure 14-1. The falling

FIGURE 14-2
Kinetic energy is transferred from a fast-moving molecule (dark) to a slow-moving molecule by means of a collision.

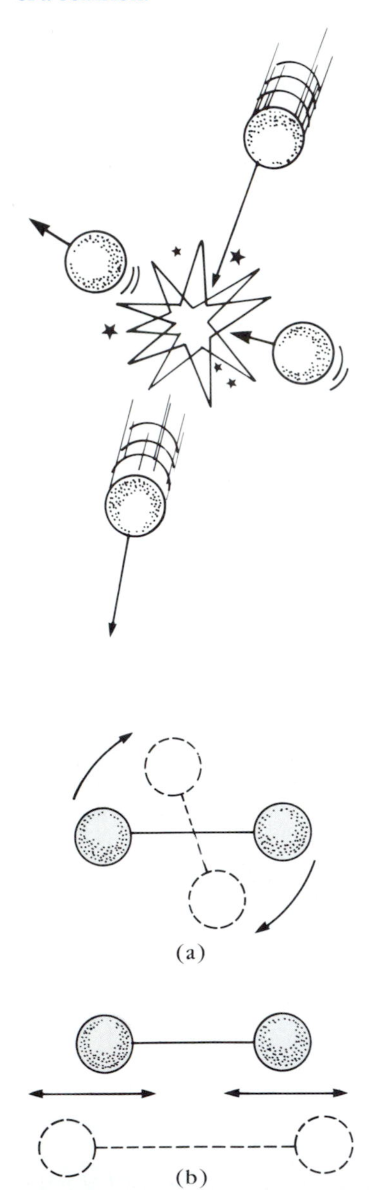

FIGURE 14-3
Molecules can have (a) rotational energy and (b) vibrational energy (atoms in a molecule vibrate back and forth within the molecule).

weight causes the paddle wheel to turn; the friction between the water and the paddle wheel causes the temperature of the water to rise. Of course, the same temperature rise could also be obtained by heating the water on a hot stove. In this and a great many other experiments (some involving electrical energy), Joule found that a given amount of work was always equivalent to a particular amount of heat. Quantitatively, 4.18 joules (J) of work was found to be equivalent to 1 calorie of heat. This is known as the *mechanical equivalent of heat*:

$$4.18\,\mathrm{J} = 1\ \mathrm{cal}, \text{ or } 4180\,\mathrm{J} = 1\ \mathrm{kcal}.$$

As a result of these and other experiments, scientists came to interpret heat not as the flow of a substance but as a transfer of energy; when heat flows from a hot object to a cooler one, it is energy that is being transferred from the hot to the cold object. Thus heat refers to *energy that is transferred from one body to another because of a difference in temperature*. The SI unit for heat, as for any form of energy, is the joule. Nonetheless, calories and kcal are still often used.

The development of kinetic theory fully supports, and indeed nicely explains, the idea of heat as a transfer of energy. Let us examine the process of heating a pot of water on a stove. According to kinetic theory, the average kinetic energy of molecules increases with temperature; thus the molecules of the stove burner have much more kinetic energy on the average than those of the cold water or the pot. When the fast-moving stove molecules collide with the slower-moving molecules of the pot, some of their kinetic energy is transferred to the pot molecules, just as a fast-moving billiard ball transfers some of its kinetic energy to a ball it collides with (Figure 14-2). The molecules of the pot gain in kinetic energy. The now faster-moving pot molecules, in turn, transfer some of their kinetic energy by collision to the slower-moving water molecules. The temperature of the water and the pot consequently increases. Thus we see how heat flow is a transfer of energy.

Whenever a hot substance is in contact with a cold substance, heat flows from the hotter one to the colder. Unless the process is interrupted, the two objects eventually reach the same temperature. When a pot is heated on the stove, however, the burner doesn't cool down very much because energy is continually being supplied to it by electricity or gas. The pot gets hotter and hotter. But the stove burner cannot heat the pot to a temperature higher than its own temperature; heat will only flow "downhill" so to speak—only from a high temperature to a low temperature.

When an object is heated, not all the energy it absorbs goes into increasing the velocity of its molecules. Some goes into rotational motion of the molecules and some into internal vibrational energy; this is illustrated in Figure 14-3. Thus the total energy of the molecules in a substance is the sum of the ordinary kinetic energy, the rotational kinetic energy, and the vibrational energy.

The sum total of all the energy of all the molecules in an object is called its **internal energy**, or **thermal energy**. We will use the two terms interchangeably. Occasionally, the term "heat content" of a body is used for this purpose. However, this is not a good term to use for it can be confused with heat itself. Heat, as we have seen, is not the energy a body contains but rather refers to the amount of energy transferred from a hot to a cold body.

Distinction Between Temperature, Heat, and Internal Energy

Using the kinetic theory, we can now make a clear distinction between temperature, heat, and internal energy. Temperature is a measure of the *average* kinetic energy of individual molecules. Thermal or internal energy refers to the *total* energy of all the molecules in the object. Thus two equal-mass hot ingots of iron may have the same temperature, but two of them have twice as much thermal energy as one does. Heat, finally, refers to a *transfer* of energy (usually thermal energy) from one object to a second which is at a lower temperature. Heat, as Count Rumford saw, can be generated indefinitely; but the thermal energy of a body is strictly limited.

Notice that the direction of heat flow between two objects depends on their temperatures, not on how much internal energy they each have. Thus, if 50 g of water at 30°C is placed in contact (or mixed) with 200 g of water at 25°C, heat flows *from* the water at 30°C *to* the water at 25°C even though the internal energy of the 25°C water is much greater because there is so much more of it.

EXPERIMENT-PROJECT

Find two kitchen pots of about the same size and made of the same material. Put 2 cups of water in one and 8 cups of water in the other—Figure 14-4(a). Put them on identical burners and, using a thermometer, heat the first to boiling, 100°C (212°F), and the second only to 75°C (167°F). Notice which pot takes longer to heat and therefore requires more heat. Remove both pots from the stove and immediately place an identical tin can (or glass) containing one cup of cold water in each pot—Figure 14-4(b). Wait a few

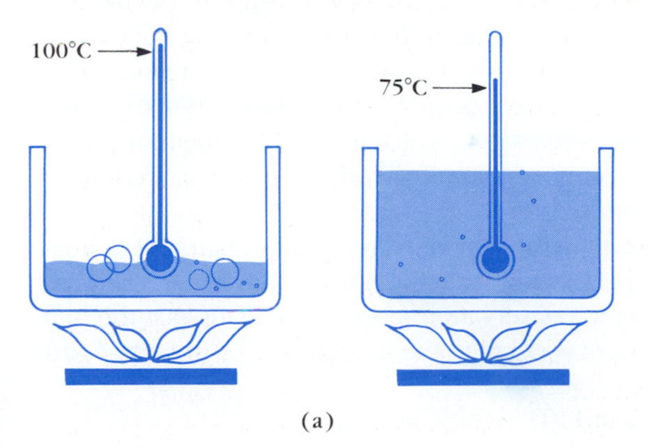

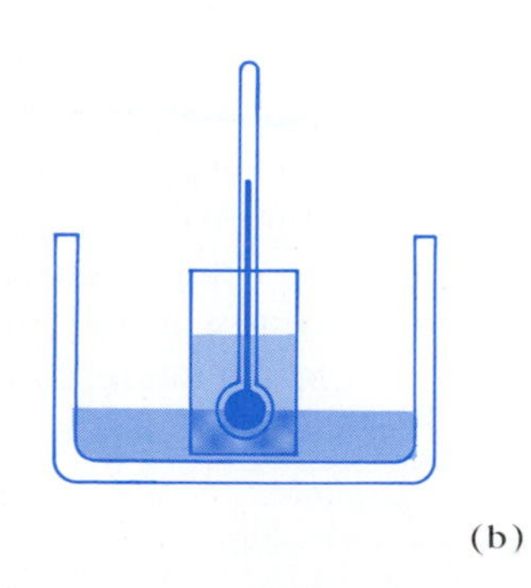

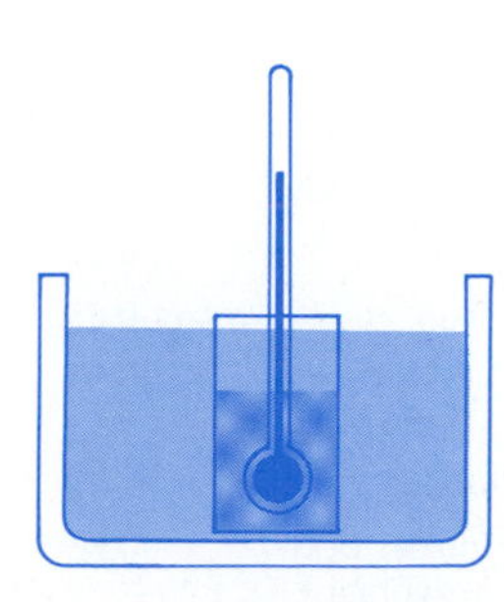

FIGURE 14-4
(a) Two pots on a stove. The pot on the left contains 2 cups of water and is heated to boiling; the pot on the right contains 8 cups and is heated only to 75°C. (b) Both pots are removed from the stove, and identical glasses or tin cans are placed in each (see experiment in text).

minutes and then check the temperature of the water in each tin can. Which can of water is heated to the higher temperature? Which pot of water, the 2 cups at 100° C or the 8 cups at 75° C, possessed the greater thermal energy? Alternatively, after removing the two pots from the stove, see how many ice cubes each can melt.

2. SPECIFIC HEAT

When an object or a substance is heated, it is found that the change in temperature is proportional to the heat added. It is also found that for a given temperature rise, the amount of heat required is proportional to the amount, or mass, of substance. For example, the more water there is in a pot the more heat it takes to heat it from, say, 20° C to 100° C. Furthermore, different materials require different amounts of heat even when the mass and temperature changes are the same. These observations are expressed in the relation

$$Q = mc\,(T - T_0)$$

where Q is the heat required to raise a mass m of the substance from a temperature T_0 to T; the symbol c is a quantity characteristic of the material called the **specific heat**. For water at 15°C, $c = 1.00$ kcal/kg · C° or 4180 J/kg · C°; this corresponds to the fact that it takes 1 kcal of heat to raise the temperature of 1 kg of water by 1 C°. Table 14-1 gives the values of specific heat for other substances.

Let us calculate how much heat is required to raise the temperature of 25 kg of iron from 20° C to 100° C. From Table 14-1, the specific heat of iron is 0.11 kcal/kg · C°. The change in temperature is 100°C − 20°C = 80 C°. Thus

$$Q = mc\,(T - T_0) = (25\text{ kg})(0.11\text{ kcal/kg}\cdot\text{C}^\circ)(80\text{ C}^\circ) = 220\text{ kcal}.$$

If the iron had been cooled from 100°C to 20°C, 220 kcal of heat would have flowed out of the iron. In other words, we can use the specific heat equation for heat flow either in or out, with a corresponding increase or decrease in temperature. If 25 kg of water had been heated from 20° C to 100° C, the heat required would have been 2000 kcal. Water has one of the highest heat capacities of all substances, which makes it an ideal substance for hot-water radiator-heating systems and other uses that require a minimal drop in temperature.

When different parts of an isolated system are at different temperatures, heat will flow from the part at higher temperature to the part at lower temperature. In this case the conservation of energy tells us that the heat lost by one part of the system is equal to the heat gained by the other part:

$$(\text{heat lost}) = (\text{heat gained}).$$

TABLE 14-1
Specific heat

SUBSTANCE	SPECIFIC HEAT kcal/kg · C°	J/kg · C°
Aluminum	0.22	900
Copper	0.093	390
Glass	0.20	840
Ice (−5°C)	0.50	2100
Iron or steel	0.11	450
Marble	0.21	860
Silver	0.056	230
Alcohol (ethyl)	0.58	2400
Water	1.00	4180
Human body (average)	0.83	3470

3. CHANGE OF STATE

When water is heated to a high enough temperature, it boils and becomes a gas. When water is cooled to a low enough temperature, it freezes and becomes a solid.

The temperature at which a liquid boils and changes to a gas is called its **boiling point**; the gas cools and condenses into a liquid at the *same* temperature (the **condensation point**). The temperature at which a solid melts into a liquid is called its **melting point**; the liquid freezes and changes back into a solid at the *same* temperature (the **freezing point**). Each of these processes is referred to as a *change of state* of the substance. The melting and boiling points for a variety of common substances are shown in Table 14-2.

TABLE 14-2
Melting and Boiling Points of Various Substances

SUBSTANCE	MELTING POINT (°C)	BOILING POINT (°C)
Helium	−271*	−269
Oxygen	−219	−183
Nitrogen	−210	−196
Alcohol (ethyl)	−114	−78
Mercury	−39	−357
Water	0	−100
Gold	1063	2933
Iron	1535	3000

* Under pressure of 26 atmospheres. (Does not condense under one atmosphere.)

Heat Is Involved in a Change of State

Let us examine the process of continuously heating a solid substance until it becomes a liquid and then a gas. We will use a concrete example, water. Suppose we have one gram of ice at, say, −40°C and we heat it at a continuous rate (Figure 14-5). As we heat it, its temperature rises. It takes about one half a calorie to raise the temperature of one gram of ice by one degree Celsius. When the temperature of the ice reaches 0°C, the ice begins to melt. But it doesn't all melt at once. As heat is added, the ice melts to water little by little and a water-ice mixture results. Throughout this process, the temperature of the ice does not change; it remains at 0°C until all the ice has melted. Heat is required just to melt ice without any increase in temperature. To melt one gram of ice at 0°C to water at 0°C requires 80 calories; this is known as the *heat of melting* for water. Once all the ice has melted to water, the temperature begins to rise again. Because it takes about one calorie to raise the temperature of one gram of water by one degree Celsius, a total of 100 calories is necessary to bring the one gram of water to the boiling point at 100°C. At this point a

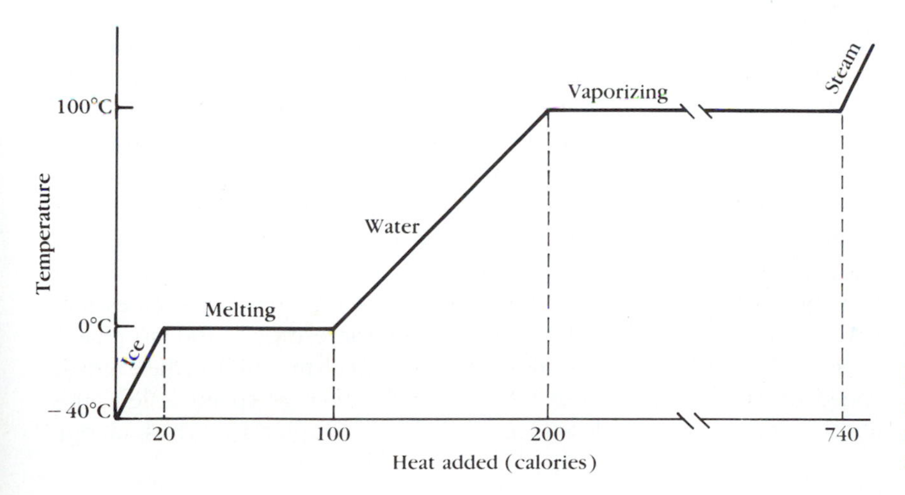

FIGURE 14-5
Graph showing the heat required to heat 1 gram of ice at −40°C to steam at 100°C.

considerable amount of energy is required to change from the liquid to the gaseous state. This is the *heat of vaporization*, which for water is 540 calories per gram. The temperature of the water remains constant until it has all been converted to steam. With further heating, the temperature of the steam increases. This whole process is graphed in Figure 14-5.

We can make use of kinetic theory to see why energy is needed to melt or vaporize a substance. At the melting point, the latent heat of fusion does not increase the kinetic energy (and the temperature) of the molecules in the solid but, instead, is used to overcome the potential energy associated with the forces between the molecules. That is, work must be done against these attractive forces to break the molecules loose from their relatively fixed positions in the solid so they can freely roll over one another in the liquid phase. Similarly, energy is required for molecules held close together in the liquid phase to escape into the gaseous phase. This process is a more violent reorganization of the molecules than is melting (the average distance between the molecules is greatly increased); hence the heat of vaporization for a given substance is generally much greater than the heat of fusion.

Evaporation and Condensation

To change water from the liquid to the gas state—which we often call water vapor—does not necessarily require that the water be heated to boiling. It happens even at room temperature, although much more slowly. This is the process of *evaporation*. If you leave a glass of water out overnight, you will find that the water level has gone down by the next morning.

The latent heat to change a liquid to a gas is needed not only at the boiling point but also in the evaporation process. As a result, when water evaporates, it cools, since the energy required (the latent heat of vaporization) comes from the water itself; so its internal energy, and therefore its temperature, must drop.

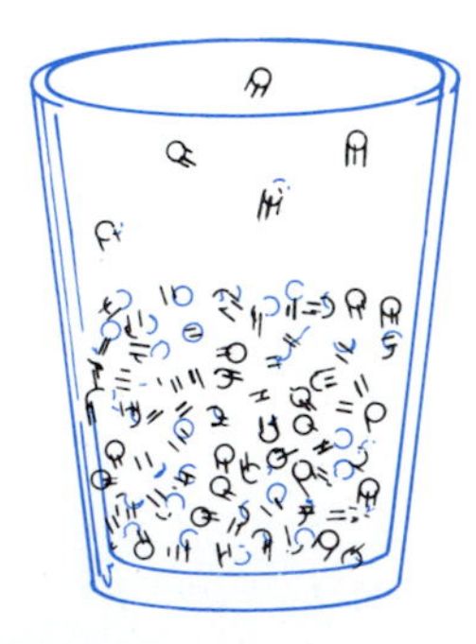

FIGURE 14-6
Evaporation. Very fast-moving molecules near the surface of the liquid can escape.

This process of evaporation can be explained on the basis of kinetic theory. The molecules in a liquid move past one another with a variety of speeds. There are strong attractive forces between these molecules, which is why they stay together in the liquid phase. A molecule in the upper regions of the liquid may, because of its speed, leave the liquid momentarily. But just as a rock thrown into the air returns to the earth, so the attractive forces of the other molecules pull the vagabond molecule back to the liquid surface—at least if its velocity is not too large. If the molecule has a high enough velocity, it will escape from the liquid entirely and become part of the gas phase (Figure 14-6). Only those molecules that have kinetic energy above a particular value can escape to the gas phase.

Since it is the fastest molecules that escape from the surface, the average speed of those remaining is less. When the average speed is less, the absolute temperature is less (see Section 3 of Chapter 13). Thus, kinetic theory predicts that evaporation is a cooling process. Experiment bears this out: you have no doubt noticed, for example, that when you step out of a warm shower and the water on your body begins to evaporate you feel cold; and after working up a sweat on a hot day, you will feel cool from even a slight breeze.

The reverse of evaporation also occurs. Gas molecules may strike the surface of the liquid and be absorbed. This is called *condensation*. Condensation occurs when gas molecules are slowed down (or cooled): for example, when water in the air condenses on the outside of a glass of ice water. Water molecules in the vicinity of the cold glass lose kinetic energy through collisions with the slow-moving molecules of the glass and the cool layer of air right next to the glass. As more and more of the water molecules slow down, the attractive forces between them can pull them together and form a drop of water on the glass.

The common phenomena of fog, clouds, and rain are also the result of condensation of water droplets when air is cooled. The cooling of air at night often results in the condensation we know as morning dew.

These explanations in terms of kinetic theory remind us that a theory is only as good as its power to explain phenomena; and kinetic theory works very well here, as elsewhere.

The Boiling Process

Let's look at the boiling process in more detail.

EXPERIMENT

Put some cold water in a pan on a stove and heat it to boiling. Notice where and how bubbles form in the water as the boiling point is approached. Examine the water carefully as it approaches the point of vigorous boiling.

As the boiling point is approached, evaporation begins to take place below the surface of the water on the bottom and sides of the pan where the temperature is highest. The molecules are moving so fast there that a group of them may enter the gaseous state and form a bubble. Any bubble that tends to form will immediately collapse unless the pressure due to the fast movement of molecules within the liquid is sufficient to balance the downward pressure of the liquid around it (Figure 14-7). As heating continues, the vapor pressure within the bubble builds up until it releases itself from the bottom of the pan and floats to the surface. Bubbles soon begin appearing throughout the liquid and rise to the top; vigorous boiling has begun.

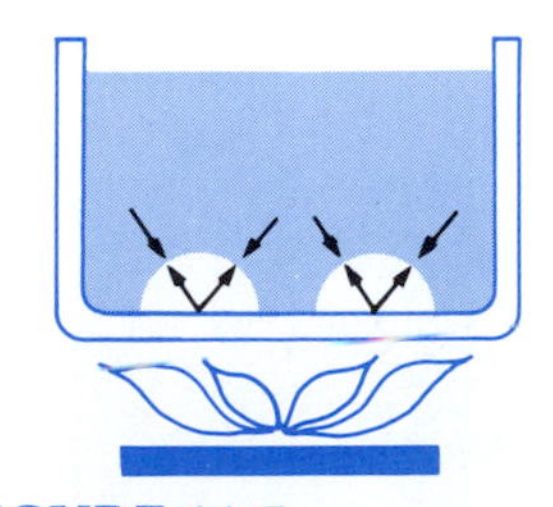

FIGURE 14-7
Bubbles of steam forming at the bottom of a heated pot just before boiling takes place.

Boiling occurs when the pressure within the bubble equals or exceeds outward pressure due to the liquid and the atmosphere above it. Thus, if the pressure at the surface of the liquid is increased, a higher temperature is necessary before the molecules are moving fast enough to form a bubble. This is how a pressure cooker works. The evaporating water builds up a high pressure within the well-sealed pot and boiling occurs at a higher than normal temperature. At this higher temperature foods cook much more quickly.

At high elevations, where the air pressure is less, it takes longer to cook foods. This is because bubbles can form at a lower temperature, since less internal vapor pressure is needed. Thus the temperature of boiling water is less at high elevations. If you apply more heat to the water in an attempt to increase its temperature, you find that its temperature won't rise any higher; instead, it evaporates, or boils, all the faster. Faster boiling does not cook food any faster. Only the temperature affects the cooking rate. Since the maximum temperature of liquid water is less at the lower pressure of high altitudes, cooking takes longer. Pressure cookers can of course be used at high altitudes to reduce cooking time.

4. HEAT TRANSFER: CONDUCTION, CONVECTION AND RADIATION

Heat can be transferred from one place to another in three different ways: by **conduction**, by **convection**, and by **radiation**.

Conduction

When you put one end of a metal rod in a fire or a silver teaspoon in a hot cup of soup, the other end becomes warm even though it is not directly in contact with the source of heat. We say that heat has been *conducted* from the hot end to the cold end (Figure 14-8).

Conduction is the transfer of energy through the collision of molecules. As one end of the object is heated, the molecules at that end move faster and faster. When they collide with their slower-moving neighbors, some of their energy is transferred to them. These molecules now speed up. When they in turn collide with other molecules farther up the rod (or spoon), they transfer energy to them. And so on.

FIGURE 14-8
Heat conduction.

Heat conduction takes place only when there is a difference in temperature, and the heat always flows from the higher temperature to the lower temperature. The rate at which the heat flows depends on the temperature difference, on the size and shape of the material, and on the kind of material it is. Some materials are good **conductors** of heat, which means that heat flows through them quickly. Other materials are poor conductors, or heat **insulators**; these materials conduct heat only very slowly.

EXPERIMENT

To demonstrate the difference between conductors and insulators, you can use a lighted candle or the burner of a stove. Be careful. Hold one end of a paper clip in your fingers and stick the other end in the flame; drop it quickly if it gets too hot. Now try a glass rod and other materials such as a nail, a silver spoon, a wooden rod (be careful it doesn't burn, though charring won't hurt), and a stone. Which materials are good conductors? Which are good insulators?

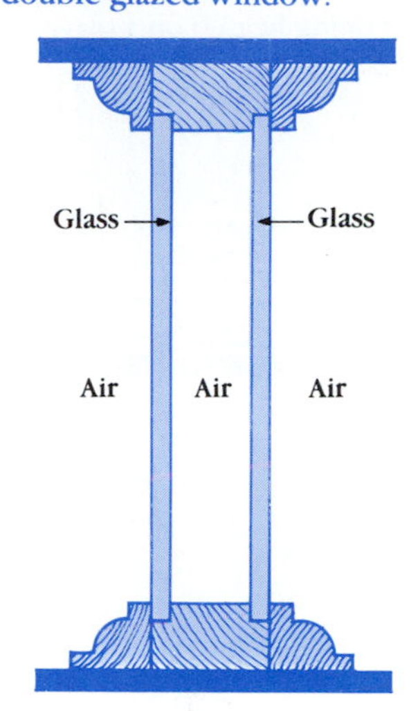

FIGURE 14-9
A double-glazed window.

Most metals are good conductors of heat, although there is wide variability. Test this statement by putting a silver spoon and a stainless steel spoon in the same cup of hot water and then touching the handles.

Even the best insulators conduct some heat. Glass, for example, is a much poorer conductor of heat than metals are. Even so, a good deal of heat is lost through the windows of a house on a cold day; the glass, though a poor conductor, is usually quite thin, so the heat doesn't have to travel far to be lost. The use of "double glazing," two panes of glass separated by a layer of air (Figure 14-9) helps considerably to reduce heat loss by conduction.

Air and other gases are excellent insulators because the molecules in a gas are so far apart on the average that they make fewer collisions with their neighbors than do the molecules in solids; hence they transfer energy much more slowly. A perfect vacuum would be the ideal insulator, because it would contain no molecules at all to transfer energy. The double-glazed window of Figure 14-9 insulates well because of the air layer between the sheets of glass.

The best insulating materials are those that trap air in pockets or between their fibers: materials like fiberglass and the various types of foam that are used to insulate houses and buildings. Wool and down are very good for insulating people; down is especially effective, because even a small amount of it fluffs up and traps a great deal of air. It is not the material our clothes are made of that insulates us, but the air trapped in the material.

Convection

Although liquids and gases are generally not very efficient conductors of heat, they can transfer heat quite rapidly by convection. **Convection** is the transfer of heat through the mass movement of molecules from one place to another. In conduction, molecules move only through very small distances and the energy is transferred by collision; but in convection, molecules move through large distances and themselves carry the energy.

A forced-air furnace, in which air is heated and is then blown by a fan into a room, is an example of **forced convection**. **Natural convection** occurs as well, and one familiar example is the fact that hot air rises. For example, the air above a radiator expands as it is heated; hence its density decreases and it rises.† Warm and cold ocean currents, such as the Gulf

† This is essentially buoyancy (see Section 2 of Chapter 12). The less dense air "floats" on top of the denser air.

FIGURE 14-10
Convection currents in a pot of water being heated on a stove.

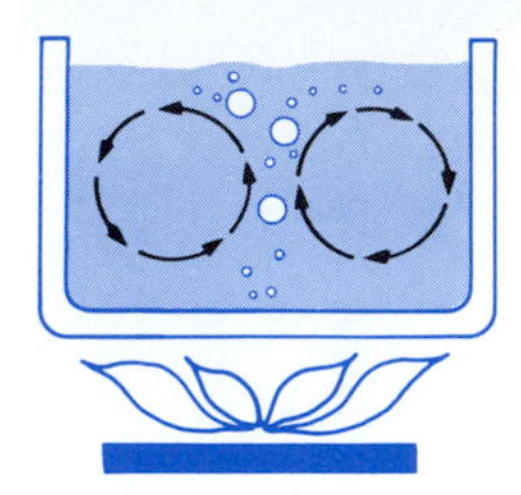

Stream, are examples of natural convection on a large scale. Wind is another example, and weather in general is the result of convective air currents.

When a pot of water is heated (Figure 14-10) convection currents are set up as the heated water at the bottom of the pot becomes less dense, rises, and is replaced by cooler water from above. This same principle is used in home heating systems like the hot-water radiator system shown in Figure 14-11. As water is heated in the furnace its temperature goes up, and the water expands and rises into the radiators. As the water circulates through the system, heat is transferred by conduction to the radiators and thence to the air; the cooled water then returns to the furnace to be heated again. The water circulates because of convection, though pumps are sometimes used to speed up its circulation. Convection also brings about the uniform heating of the rooms in which the radiators sit. As the radiators heat the air, the air rises and is replaced by cooler air, resulting in convective air currents (Figure 14-11).

Other types of furnaces also depend on convection. Hot-air furnaces with openings near the floor often depend on natural convection to bring about the distribution of heated air; the cold air returns to the furnace through other openings.

Another example of convection and its effects is given in the following excerpt from Francois Matthes's "The Winds of Yosemite Valley":†

> It happens to be so ordained in nature that the sun shall heat the ground more rapidly than the air. And so it comes that every slope or hillside basking in the morning sun soon becomes itself a source of heat. It gradually warms the air immediately over it, and the latter, becoming lighter, begins to rise. But not vertically upward, for above it is still the cool air pressing down. Up along the warm slope it ascends, much as shown by the arrows in the accompanying diagram—Figure 14-12(a). Few visitors to the valley but will remember toiling up some never-ending zigzags on a hot and breathless day, with the sun on their backs and their own dust floating upward with them in an exasperating, choking cloud. Perhaps they thought it was simply their misfortune that the dust should happen to rise on that particular day. It always does on a sun-warmed slope.
>
> But again, memories may arise of another occasion when, on coming down a certain trail the dust ever descended with the travelers, wafting down upon them from zigzag to zigzag as if with malicious pleasure. That, however, undoubtedly happened on the shady side of the valley. For there the conditions are exactly reversed. When the sun leaves a slope the latter begins at once to lose its heat by radiation, and in a short time is colder than the air. The layer next to the ground then gradually chills by contact, and, becoming heavier as

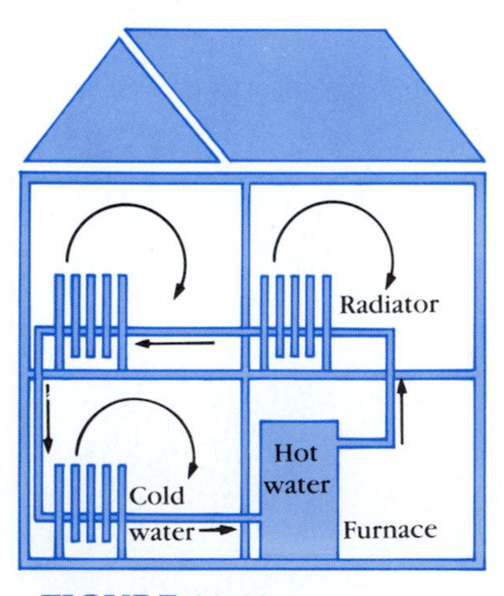

FIGURE 14-11
A hot water radiator heating system uses convection.

† Reprinted from the *Science Club Bulletin*, June 1911, pp. 91–92.

it condenses, begins to creep down along the slope—Figure 14-12 (b). There is, thus, normally a warm updraft on a sunlit slope and a cold downdraft on a shaded slope—and that rule one may depend on almost any day in a windless region like the Yosemite. Indeed, one might readily take advantage of it and plan his trips so as to have a dust-free journey.

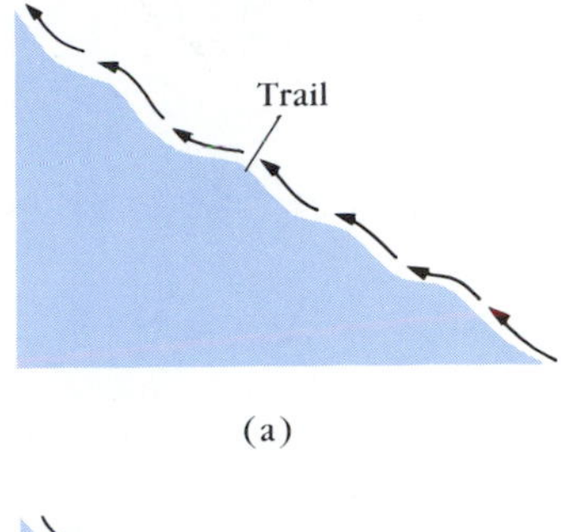

(a)

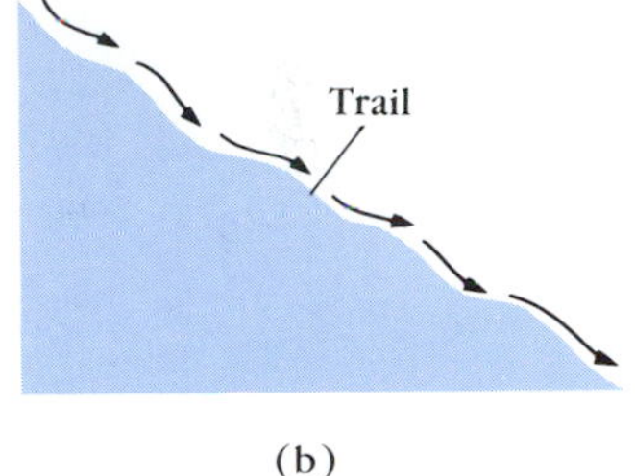

(b)

FIGURE 14-12
Convection on a hiking trail: (a) upward movement of air in the morning because it is heated. (b) downward movement in evening because it is cooled.

EXPERIMENT

On a cold day, open an outside door in a well-heated room in your house. Light a candle and hold it first near the very top of the door opening and then near the floor (Figure 14-13). Does the flame change direction? The hot air, which has expanded and risen, escapes near the top; the cold air, on the other hand, is drawn into the room near the floor and replaces the air that has escaped.

The human body generates a great deal of heat. Only about 20 percent of the food energy transformed by the body is used to do work; some 80 percent appears as heat, and this heat must somehow be transferred to the outside if the body is not to become overheated. But the body is not a very good conductor of heat. How does it get rid of this superfluous heat? The blood acts as a convective fluid and carries the heat to the surfaces of the body where it is transferred to the air by conduction. Our blood serves many purposes.

Radiation

Conduction and convection require the presence of matter. The third form of energy transfer, **radiation**, does not. All life on earth depends on energy from the sun, and this energy is carried to the earth at the speed of light by radiation. The warmth we receive from a fire reaches us mainly through radiation.

As we shall see later, radiation consists of electromagnetic waves; for now, we merely point out that the radiation from the sun includes not only visible light but many other wavelengths as well.

All objects give off radiation. But the higher the temperature of the object, the more energy it radiates. In fact, the rate of radiation of energy is proportional to the fourth power of the Kelvin temperature (T^4). Thus an object at 600 K as compared to one at 300 K has only twice the temperature but radiates $(2)^4 = 16$ times as much energy!

Bodies that glow, such as the sun, a fire, or a light bulb, are radiating energy. Indeed, any object will glow (that is, will give off *visible* radiation) if it is raised to a high enough temperature. Even at lower temperatures objects give off radiation, though it may not be visible.

Consider, for example, two bodies close to one another that are otherwise identical, except that one is cold and the other hot. Both objects emit energy, but the hot one emits more energy than the cold one. Thus there is a net transfer of radiant energy from the hot one to the cold one.

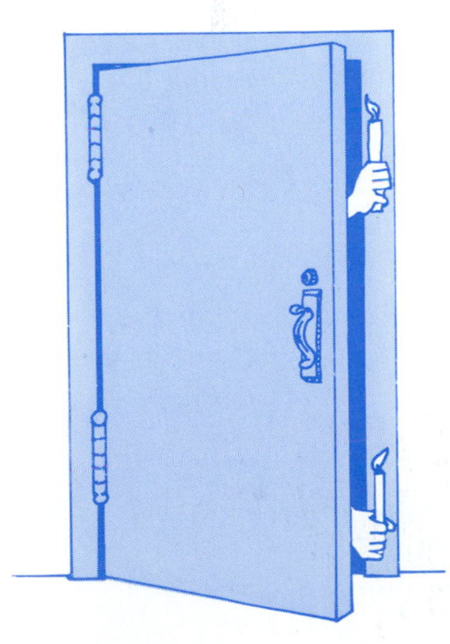

FIGURE 14-13
Detecting convection currents.

Indeed, there will be a net transfer of energy by radiation between two bodies as long as there is a difference in their temperatures.

Every object both emits energy and absorbs energy. But an object may not absorb all the energy that falls on it; it may reflect some of it. Good absorbers reflect very little of the radiant energy that falls on them and thus they appear black; that is why dark-colored objects are relatively good absorbers. Light-colored objects, on the other hand, reflect most of the energy that falls on them and so are poor absorbers. That is why light-colored clothing is more comfortable than dark clothing on a bright hot day.

It is found, too, that good absorbers are good emitters and poor absorbers are poor emitters. That is, at the same temperature, a black body will radiate energy at a faster rate than a similar white body at the same temperature. For example, hot tea in a dark-colored metal container cools more quickly as a result of radiation than does tea in a brightly polished metal container.

The fact that all objects, including human bodies, emit radiation has many practical consequences. For example, a person may be sitting in a room where the air temperature is 25°C (77°F), which you might think would be comfortable enough. But the room may feel chilly if the walls are at a somewhat lower temperature (perhaps because they are in contact with cold outside air). Human skin temperature is normally about 33° or 34°C; if the walls are at, say, 15°C, there may be a significant flow of radiant energy away from the person, and the person becomes chilled. Rooms are most comfortable when the walls and floor are warm and the air a bit less warm. Hot-water pipes or electric heating elements could be used to heat walls and floors, but such systems are rare today. Interestingly, though, 2000 years ago the Romans heated the floors of their houses with hot-water and steam pipes, even in the remote province of Britain.

Glass windows allow substantial heat loss through both conduction and radiation. They not only conduct heat out of a room but at night, when the outside temperature is low, considerable radiation from a body occurs to the outside. That's why it helps to close the curtains on a cold night.

SUMMARY

Thermal energy, or *internal energy*, refers to the total energy of all the molecules in a body. *Heat* refers to the transfer of energy from one body to another because of a difference of temperature. Heat is measured in energy units, such as joules. Heat and thermal energy are also sometimes specified in calories or kilocalories, where 1 cal = 4.18 J is the amount of heat needed to raise the temperature of 1 g of water by 1 C°.

The *specific heat*, *c*, of a substance is defined as the energy (or heat) required to change the temperature of 1 kg of substance by 1 degree. As an

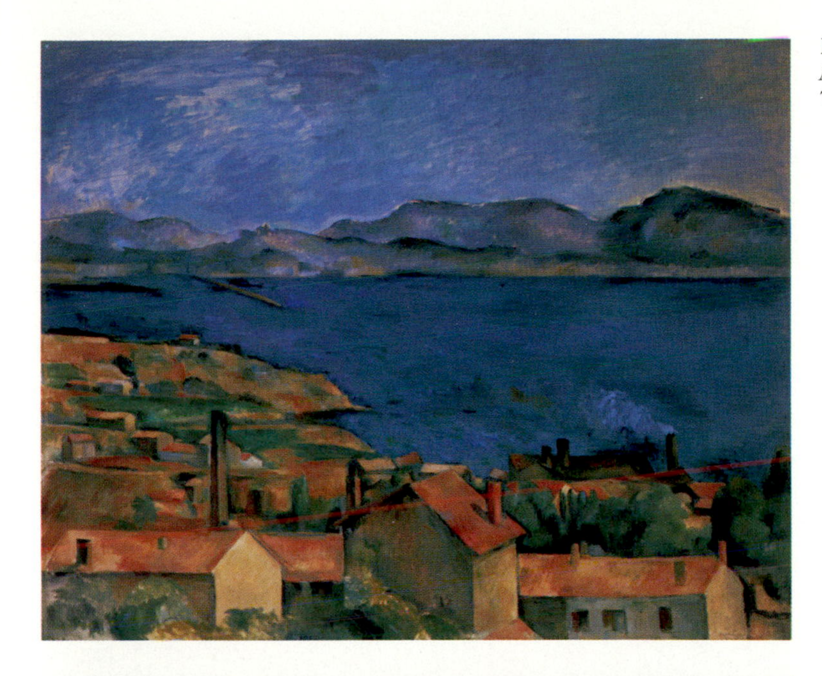

FIGURE 1-1 Paul Cézanne, *Bay of Marseilles Seen from L'Estaque*, 1886–90. The Art Institute of Chicago.

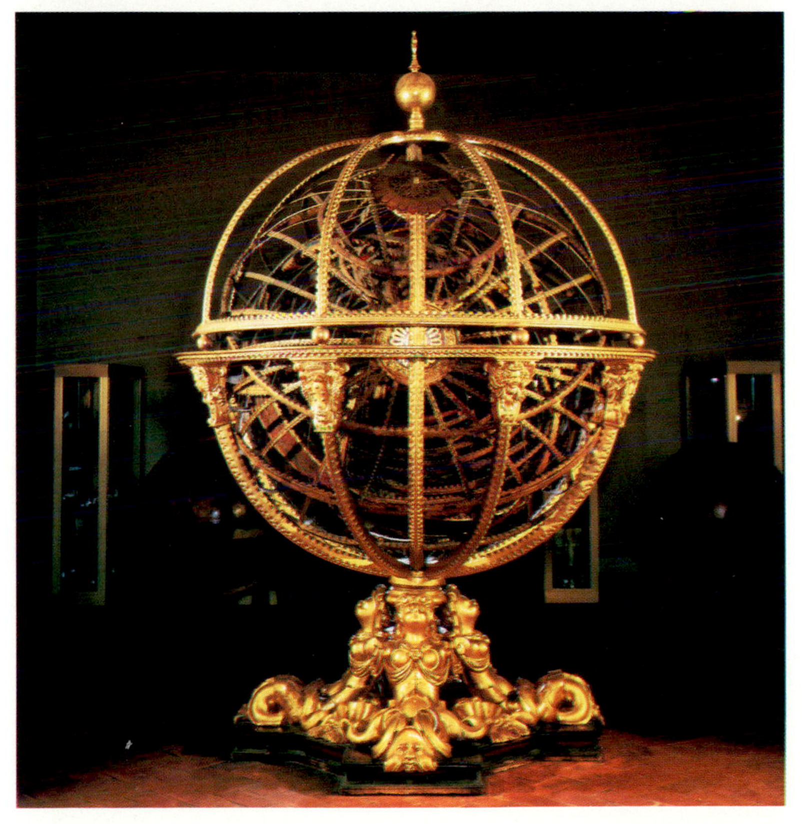

FIGURE 2-5(a) Armillary sphere. A three-dimensional geocentric model of the universe containing the spheres of the seven "wanderers," the fixed stars, and the prime mover (God), centered on the earth. The model, built by Santucci dalle Pomerance from 1588–93, is 3.5 meters ($\approx$ 12 feet) high and could be turned by a crank to show the motions of the heavenly bodies. (Courtesy of the Museum of the History of Science, Florence.)

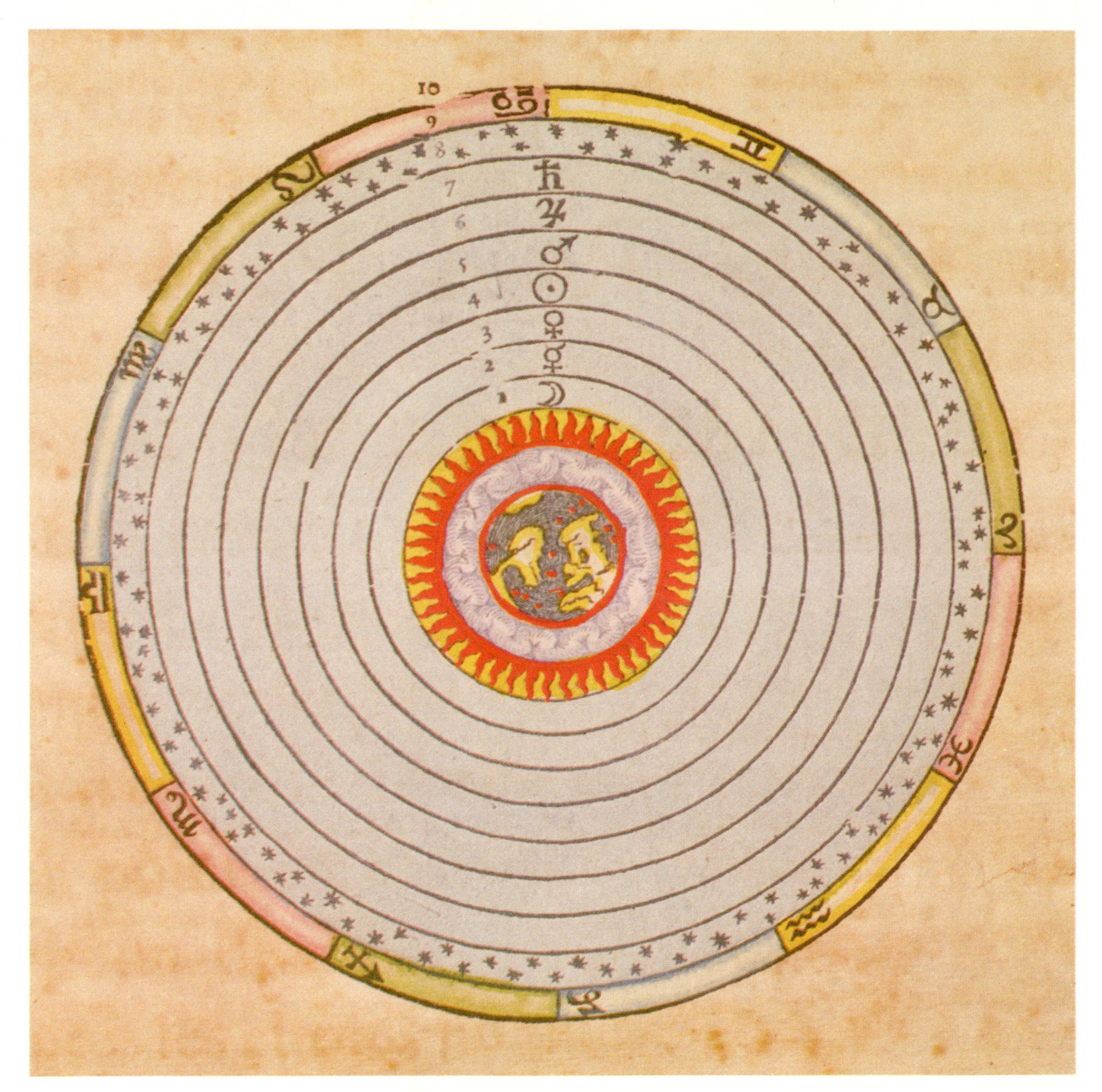

FIGURE 2-5(b) Geocentric view of the universe. Note at the center the four elements of the ancients: earth, water, air (clouds around the earth), and fire; then the circles, with symbols, for Moon, Mercury, Venus, Sun, Mars, Jupiter, Saturn, fixed stars, and signs of the zodiac. (From Jean Blaeu, *Geographie, qui est la Premiere Partie de la Cosmographie Blaviane,* 1667. Courtesy of the Museum of the History of Science, Florence.)

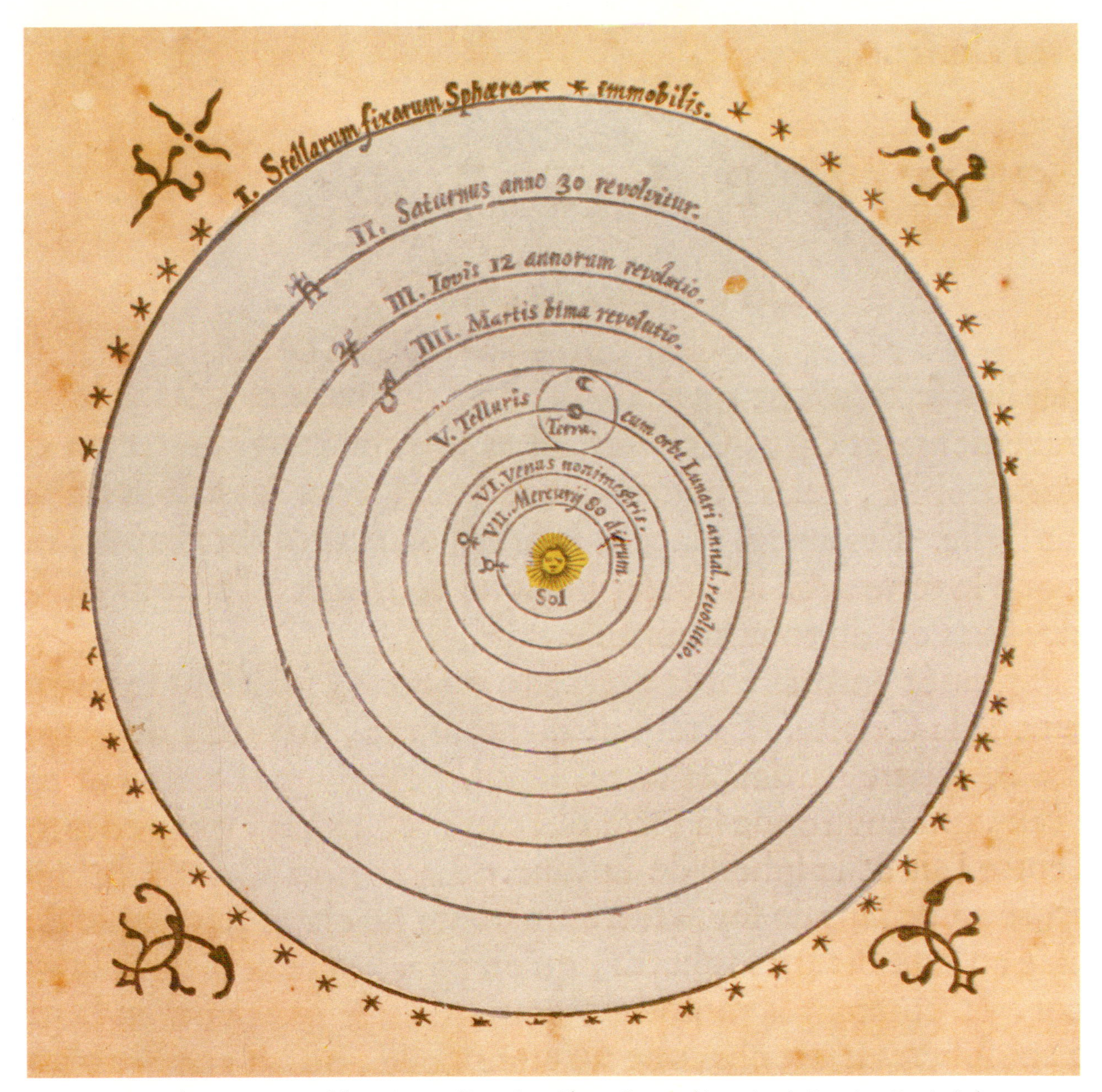

FIGURE 2-7 Heliocentric view of the universe. (From Jean Blaeu, *Geographie, qui est la Premiere Partie de la Cosmographie Blaviane,* 1667. Courtesy of the Museum of the History of Science, Florence.)

FIGURE 10-12 Greek temple (Agrigento, Sicily).

FIGURE 10-15 Roman arch (Roman Forum, Rome).

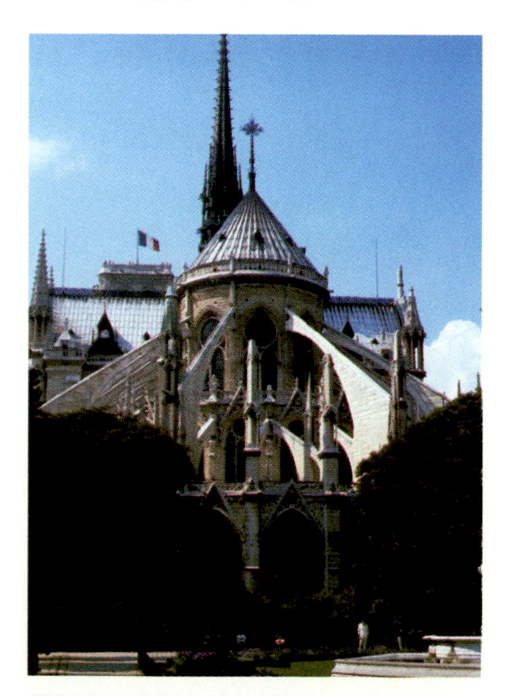

FIGURE 10-18 Flying buttresses that support the walls of Notre Dame cathedral, Paris. They are lighter than the buttressing required for churches built with round arches.

FIGURE 10-21(b) The great dome of Brunelleschi, cathedral of Florence [see also Figure 10-21(a) in the text].

FIGURE 13-1 These exquisite and sensitive thermometers, built by the Accademia del Cimento (1657–1667) in Florence, are among the earliest known. They contained alcohol, sometimes colored red for easier reading, like many thermometers today. (Courtesy of the Museum of the History of Science, Florence.)

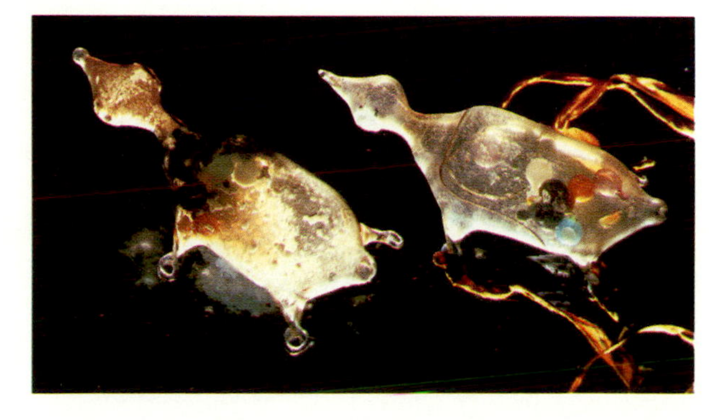

FIGURE 13-2 These clinical thermometers in the shape of a frog, also built by the Accademia del Cimento, could be tied to a patient's wrist. The small spheres suspended in the liquid each have a slightly different density; the number of spheres that sank were a measure of the patient's temperature. (Courtesy of the Museum of the History of Science, Florence.)

FIGURE 23-18(b) This seventeenth-century Galilean-type microscope contains three lenses, and could be moved up and down in its support ring. (Courtesy of the Museum of the History of Science, Florence.)

FIGURE 23-19 (a) Objective lense, mounted in a carved ivory frame, from the telescope with which Galileo made his world-shaking observations including the discovery of the moons of Jupiter (see Chapter 2). (b) Telescopes made by Galileo. (Courtesy of the Museum of the History of Science, Florence.)

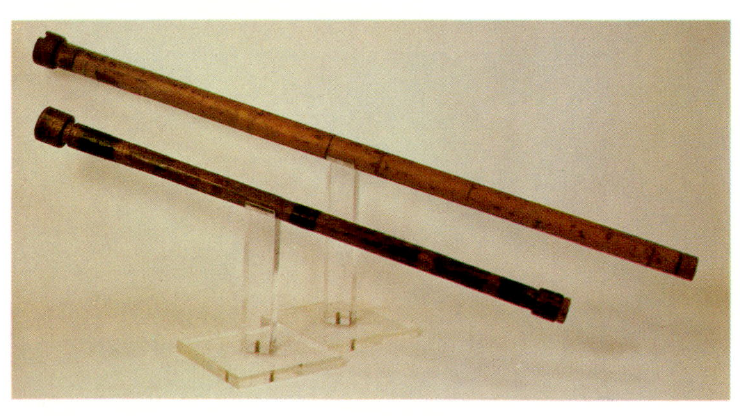

FIGURE 24-8 The spectrum of visible light. Wavelengths in mμ (millimicrons) are the same as nm (nanometers).

FIGURE 24-21 Mixing of the subtractive primaries. Here light is passed through colored filters. Each filter subtracts a portion of the spectrum; when filters overlap, additional wavelengths are subtracted out. The result is subtractive mixing, as for pigments.

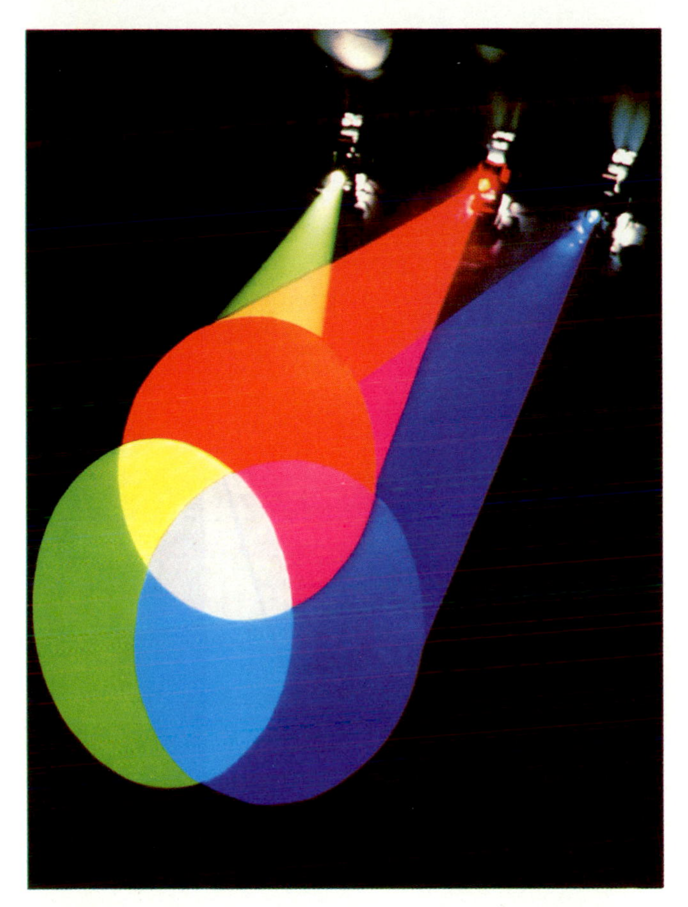

FIGURE 24-23 Mixing the additive primaries by shining colored lights on a screen.

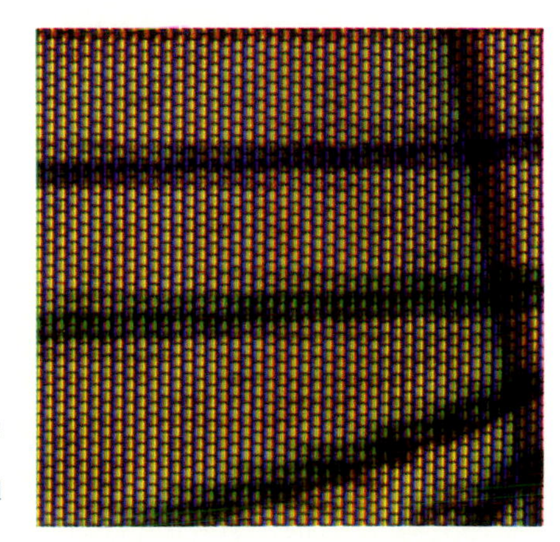

FIGURE 24-24 Close-up of a color television screen showing the vertical lines that alternate among the three primary colors: red, green, blue. From any distant point, the color we see is determined by the relative brightnesses of the three phosphors.

FIGURE 24-26 Claude Monet, *La Grenouilliere* ("The Frog Pond"), 1869. Metropolitan Museum of Art, New York.

FIGURE 32-2 Galaxies (a) in Andromeda, (b) near Cygnus and Cepheus.

equation, $Q = mc\ (T - T_0)$, where Q is the heat absorbed or given off, $(T - T_0)$ is the temperature rise or decline, and m is the mass of the substance. When heat flows within an isolated system, the heat gained by one part of the system is equal to the heat lost by the other part of the system.

An exchange of energy occurs without a change in temperature whenever a substance changes phase; this happens because the potential energy of the molecules changes as a result of the changes in the relative positions of the molecules. The *heat of fusion* is the heat required to melt 1 kg of a solid into the liquid phase; it is also equal to the heat given off when the substance changes from liquid to solid. The *heat of vaporization* is the energy required to change 1 kg of a substance from the liquid to the vapor phase; it is also the energy given off when the substance changes from vapor to liquid.

Heat is transferred from one place (or body) to another in three different ways. By *conduction*, energy is transferred from higher-KE molecules to lower-KE neighboring molecules when they collide. *Convection* is the transfer of energy by the mass movement of molecules over considerable distances. *Radiation*, which does not require the presence of matter, is energy transfer by electromagnetic waves, such as from the sun. All bodies radiate energy in an amount that is proportional to the fourth power of their Kelvin temperature. The energy radiated (or absorbed) also depends on the nature of the surface (dark and absorbing versus brightly reflecting).

QUESTIONS

1. What happens to the work done when a container of orange juice is shaken vigorously?
2. From the point of view of kinetic theory, describe how a hot oven heats food.
3. Why would you expect the water temperature at the bottom of a waterfall to be slightly higher than at the top?
4. When a hot object warms a cooler object, does temperature flow between them? Are the temperature changes of the two objects equal?
5. Does an electric fan actually cool the air? Why or why not? If not, why use it?
6. In warmer areas where the tropical plants grow yet the temperature drops below freezing a few times each winter, the destruction of sensitive plants due to freezing can be reduced by watering them in the evening. Explain.
7. The specific heat of water is quite large. Explain why this fact makes water particularly good for heating systems (that is, hot-water radiators).

8. Why does water in a canteen stay cooler if the cloth jacket surrounding the canteen is kept moist?
9. Why does water condense on the inside of windows on a cold day?
10. Why do you feel cool when you step out of a swimming pool into a warm breeze?
11. You can determine the wind direction by wetting your finger and holding it up in the air. Explain.
12. When your body overheats, it perspires. What is the value of this process?
13. Briefly describe what happens at the atomic level when alcohol at room temperature is heated to its boiling point of 78°C.
14. Alcohol evaporates more quickly than water at room temperature. What can you infer about the molecular properties of alcohol relative to those of water?
15. Explain why skin burns caused by steam are often so severe.
16. Will potatoes cook faster if the water is boiling faster?
17. A house can be kept cooler on a hot sunny day by closing the windows and curtains. Explain.
18. Heat loss occurs through windows (a) by drafts through cracks around the edges (b) through the frame, particularly if it is metal (c) through the glass panes. For each of these, which mechanisms are involved: conduction, convection, and/or radiation? Which of these will heavy curtains reduce?
19. If you hold a Kleenex tissue over a hot plate, why does the tissue wave up and down? (Try it!)
20. Sea breezes often occur at the shore of a large body of water. Explain, using the fact that as a result of the sun's radiation, the temperature of the land rises more rapidly than that of the water.
21. At the same temperature, why does a tile floor seem so much cooler to your bare feet than a carpet?
22. Down parkas and sleeping bags are often specified as so many centimeters or inches of "loft," which refers to the actual thickness of the garment when fluffed up. Why is tthis term used?
23. Why is a down parka useless if it becomes soaked?
24. Why are air temperature readings always taken with the thermometer in the shade?
25. Why do Bedouins wear several layers of clothes in the desert even when the temperature reaches 50°C (122°F)?
26. Explain why cold drafts are so uncomfortable.
27. A premature baby in an incubator can be dangerously cooled even when the air inside is warm. How can this be?

28. Why does the earth cool at night more quickly when the sky is clear than when it is cloudy?
29. The amount of heat needed to heat a room whose windows face north is much greater than is needed to heat a room whose windows face south. Explain.
30. Why does dirty snow melt faster than clean snow?
31. A thermos consists of a bottle with two glass walls with shiny silvered surfaces separated by a vacuum. Explain how this construction reduces conduction, convection, and radiation.

EXERCISES

1. How much heat (joules) is required to raise the temperature of 2.0 kg of water from 0°C to 20°C?
2. To what temperature will 9500 J of work raise 2.0 kg of water initially at 10.0°C?
3. How much work must a person do to offset eating a 400-Cal piece of cake?
4. A British thermal unit (Btu) is a unit of heat in the British system of units. One Btu is defined as the heat needed to raise 1 pound of water by 1 F°. Show that 1 Btu = 0.252 kcal = 1055 J.
5. A water heater can generate 8500 kcal/hr. How much water can it heat from 10°C to 60°C per hour?
6. How many kilocalories of heat are generated when the brakes are used to bring a 1400-kg car to rest from a speed of 80 km/hr?
7. If coal gives off 7000 kcal/kg when it is burned, how much coal will be needed to heat a house that requires 4.2×10^7 kcal for the entire winter? Assume that an additional 30 percent of the heat is lost up the chimney.
8. How many kilocalories of heat are given off when 2 kg of water cools from 40°C to 10°C?
9. Show that the specific heat capacity of water, in British units, is 1 Btu/lb·F°. (See Exercise 4.)
10. An automobile cooling system holds 12 liters of water. How much heat does it absorb if its temperature rises from 20°C to 70°C?
11. What is the specific heat capacity of a metal substance if 36 kcal of heat is needed to raise 4.0 kg of the metal from 20°C to 32°C?
12. What is the temperature rise of 100 kg of iron if 9.5×10^6 J of heat is transferred to it?
13. How many kilograms of copper will undergo the same temperature rise as 10 kg of water when the same amount of heat is absorbed?

14. How much heat is required to boil 4 liters of water at 100°C?
15. How much heat is required to heat 5 liters of water from 20°C to steam at 100°C?
16. A person gives off 100 kcal of heat by evaporation of water from the skin. How much water (in kilograms) was lost?
17. How much energy does a refrigerator have to remove from 1.5 liters of water at 20°C to make ice at −12°C?
18. If 1.70×10^5 J of energy is supplied to a flask of oxygen at −183°C, how much oxygen will evaporate?

*19. If a person eats 3000 Cal in a day, what must his average power output be (in watts) so no weight is gained?

*20. What will be the final result when equal amounts of ice at 0°C and steam at 100°C are mixed together?

THE LAWS OF THERMODYNAMICS

CHAPTER

15

Thermodynamics is the name we give to the study of processes where energy is transferred as heat and as work. In Chapter 7 we saw that work is done when energy is transferred from one body to another by mechanical means. In Chapter 14 we saw that heat is a transfer of energy from one body to a second body that is at a lower temperature. Thus heat is much like work. To distinguish them, *heat* is defined as a *transfer of energy due to a difference in temperature*; work is defined as a *transfer of energy not caused by a difference in temperature*.

In discussing thermodynamics we shall often refer to a particular system. A *system* is any object or set of objects we wish to consider. Everything else in the universe is called the "environment." There are several kinds of systems. A *closed system* is one whose mass is constant. In an *open system*, mass may enter or leave. Many systems we study in physics are closed systems; but many systems, including plants and animals, are open systems since they exchange materials with the environment (food, oxygen, waste products). A closed system is said to be *isolated* if no energy in any form passes across its boundaries.

1. THE FIRST LAW OF THERMODYNAMICS

In Chapter 14 we defined the internal energy of a system as the sum total of all the energy of the molecules of the system. We would expect the internal energy of a system to be increased either by doing work on the system or by adding heat to it, and to be decreased if heat flows out of the system or if work is done by the system on something else. Thus, it is reasonable to propose an important law: *the change in internal energy of a closed system is equal to the heat added to the system minus the work done by the system.* In symbols we can write

$$\Delta U = Q - W$$

where ΔU stands for the change in internal energy ("Δ" is the Greek delta and stands for "change in"), Q is the heat *added* to the system, and W is the work done *by* the system. Note that if work is done *on* the system, W will be negative and the internal energy will increase.† Similarly, if heat leaves the system Q is negative. This relation is known as the **first law of thermodynamics**. It is one of the great laws of physics, and its validity rests on experiments (such as Joule's) in which no exceptions have been seen. Since Q and W represent energy transferred into or out of the system, the internal energy changes accordingly; thus the first law of thermodynamics is a statement of the *law of conservation of energy*. It is worth noting that the conservation of energy law was not formulated until the nineteenth century, for it depended on the interpretation of heat as a transfer of energy.

The first law of thermodynamics as written above applies to a closed system. It also applies to an open system if we take into account the change in internal energy due to the increase or decrease in the amount of matter.

As an example of the first law, let us briefly analyze the boring-out of a cannon barrel (discussed in Section 1 of Chapter 14). The work done in turning the cutting tool goes into increasing the internal energy of the barrel. At the same time, some of the increased internal energy of the cannon flows from the cannon to the cooling water, and because this flow of energy is due to a temperature difference we say that it is a flow of heat. Thus the work done on the cannon goes partly into increasing the internal energy of the cannon and partly into heat flow out of the cannon.

2. HUMAN METABOLISM AND THE FIRST LAW

Human beings and other animals do work. Work is done when a person walks or runs, or lifts a heavy object. Work requires energy. Energy is also needed for growth—to make new cells and to replace old ones. A great many energy-transforming processes occur within an organism, and they are referred to as *metabolism*.

We can apply the first law of thermodynamics to an organism, say the human body. Work W is done by the body, resulting in a decrease in the body's internal energy (and temperature) that must be replenished. The body's internal energy is not, however, maintained by a flow of heat Q into

† Of course we could have defined W as the work done *on* the system, in which case there would be a plus sign in our equation. However, it is conventional to define W and Q as we have.

the body. Normally the body is at a higher temperature than its surroundings, so heat usually flows *out* of the body. Even on a very hot day when heat is absorbed, the body has no way of utilizing this heat to support its vital processes. What then is the source of energy? It is the internal energy (chemical potential energy) stored in foods. In a closed system the internal energy can change only as a result of heat flow or work done; in an open system, such as a living organism, internal energy itself can flow into or out of the system. When we eat food we are bringing internal energy directly into our bodies, which thus increases the total internal energy in our bodies. This energy eventually goes into work and heat flow from the body according to the first law.

The "metabolic rate" is the rate at which internal energy is transformed within the body. It is usually specified in kcal/hr or in watts. Typical metabolic rates for a variety of human activities are given in Table 15-1 for an "average" 65-kg adult. As an example, consider how much energy is transformed in 24 hr by a 65-kg person who spends 8.0 hr sleeping, 1.0 hr at moderate physical labor, 4.0 hr in light activity, and 11.0 hr working at a desk or relaxing. Table 15-1 gives the metabolic rate in watts (J/s). Since there are 3600 s in an hour, the total energy transformed is $[(8.0\ \text{hr})(70\ \text{J/s}) + (1.0\ \text{hr})(480\ \text{J/s}) + (4.0\ \text{hr})(230\ \text{J/s}) + (11.0\ \text{hr})(115\ \text{J/s})]$ $[3600\ \text{s/hr}] = 1.16 \times 10^7\ \text{J}$. Since $4.18 \times 10^3\ \text{J} = 1\ \text{kcal}$, this is equivalent to 2800 kcal; so a food intake of 2800 Cal would compensate for this energy output.

TABLE 15-1
Metabolic rate for 65-kg human being

ACTIVITY	APPROXIMATE METABOLIC RATE (WATTS)
Sleeping	70
Sitting upright	115
Light activity (eating, dressing, household chores)	230
Moderate work (tennis, walking)	480
Running (15 km/hr)	1150
Bicycling (race)	1250

3. THE SECOND LAW OF THERMODYNAMICS

The first law of thermodynamics states that energy is conserved. There are, however, many processes we can imagine that conserve energy but are not observed to occur in nature. For example, when a hot object is placed in contact with a cold object, heat flows from the hotter one to the colder one, never the reverse. If heat were to leave the colder one and pass to the hotter one, energy would still be conserved. Yet it doesn't happen. As a second example, consider what happens when you drop a rock and it hits the ground. The initial potential energy of the rock changes to kinetic as the rock falls and when the rock hits the ground this energy in turn is transformed into internal energy of the rock and the ground in the vicinity of the impact; the molecules move faster and the temperature rises slightly. But have you seen the reverse happen—a rock at rest on the ground (Figure 15-1) suddenly rise in the air because the thermal energy of molecules is transformed into kinetic energy of the rock as a whole? Energy would be conserved in this process, yet we never see it happen.

There are many other examples of processes that occur in nature but whose reverse do not. Here are two more. If you put a layer of salt in a jar and cover it with a layer of pepper, when you shake it you get a thorough mixture; no matter how long you shake it, the mixture is unlikely to

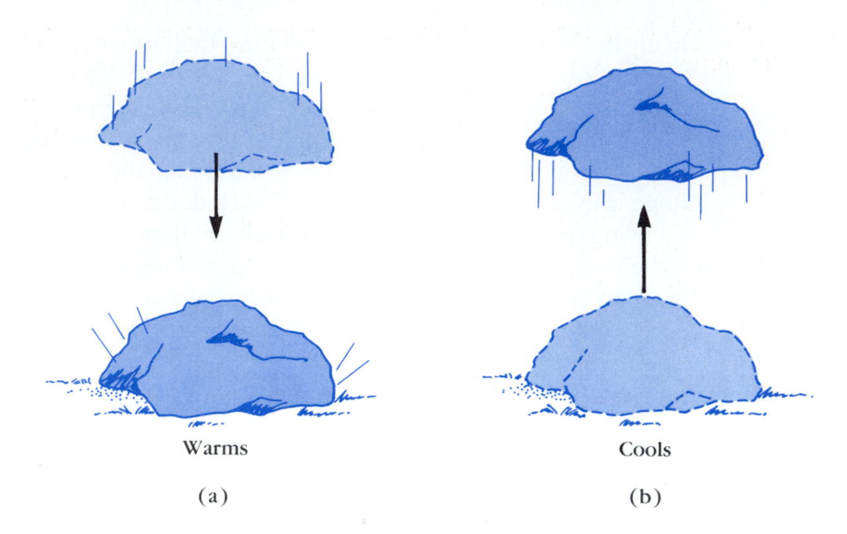

FIGURE 15-1
(a) A rock falls and its kinetic energy is transformed into thermal energy upon hitting the ground. Energy is conserved. (b) If the rock were suddenly to cool and the thermal energy be transformed into kinetic energy, the rock would rise. Energy would be conserved. Does this ever happen?

separate into two layers again. Coffee cups and glasses break spontaneously if you drop them; but they don't go back together spontaneously.

The first law of thermodynamics would not be violated if any of these processes occurred in reverse. To describe this lack of reversibility, scientists in the latter half of the nineteenth century formulated a new principle known as the **second law of thermodynamics**. This law is a statement about which processes occur in nature and which do not. It can be stated in a variety of ways, all of which are equivalent. One statement, given by R. J. E. Clausius (1822–1888), is that *heat flows naturally from a hot object to a cold object; heat will not flow spontaneously from a cold object to a hot object*. Since this statement applies to one particular process, it is not obvious how it applies to other processes. A more general statement is needed that will include other possible processes in a more obvious way. This is no easy task, for the second law of thermodynamics is a strange law, perhaps the most abstract in all of physics. Yet it is one of the most interesting and has stimulated a great deal of philosophical discussion.

In the following few sections we investigate how a general unifying statement of the second law was developed, and we examine some of the ramifications of this law.

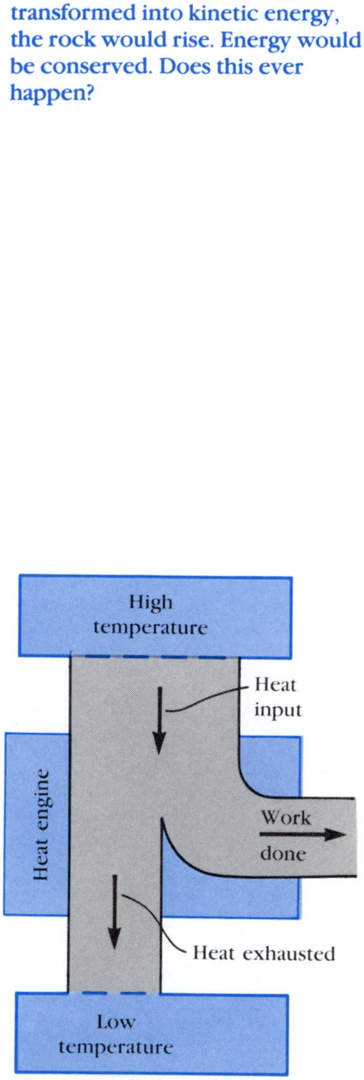

High temperature
Heat input
Heat engine
Work done
Heat exhausted
Low temperature

FIGURE 15-2
General principle behind any heat engine. When heat is allowed to flow from one substance at a high temperature to another at a low temperature, part of this heat can be turned into work.

4. HEAT ENGINES

The development of the second law of thermodynamics was prompted by studies on the changing of mechanical energy into heat and the reverse, the changing of heat or thermal energy into mechanical energy. Changing mechanical energy into heat is easily accomplished, as shown, for instance, in Figure 15-1(a), or by simply rubbing your hands together. In any process in which friction is present, mechanical energy is transformed into thermal energy. But what about the reverse process? Is it possible to change thermal energy into mechanical energy that can be used to do mechanical work? We know that this doesn't happen spontaneously, as in Figure 15-1(b); but is there some other way it can be done?

It *is* possible to get mechanical energy from thermal energy, and the devices that do so are called **heat engines**. Common examples are the steam engine and the internal combustion engine.

Early in the nineteenth century the French engineer and physicist Sadi Carnot (1796–1832) investigated the process of transforming heat into mechanical energy and recognized that there are natural limitations on the work that can be done by a heat engine. He realized that mechanical energy can be obtained from heat only when heat is allowed to flow from a substance at high temperature to one at a lower temperature; some of this heat can then be diverted into mechanical work, as diagrammed in Figure 15-2. The high and the low temperatures are called the *operating temperatures* of the engine. We will be concerned only with engines that run in a cycle, and thus run continuously.

The operation of two practical engines, the steam engine and the internal combustion engine (used in most automobiles), are illustrated in Figures 15-3 and 15-4. Steam engines are of two principal types. In the reciprocating type—Figure 15-3(a) and (b)—the heated steam passes through the intake valve and expands against a piston, forcing it to move;

FIGURE 15-3
The steam engine:
(a) and (b) reciprocating types;
(c) steam turbine.

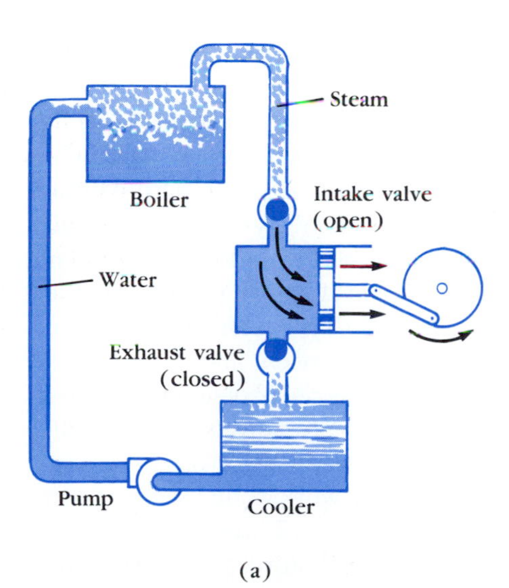

(a)

Expansion stroke. High-pressure steam expands against piston, driving wheel around.

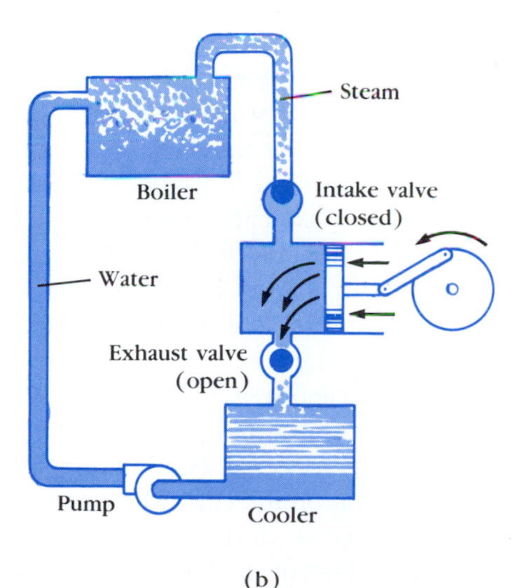

(b)

Exhaust stroke. Steam, now at low pressure, is pushed out so that new steam can enter cylinder as in (a). Expanded steam is cooled and pumped back to boiler to be heated and compressed.

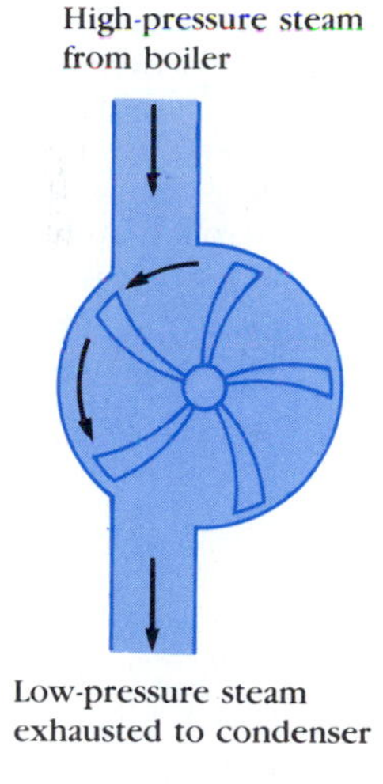

(c)

Turbine (boiler condenser not shown).

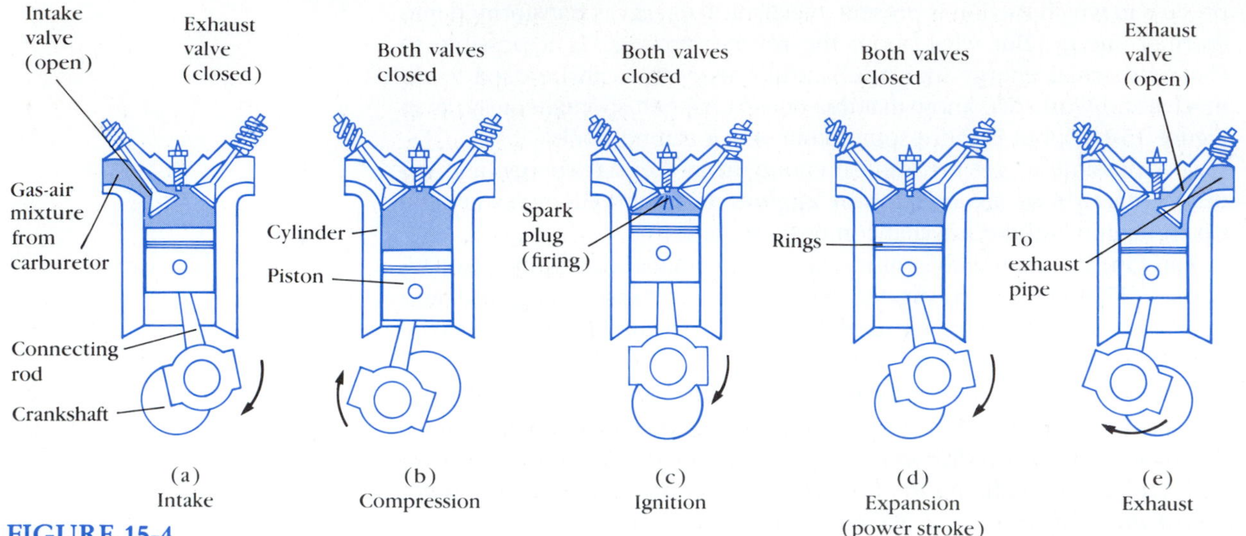

FIGURE 15-4
Four-cycle internal combustion engine. (Only one cylinder is shown.)
(a) The gasoline-air mixture flows into the cylinder as the piston moves down; (b) the piston moves upward and compresses the gas; (c) firing of the spark plug ignites the gasoline-air mixture, raising it to a high temperature; (d) the gases, now at high temperature and pressure, expand against the piston in this, the power stroke; (e) the burned gases are pushed out to the exhaust pipe. The intake valve then opens, and the whole cycle repeats.

as the piston returns to its original position, it forces the gases out the exhaust valve. In a steam turbine—Figure 15-3(c)—everything is essentially the same except that the reciprocating piston is replaced by a rotating turbine; this resembles a paddlewheel but usually has several sets of blades. Most electricity today is generated by using steam turbines.† In a steam engine, the high temperature is obtained by burning coal, oil, or other fuel to heat the steam. In an internal combustion engine, the high temperature is achieved by burning the gasoline-air mixture in the cylinder itself (ignited by the spark plug).

To see why a *temperature difference* is required to run an engine, let us examine the steam engine. In the reciprocating engine, for example Figure 15-3(a), suppose there were no condenser or pump and that the steam was at the same temperature throughout the system. This would mean that the pressure of the gas being exhausted would be the same as that on intake. Thus, although work would be done by the gas *on* the piston when it expands, an equal amount of work would have to be done *by* the piston to force the steam out the exhaust; hence no net work would be done. In a real engine, the exhausted gas is cooled to a lower temperature and condensed so that the exhaust pressure is less than the intake pressure. Thus, although the piston must do work on the gas to expel it on the exhaust stroke, it is less than the work done by the gas on the piston during the intake. So a net amount of work can be obtained—but only if there is a difference in temperature. Similarly, in the gas turbine if the gas were not cooled, the pressure on each side of the blades would be the same; by cooling the gas on the exhaust side,

† Even nuclear power plants utilize steam turbines; the nuclear fuel—uranium—merely serves as fuel to heat the steam.

the pressure on the front side of the blade is greater and therefore the turbine turns.

In any heat engine, heat must flow out to the atmosphere; this is the heat released at the lower temperature, as shown in Figure 15-2. In the internal combustion engine the heat is released in two ways: it is carried away in the exhaust by the burned gases, and it is carried away by the engine coolant which releases it to the atmosphere through the radiator. The schematic drawing of Figure 15-2 applies to all heat engines; thermal energy is taken in at a high temperature, some of it is transformed into mechanical energy, and the rest is exhausted at the lower temperature.

Efficiency of Engines

The efficiency of a heat engine is defined as the ratio of the work done to the heat input (see Figure 15-2). That is, the more work an engine can do for a given amount of heat input, the greater its efficiency. Carnot showed that the maximum efficiency of any heat engine is related to the operating temperatures of the engine. The relation he derived was:

$$\text{maximum efficiency} = (\text{work output})/(\text{heat input}) = 1 - T_L/T_H$$

where T_H is the kelvin temperature of the high-temperature source of heat and T_L is the low temperature (kelvins) of the exhaust heat. Clearly, the higher the input temperature and the lower the exhaust temperature the greater the efficiency. For example, a steam engine at a power plant may take in steam at 400°C and exhaust it at 200°C; to find the maximum possible efficiency we must first change these to kelvins: $T_H = 400 + 273 = 673$ K and $T_L = 200 + 273 = 473$ K. Then the efficiency is $1 - \frac{473\text{ K}}{673\text{ K}} = 1 - 0.70 = 0.30$; or, as a percentage (we multiply by 100) it is 30 percent efficient. The efficient given by this formula is the *maximum* that is consistent with the laws of thermodynamics. Well-designed real engines reach 60 to 80 percent of this ideal value. Thus, the engine in our example would have an actual efficiency of perhaps 18 to 25 percent.

It is quite clear from the efficiency relation (max. eff. $= 1 - T_L/T_H$) that at normal temperatures a 100-percent efficient engine is not possible. Only if the exhaust temperature, T_L, were at absolute zero could 100-percent efficiency be obtained; but this is a practical (as well as theoretical) impossibility.† Thus Carnot was able to state that *no device is possible whose sole effect is to transform a given amount of heat completely into work*. That is, there can be no perfect (100-percent efficient) heat engine such as the one diagrammed in Figure 15-5. This statement is another way of expressing the **second law of thermodynamics**.

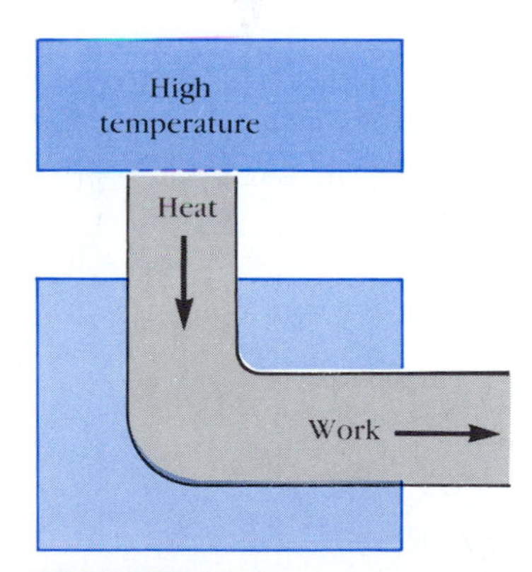

FIGURE 15-5
Schematic diagram of a hypothetical perfect heat engine in which all the heat input is used to do work. It is not possible to construct such a perfect heat engine.

† It seems, from careful experimentation, that absolute zero is unattainable. This result is known as the *third law of thermodynamics*.

FIGURE 15-6
Schematic diagram of the operation of a refrigerator or air conditioner.

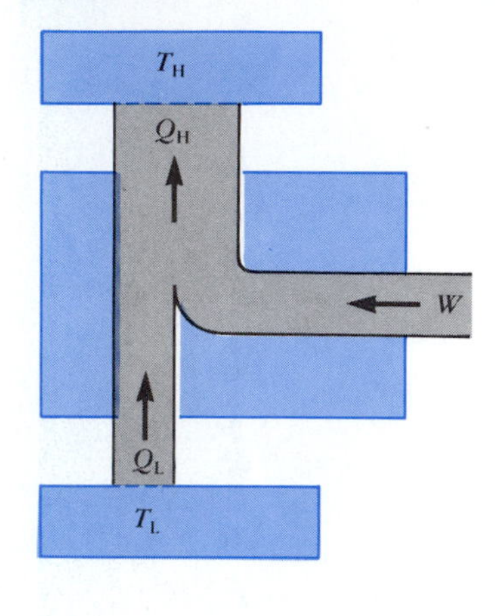

If the second law were not true, so that a perfect engine could be built, some rather remarkable things could happen. For example, if the engine of a ship did not need a low-temperature reservoir to exhaust heat into, the ship could sail across the ocean using the vast resources of the internal energy of the ocean water; indeed, we would have no fuel problems at all!

Refrigerators, Air Conditioners, and Heat Pumps

The operating principle of a **refrigerator**, or other **heat pump** (such as one to produce a flow of heat into or out of a house—the latter is called an air conditioner) is just the reverse of a heat engine. As diagrammed in Figure 15-6, by doing work W, heat is taken from a low-temperature region T_L (inside a refrigerator, say), and a greater amount of heat is exhausted at a high temperature T_H (the room). You can often feel this heat blowing out from beneath a refrigerator. The work W is usually done by a compressor motor which compresses a fluid, as illustrated in Figure 15-7. A perfect refrigerator—one in which no work is required to take heat from the low-temperature region to the high-temperature region—is not possible. This is another way to state the second law of thermodynamics. This was, in fact, implicit in our first statement of the second law of thermodynamics in Section 3: heat does not flow spontaneously from a cold object to a hot object. To accomplish such a task, work must be done. Thus, *there can be no perfect refrigerator*.

FIGURE 15-7
Typical refrigeration system.

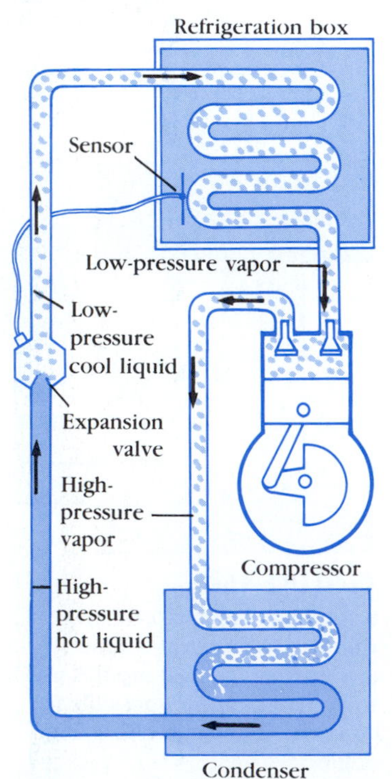

5. ORDER TO DISORDER: ENTROPY AND THE SECOND LAW OF THERMODYNAMICS

We have mentioned several different applications of the second law of thermodynamics. But we have still not found the general principle behind these examples. It was not until the end of the nineteenth century that a general statement of the **second law of thermodynamics** was finally achieved. It can be stated as follows:

> **natural processes tend to proceed toward a state of greater disorder.**

Exactly what we mean by disorder may not always be clear; to make it clear we now consider a few examples. Some of these examples will reveal how this general statement of the second law actually applies beyond what we usually consider as thermodynamics.

Let us first look at the simple processes mentioned in Section 3. A jar containing separate layers of salt and pepper is more orderly than when the salt and pepper are all mixed together. Shaking a jar containing separate layers results in a mixture, and no amount of shaking brings back the layers. The natural process is from a state of relative order (layers) to one of relative disorder (a mixture), not the reverse. That is, disorder increases. Similarly, a solid coffee cup is a more "orderly" object than the pieces of a broken cup. Cups break when they fall, but they do not spontaneously mend themselves. Again, the normal course of events is an increase of disorder.

When a hot object is put in contact with a cold object, heat flows from the high temperature to the low temperature until the two objects reach the same intermediate temperature. At the beginning of the process we can distinguish two classes of molecules: those with a high-average kinetic energy and those with a low-average kinetic energy. After the process, all the molecules are in one class with the same average kinetic energy, and we no longer have the more orderly arrangement of molecules in two classes. Order has gone to disorder. To see this more clearly, note that the separate hot and cold objects could serve as the hot- and cold-temperature regions of a heat engine and thus could be used to obtain useful work. But once the two objects are put in contact and reach the same temperature, no work can be obtained from them. Disorder has increased.

In general, we associate disorder with randomness: salt and pepper in layers is more orderly than a random mixture; a neat stack of numbered pages is more orderly than pages strewn randomly about on the floor.

The remaining example of those we discussed earlier was that of a stone falling to the ground, its kinetic energy being transformed to thermal energy. So long as the rock has kinetic energy, it is in a relatively ordered state. The molecules are all moving in the same direction, downward, and the rock is capable of doing some work—for example, driving a stake into the ground. After the rock hits the ground, the rock's energy increases the random disordered motion of molecules and the energy is no longer available to do work. Again, disorder increases.

Degradation of Energy

These last two examples of order going to disorder illustrate another very important aspect of the second law of thermodynamics—that thermal energy can be considered as a lower and less useful form of energy than mechanical energy. Mechanical energy can be used to do work. But after it is changed to thermal energy, the possibility of doing work is lost. Thus we can consider mechanical and other forms of energy as more ordered than thermal energy. In natural processes, energy is in a sense **degraded**, going from more orderly forms to less orderly forms, eventually ending up as thermal energy. This points to another important aspect of the second law of thermodynamics—that *in any natural process, some energy becomes unavailable to do useful work*.

Entropy: A Measure of Disorder

The quantitative measure of disorder is known as **entropy**. This new term enables us to state the second law of thermodynamics in yet another way:

the entropy of the universe continually increases.

Evolution and Growth

An interesting example of the increase in entropy (or disorder) relates to biological evolution and to the growth of organisms. Clearly, a human being is a highly ordered organism. The process of evolution from the early macromolecules and simple forms of life to *Homo sapiens* is a process of increasing order. So, too, the development of an individual from a single cell to a grown person is a process of increasing order. Do these processes violate the second law of thermodynamics? No, they do not. In the processes of evolution and growth, and even during the mature life of an individual, waste products are eliminated. These small molecules that remain as a result of metabolism are simple molecules without much order. Thus they represent relatively great disorder or entropy. Indeed, the total entropy of the molecules cast aside by organisms during the processes of evolution and growth is greater than the decrease in entropy associated with the order of the growing individual or evolving species.

Entropy As "Time's Arrow"

Another aspect of the second law of thermodynamics is that it tells us in which *direction* processes go. If you were to see a film being run backward, you would undoubtedly be able to tell that it *was* run backward. For you would see odd occurrences, such as a broken coffee cup rising from the floor and reassembling on a table or a torn balloon suddenly becoming whole again and filled with air. We know these things don't happen in real life; they are processes in which order increases—or entropy decreases. They violate the second law of thermodynamics. When watching a movie (or imagining that time could go backward), we are alerted to a reversal of time by observing whether entropy is increasing or decreasing. Hence, entropy has been called "time's arrow," for it tells in which direction time is going.

Heat Death

A prediction of the second law of thermodynamics is that as time goes on the universe will approach a state of maximum disorder. Matter will become a uniform mixture; heat will have flowed from high-temperature regions to low-temperature regions until the whole universe is at one temperature. No work can then be done. All the energy of the universe will have become degraded to thermal energy. All change will cease. This so-called *heat death* of the universe has been much discussed by philosophers. This final state seems an inevitable consequence of the second law of thermodynamics, although very far in the future. Yet it is based on the assumption that the universe is finite, which cosmologists are not really sure of. Furthermore, there is some question as to whether the second law of thermodynamics, as we know it, actually applies in the vast reaches of the universe. The answers are not yet in.

6. ENERGY RESOURCES AND THERMAL POLLUTION

When we speak in everyday life of energy usage, we are speaking of the transformation of energy from one form to another form that we want to use. For space heating of homes and buildings the direct burning of fuels such as gas, oil, or coal releases energy stored as potential energy of the molecules. For many transformations, however, a *heat engine* is required: to operate automobiles, aircraft, other similar vehicles, and, very importantly, to generate electricity.

We will discuss various means for producing electricity in a moment. But first we look at two types of pollution associated with heat engines: air pollution and thermal pollution. Air pollution can result from the burning of any fossil fuels (coal, oil, gas), such as industrial smelting furnaces and electric generating plants. The internal combustion engines of automobiles are especially polluting because the burning occurs so quickly that complete combustion does not take place and more noxious gases are thus produced. To help reduce air pollution, special devices must be used.

Another type of environmental pollution is **thermal pollution**. Every heat engine, from automobiles to power plants, exhausts heat to the environment (see Figure 15-2). Most electricity-producing power plants today make use of a heat engine to transform thermal energy into electricity, and the exhaust heat is generally absorbed by a coolant such as water. If the engine is run efficiently (at best, 30 to 40 percent today), the low temperature T_L (see subsection on efficiency in Section 4) must be kept as low as possible. Hence a great deal of water must flow as coolant through a power plant. The water is often obtained from a nearby river or lake, or from the ocean. As a result of the transfer of heat to the water, its temperature rises. This can cause great damage to aquatic life in the vicinity, in large part because the warmed water holds less dissolved oxygen. The lack of oxygen can adversely affect fish and other organisms, and at the same time may encourage excessive growth of other (perhaps alien) organisms such as algae, thus disrupting the ecology of an area. Another way of exhausting waste heat is to discharge it into the atmosphere by means of large cooling towers. Unfortunately this method can also have environmental effects, for the heated air can alter the weather and climate of a region. Even if careful controls eventually reduce air pollution to an acceptable level, thermal pollution cannot be avoided. All we can do, in light of the second law of thermodynamics, is to use less energy and try to build more efficient engines. One hope is that eventually we can develop an engine that can transform nuclear energy, say, directly into electricity without using a heat engine. This would greatly improve efficiency and reduce pollution, but such an engine is not yet a real possibility.

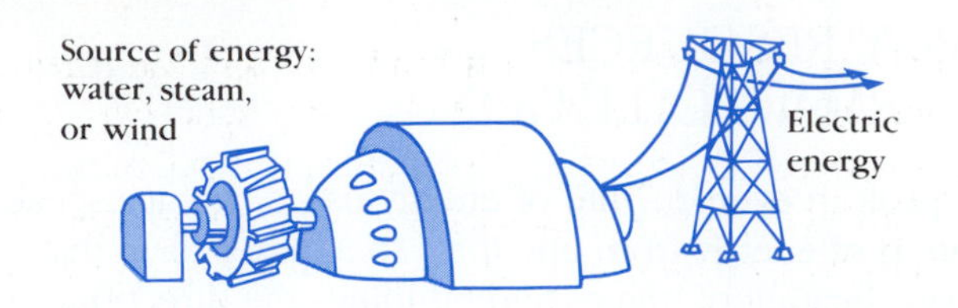

FIGURE 15-8
Basic plan of an electric generating plant.

Some Practical Aspects of Electric-Power Production

Almost all the electricity in the United States at the present time is the product of a heat engine coupled with an electric generator; there are a few other techniques in use, which we will also discuss in this section. Electric generators are devices that transform mechanical energy, usually rotational kinetic energy of turbines containing many blades (Figure 15-8), into electric energy. (How electric generators work is examined in Chapter 21.) The various means of driving the turbine will now be discussed, along with some of the advantages and disadvantages of each.

Fossil-fuel steam plants. At such a plant, coal, oil, or natural gas is burned to boil water and produce high-pressure steam that drives the turbine. The basic principles of this sort of steam engine were covered in Section 4. The advantages of such plants are that we know how to build them and they are not yet too expensive to operate. The disadvantages are that the products of combustion create air pollution; like all heat engines their efficiency is limited (30 to 40 percent is typical); the waste heat produces thermal pollution; the extraction of the raw materials can be devastating to the land, especially in coal strip mining and oil-shale recovery; accidents occur, such as oil spills in the sea; and there may not be a great deal of fossil fuel left—estimates range from a supply lasting a few centuries to only a few decades.

Nuclear power. Two nuclear processes release energy: fission and fusion. In fission, the nuclei of uranium or plutonium atoms are made to split ("fission"), releasing energy in the process. In fusion, energy is released when small nuclei, such as those of hydrogen, combine ("fuse"); these processes are described in greater detail in Chapter 30. The fission process is used in all present nuclear power plants. Nuclear energy is used to heat steam just as are fossil fuels. A nuclear power plant is thus essentially a steam engine using uranium as its fuel; it suffers from the low efficiency characteristic of all heat engines and the accompanying thermal pollution. Although nuclear power plants produce almost no air pollution (unless an accident occurs), they do present problems: radioactive substances are produced that are difficult to dispose of; a serious accident could result in the release of radioactive material into the air; there is the possibility of acquisition of nuclear material by terrorists; and the fuel supply is limited. However, the energy produced per kilogram of fuel is very large and extraction is less damaging to the land than with fossil fuels.

FIGURE 15-9
Heat from earth's interior can produce steam to run a turbine generator.

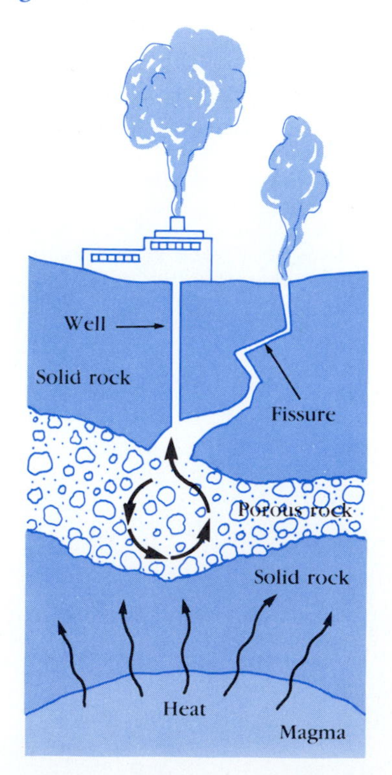

On the other hand, the fusion process has fewer disadvantages and enjoys the advantage of a vast supply of fuel—the hydrogen in the water molecules (H_2O) of the oceans. Unfortunately, this process cannot yet be controlled sufficiently, but holds promise for the future. (See Chapter 30.)

Geothermal power. Both fossil-fuel and nuclear plants heat water to steam for the steam turbine. Natural steam can be obtained from the earth itself. In many places, water beneath the ground is in contact with the hot interior of the earth and is raised to high temperature and pressure. It comes to the surface as hot springs, geysers, or steam vents. We can use not only natural vents but can also drill down to trapped steam beds (Figure 15-9). Already a large geothermal plant operates at the Geyers in northern California. A similar plant has been operating successfully for over 70 years in Italy, and a number of others are functioning in various places throughout the world. There are, however, a limited number of sites where high temperature and pressure exist naturally. Another possibility is to drill two parallel wells down to hot dry rock in contact with the earth's interior and to pass cold water under pressure down one well, the heated water (or steam) returning up the other well (Figure 15-10). Geothermal energy appears to be clean in that it produces little air pollution, although there is some (nonsteam) gas emission. There is also the thermal pollution of the spent hot water; and the mineral content of the water (often high) may not only be environmentally polluting but can be corrosive to the parts of the apparatus itself. Nonetheless, it is a reasonably inexpensive means of production, and may hold some promise.

FIGURE 15-10
A geothermal plant in which cold water is heated to steam by being passed over hot rock in the earth's interior.

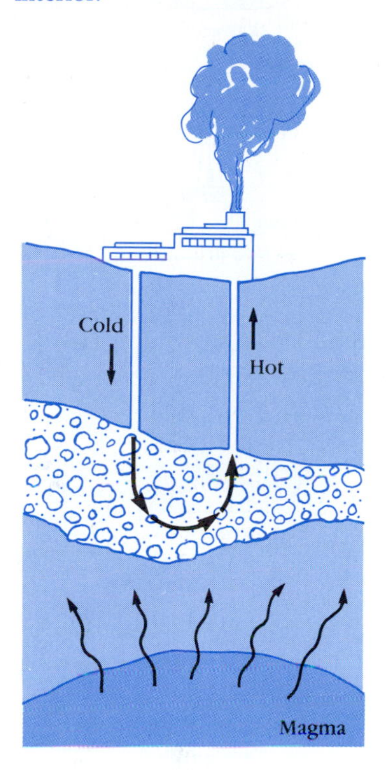

Tropical seas. There is a difference in temperature of perhaps 20°C between the surface and the depths of the sea in the tropics; although not yet tried, a number of designs have been suggested in which this temperature difference could be used to drive a heat engine. (Remember that a heat engine must exhaust heat at a lower temperature than that of the heat input.) The working fluid, of course, could not be steam; rather, it would have to be a substance with a lower boiling point that could drive the turbines. Before this method is tried, many difficulties must be overcome. The efficiency would be low because of the small temperature difference (at best, 7 percent); losses due to heat transfer must be kept very low, so the efficiency does not drop to zero. The ocean environment causes problems such as corrosion and fouling by biological organisms caught in the intake water. On a large scale, the ocean currents of the world could be affected. Yet there would be little problem with air pollution or radioactive disposal. So the advantages, if used on a limited scale, might outweigh disadvantages.

Hydroelectric power plants. Hydroelectric plants use falling water instead of steam to drive the turbines. They are usually located at the base of a dam. Hydroelectric power plants produce practically no air or water pollution; furthermore, they are nearly 100 percent efficient, since very little waste heat is produced. However, there remain few good locations at

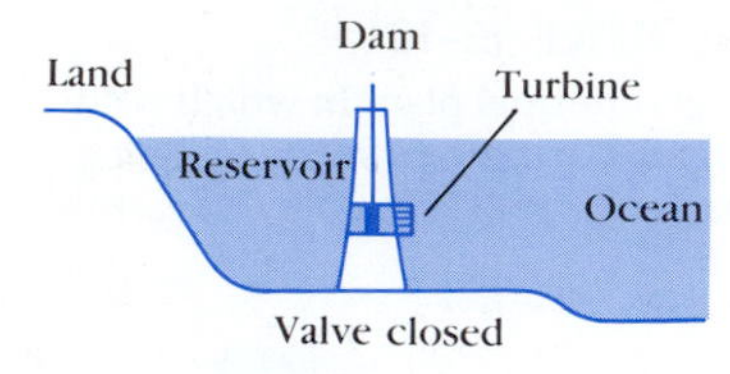

High tide

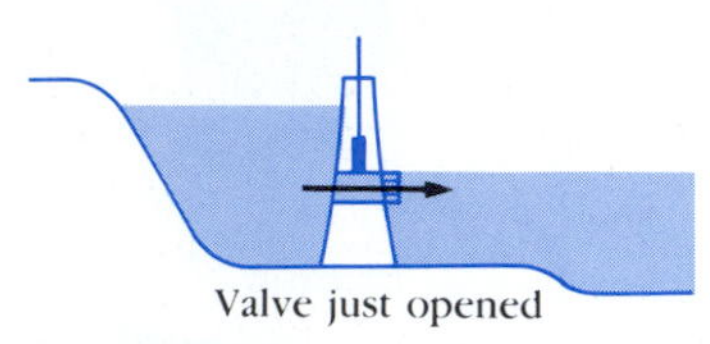

Low tide: Water is beginning to flow out of reservoir to ocean, driving turbines

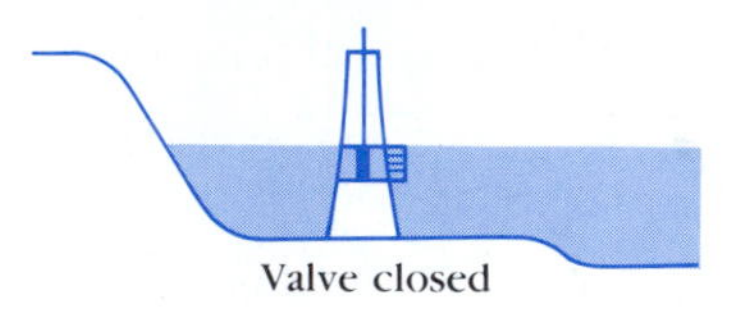

Water level equalized

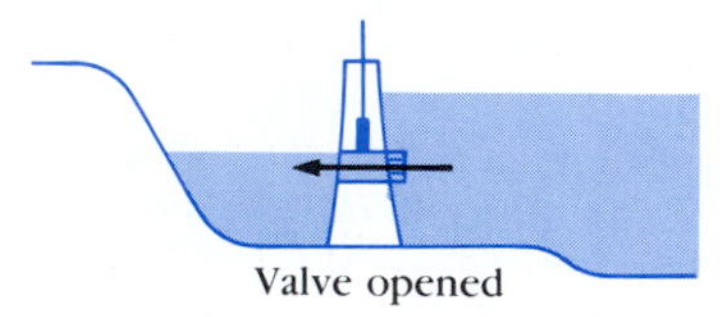

High tide: Water is allowed to flow back into reservoir, driving turbines

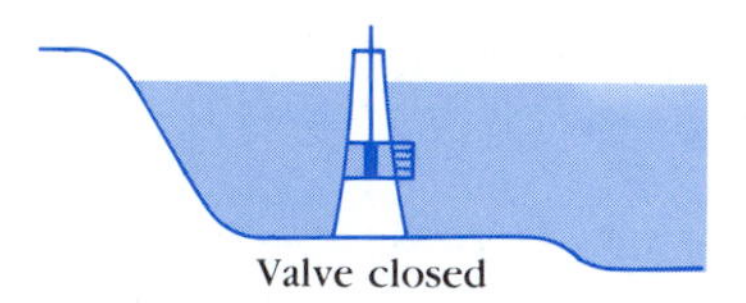

Water level equalized

FIGURE 15-11
Tidal power plant. Turbines are located inside dam.

which to build dams, and the resulting reservoirs inundate land that may be fertile or of great scenic beauty.

Tidal power. The earth's tides have been put to work at a plant in France. As shown in Figure 15-11, a basin behind a dam is filled at high tide and the water is released at low tide to drive turbines. At the next high tide the reservoir is filled again and the inrushing water also drives turbines. Good sites for tidal power (where there is a large difference between high and low tides) are not plentiful and would require large dams across natural or artificial bays. The rather abrupt changes of water level could have an effect on wildlife, but otherwise tidal power would seem to have a minimal environmental effect. Unfortunately, reasonable estimates of available sites indicate that tidal power could at best produce only a small fraction of the world's energy needs. Nonetheless, since our future energy needs may have to be met by a variety of small sources rather than mainly one as is the case now (fossil fuels), tidal power could become more important.

Wind power. Windmills were once very practical devices. Their resurrection—on a much grander scale—is a possibility as a means of turning a generator to produce electricity. One specific proposal projects some 300,000 of them, each over 800 ft high with blades 50 ft in diameter, dispersed throughout the midwestern United States where winds are strong and steady. Such an array might fullfill a good portion of United States energy needs. Of course, such a vast project might be something of an eyesore; it also might affect the weather. But wind power, on a modest scale, offers possibilities. Small projects already functioning in California show promise.

Solar power. Many kinds of solar energy are already in use: fossil fuels are the remains of plant life that grew by photosynthesis of light from the sun; hydroelectric power depends on the sun to evaporate water that later comes down as rain; and wind power relies on convection currents produced by the sun heating the atmosphere. But now let us see what we can do with direct sunrays. Figure 15-12 shows how absorption of the sun's energy by a black surface can be used to raise the temperature of water to heat a house. The sun's rays could be concentrated by large mirrors or lenses onto a small surface in order to produce temperatures high enough to produce steam to drive a turbine. Such a system could be used for a home generating system, although a backup system would probably be needed for cloudy days. On a grander scale, large areas of land would be required to collect sufficient sunlight. The principal disadvantages are that the sun does not shine brightly every day and not at all at night; thus a safe storage method would be necessary. Also, large areas of land are needed for the collectors and concentrators. Although the ubiquitous thermal pollution would exist and the climate might be affected, essentially there would be no air or water pollution, no radioactivity, and the technology would not be too difficult.

Another user of direct sunlight is the *solar cell*—or more correctly, *photovoltaic cell*—which was developed by the United States space program to convert sunlight directly into electricity. Unfortunately, they are very expensive, costing many times more per unit of power than ordinary means. However, if the cost should decrease sufficiently, thev would be very desirable, since thermal (and other) pollution would be very low. (Note that no heat engine is involved.) They might be placed on roofs for home use. On a larger scale they would again require a large land area because the sun's energy is not very concentrated. Another possibility is to place solar cells and concentrators in orbit around the earth. The sun's radiation would be greater there, before the atmosphere absorbs some of it, and the satellites would spend a minimal time in the earth's shadow. The electricity produced might be fed to a microwave generator and then transmitted to receivers on earth.

It should be clear that all forms of energy production have undesirable side effects. Some are worse than others, and not all problems can be anticipated. New forms of energy production will be required in the future as old reserves of fuel are used up. It is possible that many different methods—including those we have just discussed—will have to be used to satisfy the future needs of our energy-consuming society. The problems posed by any one form of energy production might not loom as large since there would be diverse fuel sources available. Finally, it is clear that conserving our limited fuel supplies by avoiding wasteful use of energy must be a high priority of our society.

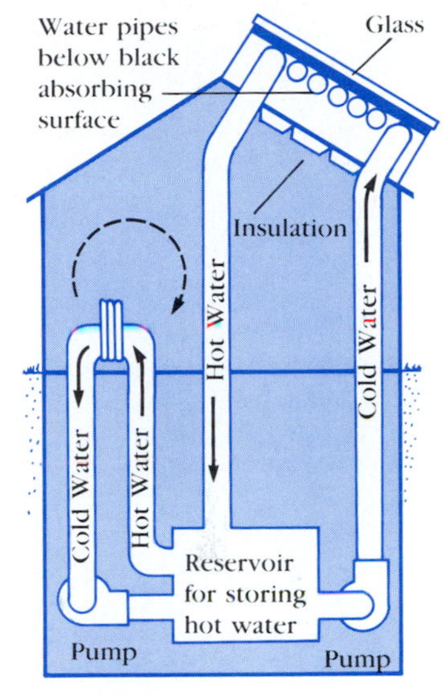

FIGURE 15-12
A solar-energy heating system for an individual house. On the roof are water-carrying tubes in contact with a large black surface that absorbs the sun's radiant energy and heats the water. The surface is covered with glass to prevent heat loss by convection and the tubes are well insulated to reduce heat loss by conduction. The heated water is then circulated to a large well-insulated reservoir, where it is stored and recirculated to heaters in the house. The reservoir can also supply hot water for other needs. A backup system is often necessary in climates with prolonged periods of cloudiness.

SUMMARY

The *first law of thermodynamics* states that the change in internal energy (ΔU) of a system is equal to the heat added to the system (Q) minus the work (W) done by the system: $\Delta U = Q - W$. This is a statement of the conservation of energy, and is found to hold for all types of processes.

A *heat engine* is a device for changing thermal energy into useful work by means of heat flow. The efficiency of a heat engine is defined as the ratio of the work done by the engine to the heat input to the engine. The upper limit of efficiency can be written in terms of the high and low operating temperatures (in kelvins) of the engine, T_H and T_L: $e = 1 - T_H/T_L$. The operation of refrigerators and air conditioners is the reverse of that of a heat engine: work is done to extract heat from a cold region and exhaust it to a region at a higher temperature.

The *second law of thermodynamics* can be stated in several equivalent ways: (1) heat flows spontaneously from a hot object to a cold object but not the reverse; (2) there can be no 100-percent efficient heat engine—that is, one that can change a given amount of heat completely into work;

(3) natural processes tend to move toward a state of greater disorder or greater entropy. *Entropy* is a quantitative measure of the disorder of a system. As time goes on, energy is degraded to less useful forms—that is, it is less available to do useful work.

Electrical energy is usually produced by a turbine that drives a generator. The turbine may be driven by wind (as by a windmill), by water (over a dam or in the tides), or most often by steam produced in a heat engine that burns fossil fuels or uses nuclear energy. The heat of the earth itself (geothermal) and the rays of the sun can also be used.

QUESTIONS

1. What happens to the internal energy of water vapor in the air that condenses on the outside of a cold glass of water? Is work done or heat exchanged? Explain.
2. Use the conservation of energy to explain why the temperature of a gas increases when it is compressed—say by pushing down on a cylinder—whereas the temperature decreases when the gas expands.
3. Is it possible for the temperature of a system to remain constant even though heat flows into or out of it? If so, give examples.
4. Is it possible to cool down a room on a hot summer day by leaving the refrigerator door open?
5. Can mechanical energy ever be transformed completely into heat or internal energy? Can the reverse happen? In each case, if your answer is no, explain why not; if yes, give examples.
6. The oceans contain a tremendous amount of thermal energy. Why, in general, is it not possible to put this energy to useful work?
7. A refrigerator takes heat from its cold interior and transfers it to the warmer exterior. Why is this not a violation of the second law of thermodynamics?
8. Give three examples, other than those mentioned in this chapter, of naturally occurring processes in which order goes to disorder. Discuss the observability of the reverse process.
9. Which do you think has the greater entropy, 1 kg of solid iron or 1 kg of liquid iron? Why?
10. What happens if you remove the lid of a bottle containing chlorine gas? Does the reverse process ever happen? Why or why not?
11. Think of several processes (other than those already mentioned) that would obey the first law of thermodynamics but, if they actually occurred, would violate the second law.
12. Suppose you collect papers that are strewn all over the floor and put them in a neat stack; does this violate the second law of thermodynamics? Explain.

13. The first law of thermodynamics is sometimes whimsically stated as, "You can't get something for nothing" and the second law as, "You can't even break even." Explain how these statements could be equivalent to the formal statements.
14. Give three examples of naturally occurring processes that illustrate the degradation of usable energy into internal energy.
15. Entropy is often called "time's arrow" because it tells us in which direction natural processes occur. If a movie film were run backward, name some processes that you might see that would tell you that time

EXERCISES

1. What is the average power output of a person who spends 10 hr each day sleeping, 4 hr at light activity, and 10 hr watching television or loafing?
2. A person decides to lose weight by sleeping $\frac{1}{2}$ hr less per day, using the time for light activity. How much weight (or mass) can this person expect to lose in 1 year assuming no change in food intake? Use the fact that 1 kg of fat stores about 9000 kcal of energy.
3. Calculate the maximum efficiency of a steam engine that operates between 500°C and 250°C.
4. If the steam engine in the above problem can reach only 60 percent of the maximum, what is its efficiency?
5. A steam engine is designed to exhaust heat at 200°C. If its efficiency is to be 40 percent, what is the minimum value for the high temperature source?
6. An engine operating between 1200°C and 600°C has an efficiency of 30 percent. What percentage is this of its maximum theoretical efficiency?
7. A heat engine produces 9500 J of heat while performing 2500 J of useful work. What is the efficiency of this engine?
8. The oceans in the tropics have a surface temperature of about 25°C, whereas below in the depths the temperature is about 5°C. The possibility of building heat engines to generate electricity using this temperature difference is being considered. What would be the maximum efficiency of such an engine? Why might such an engine be practical in spite of its low efficiency? What adverse environmental effects might occur?
9. A heat engine utilizes a heat source at 550°C and has an ideal efficiency of 30 percent. To increase the efficiency to 40 percent, what must be the temperature of the heat source?
10. An engine that operates at half its theoretical efficiency operates between 545°C and 310°C while producing work at the rate of 1000 kW. How much heat is wasted per hour?

11. A maximum-efficiency (ideal) engine performs work at the rate of 500 kW while using 960 kcal of heat per second. If the temperature of the heat source is 620°C, at what temperature is the waste heat exhausted?

12. The burning of gasoline in a car releases about 3.0×10^4 kcal/gal. If a car averages 35 km/gal when driving 90 km/hr, which requires 25 hp, what is the efficiency of the engine under those conditions?

*13. Solar cells can produce about 40 W of electricity per square meter of surface area if directly facing the sun. How large an area is needed to supply the needs of a house that requires 3.6×10^8 J/day? Would this fit on the roof of an average house? (Assume that the sun shines 12 hr/day.)

*14. Water falls 80 m over a dam at a rate of 16,000 kg/s. How many megawatts of electric power could be produced by a power plant using this energy?

*15. One way of storing energy for use during peak demand periods is to pump water to a high reservoir when the demand is low and then release it to drive turbines when needed. Suppose that water is pumped to a lake 100 m above the turbines at a rate of 1.0×10^6 kg/s for 8.0 hr during the night. (a) How much energy (joules) is needed to do this each night? (b) If all this energy is released during a 16-hr day, what is the average power output? Assume that the process is 80 percent efficient.

*16. Suppose a power plant delivers energy at 1000 MW using steam turbines. The steam goes into the turbines superheated at 520 K and deposits its unused heat in river water at 290 K. Assume that the turbine operates as an ideal maximum-efficiency engine. If the river flow rate is 40 m^3/s, calculate the temperature increase of the river water immediately downstream from the power plant.

*17. The basin of the tidal power plant at the mouth of the Rance River in France covers an area of 23 km^2. The average difference in water height between high and low tide is 8.5 m. Estimate how much work the falling water can do on the turbines per day assuming there are two high tides and two low tides per day. Assume the basin is flat; disregard any KE of water before and after its fall.

VIBRATIONS AND WAVES

CHAPTER

16

Many objects vibrate or oscillate—an object on the end of a spring, a tuning fork, the balance wheel of a watch, a pendulum, the strings of a guitar or piano. Spiders detect prey by the vibrations of their webs, cars oscillate up and down when they hit a bump, buildings and bridges vibrate when heavy trucks pass or the wind is fierce. Indeed, because most solids are elastic (see Chapter 10), most material objects vibrate (at least briefly) when given an impulse. Electrical oscillations occur in radio and television sets. And at the atomic level, atoms vibrate within a molecule, and the atoms of a solid vibrate about their relatively fixed positions.

Vibrations and wave motion are intimately related subjects. Waves—whether ocean waves, waves on a string, earthquake waves, or sound waves in air—have as their source a vibration. In the case of sound, not only is the source a vibrating object, but so is the detector—the eardrum or the membrane of a microphone. Indeed, the medium through which a wave travels itself vibrates (such as air for sound waves). In the second half of this chapter, after we discuss vibrations, we will discuss simple waves such as those on water and on a string. In Chapter 17 we will study sound waves, and in later chapters we will encounter other forms of wave motion, including light.

1. SIMPLE HARMONIC MOTION

Oscillations of a Spring

When we speak of a **vibration** or an **oscillation** we mean the motion of an object that repeats itself, back and forth, over the same path. That is, the motion is *periodic*. The simplest form of periodic motion is represented by an object oscillating on the end of a coil spring. Because many other types of vibrational motion closely resemble this system, we will look at it

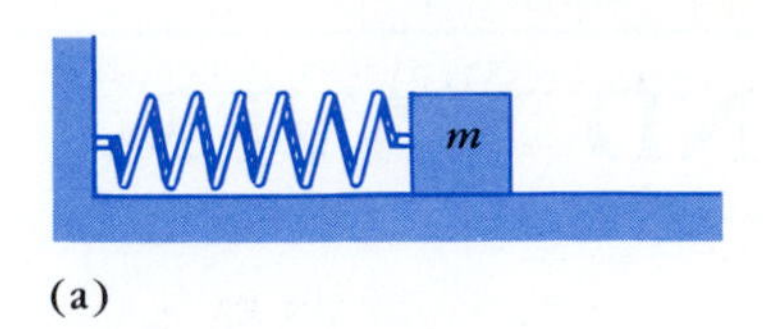

(a)

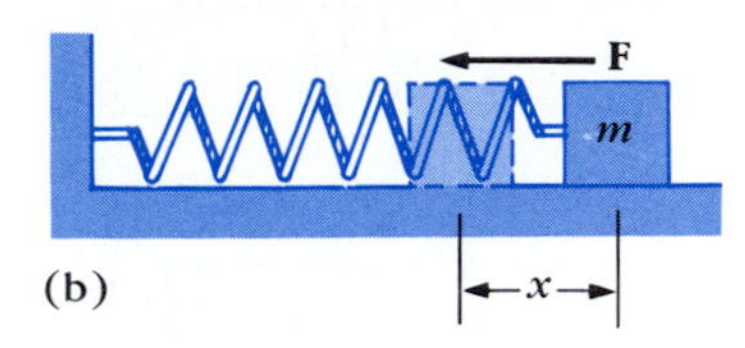

(b)

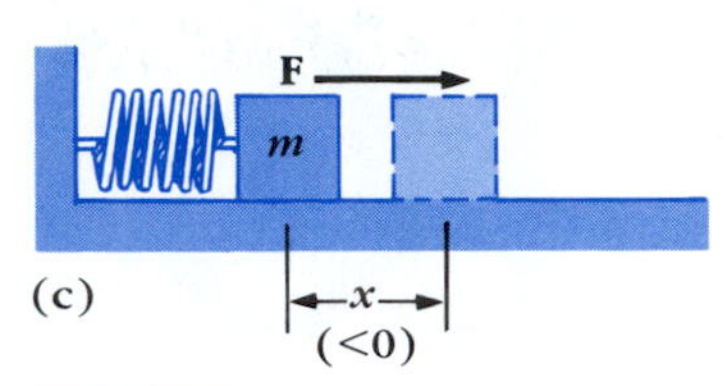

(c)

FIGURE 16-1
Mass vibrating at the end of a spring

in detail. We assume that the mass of the spring can be ignored and that the spring is mounted horizontally as shown in Figure 16-1(a), so that the object of mass m slides without friction on the horizontal surface. Any spring has a natural length at which it exerts no force on the mass m, and this is called the *equilibrium position*. If the mass is moved either to the left, which compresses the spring, or to the right, which stretches it, the spring exerts a force on the mass which acts in the direction of returning it to the equilibrium position; hence it is called a "restoring force." The magnitude of the restoring force F is found to be directly proportional to the distance x that the spring has been stretched or compressed—Figure 16-1(b) and (c):

$$F = -kx.$$

The farther the mass is pushed or pulled from its equilibrium position, the stronger the spring pulls it back. This relation, often referred to as Hooke's law, is accurate as long as the spring is not compressed to the point where the coils touch, or is stretched beyond the elastic region (see Figure 10-9). The minus sign reminds us that the restoring force is always in the direction opposite to the displacement x. For example, if the mass is displaced to the right in Figure 16-1, the direction of the restoring force is to the left. If the spring is compressed to the left, the force F acts toward the right—Figure 16-1(c). The proportionality constant k is called the "spring constant." Notice that in order to stretch the spring a distance x one has to exert an (external) force on the spring at least equal to $F = +kx$. The greater the value of k the greater the force needed to stretch a spring a given distance. That is, the stiffer the spring the greater the spring constant k.

Let us examine what happens when the spring is initially stretched a distance $x = A$ as shown in Figure 16-2(a) and then released. The spring exerts a force on the mass that pulls it toward the equilibrium position; but because the mass has been accelerated by the force it passes the equilibrium position with considerable speed. Notice that as the mass reaches the equilibrium position the force on it decreases to zero, but its speed at this point is a maximum—Figure 16-2(b). As it moves farther to the left the force on it acts to slow it down, and it stops momentarily at $x = -A$, Figure 16-2(c). It then begins moving back in the opposite direction, Figure 16-2(d), until it reaches the original starting point, $x = A$, Figure 16-2(e). It then repeats the motion.

To discuss vibrational motion, we must define a few terms. The distance x of the mass from the equilibrium point at any moment is called the *displacement*. The displacement is considered positive in one direction and negative in the opposite direction. The maximum displacement—the greatest distance from the equilibrium point—is called the **amplitude**, A. One **cycle** refers to the complete to-and-fro motion from some initial point back to that same point, say from $x = A$ to $x = -A$ back to $x = A$. The **period**, T, is defined as the time required for one complete cycle. Finally,

ders Company | PACKING LIST (CUSTOMER'S COPY)
PACKING LIST

Publishers, Inc. (G&S) Grune & Stratton, Inc.

USTOMER RELATIONS DEPT.
ENT IS:
2

Send returns to:

HRW/W.B. SAUNDERS
151 BENIGNO BLVD.
BELLMAWR, NEW JERSEY 08031

UNT NO.
X

YOUR ORDER NUMBER
542097583000544

PAGE NO.
1

SHIP TO:

MARGARET EDDIE
2105 M STREET
AURORA NE 68818

Z4

PRICE

40.00 NONE

OF PHYSICS 3ED

* * * * * * ADDITIONAL INSTRUCTIONS * * * * * *

FOB: SHIPPING PT

H | PICKER | CHECKER | OTHER SHP INSTRUCTIONS

SHP CH | 479 | 152

LLOW UNDER SEPARATE COVER. * *

0000000342

DOMESTIC RE

SCHOO

School titles cannot be returned without prior written auth

COLLEGE AND PR

College Publishing (Holt, Rinehart and Winston, Dryden Press and Saund
authorization returns from booksellers of all (100%) overstocked College
to the following terms and conditions:

1. Returns must reach our warehouse within 18 months of their invoice
Books received by us more than 18 months after their invoice date

2. To receive credit at prices billed, the return shipment must be a
each title returned quantity, author, title, invoice number, invoice
you list your SAN and the ISBN's for all titles. If you use your

3. We reserve the right to impose a 20% overstock returns limitation

4. Materials declared out-of-print are not acceptable for return or c
to stores in May and November. Titles identified as OSI (out-of-s
is available. If no stock is available your order will be cancelled in
our warehouse by the dates indicated. On those dates the boo
by us and no credit will be issued. No extensions of the time limit fo

	M
Last date titles will be sold	J
Date by which returns must reach Warehouse	Sep

TERMS: Subject to change without notice.

RNS POLICY

TLES

ation from our Regional Office which serves your state.

SSIONAL TITLES

College Publishing) and W.B. Saunders will accept without prior written
shing and W.B. Saunders books in new and unmarked condition, subject

Given this liberal time period we will not consider requests for extension.
retained by us and no credit will be issued.

panied by an invoice/packing list or customer packing list including for
price and discount billed. It speeds processing if on your packing list
packing lists please make sure the information on the form is legible.

ksellers whose returns are determined by us to be excessive.

Out-of-print titles will be identified in our Complete List which is mailed
ndefinitely) in the Complete List will be sold for a short period if stock
tely and you will be so notified. Returns of these OSI titles must reach
declared out of print and returns of these OP titles will be retained
ns can be granted.

November list
December 15
1 March 1

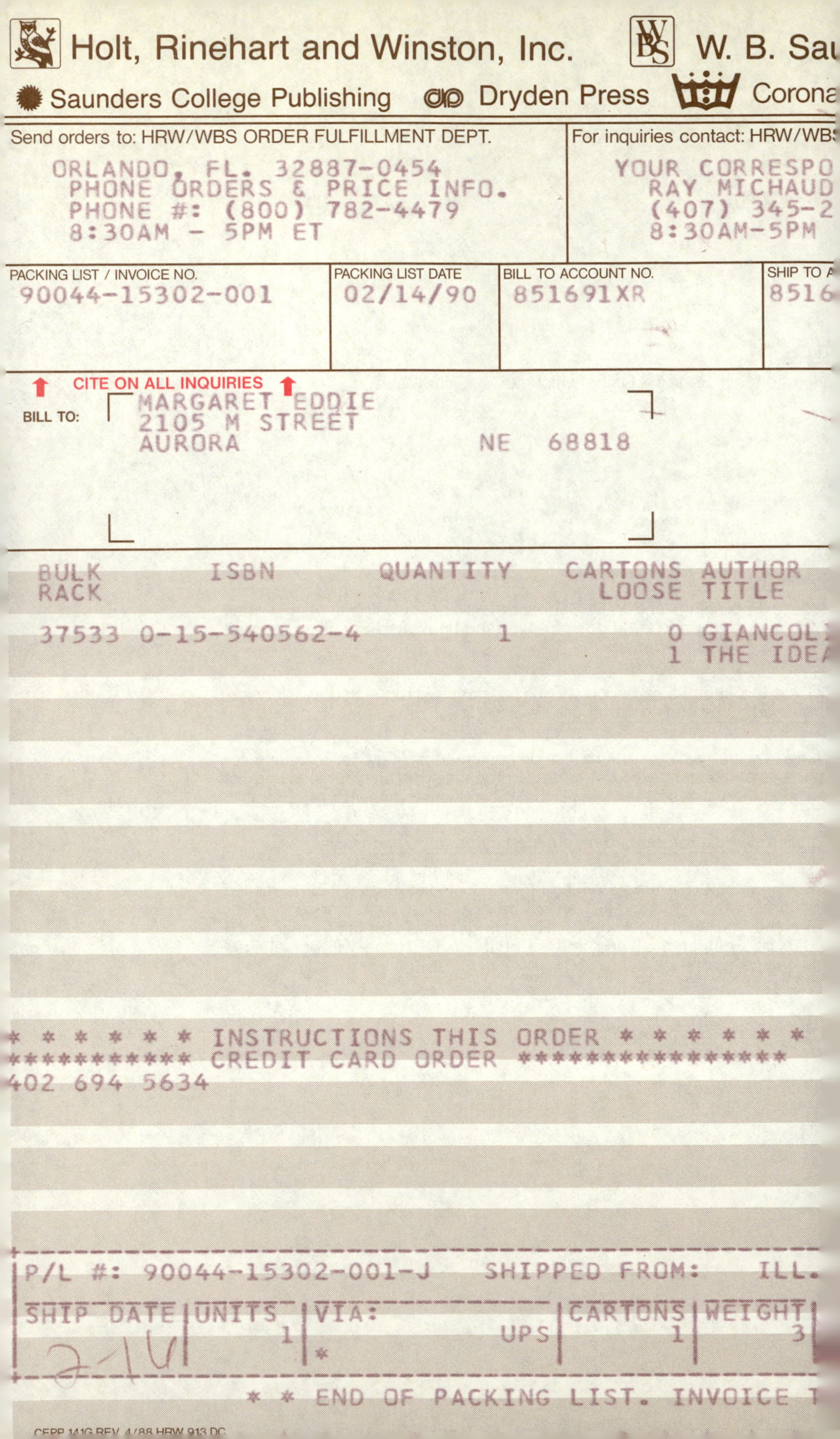

Holt, Rinehart and Winston, Inc. W. B. Sau
Saunders College Publishing Dryden Press Corona

Send orders to: HRW/WBS ORDER FULFILLMENT DEPT.
ORLANDO, FL. 32887-0454
PHONE ORDERS & PRICE INFO.
PHONE #: (800) 782-4479
8:30AM - 5PM ET

For inquiries contact: HRW/WBS
YOUR CORRESPO
RAY MICHAUD
(407) 345-2
8:30AM-5PM

PACKING LIST / INVOICE NO.	PACKING LIST DATE	BILL TO ACCOUNT NO.	SHIP TO A
90044-15302-001	02/14/90	851691XR	8516

CITE ON ALL INQUIRIES

BILL TO: MARGARET EDDIE
2105 M STREET
AURORA NE 68818

BULK RACK	ISBN	QUANTITY	CARTONS LOOSE	AUTHOR TITLE
37533	0-15-540562-4	1	0 1	GIANCOLI THE IDEA

* * * * * * INSTRUCTIONS THIS ORDER * * * * * * *
*********** CREDIT CARD ORDER *****************
402 694 5634

P/L #: 90044-15302-001-J SHIPPED FROM: ILL.

SHIP DATE	UNITS	VIA:	CARTONS	WEIGHT
2-14	1	UPS *	1	3

* * END OF PACKING LIST. INVOICE T

CFPP 141G REV. 4/88 HRW 913 DC

the **frequency**, f, is the number of complete cycles per second. Frequency is usually specified in hertz (Hz) where 1 Hz = 1 cycle per second. It is evident that

$$f = \frac{1}{T} \text{ and } T = \frac{1}{f}.$$

For example, if the frequency is 5 cycles per second, then each cycle takes $\frac{1}{5}$ s.

The oscillation of a spring hung vertically is essentially the same as that of a horizontal spring. Because of the force of gravity, the length of the vertical spring at equilibrium will be longer than when it is horizontal.

If a pen is attached to a vibrating mass and a sheet of paper is moved steadily beneath it, a curve will be drawn as shown in Figure 16-3. This is a graph of the position (displacement) of the mass as a function of time. This shape of curve is called *sinusoidal*.

Any vibrating system for which the restoring force is directly proportional to the displacement ($F \propto x$) will vibrate sinusoidally, and is said to exhibit **simple harmonic motion** (SHM). Such a system is often called a **simple harmonic oscillator** (SHO). We saw in Section 4 of Chapter 10 that most solid materials stretch or compress according to $F = kx$ as long as the displacement is not too great. Because of this, many natural vibrations are simple harmonic or close to it.

Now let us consider a numerical example. Suppose a family of four people with a total mass of 200 kg steps into their 1200-kg car and the car's springs compress 3.0 cm. What is the spring constant of the car's springs (assuming they act as a single spring) and how far would the car descend if loaded with 300 kg? To solve this, we note that the force of $(200 \text{ kg})(9.8 \text{ m/s}^2) = 1960 \text{ N}$ causes the springs to compress 3.0×10^{-2} m. Therefore, using $F = kx$, we can solve for k and obtain:

$$k = \frac{F}{x} = \frac{1960 \text{ N}}{3.0 \times 10^{-2} \text{ m}} = 6.5 \times 10^4 \text{ N/m}.$$

Next, if the car is loaded with 300 kg, $x = F/k = (300 \text{ kg})(9.8 \text{ m/s}^2)/(6.5 \times 10^4 \text{ N/m}) = 4.5 \times 10^{-2}$ m, or 4.5 cm. We could have obtained this answer without solving for k: since x is proportional to F, if 200 kg compresses the spring 3.0 cm then 1.5 times the force will compress the spring 1.5 times as much, or 4.5 cm.

FIGURE 16-2

Force on, and velocity of, mass at different positions of its oscillation.

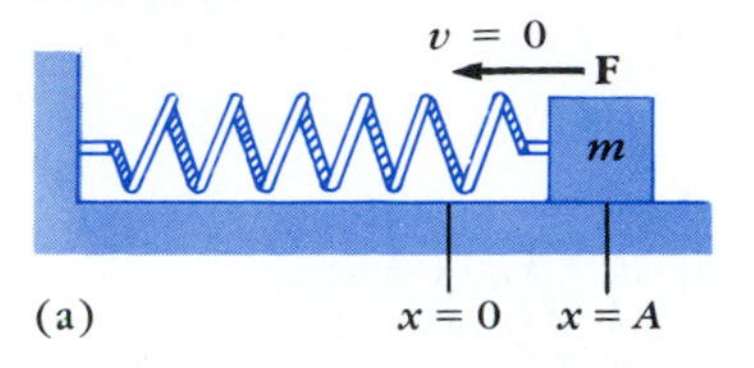

(a) $x = 0$ $x = A$

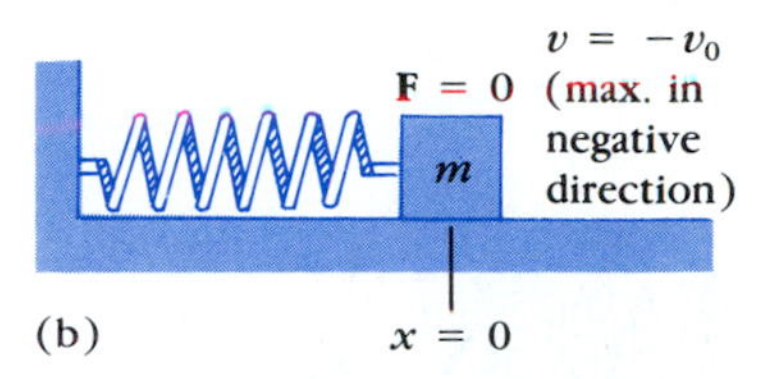

(b) $x = 0$

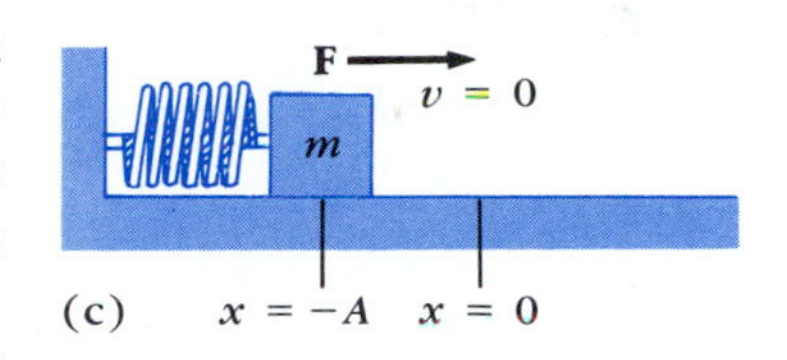

(c) $x = -A$ $x = 0$

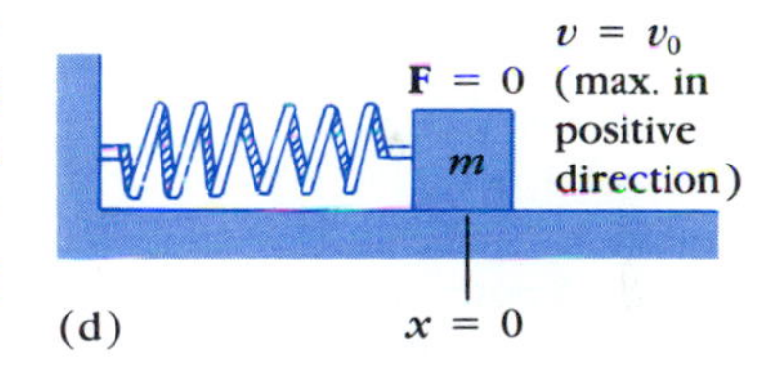

(d) $x = 0$

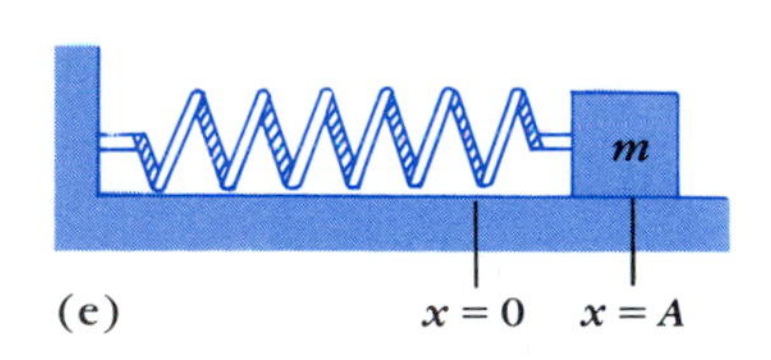

(e) $x = 0$ $x = A$

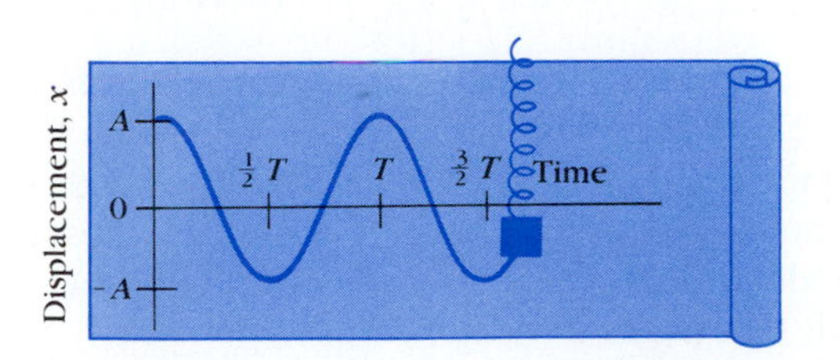

FIGURE 16-3

Sinusoidal nature of SHM as a function of time; T is the period.

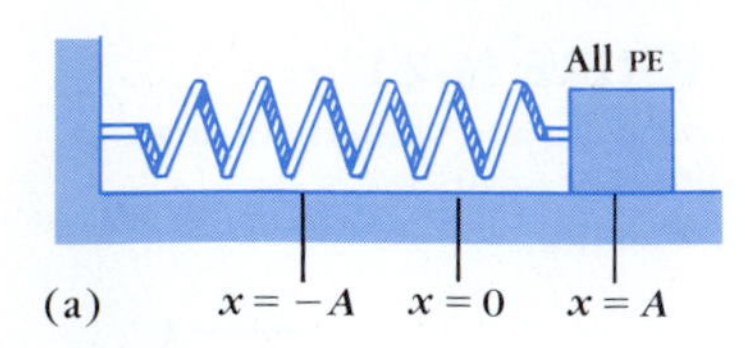

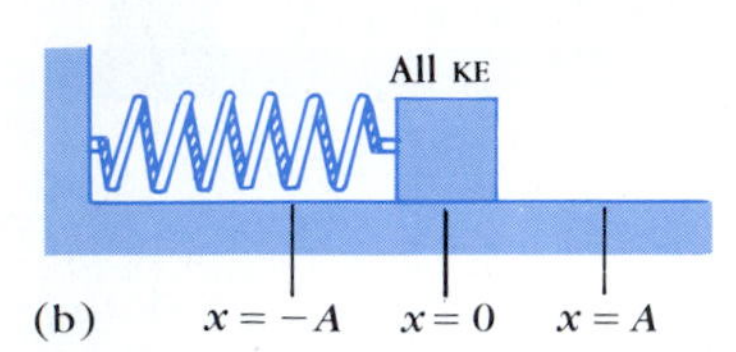

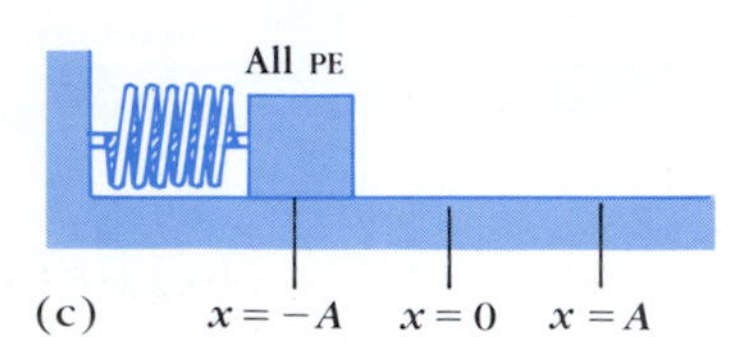

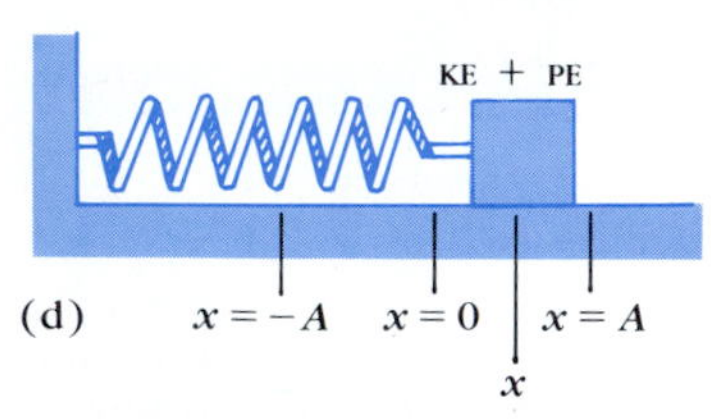

FIGURE 16-4
Energy changes from kinetic energy to potential energy and back again as the spring oscillates.

Energy in the Simple Harmonic Oscillator

To stretch or compress a spring, work must be done; hence potential energy is stored in a stretched or compressed spring. The total mechanical energy is the sum of the kinetic and potential energies. As long as there is no friction, the total mechanical energy remains constant. As the mass oscillates back and forth, the energy continuously changes from potential energy to kinetic energy and back again (Figure 16-4). At the extreme points, $x = \mathrm{A}$ and $x = -\mathrm{A}$, all the energy is potential energy (and is the same whether the spring is compressed or stretched to the full amplitude). At the equilibrium point, the energy is all kinetic. At other points, the energy is part kinetic and part potential.

The Period of SHM (optional)

As we mentioned earlier, the *period* of a vibration is the time required for a complete cycle. For the mass on a spring, Figure 16-1, this would be the time required for one complete back-and-forth motion. On what factors does the period depend? It is found that if a greater mass is hung from the spring, the period is increased. This makes sense because a greater mass means more inertia and therefore slower response. The period also depends on the stiffness of the spring: the greater k is, the shorter the period. This too makes sense since greater k means greater force and therefore quicker response. Initially you might think the period is directly proportional to the mass m and inversely proportional to the stiffness k. But this is not quite the case. Instead, the period depends on the *square root* of these quantities; that is,

$$T \propto \sqrt{\frac{m}{k}}.$$

The constant of proportionality in this relation turns out to be 2π. Thus

$$T = 2\pi\sqrt{\frac{m}{k}}.$$

As an example, let us calculate the period and frequency of the car in our earlier example when it hits a bump and begins to oscillate up and down. The mass of the car including occupants is 1400 kg, and we calculated k to be 6.5×10^4 N/m. Then

$$T = (2)(3.14)\ \sqrt{(1400\ \mathrm{kg})/(6.5 \times 10^4\ \mathrm{N/m})} = 0.92\mathrm{s},$$

or slightly less than one second. As mentioned earlier, the frequency f equals $1/T$ so $f = 1/0.92\mathrm{s} = 1.08$ cycles per second, or 1.08 Hz. Consequently, after hitting a bump the car begins to vibrate at a rate of slightly more than one up-and-down motion per second. Actually, if the shock absorbers in the car are in good condition, there will be considerable friction and the vibrations will stop quickly (see the next section).

It is interesting to note that—strange as it may seem—the period does not depend on the amplitude; you can determine this for yourself by using a watch and counting 10 or 20 cycles of a spring for a small amplitude and then for a large amplitude.

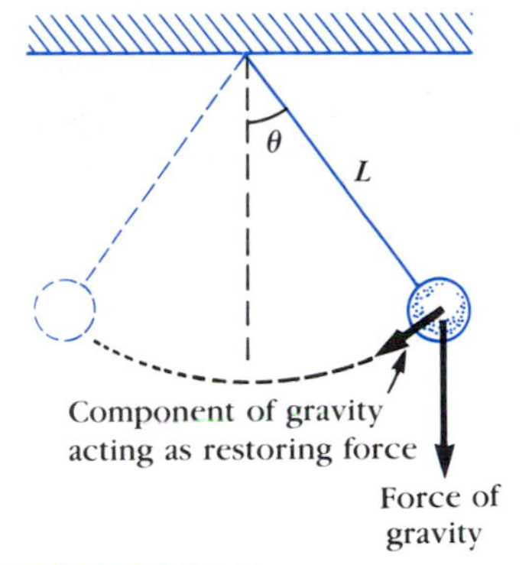

FIGURE 16-5
A simple pendulum.

The Simple Pendulum

The simple pendulum consists of a small object (the pendulum "bob") suspended from the end of a light cord (Figure 16-5). A simple pendulum moving back and forth with negligible friction resembles simple harmonic motion: it oscillates along the arc of a circle with equal amplitude on either side of its equilibrium point (where it hangs vertically), and as it passes through the equilibrium point it reaches its maximum speed.

The motion of a pendulum is close to SHM, especially when the angular amplitude (θ) is small (the only case we'll consider). The period of a pendulum is proportional to the square root of its length L: the longer the pendulum cord the longer the period. A surprising result is that the period does not depend on the mass of the pendulum bob. You may have noticed this if you pushed a small child and a large one on the same swing.

We saw earlier that the period of any SHM (and this includes a small amplitude pendulum) does not depend on the amplitude. Galileo is said to have first noted this fact while watching a swinging lamp in the cathedral at Pisa. This discovery led to the pendulum clock, the first really precise timepiece and one which became the standard for centuries.

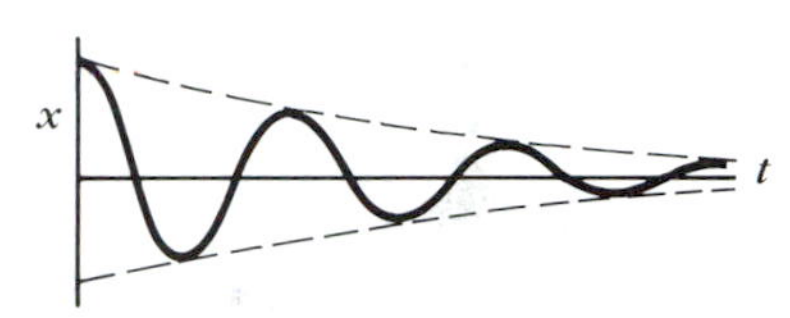

FIGURE 16-6
Damped harmonic motion.

2. DAMPING AND RESONANCE

Damped Harmonic Motion

The amplitude of any real oscillating spring or swinging pendulum slowly decreases in time until the oscillations stop altogether. Figure 16-6 shows a typical graph of the displacement as a function of time. This is called **damped harmonic motion**. The damping† is generally due to the resistance of air and to internal friction within the oscillating system. The energy that is thus dissipated to thermal energy is reflected in a decreased amplitude of oscillation.

Sometimes the damping is so large that the motion no longer resembles vibratory motion. Three common cases of heavily damped systems are shown in Figure 16-7. Curve C represents the situation when the damping is so large that it takes a long time to reach equilibrium; the system is *overdamped*. Curve A represents an *underdamped* situation in which the system makes several swings before coming to rest. Curve B represents *critical damping*; in this case equilibrium is reached the quickest. These terms all derive from the use of practical damped systems

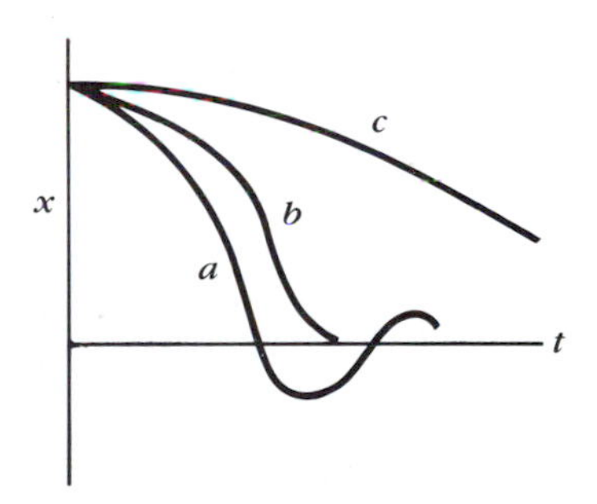

FIGURE 16-7
Underdamped (a), critically damped (b), and overdamped (c) motion.

† To "damp" means to diminish, restrain, or extinguish, as to "dampen one's spirits."

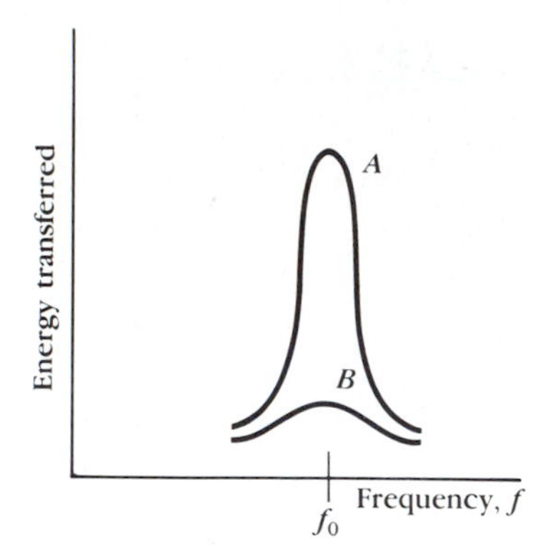

FIGURE 16-8
Resonance for (A) lightly damped, (B) heavily damped systems.

such as door-closing mechanisms and shock absorbers in cars. These are usually designed to give critical damping; but as they wear out, underdamping occurs: a door slams, and a car bounces up and down several times whenever it hits a bump. Needles on electronic instruments (voltmeters, ammeters, level indicators on tape recorders) are usually critically damped or slightly underdamped. If they were very underdamped, they would swing back and forth excessively before arriving at the correct value; if overdamped, they would take too long to reach equilibrium and rapid changes in the signal (say, recording level) would not be detected.

Forced Vibrations and Resonance

When a vibrating system is set into motion, such as a mass on the end of a spring (Section 1), it vibrates at its natural frequency. However, a system is often not left to merely oscillate on its own, but may have an external force applied to it which itself oscillates at a particular frequency. For example, we might pull the mass on the spring of Figure 16-1 back and forth at a frequency f. The mass thus vibrates at the frequency f of the external force even if this frequency is different from the **natural frequency** of the spring, which we will now denote as f_0†.

This is an example of **forced vibration**. The amplitude of vibration, and hence the energy transferred to the vibrating system, is found to depend on the difference between f and f_0. It is a maximum when the frequency of the external force equals the natural frequency of the system; that is, when $f = f_0$. The energy transferred to the system (proportional to the amplitude squared) is plotted in Figure 16-8 as a function of the external frequency f. Curve A represents light damping and curve B heavy damping. The amplitude becomes very large when $f = f_0$, particularly when the damping (friction) is small. This is known as **resonance**, and the natural vibrating frequency of a system is often called its **resonant frequency**.

FIGURE 16-9
Large amplitude oscillation of the Tacoma Narrows bridge due to heavy gusty winds immediately prior to collapse (November 7, 1940).

A simple illustration of resonance is pushing a child on a swing. A swing, like any pendulum, has a natural frequency of oscillation. If you were to close your eyes and push on the swing at a random frequency, the swing would bounce around and reach no great amplitude. But if you push with a frequency equal to the natural frequency of the swing, the amplitude increases greatly. This clearly illustrates the fact that at resonance, relatively little effort is required to obtain a large amplitude.

The great tenor Enrico Caruso was said to be able to break a goblet by singing a note of just the right frequency at full voice. This is an example of resonance, for the sound waves emitted by the voice act as a forced vibration on the glass. At resonance, the resulting vibration of the goblet

† As we saw in Section 1, the natural frequency f_0 can be written as

$$f_0 = \frac{1}{T} = \frac{1}{2\pi}\sqrt{\frac{k}{m}}.$$

may be large enough in amplitude that the glass exceeds its elastic limit and breaks.

Since material objects are in general elastic, resonance is an important phenomenon in a variety of situations. It is particularly important in building, although the effects are not always to be foreseen. For example, it has been reported that a railway bridge collapsed because a nick in one of the wheels of a passing train set up a resonant vibration in the bridge. Marching soldiers break step when crossing a bridge to avoid the possibility of a similar catastrophe. And the famous collapse of the Tacoma Narrows bridge (Figure 16-9) in 1940 was due in part to resonance of the bridge.

We will meet important examples of resonance later in this chapter and in succeeding chapters. We will also see that vibrating objects often have not one, but many resonant frequencies.

3. WAVE MOTION

When you throw a stone into a lake or pool of water, circular waves form and move outward (Figure 16-10). Waves will also travel along a cord (or a "slinky") that is stretched out straight on a table if you vibrate one end back and forth as shown in Figure 16-11. Water waves and waves on a cord are two common examples of wave motion. We will meet other kinds of wave motion later, but for now we will concentrate on these simple "mechanical" waves.

If you have ever watched ocean waves moving toward shore, you may have wondered if the waves were carrying water into the beach. This is, in fact, not the case.† Water waves move with a recognizable velocity. But each particle of the water itself merely oscillates about an equilibrium point. This is clearly demonstrated by observing leaves on a pond as waves move by. The leaves (or a cork) are not carried forward by the waves but simply oscillate about an equilibrium point because this is the motion of the water itself. Similarly, the wave on the rope of Figure 16-11 moves to the right, but each piece of the rope only vibrates to and fro. This is a general feature of waves: waves can move over large distances but the medium (the water or the rope) itself has only a limited movement. Thus, although a wave is not matter, the wave pattern can travel in matter. A wave consists of oscillations that move without carrying matter with them.

However, waves do carry energy from one place to another. Energy is given to a water wave, for example, by a rock thrown into the water or by wind far out at sea. The energy is transported by waves to shore. If you have been under an ocean wave when it breaks, you know the energy it carries. The oscillating hand in Figure 16-11 transfers energy to the rope

(a)

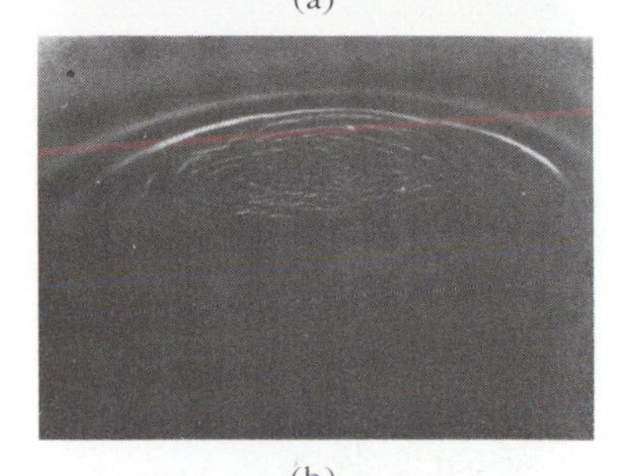

(b)

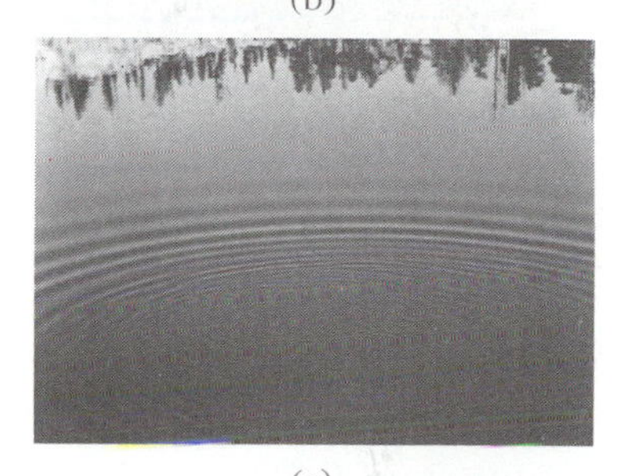

(c)

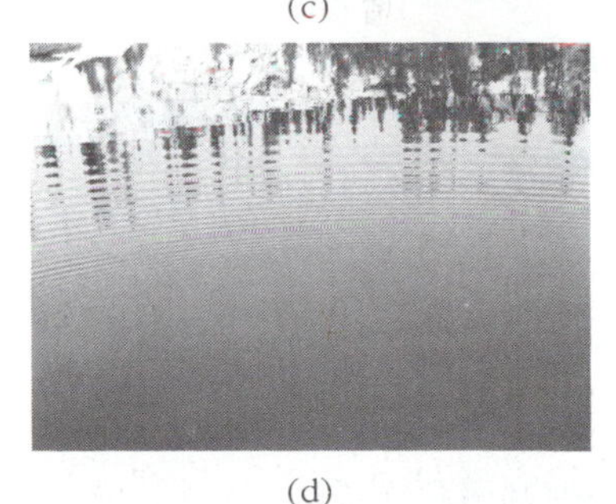

(d)

FIGURE 16-10
Water waves spreading outward from a source.

† Do not be confused by the "breaking" of ocean waves that occurs when the waves interact with the bottom in shallow water and hence are no longer simple waves.

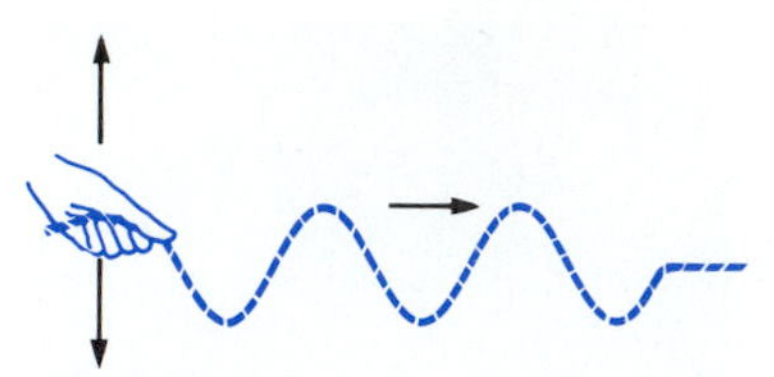

FIGURE 16-11
Wave traveling on a rope.

that is then transported down the rope and can be transferred to an object at the other end. All forms of wave motion transport energy.

EXPERIMENT-PROJECT

Fill a tub with water. Float a cork at each end. Wait until the water is smooth and then move *one* of the corks up and down. Does this produce travelling waves? Does the second cork respond? Is work done on the second cork? If so, how was the energy transferred?

Let us look a little more closely at how a wave is formed and how it comes to "travel." We first look at a single wave bump or *pulse*. A single pulse can be formed on a rope by a quick up-and-down motion of the hand (Figure 16-12). The hand pulls up on one end of the rope, and because the end piece is attached to adjacent pieces these respond to an upward force and also begin to move upward. As each succeeding piece of rope moves upward, the wave crest moves outward along the rope. Meanwhile, the end piece of rope has been returned to its original position by the hand, and as each succeeding piece of rope reaches its peak position it, too, is pulled down again. Thus the source of a traveling wave pulse is a disturbance, and cohesive forces between adjacent pieces of rope cause the pulse to travel outward. Waves in other media are created and propagate outward in a similar fashion.

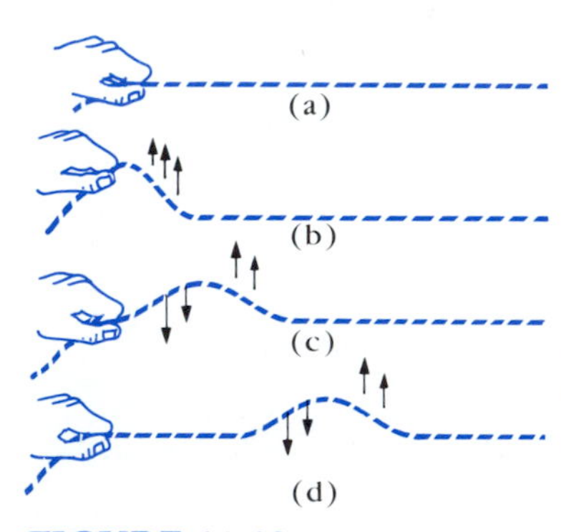

FIGURE 16-12
Motion of a wave pulse. Arrows indicate velocity of rope particles.

A *continuous* or *periodic wave*, such as that shown in Figure 16-11, has as its source a disturbance that is continuous and oscillating; that is, the source is a *vibration* or *oscillation*. In Figure 16-11, a hand oscillates one end of the rope. Water waves may be produced by any vibrating object (such as your hand) placed at the surface, or the water itself may be made to vibrate when wind blows across it or a rock is thrown into it. A vibrating tuning fork or drum membrane gives rise to sound waves in air; we will see later that oscillating electric charges give rise to light waves. Indeed, almost any vibrating object sends out waves.

The source of any wave, then, is a vibration. And it is the *vibration* that propagates outward and thus constitutes the wave. If the source vibrates sinusoidally in SHM, then the wave itself—if the medium is perfectly elastic—will have a sinusoidal shape.

4. CHARACTERISTICS AND TYPES OF WAVES

Wave Properties

FIGURE 16-13
Characteristics of a continuous wave.

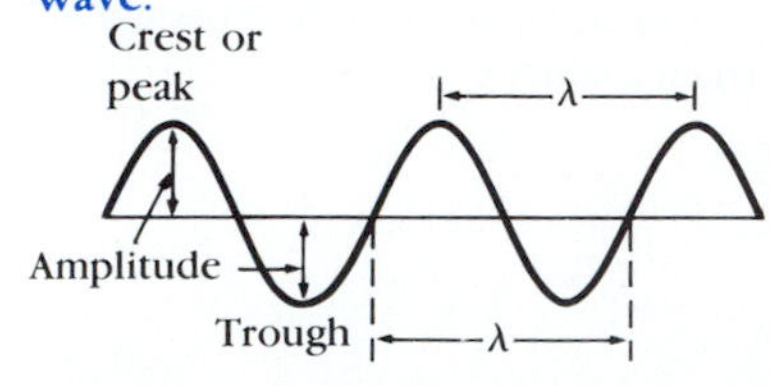

Some of the important quantities used to describe a periodic sinusoidal wave are shown in Figure 16-13. The high points on a wave are called crests, the low points troughs. The **amplitude** is the maximum height of a crest or depth of a trough relative to the normal (or equilibrium) level; the total swing from a crest to a trough is twice the amplitude. The distance between two successive crests is called the **wavelength**, λ (the Greek letter lambda). The wavelength is also equal to the distance between *any*

two successive identical points on the wave. The **frequency**, f, is the number of crests—or complete cycles—that pass a given point per unit time. The period, T, of course, is just $1/f$. For example, if three wave crests pass a small rock in a pond every second, the frequency of the wave is three cycles per second, or 3 Hz. If ocean waves pass a pier at the rate of one every ten seconds, the frequency is 1/10 Hz. Because a wave is caused by a vibrating source of some kind, the frequency of any wave is equal to the frequency of the source that produces it.

The **wave velocity**, v, is the velocity at which wave crests appear to move. (The wave velocity must be distinguished from the velocity of a particle of the medium itself. For example, for a wave traveling along a string as in Figure 16-11, the wave velocity is to the right and along the string, whereas the velocity of a particle of the string is perpendicular to it.) Since a wave crest travels a distance of one wavelength, λ, in one period, T, the wave velocity $v = \lambda/T$, or (since $1/T = f$)

$$v = \lambda f$$

wave velocity = wavelength × frequency.

For example, suppose a wave has a wavelength of 5 m and a frequency of 3 Hz. Since 3 crests pass a given point per second, and the crests are 5 m apart, the first crest (or any other part of the wave) must travel a distance of 15 m during the 1 s; so its speed is 15 m/s.

EXPERIMENT-PROJECT

Throw a ball into a still pond or swimming pool and observe the waves created. Note the following:

1. Is one crest formed? Or many? Is the wave continuous, a single pulse, or something in between?
2. What caused the wave? Was it a vibration? Of what? To answer this you must observe very carefully.
3. Estimate the wavelength, the frequency, and the velocity of the wave produced. Does the velocity equal frequency times wavelength? To estimate the frequency, count the number of waves that pass a stick or a small rock within a particular time interval measured by your watch.
4. By using different-sized rocks or by throwing a rock into the water in different places, or by some other means, can you change the amplitude, wavelength, or velocity of the resulting waves?
5. What happens when the wave strikes a barrier such as a rock or the shore?

Wave Types: Transverse and Longitudinal

We saw earlier that although waves may travel over long distances, the particles of the medium vibrate only over a limited region of space. When a wave travels down a rope, say from left to right, the particles of the rope vibrate up and down in a direction transverse (or perpendicular) to the motion of the wave itself. Such a wave is called a **transverse wave**. There exists another type of wave known as a **longitudinal wave**. In a longitudinal wave, the vibration of the particles of the medium is along the *same*

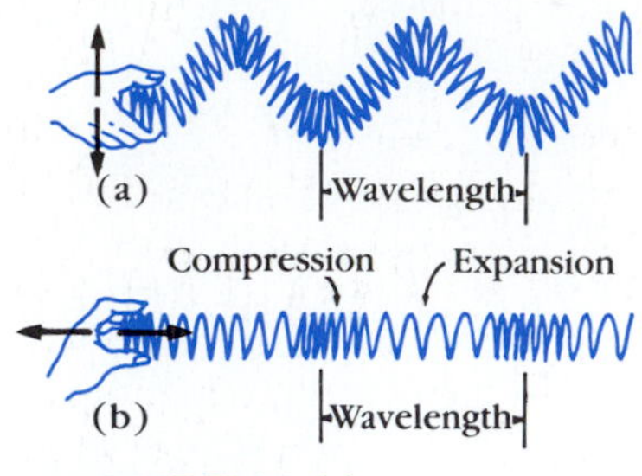

FIGURE 16-14
(a) Transverse wave; (b) Longitudinal wave.

direction as the motion of the wave. Longitudinal waves are readily formed on a stretched spring or "slinky" by alternately compressing and expanding one end. This is shown in Figure 16-14(b) and can be compared to the transverse wave in Figure 16-14(a). A series of compressions and expansions propagate along the spring. The *compressions* are those areas where the coils are momentarily close together. *Expansions* (sometimes called *rarefactions*) are regions where the coils are momentarily far apart. Compressions and expansions correspond to the crests and troughs of a transverse wave.

An important example of a longitudinal wave is a sound wave in air. A vibrating drumhead, for example, alternately compresses and rarefies the air and produces a longitudinal wave that travels outward in the air as shown in Figure 16-15.

As in the case of transverse waves, each section of the medium in which a longitudinal wave passes oscillates over a very small distance whereas the wave itself can travel large distances. Wavelength, frequency, and wave velocity all have meaning for a longitudinal wave. The wavelength is the distance between successive compressions (or between successive expansions) and frequency is the number of compressions that pass a given point per second. The wave velocity is the velocity with which each compression appears to move and is equal to the product of wavelength and frequency.

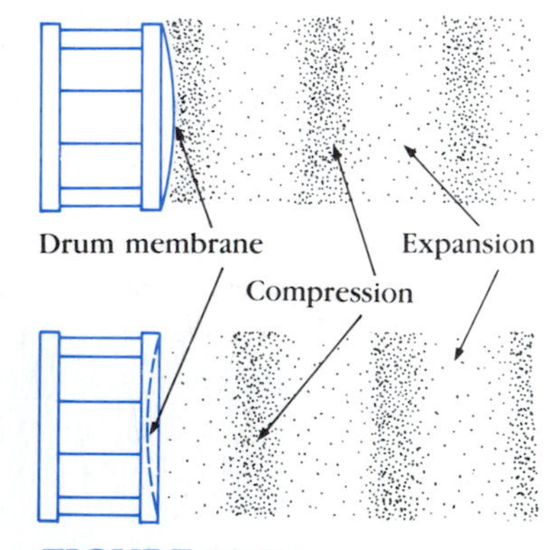

FIGURE 16-15
Production of a sound waves which is longitudinal.

A graphical representation of a longitudinal wave can be made by graphing the density of air molecules (or coils of a slinky) versus position (Figure 16-16): notice that the graphical representation looks just like a transverse wave.

5. BEHAVIOR OF WAVES: REFLECTION, REFRACTION, INTERFERENCE, AND DIFFRACTION

Reflection

When a wave strikes an obstacle or comes to the end of the medium it is traveling in, at least a part of the wave is reflected. You have probably seen water waves reflect off a rock or the side of a swimming pool. And you may

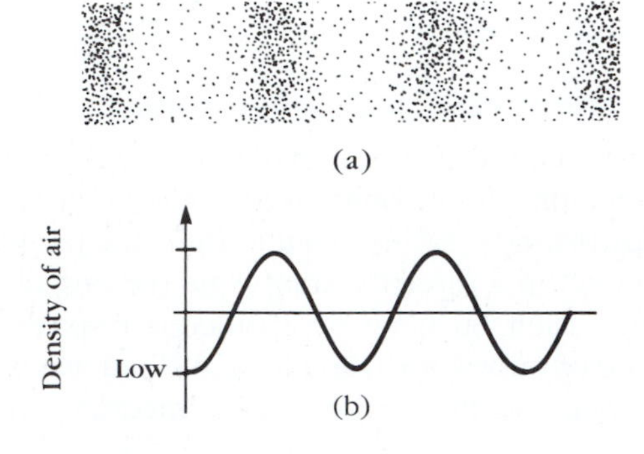

FIGURE 16-16
(a) Longitudinal wave. (b) Its graphical representation.

have heard a shout reflected from a distant cliff—which we call an "echo." A wave pulse traveling down a rope (or slinky) is reflected as shown in Figure 16-17. You can observe for yourself that the reflected pulse is inverted as shown in part (a) if the end of the rope is fixed, and returns right side up, as in part (b), if the end is free. In the case of two-dimensional waves (for example, waves on the surface of water) or three-dimensional waves, we are concerned with *wave fronts*, by which we mean the whole width of a wave crest. A line drawn in the direction of motion, perpendicular to the wave front, is called a *ray*. As shown in Figure 16-18, the angle that the incoming or *incident* wave makes with the reflecting surface is equal to the angle made by the reflected wave. That is, *the angle of reflection equals the angle of incidence*; the "angle of incidence" is defined as the angle the incident ray makes with a perpendicular to the reflecting surface (or the angle the wave front makes with the surface) and the "angle of reflection" is the corresponding angle for the reflected wave.

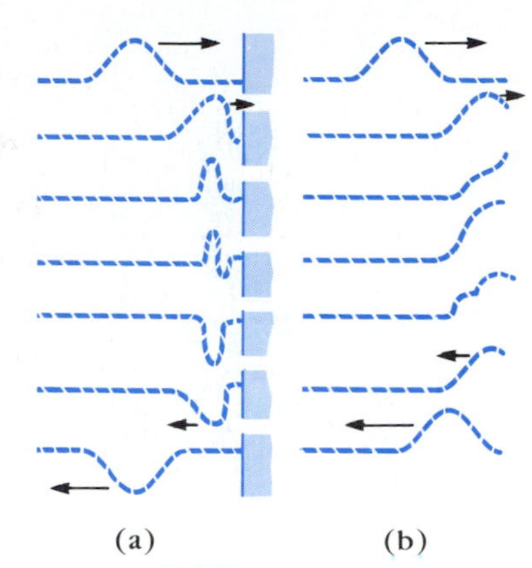

FIGURE 16-17
Reflection of a wave pulse on a rope when the end of the rope is (a) fixed and (b) free.

Refraction

When any wave strikes a boundary, some of the energy is reflected and some is transmitted or absorbed. When a two- or three-dimensional wave traveling in one medium crosses a boundary into a medium where its velocity changes, the transmitted wave may move in a different direction than the incident wave, as shown in Figure 16-19. This phenomenon is known as **refraction**. One example is a water wave; the velocity decreases in shallow water and the waves refract (Figure 16-20). We will examine refraction (as well as other wave properties) in more detail when we study light waves.

FIGURE 16-18
Law of reflection.

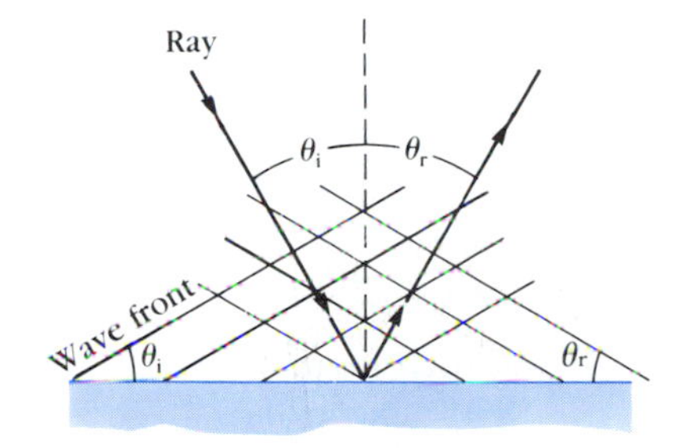

Interference

Interference refers to what occurs when two waves pass through the same region of space at the same time. Suppose two different waves of equal

FIGURE 16-19
Refraction of waves passing a boundary.

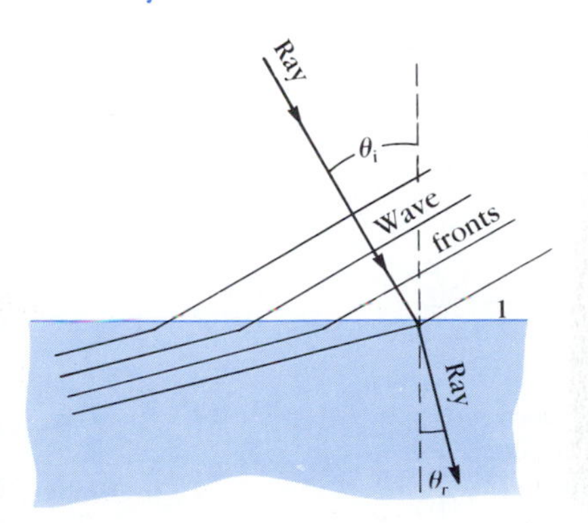

FIGURE 16-20
Water waves refracting.

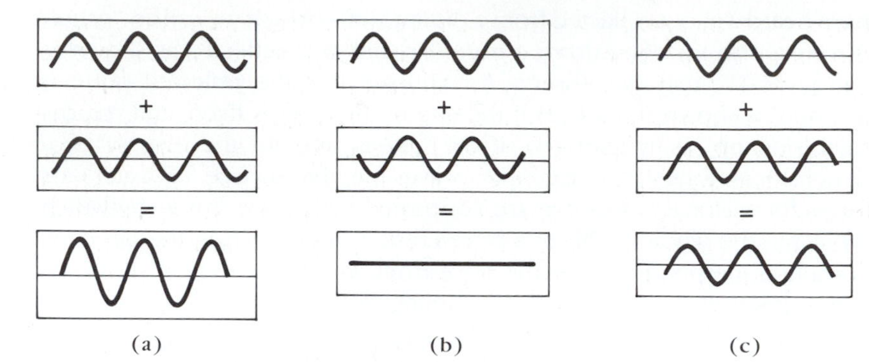

FIGURE 16-21
Two waves interfere: (a) constructively, (b) destructively, (c) partially destructively.

amplitude approach each other. They could be waves begun from opposite ends of a rope or they could be two sets of circular waves formed by throwing two rocks into a pond. When the two sets of waves meet, they are neither reflected nor absorbed; they pass right through each other. At those points where the waves overlap, the net amplitude of the combined wave will be the algebraic sum of the displacements of the two separate waves.† If at a given point the crests of the two waves arrive at the same time and the troughs arrive at the same time, then the combined wave is larger then either of the separate waves—Figure 16-21(a). This is called **constructive interference**. On the other hand, if the crests of one wave arrive at the same time as the troughs of the other, the net amplitude is zero, as shown in Figure 16-21(b). This is known as **destructive interference**. There is no wave motion at all at this point. When the two

† This is termed the *principle of superposition*.

FIGURE 16-22
Interference of water waves.

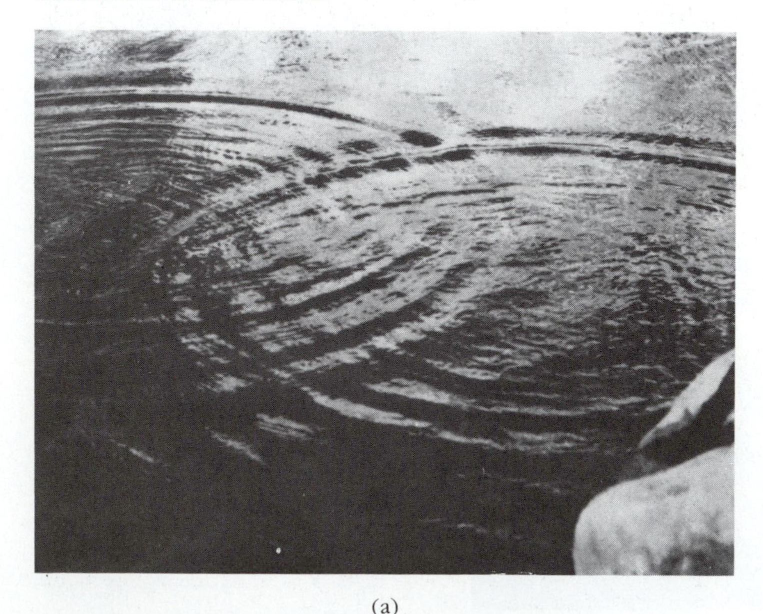

(a)

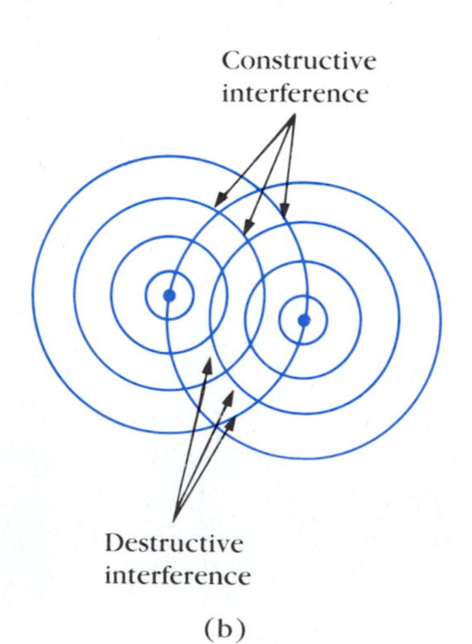

(b)

waves meet in a manner somewhere between these two extremes, as shown in Figure 16-21(c), the result is called **partially destructive interference**. If the two waves have different amplitudes and/or different wavelengths, again partially destructive interference occurs.

You can demonstrate interference between water waves by throwing two rocks into a still body of water at exactly the same time, as shown in Figure 16-22. The pattern of the two waves crossing each other is called an "interference pattern." Other types of waves, such as sound waves, also produce interference patterns.

EXPERIMENT-PROJECT

Simultaneously throw two rocks or tennis balls a few feet apart into a still pond. Locate regions of constructive and of destructive interference. The regions of constructive interference are those where the water moves up and down with the greatest amplitude, and the regions of destructive interference are those where the water remains almost still. These regions are easier to see if you float a few leaves on the water's surface. Diagram the waves produced, indicating the regions of constructive and destructive interference. Repeat the experiment with either a larger or a smaller distance between the rocks or tennis balls: again diagram the waves and note differences in the interference pattern. You may have to perform this experiment several times before you are able to discern the regions of constructive and destructive interference.

Diffraction

Waves exhibit another important characteristic known as **diffraction**. This refers to the fact that waves spread as they travel, and when they encounter an obstacle they bend around it somewhat and pass into the region behind, as shown in Figure 16-23 for water waves. The amount of diffraction depends on the wavelength of the wave and on the size of the obstacle. This is shown in Figure 16-24. If the wavelength is much larger than the object, such as the grass blades of Figure 16-24(a), the wave bends around them almost as if they were not there. For larger objects, parts (b) and (c), there is more of a "shadow" region behind the obstacle. But notice in part (d), where the obstacle is the same as in part (c) but the wavelength is longer, that there is more diffraction into the shadow region. As a rule of thumb, only if the wavelength is less than the size of the object will there be a significant shadow region. It is worth noting that this rule applies to *reflection* from the obstacle as well. Very little of the wave is reflected unless the wavelength is less than the size of the obstacle.

The fact that waves can bend around obstacles, and thus can carry energy to areas behind obstacles, is in clear distinction to energy carried by material particles. A clear example is the following: if you are standing behind a wall, you can't be hit by a baseball thrown from the other side but you can hear a shout or other sound because the sound waves diffract around the edges.

FIGURE 16-23
Wave diffraction. The waves come from the upper right. Note how they bend around and behind the headland. Note also how they bend around and behind the rock outcrop on both sides, as diagrammed in Figure 16-24 (b). (The headland and rock outcrop can be considered as forming a slit through which the waves pass and spread out.)

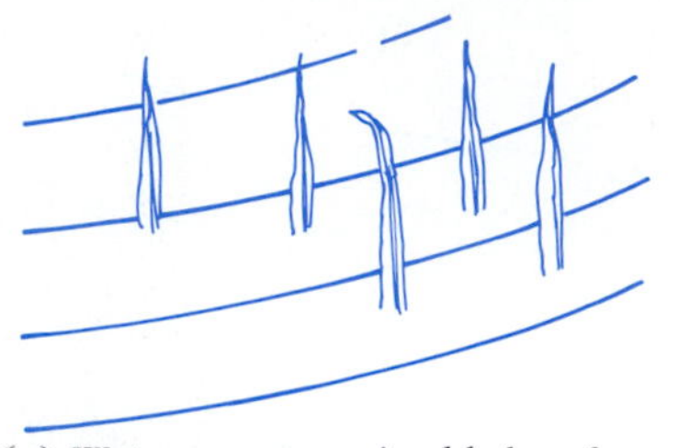

(a) Water waves passing blades of grass

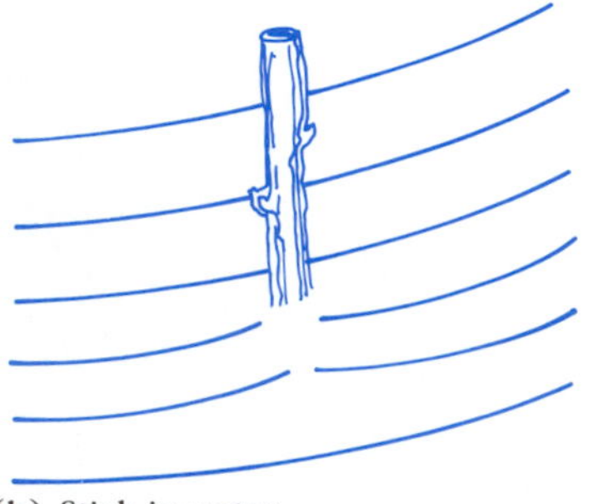

(b) Stick in water

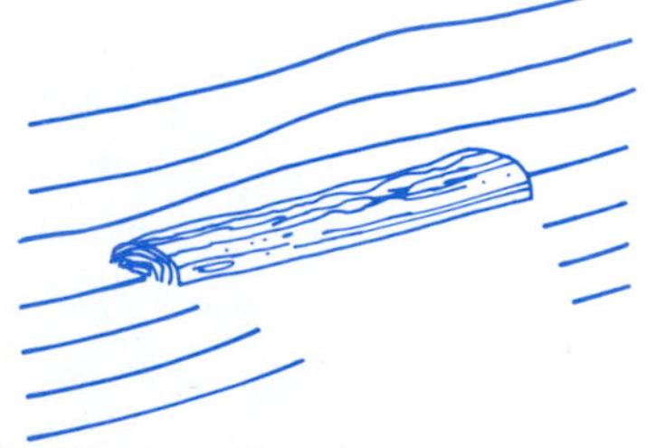

(c) Short wavelength waves passing log

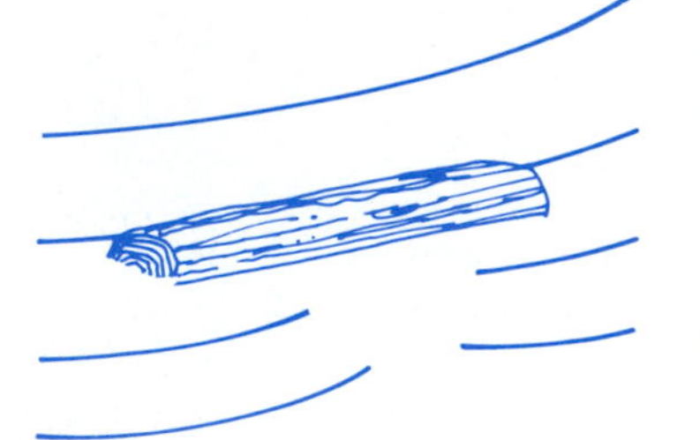

(d) Long wavelength waves passing log

FIGURE 16-24

Water waves passing objects of various sizes. Note that the larger the wavelength compared to the size of the object, the more diffraction there is into the "shadow region."

Both interference and diffraction occur only for energy carried by waves and not for energy carried by material particles. This distinction is important for understanding the nature of light, and of matter itself, as we shall see in later chapters.

6. STANDING WAVES

If you shake one end of a rope (or slinky) and the other end is kept fixed, a continuous wave will travel down to the fixed end and be reflected back. As you continue to vibrate the rope there will be waves traveling in both directions, and the wave traveling down the rope will interfere with the reflected wave coming back. Usually, there will be quite a jumble. But if you vibrate the rope at just the right frequency, these two waves will interfere in such a way that a large-amplitude **standing wave** will be produced (Figure 16-25). It is called a "standing wave" because it does not appear to be moving. The points of destructive interference, called **nodes**, and of constructive interference, called **antinodes**, remain in fixed positions. Standing waves occur at more than one frequency. The lowest frequency of vibration that produces a standing wave gives rise to the pattern shown in Figure 16-25(a). The standing waves shown in parts (b) and (c) are produced at precisely twice and three times the lowest frequency (assuming that the tension in the rope is the same). The rope can also vibrate with four loops at four times the lowest frequency, and so on.

The frequencies at which standing waves are produced are the **natural frequencies** or **resonant frequencies** of the rope, and the different standing-wave patterns shown in Figure 16-25 are different "resonant modes of vibration." For although a standing wave is the result of the interference of two waves traveling in opposite directions, it is also an example of a vibrating object at resonance (Section 2). When a standing wave exists on a rope, the rope is vibrating in place; and at the frequencies at which resonance occurs, little effort is required to achieve a large amplitude. Standing waves then represent the same phenomenon as the resonance of a vibrating spring or pendulum which we discussed earlier. The only difference is that a spring or pendulum has only one resonant frequency whereas the rope has an infinite number of resonant frequencies, each of which is a whole-number multiple of the lowest frequency.

Now let us consider a string stretched between two supports that is plucked like a guitar string—Figure 16-26(a). Waves of a great variety of frequencies will travel in both directions along the string, will be reflected at the ends, and will travel back in the opposite direction. Most of these waves interfere in a random way with each other and quickly die away. However, those waves that correspond to the resonant frequencies of the string will persist. The ends of the string, since they are fixed, will be nodes. There may be other nodes as well. Some of the possible resonant modes of vibration (standing waves) are shown in Figure 16-26(b). Generally, the motion will be a combination of these different resonant modes; but only those frequencies that correspond to a resonant frequency will be present.

To determine the resonant frequencies, we first note that the wavelengths of the standing waves bear a simple relationship to the length L of the string. The lowest frequency, called the **fundamental frequency** (or **first harmonic**), corresponds to one antinode (or loop); and, as can be seen in Figure 16-26(b), the entire length corresponds to one-half wavelength. Thus $L = \frac{1}{2}\lambda_1$, where λ_1 stands for the wavelength of the fundamental. The next mode has two loops and is called the **second harmonic**†; the length L of the string corresponds to one complete wavelength: $L = \lambda_2$. For the third and fourth harmonics, $L = \frac{3}{2}\lambda_3$ and $L = 2\lambda_4$, respectively, and so on. In general, we can write

$$L = \frac{n\lambda_n}{2}, \text{ where } n = 1, 2, 3 \ldots$$

The integer n labels the number of the harmonic; $n = 1$ for the fundamental; $n = 2$ corresponds to the second harmonic, and so on. (The second harmonic is also called the first *overtone*, the third harmonic the second overtone, and so on.) We can write this equation as

$$\lambda_n = \frac{2L}{n}, \text{ where } n = 1, 2, 3 \ldots$$

Since the frequency f is inversely proportional to the wavelength ($f = v/\lambda$, as discussed earlier), we see that the frequency is proportional to

† The term "harmonic" derives from music because whole-number multiples of frequencies harmonize.

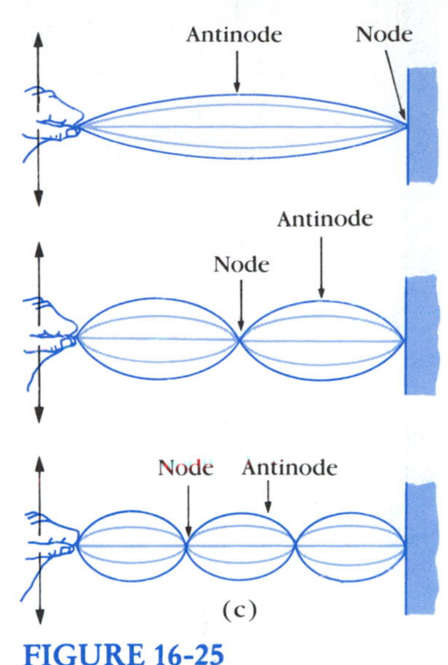

FIGURE 16-25
Standing waves corresponding to three resonant frequencies.

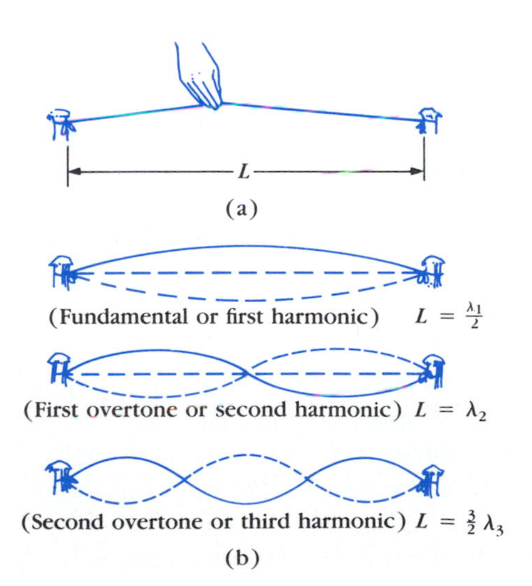

FIGURE 16-26
(a) A string is plucked. (b) Only standing waves corresponding to resonant frequencies persist for long.

n and inversely proportional to L: $f = v/\lambda = nv/2L$. The concept of wave speed, v, still makes sense for a standing wave because a standing wave is equivalent to two traveling waves moving in opposite directions.

As an example, suppose waves travel along a 0.60-m long guitar string at a speed of 420 m/s. The wavelength of the fundamental ($n = 1$) is $2L =$ 1.20 m. The frequency of the fundamental is then $f = v/\lambda = (420 \text{ m/s})/(1.20 \text{ m}) = 350$ Hz. The second harmonic or first overtone ($n = 2$) will have a wavelength $\lambda_2 = 2L/2 = 0.60$ m, so its frequency will be $f = v/\lambda = (420 \text{ m/s})/(0.60 \text{ m}) = 700$ Hz, which is double the fundamental frequency. Similarly, the third harmonic will have three times the frequency of the fundamental, or 1050 Hz, and so on.

Standing waves are produced not only on strings but on any object that is set into vibration. Even when a rock or a piece of wood is struck with a hammer, standing waves are set up that correspond to the natural resonant frequencies of that object. In general, the resonant frequencies depend on the dimensions of the object, just as for a string they depend on its length. For example, a small object does not have as low resonant frequencies as a large object. All musical instruments depend on standing waves to produce their musical sounds: from string instruments to wind instruments (in which a column of air vibrates as a standing wave) to drums and other percussion instruments. We will look at this in more detail in the next chapter.

SUMMARY

A vibrating object undergoes *simple harmonic motion* (SHM) if the restoring force is proportional to the displacement (it obeys Hooke's law). The force constant k is the ratio of restoring force to displacement, $k = F/x$. The maximum displacement is called the *amplitude*. The *period* T is the time required for one complete cycle (back and forth) and the *frequency* f is the number of cycles per second; they are related by $f = 1/T$.

During a vibration, the energy continually alternates between kinetic and potential. When friction is present, the motion is said to be *damped*: the displacement decreases in time and the energy is eventually all transformed to heat. When an oscillating force is applied to a system capable of vibrating, the amplitude of vibration is very large if the frequency of the applied force equals (or nearly equals) the natural frequency of vibration for the object. This is called *resonance*.

A vibrating object can give rise to waves that travel outward. Waves on water or on a rope are simple examples. The wave may be a pulse (a single crest) or it may be continuous (many crests and troughs). The *wavelength* of a continuous wave is the distance between two adjacent crests; and its *frequency* is equal to the number of crests that pass a given point per second. The *velocity* of a wave (how fast a crest moves) is equal to the product of wavelength and frequency: $v = \lambda f$. The *amplitude* of a wave is defined as the height of a crest or depth of a trough. Waves carry energy from place to place without matter being carried. In a *transverse wave*, the oscillations are perpendicular to the direction in which the wave travels. In a *longitudinal wave*, the oscillations are along the line of travel; sound is an example of a longitudinal wave.

Waves reflect off objects in their path. When a *wave front* strikes an object obliquely, the angle of reflection equals the angle of incidence. When a wave strikes a boundary between two materials in which it can travel, part of the wave is reflected and part is transmitted. A transmitted wave front may undergo *refraction* or bending. When two waves pass through the same region at the same time, they *interfere*. The resultant displacement or amplitude is the sum of their separate displacements; this can result in *constructive interference, destructive interference*, or something in between, depending on the amplitude and relative phases of the waves. Waves also undergo *diffraction*, which means that they tend to spread as they travel and bend around and behind objects in their path.

Waves traveling on a string (or other medium) of fixed length can interfere with waves that have reflected off the end and are traveling in the opposite direction. At certain frequencies *standing waves* can be produced in which the waves seem to be standing still instead of traveling. The string (or other medium) is vibrating as whole; this is essentially a resonance phenomenon and the frequencies at which standing waves occur are called *resonant frequencies*. The points of complete destructive interference (no vibration) are called *nodes*. Points of maximum vibrational amplitude are called *antinodes*.

QUESTIONS

1. Give some examples of everyday vibrating objects. Which follow SHM, at least approximately?
2. Is the motion of a piston in an automobile engine simple harmonic? Explain. (See Figure 15-4.)
3. Is the acceleration of a simple harmonic oscillator ever zero? If so, where?
4. A 10-kg fish is attached to the hook of a vertical spring scale and is then released. Describe the scale reading as a function of time.
5. How could you double the maximum speed of a SHO?
6. Why do you suppose circular water waves decrease in amplitude as the radius increases?
7. What is the difference between a transverse and a longitudinal wave?
8. What kind of waves will travel along a horizontal rod if you strike it on its end? What if you strike it from above?

9. What do we call (a) the number of wave crests passing a given point per second, and (b) the distance from crest to crest of a wave?
10. Is the frequency of a simple continuous wave the same as the frequency of its source?
11. Explain the difference between the speed of a transverse wave traveling down a rope and the speed of a tiny piece of the rope.
12. The fact that harmonic motion is damped is said to be another example of the second law of thermodynamics. Explain.
13. Why can you make water slosh back and forth in a pan only if you shake the pan at a certain frequency?
14. What do you hear when you put a seashell to your ear? Explain. If you don't have a seashell, try an empty can. Listen to the sounds from a small shell (or can) and from a large one. Why is the sound higher in pitch from the smaller one and lower from the larger one?
15. A tuning fork of natural frequency 264 Hz sits on a table at the front of a room. At the back of the room, two tuning forks, one of natural frequency 260 Hz and one of 420 Hz, are initially silent. When the tuning fork at the front of the room is set into vibration, the 260-Hz fork spontaneously begins to vibrate but the 420-Hz fork does not. Explain.
16. Give several everyday examples of resonance.
17. Is a rattle in a car ever a resonance phenomenon? Explain.

*18. Over the years, buildings have been built out of lighter and lighter materials. How has this affected the natural vibration frequencies of buildings and the problems of resonance due to passing trucks, airplanes, or natural sources of vibration?

19. AM radio signals can usually be heard behind a hill, but FM often cannot. That is, AM signals bend more than FM. Explain. (Radio signals, as we shall see, are carried by electromagnetic waves whose wavelength for AM is typically 200 to 600 m and for FM about 3 m.)
20. If a string is vibrating in three segments, are there any places one can touch it with a knife blade without disturbing the motion?

EXERCISES

1. A piece of rubber is 45 cm long when a weight of 8.0 kg hangs from it and is 58 cm long when a weight of 12.5 kg hangs from it. What is the "spring" constant of this piece of rubber?
2. If a particle undergoes SHM with amplitude A, what is the total distance it travels in one period?

3. A watch spring oscillates with a frequency of 3.22 Hz. How long does it take to make 100 vibrations?
4. An unloaded spring is 38 cm long and its spring constant is 62 N/m. How much will it stretch when 0.50 kg is hung from the end of it?
*5. When an 80-kg person climbs into an 1100-kg car, the car's springs compress vertically by 1.2 cm. What will be the frequency of vibration when the car hits a bump? (Ignore damping.)
*6. A spring vibrates at 3.0 Hz when a weight of 0.82 kg is hung from it. What is the spring constant?
*7. A fisherman's scale stretches 2.5 cm when a 2.1-kg fish hangs from it. What is the spring constant and what will be the frequency of vibration if the fish is pulled down and released so that it vibrates up and down?
*8. A small cockroach of mass 0.30 g is caught in a spider's web. The web vibrates predominately with a frequency of 15 Hz. (a) What is the value of the spring constant k for the web? (b) At what frequency would you expect the web to vibrate if an insect of mass 0.10 g were trapped?
9. A fisherman notices that wave crests pass the bow of his anchored boat every 6.0 s. He measures the distance between two crests to be 20 m. How fast are the waves traveling?
10. A water wave has a wavelength of 12 cm and a speed of 50 cm/s. What is its frequency?
11. A sound wave in air has a frequency of 262 Hz and travels with a speed of 330 m/s. How far apart are the wave crests (compressions)?
12. AM radio signals have frequencies between 550 kHz and 1600 kHz (kilohertz) and travel with a speed of 3.0×10^8 m/s. What are the wavelengths of these signals? On FM the frequencies range from 88 MHz to 108 MHz (megahertz) and travel at the same speed. What are their wavelengths?
13. What is the velocity of waves on a violin string 0.40 m long whose fundamental frequency is 262 Hz?
14. An earthquake wave travels with a speed of 5.0 km/s. What is its wavelength if its frequency is 50 Hz?
15. A sailor produces a sound by striking the hull of his ship beneath the waterline and hears the echo of the wave reflected from the ocean bottom 1.5 s later. How deep is the ocean at this point? The speed of sound in water is approximately 1440 m/s.
16. The two pulses shown in Figure 16-27 are moving toward each other. (a) Sketch the shape of the string at the moment they directly overlap. (b) Sketch the shape of the string a few moments later.
17. A violin string vibrates at 294 Hz as its fundamental frequency. What are the frequencies of the first three overtones?

FIGURE 16-27

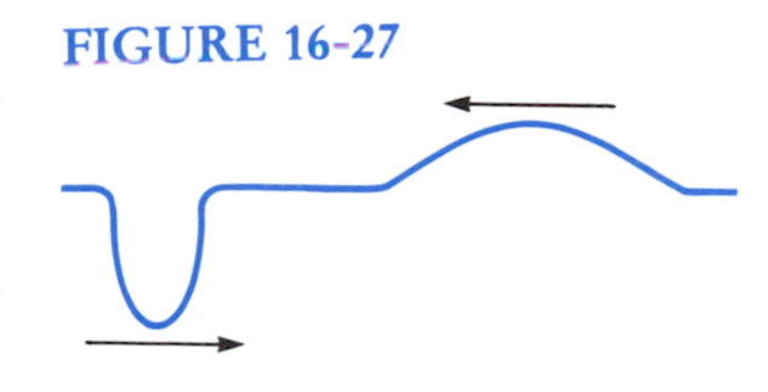

18. A particular string vibrates in four segments at a frequency of 120 Hz. What is the fundamental frequency of this string?

19. The velocity of waves on a string is 400 m/s. If the frequency of standing waves is 380 Hz, how far apart are the nodes?

20. When you slosh the water back and forth in a tub at just the right frequency, the water alternately rises and falls at each end. Suppose the frequency to produce such a standing wave in a 50-cm wide tub is 0.85 Hz. What is the speed of the water wave?

*21. The ripples in a certain groove 12 cm from the center of a 33-rpm phonograph record have a wavelength of 2.4 mm. What will be the frequency of the sound emitted?

SOUND

CHAPTER

17

Sound is associated with our sense of hearing and, therefore, with the physiology of our ears and the psychology of our brain which interprets the sensations that reach our ears. The term *sound* also refers to the physical sensation that stimulates our ears: namely, longitudinal waves.

We can distinguish three aspects of any sound. First, there must be a *source* for a sound; and as with any wave, the source of a sound wave is a vibrating object. Second, the energy is transferred from the source in the form of longitudinal sound *waves*. And third, the sound is *detected* by an ear or an instrument. We will discuss sources and detectors of sound, as well as some important applications to fields such as medicine, later in this chapter. But first we will look at certain aspects of sound waves themselves.

1. CHARACTERISTICS OF SOUND

We already saw in Chapter 16, Figure 16-15, how a vibrating drumhead produces a sound wave in air. Indeed, we usually think of sound waves traveling in the air, for normally it is the vibrations of the air that force our eardrums to vibrate. But sound waves can also travel in other materials. Two stones struck together under water can be heard by a swimmer beneath the surface, for the vibrations are carried to the ear by the water. When you put your ear flat against the ground, you can hear an approaching train or truck. In this case the ground does not actually touch your eardrum but the longitudinal wave transmitted by the ground is called a sound wave just the same. Indeed, longitudinal waves traveling in any material medium are often referred to as sound waves. Clearly sound cannot travel in the absence of matter. For example, a bell ringing inside an evacuated jar cannot be heard.

Speed of Sound

In air at 0°C and atmospheric pressure, a sound wave travels with a speed of 331 m/s. The speed of sound in air varies somewhat with temperature: it increases about 0.60 m/s for each Celsius degree increase in temperature. Thus, at 20°C the speed of sound is (0.60)(20) m/s = 12 m/s faster than at 0°, or 331 m/s + 12 m/s = 343 m/s.

The speed of sound is quite low compared with the speed of light. Consequently, a distant event is seen before it is heard. For example, if you have sat in the balcony of a concert hall several hundred feet from the musicians, you may have noticed a time lag of a fraction of a second between the time a musician begins to play and the time you hear the sound. The time lag for more distant events is even greater. Thunder is often heard many seconds after the flash of lightning was seen. By measuring the time difference we can determine how far away the lightning flashed. Because the speed of light is so great, we see events on earth almost instantaneously. But sound has a speed of only about 340 m/s (1100 ft/s), and therefore it takes about 3 seconds for sound to travel a kilometer (5 seconds to travel a mile). Thus, if thunder is heard 6 seconds after a lightning flash was seen, the lightning must have struck 2 km away.

The speed of sound in different materials is different. In helium it is about 1000 m/s, in water about 1500 m/s (almost 5 times its speed in air), and in iron or aluminum about 5000 m/s.

Loudness and Pitch

Two aspects of any sound are immediately evident to a human listener. These are "loudness" and "pitch," and each refers to a sensation in the consciousness of the listener. But to each of these subjective sensations there corresponds a physically measurable quantity. **Loudness** is related to the energy in the sound wave, and we shall discuss it in the next section.

The **pitch** of a sound refers to whether it is high, like the sound of a piccolo or violin, or low, like the sound of a bass drum or string bass. The physical quantity that determines pitch is the frequency, a fact that was first noted by Galileo. The lower the frequency, the lower the pitch; the higher the frequency, the higher the pitch.[†]

EXPERIMENT

Hold a card against the spokes of a bicycle wheel and spin the wheel slowly; then increase the speed. What do you notice?

OR

Play a 33- or 45-rpm record at the wrong speed on a phonograph. What happens to the sound? Explain.

[†] Although pitch is determined mainly by frequency, it also depends to a slight extent on loudness; for example, a very loud sound may seem slightly lower in pitch than a quiet sound of the same frequency.

The human ear responds to frequencies in the range from about 20 Hz to about 20,000 Hz. This is called the **audible range**. These limits vary somewhat from one individual to another. One general trend is that as people age they are less able to hear the high frequencies, so that the high-frequency limit may be 10,000 Hz or less.

Sound waves whose frequencies are outside the audible range may reach the ear, but we are not generally aware of them. Frequencies above 20,000 Hz are called **ultrasonic** (do not confuse with *supersonic*, which means a speed faster than the speed of sound). Many animals can hear ultrasonic frequencies; dogs, for example, can hear sounds as high as 50,000 Hz and bats can detect frequencies as high as 100,000 Hz. Ultrasonic waves have a number of applications in medicine and other fields, which we will discuss later in this chapter.

Sound waves whose frequencies are below the audible range (that is, less than 20 Hz) are called **infrasonic**. Sources of infrasonic waves are earthquakes, thunder, volcanoes, and waves produced by vibrating heavy machinery. This last source can be particularly troublesome to workers, for infrasonic waves—even though inaudible—can cause damage to the human body. These low-frequency waves act in a resonant fashion, causing considerable motion and irritation of internal organs of the body.

2. SOUND INTENSITY

Loudness, like pitch, is a sensation in the human consciousness. It is therefore a subjective quality, yet it, too, is related (though not in a simple mathematical way) to a physically measurable quantity, *intensity*. The **intensity** of a sound refers to the rate energy is carried across unit area by a sound wave (watts/m^2) and is numerically proportional to the square of the amplitude of the wave. The greater the amplitude of a sound wave, the greater the intensity; and the greater the intensity, the louder it sounds.

The human ear can detect sounds with an intensity as low as 10^{-12} W/m^2 and as high as 1 W/m^2 (and even higher, although above this it is painful). This is an incredibly wide range of intensity, spanning a factor of 10^{12} from lowest to highest. Presumably because of this wide range, what we perceive as loudness is not directly proportional to the intensity. True, the greater the intensity, the louder the sound. But to produce a sound that sounds about twice as loud requires a sound wave that has about 10 times the intensity. For example, a sound wave of intensity 10^{-9} W/m^2 sounds to an average human being as if it is about twice as loud as one whose intensity is 10^{-10} W/m^2; and an intensity of 10^{-2} W/m^2 sounds about twice as loud as 10^{-3} W/m^2 and four times as loud as 10^{-4} W/m^2.

Because the relationship between the subjective sensation of loudness and the physically measurable quantity intensity is not a linear one, we do not use a linear scale; instead, we use a compressed, or multiplicative, scale known as *logarithmic*. The zero point on this scale does not correspond to zero amplitude. Rather, the zero point is usually chosen to be at the threshold of human hearing—at the lowest-intensity sound that the average human ear can detect. The unit on this scale is a *bel*, or more commonly, the *decibel* (dB), which is $\frac{1}{10}$ bel. Thus, the **intensity level**, β, of any sound is defined in terms of its intensity, I, as follows:

$$\beta \text{ (in dB)} = 10 \log \frac{I}{I_0}$$

where I_0 is the intensity of the reference level, normally taken as the minimum intensity audible to an average person, the "threshold of hearing," which is $I_0 = 1.0 \times 10^{-12}\,\text{W/m}^2$. The intensity level of a sound whose intensity, for example, is $I = 1.0 \times 10^{-10}\,\text{W/m}^2$ will be (logarithms are discussed in Appendix B):

$$\beta = 10 \log \frac{10^{-10}}{10^{-12}} = 10 \log 100 = 10 \times 2 = 20\text{ dB},$$

since log 100 is equal to 2.0. Notice that the intensity level at the threshold of hearing is 0 dB; that is, $\beta = 10 \log (10^{-12}/10^{-12}) = 10 \log 1 = 0$. Notice, too, that an increase in intensity by a factor of 10 corresponds to a level increase of 10 dB. An increase in intensity by a factor of 100 corresponds to a level increase of 20 dB. Thus a 50-dB sound is 100 times more intense than a 30-dB sound.

TABLE 17-1 Intensity of Various Sounds

SOURCE OF THE SOUND	INTENSITY LEVEL (dB)	INTENSITY (W/m^2)
Jet plane at 30 m	140	100
Threshold of pain	120	1
Loud indoor rock concert	120	1
Siren at 30 m	100	1×10^{-2}
Auto interior, moving at 90 km/h	75	3×10^{-5}
Busy street traffic	70	1×10^{-5}
Ordinary conversation at 50 cm	65	3×10^{-6}
Quiet radio	40	1×10^{-8}
Whisper	20	1×10^{-10}
Rustle of leaves	10	1×10^{-11}
Threshold of hearing	0	1×10^{-12}

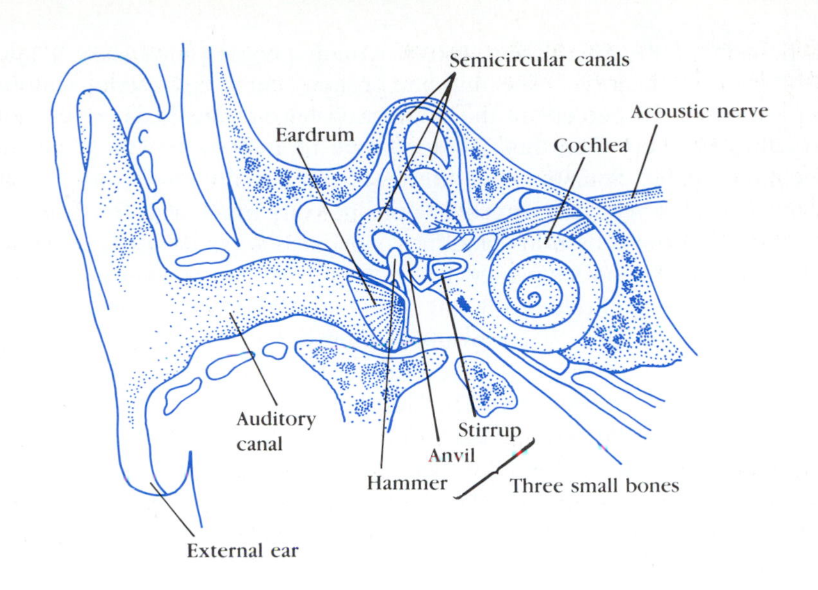

FIGURE 17-1
Diagram of the human ear.

The intensities and intensity levels for a number of common sounds are listed in Table 17-1.

3. THE EAR AND ITS RESPONSE; LOUDNESS

The human ear is a remarkably sensitive detector of sound. Mechanical detectors of sound, namely microphones, can barely match the ear in detecting low-intensity sounds.

The function of the ear is to efficiently transform the vibrational energy of sound waves into electrical signals that are carried to the brain by way of nerves. A microphone performs a similar task (discussed in Chapter 21).

A human ear is diagrammed in Figure 17-1. Sound waves enter the ear and strike a diaphragm called the *eardrum*. The waves set the eardrum into motion, which then vibrates with the same frequency or frequencies as the sound waves themselves. Three small bones connected to the eardrum transmit the vibrations to a fluid in the inner ear, the *cochlea*. The motion of the fluid is detected by tiny hairs connected to nerves; at this point the vibrations are transformed into electrical signals and are carried by the nerves to the brain, where the sensation of sound is realized. Sound waves can also reach the inner ear by traveling directly through the bones of the skull, as when you tap your head.

The ear is not equally sensitive to all frequencies. The lowest curve in Figure 17-2 represents the intensity level, as a function of frequency, in the softest sound that is just audible. As can be seen, the ear is most sensitive to sounds of frequency between 2000 and 3000 Hz. Whereas a 1000-Hz sound is audible at a level of 0 dB, a 100-Hz sound must be at least 30 dB to be heard. This lowest (solid) curve represents a very good ear; only about

FIGURE 17-2
Sensitivity of the human ear as a function of frequency (see text). Note that the frequency scale is "logarithmic" in order to cover a wide range of frequencies.

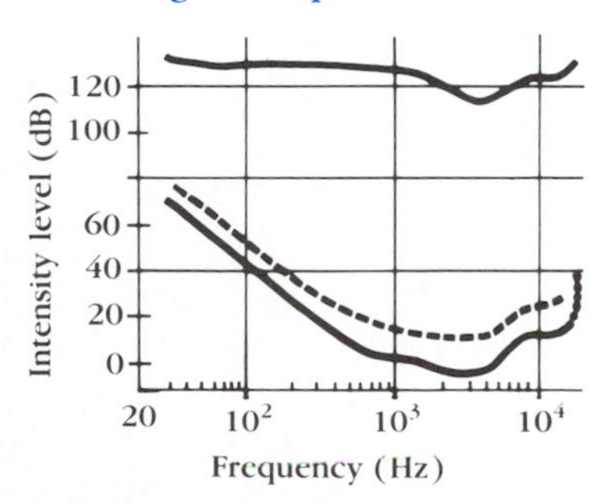

1 percent of the population, mostly young people, have such a low "threshold of hearing." The middle (broken) curve represents a more typical curve; 50 percent of the population have a threshold of hearing equal to or better than this. The top curve represents the "threshold of feeling or pain." Sounds above this level can actually be felt and cause pain. As can be seen, the threshold does not vary much with frequency.

Our subjective sensation of loudness obviously depends not only on the intensity but also on frequency. For example, as seen in Figure 17-2, an average person will detect a 30-dB sound at 1000 Hz as reasonably loud but a 30-dB sound at 50 Hz would not be heard at all. Thus at lower intensity levels, our ears are less sensitive to the high and low frequencies relative to middle frequencies. The "loudness" control on stereo and hi-fi systems is intended to compensate for this. As the volume is turned down, the loudness control boosts the high and low frequencies relative to the middle frequencies so that the sound will be more uniform. Many listeners, however, find the sound more pleasing or natural without the loudness control.

4. SOURCES OF SOUND: VIBRATING STRINGS AND AIR COLUMNS

The source of any sound is a vibrating object. Almost any object can vibrate and hence be a source of sound. We shall now discuss some simple sources of sound, particularly as they apply to those most pleasant of sounds, music. We will also discuss briefly nonmusical sound—that is, noise.

Musical instruments make use of a variety of vibrating materials to produce sound. A drum has a stretched membrane that vibrates. Xylophones and marimbas have metal or wood resonators that can be set into vibration. Bells, cymbals, and the gong also make use of a vibrating metal. The most widely used instruments make use of vibrating strings, such as the violin, guitar, and piano, or make use of vibrating columns of air, such as the flute, trumpet and pipe organ.

In each musical instrument the source is set into vibration by striking, plucking, bowing, or blowing. Standing waves are produced and the object vibrates at its natural resonant frequencies.

Strings

We saw in Chapter 16, Figure 16-26, how standing waves are established on a string. This is the basis for all stringed instruments. The pitch of a note is normally determined by the lowest resonant frequency, the **fundamental**, which corresponds to nodes occurring only at the ends. As we saw in Figure 16-26, the wavelength of the fundamental is equal to twice the length of the string; therefore, the fundamental frequency $f = v/\lambda = v/2L$, where v is the velocity of the wave on the string. When a finger is placed on the string of, say, a guitar or violin, the effective length of the string is shortened; when it is plucked or a bow is pulled across it, the pitch is higher since the wavelength of the fundamental is now shorter (Figure 17-3). The strings on a guitar or violin are all the same length when unfingered. They each sound at a different pitch because the strings have different mass; and the velocity of the waves generated by the heavier strings is less than the velocity of the waves generated by the lighter strings. This makes sense because the heavier strings offer more inertia to a traveling wave and we would therefore expect the wave to travel more slowly.† Therefore, since frequency = velocity/wavelength, the frequency or pitch of a heavy string will be lower than that of a light string of equal length.

The wide variety of musical notes that can be played on a violin or guitar thus arises from the different thicknesses of the strings and from the changing of the effective length of each string by fingering. The piano and harp, unlike other stringed instruments, have strings of different lengths for each note and thus require no fingering.

Stringed instruments would not be very loud if they relied on their vibrating strings to produce the sound waves. The strings on a stringed instrument are simply too thin to compress and expand very much air. Because of this, all stringed instruments make use of a kind of mechanical amplifier known as a *sounding board* or *sounding box*, which is sometimes called a *resonator*. The strings on a piano are attached to a large wooden board—the sounding board. On a guitar, a violin, and a cello, they are attached through the "bridge" to the sounding box (Figure 17-4). When the strings are set into vibration, the sounding board or box is set into vibration as well. Since it has much greater area in contact with the air it can produce a much stronger sound wave, and thus acts to amplify the sound. On an electric guitar the sounding box is not so important, since the vibrations of the strings are amplified electrically.

FIGURE 17-3

The wavelength of a fingered string (b) is shorter than that of an unfingered string (a). Hence, the frequency of the fingered string is higher. Only the simplest standing wave, the fundamental, is shown.

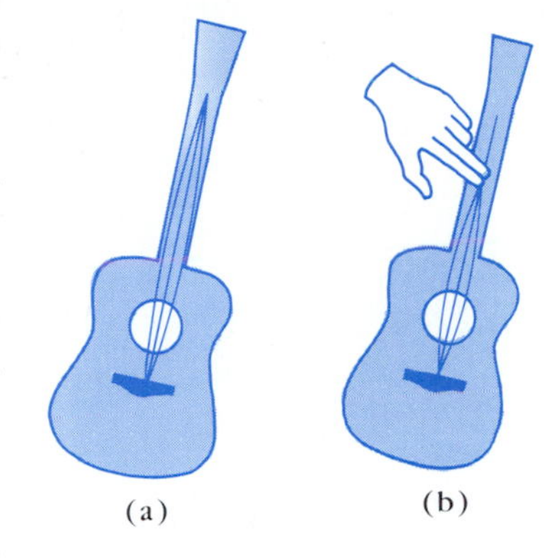

EXPERIMENT

Examine the inside of a piano. Note the different diameters and lengths of the strings. Do you discern a pattern? How do the strings come into contact with the sounding board? Why is it that for some notes the hammer strikes a single string and for others two or three identical strings?

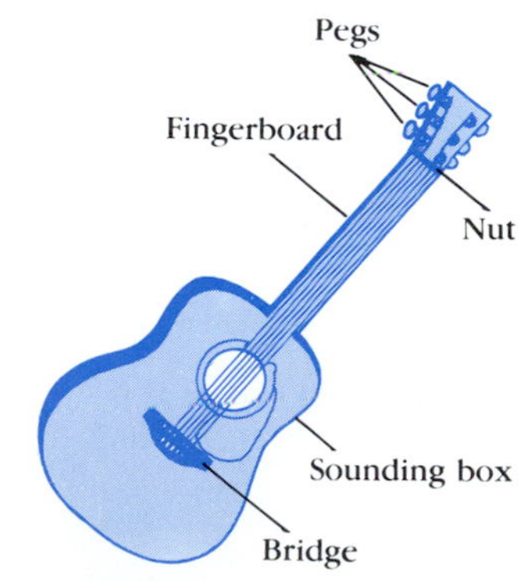

FIGURE 17-4

Stringed instruments have a sounding board or chamber to amplify the vibrations.

Wind Instruments: Vibrating Columns of Air

Many instruments produce sound from the vibrations of standing waves in a column of air within the tube or pipes of the instrument. These include

---

that the velocity of the (standing) waves on a string is different from the velocity of sound waves in air. However, the frequency of the vibrating string is the same as the frequency of the sound wave it produces.

the woodwinds, the brasses, and the pipe organ. The amplification by the sounding box of a violin or guitar and that of the human voice also depends to some extent on vibrations of the air within the box or oral cavity.

Standing waves can occur in the air of any cavity, but the frequencies are difficult to calculate for any but very simple shapes, such as a long narrow tube. Fortunately, and for good reason, this is the situation for most wind instruments.

In some instruments, such as some woodwinds and brasses, a vibrating reed or the vibrating lip of the player helps to set up the vibrations of the air column. In other instruments, such as the flute and the organ, a stream of air is directed against one edge of the opening or mouthpiece; the deflection of the air leads to turbulence, which directly sets up vibrations in the column of air. Because of the disturbance, whatever its source, the air within the tube vibrates with a variety of frequencies, but only certain frequencies persist, which correspond to standing waves. For a string fixed at both ends, we saw in Section 6 of Chapter 16 that the standing waves have nodes (no movement) at the two ends and one or more antinodes (large amplitude of vibration) in between; a node also separates each antinode. The lowest-frequency standing wave, the ***fundamental***, corresponds to a single antinode. The higher-frequency standing waves are the *overtones* or *harmonics*; specifically, the first harmonic is the fundamental, the second harmonic has twice the frequency of the fundamental, and so on (see Figure 16-26).

The situation is similar for a column of air, but we must remember that it is now air itself that is vibrating. Thus the air at the closed end of a tube must be a node, since the air is not free to move there, whereas at the open end of a tube there will be an antinode since the air can move freely. The air within the tube vibrates in the form of longitudinal standing waves. The possible modes of vibration for a tube open at both ends (called an *open tube*) and for one that is open at one end but closed at the other (called a *closed tube*) are shown graphically in Figure 17-5. The graphs represent the amplitude of motion of the air molecules within the tube. Let us look first at the open tube—Figure 17-5(a). An open tube has antinodes at both ends. Notice that there must be at least one node within an open tube if there is to be a standing wave at all. This corresponds to the fundamental frequency of the tube; since the distance between two successive nodes, or between two successive antinodes, is $\frac{1}{2}\lambda$, there is one-half a wavelength within the length of the tube in this case: $L = \frac{1}{2}\lambda$, so the fundamental frequency is $f_1 = v/\lambda = v/2L$, where v is the velocity of sound in air. The standing wave with two nodes is the first overtone or second harmonic and has half the wavelength ($L = \lambda$), and therefore twice the frequency. Indeed, the frequency of each overtone is a whole-number multiple of the fundamental frequency. This is just what is found for a string.

For a closed tube—Figure 17-5(b)—there is always a node at the closed end and an antinode at the open end. Since the distance between a

FIGURE 17-5
Modes of vibration (standing waves) for (a) an open tube, (b) a closed tube.

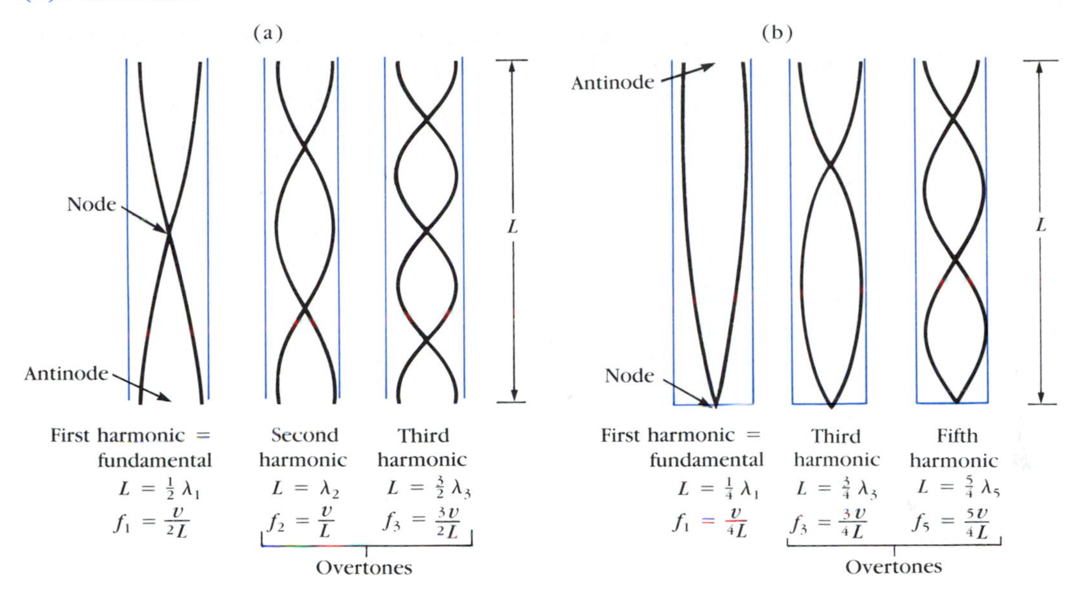

node and the nearest antinode is $\frac{1}{4}\lambda$, we see that the fundamental in this case corresponds to only one-fourth of a wavelength within the length of the tube: $L = \lambda/4$. The fundamental frequency is thus $f_1 = v/4L$, or half what it is for an open tube of the same length. There is another difference, for as we can see from Figure 17-5(b) only the odd harmonics are present in a closed tube. That is, the overtones have frequencies equal to 3, 5, 7 . . . times the fundamental frequency. There is no way for a wave with 2, 4 . . . times the fundamental to have a node at one end and an antinode at the other; thus they cannot exist as standing waves in a closed tube.

Organs make use of both open and closed pipes. Notes of different pitch are sounded using different pipes which vary in length from a few centimeters to 5 m or more. Other musical instruments can act like a closed tube or like an open tube. A flute, for example, is an open tube, for it is open not only where you blow into it but also at the opposite end. The different notes on a flute and many other instruments are obtained by shortening the length of the tube—that is, by uncovering holes along its length. In a trumpet, on the other hand, the pushing down of the valves opens additional lengths of tube. In all these instruments, the longer the length of the vibrating air column the lower the pitch.

As a numerical example, let us calculate the fundamental frequencies and first three overtones for a 26-cm-long organ pipe at 20°C if it is (a) open and (b) closed. At 20°C, the speed of sound in air is 343 m/s

(Section 1). (a) For the open pipe, the fundamental frequency is:

$$f_1 = \frac{v}{2L} = \frac{343\ \text{m/s}}{2(0.26\ \text{m})} = 660\ \text{Hz}.$$

The overtones, which include all harmonics, are 1320 Hz, 1980 Hz, 2640 Hz, and so on. (b) Referring to Figure 17-5(b) we see that

$$f_1 = \frac{v}{4L} = \frac{343\ \text{m/s}}{4(0.26\ \text{m})} = 330\ \text{Hz}.$$

But only the odd harmonics will be present, so the first three overtones will be 990 Hz, 1650 Hz, and 2310 Hz.

Since the speed of sound depends on temperature, if the temperature varies the frequency also varies for a wind instrument. As we saw in Section 1, the speed of sound in air at 0°C is 331 m/s and increases 0.60 m/s for each Celsius degree increase. Thus at 10°C, $v = [331 + (0.60)(10)]$ m/s = 337 m/s. At 10°C, the organ pipe discussed just above in (a) would have a fundamental frequency of $f = v/2L = (337\ \text{m/s})/2(0.26\ \text{m}) = 648$ Hz, and would thus be "out of tune."

This is the reason players of wind instruments take time to "warm up" their instruments—so they will be in tune. The effect of temperature on stringed instruments is much smaller.

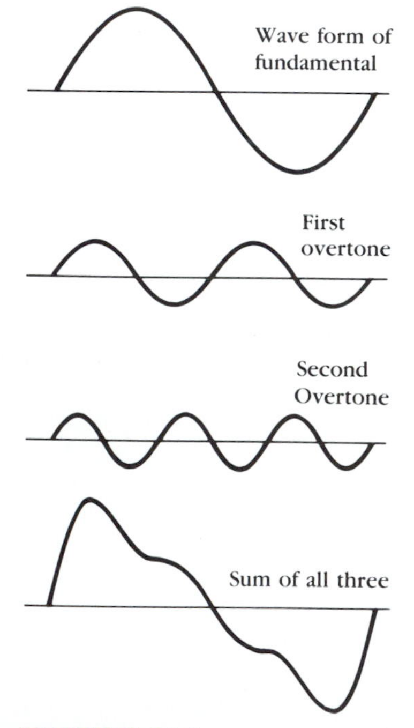

FIGURE 17-6
The amplitudes of the fundamental and first two overtones are added at each point to get the "sum," or composite waveform.

5. QUALITY OF SOUND

Whenever we hear a sound, particularly a musical sound, we are aware of its loudness, its pitch, and also of a third aspect called *quality*. For example, when a piano and then an oboe play a note of the same loudness and the same pitch (say middle C), there is a clear difference in the overall sound. We would never mistake a piano for an oboe. This is what is meant by the *quality* of a sound; for musical instruments the terms *timbre* or *tone color* are also used.

Just as loudness and pitch can be related to physically measurable quantities, so too can quality. The quality of a sound depends on the presence of overtones—their number and their relative amplitudes. Generally, when a note is played on a musical instrument, the fundamental as well as overtones are present simultaneously. Figure 17-6 illustrates how the superposition of three wave forms, in this case the fundamental and first two overtones (with particular amplitudes), would combine to give a composite *wave form*. Of course more than two overtones are usually present.

The relative amplitudes of the various overtones are different for different musical instruments, and this is what gives each instrument its characteristic quality or timbre. String and wind instruments produce overtones that are whole-number multiples of the fundamental. The overtones on a drum membrane, on the other hand, are not simple multiples of the fundamental; that is why a drum does not sound as harmonious as do string or wind instruments.

EXPERIMENT

Hum into a bottle (Figure 17-7); start at a low pitch and slowly raise the pitch. At what pitches does the bottle respond? Do you hear "resonance" at different pitches? Explain.

Normally, the fundamental has a greater amplitude than any of the overtones. It is this predominant frequency, the fundamental, that we hear as *pitch*. The quality of a musical sound is governed by the mixture of overtones present. Different instruments emphasize different overtones, and this is what gives rise to their different quality. For example, a note played on a piano contains a large number of overtones, whereas a note played on a flute contains very few overtones; on a clarinet the lower overtones are not as strong as some of the higher ones.

The manner in which an instrument is played influences the mixture of overtones. For example, plucking a violin string results in a different sound quality than drawing a bow across it. Even the pitch of a musical note can be changed in some instruments by the manner in which the vibrations are produced. For example, the lips of a flutist can be shaped so that the air is blown more into the hole than directly across it. When this is done the fundamental is excited hardly at all. Thus the first overtone, whose frequency is twice that of the fundamental, predominates; and the pitch, in musician's terms, is an octave higher.

The precise physical shape of an instrument and the kind of material of which it is made influence the particular overtones it will produce. Often small changes in the construction of a musical instrument will have a marked effect on the quality of sound it produces. This is especially true of stringed instruments with sounding boards; not only does the sounding board amplify the sound but because it has its own resonant frequencies it helps determine which overtones will be emphasized. A good violin maker carefully chooses and shapes the wood that goes into the instrument so that the quality of sound will be pleasing; even the age of the wood is important. To construct a violin with an exceptionally pleasing tone like a Stradivarius is an art.

An ordinary noise, like the sound of a hammer striking a nail, has a definite quality but usually lacks any distinct pitch. Such noises are mixtures of a great variety of frequencies that bear little relation to one another. This is because a complex object like a nail or a hammer, unlike a simple stretched string or a narrow air column in a pipe, has a complicated set of resonant frequencies. No one frequency predominates, and those that are present are not simply related to one another as they are in a musical instrument. Such a sound we call "noise" in comparison with the more harmonious sounds which contain frequencies that are simple multiples of the fundamental.

FIGURE 17-7
Humming into a bottle to detect resonant frequencies.

TABLE 17-2 Diatonic C Major Scale‡

NOTE	LETTER NAME	FREQUENCY (cps)	FREQUENCY RATIO	INTERVAL
do	C	264		
			9/8	whole
re	D	297		
			10/9	whole
mi	E	330		
			16/15	half
fa	F	352		
			9/8	whole
sol	G	396		
			10/9	whole
la	A	440		
			9/8	whole
ti	B	495		
			16/15	half
do'	C'	528		

‡ Only one octave is included.

TABLE 17-3 Diatonic D Major Scale‡

NOTE	LETTER NAME	FREQUENCY (cps)	FREQUENCY RATIO	INTERVAL
do	D	297		
			9/8	whole
re	E	334		
			10/9	whole
mi	F#	371		
			16/15	half
fa	G	396		
			9/8	whole
sol	A	445		
			10/9	whole
la	B	495		
			9/8	whole
ti	C#	557		
			16/15	half
do	D'	594		

‡ Only one octave is included.

TABLE 17-4 The Equally Tempered Chromatic Scale‡

NOTE	FREQUENCY (cps)	FREQUENCY RATIO	INTERVAL
C	262		
		1.06	half
C# or D♭	277		
		1.06	half
D	294		
		1.06	half
D# or E♭	311		
		1.06	half
E	330		
		1.06	half
F	349		
		1.06	half
F# or G♭	370		
		1.06	half
G	392		
		1.06	half
G# or A♭	415		
		1.06	half
A	440		
		1.06	half
A# or B♭	466		
		1.06	half
B	494		
		1.06	half
C'	524		

‡ Only one octave is included.

6. MUSICAL SCALES

Nearly all music makes use of notes or tones that have a definite pitch relationship to one another. The basic set of notes that are normally played is called a musical *scale*. Many different scales have been known throughout the ages and in different cultures. The simplest one in Western music is the *diatonic major scale*. This is the scale most of us learned in school as "do-re-mi-fa-sol-la-ti-do." Each of the notes on the diatonic scale is named by a letter from A to G and each corresponds to a definite frequency. The difference in pitch between each note on the diatonic scale is called a *whole-interval*, except for the intervals between "mi" and "fa" and between "ti" and "do" which have only about half the difference in pitch of the others and are called *half-intervals*. The pitch of the first "do" can be chosen arbitrarily, but the pitch of the subsequent notes conform to a regular pattern. The diatonic scale beginning on middle C as "do," taken to be 264 cycles per second, is shown in Table 17-2 along with the frequencies of each note and the intervals between them. This is the C major scale.

The interval from middle C to C above middle C (or C′) is called an *octave*, from the Latin word for "eight," since there are eight notes in that interval counting both Cs. Notice that C′ has twice the frequency of middle C. This is always the case as one goes farther up the scale. Each C has twice the frequency of the preceding C, each A has twice the frequency of the preceding A, and so on. Thus the frequency of each note doubles with each octave.

If the diatonic major scale begins on D, let us say, the intervals between successive notes must maintain the same relationships of whole- and half-intervals. Some of the notes must therefore be raised or lowered a half-interval as shown in Table 17-3; when a note is raised by a half-interval it is said to be a "sharp" (♯) and if it is lowered by a half-interval it is a "flat" (♭). Sharps and flats are also necessary when one of the *minor* scales is used in which the intervals between notes are ordered slightly differently than for the major scale.

A scale that includes all the sharps and flats is called a "chromatic scale." Table 17-4 illustrates the "equally tempered chromatic scale," the most common scale for Western music. On this scale C♯ and D♭, for example, are taken to have exactly the same frequency. Originally the frequencies of C♯ and D♭ were slightly different, and solo violin and other string players often play them that way still. The equally tempered scale has a great advantage for fixed note instruments like the piano and flute since the number of keys or stops necessary is reduced; that is, the same key is used to play both C♯ and D♭. If C♯ and D♭, and other like pairs, were different,[†] there would have to be many more keys on a piano. The interval

[†] They are slightly different on the diatonic scale.

between each note on the equally tempered chromatic scale is a half-interval, and each succeeding note has a frequency that is about 1.06 times larger than the preceding note.

7. INTERFERENCE OF SOUND WAVES AND BEATS

We saw in Chapter 16 that when two waves pass through the same region of space simultaneously, they interfere with one another. Since this is true of any kind of wave, we should expect that interference will occur with sound waves, and indeed it does.

An interesting and important example of interference of sound waves occurs in the phenomenon known as *beats*. This is the phenomenon that occurs if two sources of sound—say, two tuning forks or two identical strings on a piano—are close in frequency but not exactly the same. Sound waves from the two sources interfere with each other and the sound level alternately rises and falls; the regularly spaced intensity changes are called beats and often sound eerie.

To see how beats arise, consider two equal-amplitude† sound waves of frequency $f_1 = 50$ Hz and $f_2 = 55$ Hz, respectively. In 1.00 s, the first source makes 50 vibrations whereas the second makes 55. The wave forms for each wave as a function of time are shown on the first two lines of Figure 17-8; the third line shows the sum of the two waves. At time $t = 0$ the two waves are shown to be in phase and interfere constructively. Because the two waves vibrate at different rates, at time $t = 0.10$ s they are completely out of phase and destructive interference occurs as shown in the figure. At $t = 0.20$ s they are again in phase and the resultant amplitude is again large. Thus the resultant amplitude is large every 0.20 s and in between it drops drastically. This rising and falling of the intensity is what is heard as beats. In this case the beats are 0.20 s apart. That is, the *beat frequency* is five per second or 5 Hz. This result, that the beat frequency equals the difference in frequency of the two waves, is valid in general.

The phenomenon of beats can occur with any kind of wave and is a very sensitive method for comparing frequencies. For example, to tune a piano,

† Beats will be heard even if the amplitudes are not equal, providing the difference is not great.

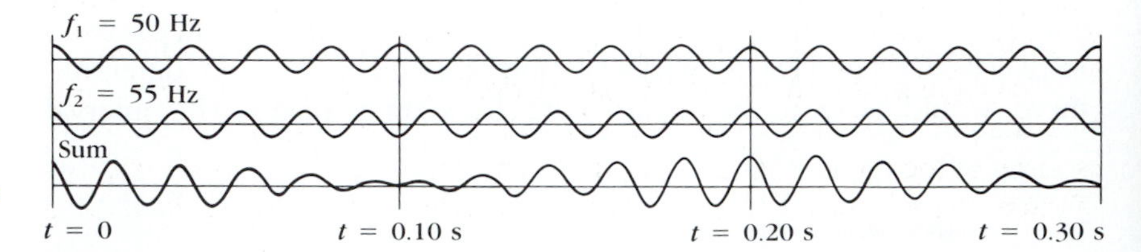

FIGURE 17-8

Beats occur as a result of the superposition of two sound waves of slightly different frequency.

a piano tuner listens for beats produced between his standard tuning fork and that of a particular string on the piano, and knows it is in tune when the beats disappear; he also listens for beats between the different piano strings. The members of an orchestra tune up by listening for beats between their instruments and that of a standard tone (usually A above middle C at 440 Hz) produced by a piano or an oboe.

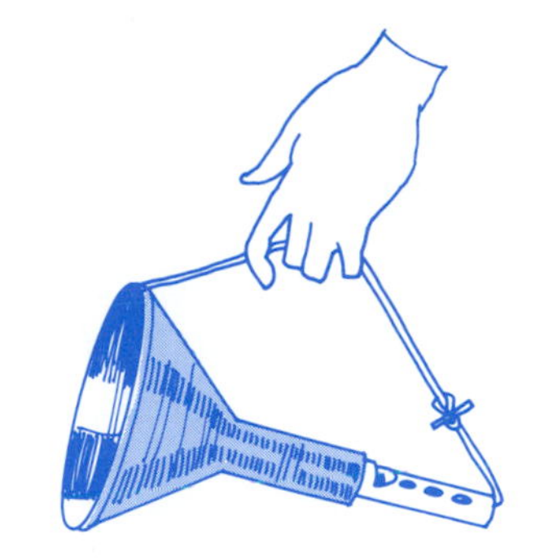

8. THE DOPPLER EFFECT AND THE SONIC BOOM

The Doppler Effect: Pitch Changes if Source or Listener is Moving

You may have noticed that the pitch of the siren on a speeding firetruck drops abruptly as it passes you. Or you may have noticed the change in pitch of a blaring horn on a fast-moving car as it passes by. The pitch of the sound from the engine of a race car changes as it passes an observer. When a source of sound is moving toward an observer, the pitch is higher than when the source is at rest; and when the source is traveling away from the observer, the pitch is lower. This phenomenon is known as the **Doppler effect** and occurs for all types of waves.

FIGURE 17-9
A funnel-whistle used to detect the Doppler effect.

EXPERIMENT

Find a small round whistle that will fit into a funnel, as shown in Figure 17-9. Tie a string as shown and have someone rotate the funnel-whistle in a horizontal circle. Notice how the pitch changes as the whistle approaches and retreats from you. (Try to distinguish between loudness and pitch changes).

To understand the origin of the Doppler effect, let us consider the siren of a firetruck at rest and emitting sound of a particular frequency in all directions as shown in Figure 17-10(a). If the firetruck is moving, the siren emits sound at the same frequency. But the sound waves it emits forward are closer together than normal as shown in Figure 17-10(b). This is because the fire engine, as it moves, is "catching up" with the previously emitted waves. Thus an observer on the sidewalk will detect more wave crests passing per second, so the frequency is higher. The waves emitted behind the truck are, on the other hand, farther apart than normal because the truck is speeding away from them. Hence, fewer wave crests per second pass by an observer behind the truck and the pitch is lower.

To calculate the change in frequency, we make use of Figure 17-11, and we assume the air (or other medium) is still. In Figure 17-11(a), the source of the sound (say a siren) is at rest; two successive wave crests are shown, the second of which is just in the process of being emitted. The

FIGURE 17-10
(a) Both observers on the sidewalk hear the same frequency from the firetruck at rest.
(b) Doppler effect: observer toward whom the firetruck moves hears a higher-frequency sound and observer behind the firetruck hears a lower frequency.

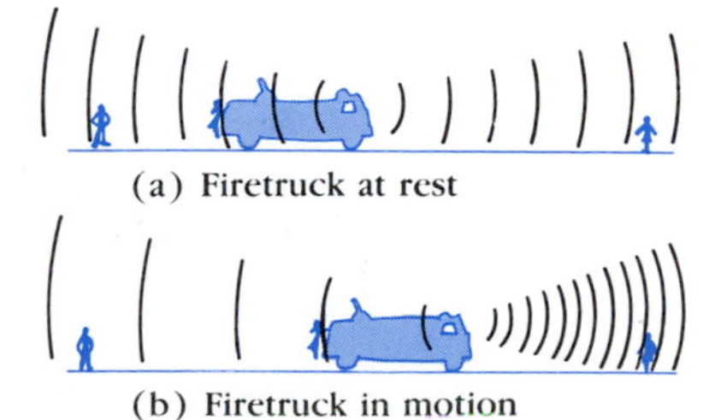

distance between these crests is λ, the wavelength. If the frequency of the source is f, then the time between emissions of wave crests is

$$T = \frac{1}{f}.$$

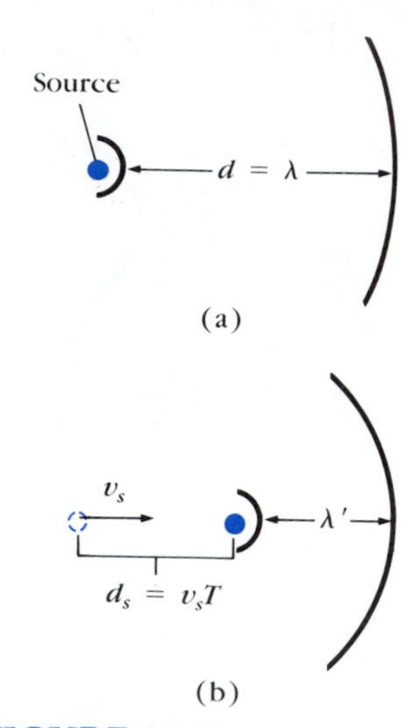

FIGURE 17-11
Determination of frequency change in the Doppler effect.

In Figure 17-11(b), the source is moving with a velocity v_s. In a time T (as just defined) the first wave crest has moved a distance $d = vT$ where v is the velocity of the sound wave in air (which is the same whether the source is moving or not). In this same time, the source has moved a distance $d_s = v_sT$. Then the distance between successive wave crests, which is the new wavelength λ', is

$$\lambda' = d - d_s = vT - v_sT = (v - v_s)T = (v - v_s)\frac{1}{f},$$

since $T = 1/f$. The frequency f' of the wave is

$$f' = \frac{v}{\lambda'} = \left(\frac{v}{v - v_s}\right) f = \frac{1}{(1 - v_s/v)} f.$$

Because the denominator is less than one, $f' > f$. For example, if a source emits a sound of frequency 400 Hz when at rest, then when the source moves toward a fixed observer with a speed of 30 m/s the observer hears a frequency (at 20°C) of

$$f' = \frac{400 \text{ Hz}}{\left(1 - \dfrac{30 \text{ m/s}}{343 \text{ m/s}}\right)} = 438 \text{ Hz}.$$

For a source that is moving *away* from the observer at a speed v_s, the new wavelength will be

$$\lambda' = d + d_s,$$

and the frequency f' will then be

$$f' = \frac{1}{(1 + v_s/v)} f.$$

In this case, an observer behind the fire truck traveling at 30 m/s whose siren vibrates at 400 Hz would hear a frequency of about 368 Hz.

The Doppler effect also occurs when the listener moves toward or away from a stationary source of sound. If he moves toward the source, the wave crests will pass him at a faster rate, so the pitch is higher. If he moves away from the source, the wave crests take a little longer to catch up with him; therefore the frequency with which they pass him will be less, and the pitch will be lower. [The formula for the frequency in these cases is $f' = (1 \pm v_{0bs}/v)f$.]

The Astronomical "Redshift"

The Doppler effect occurs for other types of waves as well. Light (and other types of electromagnetic waves) exhibit the Doppler effect; although the formulas for the frequency shift are not identical to those for sound, the effect is the same. An important application is found in astronomy where, for example, the velocities of distant galaxies can be determined from the Doppler shift. Light from such galaxies is shifted toward lower frequencies (this is called the "redshift" because red has the lowest frequency of visible light), indicating the galaxies are moving away from us. The greater the frequency shift, the greater the velocity of recession. It is found that the farther the galaxies are from us, the faster they move away. This observation is the basis for the idea that the universe is expanding (see Chapter 33).

On a less lofty plane, police radar "speed traps" make use of the Doppler effect (also of electromagnetic waves) to measure the speeds of approaching cars.

Shock Waves and the Sonic Boom

When an aircraft exceeds the speed of sound, it is said to have a **supersonic** speed.† Supersonic speeds are sometimes rated in "Mach numbers," named after the physicist-philosopher Ernst Mach. The Mach number is defined as the ratio of the speed of the object to the speed of sound in air. Thus "Mach 2" means the object is moving at twice the speed of sound, and "Mach 0.5" means it is moving at half the speed of sound.

We have seen that a moving source of sound gives rise to changes in pitch known as the Doppler effect [see Figure 17-12(a) and (b)]. In a sense,

† Do not confuse supersonic with *ultrasonic*; the latter refers to a frequency greater than audible frequencies.

FIGURE 17-12

Sound waves emitted by an object at rest (a) or moving (b, c, d). If the object's velocity is less than the velocity of sound, the Doppler effect occurs (b); if its velocity is greater than the velocity of sound, a shock wave is produced (d).

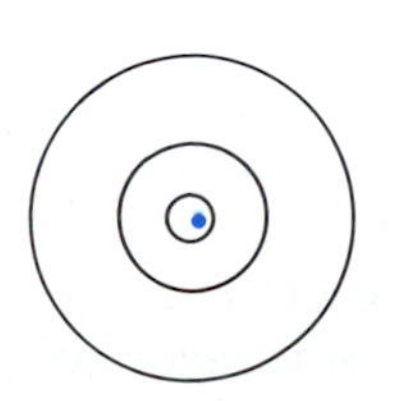

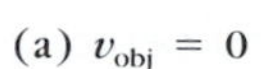

(a) $v_{obj} = 0$

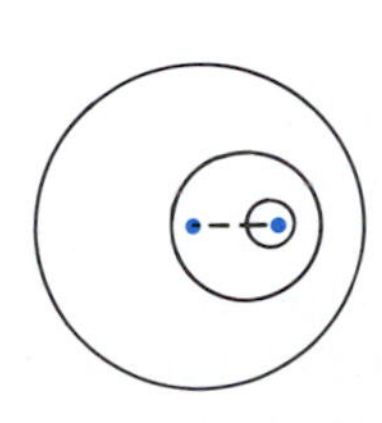

(b) $v_{obj} < v_{snd}$

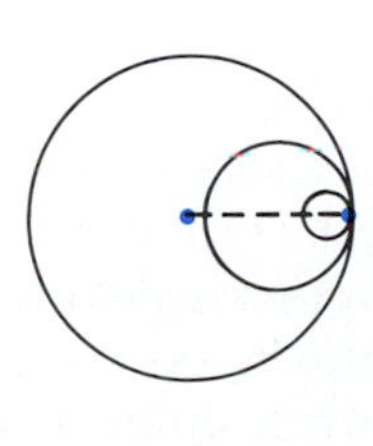

(c) $v_{obj} = v_{snd}$

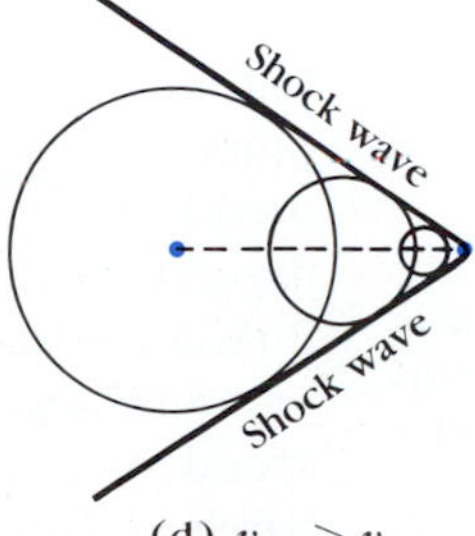

(d) $v_{obj} > v_{snd}$

FIGURE 17-13

Bow waves of a boat are analogous to shock waves.

the source is chasing the sound waves moving out in front of it. If the moving source exceeds the speed of sound, far more dramatic effects occur. In this case the source is actually "outrunning" the waves it produces. As shown in Figure 17-12(c), when the source is traveling at the speed of sound the waves it emits in the forward direction "pile up" directly in front of it. When the object moves at a supersonic speed the waves pile up on one another along the sides as shown in Figure 17-12(d). The different wave crests overlap one another and form a single very large crest which is the shock wave. Behind this very large crest there is usually a very large trough. A shock wave is essentially due to the constructive interference of a large number of waves. A shock wave in air is analogous to the bow wave of a boat traveling faster than the speed of the water waves it produces (Figure 17-13).

When an airplane travels at supersonic speeds, the noise it makes and its disturbance of the air form into a shock wave containing a tremendous amount of sound energy. When the shock wave passes a listener, it is heard as a loud "sonic boom." A sonic boom lasts only a fraction of a second, but the energy it contains is often sufficient to break windows and cause other damage. Actually the sonic boom from a supersonic aircraft is a double boom since a shock wave forms at both the front and the rear of the aircraft (Figure 17-14).

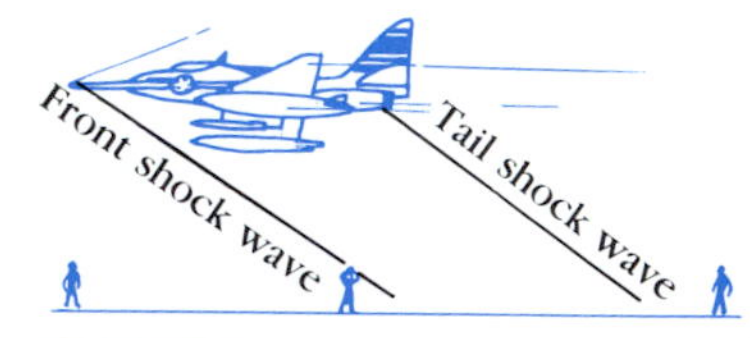

FIGURE 17-14

The (double) sonic boom has already been heard by the person on the right; it is just being heard by the person in the center; and it will shortly be heard by the person on the left.

When an aircraft approaches the speed of sound, it encounters a barrier of sound waves in front of it—Figure 17-12(c). In order to exceed the speed of sound, extra thrust is needed to pass through this "sound barrier." This is called "breaking the sound barrier." Once a supersonic speed is attained, this barrier no longer impedes the motion. It is sometimes erroneously thought that a sonic boom is produced only at the moment an aircraft is breaking through the sound barrier. Actually, a shock wave follows the aircraft at all times it is traveling at supersonic speeds. A series of observers on the ground will each hear a loud "boom" as the shock wave passes (Figure 17-14).

9. ULTRASOUND: SONAR AND MEDICAL APPLICATIONS

The reflection of sound is used in many applications to determine distance. The *sonar* or pulse-echo technique is used to locate underwater objects. A transmitter sends out a sound pulse through the water, and a detector receives its reflection, or echo, a short time later. This time is carefully measured, and from it the distance to the reflecting object can be determined because the speed of sound in water is known. The depths of the sea and the location of reefs, sunken ships, submarines, or schools of fish can be determined in this way. The interior structure of the earth is studied in a similar way by detecting reflections of waves traveling through the earth, waves whose source was a deliberate explosion (called "soundings"). An analysis of waves reflected from various structures and boundaries within the earth reveals characteristic patterns that are also useful in the exploration for oil and minerals.

Sonar generally makes use of *ultrasonic* frequencies: that is, waves whose frequencies are above 20 kHz, beyond the range of human detection. For sonar, the frequencies are typically in the range 20 kHz to 100 kHz. One reason for using ultrasound, other than the fact that its frequencies are inaudible, is that for shorter wavelengths there is less diffraction. As we saw in Chapter 16, particularly in Figure 16-24, an obstacle intercepts and reflects a portion of a wave significantly only if the wavelength is less than the size of the object. Indeed, the smallest-sized objects that can be detected are on the order of the wavelength of the wave used. With the higher frequencies of ultrasound, the wavelength is smaller, so smaller objects can be detected.

In medicine, ultrasonic waves are used both in diagnosis and in treatment. Treatment involves destruction or reduction of unwanted tissue or objects in the body (such as tumors or kidney stones) using ultrasonic waves of very high intensity focused on the subject material.

The diagnostic use of ultrasound in medicine is a more complicated and very interesting application of physical principles. A *pulse-echo technique* is used, much like sonar. A high-frequency sound pulse is directed into the body, and its reflection from boundaries or interfaces between organs and other structures and lesions in the body are then detected. Using this technique, tumors and other abnormal growths, or pockets of fluid, can be distinguished; the action of heart valves and the

FIGURE 17-15
(a) An ultrasound pulse passes through the abdomen and interior organs, reflecting from surfaces in its path. (b) Reflected pulses are plotted as a function of time (A-scan) when received by the transducer; time is proportional to the distance of travel. The vertical dashed lines simply point out which reflected pulse goes with which surface. (c) B-scan mode display for the same echoes: the brightness of each dot is related to the signal strength.

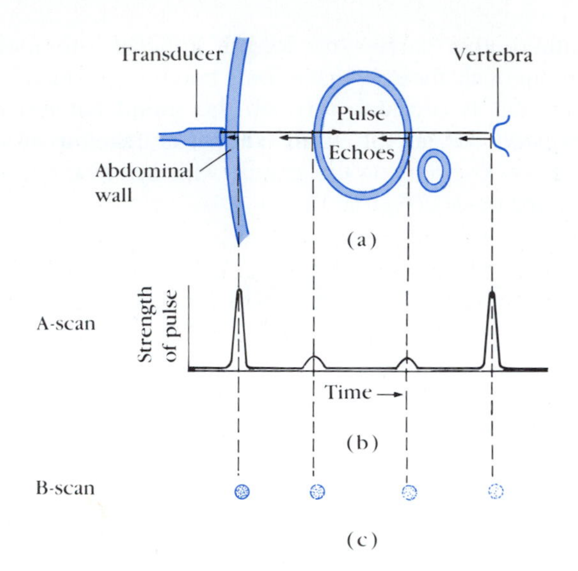

FIGURE 17-16
(a) Ten traces are made across the abdomen by moving the transducer or by using an array of transducers. (b) The echoes are plotted in the B-scan mode to produce the image. More closely spaced traces would give a more detailed image.

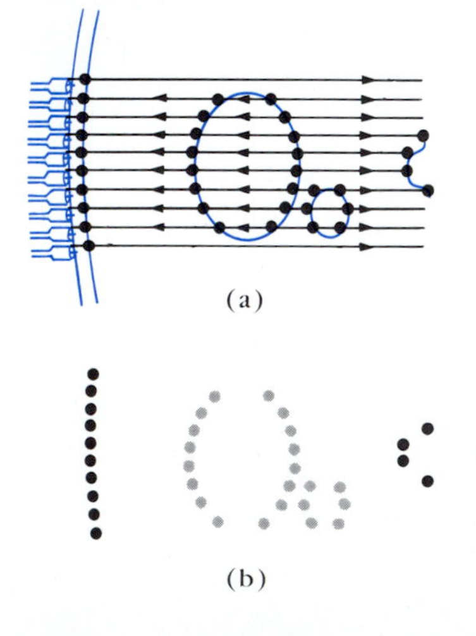

development of a fetus can be examined; and information about various organs of the body, such as the brain, heart, liver, and kidneys, can be obtained. Although ultrasound does not replace X rays, for certain kinds of diagnosis it is more helpful. It is also possible to have "real-time" ultrasound images, as if one were watching a movie of a section of the interior of the body. Furthermore, at the low levels used for diagnosis no adverse affects have been reported; so ultrasound is considered a noninvasive method for probing the body.

The frequencies used in ultrasonic diagnosis are in the range 1 to 10 MHz (1 megahertz = 10^6 Hz). The speed of sound waves in the tissues of the human body averages about 1500 m/s (close to that for water); so the wavelength of a 1-MHz wave is about $\lambda = v/f = (1500\ \text{m/s})/(10^6\ \text{s}^{-1}) \approx 1.5 \times 10^{-3}\ \text{m} = 1.5\ \text{mm}$, and this sets the limit to the smallest-sized objects that can be detected.

The technique works as follows. A brief pulse of ultrasound is emitted by a transducer (similar to a loudspeaker, Chapter 20). The same transducer detects the pulses reflected from various structures in the body, as shown in Figure 17-15(a) for pulses made to pass through the abdomen. The time between sending the pulse and when each reflection is received is a measure of the distance to each reflecting surface inside the body. Figure 17-15(b) shows a plot of the reflected pulses as a function of the time they are received by the transducer. The reflected pulses can be displayed as a set of dots, as shown in part (c) (this is called a B-scan). If the transducer is moved across the body, as shown in Figure 17-16, or many transducers are used so that a set of B-scan echos are obtained, an image of a cross-section of the body is obtained as shown in Figure 17-16(b). The display can be done on the screen of a cathode-ray tube (see Chapter 21). Only ten lines are shown in Figure 17-16, so the image is crude. More lines would give a more precise image.† A photograph of an ultrasound image is shown in Figure 17-17.

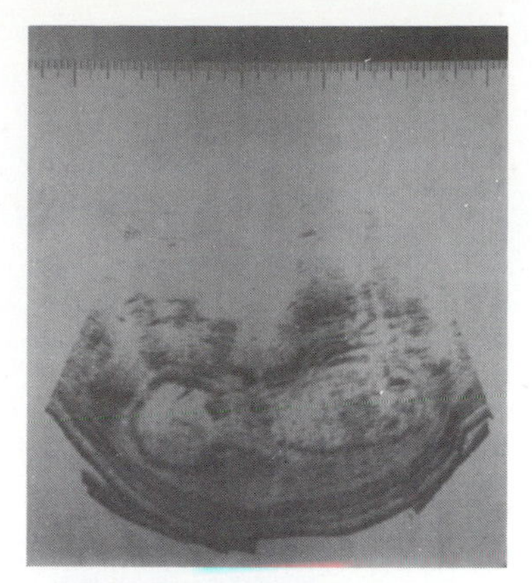

FIGURE 17-17 Ultrasound image of a 14-week-old human fetus.

SUMMARY

Sound travels as a longitudinal wave in air and other materials. In air, the speed of sound increases with temperature; at 20°C it is about 343 m/s.

The *pitch* of a sound is determined by the frequency; the higher the frequency, the higher the pitch. The *audible range* of frequencies is roughly 20 to 20,000 Hz (1 Hz = 1 cycle per second). The *loudness* or *intensity* of a sound is related to the amplitude of the wave. Because the human ear can detect sound intensities over an extremely wide range, intensity levels are specified on a logarithmic scale: the *intensity level* β, specified in decibels, is $\beta = 10 \log (I/I_0)$, where the reference intensity I_0 is usually taken to be $10^{-12}\,\mathrm{W/m^2}$. An increase in intensity by a factor of 100, for example, corresponds to a level increase of 20 dB.

Musical instruments are simple sources of sound in which standing waves are produced. The strings of a stringed instrument may vibrate as a whole with nodes only at the ends; the frequency at which this occurs is called the *fundamental*. The string can also vibrate at higher frequencies called *overtones* or *harmonics*, in which there are one or more additional nodes. The frequency of each harmonic is a whole-number multiple of the fundamental. In wind instruments, standing waves are set up in the column of air within the tube. The vibrating air in an open tube (open at both ends) has antinodes at both ends; the fundamental frequency corresponds to a wavelength equal to twice the tube length. The harmonics have frequencies that are 2, 3, 4,... times the fundamental frequency. For a closed tube (closed at one end) the fundamental corresponds to a wavelength four times the length of the tube; only the odd harmonics are present, equal to 1, 3, 5, 7,... times the fundamental frequency.

Sound waves from different sources can interfere with each other. If two sounds are at slightly different frequencies, *beats* can be heard at a frequency equal to the difference in frequency of the two sources.

The *Doppler effect* refers to the change in pitch of a sound due to the motion either of the source or of the listener. If they are approaching each other, the pitch is higher; if they are moving apart, the pitch is lower.

Shock waves and *sonic booms* are large-amplitude *wave pulses* that occur when the speed of the source of vibration exceeds the speed of sound in the medium in which it is moving.

Ultrasound (frequencies higher than the audible range) pulses are used in sonar, for medical treatment, and for diagnostic imaging.

† *Radar* used for aircraft involves a similar pulse-echo technique except that it uses electromagnetic (EM) waves which, like light, travel with a speed of 3×10^8 m/s.

QUESTIONS

1. What is the evidence that sound travels as a wave?
2. Why is it difficult for a large group of people to sing in unison?
3. What is an "echo"?
4. The voice of a person who has inhaled helium sounds very high pitched. Why?
5. In some concert halls there are locations where it is difficult to hear the performers, yet in adjacent locations it is easy to hear. Explain.
6. What is the evidence that sound is a form of energy?
7. What is the lowest intensity level that can be heard by an average ear (middle curve of Figure 17-2)?
8. What are the lowest and highest frequencies that the average ear (middle curve in Figure 17-2) can hear when the intensity level is 25 dB?
9. What is the reason that catgut strings on some musical instruments are wrapped with fine wire?
10. Whistle through your lips and explain how you control the pitch of the sound. Are there any similarities between the interior of your mouth and the length of an organ pipe or an air column in a woodwind instrument?
11. By very lightly touching a guitar string at its midpoint, a very pure tone can be obtained which is one octave above the fundamental for that string. Explain.
12. Why are the frets on a guitar closer together as you move up the neck? Remember that on the equally tempered chromatic scale the ratio of frequencies of any two successive notes is the same all along the scale.
13. How will the air temperature in a room affect the pitch of organ pipes?
14. When a sound wave passes into water, where its speed is higher, does its frequency or its wavelength change?
15. If a string is vibrating in three segments, are there any points where you can touch it and not disturb its motion?
16. Noise control is an important goal today. One mode of attack is to reduce the area of vibration of noisy machinery, for example by keeping it as small as possible or isolating it (acoustically) from the floor and walls. A second method is to make the surface out of a thicker material. Explain how each of these can reduce the noise level.
17. Why does high C played on the piano sound different from high C played on a violin?
18. Standing waves can be said to be due to "interference in space," whereas beats can be said to be due to "interference in time." Explain.

19. A sonic boom sounds much like an explosion. Explain the similarity.
20. Will the frequency of a sound heard by a listener at rest with respect to the source be altered if a wind is blowing? Will the velocity or wavelength be changed?

EXERCISES

1. Audible sound waves range from 20 to 20,000 cycles per second and the speed of sound is about 340 m/s. What, therefore, is the range of wavelengths for audible sounds? Remember, wavelength = velocity/frequency.
2. What is the speed of sound in air (20°C) in km/hr? In mi/hr?
3. A hiker wants to determine the length of a lake by shouting and listening for the echo from a cliff at the far end. If the echo returns after 2 seconds, how long is the lake?
4. The speed of sound in iron is about 5000 m/s. If you put your ear to a rail, you can hear the sound of an approaching train. (a) How long does it take the sound to reach you if the train is 1 km away? (b) If the train is traveling 100 km/hr, how soon will the train reach you?
5. Bats emit sounds whose frequencies are as high as 100,000 Hz. What is the wavelength of such waves?
6. What is the frequency of a sound whose wavelength is 0.50 meters in air at 20°C?
7. A sound wave in air at 20°C has a frequency of 524 Hz. How far apart are successive compressions?
8. If the frequency of a sound wave in air is 340 Hz, what is the distance between each compression and the adjacent expansion?
9. What is the intensity level of a sound with an intensity of 1 W/m^2?
10. What is the intensity level of a sound whose intensity is 1.0×10^{-8} W/m^2?
11. What is the intensity of a sound whose intensity level is 50 dB?
12. A stereo tape recorder is said to have a signal-to-noise ratio of 62 dB. What is the ratio of intensities of the signal and the background noise?
13. A single mosquito 10 m from a person makes a sound close to the threshold of human hearing (0 dB). What will be the intensity level of 1000 such mosquitoes?

*14. A high-quality loudspeaker is advertised to reproduce, at full volume, frequencies from 30 Hz to 18,000 Hz with uniform intensity ±3 dB. That is, over this frequency range the intensity level does not vary by more than 3 dB from the average. By how much does the intensity change for the maximum intensity-level change of 3 dB?

15. The frequency of A above middle C is 440 cycles per second. What is its wavelength in air? How long would an organ pipe, open at one end, have to be in order to produce this sound as its fundamental?
16. An unfingered guitar string is 0.70 m long and is tuned to play E above middle C (330 Hz). How far from the end of this string must the finger be placed to play A above middle C (440 Hz)?
17. A violin string vibrates at 294 Hz when unfingered. At what frequency will it vibrate when fingered one-third of the way from one end?
18. What is the fundamental frequency of a 90-cm-long organ pipe closed at one end? What if it is open at both ends?
19. An organ pipe is 80 cm long. What are the fundamental and first three audible overtones (a) if the pipe is closed at one end and (b) if it is open at both ends?
20. Determine the length of a closed organ pipe that emits middle C (264 Hz) when the temperature is 15°C.
21. An organ is in tune at 20°C. By what fraction will the frequency be off at 0°C?
22. Calculate the resonant frequency of the column of air in the outer ear of a human being, which is about 2.5 cm long. Does this correspond to a region of high sensitivity in the ear? Explain.
23. A flute is designed to play middle C (264 Hz) at 20°C when all holes are covered. Approximately how long should the flute be? A flute has both ends open.
24. If the hole nearest the mouthpiece in the flute in the above problem were exactly halfway to the end, what note would it play? How far from the end should a hole be to play A above middle C?
25. Two horns emitting sounds of frequency 665 and 671 Hz, respectively, will produce beats of what frequency?
26. Two flutes are slightly out of tune when they play middle C so that beats are heard every 2 s. By how much do the two flutes differ in frequency?
27. What will be the "beat frequency" if middle C (262 Hz) and C# (277 Hz) are played together? Will this be audible? What if each is played two octaves lower (each frequency reduced by a factor of 4)?
28. A police car siren emits a sound of 1400 Hz. If it travels away from an observer on the sidewalk at 30 m/s, what frequency does the observer hear?
29. A factory whistle emits a sound at 380 Hz. What frequency will be heard by a person driving 90 km/hr (25 m/s) (a) toward the source (b) away from it?
30. How fast must a person be traveling so that the frequency of a sound from a source at rest is (a) twice as high as normal (b) half as high?

ELECTRICITY AT REST

CHAPTER

18

The word "electricity" may evoke an image of complex modern technology: computers, lights, motors, electric power. But the electric force plays an even deeper role in our lives, since—according to atomic theory—the forces that act between atoms and molecules to hold them together to form liquids and solids are electrical forces. Similarly the electric force is responsible for the metabolic processes that occur within our bodies. Even ordinary pushes and pulls are the result of the electric force between the molecules of your hand and those of the object being pushed or pulled. Indeed, most of the forces we have dealt with so far, such as elastic forces and contact forces, are now considered to be electric forces acting at the atomic level. This does not include gravity, however, which is a separate force.

In this century physicists came to recognize only four different forces in nature: (1) gravitational force, (2) electromagnetic force (we will see later that electric and magnetic forces are intimately related), (3) strong nuclear force and (4) weak nuclear force. The last two forces operate at the level of the nucleus of an atom and although they manifest themselves in such phenomena as radioactivity and nuclear energy, they are much less obvious in our daily lives.

The earliest studies on electricity date back to the ancients, but it has been only in the past two centuries that electricity was studied in detail. We will discuss the development of ideas about electricity, including practical devices, as well as the relation to magnetism, in the next four chapters.

1. ELECTRIC CHARGE

Static Electricity and Charge

The word *electricity* comes from the Greek word *elektron*, which means "amber." Amber is petrified tree resin, and the ancients knew that if you

FIGURE 18-1
Bringing a plastic ruler up to a few tiny pieces of paper.

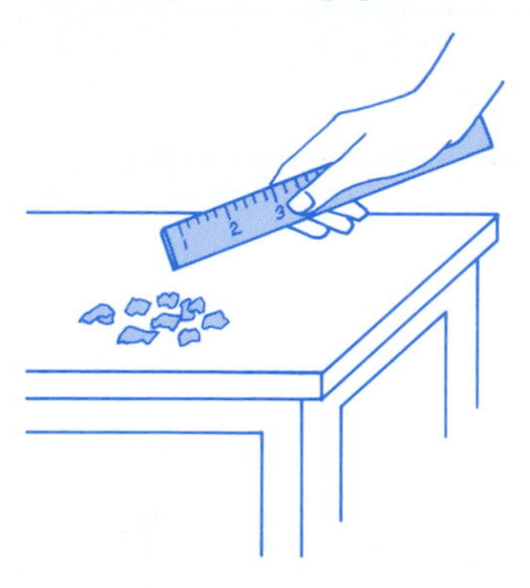

rub an amber rod with a piece of cloth the amber attracts small pieces of leaves or dust. A piece of hard rubber, a glass rod, or a plastic ruler rubbed with a cloth will also display this "amber effect," or **static electricity** as we call it today. You can demonstrate the amber effect with readily available materials.

EXPERIMENT

Place some tiny pieces of paper (1 cm square or so) on a table. Hold a plastic ruler or comb directly above and close to the bits of paper, Figure 18-1. Does anything happen? Now rub the plastic ruler or comb with a clean, dry paper towel or piece of wool and again place it directly above and close to the bits of paper. What happens now? Note that the plastic must be rubbed often if it is to maintain its ability to attract.

You have probably experienced static electricity when combing your hair or upon taking a synthetic blouse or shirt from a clothes dryer. And you may have felt a shock when you touched a metal door knob after sliding across a car seat or walking across a nylon carpet. In each case, an object becomes "charged" due to a rubbing process and is said to possess an **electric charge**.

There Are Two Kinds of Electric Charge

Is all electric charge the same or is it possible that there is more than one type? In fact, there are two types of electric charge as the following simple experiments show. A plastic ruler is suspended by a thread and rubbed vigorously to charge it; when a second ruler, which has also been charged by rubbing, is brought close to the first, it is found that the one ruler *repels* the other. This is shown in Figure 18-2(a). Similarly, if a rubbed glass rod is brought close to a second charged glass rod, again a repulsive force is seen to act—Figure 18-2(b). However, if the charged glass rod is brought close to the charged plastic ruler, it is found that they *attract* each other—Figure 18-2(c). The charge on the glass must therefore be different from that on the plastic. Indeed, it is found experimentally that all charged objects fall into one of two categories. Either they are attracted to the plastic and repelled by the glass, just as glass is; or they are repelled by the plastic and attracted to the glass, just as the plastic ruler is. Thus there seem to be two, and only two, types of electric charge. Each type of charge repels the same type but attracts the opposite type. That is

unlike charges attract; like charges repel.

These two types of charge were referred to as **positive** and **negative** by the American statesman, philosopher, and scientist Benjamin Franklin (1706–1790). The choice of which name went with which type of charge was of course arbitrary. Franklin's choice sets the charge on the rubbed glass rod to be a positive charge; so the charge on a rubbed plastic ruler

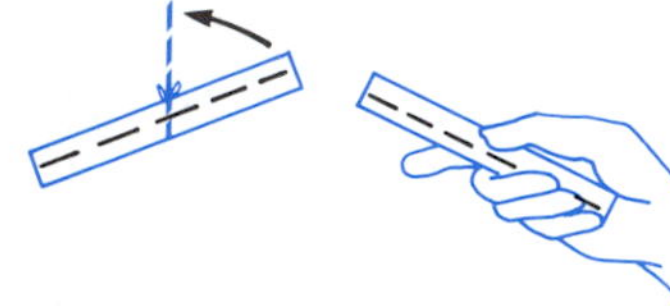

(a) Two plastic rulers repel

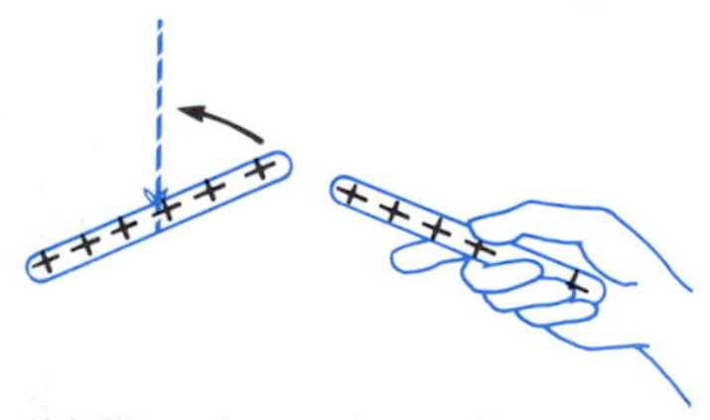

(b) Two glass rods repel

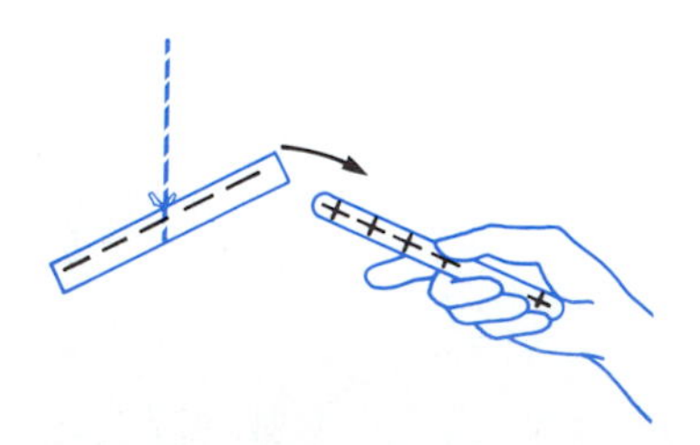

(c) Glass rod attracts plastic ruler

FIGURE 18-2
Unlike charges attract, whereas like charges repel one another.

(or amber) is called a negative charge. We still follow this convention today.

Electric Charge Is Conserved

Franklin's theory of electric charge was actually a "single-fluid" theory that viewed a positive charge as an excess of the electric fluid beyond an object's normal content of electricity, and a negative charge as a deficiency. Franklin argued that whenever a certain amount of charge is produced on one body in a process, an equal amount of the opposite type of charge is produced on another body. The names positive and negative are to be taken *algebraically*, so that during any process the net change in the amount of charge produced is zero. For example, when a plastic ruler is rubbed with a paper towel, the plastic acquires a negative charge and the towel an equal amount of positive charge; the charge is separated, but the sum of the two is zero. This is an example of a law that is now well established: the **law of conservation of electric charge**, which states that

the net amount of electric charge produced in any process is zero.

No violations have ever been found, and this conservation law is as firmly established as those for energy and momentum.

Electric Charge Originates in the Atom

Only within the past century has it become clear that electric charge has its origin within the atom itself. In later chapters we will discuss atomic structure and the ideas that led to our present view of the atom in more detail; but it will help our understanding of electricity if we discuss it briefly now.

Today's view, slightly simplified, shows the atom as having a heavy, positively charged nucleus surrounded by one or more negatively charged electrons (Figure 18-3). In its normal state, the positive and negative charges within the atom are equal and the atom is electrically neutral. Sometimes, however, an atom may lose one or more of its electrons, or may gain extra electrons. In this case the atom will have a net positive or negative charge, and is called an *ion*.

The nuclei in a solid material can vibrate but they tend to remain close to fixed positions, whereas some of the electrons move quite freely. The charging of an object by rubbing is explained by the transfer of electrons or ions from one material to the other. When a plastic ruler becomes negatively charged by rubbing with a paper towel, the transfer of charged particles from the one to the other leaves the towel with a positive charge equal in magnitude to the negative charge acquired by the plastic.

Normally when objects are charged by rubbing, they hold their charge only for a limited time and eventually return to the neutral state. Where does the charge go? In some cases it is neutralized by charged ions in the air (formed by cosmic rays, etcetera). Often more importantly, the charge

FIGURE 18-3
Simple view of the atom.

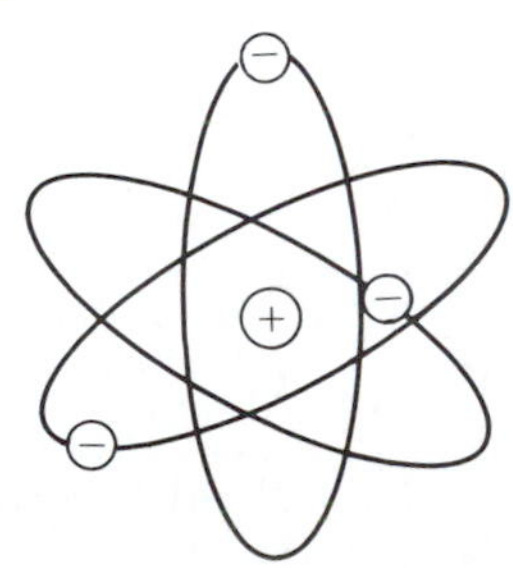

FIGURE 18-4
Diagram of a water molecule. Because it has opposite charges on different ends, it is called a "polar" molecule.

can "leak off" onto water molecules in the air. This is because water molecules are *polar*—that is, even though they are neutral their charge is not distributed uniformly (Figure 18-4). Thus the extra electrons on, say, a charged plastic ruler can "leak off" into the air because they are attracted to the positive end of water molecules. A positively charged object, on the other hand, can be neutralized by transfer of loosely held electrons from water molecules in the air. On dry days, static electricity is much more noticeable since the air contains fewer water molecules to allow leakage. On humid or rainy days, it is difficult to make any object hold its charge for long.

2. INSULATORS AND CONDUCTORS

Suppose we have two metal spheres, one highly charged and the other electrically neutral. If we now place an iron nail so that it touches both the spheres, it is found that the previously uncharged sphere quickly becomes charged. If, however, we had connected the two spheres together with a wooden rod or a piece of rubber, the uncharged sphere would not have become noticeably charged. Materials like the iron nail are said to be **conductors** of electricity, whereas wood and rubber are called **non-conductors** or **insulators**.

Metals are generally good conductors, whereas most other materials are insulators (although even insulators conduct electricity very slightly). It is interesting that nearly all natural materials fall into one or the other of these two quite distinct categories. There are a few materials, however, (notably silicon, germanium, and carbon) that fall into an intermediate (but distinct) category known as **semiconductors**.

From the atomic point of view the electrons in an insulating material are bound very tightly to the nuclei, whereas in a conductor many of them are bound very loosely and can move about freely within the material. When a positively charged object is brought close to or touches a conductor, the free electrons move quickly toward this positive charge. On the other hand, the free electrons move swiftly away from a negative charge that is brought close. In a semiconductor there are very few free electrons and in an insulator, almost none.

3. THE ELECTRIC FORCE AND COULOMB'S LAW

When one charged body attracts or repels a second charged body, as in Figure 18-2, some kind of force is clearly involved. We call this force the **electric force**. When a positively charged body attracts a negatively charged body, the electric force is said to be *attractive*; each of the bodies exerts an attractive force on the other—Figure 18-5(a). If the two bodies have the same charge they each exert a *repulsive* force on the other—Figure 18-5(b) and (c). Electric forces act even though the two bodies do not touch. As in the case of the gravitational force, the electric force acts "at a distance." In the 1780s, the French physicist Charles Coulomb (1736–1806) undertook the task of investigating these electric forces quantitatively.

Using the apparatus shown schematically in Figure 18-6, Coulomb found that the force exerted by one charged sphere on a second one is directly proportional to the charge on each of them. Thus, when the charge on one body was doubled, the force was doubled. When the charge on both bodies was doubled, the force was four times as great. If either charge was zero, the force was zero. If we use the symbol q_1 to represent the amount of charge on one body and q_2 to represent the amount of charge on the second body, then this result can be written mathematically as $F \propto q_1 q_2$, where, as usual, "$\propto$" means "is proportional to."

The above results hold true as long as the distance between the two spheres is not changed. If the distance is changed, the force changes as well. To determine just how the force varies with distance, Coulomb measured the strength at various distances while keeping the charge on each sphere unchanged. He found that as the distance between the spheres increases, the force diminishes quite rapidly. Quantitatively, if the distance between two small spheres doubles, the force is only $\frac{1}{4}$ as strong. If the distance between them triples, the force is only $\frac{1}{9}$ of the original force. Noting that $\frac{1}{2^2} = \frac{1}{4}$ and $\frac{1}{3^2} = \frac{1}{9}$, Coulomb came to the conclusion that the force between two small charged spheres varies inversely as the square of the distance between them. Mathematically, $F \sim \frac{1}{d^2}$, where d represents the distance between the two charged spheres.

The dependence of the force on the three variables—the charge on each of the two bodies and the distance between them—can be written as a single equation: $F \propto \frac{q_1 q_2}{d^2}$. This will be an equality if we insert a constant of proportionality, call it k:

$$F = k\frac{q_1 q_2}{d^2}.$$

This is known as **Coulomb's law**.

This equation gives the magnitude of the force that one charged object exerts on another. The direction of the force is always along the line that joins the two charges; it points away from the other charge if both charges have the same signs, and toward the other charge if they have opposite signs (Figure 18-5). Notice that the force one charge exerts on the second is equal but opposite to that exerted by the second on the first; this is in accord with Newton's third law.

The most commonly used unit of measurement for electric charge is called the coulomb, in honor of Charles Coulomb. It is the SI unit and is always used in conjunction with other metric units: meters for distance and newtons for force. The constant of proportionality, k, has the value

FIGURE 18-5

(a) Unlike charges attract; (b) and (c): like charges repel. Arrows represent the force on each charge due to the other charge.

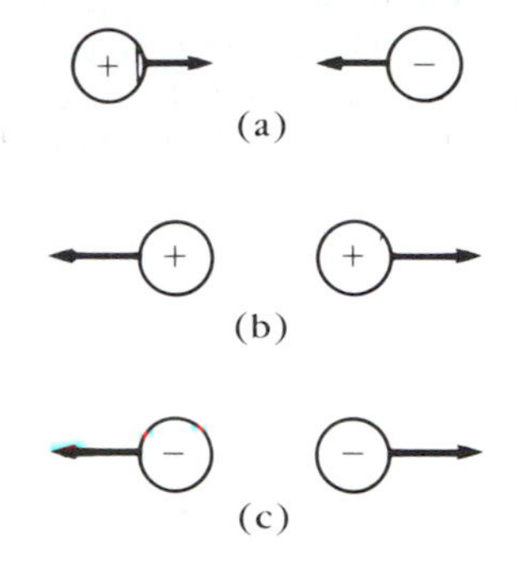

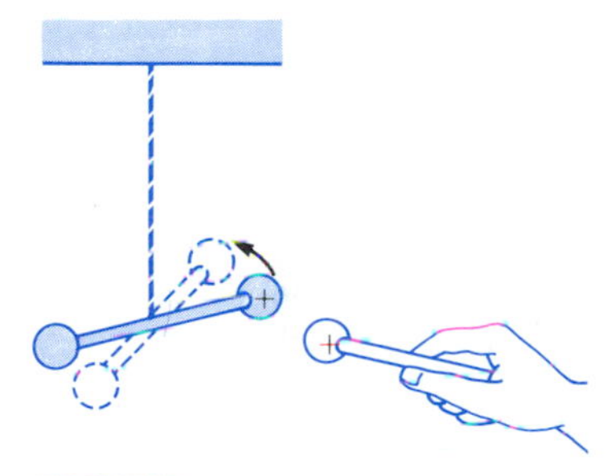

FIGURE 18-6

Schematic diagram of Coulomb's apparatus. It is similar to Cavendish's, which was used to measure the gravitational force. When a charged sphere is placed close to the one on the suspended bar, the bar rotates slightly. The suspending fiber resists the twisting motion and the angle of twist is proportional to the force applied. By use of this apparatus, Coulomb was able to investigate how the electric force varied as a function of the magnitude of the charges and of the distance between them.

9×10^9 newton-meters²/coulomb². Thus, 1 C is that amount of charge which, if it exists on each of two tiny objects placed 1 m apart, will result in each object exerting a force of about $(9.0 \times 10^9\ \text{N} \cdot \text{m}^2/\text{C}^2)(1.0\ \text{C})(1.0\ \text{C})/(1.0\ \text{m})^2 = 9.0 \times 10^9\ \text{N}$ on the other.

Charges produced by rubbing ordinary objects (such as a comb or plastic ruler) are typically a microcoulomb ($1\ \mu\text{C} = 10^{-6}\ \text{C}$) or less. The magnitude of the charge on one electron, on the other hand, is measured to be about 1.6×10^{-19} C (and is negative). This is the smallest known charge,† and because of its fundamental nature it is given the symbol e and is often referred to as the *elementary charge*:

$$e = 1.602 \times 10^{-19}\ \text{C}.$$

Since an object cannot gain or lose a fraction of an electron, the net charge on any object must be an integral multiple of this charge. Electric charge is thus said to be *quantized* (existing only in discrete amounts); because e is so small, however, we normally don't notice this discreteness in macroscopic charges ($1\ \mu\text{C}$ requires about 10^{13} electrons) which thus seem continuous.

Coulomb's law looks much like the universal law of gravitation, $F = Gm_1m_2/d^2$. Charge plays the same role for the electric force that mass does for the gravitational force. These two forces, however, are completely independent phenomena. Despite the similarity of the two force laws there are some very important differences between them. The gravitational force is always an attractive force; the electric force can be both attractive and repulsive. Furthermore, the electric force is generally much stronger than the gravitational force. For example, the electric force between two electrons that have a charge of 1.6×10^{-19} coulombs is 10^{40} times greater than the gravitational force between them. When dealing with charged objects, gravitational forces can usually be ignored compared to the electric forces. Only in the case of a massive body like the earth or the sun is the gravitational force significant enough to be noticed. At the atomic level, the gravitational force can almost always be ignored compared to the much stronger electrical force.

An Example (optional)

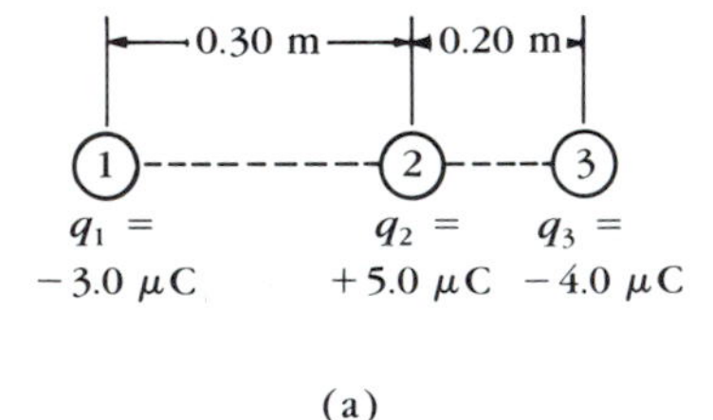

FIGURE 18-7
Using Coulomb's law to determine the net electric force on particle 3 due to particles 1 and 2.

As an example of the use of Coulomb's law in a more complicated situation, let us determine the net electric force on particle number 3 in Figure 18-7(a) due to the other two charged particles shown. The net force on particle 3 ($q_3 = -4.0 \times 10^{-6}\ \text{C}$) will be the sum of the force $\mathbf{F}_{31}$ exerted by particle 1 ($q_1 = -3.0 \times 10^{-6}\text{C}$) and the force $\mathbf{F}_{32}$ exerted by particle 2 ($q_2 = 5.0 \times 10^{-6}\text{C}$). The magnitudes of these two forces are

† Elementary-particle physicists have theorized the existence of smaller particles, called quarks, that would have a smaller charge equal to $\frac{1}{3}$ or $\frac{2}{3}e$. They have not been detected for certain experimentally, and theory indicates that free quarks may not be detectable (see Chapter 31).

$$F_{31} = \frac{(9.0 \times 10^9\ \text{N} \cdot \text{m}^2/\text{C}^2)(-4.0 \times 10^{-6}\ \text{C})(-3.0 \times 10^{-6}\ \text{C})}{(0.50\ \text{m})^2} = 0.43\ \text{N},$$

$$F_{32} = \frac{(9.0 \times 10^9\ \text{N} \cdot \text{m}^2/\text{C}^2)(5.0 \times 10^{-6}\ \text{C})(-4.0 \times 10^{-6}\ \text{C})}{(0.20\ \text{m})^2} = -4.5\ \text{N}.$$

$\mathbf{F}_{31}$ is repulsive and $\mathbf{F}_{32}$ is attractive, so the direction of the forces is as shown in Figure 18-7(b). The net force on particle 3 is then

$$F = F_{32} + F_{31} = -4.5\ \text{N} + 0.4\ \text{N} = -4.1\ \text{N}.$$

The magnitude is 4.1 N, and it points to the left. (Notice that the charge in the middle (q_2) in no way blocks the effect of the other one on the end (q_1); it does exert its own force, of course.)

FIGURE 18-8

Charging by induction. (a) A neutral metal rod. (b) A negatively charged body brought close to but not touching the neutral metal rod induces a separation of charge in the latter because some electrons are repelled to the opposite end.

(a) Neutral metal rod

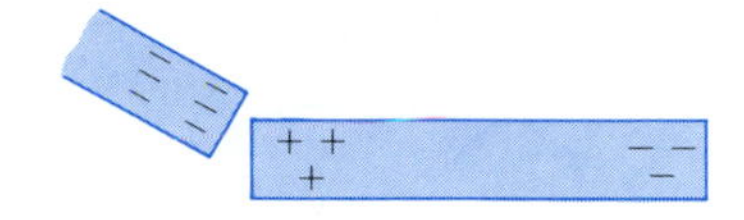

(b) Metal rod remains neutral but undergoes a separation of charge

4. INDUCED CHARGE AND THE ELECTROSCOPE

Charges Can Be Induced

When a negatively charged metal object touches a neutral metal object, some free electrons flow from the charged object into the previously neutral one, and the latter also becomes negatively charged. This is called "charging by conduction."

Now suppose a negatively charged object (say, a charged plastic ruler) is brought close to a neutral metal rod but does not touch it. Although the electrons of the metal rod do not leave the rod, they still move within the metal and away from the negatively charged ruler; this leaves a positive charge at the opposite end—Figure 18-8(b). A charge is said to have been **induced** at the two ends of the metal rod. Of course no charge has been created; it has merely been *separated*; the net charge on the metal rod is still zero. However, if the metal were now cut in half we would have two charged objects, one charged positively and one charged negatively.

Figure 18-8(b) illustrates how a charged object can attract a neutral object—such as the plastic ruler discussed at the beginning of this chapter (Figure 18-1), charged by rubbing, that attracts bits of neutral paper. In Figure 18-8(b), the neutral metal rod has zero net charge. You might therefore expect that the net electric force on it would be zero. However, there is a separation of charge, and we must remember the distance factor in Coulomb's law ($F \propto 1/d^2$). Referring to Figure 18-8(b) again, we see that the negatively charged ruler on the left must exert an attractive force on the positive end of the metal rod and a repulsive force on the negative end. Because the positively charged end is closer to the ruler, Coulomb's law tells us that the attractive force must be greater than the repulsive force. Thus the net force on the rod is attractive.

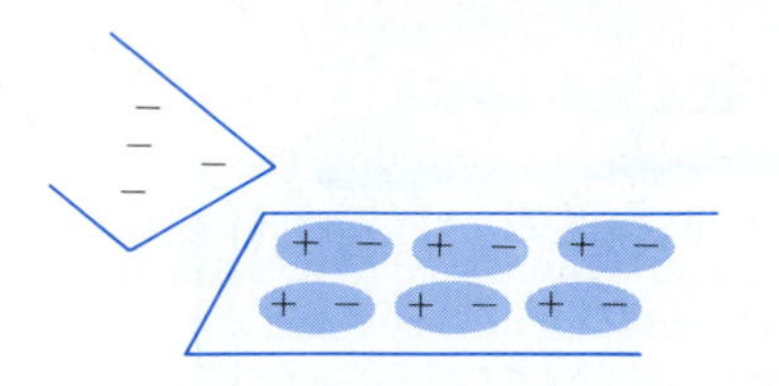

FIGURE 18-9
In a nonconductor the charged rod induces a separation of charge within each molecule, and a net attractive force results.

In nonconducting materials in which there are practically no free electrons, the electrons are still free to reorient themselves within each molecule. If a negatively charged rod is brought close to such a material, the electrons within each atom will be repelled by the rod and a separation of charge within each atom will occur, as shown in Figure 18-9. The positive part of each atom will be closest to the rod, and again a net attraction will occur. This is the situation for the charged plastic ruler of Figure 18-1 attracting bits of paper or leaves.

Figure 18-8 shows how a charge can be induced in a neutral metal rod. Another way to induce a net charge on a metal object is to connect it with a conducting wire to the ground (or a pipe leading into the ground) as shown in Figure 18-10(a). (⏚ means "ground"). The object is then said to be "grounded" or "earthed." Now the earth, being so large, can easily accept or give up electrons; hence it acts like a reservoir for charge. If a charged object—let's say negative this time—is brought up close to the metal, free electrons in the metal are repelled and many of them move down the wire into the earth—Figure 18-10(b). This leaves the metal positively charged. If the wire is now cut, the metal will have a positive induced charge on it. If the wire were cut after the negative object is moved away, the electrons would all have moved back into the metal and it would be neutral.

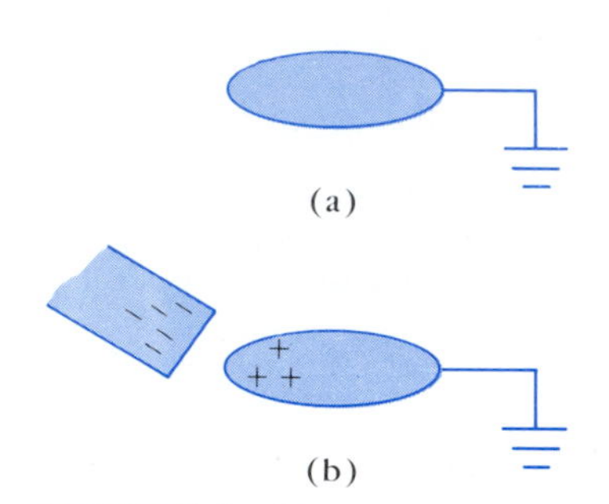

FIGURE 18-10
Inducing a charge on an object connected to ground.

The Electroscope

An *electroscope* (or simple *electrometer*) is a device that can be used for detecting charge. As shown in Figure 18-11, it consists of a case that contains two movable leaves, often made of gold. (Sometimes only one leaf is movable.) The leaves are connected by a conductor to a metal ball on the outside of the case, but are insulated from the case itself. If a charged object is brought close to the knob, a separation of charge is induced on it, Figure 18-12(a); the two leaves become charged and repel each other as shown. If, instead, the knob is charged by conduction, the entire apparatus becomes charged as shown in Figure 18-12(b). In either case, the greater the amount of charge the greater the separation of the leaves.

Note, however, that you cannot tell the sign of the charge in this way because a negative charge will cause the leaves to separate just as much as an equal magnitude positive charge—in either case the two leaves repel each other. An electroscope can, however, be used to determine the sign of the charge if it is first charged by conduction, say negatively, as in Figure 18-13(a). Now if a negative object is brought close, as in Figure 18-13(b), electrons are induced to move further down into the leaves which separate further. On the other hand, if a positive object is brought close, the electrons are induced to flow upward, leaving the leaves less negatively charged and their separation reduced—Figure 18-13(c).

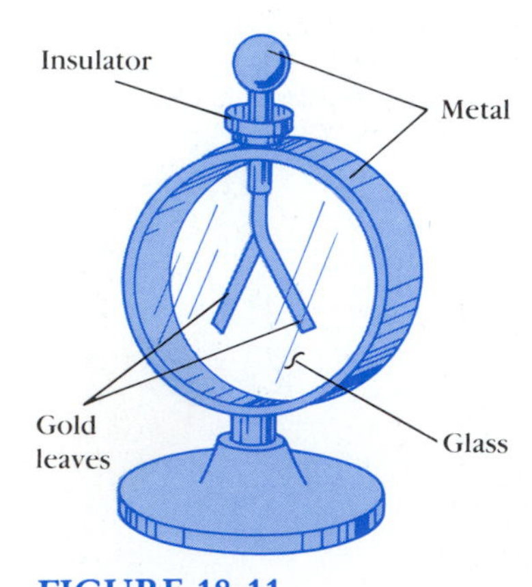

FIGURE 18-11
Electroscope.

The electroscope was much used in the early days of electricity. The same principle, aided by some electronics, is used in much more sensitive modern *electrometers*.

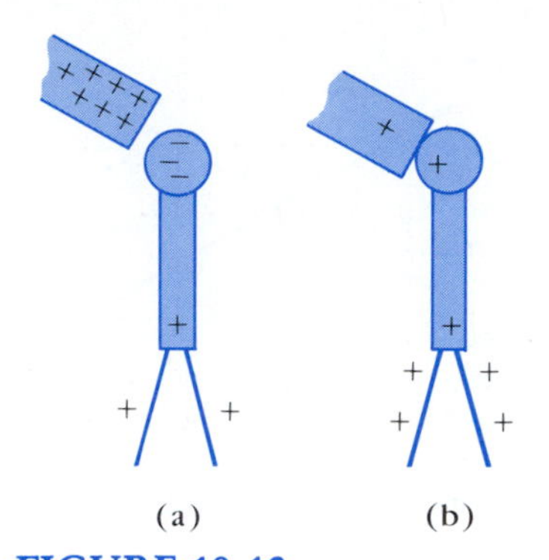

FIGURE 18-12
Electroscope charged (a) by induction, (b) by conduction.

5. ELECTRIC FIELD

Most of the forces we are familiar with in daily life could be called "contact" forces. Contact is required between that which exerts the force and that which feels the force, as when a person *pushes* (exerts a force on) a lawnmower or a tennis racket hits (exerts a force on) a tennis ball. In other words, when an object feels a force, something has to be "there" to exert the force.

But we have encountered two forces, the gravity force and the electric force, that act over a distance; that is, they act even when the bodies involved are not touching. This idea of a force "acting at a distance" was difficult for early thinkers to accept, and even today it is a difficult concept for many people.

This conceptual difficulty can be overcome with the idea of a *field*, which was developed by the British scientist Michael Faraday (1791–1867). We visualize an electric charge as giving rise to an **electric field** that extends outward from the charge and permeates all of space—Figure 18-14(a). Consequently, the force that one electric charge exerts on a second charge can be described as the interaction between one charge and the field set up by the other. In Figure 18-14(a), an electric field exists at all points in space due to the charge q_1. A second charge q_2 is placed near it, Figure 18-14(b); the charge q_2 feels a force due to the electric field of q_1. Thus, an electric charge feels a force because an electric field is there to exert the force. It must be emphasized however, that a field is *not* some form of matter.

Mathematically, the magnitude of the electric field is defined as force per unit charge. In other words, if a tiny[†] "test" charge q undergoes an electric force **F** at some point in space, then the electric field at that point in space is defined as

$$\mathbf{E} = \frac{\mathbf{F}}{q}.$$

The electric field, like force, has both magnitude and direction, and thus is a vector.

As an example, let us recall that the net force on charge number 3 in Figure 18-7 was 4.1 N, and pointed to the left. Whether charge 3 is

[†] We use a tiny electric charge for "testing," so the electric field of this charge can be ignored compared to the electric field already there.

FIGURE 18-13
A previously charged electroscope can be used to determine the sign of a given charge.

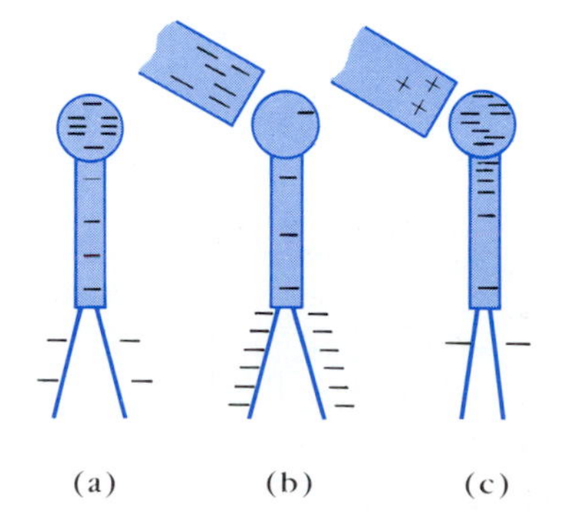

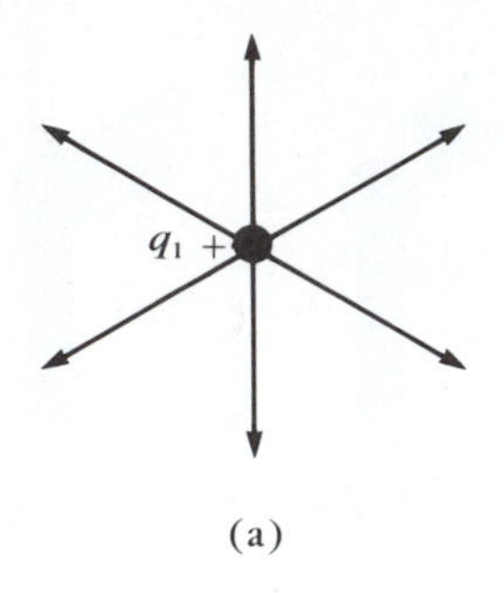

(a)

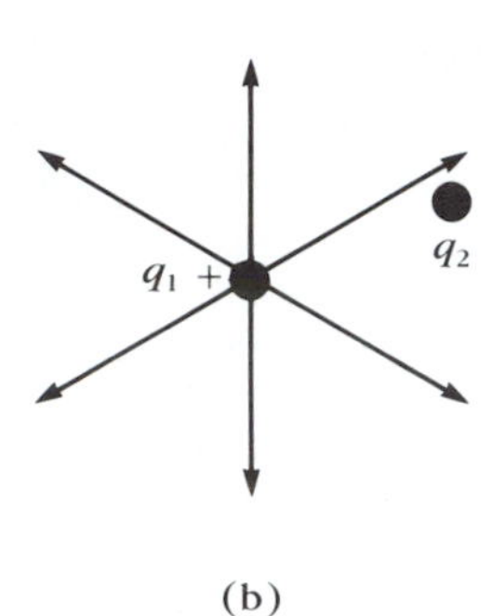

(b)

FIGURE 18-14
(a) Charge q_1 gives rise to an electrical field. (b) Charge q_2 feels a force due to the field of charge q_1.

present or not (it has a charge of magnitude $q_3 = 4.0\ \mu C$), the electric field at its position due to the charges q_1 and q_2 is $E = (4.1\ N)/(4.0 \times 10^{-6} C) = 1.0 \times 10^6\ N/C$. Notice that the units for electric field are newtons/coulomb.

Lines of Force

It is very useful to represent pictorially electric fields by drawing lines to indicate the direction of the electric field, as in Figure 18-14. The **electric field lines** or **lines of force** are drawn to indicate the direction of the force on a small positive charge placed in the field. The lines then indicate the path that a tiny positive charge would follow if released in this field. For example, the lines of force around a single positive point charge are directed outward, as in Figure 18-15(a); this is because a positive test charge would be repelled and move away in a straight line. Similarly, the direction of the field around a single negative point charge points inward—Figure 18-15(b)—since a positive charge would be attracted toward it. Notice in Figures 18-15(a) and (b) that near the charge the lines of force are closest together. This is just the region where the force on a second charge, and hence the electric field, is strongest. This is a general property of lines of force: *the closer together the lines are, the stronger the electric field in that region*.

Figure 18-15(c) shows the electric field surrounding two charges of opposite sign; Figure 18-15(d) shows the field between two oppositely charged parallel plates. Notice in each case that a small positive test charge would be repelled by the positive charge and attracted by the negative charge. The electric field lines always start on positive charges and are directed toward negative charges where they end.

The concept of electric field gives us a better sense of how one electric charge attracts another. But note in Figures 18-14 and 18-15 that only a few of the possible field lines are drawn in each case. Note too that the electric field is not confined to these lines but is continuous and exists between the lines as well.

The concept of a field can also be used for the gravitational force; it too acts "at a distance." A **gravitational field** surrounds every material object.

FIGURE 18-15
Electric field lines due to different arrangements of electric charges.

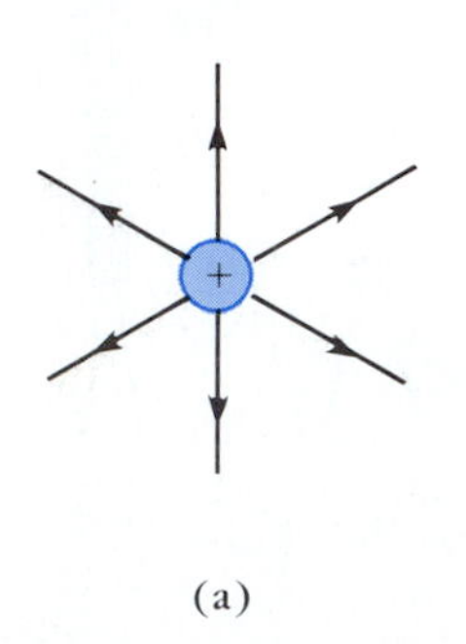

(a)

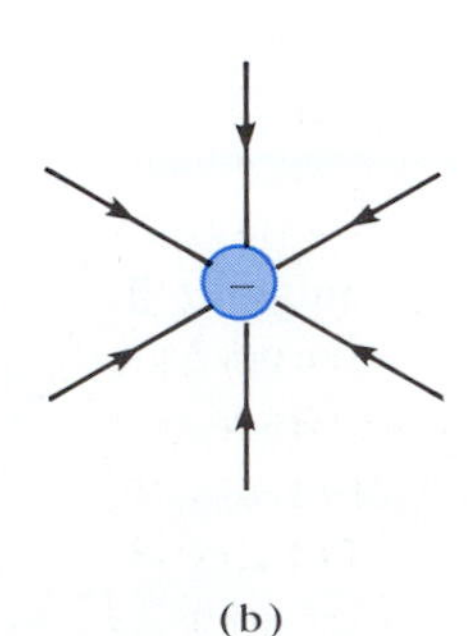

(b)

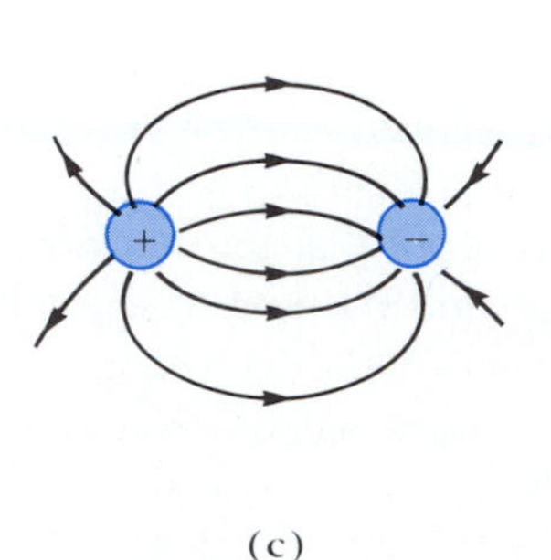

(c)

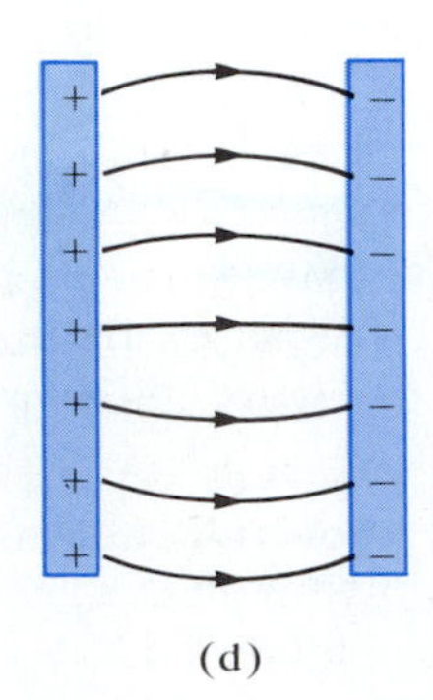

(d)

One object attracts another by virtue of its gravitational field (Figure 18-16). When we drop an object it falls in the gravitational field of the earth. Similarly the earth feels the effect of the gravitational field of the sun.

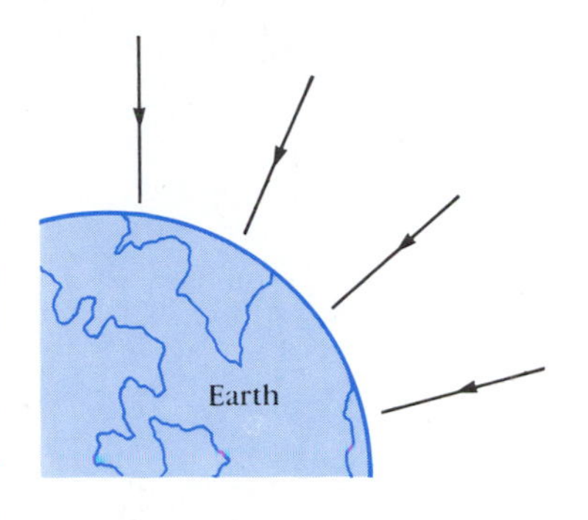

FIGURE 18-16
Gravitational field lines due to the earth.

6. ELECTRIC POTENTIAL

A body placed in a gravitational field feels a force. If it is allowed to fall, the body will gain kinetic energy as it loses potential energy. A particle has potential energy because of its *position* in a gravitational field and the total energy, kinetic plus potential, is conserved (see Chapter 7). An electrically charged body placed in an electric field also has potential energy by virtue of its position in that field. Take for example the case of two oppositely charged parallel plates, as in Figure 18-17. A positively charged particle released near the positive plate will, since a force is exerted on it, be accelerated to the right toward the negative plate. As it moves to the right its velocity, and hence its kinetic energy, increases: conservation of energy tells us that as the kinetic energy increases, the potential energy decreases. A positively charged particle thus has more potential energy when it is near the positive plate than when it is near the negative plate. This is in accord with the idea that potential energy is the ability (or "potential") to acquire kinetic energy, or to do work.

We defined the electric field (Section 5) as the force per unit charge; similarly, it is useful to define the **electric potential** (or simply the **potential** when "electric" is understood) as the *potential energy per unit charge*. Electric potential is given the symbol V; so we define the electric potential, V, at some point as

$$V = \text{electric potential} = \frac{\text{electric potential energy}}{\text{electric charge}}.$$

The unit of measurement of electric potential is joules/coulomb. It is called a volt (abbreviated V) in honor of the Italian scientist who invented the battery, Alessandro Volta (1745–1827): 1 volt = 1 joule/coulomb.

Electric potential is sometimes referred to as **voltage**. For example, a 6-volt battery is one whose electric potential difference between the + and − terminals is 6 volts. A battery of twice the voltage is able to give twice as much energy to electrons that flow in any wire or device connected to the terminals.

Since electric potential is defined as electric potential energy per unit charge (in symbols, $V = \text{PE}/q$), to find how much potential energy a certain amount of charge possesses we merely multiply the electric potential by its charge (in symbols, $\text{PE} = q \times V$). For example, if the potential difference between the two plates in Figure 18-17 is 6 V, then if a 1-C charge is moved (say by an external force) from B to A it will gain (1 C)(6 V) = 6 J of electric potential energy. (And it will lose 6 J of electric PE as it moves from A to B.)

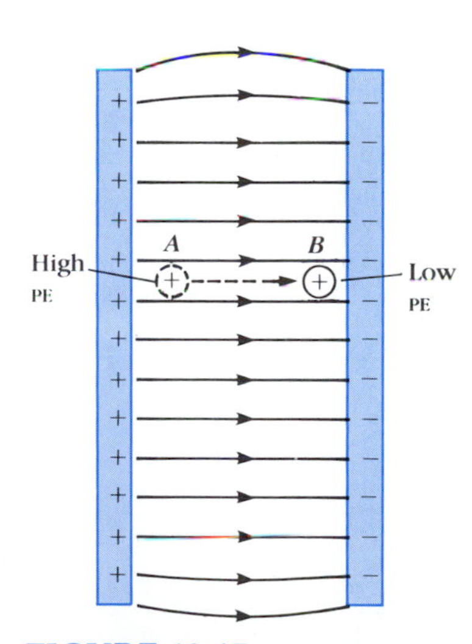

FIGURE 18-17
A positive charge is moved (by the electric field) from the positive toward the negative plate; the charge loses PE but gains in KE.

Similarly, a 2-C charge will gain 12 J, and so on. Thus, electric potential is a measure of how much energy an electric charge can acquire in a given situation. And, since energy is the ability to do work, the electric potential is also a measure of how much work a given charge can do. The exact amount depends both on the potential and on the charge.

To better understand electric potential, let's make a comparison to the gravitational case when a rock falls from the top of a cliff. The greater the height of a cliff, the more potential energy the rock has and the more kinetic energy it will have when it reaches the bottom. The actual amount of kinetic energy it will acquire and the amount of work it can do depend both on the height of the cliff and the mass of the rock. Similarly in the electrical case: the potential energy change and the work that can be done depend both on the potential difference (corresponding to the height of the cliff) and on the charge (corresponding to mass): $\text{PE} = q \times V$.

Practical sources of electrical energy such as batteries and electric generators are intended to maintain a particular potential difference; the actual amount of energy used or transformed depends on how much charge flows. For example, consider an automobile headlight connected to a 12-V battery. The amount of energy transformed (into light, and of course heat) is proportional to how much charge flows, which in turn depends on how long the light is on. If over a given period 5.0 C of charge flows through the light, the total energy transformed is (5.0 C)(12 V) = 60 J. If the headlight is left on twice as long, 10.0 C of charge will flow and the energy transformed is (10.0 C)(12 V) = 120 J.

SUMMARY

There are two kinds of *electric charge*, positive and negative. These designations are to be taken algebraically—that is, any charge is plus or minus so many coulombs (C). Electric charge is *conserved*: if a certain amount of one type of charge is produced in a process, an equal amount of the opposite type is also produced on the same body or on a different body, so the *net* charge produced is zero. According to the atomic theory, electric charge originates in the atom, which consists of a positively charged nucleus surrounded by negatively charged electrons. Each electron has a charge $-e = -1.6 \times 10^{-19}$ C. An object is negatively charged when it has an excess of electrons and positively charged when it has fewer than its normal number of electrons. Electrical conductors are those materials in which many electrons are relatively free to move, whereas electric insulators are those in which very few electrons are free to move.

An object can become charged in three ways: by rubbing, in which electrons are transferred from one material to another; by conduction, which is the transfer of charge from one charged object to another by touching; or by induction, the separation of charge within an object because of the close approach of another charged object but without touching.

Electric charges exert a force on each other. If two charges are of opposite types, one positive and one negative, they each exert an attractive force on the other. If the two charges are the same type, each repels the other. The magnitude of the force one point charge exerts on another is proportional to the product of their charges and inversely proportional to the square of the distance between them:

$$F = k\frac{q_1 q_2}{d^2};$$

this is *Coulomb's law*.

An *electric field* is imagined to exist in space due to any charge or group of charges. The force on another charged object is then considered to be due to the electric field present at its location. The *electric field*, $\mathbf{E}$, at any point in space due to one or more charges is defined as the force per unit charge that would act on a test charge q placed at that point: $\mathbf{E} = \mathbf{F}/q$. Electric fields are represented by *lines of force* which start on positive charges and end on negative charges. Their direction indicates the direction the force would be on a tiny positive test charge placed at a point; the closer the lines are to each other in a region of space, the greater the electric field.

A charged body in an electric field possesses potential energy by virtue of its position in that field. The potential energy per unit charge is called the *electric potential*, V, where $V = \text{PE}/q$; V is also called *voltage*, and is measured in volts.

QUESTIONS

1. The magnitude of the electric force between two bodies depends on what three factors?
2. If you charge a pocket comb by rubbing it with a silk scarf, how can you determine if the comb is positively or negatively charged?
3. Why does a shirt or blouse taken from a clothes dryer sometimes cling to your body?
4. Why does a phonograph record attract dust just after it has been wiped clean?
5. Explain why fog or rain droplets tend to form around ions or electrons in the air.
6. Can you guess why trucks carrying flammable fluids drag a chain along the ground? (*Hint*: Have you ever experienced a slight shock when getting out of a car?)

7. How do we know that an amber rod rubbed with a piece of fur acquires a negative charge rather than a positive charge?
8. Why do you sometimes feel a slight shock when you touch a metal object after walking across a rug?
9. A positively charged rod is brought close to a neutral piece of paper, which it attracts. Draw a diagram showing the separation of charge and explain why attraction occurs.
10. Name three good conductors of electricity and three good insulators. How well do these materials conduct heat? Do you think there might be a connection between the conduction of heat and the conduction of electricity?
11. Why does a plastic ruler that has been rubbed with a cloth have the ability to pick up small pieces of paper? Why is this difficult to do on a humid day?
12. Contrast the *net charge* on a conductor to the "free charges" in the conductor.
13. The form of Coulomb's law is very similar to that for Newton's law of universal gravitation. What are the differences between these two laws? Compare also gravitational mass and electric charge.
14. We are not normally aware of the gravitational or electrical force between two ordinary objects. What is the reason in each case? Give an example where we are aware of each one and why.
15. When a charged ruler attracts small pieces of paper, sometimes a piece jumps quickly away after touching the ruler. Explain.
16. Figure 18-8 shows how a charged rod placed near an uncharged metal object can attract (or repel) electrons. There are a great many electrons in the metal, yet only some of them move as shown. Why not all of them?
17. When an electroscope is charged, the two leaves repel each other and remain at an angle. What balances the electric force of repulsion so that the leaves don't separate further?
18. If a negatively charged particle is placed halfway between two charged parallel plates, one positive and the other negative as in Figure 18-17, in what direction will the particle move?
19. A positively charged object is placed to the left of a negatively charged object, as in Figure 18-15(c). Consider the electric field close to the positive charge. On which side of the positive charge—top or bottom, left or right—is the electric field the greatest? The smallest?
20. Why can lines of force never cross one another?
21. If one body has half the electric potential energy of another, does the first necessarily have half the electric potential? Explain.

22. If a negative charge is initially at rest in an electric field, will it move toward a region of higher potential or lower potential? What about a positive charge? How does the potential energy of the charge change in each case?

EXERCISES

1. How many electrons make up a charge of 100 μC?
2. Two charged bodies exert a force of 480 mN on each other. What will be the force if they are moved so they are only one-eighth as far apart?
3. How far apart must two electrons be if the force between them is to be 1.0×10^{-12} N?
4. What is the magnitude of the force of a 10-μC charge exerts on a 3.0-mC charge 2.0 m away? ($1\ \mu C = 10^{-6}$ C, $1\ mC = 10^{-3}$ C.)
5. Two small objects have the same positive charge. The force between them is 12 N when they are 2 cm apart. What will be the force (a) when they are 6 cm apart (b) when they are 1 cm apart (c) when one charge is tripled (d) when both charges are tripled?
6. A hydrogen atom consists of an electron which orbits a proton (charge $+e$) at an average distance of 0.53×10^{-10} m. Calculate the electrical and gravitational forces of attraction between them. Comment on the relative significance of electrical and gravitational forces at the atomic level. (The proton's mass is 1.67×10^{-27} kg and the electron's is 9.1×10^{-31} kg.)

*7. Particles of charge +86, +48, and −90 μC are placed in a line. The center one is 1.5 m from each of the others. Calculate the net force on each due to the other two.

8. A force of 2.4 N is exerted on a −1.8-μC charge in a downward direction. What is the magnitude and direction of the electric field at this point?
9. An electron in an electric field experiences a force of 8.0×10^{-16} N. What is the magnitude and direction of the electric field at this point?
10. What is the magnitude of the force on an electron in an electric field of 600 N/C?
11. What is the magnitude and direction of the electric field 12.0 m directly above a 13.0×10^{-6} C charge?
12. What is the magnitude and direction of the electric field at a point midway between a −20-μC and a +60-μC charge 40 cm apart?
13. A proton ($m = 1.67 \times 10^{-27}$ kg) is suspended at rest in a uniform electric field **E**. Take into account gravity and determine **E**.
14. Measurements indicate that there is an electric field surrounding the

earth. Its magnitude is about 100 N/C at the earth's surface and points inward toward the earth's center. What is the magnitude of the electric charge on the earth? Is it positive or negative? (*Hint*: The electric field due to a uniformly charged sphere is the same as if all the charge were concentrated at its center.)

15. How much energy does a 5-coulomb charge acquire when accelerated through a potential difference of 1000 volts?
16. How much energy would an electron acquire if it moved through a potential difference of 100 kV (kV = kilovolts)?
17. An electron is accelerated by a potential difference of 100 volts. How much greater would its final speed be if it were accelerated by 400 volts?
18. How much work is needed to move a $-3.0\text{-}\mu C$ charge from ground (0 V) to a point whose potential is +60 V?
19. An electron acquires 6.4×10^{-16} J of kinetic energy when it is accelerated by an electric field from plate *A* to plate *B*. What is the potential difference between the plates, and which plate is at the higher potential?
20. An electron in the picture tube of a TV set is accelerated from rest through a potential difference $V_{ab} = 5000$ V. (a) What is the change in potential energy of the electron? (b) What is the speed of the electron as a result of this acceleration?
21. A lightning flash transfers 30 C of charge to earth through a potential difference of 3.5×10^{7} V. (a) How much energy is dissipated? (b) How much water at 0°C could be brought to boiling?

*22. What is the total charge on all the electrons in 1.0 kg of H_2O? Why don't we notice such a large charge?

*23. Calculate the force on charge q_3 shown in Figure 18-18 due to the charges q_1 and q_2. (*Hint*: Recall how to add vectors, Chapter 5.)

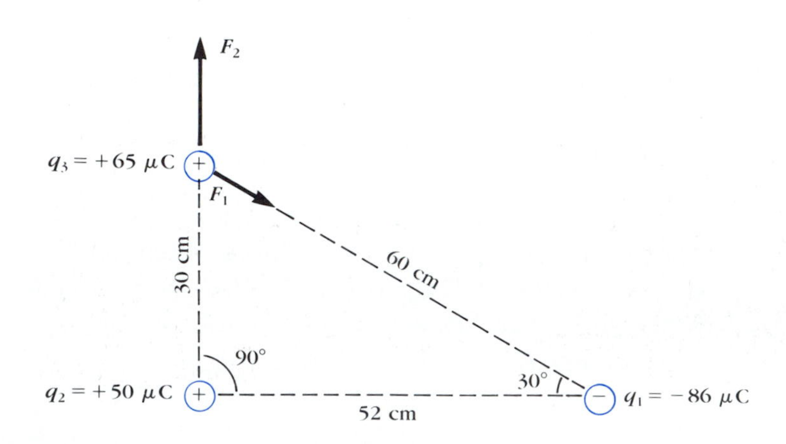

FIGURE 18-18

ELECTRIC CURRENTS

CHAPTER

19

Until the year 1800, the technical development of electricity consisted mainly of producing a static charge by friction. In the preceding century a number of machines had been built that could produce rather large potential by frictional means; one type of such apparatus is shown in Figure 19-1. Large sparks could be produced by these machines, but they had little practical value.

In nature itself there were grander displays of electricity such as lightning and "St. Elmo's fire," which is a glow that appeared around the yardarms of ships during storms. The fact that these phenomena were electrical in origin was not recognized until the eighteenth century. For example, it was only in 1752 that Benjamin Franklin, in his famous kite experiment, showed that lightning was an electric discharge—a giant electric spark.

Finally, in 1800 an event of great practical importance occurred: Alessandro Volta (1745–1827) invented the electric battery, and with it produced the first steady flow of electric charge—that is, a steady electric

FIGURE 19-1
Early electrostatic generator.

current. This discovery opened a new era that transformed our civilization, for today's sophisticated technology is based on electric current.

1. THE ELECTRIC BATTERY

The events that led to the discovery of the battery are interesting; for not only was this an important discovery but it also gave rise to a famous scientific debate between Volta and Luigi Galvani (1737–1798), eventually involving many others in the scientific world.

In the 1780s Galvani, a professor at the University of Bologna (thought to be the world's oldest university still in existence)[†], carried out a long series of experiments on the contraction of a frog's leg muscle through electricity produced by a static-electricity machine. In the course of these investigations, Galvani found, much to his surprise, that contraction of the muscle could be produced by other means as well: when a brass hook was pressed into the frog's spinal cord and then hung from an iron railing that also touched the frog, the leg muscles again would contract. Upon further investigation, Galvani found that this strange but important phenomenon occurred for other pairs of metals as well.

What was the source of this unusual phenomenon? Galvani believed that the source of the electric charge was in the frog muscle or nerve itself and that the wire merely transmitted the charge to the proper points. When he published his work in 1791, he termed it "animal electricity." Many wondered, including Galvani himself, if he had discovered the long-sought "life-force."

Volta, at the University of Pavia 200 km away, was at first skeptical of Galvani's results; but at the urging of his colleagues he soon confirmed and extended those experiments. But Volta doubted Galvani's idea of animal electricity. Instead he came to believe that the source of the electricity was not in the animal but rather in the *contact between the two metals*. Volta made public his views and soon had many followers, although others still sided with Galvani.

Volta was both a strong theoretician and a careful and skillful experimenter. He soon realized that a moist conductor, such as a frog muscle or moisture at the contact point of the two dissimilar metals, was necessary if the effect was to occur. He also saw that the contracting frog muscle was a sensitive instrument for detecting electric "tension" or "electromotive force" (his words for what we now call potential or voltage), in fact more

[†] As long ago as the thirteenth century it had 10,000 students and many women professors. A thirteenth-century historian reported that one of the women professors, Novella d'Andrea, was so beautiful that she had to deliver her lectures from behind a curtain to keep from distracting the students.

sensitive than the best available electroscopes that he and others had developed. Most important, he recognized that a decisive answer to Galvani could be given only if the sensitive frog leg was replaced by an inorganic detector; that is, to cement his view that it was the contact of two dissimilar metals that caused the frog muscle to contract, he would have to connect the two dissimilar metals directly to an electroscope and observe a separation of the leaves representing a potential difference. This proved difficult since his most sensitive[†] electroscopes were much less sensitive than the frog muscle. However, the eventual success of this experiment vindicated Volta's theory.

Volta's research showed that certain combinations of metals produced a greater effect than others, and, using his measurements, he listed them in order of effectiveness. (This "electrochemical series" is still used by chemists today.) And he further found that carbon could be used in place of one of the metals.

Volta then conceived his greatest contribution to science. Between a disc of zinc and one of silver he placed a piece of cloth or paper soaked in salt solution or dilute acid and piled a "battery" of such couplings, one on top of another, as shown in Figure 19-2; this "pile" or "battery" produced a much increased potential difference. Indeed, when strips of metal connected to the two ends of the pile were brought close, a spark was

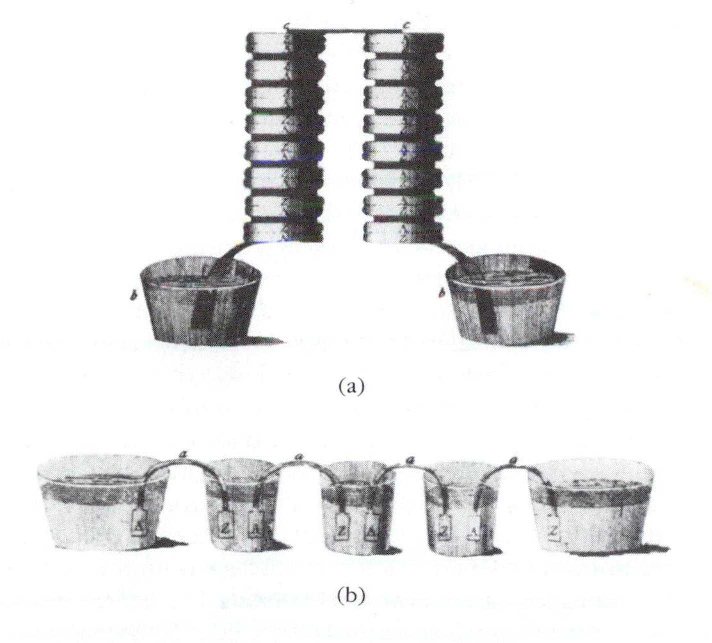

(a)

(b)

FIGURE 19-2
Two types of voltaic battery: (a) a pile, (b) "crown of cups." Z stands for zinc and A for silver (*argentum* in Latin). Taken from Volta's original publication.

[†] Volta's most sensitive electroscope measured about 40 V per degree (of leaf separation). Nonetheless, he was finally able to estimate the potential differences produced by dissimilar metals in contact: for a silver-zinc contact he got about 0.7 V, remarkably close to today's value of 0.78 V.

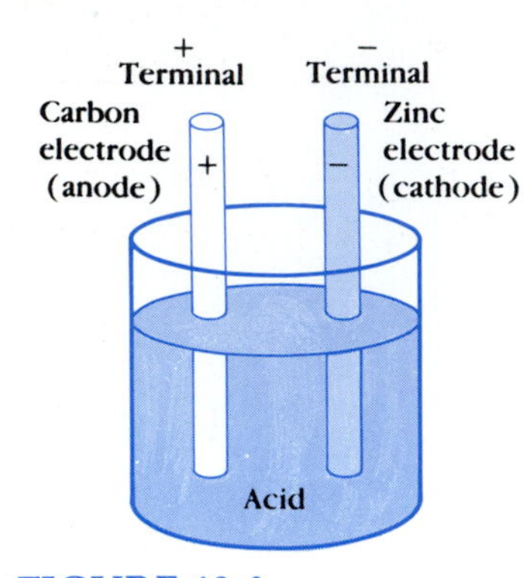

FIGURE 19-3
Simple electric cell.

produced. Volta had designed and built the first electric battery. A second design, known as the "crown of cups," is also shown in Figure 19-2. Volta made public this great discovery in 1800.

The potential produced by Volta's battery was still weak compared to that produced by the best friction machines of the time, although it could produce considerable charge. (The electrostatic machines were high-potential, low-charge devices.) But it had a great advantage: it was "self-renewing"—it could produce a flow of electric charge continuously for a relatively long period of time. It was not long before even more powerful batteries were constructed.

After Volta's discovery of the electric battery, it was eventually recognized that a battery produces electricity by transforming chemical energy into electrical energy. Today a great variety of electric cells and batteries are available, from flashlight batteries (sometimes called "dry cells") to the storage battery of a car. The simplest batteries contain two plates or rods made of dissimilar metals (one can be carbon) called *electrodes*. The electrodes are immersed in a solution, such as a dilute acid, called the *electrolyte*. (In a dry cell, the electrolyte is absorbed in a powdery paste.) Such a device is properly called an *electric cell*, and several cells connected together make a battery. The chemical reactions involved in most electric cells are quite complicated. Here we describe how one very simple cell works.

The simple cell shown in Figure 19-3 uses dilute sulfuric acid as the electrolyte. One of the electrodes is made of carbon, the other of zinc. That part of each electrode remaining outside the solution is called the *terminal*, and connections to wires and circuits are made here. The acid attacks the zinc electrode and tends to dissolve it. But each zinc atom leaves two electrons behind, so it enters the solution as a positive ion. The zinc electrode thus acquires a negative charge. As more zinc ions enter the solution, the electrolyte becomes increasingly positively charged. Because of this, and through other chemical reactions, electrons are pulled off the carbon electrode. Thus the carbon electrode becomes positively charged. (The positive electrode is called the *anode*; the negative electrode is called the *cathode*.) Because there is an opposite charge on the two electrodes, there is a potential difference between the two terminals. In a cell whose terminals are not connected, only a small amount of the zinc is dissolved; for as the zinc electrode becomes increasingly negative, any new positive zinc ions produced are attracted back to the electrode. Thus a particular potential difference or voltage is maintained between the two terminals. If charge is allowed to flow between the terminals, say through a wire (or a light bulb), then more zinc can be dissolved. The carbon too suffers disintegration. After a time, one or the other electrode is used up and the cell becomes "dead."

The voltage that exists between the terminals of a battery depends on what the electrodes are made of and their relative ability to be dissolved or give up electrons. The voltage of typical cells is 1.0 to 2.0 V. When two or more cells are connected so that the positive terminal of one is connected to the negative terminal of the next, they are said to be connected in *series* and their voltages add up. Thus the voltage between the ends of two flashlight batteries so connected is 3.0 V, while the six 2-V cells of an automobile storage battery give 12 V.

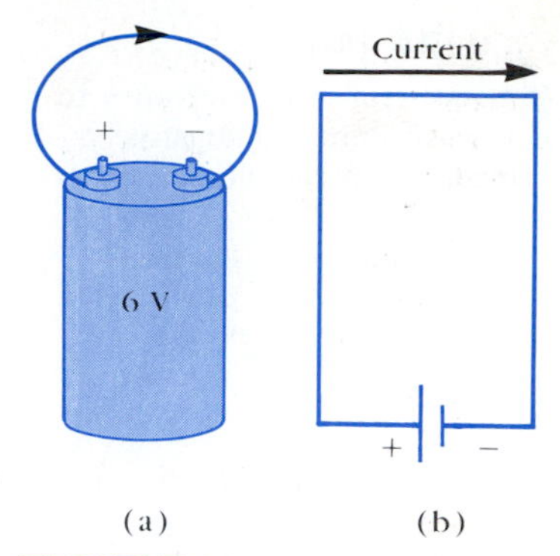

FIGURE 19-4
(a) Very simple electric circuit. (b) Schematic drawing of the circuit in part (a).

2. ELECTRIC CURRENT

Electric Charges Can Flow in Matter

When a continuous conducting path, such as a wire, is connected to the terminals of a battery, we have an electric **circuit**, Figure 19-4(a). On a diagram of a circuit, as in Figure 19-4(b), a battery is represented by the symbol "⊣⊢" (the longer line on this symbol represents the positive terminal and the shorter line the negative terminal). When such a circuit is formed, charge can flow through the circuit from one terminal of the battery to the other. A flow of charge such as this is called an **electric current**.

We speak of an electric current in much the same way we speak of a river current. In a river a mass of water flows; the more water and the faster it flows the greater is the current. An electric current, on the other hand, is a flow of electric *charge*; the more charge and the faster it flows the greater is the electric current. The magnitude of an electric current is defined to be the amount of charge flowing past a given point per second; it is usually given the symbol I:

$$I = \text{electric current} = \frac{\text{charge}}{\text{time}}.$$

Electric current is measured in coulombs/second, and this unit is given the special name "ampere" (abbreviated A or amp) in honor of the French physicist André M. Ampère (1775–1836): 1 ampere = 1 coulomb/second (1A = 1C/s). When one ampere flows through a wire, it means that one coulomb of charge flows by any point of the wire every second. When a current of two amperes flows in a wire, two coulombs of charge pass any point of the wire every second (Figure 19-5).

FIGURE 19-5
A section of wire in which a two ampere current is flowing.

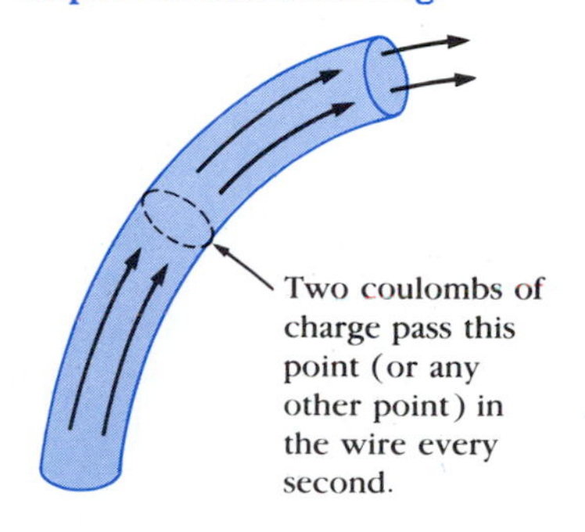

We saw in Chapter 18 that conductors contain many free electrons; thus, when a conducting wire is connected to the terminals of a battery as in Figure 19-4, it is actually the negatively charged electrons that flow in the wire. When the wire is first connected, free electrons at one end of the wire are attracted into the positive terminal. At the same time, electrons leave the negative terminal of the battery and enter the wire at the other end. Thus there is a continuous flow of electrons through the wire that begins as soon as the wire is connected to *both* terminals. However, when the conventions of positive and negative charge were invented two centuries ago, it was assumed positive charge flowed in a wire. Actually, for

FIGURE 19-6
Conventional current from + to − is equivalent to a negative (electron) current flowing from − to +.

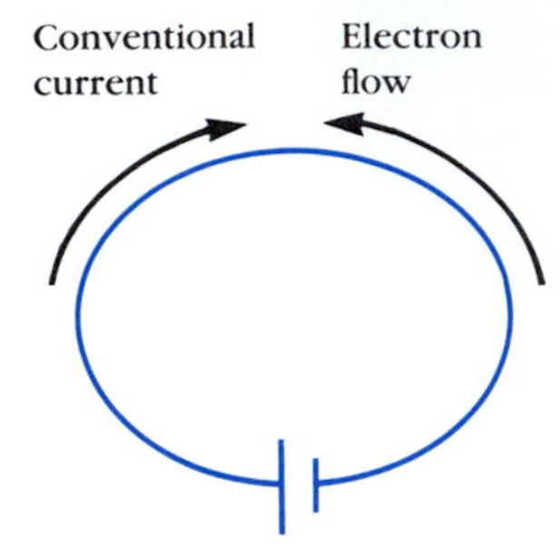

nearly all purposes, positive charge flowing in one direction is exactly equivalent to negative charge flowing in the opposite direction (Figure 19-6). Today, we still use the historical convention of positive current flow when discussing the direction of a current. This is sometimes referred to as *conventional current*. When we want to speak of the direction of electron flow, we will specifically state that it is the electron current. In liquids and gases, both positive and negative charges (ions) can move: together these comprise the electric current.

It is important to make a distinction between charge at rest and charge in motion. When there is a current in a wire, there are charges moving down the wire. Yet the wire has very little or no *net* charge on it. Whenever an electron enters one end of the wire, an electron leaves the wire at the other end. The wire remains electrically neutral even though a current flows through it.

Electric Sparks and Lightning

A dramatic example of a flow of charge is the electric spark formed by lightning. If the potential difference between two close objects—in this case a cloud and the earth or between two clouds—is great enough, the air, though it is usually a good insulator, breaks down. The strong electric field present pulls electrons off the air molecules, and the air becomes a conductor. This is known as an electric discharge. Electrons flow between the cloud and earth very quickly, eventually reducing the separation of charge and the difference of potential. We see a bright flash because electrical energy is transformed to light energy as electrons recombine with atoms. A streak of lightning indicates the path of the electron flow from a highly charged cloud to the earth.

3. OHM'S LAW: THE RELATION BETWEEN CURRENT AND VOLTAGE

Potential difference is the driving force behind electric current. Generally, the greater the potential difference the greater the current.

When we put electricity to practical use—in light bulbs, toasters, television sets, and other electric appliances—it is important to know the relationship between current and potential difference.

It was Georg Simon Ohm (1787–1854) who discovered the experimental fact that the current in a wire is directly proportional to the voltage (or potential difference) between its two ends:

$$\text{current} \propto \text{voltage},$$

or in symbols

$$I \propto V.$$

This is known as **Ohm's law**. If, for example, we connect a wire to a 6-V battery, the current flow will be twice what it would be if the wire were connected to a 3-V battery.

It is helpful to compare an electric current to the flow of water in a river or a pipe. If the pipe (or river) is nearly level, the flow rate is small. But if one end is somewhat higher than the other, the flow rate—or current—is much greater. The greater the difference in height, the greater the current. We saw in Chapter 18 that electric potential is analogous, in the gravitational case, to the height of a cliff; and this applies in the present case to the height through which the fluid flows. Just as an increase in height causes a greater flow of water, so a greater electric potential difference, or voltage, causes a greater current flow.

Exactly how much current flows in a wire depends not only on the voltage but also on the resistance the wire offers to the flow of electrons. The walls of a pipe, or the banks of a river and rocks in the middle, offer resistance to the flow of current. Similarly, electrons are slowed because of interactions with the atoms of the wire. The property of a wire that causes it to resist the flow of current is known as **resistance** and is denoted by the symbol R. A high-resistance wire will allow less current to flow than will a low-resistance wire. Current is thus inversely proportional to the resistance. Combining this fact with the above proportionality, we write Ohm's law in its final form as

$$\text{current} = \frac{\text{voltage}}{\text{resistance}}.$$

We can turn this relationship around to read *voltage* = *current* × *resistance* or, in symbols (this equality is also known as **Ohm's law**):

$$V = IR.$$

The resistance of a wire depends on the arrangement of the atoms in the material—and therefore on the kind of material it is—as well as on the size and shape of the wire. A long wire will offer more resistance than a short one, since there will be more opportunities for collisions to occur in the longer one. A wire with a large cross-sectional area will offer less resistance than a wire with a small cross-sectional area, because there is more area for the electrons to flow through.

The unit for resistance is called the *ohm* and is abbreviated Ω (Greek capital omega). Because $R = V/I$, we see that 1.0 Ω is equivalent to 1.0 V/A As a numerical example, suppose the plate on the bottom of a small cassette recorder specifies that it should be connected to 6.0 V and will draw 300 mA. (a) What is the net resistance of the recorder? (b) If the voltage dropped to 5.0 V, how would the current change? To answer (a) we write Ohm's law in the form $R = V/I = 6.0\ V/0.30\ A = 20\,\Omega$. For (b), if the resistance remained constant the current would be approximately

$I = V/R = 5.0\text{ V}/20\ \Omega = 0.25\text{ A}$, or 250 mA, which is a drop of 50 mA. Actually, resistance depends on temperature, so this is only an approximation.

In order for a current to flow in a circuit, it is important to remember that there must be a continuous conducting path. If a wire is cut, or is not properly connected in a circuit, or there is a switch that is open, the current will not flow and the circuit is said to be "open." If there are no interruptions in the path to stop the current flow, the circuit is "closed."

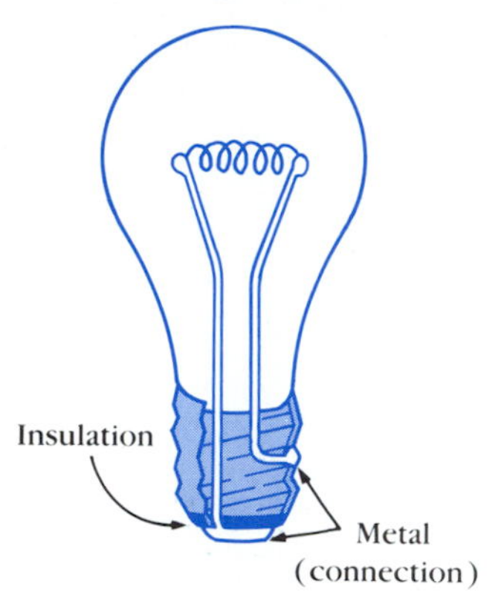

FIGURE 19-7
Diagram of an ordinary incandescent light bulb.

EXPERIMENT

Take a flashlight apart and figure out how the complete circuit is formed. If it's a metal flashlight, the casing may serve as a conductor. Examine the interior of the bulb, including the filament, with a magnifying glass. Is there a continuous conducting path within the bulb? The two connections to the bulb are made at the conductor on the very bottom and on the threads (see Figure 19-7). Connect a circuit consisting of the bulb, one battery, and one or two wires. You can use tape to hold the connections but secure them as tightly as possible. Note how brightly the bulb glows. Reconnect the circuit using two batteries connected end to end; you might lay them in the fold of an open book to keep them in line. Why does the bulb glow more brightly with two batteries than with one?

4. SERIES AND PARALLEL CIRCUITS

When two or more light bulbs or electric devices are connected in one circuit, they can be arranged in *series* or in *parallel*. When they are arranged in **series**, the devices are connected in a single conducting path—Figure 19-8(a). When they are arranged in **parallel**, each device is on a separate path, so that electrons that flow through one do not flow through the other—Figure 19-8(b).

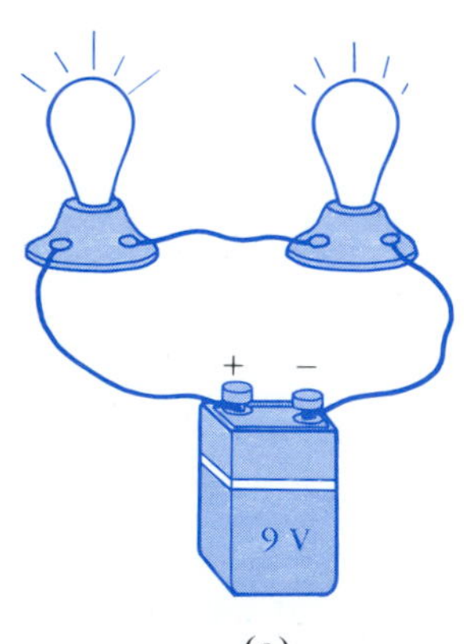

(a)

(b)

FIGURE 19-8
Two light bulbs connected (a) in series and (b) in parallel.

EXPERIMENT

Obtain two flashlight bulbs, a battery, and several lengths of electrical wire. Connect one bulb to the battery with two pieces of wire. You can make the connections by wrapping the wire around the base of the bulb and securing it with sturdy tape. Next, connect the second bulb into the circuit so that the two bulbs are (a) in series and then (b) in parallel.

Series and parallel circuits have different characteristics. You may have noticed in this experiment that the two bulbs in the parallel circuit glowed nearly as brightly as when only one bulb was used (see previous experiment); but the two bulbs in the series circuit glowed somewhat less brightly. To see why, let's look at the differences between these two types of circuit.

In the **series** circuit, the same current is flowing in all parts of the circuit (what flows in one end flows out the other). The current through the bulbs in this case is less than it would be if only one bulb were connected to the battery, because each bulb offers resistance to the flow of current. In a series circuit, the total resistance is *the sum of the individual resistances*. If the two bulbs have resistances R_1 and R_2, then the total resistance of the circuit is

$$R = R_1 + R_2. \qquad \text{[series circuit]}$$

For three or more devices in series, we just add more terms: $R = R_1 + R_2 + R_3 + \ldots$. If the two bulbs in series of Figure 19-8(a) are identical, the total resistance is double the resistance offered by each. And so the current will be only half what it would be through a single bulb connected to the same battery. This is why the two bulbs in the series circuit glowed less brightly than a single bulb in the circuit.

To look at this in more detail, suppose the two bulbs each have resistance $R_1 = R_2 = 25\,\Omega$. If only one were connected to the 9.0-V battery, the current through it would be $I = V/R = 9.0\text{ V}/25\,\Omega = 0.36\text{ A}$. But when both are connected in series, as in Figure 19-8(a), the total resistance of the circuit is $R = 25\,\Omega + 25\,\Omega = 50\,\Omega$. In this case, the current will be $I = V/R = 9.0\text{ V}/50\,\Omega = 0.18\text{ A}$, or half what it was for a single bulb. Notice also that when the bulbs are connected in series, the voltage is divided between them: the voltage across each bulb is $V = IR = (0.18\text{ A})(25\,\Omega) = 4.5\text{ V}$. Of course the sum of the voltage across the two bulbs in series adds up to the full 9.0 volts of the battery.

What happens in a **parallel** circuit? Here the current leaving the battery splits into two (or perhaps more) branches. Each of the bulbs in Figure 19-8(b) has the full voltage of the battery across it, since each is connected directly (by nearly resistanceless wire) to the battery. Thus the current through each bulb will be essentially the same as if only one bulb were in the circuit. Hence bulbs in parallel each burn as brightly as a single bulb. Since the current flowing from the battery is following two paths, the *total* current is the sum of the individual currents and is thus greater than the current drawn by a single bulb. Thus, arranging the bulbs in parallel actually reduces the net resistance of the circuit. This may seem surprising; but remember that when you add another path in parallel, you give the electrons another channel to flow in; it's like making the wire fatter, so there must be less net resistance. The same is true of water in a pipe or a river—if you give it an additional channel to flow through, more water will flow.

To determine the net resistance when two or more resistances are arranged in *parallel*, we note that the total current I is the sum of the individual currents through each resistance: I_1, I_2, and so on. That is

$$I = I_1 + I_2 + \ldots.$$

Since the voltage V across each resistor is the same, we have

$$I = \frac{V}{R_1} + \frac{V}{R_2} + \cdots = \frac{V}{R},$$

FIGURE 19-9
Connection of household appliances.

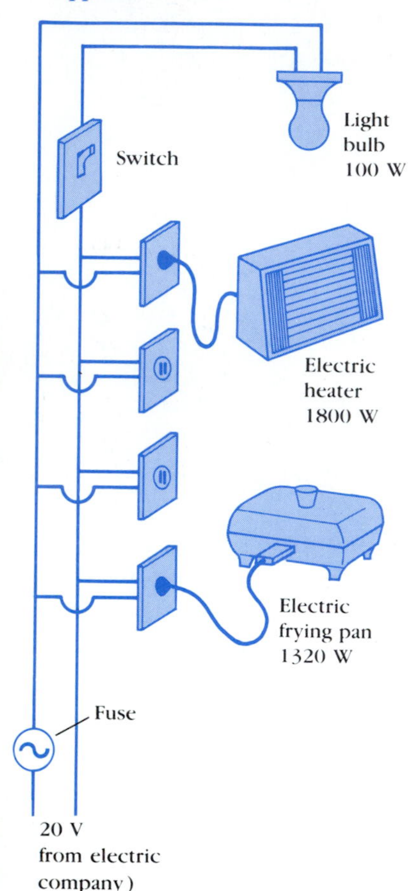

where $R_1, R_2 \ldots$ are the individual resistances and R is the "effective" or net resistance of the whole combination. Thus,

$$\frac{1}{R} = \frac{1}{R_1} + \frac{1}{R_2} + \ldots. \qquad \text{[parallel circuit]}$$

If the two bulbs in Figure 19-8(b) each have resistance $R_1 = R_2 = 25\,\Omega$, then the net resistance of this parallel circuit is found from

$$\frac{1}{R} = \frac{1}{25\,\Omega} + \frac{1}{25\,\Omega} = \frac{2}{25\,\Omega};$$

this gives $1/R$; R itself is the reciprocal:

$$R = \frac{25\,\Omega}{2} = 12.5\,\Omega$$

This parallel circuit has a total resistance equal to half the resistance of each bulb individually. The total current leaving the battery will be $I = V/R = 9.0\text{ V}/12.5\,\Omega = 0.72\text{ A}$. This is twice the current one bulb draws from the battery (0.36 A). This of course makes sense since both bulbs are attached directly to the battery and each draws 0.36 A for a total of 0.72 A.

A series circuit has one disadvantage that a parallel circuit does not. If one of the lights burns out or comes loose in the socket, the circuit is broken, the current stops flowing, and the other bulb goes out too. Some strings of Christmas-tree lights are connected in series and when one light burns out, they all go out. In a parallel circuit, by contrast, if one bulb goes out the others are unaffected and continue to glow.

Houses are always wired in parallel to ensure that each device receives the same voltage, usually 120 V in the United States.[†] Two lead-in wires are connected to each outlet in parallel, as shown in Figure 19-9. Thus you can turn one light on or off without affecting the others.

5. ELECTRIC POWER

Electric Current Gives Rise to Heat and Light

Electric energy can be easily transformed into other forms of energy. How is this done? A difference of potential gives rise to an electric current, a flow of charge. The potential energy that each charge loses as it moves through this potential difference is equal to the amount of work it can do. For example, each coulomb of charge that moves through a potential difference of 10 volts is capable of doing 10 joules of work.

[†] Most houses in the United States receive 240 V. This full voltage is used for such appliances as electric stoves, but it is split in half for other appliances. In many other countries, 240 V is standard for all appliances.

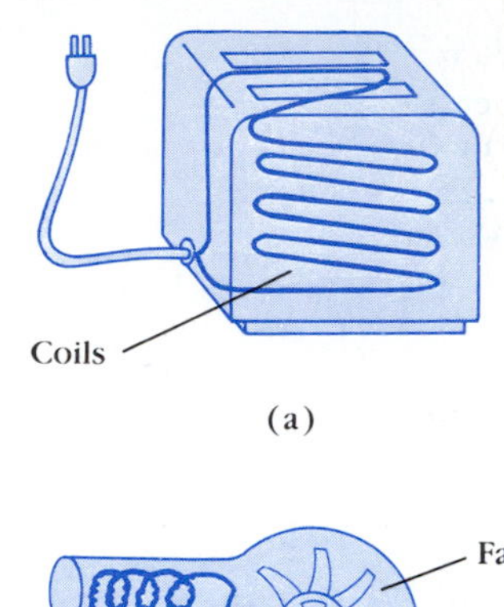

(a)

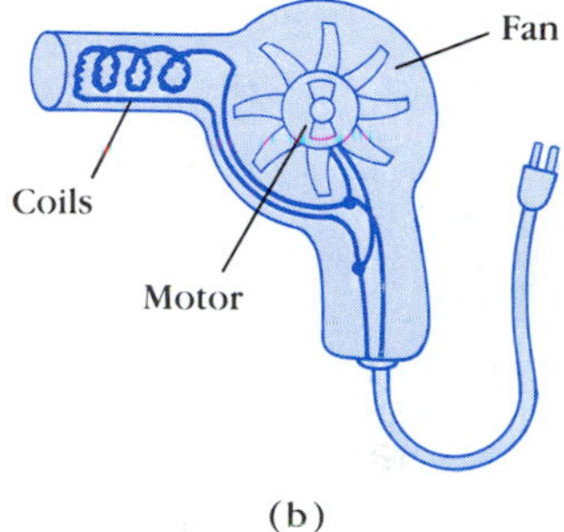

(b)

FIGURE 19-10
Electric toaster (a) and hair dryer (b), showing heating elements in each.

In devices such as electric heaters, stoves, toasters, and hair dryers (Figure 19-10), electric energy is transformed into thermal energy in a wire resistance known as a "heating element." And in an ordinary light bulb, the tiny wire filament (Figure 19-7) becomes so hot it glows. Only a few percent of the energy is transformed into light: the rest, over 90 percent, is transformed into thermal energy. Light-bulb filaments and heating elements in household appliances have a resistance typically of a few ohms to a few hundred ohms.

Electric energy is transformed into thermal energy or light in such devices because the current is usually rather large, and there are many collisions between the moving electrons and the atoms of the wire. In each collision, part of the electron's kinetic energy is transferred to the atom with which it collides. As a result, the kinetic energy of the atoms increases and hence the temperature of the wire element increases. The increased thermal energy (internal energy) can be transferred as heat by conduction and convection to the air in a heater or to food in a pan, by radiation to toast in a toaster, or radiated as visible light.

We shall see in the next chapter that electric energy can be transformed into mechanical energy by an electric motor and into sound by a loudspeaker.

Calculating the Power

How much energy an electric appliance transforms (it is energy you pay for on your electric bill) depends on how long you use the appliance and at what rate it uses energy. A radio left on for two hours uses twice as much energy as one left on for only one hour. The characteristic of any given appliance is the *rate* at which it uses energy, which we call *power*:

$$\text{power} = \frac{\text{energy}}{\text{time}}.$$

As we saw in Chapter 7, the unit of power is the watt; 1 watt = 1 joule/second. The total amount of energy used by any appliance is simply the product of its power consumption an the length of time it is on. A 150-watt light bulb uses up 150 joules of energy every second it is on. If it is on for an hour, a total of 3600 seconds, it uses a total of $150 \times 3600 = 540{,}000$ joules. Electric companies use a much larger unit for measuring energy—the "kilowatt-hour." One kilowatt is a thousand watts. One kilowatt-hour (kWh) is the energy consumed at the rate of 1000 watts for one hour. Therefore 1 kWh = 1000 joules/s × 3600 s/hr × 1 hr = 3,600,000 joules. Electric energy costs the consumer a few cents per kilowatt-hour, depending on the locale. It is worthwhile to be able to calculate how much it costs to run an electric applance. At $.10/kWh, a 100-watt light bulb (100 watts = 0.1 kilowatt) burning for 20 hours would cost: 0.1 kW × 20 hr × $.10/kWh = $.20. A 1300-watt electric frying pan used for 2 hours would cost: 1.3 kW × 2 hr × $.10/kWh = $.26.

Calculate how much energy (in kilowatt-hours) you use in an average day at your residence. You will have to find out the power consumption or "wattage" of each device used; it is usually printed somewhere on the appliance. If only the current rating is given, you can use the power relationship below to determine the power. To find the energy used, you must also estimate how long each appliance is on during an average day. Inquire from your electric company how much a kilowatt-hour costs and determine your electricity costs per day and per month. (Does this match your electric bill?) You can also check your calculation by reading the electric meter.

It is useful to know the relationship between power, voltage, and current. In an electric circuit when an amount of charge q moves through a potential difference V, an amount of energy equal to $q \times V$ can be transformed into work or other forms of energy. Then the power P transformed must be

$$P = \text{power} = \frac{\text{energy}}{\text{time}} = \frac{q \times V}{t}.$$

The charge per unit time, q/t, is simply the current I; so

$$\text{power} = I \times V = \text{current} \times \text{voltage}.$$

When the current is in amperes and the voltage is in volts, the power will be in watts. The power relationship can be written in various other ways using Ohm's law ($V = IR$):

$$\begin{aligned} P &= IV \\ &= I^2R \ (\text{since } P = IV = I \times IR = I^2R) \\ &= \frac{V^2}{R} \left(\text{since } P = IV = \frac{V}{R} \times V = \frac{V^2}{R}\right). \end{aligned}$$

As a simple example, let us calculate the resistance of a 60-watt light bulb. We use $P = V^2/R$ and solve for R, which gives us $R = V^2/P =$ (120 volts)^2/(60 watts) = 240 ohms. The current through the bulb can be calculated by using the relation $P = IV$ and solving for I: $I = P/V = 60$ watts/120 volts = 0.5 amp; the same result can be obtained by using Ohm's law and the value of R just calculated: $I = V/R =$ (120 volts)/(240 ohms) = 0.5 amp.

The electric wires that carry electricity to lights and other electric appliances have some resistance, although usually it is quite small. Nonetheless, if the current is large enough the wires will heat up and produce heat equal to I^2R, where R is the wires' resistance. One possible hazard is that the current-carrying wires in the wall of a building may become so hot that they will start a fire. Thicker wires offer less resistance and thus can carry more current without becoming too hot. When a wire carries more current than is safe, it is said to be "overloaded." To prevent over-loading, fuses or circuit breakers are installed in circuits. They are basically switches (Figure 19-11) that open the circuit when the current exceeds a particular value. A 20-A fuse or circuit breaker, for example, opens when the current passing through it exceeds 20 A.

When wiring is installed in a building, the fuses are chosen to match the particular wire being used. For example, if the wire can take currents up to 20 amperes safely, a 20-A fuse is used. If the fuse is replaced with a 30-A fuse, overload protection is lost and serious fire danger may result.

If a circuit in your house repeatedly burns out a fuse or opens a circuit breaker, there are two possibilities: (a) you have too many appliances drawing current in that circuit or (b) there is a fault somewhere, such as a "short." A short, or "short circuit," means that two wires have crossed, perhaps because the insulation has worn down; so the path of the current is shortened. The resistance of the circuit is then very small, so the current will be very large. Short circuits should, of course, be remedied immediately.

Here's an example: suppose you blow a fuse whenever you use a 100-watt overhead light, an electric heater, and an electric frying pan (Figure 19-9). On each of these appliances you find a plate giving its wattage (although sometimes the current drain is given directly). Suppose the wattage is 1320 watts for the frying pan and 1800 watts for the heater, and they are plugged into a standard 120-volt outlet. How much current is being drawn by each? We use the formula $P = IV$ and solve for I, which gives $I = \frac{P}{V}$. The light bulb draws a current $I = \frac{P}{V} = \frac{100 \text{ watts}}{120 \text{ volts}} = 0.8$ amperes; the frying pan draws $I = \frac{1320 \text{ watts}}{120 \text{ volts}} = 11$ amperes; and the heater draws $I = \frac{1800 \text{ watts}}{120 \text{ volts}} = 15$ amperes. The total current drawn is $15 + 11 +$

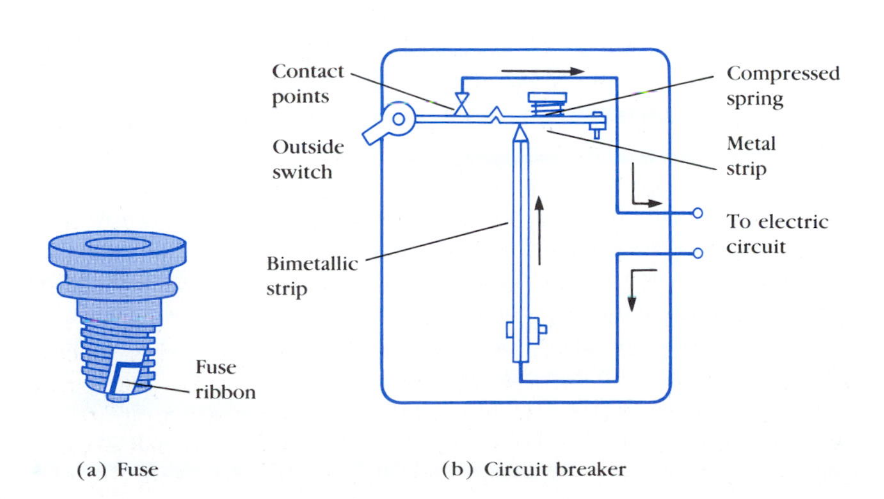

FIGURE 19-11
(a) A fuse. When the current exceeds a certain value, the ribbon melts and the circuit opens. Then the fuse must be replaced. (b) A circuit breaker. Electric current passes through a bimetallic strip. When the current is great enough, the bimetallic strip heats sufficiently to bend so far to the left that the notch in the spring-loaded metal strip drops down over the end of the bimetallic strip. The circuit then opens at the contact points (one is attached to the metal strip) and flips the outside switch. As the device cools down, it can be reset using the outside switch.

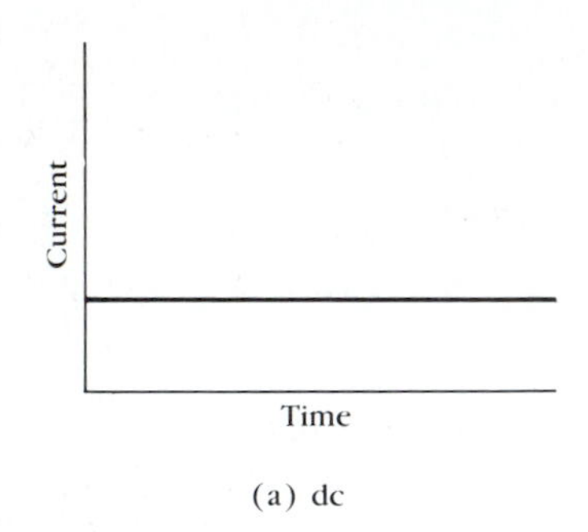

(a) dc

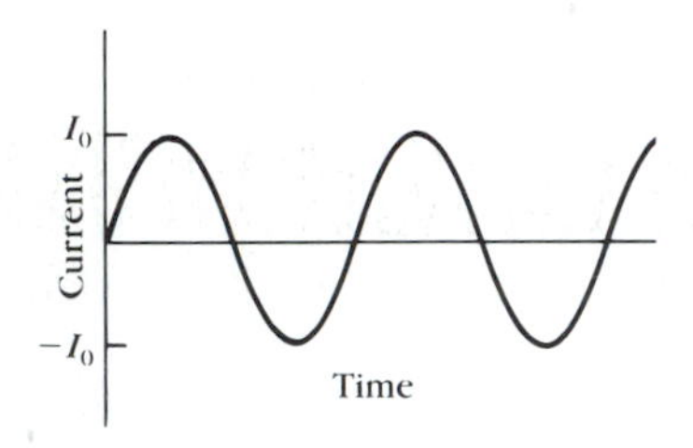

(b) ac

FIGURE 19-12
(a) Direct current. (b) Alternating current.

0.8 ≈ 27 A. If the circuit has a 20-A fuse, no wonder it blows! If it has a 30-A fuse it should not blow, so we would look for a short. The most likely place is in the cord of one of the appliances.

6. AC AND DC

If you look at the information plate on an electric appliance, you may see "110–120 V ac" or "9 V dc," as well as the wattage or current requirements. The "110 V" or "9 V" refers to the voltage required. In the former case, you use ordinary house current; you plug it in. In the latter case, a 9-volt battery is needed. But what do "ac" and "dc" mean?

Dc stands for "direct current." A direct current is one that stays at a constant value—Figure 19-12(a). The electrons move in one direction only. Direct current is produced by batteries and by some generators known as dc generators.

Ac stands for "alternating current." An alternating current reverses direction many times a second—Figure 19-12(b). The electrons move first in one direction and then in the other. This is accomplished by applying an alternating voltage produced by an ac generator (discussed in Chapter 21). House current in nearly all of the United States and in many other countries is ac. In the United States, the current alternates back and forth 60 times per second and is commonly referred to as "60 cycle," or as having a frequency of 60 Hz. The electric company usually maintains the frequency very close to the 60 cycles per second, independent of the voltage or current drawn. In some locales a different frequency is used; for example, 50 Hz is used in some countries.

7. ELECTRIC HAZARDS

It is primarily current and not voltage that causes shocks. An electric shock can cause damage to the body and can even be fatal. The severity of a shock depends on the magnitude of the current, how long it acts, and through what part of the body it passes. A current that flows from the thumb to a finger on the same hand is much less damaging than one that flows from one hand to the other and passes through the heart. Indeed, a current that passes through vital organs such as the heart and brain is especially serious, for it can interfere with their operation. A current heats tissues and can cause burns, particularly on the skin, where the resistance is high. A current also stimulates the nerves and muscles of the body. We feel a "shock" because our muscles contract.

Most people can feel a current of about 1 mA (1 mA = 1 milliamp = 1/1000 A). Currents of a few mA cause pain but rarely much damage in a healthy person. However, currents above 10 mA cause severe contraction of the muscles; in this case a person may not be able to release the source of the current (say a faulty appliance or wire). Death from paralysis of the respiratory system can then occur; artificial respiration, however, can often revive a victim. If a current above about 70 mA passes across the torso so that a portion passes through the heart for a second or more, the heart muscle begins to contract irregularly and blood is no longer pumped properly; it is difficult for the heart to recover from this state, which is why death is common. Strange as it may seem, however, if the current is larger than about 1 A, the damage may be less and death less likely. Apparently such large currents bring the heart to a complete standstill; with release of the current, the heart may then resume normal rhythm.

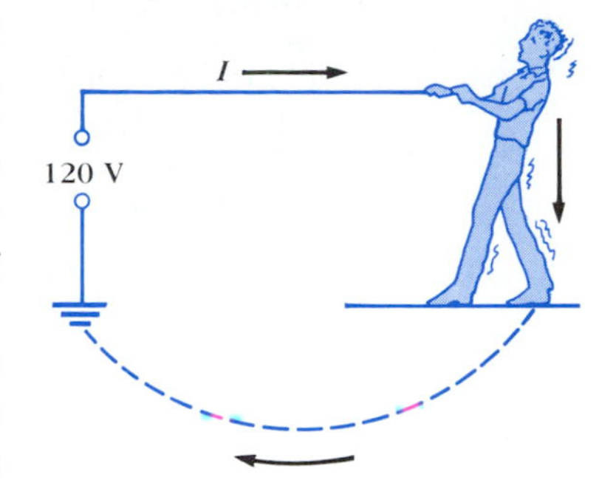

FIGURE 19-13
A person receives an electric shock when the circuit is completed.

The seriousness of a shock depends on the effective resistance of the body. Living tissue has quite low resistance since the fluid of cells contains ions that can conduct quite well. However, the outer layers of skin, when dry, offer much resistance. The effective resistance between two points on opposite sides of the body when the skin is dry is in the range of 10^4 to $10^6\,\Omega$. However, when the skin is wet the resistance may be 1000 Ω or less. A person in good contact with the ground who touches a 120-V dc line with wet hands can suffer a current

$$I = \frac{120\text{ V}}{1000\,\Omega} = 120\text{ mA}.$$

As we saw earlier, this could be lethal.

Figure 19-13 shows how the circuit is completed when a person touches an electric wire. One side of a 120-V source is connected to ground by a wire connected to a buried conductor (say, a water pipe). Thus the current passes from the high-voltage wire through the person to the ground; it passes through the ground back to the other terminal of the source to complete the circuit. If the person in Figure 19-13 stands on a good insulator—thick-soled shoes or a dry wood floor—there will be much more resistance in the circuit and consequently much less current will flow. However, if the person stands barefooted on the ground or sits in a bathtub, there is considerable danger. In a bathtub, not only are you wet and thus your resistance low, but the water is in contact with the drain pipe that leads to the ground. That is why it is strongly recommended that you touch nothing electrical in such a situation.

A principal danger lies in touching a bare wire whose insulation has worn away or touching a bare wire inside an appliance when you tinker with it. (Always unplug an electrical device before investigating its insides). Sometimes a wire inside a device breaks or loses its insulation and comes in contact with the case. If the case is metal, it will conduct electricity. A person could then suffer a severe shock merely by touching the case. To prevent an accident, most cases are supposed to be connected to a ground. Then if a "hot" wire touches the case, a short circuit to a ground immediately occurs and the fuse or circuit breaker opens the circuit.

SUMMARY

By transforming chemical energy into electric energy, an electric battery serves as a source of potential difference or voltage. A simple battery consists of two electrodes made of different metals immersed in a solution or paste known as an electrolyte.

Electric current, I, refers to the rate of a flow of electric charge and is measured in amperes (A): 1 A equals a flow of 1 C/s past a given point. A continuous conducting path between the terminals of a source of electrical energy such as a battery is called a *circuit*.

The difference in electric potential, or voltage, produced by batteries or generators can produce electric currents. *Ohm's law* states that the current in a good conductor is proportional to the potential difference V applied to its two ends and inversely proportional to the *resistance* R of the material, so $V = IR$. The unit of resistance is the ohm (Ω), where $1\ \Omega = 1$ V/A.

When resistances are connected in series, the net resistance is the sum of the individual resistances. When connected in parallel, the reciprocal of the total resistance equals the sum of the reciprocals of the individual resistances; in a parallel connection, the net resistance is less than any of the individual resistances.

The rate at which energy is transformed in a resistance R from electric to other forms of energy (such as heat and light) is equal to the product of current and voltage; that is, the power transformed, measured in watts, is given by $P = IV$ and can be written with the help of Ohm's law as $P = I^2R = V^2/R$. The total electric energy transformed in any device equals the product of power and the time during which the device is operated. In SI units, energy is given in joules ($1\ \text{J} = 1\ \text{W} \cdot \text{s}$), but electric companies use a larger unit, the kilowatt-hour ($1\ \text{kWh} = 3.6 \times 10^6\ \text{J}$).

Electric current can be direct (dc), in which the current is steady in one direction; or it can be alternating (ac), in which the current reverses direction at a particular frequency, typically 60 Hz.

Electric shocks are caused by current passing through the body. To avoid shocks, the body must not become part of a circuit by allowing a current to flow through it.

QUESTIONS

1. When you turn on a water faucet, the water usually flows immediately. You don't have to wait for water to flow from the faucet valve to the spout. Why not? Is the same thing true when you connect a wire to the terminals of a battery?

2. Car batteries are often rated in ampere-hours (A • hr). What does this rating mean?
3. When an electric cell is connected to a circuit, electrons flow away from the negative terminal in the circuit. But within the cell, electrons flow *to* the negative terminal. Explain.
4. A 9-volt transistor radio battery may be smaller in size than a $1\frac{1}{2}$-volt flashlight battery. What, besides their voltage, do you think might be different between these two kinds of batteries?
5. Does a wire carrying an electric current necessarily have a net charge on it?
6. Show on a diagram how to make an alarm system that will ring whenever someone opens the door to your room. You have at your disposal a battery, a bell that rings when connected to the battery, a push-button switch (off when the button is "in"), and several lengths of wire.
7. Design a circuit in which two different switches of the type shown in Figure 19-14 can be used to operate the same light bulb from opposite sides of a room.
8. Is current used up in a resistance?
9. Develop an analogy between blood circulation and an electrical circuit. Discuss what plays the role of the heart for the electric case, and so on.
10. What is the difference between a volt and an ampere?
11. Discuss the advantages and disadvantages of Christmas tree lights connected in parallel versus those connected in series.
12. If all you have is a 120-V line, would it be possible to light several 6-V lamps without burning them out? How?
13. Two light bulbs of resistance R_1 and R_2 ($> R_1$) are connected in series. Which is brighter? What if they are connected in parallel?
14. The equation $P = V^2/R$ indicates that the power dissipated in a resistor decreases if the resistance is increased whereas the equation $P = I^2R$ implies the opposite. Is there a contradiction here? Explain.
15. What happens when a light bulb burns out?
16. Which draws more current, a 100-W light bulb or a 75-W bulb?
17. Electric power is transferred over large distances at very high voltages. Explain how the high voltage reduces power losses in the transmission lines.
18. Why is it dangerous to replace a 15-A fuse (that blows repeatedly) with a 25-A fuse?
19. Electric lights operated on low frequency ac (say 10 Hz) flicker noticeably. Why?

FIGURE 19-14

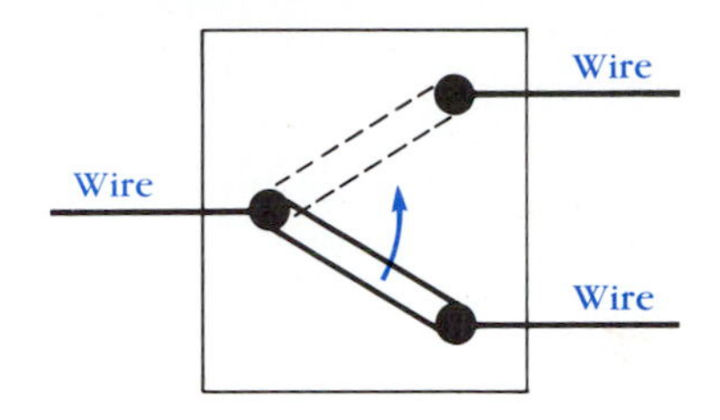

20. In a car, one terminal of the battery is said to be connected to "ground." Since it is not really connected to the ground, what is meant by this expression?
21. Why is it more dangerous to turn on an electric appliance when you are standing outside in bare feet than when you are inside wearing shoes with thick soles?

EXERCISES

1. A steady current of 2.5 A flows in a wire connected to a battery. But after 4.0 min, the current suddenly ceases because the wire is disconnected. How much charge passed through the circuit?
2. A current of 1 ampere flows in a wire. How many electrons are flowing past any point in the wire per second? The charge on one electron is 1.6×10^{-19} coulombs.
3. Automobile batteries are often rated in "ampere·hours," which is a unit of charge; from the definition of an ampere, determine how many coulombs of charge there are in a new 70-ampere-hour battery.
4. A service station charges a battery using a current of 5.5 A for 6.0 hr. How much charge passes through the battery?
5. An ordinary flashlight uses two $1\frac{1}{2}$-volt batteries. What is the voltage applied to the bulb?
6. How would the current in a circuit be changed if you doubled both the resistance and the voltage?
7. If you reduce the voltage to a heater by half, what happens to the current?
8. What is the resistance of a toaster that uses 5 amps on a 120-volt line?
9. What is the current in a 200,000-Ω resistor if the potential difference across it is 120 volts?
10. If a 500-Ω resistor is to have a current of 10 mA passing through it, what voltage battery should be connected to it?
11. Two 1.5-V batteries are connected together in series to a light bulb whose resistance is 10Ω. How many electrons per minute leave each battery?
12. What is the current through six 75-Ω resistors arranged in series if the voltage is 120 volts?
13. What is the net resistance of four 600-Ω resistors arranged in parallel? How much voltage would be required if each resistor were to have 1 ampere flowing in it?
14. Suppose you have a 500-Ω, an 800-Ω, and a 1.20 kΩ resistor. What is (a) the maximum and (b) the minimum resistance you can obtain by combining these?

15. Two 100-Ω and two 300-Ω resistors are arranged in parallel. How much total current passes through the circuit if the voltage is 12 volts?
16. How many 10-Ω resistors are needed to allow 2 amperes of current to flow when connected to a 120-volt line? Are they arranged in series or in parallel?
17. How many 100-Ω resistors are needed so that a total of 5 amperes flows when the resistors are connected to a 125-volt line? How should they be connected?
18. A 12-ampere current passes through a circuit of 60-Ω resistance. How much resistance should be added to the circuit to reduce the current to 8 amperes? Will you arrange this extra resistance in series or in parallel?
19. A circuit contains only one lamp. By what factor does the current change if two more lamps are added in series? What happens if they are added in parallel?
20. Calculate the resistance of a 40-W automobile headlight designed for 12 V.
21. How much heat is produced by a 900-watt iron in 10 minutes?
22. The element of an electric oven is designed to produce 3.0 kW of heat when connected to a 240-V source. What must be the resistance of the element?
23. An automobile starter motor draws 150 A from the 12-V battery. How much power is this?
24. What is the maximum power consumption of a 9.0-V transistor radio that draws a maximum of 400 mA of current?
25. At $0.080 per kWh, what does it cost to leave a 25-W light on all day for a year?
26. What will it cost to run a 1300-watt frying pan, six 100-watt bulbs, and a heater requiring 12 amperes at 110 volts for a total of 4 hours at $0.10 per kWh?

*27. What is the total amount of energy stored in a 12-V, 50-A • hr car battery when it is fully charged?

*28. How much resistance must be added to a 60-Ω circuit to reduce the power from 200 watts to 100 watts?

MAGNETISM

CHAPTER

20

Today it is clear that magnetism and electricity are closely related. This relationship was not discovered, however, until the nineteenth century. The history of magnetism begins much earlier with the ancient civilizations in Asia Minor. It was in a region of Asia Minor known as Magnesia that rocks were found that would attract each other. These rocks were called "magnets" after their place of discovery.

1. MAGNETS AND MAGNETIC FIELDS

Behavior of Magnets: Unlikes Attract, Likes Repel

We are all familiar with the fact that a magnet will attract paper clips, nails, and other objects made of iron. Any magnet, whether it is in the shape of a bar or a horseshoe, has two ends or faces called *poles*; this is where the magnetic effect is strongest. If a magnet is suspended from a fine thread, it is found that one pole of the magnet will always point toward the north. It is not known for sure when this fact was discovered but it is known that the Chinese were making use of it as an aid to navigation by the eleventh century and perhaps earlier. This is, of course, the principle of a compass. A compass needle is simply a magnet that is supported at its center of gravity so it can rotate freely. That pole of a freely suspended magnet which points toward the north is called the **north pole** of the magnet. The other pole points toward the south and is called the **south pole**.

It is a common observation that when two magnets are brought near one another, each exerts a force on the other. The force can be either attractive or repulsive, and can be felt even when the magnets don't touch. If the north pole of one magnet is brought near the north pole of a second magnet, the force is repulsive. Similarly, if two south poles are brought close, the force is repulsive. But when a north pole is brought near a south pole, the force is attractive (Figure 20-1). This is reminiscent of the force between electric charges; *like poles repel* and *unlike poles attract*. Confirm this by performing the following experiment. (But do not confuse magnetic poles with electric charge; they are not the same thing.)

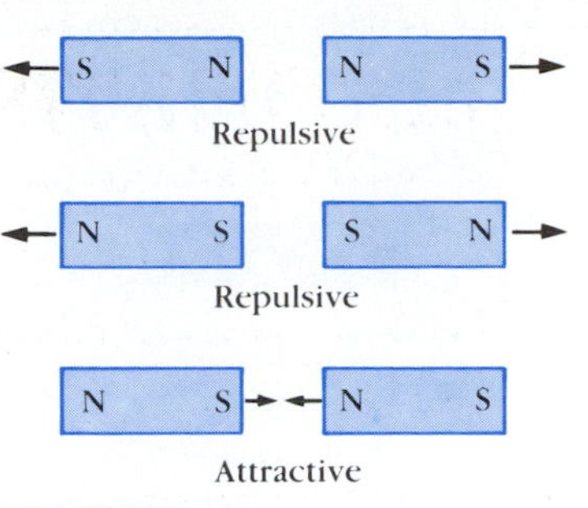

FIGURE 20-1
Like poles of a magnet repel; unlike poles attract.

EXPERIMENT

Obtain two bar magnets (compass needles may work). Suspend each magnet from a separate thread and determine the north poles of each by noting the pole that points toward the north. Now bring the north poles toward each other and determine the direction of the force. Do the same for the two south poles, and for the north and south poles.

Only iron and a few other materials such as cobalt, nickel, and gadolinium exhibit strong magnetic effects. They are said to be *ferromagnetic* (from the Latin word *ferrum* for iron). All other materials show a slight magnetic effect.

Magnetic Fields

We found it useful to speak of an electric field surrounding an electric charge. In the same way we can imagine a **magnetic field** surrounding a magnet. The force one magnet exerts on another can then be described as the interaction between one magnet and the magnetic field of the other. Just as we drew electric field lines, we can also draw magnetic field lines.

The direction of the magnetic field at a given point is defined as the direction that the north pole of a compass needle would point when placed at that point. Figure 20-2 shows how one magnetic field line around a bar magnet is found using compass needles. The magnetic fields determined in this way for a horseshoe magnet and a bar magnet are shown in Figure 20-3. Notice that because of our definition, the lines always point from the north toward the south pole of a magnet (the north pole of a magnetic compass needle is attracted to the south pole of another magnet).

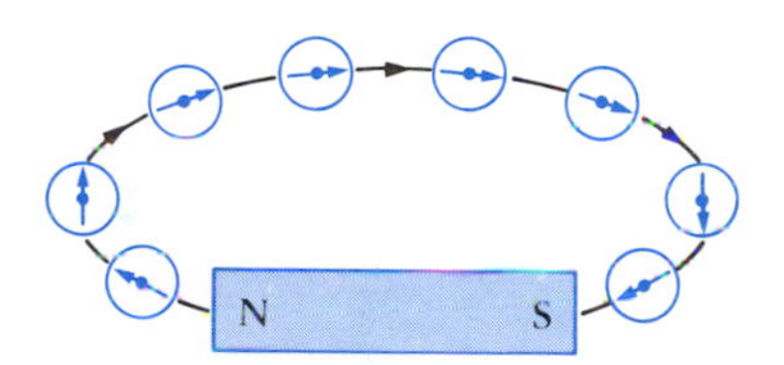

FIGURE 20-2
Plotting a magnetic field line of a bar magnet.

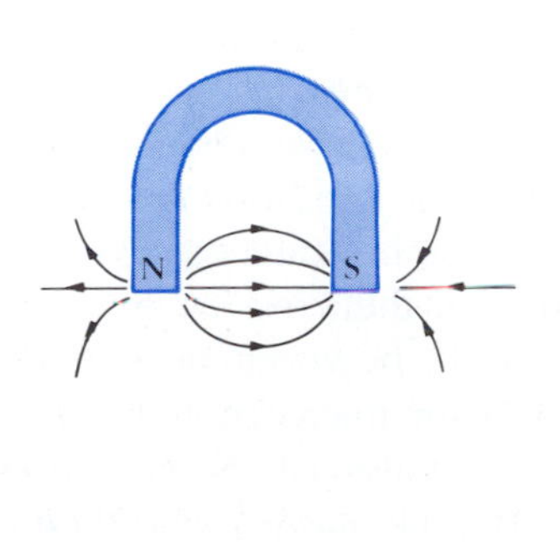

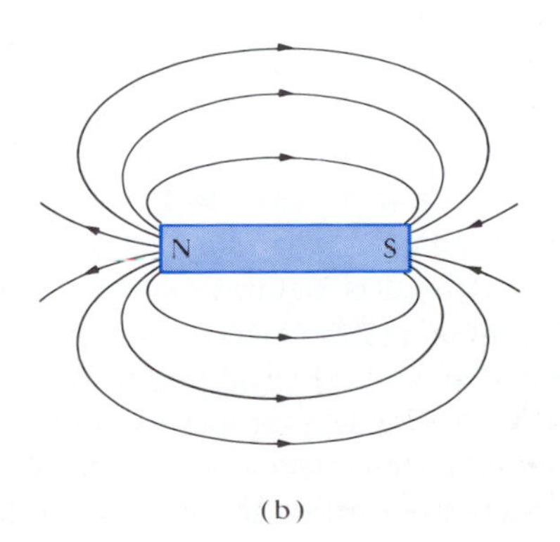

FIGURE 20-3
Magnetic field lines of (a) a horseshoe magnet and (b) a bar magnet.

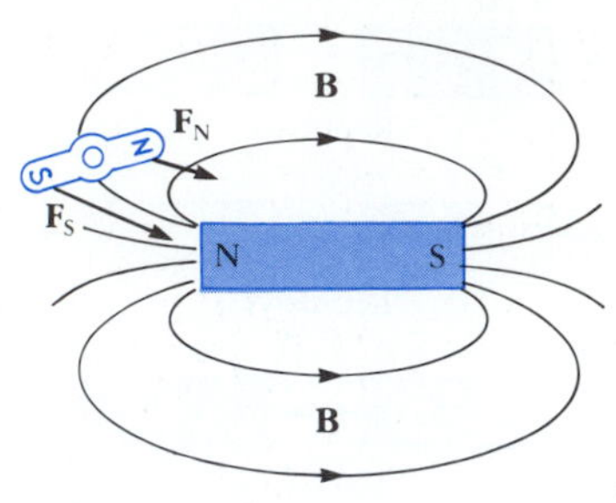

FIGURE 20-4
Forces on a compass needle that produce a torque to orient it parallel to the magnetic field lines. The torque will be zero when the needle is parallel to the magnetic field line at that point.

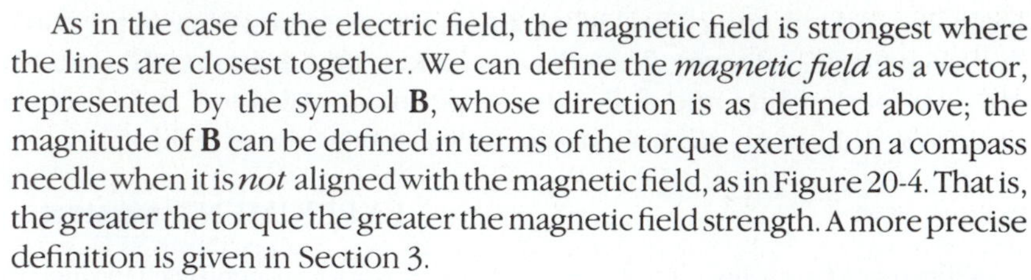

As in the case of the electric field, the magnetic field is strongest where the lines are closest together. We can define the *magnetic field* as a vector, represented by the symbol **B**, whose direction is as defined above; the magnitude of **B** can be defined in terms of the torque exerted on a compass needle when it is *not* aligned with the magnetic field, as in Figure 20-4. That is, the greater the torque the greater the magnetic field strength. A more precise definition is given in Section 3.

When iron filings are placed near a magnet, they align themselves in a pattern following the magnetic field lines (Figure 20-5). Each iron filing in the magnetic field acquires a weak north and south pole in much the same way that an electric field induces a charge separation in an otherwise neutral body (see Figure 18-9).

The Earth's Magnetic Field

The earth's magnetic field is shown in Figure 20-6. Since the north pole of a compass needle points north, the magnetic pole which is in the geographic north is magnetically a south pole (remember that the north pole of one magnet is attracted to the south pole of a second). And the earth's magnetic pole near the geographic south pole is magnetically a north pole. The earth's magnetic poles do not, however, coincide with the geographic poles (which are on the earth's axis of rotation). The magnetic south pole, for example, is in northern Canada, about 1500 km from the geographic north pole. This must be taken into account when using a compass. The angular difference between magnetic north and true (geographical) north is called the "magnetic declination." In the United States it varies from 0° to about 25°, depending on location.

Magnets Are Made Up of Domains

The forces between the like and unlike poles of a magnet are much like the forces between positive and negative electric charges; that is, unlikes attract, likes repel. And just as we can isolate positive and negative charges, we might expect that by cutting a magnet in half we could isolate the N and S poles. But we can't. If we were to cut a magnet in half we would find that we have two new magnets, each with a north and a south pole; the two new poles appear at the cut (Figure 20-7). If we were to then cut each of the new magnets in half, we would get four magnets, each with a north and south pole. We could repeat this operation many times and end up with more and more magnets, each with a north and south pole. Physicists have tried various ways to isolate a single magnetic pole, and this is an active research field today since certain theories suggest they ought to exist. But so far there is no firm experimental evidence for their existence.

FIGURE 20-5
Photograph showing how iron filings line up along magnetic field lines.

These attempts illustrate an important point, however: an ordinary magnet can be considered to be made up of many microscopic magnets. Indeed, carefully prepared samples of magnetic materials, when viewed under an electron microscope, reveal tiny regions known as *domains*, which are at most about 1 mm in length or width. Each domain behaves like a tiny magnet with a north and a south pole. In an unmagnetized piece of iron, these domains are arranged randomly as shown in Figure 20-8(a). The magnetic effects of the domains cancel each other out, so this piece of iron is not a magnet. In a magnet, the domains are preferentially aligned in one direction as shown in Figure 20-8(b) (downward in this case). A magnet can be made from an unmagnetized piece of iron by placing it in a strong magnetic field. (You can make a needle magnetic, for example, by stroking it with one pole of a strong magnet.) Careful observations show that domains may actually rotate slightly so they are more nearly parallel to the external field, or, more commonly, the borders of domains move so that those domains whose magnetic orientation is parallel to the external field grow in size at the expense of other domains. This can be seen by comparing Figure 20-8(a) and (b). This explains how a magnet can pick up unmagnetized pieces of iron like paper clips or bobby pins. The magnet's field causes a slight alignment of the domains in the unmagnetized object so that the object becomes a temporary magnet with its north pole facing the south pole of the permanent magnet, and vice versa; thus attraction results. In the same way, enlongated iron filings will arrange themselves in a magnetic field just as a compass needle does, and will reveal the shape of the magnetic field (Figure 20-5).

An iron magnet can remain magnetized for a long time, and thus it is referred to as a "permanent magnet." However, if you drop a magnet on the floor or strike it with a hammer, you may jar the domains into randomness; the magnet can thus lose some or all of its magnetism. Heating a magnet too can cause a loss of magnetism; for raising the temperature increases the random thermal motion of the atoms which tends to randomize the domains.

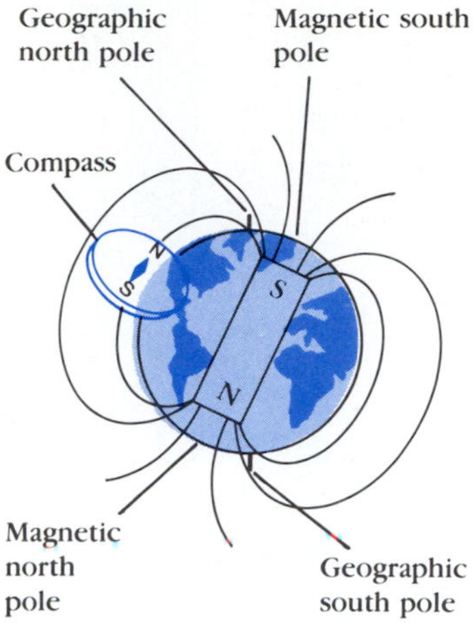

Geographic north pole

Magnetic south pole

Compass

Magnetic north pole

Geographic south pole

FIGURE 20-6

The earth acts like a huge magnet with its magnetic south pole near the geographic north pole.

2. ELECTRIC CURRENTS PRODUCE MAGNETISM

Magnets are made up of tiny domains. But what are domains made of? Or, to say it another way, what really produces magnetism?

Let's perform an experiment that requires only a compass, a length of wire, and a battery. A battery with screw terminals is best, since the wire must be firmly connected to the battery. However, a good-quality flashlight battery will do, in which case the connection between the wire and the battery can be made by holding them *firmly* together with your fingers. The experiment will work best if a thick wire is used—one that provides low resistance and ample current—although ordinary lamp cord may be satisfactory.

FIGURE 20-7

If you break a magnet in half, you do not obtain isolated north and south poles; instead, two new magnets are produced, each with a north and south pole.

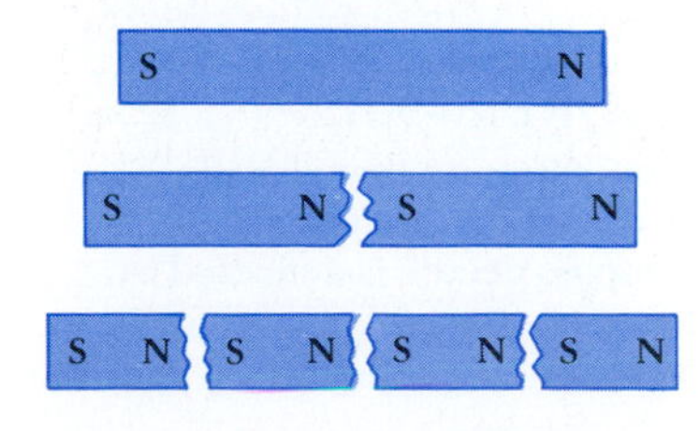

EXPERIMENT-PROJECT

Place a compass on the edge of a table (Figure 20-9). Attach one end of the wire to one terminal of the battery. Bring a straight vertical section of the wire as close as possible to the compass, Figure 20-9(a), keeping the rest of

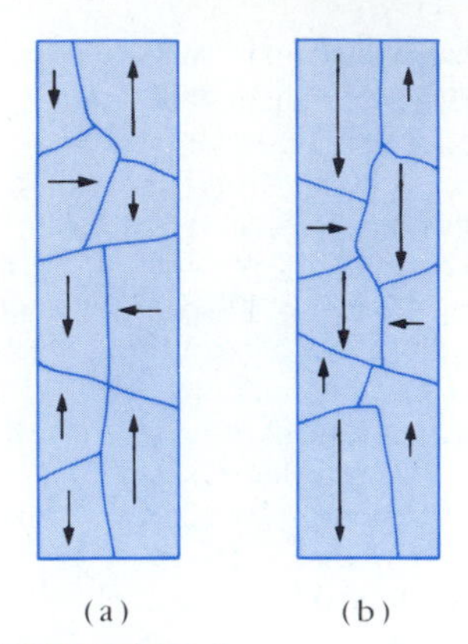

(a) (b)

FIGURE 20-8

(a) Unmagnetized pieces of iron are made up of domains that are randomly arranged. (b) In a magnet, the domains are preferentially aligned in one direction. (The tip of each arrow represents the north pole of the domain.)

the wire away from the compass. Now touch the free end of the wire to the other terminal of the battery, so that an electric current flows in the wire, Figure 20-9(b). What do you observe?

Presumably you saw the compass needle move. And what causes a compass needle to move? A magnetic field, of course! The experiment clearly shows that **an electric current produces a magnetic field**.

During the eighteenth century many natural philosophers sought to find a connection between electricity and magnetism. The first to uncover a significant connection was Hans Christian Oersted (1777–1851) in 1820. Oersted believed for a long time in the unity of nature. Philosophically, he felt there ought to be a connection between magnetism and electricity. However, a stationary electric charge and a magnet had been shown not to have any influence on each other. But Oersted found that when a compass needle is placed near an electric wire, the needle deflects as soon as the wire is connected to a battery and a current flows. Oersted was thus the first to perform the experiment you just performed (Figure 20-9). Since, as we have seen, a compass needle can be deflected by a magnetic field, what Oersted had shown was that an electric current produces a magnetic field. He had found a connection between electricity and magnetism.

Any electric current is surrounded by a magnetic field. An electric current consists of moving electric charges. Therefore, whereas a stationary electric charge fills space with an electric field, a moving charge fills space with both electric *and* magnetic fields.

Magnetic Field Due to Current in a Wire

The orientation of the magnetic field set up by the current in a wire can be easily determined. A compass needle placed in a magnetic field will align

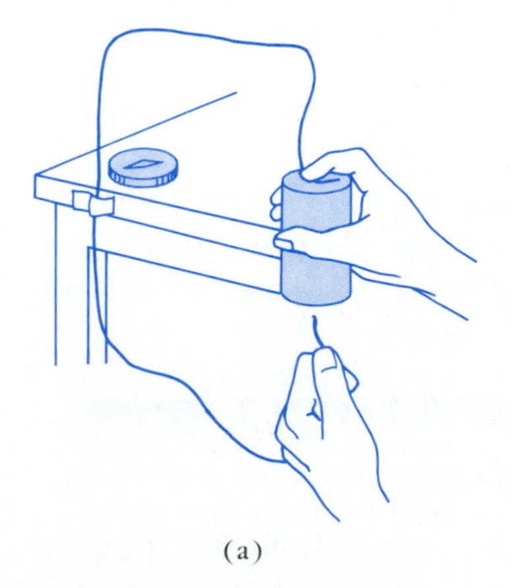

(a)

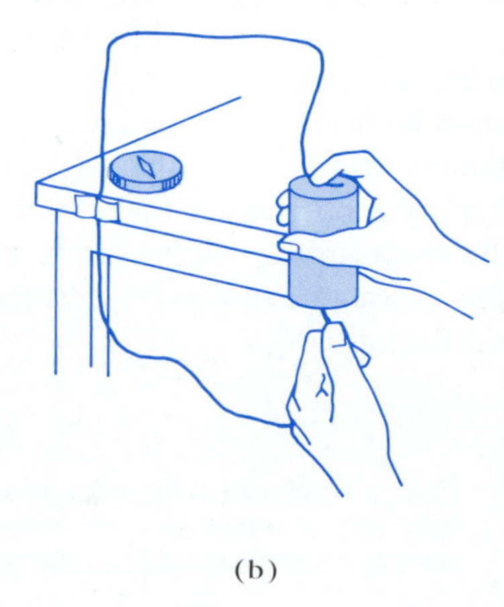

(b)

FIGURE 20-9

Demonstration that electric currents cause magnetism. Compass needle is deflected by the magnetic field of an electric current.

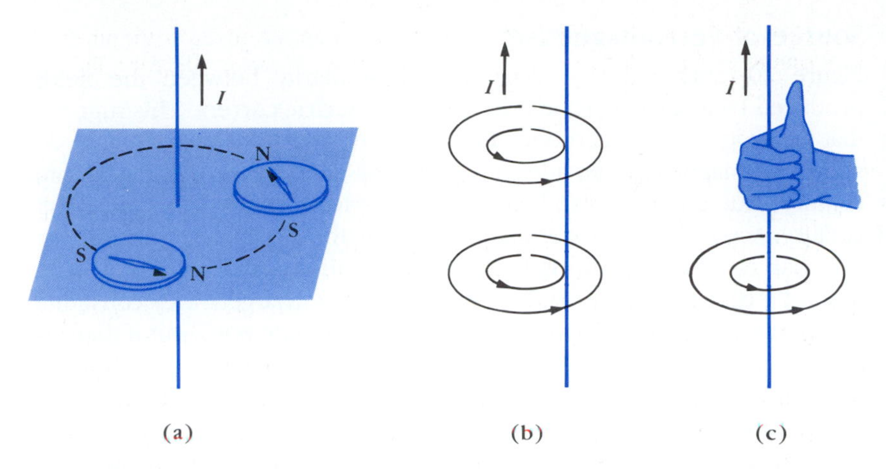

FIGURE 20-10

(a) Deflection of a compass needle near a current-carrying wire, showing the presence and direction of the magnetic field. (b) Magnetic field lines around a straight wire. (c) Right-hand rule for remembering the direction of the magnetic field: when the thumb points in the direction of the conventional current, the fingers wrapped around the wire point in the direction of the magnetic field.

itself parallel to the magnetic field lines, with its north pole pointing in the direction of the field lines at that point.†

A compass needle placed near a straight section of wire aligns itself so that it is tangent to a circle drawn around the wire, Figure 20-10(a). Thus the magnetic field lines of the wire are in the form of circles, with the wire at their center, Figure 20-10(b). The direction of these lines is indicated by the north pole of the compass in Figure 20-10(a). There is a simple way to remember the direction of the magnetic field lines in this case. It is called a *right-hand rule.* You grasp the wire with your right hand so that your thumb points in the direction of the conventional (positive) current; then your fingers will encircle the wire in the direction of the magnetic field, Figure 20-10(c).

The magnetic field lines surrounding a circular loop of current-carrying wire are shown in Figure 20-11(a). Notice that the pattern of lines is very similar to that of a bar magnet—Figure 20-11(b).

† If the current is not large, the magnetic field of the earth will be as large or larger than the field caused by the current, and a compass needle will point in the direction of the resultant *sum* of these two fields.

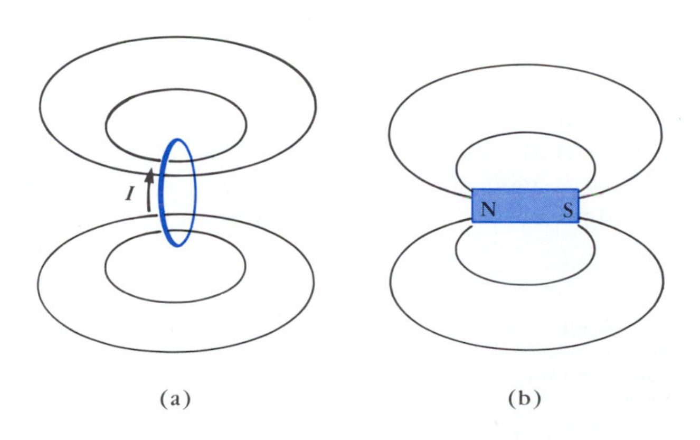

FIGURE 20-11

Magnetic field due to (a) an electric current in a circular loop of wire, (b) a simple bar magnet. Note similarities in the pattern of magnetic field lines.

Source of Ferromagnetism

Figure 20-11(a) and (b) show striking similarity between the fields produced by a bar magnet and by a loop of electric current. This suggests that the magnetic field produced by a current may have something to do with ferromagnetism, an idea proposed by Ampere in the nineteenth century; but an understanding was not arrived at until the twentieth century. According to modern atomic theory, the atoms that make up any material can be roughly visualized as containing electrons that orbit around a central nucleus. Since the electrons are charged, they constitute an electric current and therefore produce a magnetic field. But if there is no external field, the electron orbits in different atoms are arranged randomly so the magnetic effects due to the orbits in all the atoms in a material will cancel out. However, electrons produce an additional magnetic field, almost as if they were spinning about their own axes. It is the magnetic field due to electron spin[†] that is believed to produce ferromagnetism. In most materials, the magnetic fields due to electron spin cancel out because they are oriented at random. But in iron and other ferromagnetic materials, a complicated cooperative mechanism operates; the result is that the electrons contributing to the ferromagnetism in a domain "spin" in the same direction. Thus the tiny magnetic fields due to each of the electrons add up to give the magnetic field of a domain. And when the domains are aligned, as we have seen, a strong magnet results.

Today it is believed possible that *all* magnetic fields are caused by electric currents. This would explain why it has proved difficult to find a single magnetic pole. There is no way to divide up a current and obtain a single magnetic pole. Of course, if an isolated pole is found we will have to amend the idea that all magnetic fields are produced by currents.

3. MAGNETIC FIELDS EXERT FORCES ON ELECTRIC CURRENTS AND MOVING CHARGES

Magnetic Force on a Current

The interrelatedness of electric and magnetic effects is even more profound than we have seen so far.

We have seen that a current-carrying wire exerts a force on a magnet such as a compass needle; this is what convinced us that currents produce magnetism. By Newton's third law we might expect the reverse to be true as well, and this is indeed the case: **A magnet exerts a force on a current-carrying wire**. This effect was also discovered by Oersted.

[†] The name "spin" comes from the early suggestion that the additional magnetic field arises from the electron "spinning" on its axis (as well as "orbiting" the nucleus), and this additional motion of the charge was thought to produce the extra field. However, this view of an electron as actually spinning is very oversimplified.

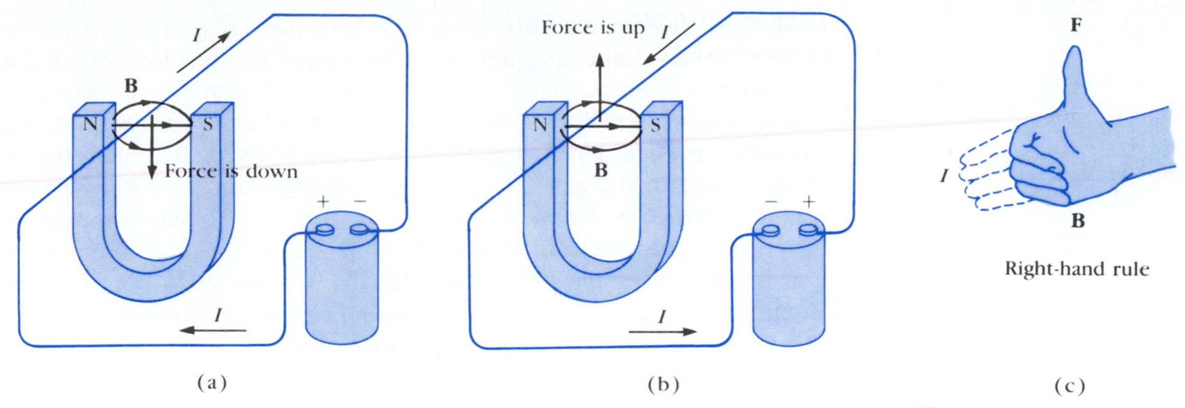

FIGURE 20-12
Force on a current-carrying wire placed in a magnetic field **B**.

Let us look at the force exerted on a wire in detail. Suppose that a straight wire is placed between the pole pieces of a magnet as shown in Figure 20-12. When a current flows in the wire, a force is exerted on the wire, but this force is *not* toward one or the other poles of the magnet. Instead, the force is directed *at right angles to the magnetic field direction*. If the current is reversed in direction, the force is in the opposite direction. It is found that the direction of the force is always perpendicular to the direction of the current I and also perpendicular to the direction of the magnetic field, **B**. Depending on the relative orientation of I and **B**, the force can be either up or down as shown in Figure 20-12(a) and (b). To be more precise, the direction is found using the right-hand rule shown in Figure 20-12(c): you orient your hand so the outstretched fingers point in the direction of the conventional (positive) current; when the fingers are bent, they point in the direction of the magnetic field **B** (**B** points from a north to a south pole). When your hand is so oriented, your thumb points in the direction of the force on the wire. Figure 20-12(c) relates to Figure 20-12(b).

Calculating the Force on a Current (optional)

The *magnitude* of the force exerted by a magnetic field on a current-carrying wire has been found experimentally to be directly proportional to the current I in the wire, to the length ℓ of wire in the magnetic field (assumed uniform), and to the magnetic field B. The force also depends on the angle θ between the wire and the magnetic field. When the wire is perpendicular to the field lines, the force is strongest and is given by

$$F = I\ell B. \qquad [I \perp B]$$

The magnetic force is greatest when the electric current is at right angles to the magnetic field. At other angles the magnitude of the force diminishes; and if the current-carrying wire is parallel to the magnetic field lines, there is no force at all. The direction of the force is always perpendicular to the magnetic field lines *and* to the direction of the current.

The relation above, $F = I\ell B$, is the basis for a precise definition of the magnetic field strength. That is, the magnetic field strength B in any region of space is defined as $B = F/I\ell$, where F is the maximum force measured to act on a wire of length l carrying a current I (when the wire is perpendicular to the field direction).

The SI unit for magnetic field strength B is the *tesla* (T). Thus, for example, if the force on a 20-cm segment of wire carrying 30 A perpendicular to a uniform magnetic field B is 4.8 N, then the magnetic field must have magnitude $B = F/I\ell = (4.8\ \mathrm{N})/(30\ \mathrm{A})(0.20\ \mathrm{m}) = 0.80$ tesla.

Magnetic Force on a Moving Charge

Since a current-carrying wire experiences a force when placed in a magnetic field and since a current in a wire consists of moving electric charges, we might expect that freely moving charged particles (not in a wire) would also experience a force when passing through a magnetic field. Indeed, this is the case.

Calculating the Force on a Moving Charge (optional)

From what we already know, let us calculate the force on a single electric charge which we assume is moving with speed v perpendicular to the magnitude field **B**. If n such particles of charge q pass by a given point in time t, they consitute a current I = charge/time = nq/t. We let t be the time for a charge q to travel a distance ℓ in a magnetic field B; the $\ell = vt$, where v is the velocity of the particle. Using the relation $F = I\ell B$, the force on these n particles is $F = I\ell B = (nq/t)(vt)B = nqvB$. The force on *one* of the n particles is then

$$F = qvB. \qquad [\mathbf{v} \perp \boldsymbol{B}]$$

This equation gives the magnitude of the force on a particle of charge q moving with velocity v perpendicular to a magnetic field of strength B. The force is greatest when the charged particles are moving at right angles to the magnetic field. At other angles the magnitude of the force diminishes, and the force drops to zero if the charged particles move parallel to the magnetic field lines. When the force is not zero, the direction of force is always perpendicular to the magnetic field lines and to the direction of motion of the charged particles; this is true even when the particles are not moving perpendicularly to the magnetic field. The direction of the force is opposite for positive and negative particles, as shown in Figure 20-13. The direction of the force on a positively charged particle is given by a right-hand rule: You orient your right hand so that your outstretched fingers point along the direction of motion of the particle and when you bend your fingers they must point along the direction of **B**; then your thumb will point in the direction of the force. This is true only for *positively* charged particles, and will be "down" for the situation of Figure 20-13. For negatively charged particles the force is in exactly the opposite direction ("up" in Figure 20-13).

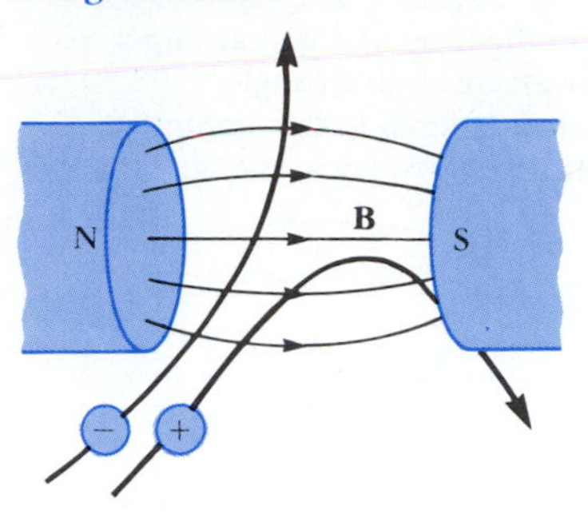

FIGURE 20-13
Force on charged particles due to a magnetic field.

Motion of a Particle in a Magnetic Field (optional)

As an example, suppose an electron ($q = 1.6 \times 10^{-19}$ C) moves with a speed $v = 1.0 \times 10^7$ m/s perpendicular to a uniform magnetic field of 0.020 tesla. What is the magnitude of the force on the electron and what is the path of the electron? The magnitude of the force is

$$\begin{aligned} F &= qvB \\ &= (1.6 \times 10^{-19}\ \text{C})(1.0 \times 10^7\ \text{m/s})(0.020\ \text{tesla}) \\ &= 3.2 \times 10^{-14}\ \text{N}. \end{aligned}$$

The force acts perpendicular to the velocity of the electron and therefore pulls it into a curved path. Since the force is constant in magnitude and always is perpendicular to the path at any moment, the path must be a *circle* (Figure 20-14). The force is directed toward the center of this circle at all points. Note that the electron moves clockwise in Figure 20-14. A positive particle would feel a force in the opposite direction and would thus move in a counterclockwise circle. The magnetic force acts as a centripetal force (see Section 1 of Chapter 6). Newton's second law tells us that $F = ma$, where the centripetal acceleration $a = v^2/r$; r is the radius of the circle and m is the mass of the electron. Thus

$$F = ma$$

$$qvB = m\frac{v^2}{r}.$$

We solve for r and find

$$r = \frac{mv}{qB}.$$

We put in numbers for our case and find

$$r = \frac{(9.1 \times 10^{-31}\ \text{kg})(1.0 \times 10^7\ \text{m/s})}{(1.6 \times 10^{-19}\ \text{C})(2.0 \times 10^{-2}\ \text{tesla})} = 2.8 \times 10^{-3}\ \text{m}.$$

The electron's path is a circle of radius 2.8 mm.

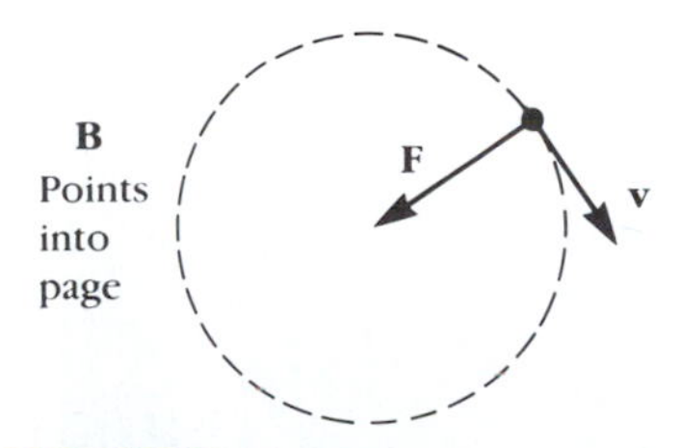

FIGURE 20-14
Force exerted by a uniform magnetic field on a moving charged particle (in this case an electron) produces a curved path.

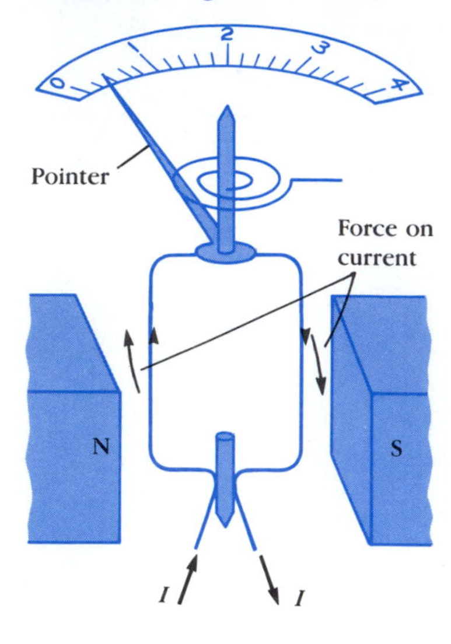

FIGURE 20-15
Galvanometer.
The magnetic field exerts a force on the loop of wire causing it to twist through an angle proportional to the amount of current flowing in the wire.

4. APPLICATIONS: METERS, MOTORS, LOUDSPEAKERS, SOLENOIDS

A number of important practical devices make use of the force that exists between a current and a magnetic field. In most of these devices, the current flows in a coil of wire.

Galvanometer

The basic component of most meters, including ammeters and voltmeters, is a *galvanometer*. As shown in Figure 20-15, a galvanometer consists of a loop or coil of wire suspended in the magnetic field of a permanent magnet. When current flows through the loop, the magnetic field exerts a force on the vertical sections of wire as shown. [Notice that, by the right-hand rule (Figure 20-12), the force on the upward current on the left is inward, whereas that on the descending current on the right is outward.] These forces give rise to a net torque that tends to rotate the coil about its vertical axis. A small coil spring resists this motion to a certain extent, so the size of the angle through which the pointer turns will be proportional to the magnetic force. The magnetic force, in turn, is proportional to the current in the wire. Thus the deflection of the pointer is a measure of the current flowing through the coil. With the addition of an appropriate resistance in parallel, any range of current can be measured; such a device is called an "ammeter." With the addition of a large resistance in series, the deflection of a galvanometer can be calibrated to measure voltage; such a device is called a "voltmeter."

Electric Motor

An *electric motor* changes electric energy into (rotational) mechanical energy. Motors are used to turn fans, grinding wheels, and washing-machine tubs, and they are found in a great many other devices such as refrigerators, hair dryers, mixers, pumps, and electric cars.

A motor works on the same principle as a galvanometer, except that the coil is larger and is mounted on a large cylinder called the *rotor* or *armature* (Figure 20-16). Actually, there are several coils, although only one is indicated in the figure. The armature is mounted on a shaft or axle. Unlike a galvanometer, a motor must turn continuously in one direction. This presents a problem: when the coil, which is rotating clockwise in Figure 20-16, passes beyond the vertical position the forces would then act to return the coil back to vertical. Thus alternation of the current is necessary if a motor is to turn continuously in one direction. This can be achieved in a *dc motor* with the use of *commutators* and *brushes*. As shown in Figure 20-17, the brushes are stationary contacts that rub against the conducting commutators mounted on the motor shaft. Every half

FIGURE 20-16
Diagram of a simple motor.

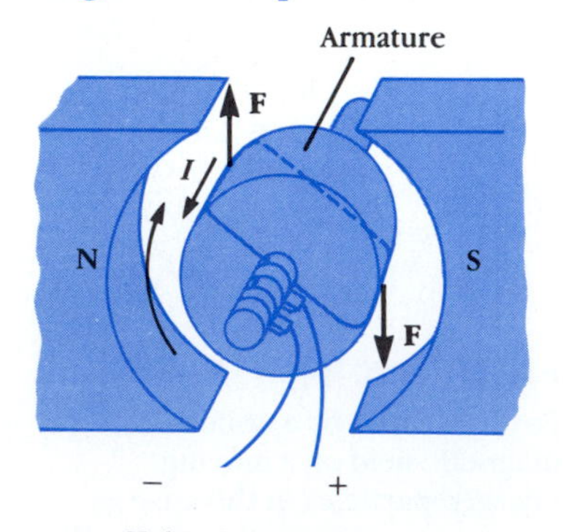

revolution each commutator changes its connection to the other brush. Thus the current in the coil reverses every half revolution, as required for continuous rotation. Most motors contain several coils, called "windings," each located in a different place on the armature (Figure 20-18). Current flows through each coil only during a small part of a revolution, at the time when its orientation results in the maximum torque. In this way, a motor produces a much steadier torque than can be obtained from a single coil.

In larger motors an "electromagnet" is used instead of a permanent magnet. An electromagnet is simply a coil of wire wrapped around a piece of magnetizable material such as iron. The magnetic field due to a current in the coil will magnetize the iron. The total magnetic field is due to that of the iron plus that of the current in the coil. Thus the magnetic field of an electromagnet can be made much larger than that of a permanent magnet alone. The design of most practical motors is more complex than described here, but the general principles remain the same.

FIGURE 20-17
This commutator-brush arrangement in a dc motor assures alternation of current in the armature to keep rotation continuous. The commutators are attached to the motor shaft and turn with it; the brushes remain stationary.

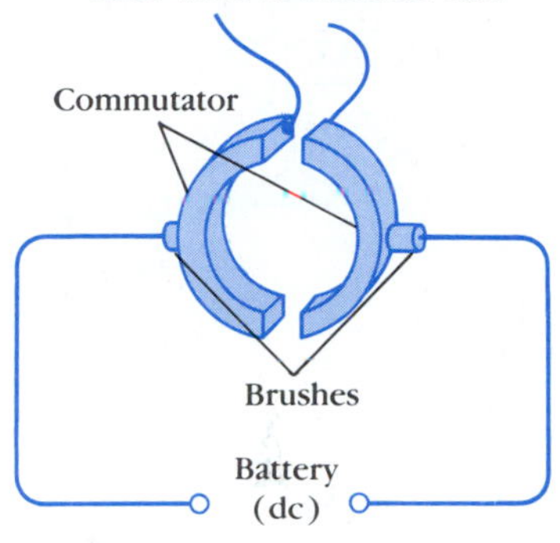

Loudspeaker

A high-fidelity *loudspeaker* also works on the principle that a magnet exerts a force on a current-carrying wire. The electrical output of a radio or TV set is connected to the wire leads of the speaker. The speaker leads are connected internally to a coil of wire, which is itself attached to the speaker cone (Figure 20-19). The speaker cone is usually made of stiffened cardboard and is mounted so that it can move back and forth freely. A permanent magnet is mounted directly in line with the coil of wire. When the alternating current of an audio signal flows through the wire coil, the coil and the attached speaker cone undergo a force due to the magnetic field of the magnet. As the current alternates at the frequency of the audio signal, the speaker cone moves back and forth at the same frequency causing alternate compressions and rarefactions of the adjacent air, and sound waves are produced. A speaker thus changes electrical energy into sound energy, and the frequencies of the emitted sound waves are an accurate reproduction of the electrical input.

Solenoid

A *solenoid* is a long coil of wire with a cylindrical piece of iron inserted partially into it. One common use of the solenoid is as a doorbell (Figure 20-20). When the circuit is closed by pushing the button, the coil effectively becomes a magnet and exerts a force on the iron rod. The rod is pulled into the coil and strikes the bell. A larger solenoid is used in the starters of cars; when you push the starter button, you are closing a circuit that not only turns the starter motor but activates a solenoid that first moves the starter into contact with the engine. Solenoids are used as switches in many other devices, such as tape recorders. They have the advantage of moving mechanical parts quickly and accurately.

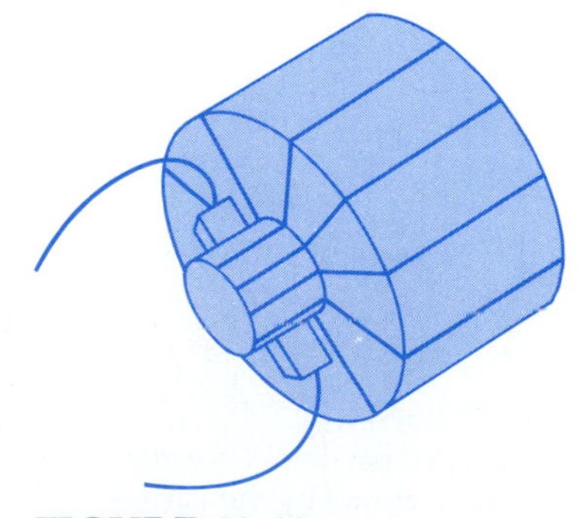

FIGURE 20-18
Motor with many windings.

5. EARTH'S MAGNETIC FIELD AND COSMIC RAYS; THE AURORA

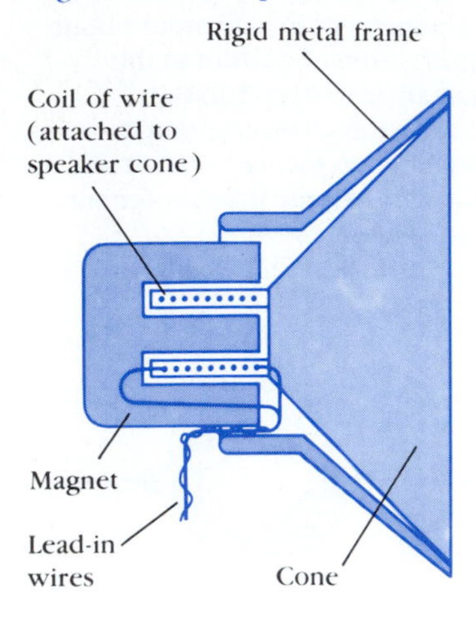

FIGURE 20-19
Diagram of a loudspeaker.

Van Allen Radiation Belts and the Aurora Borealis

The magnetic field of the earth (Figure 20-6) exerts a force on charged particles that impinge on the outer atmosphere. Because of their extraterrestrial origin, these charged particles are known as "cosmic rays." They come from the sun, stars, and other regions of the universe. The cosmic rays that come from the sun include hydrogen atoms that have been ionized by the sun's high temperatures. Huge magnetic storms, disturbances in the magnetic field of the sun, cause these charged particles to be expelled. Those that pass near the earth are deflected by the earth's magnetic field, and many are trapped by it. The trapped particles move back and forth between the earth's magnetic poles, following spiral paths around the magnetic field lines of the earth—Figure 20-21(a). The trapped radiation is largely confined to two dough-nut-shaped regions surounding the earth. These are known as the Van Allen radiation belts, Figure 20-21(b), named after James Van Allen, who determined their existence in 1958 from data gathered by the Explorer 1 satellite.

At the poles some of the charged particles descend into the earth's atmosphere, ionizing the atoms of the atmosphere and causing them to fluoresce (see Chapter 28); this action produces the beautiful Northern Lights, or Aurora Borealis.

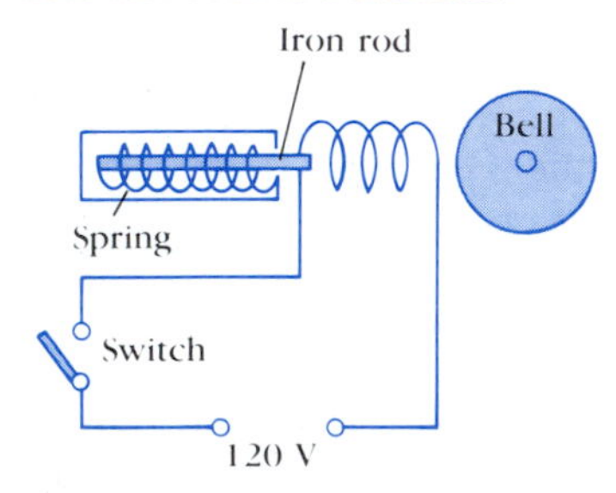

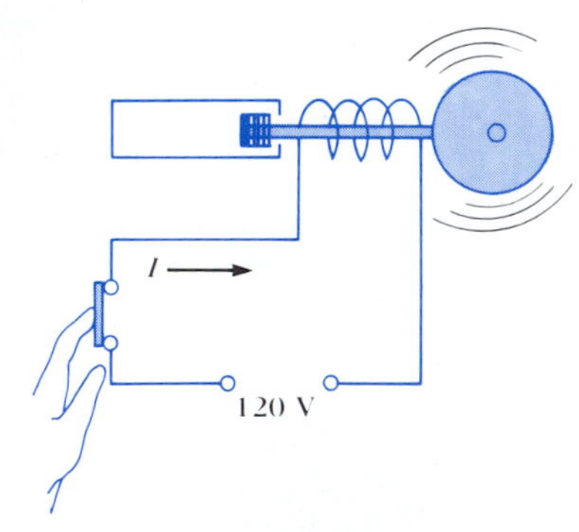

FIGURE 20-20
Solenoid used as a doorbell.

The Earth's Magnetic Field Protects Us from Radiation

Although some cosmic rays penetrate to the earth's surface, most of them are trapped in the radiation belts by the earth's magnetic field. It is fortunate that the earth has a magnetic field that protects us from this onslaught. Fast-moving charged particles cause atoms, including those in biological tissue, to ionize. Hence they can cause biological damage, especially to the genetic material DNA, and may result in mutations or alterations of the genetic makeup of the individual. While some mutations

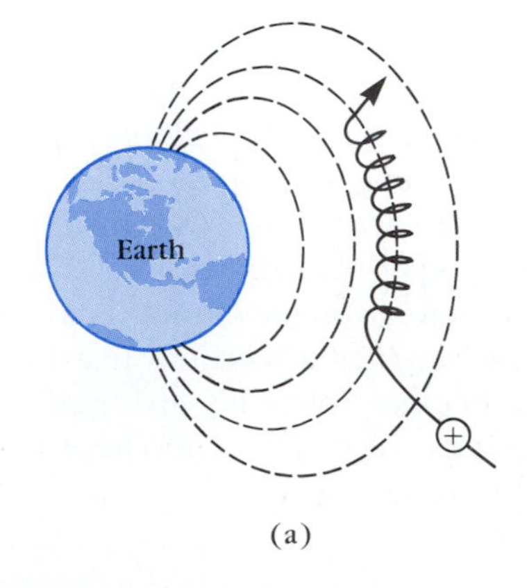

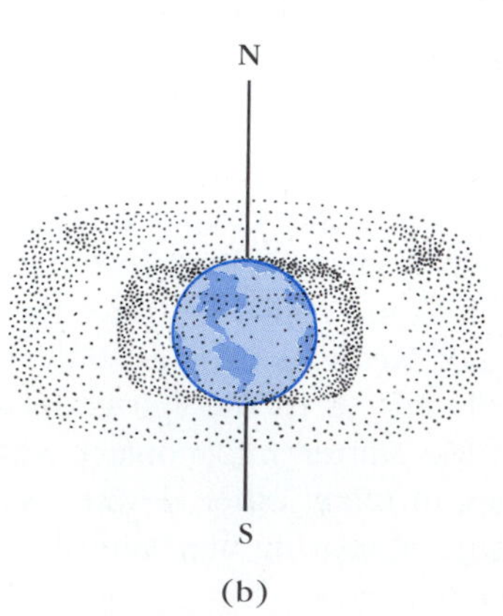

FIGURE 20-21
(a) Charged particles (cosmic rays) are trapped by the earth's magnetic field. (b) Van Allen radiation belts surround the earth in doughnut shaped regions.

lead to an improvement and better adaptability of organisms, the vast majority of mutations are harmful.

SUMMARY

A magnet has two *poles*, north and south. The north pole is that end which points toward the north when the magnet is freely suspended. Unlike poles of two magnets attract each other, whereas like poles repel. Permanent magnets are made up of tiny *domains*—each a tiny magnet—which are aligned. In an unmagnetized piece of iron, the domains are randomly oriented.

We can apply the field concept to magnetism and imagine that a *magnetic field* surrounds every magnet. The SI unit for magnetic field is the tesla (T). The force one magnet exerts on another is said to be an interaction between one magnet and the magnetic field produced by the other.

Electric currents produce magnetic fields. For example, the lines of magnetic field due to a current in a straight wire form circles around the wire and the field exerts a force on magnets placed near it.

A magnetic field exerts a force on an electric current. For a straight wire of length ℓ carrying a current I perpendicular to a magnetic field **B**, the force has magnitude $F = I\ell B$. The direction of the force is perpendicular to the wire and to the magnetic field, and is given by the right-hand rule. Similarly, a magnetic field exerts a force on a charge q moving with velocity **v** perpendicular to a magnetic field **B** whose magnitude is $F = qvB$ The direction of **F** is perpendicular to **v** and to **B**. The path of a charged particle moving perpendicular to a uniform magnetic field is a circle.

The force exerted on a current-carrying wire by a magnetic field is the basis for operation of many devices, such as meters, motors, and loudspeakers.

QUESTIONS

1. A compass needle is not always balanced parallel to the earth's surface; one end may dip downward. Explain.
2. How do you suppose the first magnets found in Magnesia were formed?
3. Why will either pole of a magnet attract an unmagnetized piece of iron?
4. Suppose that you have three iron rods, two of which are magnetized and one that is not. How would you determine which two are the magnets without using any additional objects?

5. Will a magnet attract any metallic object or only those made of iron? (Try it and see.) Why is this so?
6. An unmagnetized nail will not attract an unmagnetized paper clip. However, if one end of the nail is in contact with a magnet, the other end *will* attract a paper clip. Explain.
7. Draw the magnetic field lines around a straight section of wire carrying a current horizontally to the left.
8. In what direction are the magnetic field lines surrounding a straight wire carrying a current that is moving directly toward you?
9. Although each iron atom is a tiny magnet, not every piece of iron is a magnet. Why not?
10. Explain the existence of a permanent magnet in terms of tiny atomic currents.
11. The magnetic field due to current in wires in your home can affect a compass. Discuss the problem in terms of currents, depending on whether they are ac or dc.
12. What kind of field or fields surround a moving electric charge?
13. A horseshoe magnet is held vertically with the north pole on the left and south pole on the right. A wire passing perpendicularly between the poles carries a current toward you. In what direction is the force on the wire?
14. Can you set a resting electron into motion with a magnetic field? With an electric field?
15. A beam of electrons is directed toward a horizontal wire carrying a current from left to right. In what direction is the beam deflected?
16. Two parallel wires each carry an electric current. Is there a force between the two wires? Why?
17. A charged particle is moving in a circle under the influence of a uniform magnetic field. If an electric field is turned on that points along the same direction as the magnetic field, what path will the charged particle take?
18. How is a motor similar to a compass?
19. A current-carrying wire is placed in a magnetic field. How must it be oriented so the force on it is zero? So it is maximum?
20. A loop of wire is suspended between the poles of a magnet with its plane parallel to the pole faces. What happens if a direct current is put through the coil? What if an alternating current is used instead?
21. Each of the right-hand rules you learned in this chapter can be changed to *left-hand rules* if you are specifying the direction of movement of *negative* particles, such as electrons in a wire. Show, for each right-hand rule, that the same operations using the left hand give the same results if the direction of charge flow is for negative charges.

22. Why does more cosmic radiation strike residents of Norway than residents of Egypt?

EXERCISES

1. What is the force per meter of length on a wire carrying a 0.5-A current due to a 0.5 tesla-magnetic field?
2. The force on a wire carrying 2.0 A in a magnetic field is 3.0 N per meter of length. How large is the field strength?
3. The force on a 10-m-long electrical wire in a 0.025 tesla magnetic field is 0.66 N. How much current is flowing in the wire?
4. Calculate the force on a 0.5-coulomb object traveling 10 m/s perpendicular to the earth's magnetic field whose magnitude is about 5×10^{-5} tesla.

*5. What is the radius of curvature of the object's path in the above problem if its mass is 15 grams?

6. The force on a wire carrying 20.0 A is a maximum of 3.60 N when placed between the pole faces of a magnet. If the pole faces are 15.0 cm in diameter, what is the approximate strength of the magnetic field?
7. How much current is flowing in a wire 2.00 m long if the force on it is 0.700 N when placed in a uniform 0.0300-T field?
8. Determine the magnitude and direction of the force on an electron traveling 3.24×10^5 m/s horizontally to the east in a vertically upward magnetic field of strength 1.50 T.

*9. Calculate the force on a proton and the radius of its circular path when moving in a magnetic field of 0.031 tesla at a speed of 8.0×10^5 m/s. The mass of a proton is 1.7×10^{-27} kg and its charge is $+ 1.6 \times 10^{-19}$ coulomb.

10. An electron experiences the greatest force as it travels 2.1×10^5 m/s in a magnetic field when it is moving southward. The force is upward and of magnitude 5.6×10^{-13} N. What is the magnitude and direction of the magnetic field?
11. A proton having a speed of 5.0×10^6 m/s in a magnetic field feels a force of 8.0×10^{-14} N toward the west when it moves vertically upward. When moving horizontally in a northerly direction, it feels zero force. What is the magnitude and direction of the magnetic field in this region?
12. A particle of charge q moves in a circular path of radius r in a uniform magnetic field **B**. Show that its momentum is $p = qBr$.
13. The straight sections of wire on the armature of a motor are 10 cm

long. If the magnetic field is 0.045 tesla and the current through the wire is 3.3 A, what is the force on each section of the wire?

*14. If the current to a motor drops by 10 percent, by what factor does the output torque change?

*15. An electron is accelerated through a potential difference of 5000 V. What is the strength of the magnetic field if the radius of its path is 3.4 mm?

*16. An electron is accelerated from rest by a potential difference of 2000 volts. What will be the radius of its path in a field $B = 1.1 \times 10^{-4}$ tesla?

ELECTROMAGNETISM

CHAPTER

21

In Chapter 20 we discussed two ways in which electricity and magnetism are related: (1) an electric current produces a magnetic field and (2) a magnetic field exerts a force on an electric current or moving electric charge. These discoveries were made in 1820–1821. Scientists then began to wonder: if electric currents produce a magnetic field, is it possible that a magnetic field can produce an electric current? Ten years later the American Joseph Henry (1797–1878) and the Englishman Michael Faraday (1791–1867) independently found that it was possible. Henry actually made the discovery first; but Faraday published his results earlier and investigated the subject in more detail. We now discuss this phenomenon and some of its world-changing applications.

1. ELECTROMAGNETIC INDUCTION

A Changing Magnetic Field Can Induce A Voltage

In his attempt to produce an electric current from a magnetic field, Faraday used the apparatus shown in Figure 21-1. A coil of wire, *X*, was connected to a battery. The current that flowed through *X* produced a

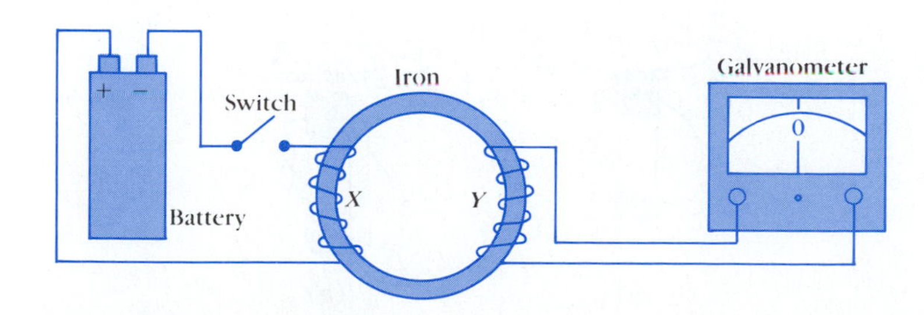

FIGURE 21-1
Faraday's experiment to induce a voltage.

magnetic field that was intensified by the iron core. Faraday hoped that by using a strong enough battery, a steady current in X would produce a great enough magnetic field to produce a current in a second coil Y. This second circuit, Y, contained a galvanometer to detect any current but contained no battery. He met no success with steady currents. But the long-sought effect was finally observed when Faraday saw the galvanometer in circuit Y deflect strongly at the moment he closed the switch in circuit X. And the galvanometer deflected strongly in the opposite direction when he opened the switch. A *steady* current in X had produced *no* current in Y. Only when the current in X was starting or stopping was a current produced in Y.

Faraday concluded that although a steady magnetic field produced no current, a *changing* magnetic field can produce an electric current!

Faraday did further experiments on **electromagnetic induction**, as this phenomenon is called. For example, Figure 21-2 shows that if a magnet is moved quickly into a coil of wire, a current is induced in the wire. If the magnet is quickly removed, a current is also induced but in the opposite direction. Furthermore, if the magnet is held steady and the coil of wire is moved toward or away from the magnet, again a current is induced. Motion or change is required to induce a current. It doesn't matter whether the magnet or the coil moves.

We know that a current flows in a wire only when a voltage is present. Therefore, the changing magnetic field, whether the change is due to the motion of the magnet or the motion of the wire loop, causes an *induced voltage*. The principle of electromagnetic induction can be stated simply as follows:

> **A voltage is induced in a loop of wire when the magnetic field through the loop changes.**

In its more general form, this is known as *Faraday's law*.

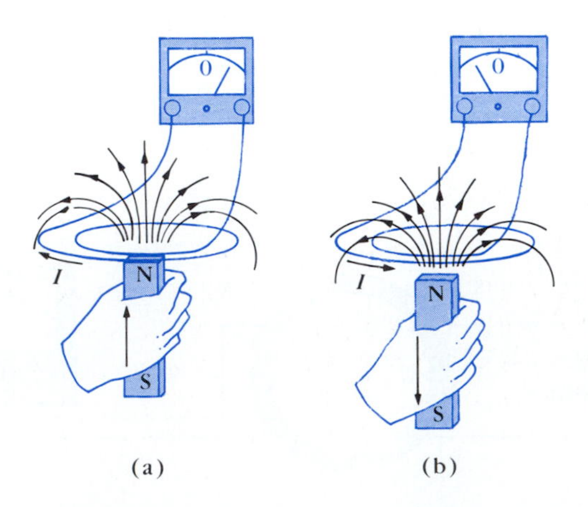

FIGURE 21-2
A current is induced when a magnet is moved toward a coil (a). The induced current is opposite when the magnet is removed (b). Note that the galvanometer zero is in the center of the scale and deflects left or right, depending on direction of the current.

Faraday's Law

Faraday investigated quantitatively what factors influence the magnitude of the voltage induced. He found first of all that it depends on time: the more rapidly the magnetic field changes, the greater the induced voltage. But the voltage is not simply proportional to the rate of change of the magnetic field, **B**, but rather to the product of the magnetic field and the area of the loop through which the field **B** passes. For example, as shown in Figure 21-3, a voltage can be induced in a loop of wire in a magnetic field if the loop is turned in the field so the effective area through which the field passes is changed, as in Figure 21-3(a), or the area is physically changed, as in Figure 21-3(b). (Note in Figure 21-3 that the **x**'s represent magnetic field lines directed *into* the paper; the **x** is meant to resemble the tail of an arrow going away.) Thus a voltage can be induced in two ways: (1) by changing the magnetic field B, or (2) by changing the area A of the loop or its orientation with respect to the field. Figures 21-1 and 21-2 illustrated case (1). Examples of case (2) are illustrated in Figure 21-3.

The product of magnetic field B and area A through which B passes is called the *flux*, and is given the symbol Φ. In symbols, $\Phi = BA$. (If **B** is not perpendicular to the loop's face, then we must use the vector component of **B** that is perpendicular to the face.)

Faraday found that the magnitude of the induced voltage in a loop of wire is equal to the rate of change of flux through the coil. For a coil of

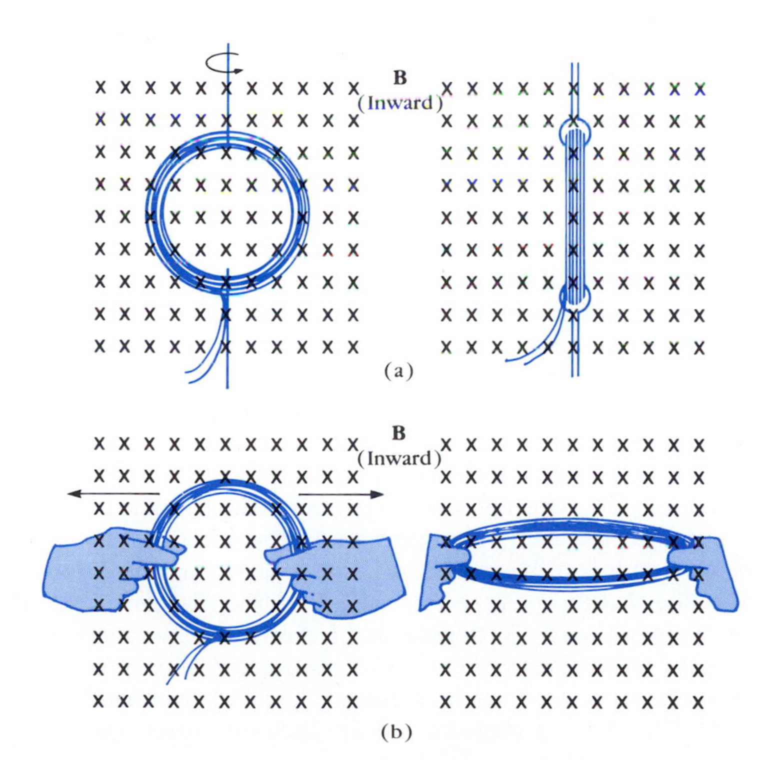

FIGURE 21-3
A current can be induced (a) by rotating the coil in the magnetic field or (b) by changing the area of the coil. In both cases the flux through the coil is reduced. The **x**'s in this figure represent magnetic field lines passing directly into the paper.

wire containing many loops, each loop receives the same induced voltage, so the total induced voltage is proportional to the number of loops N. Thus

$$\text{induced voltage} = \text{no. of loops} \times \frac{\text{change in flux}}{\text{time}}.$$

This is known as **Faraday's law of induction**. The flux, which is the product of magnetic field and area of the loop ($\Phi = BA$) can be thought of as *the total number of field lines passing through the coil.* Thus in Figure 21-2, there is an induced voltage because the number of field lines passing through the coil increases or decreases. This is also true in Figure 21-3(b). In Figure 21-3(a), many lines pass through the coil when positioned as shown on the left, but none pass through the coil as positioned on the right. Faraday's law can be stated picturesquely as: the greater the change in the number of lines through the coils per second, the greater the induced voltage.

A Numerical Calculation (optional)

As an example, suppose the circular coil in Figure 21-3(a) has a radius of 5.0 cm and contains 100 loops of wire; if it is positioned perpendicular to a 0.60-T magnetic field and then quickly rotated in 0.10 s so its face is parallel to the field (as shown on the right), what is the average induced voltage during this short time? The area of the coil is $A = \pi r^2 = (3.14)(0.050\ \text{m})^2 = 7.9 \times 10^{-3}\ \text{m}^2$. The flux is initially $\Phi = BA = (0.60\ \text{T})(7.9 \times 10^{-3}\ \text{m}^2) = 4.7 \times 10^{-3}\ \text{T} \cdot \text{m}^2$. After 0.10 s the flux is zero (no field lines pass through the loop). Therefore, the average induced voltage is

$$V = (100)\frac{4.7 \times 10^{-3}\ \text{T} \cdot \text{m}^2 - 0}{0.10\ \text{s}} = 4.7\ V.$$

Lenz's Law (optional)

We have seen that a current is induced in a loop of wire when the magnetic flux through the loop changes. But in what direction is the induced current and voltage? Experiment shows that *an induced voltage always gives rise to a current whose magnetic field opposes the original change in flux*. This is known as **Lenz's law**. Let us apply it to the case of relative motion between a magnet and a coil (Figure 21-2). The changing flux induces an emf, which produces a current in the coil; and this induced current produces its own magnetic field. In Figure 21-2(a) the distance between the coil and the magnet decreases; so the magnetic field, and therefore the flux through the coil, increases. The magnetic field of the magnet points upward. To oppose this upward increase, the field produced by the induced current must point *downward*. Thus Lenz's law tells us that the current must move as shown (use the right-hand rule). In Figure 21-2(b) the flux *decreases*, so the induced current produces an *upward* magnetic field that is "trying" to maintain the status quo. Thus the current must be as shown.

Let us consider what would happen if Lenz's law were not true but was just the reverse. The induced current in this imaginary situation would produce a flux in the same direction as the original change; this greater change in flux would produce an even larger current followed by a still greater change in flux, and so on. The current would continue to grow indefinitely, producing power even after the original stimulus ended. This would violate the conservation of energy. Such "perpetual motion" devices do not exist. Thus Lenz's law as stated above (and not its opposite) is consistent with the law of conservation of energy.

Phonograph Cartridges and Magnetic Microphones

Many microphones work on the principle of induction. In one form, a microphone is just the inverse of a loudspeaker (see Section 4 of Chapter 20). A small coil connected to a membrane is suspended close to a small permanent magnet. The coil moves in the magnetic field when sound waves strike the membrane; the frequency of the induced voltage will be just that of the impinging sound waves. In a "ribbon" microphone, a thin metal ribbon is suspended between the poles of a permanent magnet. The ribbon vibrates in response to sound waves and the voltage induced in the ribbon is proportional to its velocity.

In one type of phonograph cartridge, the needle which follows the grooves of a phonograph record is connected to a tiny magnet inside the cartridge in the arm of the record player. The magnet is suspended inside a tiny coil of wire and a voltage is induced in the coil according to the motion of the magnet. This tiny voltage, at frequencies corresponding to the wavy shape of the record grooves, is amplified electronically before sending it to a loudspeaker. Most good-quality cartridges are of the magnetic type.

Microphones and phonographs are two types of *transducer* (by which we mean a device that transforms one type of energy into another); both convert a mechanical signal (moving air, moving stylus) into an electrical one. Another transducer is the head of a tape recorder. Recorded tape has been magnetized along its length at different intensities corresponding to the program material. As tape passes the head, a tiny voltage is induced in a coil due to the changing magnetic field passing the head. The signal is amplified before going to the speaker.

2. ELECTRIC GENERATORS AND ALTERNATORS

Probably the most important practical result of Faraday's great discovery was the development of the **electric generator** or **dynamo**. A generator transforms mechanical energy into electric energy. This is just the opposite of what a motor does. Indeed, a generator is basically the inverse of a

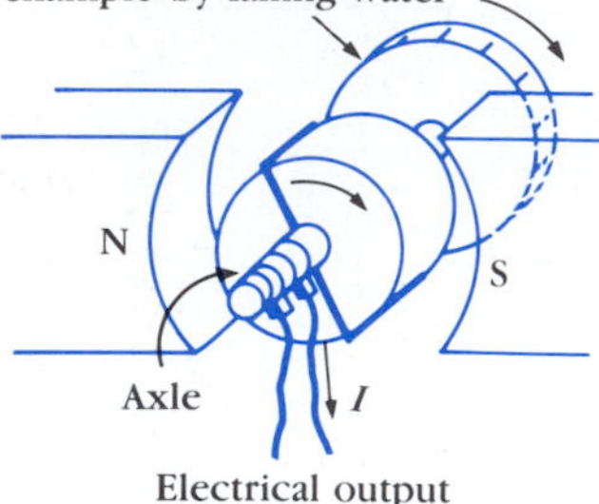

FIGURE 21-4
An electric generator. In principle it is just the opposite of an electric motor. Real generators have many loops of wire, but only one is shown here.

motor.[†] A simplified diagram of an ac generator is shown in Figure 21-4. A generator consists of many coils of wire (only one is shown) wound on an armature that can rotate in a magnetic field. The axle is turned by some mechanical means, such as high-pressure steam or water falling over a dam and striking the blades of a turbine attached to the axle. Since the rapidly turning coil is situated in a magnetic field, a current is continuously induced in it. At the instant shown in Figure 21-4, the number of magnetic field lines passing through the coil is increasing and the induced current is as shown. After the coil passes through the vertical position, the number of field lines passing through it will begin to decrease and the current will flow in the opposite direction. The current changes direction every half revolution. It is thus an alternating current, and this generator is called an ac generator. Ac generators produce nearly all the electrical energy we use.

For certain applications a dc generator is used. A dc generator contains a commutator and brushes just as a dc motor does, and a direct current is obtained as the output.

In the past, automobiles used dc generators. More common now, however, are ac generators or **alternators** (Figure 21-5). The ac output is changed to dc for charging the battery by the use of diodes (Chapter 28). Most alternators differ from the generator discussed above in that the magnetic field is made to rotate within a stationary armature called a *stator*. The brushes press against a continuous ring instead of a slotted commutator and are on the input instead of the output; so large output currents are carried in solid conductors rather than through sliding-ring commutators, which are subject to wear and arcing. In a car the rotation speed can be faster so the battery can be charged even at idling speed, and there will still be no problem with electrical arcing across the rings when the car travels at high speeds.

FIGURE 21-5
Simplified diagram of an alternator. Electromagnet current connected through continuous slip rings is preferable to the output of a dc generator through split commutators (see Figure 20-17), where poor connection or arcing can occur at high speeds. Sometimes the electromagnet is replaced by a permanent magnet.

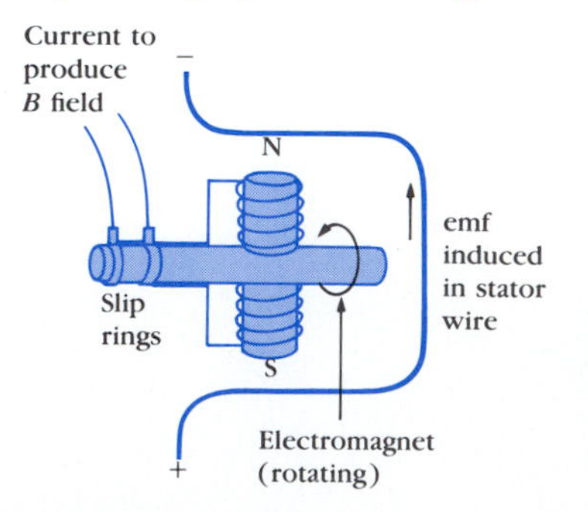

Had no one discovered electromagnetic induction and its practical application to the generator, life today would be totally different. Technology could not have reached the level it has without them (for good or bad). Even before the Industrial Revolution it had been possible to change the kinetic energy of running water into useful work by means of a waterwheel—for example, to turn the wheel of a mill that grinds grain into flour. But the mill had to be situated right on the river. With the development of electric generators and motors, the kinetic energy of moving water could be transformed into electric energy, and the electric energy could then be transported by wires to cities and transformed back into mechanical work. Thus the mill no longer had to be situated at the source of energy, the flowing water. It could be many miles away from the river; it could be built where consumers were located. With the building of high dams and large turbines to increase the kinetic energy of the falling

[†] You can, for example, actually run a car generator backward as a motor by connecting its output terminals to a battery.

water, vast amounts of energy could be produced. Where large dams were unfeasible, steam plants were built to release the chemical energy stored in coal or other fossil fuels by burning and to use the heat to make the steam that drives the turbines.

Various means of turning a generator to produce electric energy—from fossil fuel, nuclear material, geothermal sources, wind, and other energy sources—were discussed in Chapter 15.

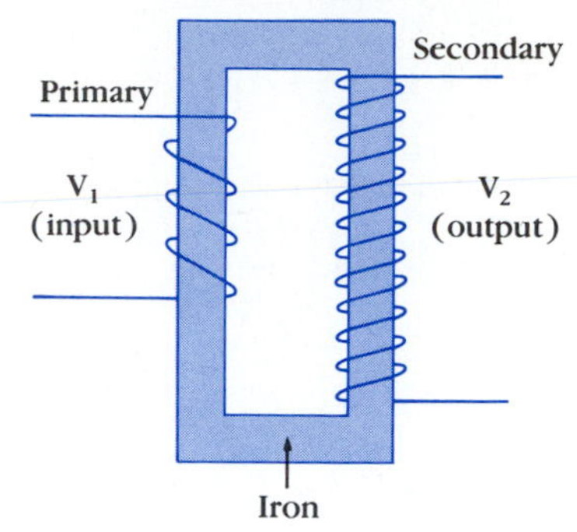

FIGURE 21-6
Step-up transformer ($N_1 = 4$, $N_2 = 12$).

3. THE TRANSFORMER

A **transformer** is a device for increasing or decreasing an ac voltage. A transformer consists of two coils of wire known as the *primary* and *secondary* coils. The two coils can be interwoven; or they can be linked by an iron core as shown in Figure 21-6. An alternating current is applied to the primary coil. As the primary current changes, the magnetic field it produces changes with it. The changing magnetic field passes through the secondary coil, and, as we know from Faraday's law, this changing magnetic field induces a current in the secondary coil.

Transformers work only with ac. A constant dc voltage applied to the primary coil gives rise to a constant magnetic field, and no voltage can be induced in the secondary. A voltage can be induced only by a *changing* magnetic field—that is, only by alternating current.

The main use of a transformer is to change the magnitude of an ac voltage, to make it either higher or lower. If the primary has the same number of turns (or loops) as the secondary, the voltage induced in the secondary is the same as that induced in the primary. But if the secondary coil contains twice as many turns as the primary, the secondary voltage will be twice that in the primary because the total voltage induced in the secondary is the sum of the voltages induced in each secondary loop. If the number of secondary turns is doubled, the total induced voltage is doubled. In general, the ratio of the induced secondary voltage, V_2, to the applied primary voltage, V_1, is equal to the ratio of the number of turns in the two coils, call them N_1 and N_2, respectively:

$$V_2/V_1 = N_2/N_1.$$

If the secondary coil has more turns than the primary, the secondary voltage will be greater than the applied primary voltage. This is called a *step-up transformer*. If the secondary has fewer turns than the primary, the secondary voltage will be less than the primary voltage. This is a *step-down* transformer. Of course we don't get something for nothing with a transformer. True, voltage can be increased with a step-up transformer, but what is gained in voltage is lost in current. The output current of a step-up transformer is smaller by at least the same factor as that by which the voltage is larger. Thus the product voltage × current, which is power, remains the same. (Actually, it is reduced because some of the energy goes into heat.)

Transformers are indispensable for the transmission of electric power. In order to avoid great losses of power during transmission over long distances, high voltage must be used. At a power station the output of the generators is boosted by a step-up transformer to 120,000 volts or more for transmission. Step-down transformers are utilized at the receiving end to reduce the voltage, usually in several stages, to 240 or 120 volts for home and industrial use. The boxes you see on telephone poles contain step-down transformers.

Transformers are common in everyday life. A step-down transformer is the essential component of a converter that allows a battery-operated transistor radio or casette tape recorder to be used on household voltage. The ignition coil of an automobile is a step-up transformer; it transforms the twelve volts of a car battery to the many thousands of volts necessary to cause the spark plugs to spark and ignite the gas-air mixture in the cylinders. However, a 12-volt dc car battery alone will not operate a transformer since alternating current is required. Therefore the automobile battery is connected to a switch (which can be mechanical or electronic "ignition points"), which opens and closes the primary circuit. In this way the steady battery voltage alternates between +12 volts and zero volts. This changing voltage can then induce the required high voltage in the secondary of the coil.

An Example (optional)

As an example, consider a transformer for a transistor radio that reduces 120 V ac to 9.0 V ac. (Such a device also contains diodes to change the 9.0 V ac to dc; see Chapter 28.) The secondary contains 30 turns and the radio draws 400 mA. Let us calculate (a) the number of turns in the primary (b) the current in the primary and (c) the power transformed. (a) This is a step-down transformer, and from the preceding equation we have

$$N_1 = N_2 \frac{V_1}{V_2} = (30)(120\ \text{V})/(9.0\ \text{V}) = 400\ \text{turns}.$$

(b) If we assume there are no losses, the input power equals the output power. Thus $I_1V_1 = I_2V_2$, so

$$I_1 = \frac{I_2V_2}{V_1} = \frac{(0.40\ \text{A})(9.0\ \text{V})}{(120\ \text{V})} = 0.030\ \text{A}.$$

(c) The power transformed is

$$P = I_2V_2 = (9.0\ \text{V})(0.40\ \text{A}) = 3.6\ \text{W},$$

which is, assuming 100 percent efficiency, the same as the power in the primary, $P = (120\,\text{V})(0.030\,\text{A}) = 3.6\,\text{W}$.

4 ELECTROMAGNETIC WAVES

Faraday's law of induction tells us that a changing magnetic field induces an electric current. Now an electric current is a flow of electric charge, and electric charges flow when an electric field acts on them. Consequently Faraday's law tells us that

A changing magnetic field gives rise to an electric field.

It is this *induced* electric field that gives rise to the induced voltage and current in a wire. The electric field is induced, however, even if no wire or other matter is present in the region of space in which the magnetic field is changing. This is the deepest essence of Faraday's discovery. With this insight, Faraday's law takes on even greater power.

A Changing Electric Field Gives Rise to a Magnetic Field

The symmetry of nature is truly remarkable. In 1860 the great physicist James Clerk Maxwell (1831–1879) postulated that the exact inverse of Faraday's law is also true:

A changing electric field gives rise to a magnetic field.

This second induction principle should not come as a surprise to us for the following reason. We know that an electric current sets up a magnetic field—this was Oersted's discovery. Now an electric current is simply electric charges in motion. Consider just one electric charge in motion. In Figure 21-7(a) we see the charge with its electric field. A moment later the charge has moved—Figure 21-7(b)—and has carried its electric field with it; the electric field at any point in space, say at P in Figure 21-7, is changing. Thus we can reinterpret our statement that moving charges or currents cause magnetic fields so that it says: a changing electric field induces a magnetic field.

This second induction principle, which is another important relation between electricity and magnetism, enabled Maxwell to tie all of electricity and magnetism together in a mathematical theory of great beauty. It is one of the great triumphs in all of physics.

We will not go into the mathematics of Maxwell's electromagnetic theory. It is essentially a formulation in mathematical terms of the facts we have already discussed. But we can discuss a startling prediction of Maxwell's theory: *accelerating electric charges or the equivalent, changing electric currents, give rise to electric and magnetic fields that move through space*. We call these rapidly moving electric and magnetic fields **electromagnetic waves.**

FIGURE 21-7
Electric field lines of a moving electric charge. Note the strength of the electric field at point P increases as the charge comes closer.

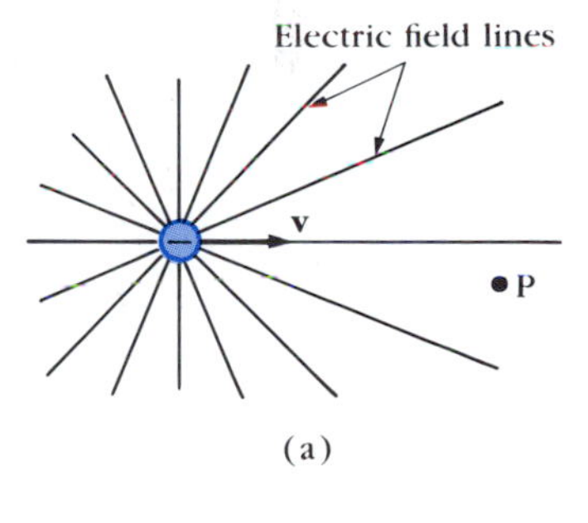

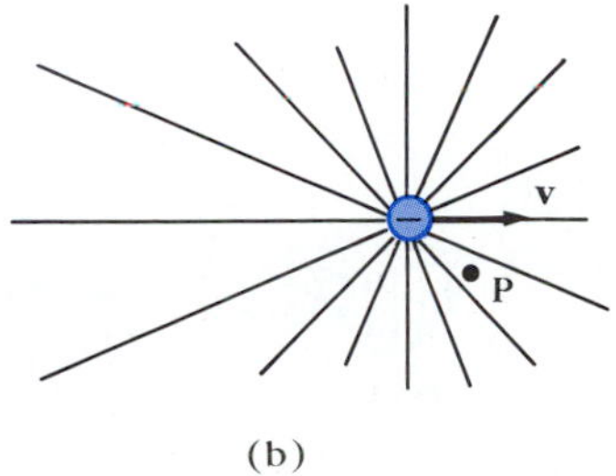

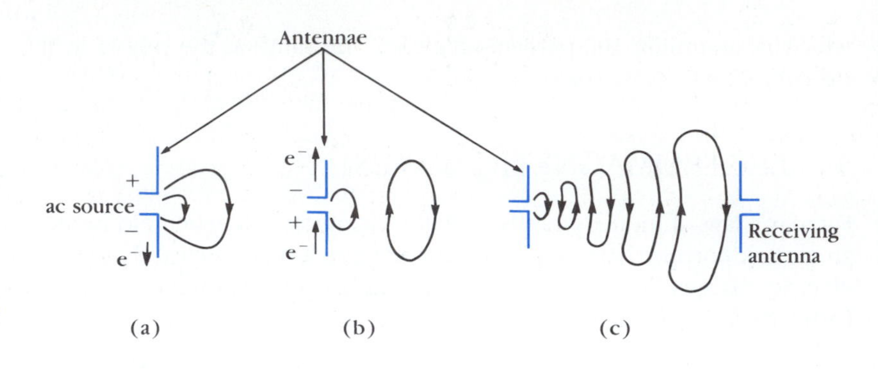

FIGURE 21-8
Production of electromagnetic waves by application of ac voltage to a wire antenna. The diagram shows the electric field lines at various moments as the ac voltage on the antenna changes. The magnetic field lines are not shown; they are perpendicular to the electric field lines and therefore point into and out of the page. In (c) a receiving antenna (on a radio or TV) is also shown.

Electromagnetic Waves Are the Result of the Two Induction Principles

Let us see how electromagnetic waves are produced. Figure 21-8 shows a wire *antenna*. The upper and lower arms of the antenna are connected to opposite terminals of an ac source of voltage. At any given instant the electrons in the two arms of the antenna are moving in the same direction, as shown in Figure 21-8(a). This is true because the electrons are attracted to the positive terminal in one arm and repelled by the negative terminal in the other. A moment later, when the voltage changes sign, the electrons change their direction and start moving in the opposite direction, Figure 21-8(b). The flow of electrons in the antenna is an electric current that changes direction at the same frequency as the ac source. It thus sets up a magnetic field which itself is continuously changing. This changing or oscillating magnetic field gives rise to an electric field, as Faraday's law of induction tells us. The changing electric field itself produces a changing magnetic field; and so on. But these are not two separate processes; they occur together. The oscillating electric and magnetic fields are everywhere linked together.

The induced electric and magnetic fields, as produced at any instant, do not exist all the way to infinity. If they did it would mean that the fields could get from the antenna to distant points in zero time, which is not possible. Instead the electric and magnetic fields are set up in the vicinity of the antenna and travel outward at a finite velocity. As they move outward, they make room next to the antenna for the oppositely directed electric and magnetic fields that will be set up a moment later when the current changes direction. This is shown in Figure 21-8(b). As the current in the antenna continues to reverse itself, the newly produced electric and magnetic fields appear next to the antenna, and the "old" fields move outward—Figure 21-8(c). These electric and magnetic fields which travel outward are known as **electromagnetic waves**. They are waves because the direction of the electric and magnetic fields alternate in an undulating fashion: Figure 21-9 is a diagram of the strength of the propagating electric

FIGURE 21-9
An electromagnetic wave consists of alternating electric and magnetic fields at right angles to each other. The entire pattern moves in a direction perpendicular to both the electric and magnetic fields at the speed of light.

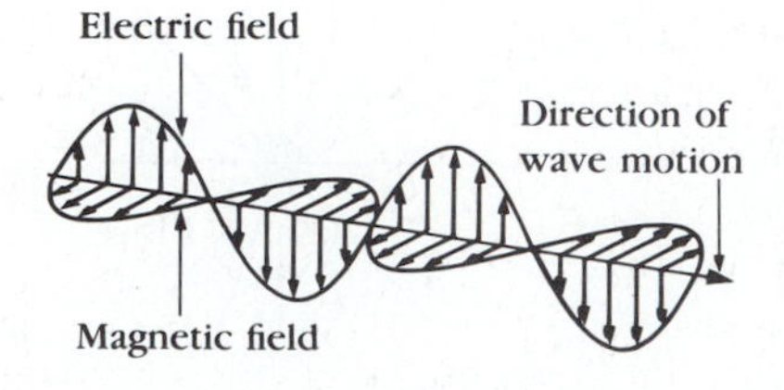

and magnetic fields at a particular instant. The electric and magnetic fields are perpendicular to each other, and perpendicular to the direction of propagation.

Maxwell first predicted the existence of electromagnetic waves as an outcome of his theory of electricity and magnetism. It was not until 1885, after Maxwell's death, that Heinrich Hertz (1857–1894) showed experimentally that electromagnetic waves do indeed exist and that they behave in just the fashion that Maxwell had predicted.

Light Is an Electromagnetic Wave

The frequency, or number of oscillations per second, of an electromagnetic wave can be a few cycles per second to hundreds of millions of cycles per second and even higher, depending on the frequency of the ac source.

Maxwell's theory predicted that all electromagnetic waves, regardless of their frequency, would travel at a velocity of 3×10^8 meters/second. This happens to be exactly the speed of light in vacuum. For a long time the nature of light had completely escaped scientists. Light was known to travel at a finite speed, and that speed was known to be 3×10^8 m/s; but what light "really is" was a great mystery. The only thing known about light, besides its speed, was that it was some sort of wave motion (see Chapter 24); but just what oscillated up and down was not known. Maxwell had at last provided a solution to the mystery of light. Light was simply an electromagnetic wave!

Other Kinds of Electromagnetic Wave: The Electromagnetic Spectrum

What distinguishes light from other electromagnetic waves is its frequency or wavelength. Our eyes are sensitive to electromagnetic waves only if they have a frequency between about 4×10^{14} and 7.5×10^{14} Hz (vibrations per second), which corresponds to wavelengths between 4×10^{-7} and 7×10^{-7} meters. (Remember from Chapter 16 that the velocity v of waves equals the product of wavelength λ and frequency f; that is, $v = \lambda f$.) Electromagnetic waves in this region are called *visible light*. Electromagnetic waves whose wavelengths are just shorter or just longer than those of visible light are called *ultraviolet light* (UV) and *infrared light* (IR), respectively. Waves with much longer wavelengths that are used for radio and TV transmission are known as *radio waves*. *X rays* are another form of electromagnetic wave, but of extremely high frequency. The classification of electromagnetic waves according to frequency is known as the **electromagnetic spectrum**. It is shown in Figure 21-10 along with the names given to the various frequency and wavelength regions. Electromagnetic waves of different frequency differ widely in their behavior. For example, as mentioned above, only waves in a certain limited frequency range can be detected by our eyes, whereas only waves of very high frequency are penetrating enough to produce medical x-ray photographs.

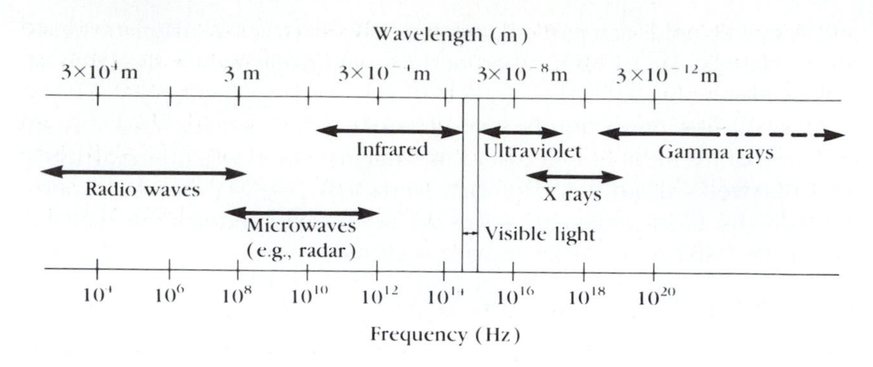

FIGURE 21-10
Electromagnetic spectrum.

Yet all electromagnetic waves travel through space at the same speed and have the same essential nature. Only the frequency and wavelength are different.

5. RADIO AND TELEVISION

Electromagnetic waves offer the possibility of transmitting information over long distances. Among the first to realize this and put it in practice was Guglielmo Marconi (1874–1937) who, in the 1890s, invented and developed the wireless telegraph. With it, messages could be sent hundreds of kilometers at the speed of light without the use of wires. The first signals were merely long and short pulses that could be translated into words by a code, such as the "dots" and "dashes" of the Morse code. Out of this early work radio and television were born, which we now discuss.

Radio and TV Transmission

The process by which a radio station transmits information (words and music) is outlined in Figure 21-11. The audio (sound) information is changed into an electrical signal of the same frequencies by a transducer such as a microphone, phonograph cartridge, or tape recorder head. This electrical signal is called an audio-frequency (AF) signal, since the frequencies are in the audio range (20 to 20,000 Hz). The signal is amplified electronically and mixed with a radio-frequency (RF) wave

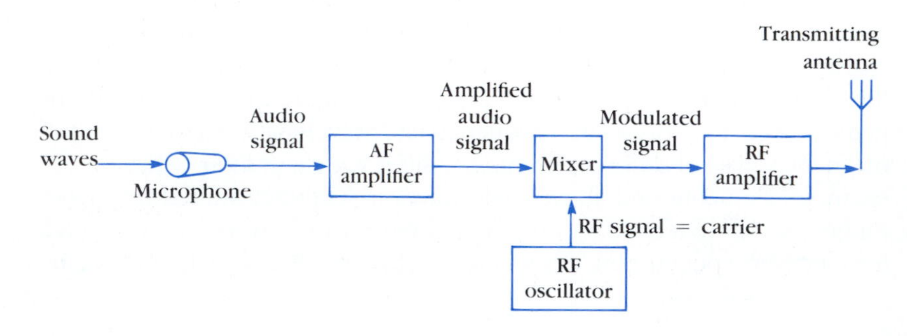

FIGURE 21-11
Block diagram of a radio transmitter.

known as the *carrier frequency*. A particular carrier frequency is assigned each station. AM radio stations have carrier frequencies between 550,000 Hz to 1,600,000 Hz (550 kHz to 1600 kHz). For example, "710 on your dial" means a station whose carrier frequency is 710 kHz. FM radio stations have much higher frequencies, between 88 and 108 MHz. The carrier frequencies for TV stations in the United States lie between 54 and 88 MHz for channels 2 through 6 and between 174 and 216 MHz for channels 7 through 13; UHF (ultra-high-frequency) stations have even higher carrier frequencies, between 470 and 890 MHz.

The mixing of the audio and carrier frequencies can be done in two ways. In amplitude modulation (AM) the amplitude of the audio signal is combined with that of the much higher carrier frequency, as shown in Figure 21-12. It is called "amplitude modulation" because the *amplitude* of the carrier is altered ("modulate" means to change or alter). In frequency modulation (FM), the *frequencies* of the audio and carrier signals are combined. Thus the *frequency* of the carrier wave is altered by the audio signal, as shown in Figure 21-13.

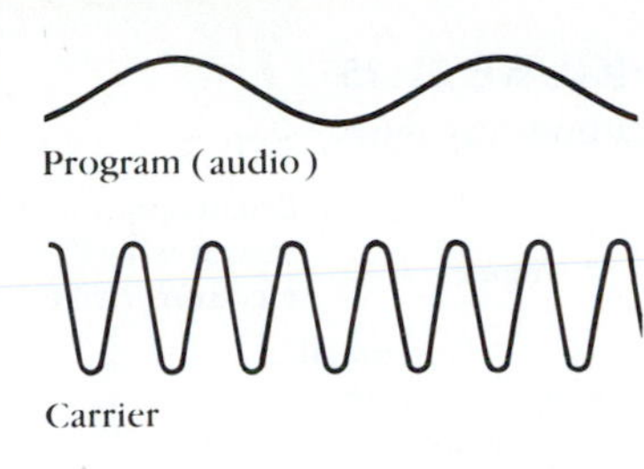

FIGURE 21-12
In amplitude modulation (AM), the amplitudes of the audio and carrier signals are combined.

The mixed signal is amplified further (since the signal contains radio frequencies, this is called an RF amplifier) and is then sent to the antenna where the complex mixture of frequencies is sent out in the form of electromagnetic (EM) waves.

A television transmitter works in a similar way, using frequency modulation, except that both audio and video (visual) signals are mixed with carrier frequencies.

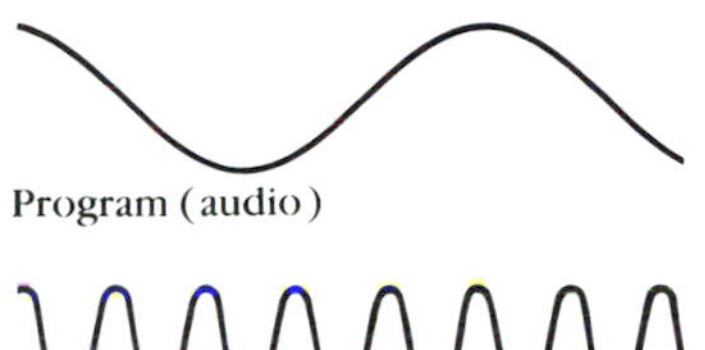

Total signal (FM)

FIGURE 21-13
In FM, the frequencies of the audio and carrier signals are combined. This method is used by FM radio and television.

Receivers

A simple radio receiver is diagrammed in Figure 21-14. The EM waves sent out by all stations are received by the antenna. One kind of antenna consists of one or more conducting rods; the electric field in the EM waves exert a force on the electrons in the conductor, causing them to move back and forth at the frequencies of the impinging EM waves. A second type of antenna, often found in AM radios, consists of a tubular coil of wire. This type of antenna detects the magnetic field of the wave, for the changing **B** field induces voltage and current in the coil. The signal from the antenna is very small and contains frequencies from many different stations. When you tune in a particular station, you are setting the receiver to select a particular RF frequency (actually a narrow range of frequencies) corresponding to the particular station's carrier frequency. The signal, containing both audio and carrier frequencies, goes to the *detector* (Figure 21-14)

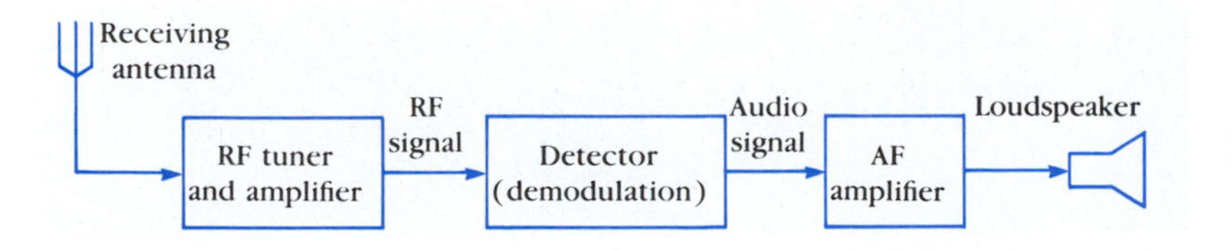

FIGURE 21-14
Block diagram of a simple radio receiver.

FIGURE 21-15
Cathode ray tube.

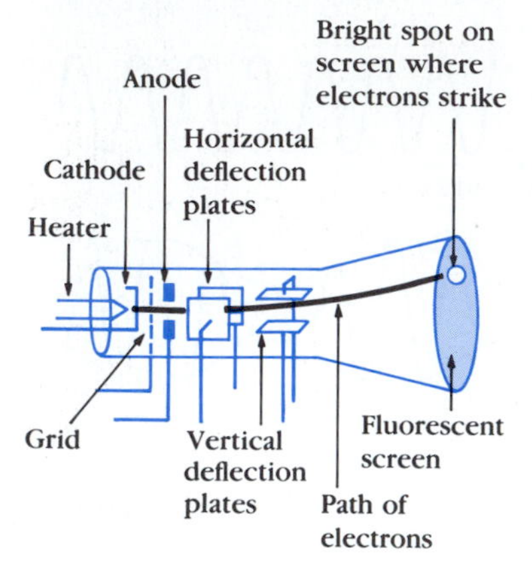

where "demodulation" takes place; that is, the RF carrier frequency is separated from the audio signal. The audio signal is then amplified and sent to a loudspeaker or headphones.†

A television receiver does similar things to both the audio and the video signals. The audio signal goes finally to the loudspeaker and the video signal to the picture tube or *cathode ray tube* (CRT), whose operation we now discuss.

6. CATHODE RAY TUBE

The picture tube of a TV set is called a cathode ray tube (CRT; "cathode ray" was an early term for the electron). Cathode ray tubes are also used in computer display terminals and in the oscilloscope, a device for visualizing electrical signals such as heart beats (ECG) and brain waves (EEG).

A simple cathode ray tube is diagrammed in Figure 21-15. Several electrodes (conductors) are inserted inside an evacuated glass tube. A heating element is made hot by a current passing through it and heats the electrode known as the cathode. When heated in this way, some of the electrons in the cathode acquire sufficient kinetic energy to escape from the metal. These electrons are accelerated by the high positive voltage on the anode (5000 to 50,000 volts). When the electrons emerge from this "electron gun" through a small hole in the anode, they speed toward the face of the tube. The tube face is coated with a fluorescent substance that glows briefly when struck by electrons. It is this glow that constitutes the visual image. Two vertical and two horizontal plates deflect the beam of negatively charged electrons whenever a voltage is applied across them. The electrons are of course attracted to that plate that at any given moment is positive. In an oscilloscope, the horizontal deflection plates cause the beam of electrons to go from left to right (repeatedly, in most cases); and the signal to be viewed, for example an ECG (Figure 21-16) is amplified and applied to the vertical deflection plates.

† For *FM stereo broadcasting*, two signals are carried by the carrier wave. One of these contains frequencies up to about 17,000 Hz, which includes most audio frequencies; the other signal includes the same range of frequencies, but 21,000 Hz is added to it. A stereo receiver subtracts this 21,000-Hz signal and distributes the two signals to the left and right channels. The first signal actually consists of the sum of left and right channels ($L + R$), so mono radios detect all the sound; the second signal is the difference between left and right ($L - R$). Hence the receiver must add and subtract the two signals to get pure left and right signals for each channel.

FIGURE 21-16
An ECG trace.

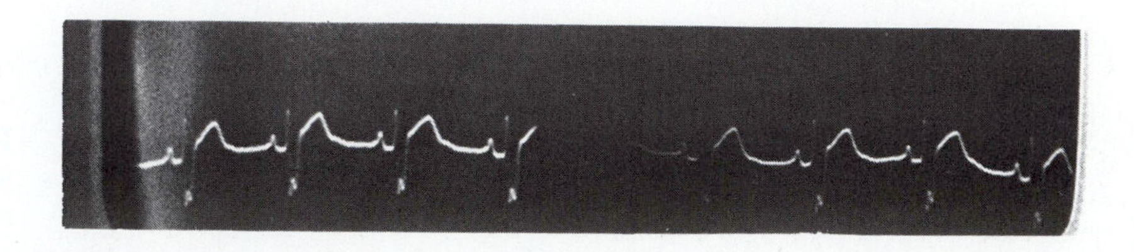

In the picture tube of a television set, magnetic deflection coils are usually used instead of electric plates. But the effect is the same. The electron beam is made to sweep over the screen in the manner shown in Figure 21-17. The beam is swept horizontally by the horizontal deflection plates or coils. When the horizontal deflecting field is maximum in one direction, the beam is at one edge of the screen. As the field decreases to zero the beam moves to the center; and as it increases to a maximum in the opposite direction the beam approaches the opposite edge. When it reaches this edge, the voltage or current abruptly changes to return the beam to the opposite side of the screen. Simultaneously, the beam is deflected downward slightly by the vertical deflection plates or coils, and then another horizontal sweep is made. In the United States 525 lines constitute a complete sweep over the entire screen. European systems use more lines, providing greater sharpness.) The complete picture of 525 lines is swept out in $\frac{1}{30}$ s. Actually, a single vertical sweep takes $\frac{1}{60}$ s and involves every other line; the lines in between are then swept out over the next $\frac{1}{60}$ s. We see a picture because the image is retained by the fluorescent screen and by our eyes for about $\frac{1}{20}$ s. The picture we see consists of the varied brightness of the spots on the screen. The brightness at any point is controlled by the grid (a "porous" electrode, such as a wire grid, that allows passage of electrons), which can limit the flow of electrons by means of the voltage applied to it: the more negative this voltage, the fewer electrons pass through. The voltage on the grid is determined by the video signal sent out by the station and received by the set. Accompanying this signal are signals that synchronize the grid voltage to the horizontal and vertical sweeps.

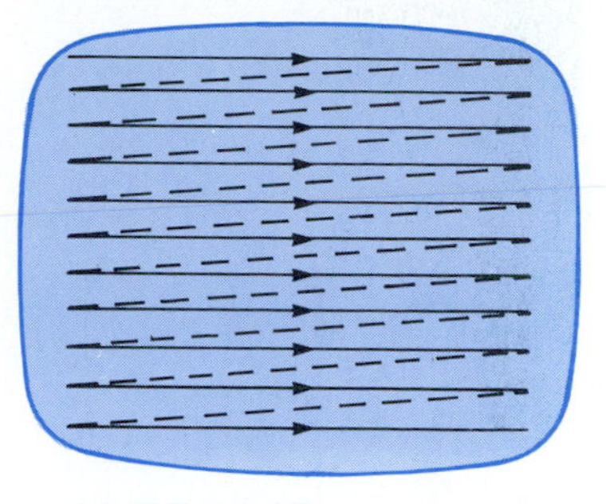

FIGURE 21-17
Television screen showing how electrons sweep across the screen in a succession of horizontal lines. (The dotted lines represent a very rapid, unnoticeable, motion.)

A color television screen contains an orderly array of tiny phosphor lines, alternating among the three primary colors: red, green, and blue. When a line at a given location is struck by the electron beam, it glows either red, green, or blue, or an intermediate color if more than one phosphor is struck (Figure 24-24, Color Plate VII). Except for the more complicated color video signal and the circuitry needed to direct the electron beam to the correct phosphors across the screen, a color TV receiver is similar to a black-and-white receiver.

SUMMARY

A changing magnetic field can induce current and voltage in a circuit. *Faraday's law* says that the induced voltage in a loop of wire is proportional to the rate of change of magnetic flux through the wire. The magnetic flux is the product of the magnetic field B and the area of the loop (assumed perpendicular to each other). The induced voltage produces a current whose magnetic field opposes the original change in flux

(*Lenz's law*). Faraday's law also tells us that a changing magnetic field produces an electric field.

An electric *generator* changes mechanical energy into electrical energy. Its operation is based on Faraday's law: a coil of wire is made to rotate by mechanical means in a magnetic field, and the changing flux through the coil induces an alternating current, which is the output of the generator.

A *transformer*, which is a device to change the magnitude of an ac voltage, consists of a primary and a secondary coil. The changing flux due to an ac voltage in the primary induces an ac voltage in the secondary. In a 100 percent efficient transformer, the ratio of output to input voltages (V_2/V_1) equals the ratio of the number of turns N_2 in the secondary to the number N_1 in the primary: $V_2/V_1 = N_2/N_1$.

Maxwell postulated that the inverse of Faraday's law is also true: a changing electric field gives rise to a magnetic field. This enabled Maxwell to tie all of electricity and magnetism together into one theory. Maxwell's theory predicted that accelerating electric charges give rise to electric and magnetic fields that move through space at the speed of light. These moving fields are called *electromagnetic* (EM) *waves*. Visible light is composed of electromagnetic waves in a particular range of frequencies. Other ranges of frequencies of electromagnetic waves include infrared, ultraviolet, microwaves, X rays, and radio waves.

Radio and TV signals are transmitted via EM waves. The audio (and video) signal is combined with the carrier frequency either by amplitude modulation (AM) or by frequency modulation (FM).

A cathode ray tube (CRT) is used as the picture tube in television sets and in oscilloscopes. Thermionically emitted electrons are accelerated by a high voltage within the tube; the resulting electron beam strikes the fluorescent screen on the inside face of the tube, which glows at that point. The electron beam is made to move about the screen by electric deflecting plates or magnetic deflecting coils.

QUESTIONS

1. What would be the advantage in Faraday's experiments (Figure 21-1) of using coils with many turns?
2. What is the difference between magnetic flux and magnetic field?
3. Suppose you are holding a circular piece of wire and suddenly thrust a magnet, south pole first, toward the center of the circle. Is a current induced in the wire? Is a current induced when the magnet is held steady within the loop? Is a current induced when you withdraw the magnet? In each case, if your answer is yes, specify the direction.
4. Since a magnetic microphone is basically like a loudspeaker, could a loudspeaker actually serve as a microphone? That is, could you speak

into a loudspeaker and obtain an output signal that could be amplified? Explain. Discuss, in light of your response, how a microphone and loudspeaker differ in construction.

5. Would a magnet rotating inside a coil of wire induce a voltage in the coil?
6. What are the similarities between a motor and a generator? What are the differences?
7. Does the voltage output of a generator change if its speed of rotation is increased? Explain.
8. In some early automobiles, the starter motor doubled as a generator to keep the battery charged once the car was started. Explain how this might work.
9. What does a transformer "transform"? Why is an alternating current needed?
10. An enclosed transformer has four wire leads coming from it. How could you determine the ratio of turns on the two coils without taking the transformer apart? How would you know which wires paired with which?
11. Is sound an electromagnetic wave? If not, what kind of wave is it?
12. Can EM waves travel through a perfect vacuum? Can sound waves?
13. How are light and sound alike? How are they different?
14. In the electromagnetic spectrum, what type of EM wave would have a wavelength of 10^3 km? 1 km? 1 m? 1 cm? 1 mm? 1 μm?
15. Are the wavelengths of radio and television signals longer or shorter than those detectable by the human eye?
16. A lost person may signal by flashing a flashlight on and off using Morse code. This is actually a modulated EM wave. Is it AM or FM? What is the frequency of the carrier, approximately?
17. Can two radio or TV stations broadcast on the same carrier frequency? Explain.
18. Why does a television set use an antenna?
19. The carrier frequencies of FM broadcasts are much higher than for AM broadcasts. On the basis of what you learned about diffraction in earlier chapters, explain why AM signals can be detected more readily behind low hills or buildings.
20. Bringing a magnet close to a television screen will distort the picture. Why? (*Do not do this to a color set; permanent damage may result.*)
21. Why, at the beginning of Chapter 18, did we speak of the electromagnetic force as one of the four forces in nature rather than considering the electric force and the magnetic force as two separate forces in nature?

EXERCISES

1. A 10-cm-diameter circular loop of wire is in a 0.50-T magnetic field. It is removed from the field in 0.10 s. What is the average induced voltage?
2. A square wire loop, 6.0 cm on a side, lies between the pole faces of an electromagnet. Calculate the average voltage induced in the wire when the electromagnet is turned on if its field increases from zero to its maximum of 1.5 T in $\frac{3}{4}$ s.
3. The magnetic field perpendicular to a circular loop of wire 12 cm in diameter is changed from $+0.35$ T to -0.15 T in 90 ms, where + means the field points away from an observer and − toward the observer. (a) Calculate the induced voltage. (b) In what direction does the induced current flow?
4. Suppose you are looking along a line through the centers of two circular (but separate) wire loops, one behind the other. A battery is suddenly connected to the front loop, establishing a clockwise current. (a) Will a current be induced in the second loop? (b) If so, when does this current start? (c) When does it stop? (d) In what direction is this current? (e) Is there a force between the two loops? (f) If so, in what direction?
5. A generator produces 100 volts when rotated at a certain speed. If its speed of rotation is doubled, what will be the output voltage?
6. A car generator produces 12 V when the armature turns at 600 rev/min. What will be its output at 1500 rev/min, assuming that nothing else changes?
7. A transformer changes 12 V to 18,000 V, and there are 6000 turns in the secondary. How many turns are there in the primary?
8. A transformer has 78 turns in the primary and 158 in the secondary. What kind of transformer is this and by what factor does it change the voltage?
9. Neon signs require 12 kV for their operation. To operate from a 120-V line, what must be the ratio of secondary to primary turns of the transformer? What would the voltage be if the transformer were connected backward?
10. Show that the ratio of the current in the secondary and primary of a transformer equals the inverse ratio of number of turns:

$$\frac{I_2}{I_1} = \frac{N_1}{N_2}.$$

11. A step-up transformer increases 30 V to 120 V. What is the current in the secondary as compared to the primary? Assume 100 percent efficiency.

12. A step-down transformer reduces 120 volts to 40 volts. What is the current in the secondary as compared to that in the primary?

*13. A transformer has 1500 primary turns and 120 secondary turns. The input voltage is 120 V and the output current is 8.0 A. What is the secondary voltage and primary current?

14. Describe a transformer that could be used to light a 6-V light bulb from a 120-V 60-Hz source. What if the source were 24 Vdc?

15. What is the frequency of a microwave whose wavelength is 1.0 cm?

16. Calculate the wavelength (a) of a 60-Hz EM wave (b) of a 1240-kHz AM radio wave.

17. When you speak on the telephone in Los Angeles to a friend in New York 4000 km away, how long does it take your voice to travel? (*Hint*: EM waves on transmission lines travel at the speed of light.)

18. An FM station broadcasts at 91.5 MHz. What is the wavelength of this wave?

19. What is the wavelength of an AM station at 1500 on the dial?

20. Compare 940 on the AM dial to 94 on the FM dial. Which has the longer wavelength, and by what factor is it larger?

21. A useful radio antenna is one whose length is such that standing waves can be set up in it. Explain why an optimum FM antenna is $1\frac{1}{2}$ meters long (5 feet). The carrier frequency for FM stations is around 10^8 cycles/s and the speed of light is 3×10^8 meters/s.

22. Who will hear the voice of a singer first, a person in the balcony 50 m away from the stage or a person 3000 km away at home whose ear is next to the radio? How much sooner? Assume that the microphone is a few centimeters from the singer and the temperature is 20°C.

*23. An average of 120 kW of electric power is sent to a small town from a power plant 10 km away. The transmission lines have a total resistance of 0.40 Ω. Calculate the power loss if the power is transmitted at (a) 240 V and (b) 24,000 V.

LIGHT: RAY OPTICS

CHAPTER 22

The sense of sight is extremely important to us. Our eyes are not only a source of pleasure, but the sense of sight provides us with a large part of our information about the world and the people around us. How do we see? What is this something called *light* that enters our eyes and causes the sensation of sight? And how does light behave so that we can see the great range of phenomena that we do? This subject of light will occupy us for the next three chapters, and we will also return to it in later chapters.

1. CHARACTERISTICS OF LIGHT; THE RAY MODEL

We see an object in one of two ways. First we can detect light emitted directly from a source such as when we look at a light bulb, a flame, or the stars. The second and more common way in which we see an object is by light reflected from it; we see trees, furniture, books, and most other everyday objects because of light that has reflected off of them into our eyes. The light may have originated from the sun or from some other source such as a lamp. An understanding of how bodies emit light was not achieved until the 1920s, and this will be discussed in Chapters 26 and 27. How light is reflected from objects was understood much earlier, and we shall discuss this in the next section.

The Ray Model

A great deal of evidence suggests that light travels in straight lines. A source of light like the sun casts distinct shadows. We can hear sounds from around the side of a wall or building but we cannot see around it. The beam of a flashlight appears to be a straight line. Indeed, we infer the positions of objects in our environment by assuming that light moves from the object to our eyes in straight-line paths. Our whole orientation to the physical world is based on this assumption.

This reasonable assumption has led to a model for light known as the **ray model**. This model assumes that light travels in straight-line paths called light **rays**. When we see an object, we are aware that it occupies some space and that light reaches our eyes from each **point** on the object—Figure 22-1(a). Although light rays leave each point in many different directions, normally only a small bundle of these rays can enter an observer's eye as shown in Figure 22-1(b). If the person's head moves to one side, a different bundle of rays will enter the eye from each point.

We saw in Chapter 21 that light can be considered as an electromagnetic wave. Although the ray model of light does not deal with this aspect of light (we discuss the wave nature of light in Chapter 24), it has been very successful in dealing with many aspects of light such as reflection, refraction, and the formation of images by mirrors and lenses. Because the explanations involve straight-line rays at various angles, this subject is referred to as *ray optics* or *geometric optics*.

The Speed of Light and Index of Refraction

Galileo attempted to measure the speed of light by trying to measure the time required for light to travel a known distance between two hilltops. He stationed an assistant on one hilltop, himself on another, and ordered the assistant to lift the cover from a lamp the instant he saw a flash from Galileo's lamp. Galileo measured the time between the flash of his lamp and when he received the light from his assistant's lamp. The time was so short that Galileo concluded it merely represented human reaction time and that the speed of light must be extremely high.

The first successful determination that the speed of light is finite was made by the Danish astronomer Ole Roemer (1644–1710). Roemer had noted that the carefully measured period of one of Jupiter's moons (Io, with an average period of 42.5 hr) varied slightly, depending on the relative motion of Earth and Jupiter. When Earth was moving away from Jupiter, the period of the moon was slightly longer; when Earth was

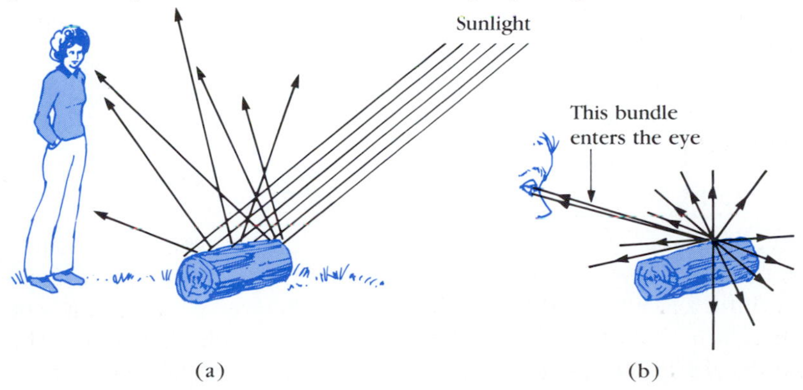

FIGURE 22-1
(a) We see an object when light is reflected off that object into our eyes. (b) Light rays radiate from each single point on an object. A small bundle of rays leaving one point is shown entering a person's eye.

moving toward Jupiter, the period was slightly shorter. He attributed this variation to the extra time needed for light to travel the increasing distance to Earth when Earth is receding, or to the shorter travel time for the decreasing distance when the two planets are approaching one another. Roemer concluded that the speed of light—though great—is finite.

Since then a number of techniques have been used to measure the speed of light. Among the most important were those carried out by the American Albert Michelson (1852–1931). Michelson used the rotating mirror apparatus diagrammed in Figure 22-2 for a series of high-precision experiments carried out from 1880 to the 1920s. Light from a source was directed at one face of a rotating eight-sided mirror. The reflected light traveled to a stationary mirror a large distance away and back again as shown. If the rotating mirror was turning at just the right rate, the returning beam of light would reflect from one face of the mirror into a small telescope through which the observer looked. At a different speed of rotation, the beam would be deflected to one side and would not be seen by the observer. From the required speed of the rotating mirror and the known distance to the stationary mirror, the speed of light could be calculated. In the 1920s, Michelson set up the rotating mirror on the top of Mt. Wilson in southern California and the stationary mirror on Mt. Baldy (Mt. San Antonio) 35 km away. He later measured the speed of light in vacuum using a long evacuated tube.

The accepted value today for the speed of light, c, in vacuum is

$$c = 2.99792458 \times 10^8 \text{ m/s}.$$

We usually round this off to

$$c = 3.00 \times 10^8 \text{ m/s}$$
$$= 300{,}000 \text{ km/s}.$$

(Note that the speed of light in vacuum is given the special symbol c

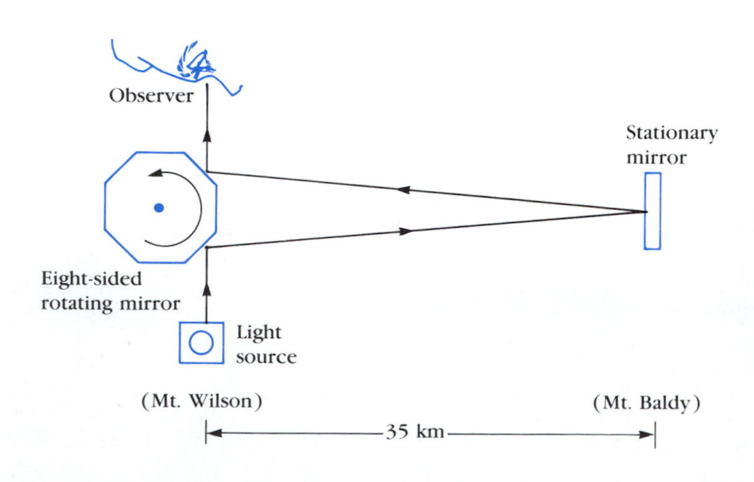

FIGURE 22-2
Michelson's speed-of-light apparatus.

instead of v.) In air the speed is only slightly less. In other transparent materials such as glass and water, the speed is always less than that in vacuum. For example, in water it travels at about $\frac{3}{4}c$. The ratio of the speed of light in vacuum to the speed v in a given material is called the *index of refraction*, n, of that material:

$$n = \frac{c}{v}.$$

The index of refraction for various materials is given in Table 22-1. For example, since $n = 2.42$ for diamond, the speed of light in diamond is

$$v = c/n = (3.00 \times 10^8\ \mathrm{m/s})/2.42 = 1.24 \times 10^8\ \mathrm{m/s}.$$

TABLE 22-1
Indices of Refraction

Material	$n = c/v$
Air (at STP)	1.0003
Water	1.33
Ethyl Alcohol	1.36
Glass	
Fused quartz	1.46
Crown glass	1.52
Light flint	1.58
Lucite or Plexiglas	1.51
Sodium chloride	1.53
Diamond	2.42

2. REFLECTION OF LIGHT; MIRRORS

When light strikes the surface of an object, some of the light is reflected. The rest is either absorbed by the object (and transformed to heat) or, if the object is transparent like glass or water, part of it is transmitted through. For a very shiny object such as a silvered mirror, over 95 percent of the light may be reflected.

The Law of Reflection

When a narrow beam of light strikes a flat surface we define the **angle of incidence,** θ_i, to be the angle an incident ray makes with the perpendicular to the surface and the **angle of reflection,** θ_r, to be the angle the reflected ray makes with the normal. For flat surfaces, it is found that the incident and reflected rays lie in the same plane with the perpendicular to the surface, and that:

the angle of incidence equals the angle of reflection.

This is the **law of reflection**, and is indicated in Figure 22-3. It was known to the ancient Greeks, and you can confirm it yourself by shining a narrow flashlight beam at a mirror in a darkened room.

EXPERIMENT

In a darkened room, put a small mirror flat on a table. Shine a flashlight beam on the mirror and notice where the reflected light beam hits the ceiling. Try various angles of incidence and note the angle of reflection. You can see the beams better if you sprinkle a little talcum powder in the air. Do your observations confirm the law of reflection?

AND

Have someone shine a beam of light on the mirror at an angle. Move your head around and note that the reflected light enters your eye only at a certain angle. Now have the person shine the light on a flat piece of white paper. Now the reflected light enters your eye from almost any angle. Why?

FIGURE 22-3
Law of reflection.

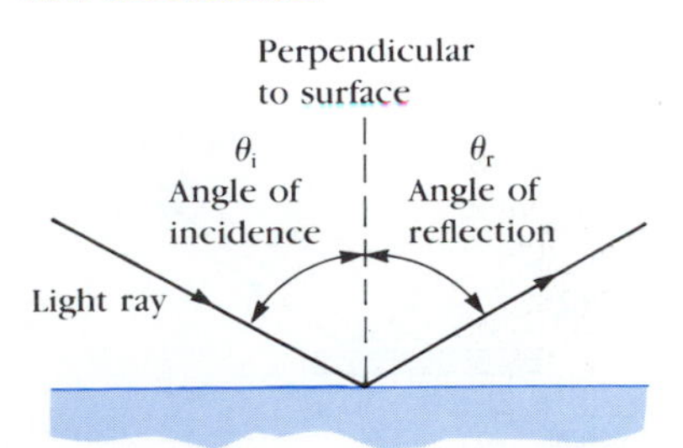

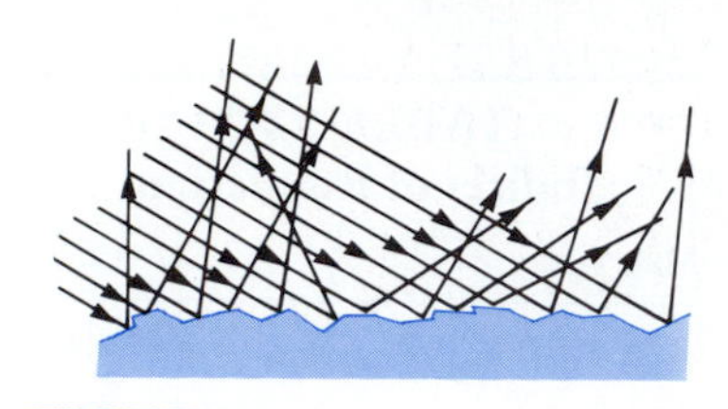

FIGURE 22-4
Diffuse reflection from a rough surface.

When light is incident upon a rough surface, even microscopically rough such as this page, it is reflected in many directions (Figure 22-4). This is called *diffuse reflection*. The law of reflection still holds, however, at each small section of the surface. Because of diffuse reflection in all directions, an ordinary object can be seen from many different angles. When you move your head to the side, a different bundle of reflected rays reaches your eye from each point on the object—Figure 22-5(a). But when a narrow beam of light is shone on a mirror, the light will not reach your eye unless it is placed at just the right place where the law of reflection is satisfied, as shown in Figure 22-5(b). This is what gives rise to the unusual properties of mirrors. (Galileo, using similar arguments,† showed that the moon must have a rough surface rather than a highly polished surface like a mirror, as some people thought.)

Image Formed by a Plane Mirror

When you look straight into a mirror, you see what appears to be yourself as well as various objects around you. Your face and the other objects look as if they are in front of you, beyond the mirror; but, of course, they are not. What you see in the mirror is an **image** of the objects.

Figure 22-6 shows how an image is formed by a plane (that is, flat) mirror. Rays from two different points on an object are shown. Rays leave each point on the object going in many directions, but only those that enclose the bundle of rays that reach the eye from the two points are shown. The diverging rays that enter the eye appear to come from behind the mirror as shown by the dashed lines. (Our eyes and brain interpret any rays that enter an eye as having traveled a straight-line path.) The point from which each bundle of rays seems to come is one point on the image. For each point on the object there is a corresponding image point. Let us concentrate on the two rays that leave the point A on the object and strike the mirror at points B and B'. The angles ADB and CDB are right angles. And angles ABD and CBD are equal because of the law of reflection. Therefore, the two triangles ABD and CDB are the same and the length $AD = CD$. That is, the image is as far behind the mirror as the object is in front: the *image distance*, d_i (distance from mirror to image, Figure 22-6), equals the *object distance*, d_o. From the geometry, we also see that the height of the image is the same as that of the object.

The light rays do not actually pass through the image itself. It merely *seems* like the light is coming from the image because our brains interpret any light entering our eyes as coming from in front of us. Since the rays do not actually pass through the image, a piece of white paper or film placed at the image would not detect the image. Therefore, it is called a *virtual image*. This is to distinguish it from a *real image* in which the

FIGURE 22-5
A beam of light from a flashlight is shined on (a) white paper, (b) a small mirror. In part (a) you can see the white light reflected at various points because of diffuse reflection. But in part (b) you see the reflected light only when your eye is placed correctly ($\theta_r = \theta_i$).

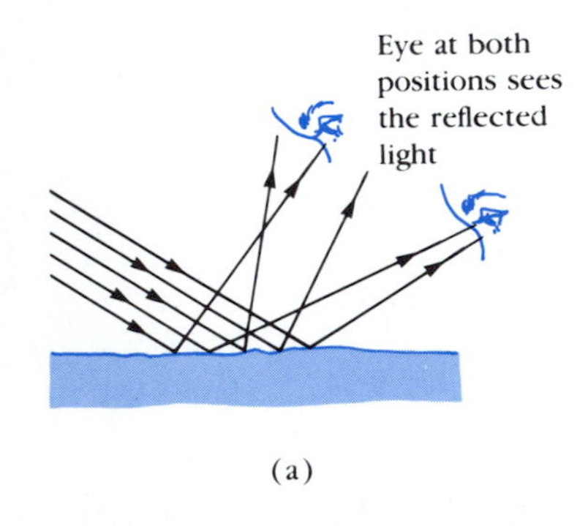
Eye at both positions sees the reflected light

(a)

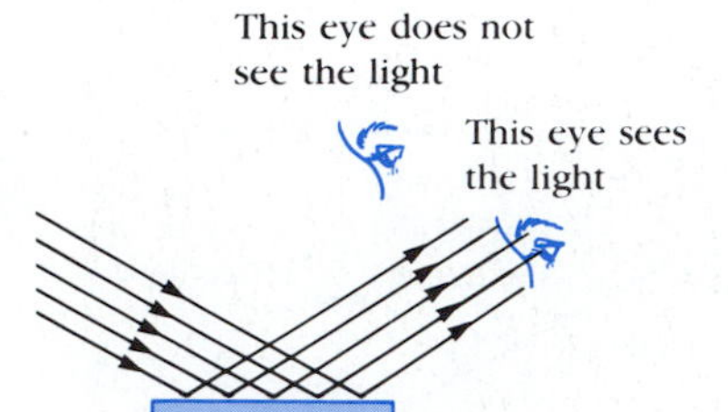
This eye does not see the light
This eye sees the light

(b)

† Galileo Galilei, *Dialogue Concerning the Two Chief World Systems*, trans. Stillman Drake (Berkeley: University of California Press, 1967). See Day 1.

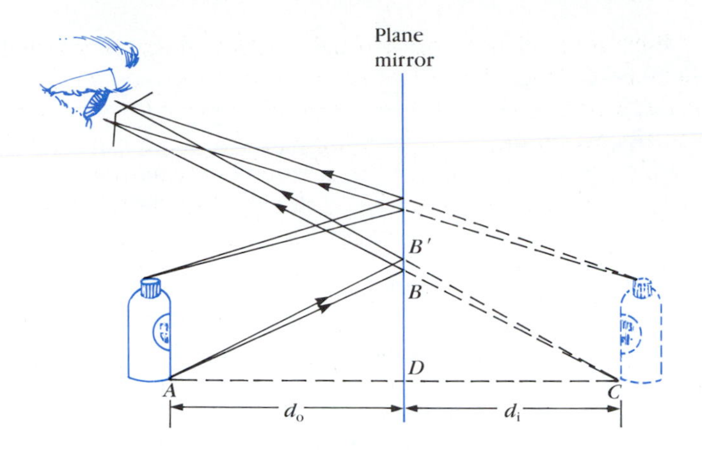

FIGURE 22-6
Formation of a virtual image by a plane mirror.

light does pass through the image, and which therefore could appear on paper or film placed at the image position. We will see that curved mirrors and lenses can form real images.

EXPERIMENT-PROJECT

Find a mirror long enough for you to see an image of your entire self, from the top of your head to your toes, when you are standing erect several feet away from it. What is the minimum-length mirror required for you to see a full-length image of yourself? To determine this, cover the top and bottom of your mirror with paper if necessary. You will find that the mirror has to be only half as long as you are. Draw a ray diagram, like that of Figure 22-6, to show why this is so. Does your distance from the mirror influence the size of mirror required? Try various distances and see.

Images Formed by Spherical Mirrors (optional)

Reflecting surfaces do not have to be flat. The most common *curved* mirrors are *spherical*, which means they form a section of a sphere. A spherical mirror is called *convex* if the reflection takes place on the outer surface of the spherical shape so that the center of the mirror bulges out toward the viewer; it is called *concave* if the reflecting surface is on the inner surface of the sphere so that the center of the mirror is farther from the viewer than the edges ("caved in" in the center). Concave mirrors find use as shaving or makeup mirrors; convex mirrors are sometimes used on vehicles and in shops (to watch for thieves), since they take in a wide field of view (Figure 22-7).

FIGURE 22-7
Looking into a convex mirror.

To see how spherical mirrors form images, we first consider an object that is very far from a concave mirror; in this case the rays from each point on the object that reach the mirror will be nearly parallel, as shown in Figure 22-8. *For an object infinitely far away* (the sun and stars approach this) *the rays would be precisely parallel*. Now consider such parallel rays falling on a concave mirror as in Figure 22-9. The law of reflection holds

FIGURE 22-8
If the object's distance is large compared to the size of the mirror, the rays are nearly parallel.

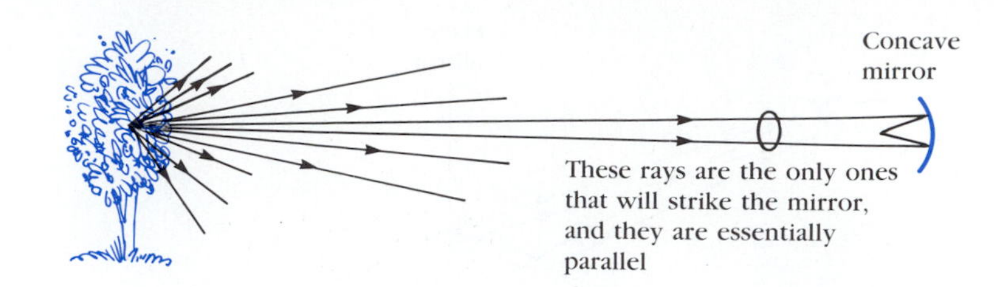

for each of these rays at the point each strikes the mirror. If the mirror width is small compared to its radius of curvature so that the reflected rays make only a small angle upon reflection, the rays will cross each other at nearly a single point, or come to a *focus*, as shown in Figure 22-9. The point F, where rays parallel to the principal axis come to a focus, is called the **focal point** of the mirror; the distance between F and the center of the mirror, length FA, is called the **focal length**, f, of the mirror. Another way to describe the focal point is to say that it is the *image point for an object infinitely far away*. The image of the sun, for example, would be at F.

From Figure 22-9 we can see that the focal length, f, is equal to half the radius of the mirror's surface. We consider a ray that strikes the mirror at B in Figure 22-9. The point C is the center of curvature of the mirror (the center of the sphere of which the mirror is a part.) So the dashed line CB is equal to r, the radius of curvature, and CB is perpendicular to the surface at B. From the law of reflection and the geometry, the three angles labeled θ are equal. The triangle CBF is isosceles because two of its angles are equal; thus length $CF = FB$. We assume the mirror is small compared to its radius of curvature, so the angles are small and the length FB is nearly equal to length FA. In this approximation, $FA = FC$. But $FA = f$, the focal length, and $CA = 2FA = r$. Thus the focal length is half the radius of curvature:

$$f = r/2.$$

Hence for an object at infinity the image is located at the focal point of a

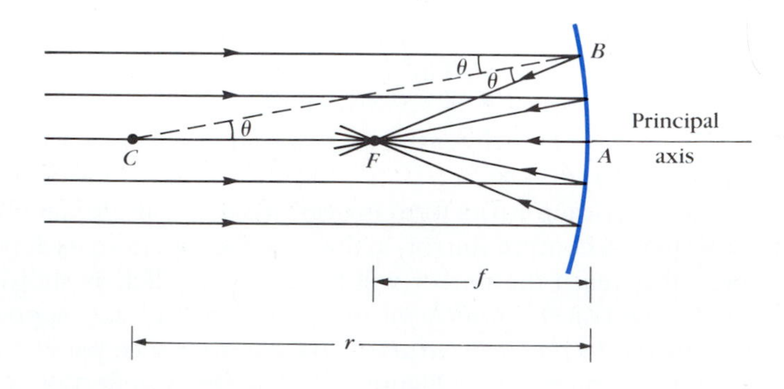

FIGURE 22-9
Rays parallel to the principal axis of a spherical mirror come to a focus at F, called the focal point, as long as the mirror is small in extent as compared to its radius of curvature r.

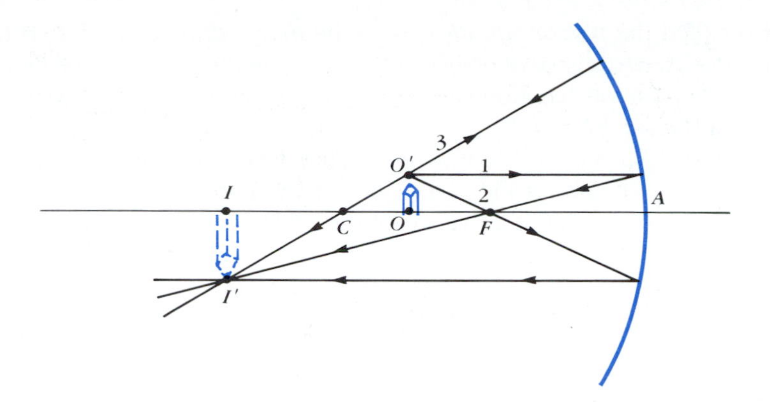

FIGURE 22-10
Rays from object at O form an image at I.

concave spherical mirror, where $f = r/2$. But where does the image lie for an object not at infinity?

To answer this, let us consider Figure 22-10. An object (OO') is placed between F and C. For a particular point O' on the object, let us find where its image point will be. To do this we can draw several rays and make sure these reflect from the mirror such that the reflection angle equals the incidence angle. This can involve much work and our task is simplified if we deal with three particularly simple rays. These are the rays labeled 1, 2, and 3 in the diagram. Ray 1 is drawn parallel to the axis; therefore it must pass through F after reflection (as in Figure 22-9). Ray 2 is drawn through F; therefore it reflects so it is parallel to the axis. Ray 3 is drawn so that it passes through C, the center of curvature, and thus is along a radius of the spherical surface; so it is perpendicular to the mirror and thus will be reflected back on itself. The point at which these rays cross (come to a focus) is the image point I'. All other rays from the same object point will pass through this image point. To find the image point for any object point, only these three types of rays need be used. Actually, only two of these rays are necessary, but the third serves as a check.

We have only shown the image point in Figure 22-10 for a single point on the object. Other points on the object are imaged nearby so a complete image of the object is formed, as shown by the dashed outline. Because the light actually passes through the image itself, this is a *real image*. This can be compared to the virtual image formed by a plane mirror.

The Mirror Equation (optional)

Image points can always be determined by drawing the three rays as described earlier. However, it is possible to derive an equation that gives the image distance if the object distance and radius of curvature are known. This equation is

$$\frac{1}{d_i} + \frac{1}{d_o} = \frac{1}{f}$$

and is called the *mirror equation*; d_i is the *image distance* and d_o is the object distance (distance of object and image from the mirror—see Figure 22-11). The mirror equation can be derived using the rays and triangles shown in Figure 22-11.

The *magnification m* of a mirror is defined as the height of the image divided by the height of the object: $m = h_i/h_o$. From the pair of similar triangles shown in Figure 22-11 we can see that

$$\text{magnification} = m = \frac{h_i}{h_o} = \frac{d_i}{d_o}.$$

In Figures 22-10 and 22-11, we see that the image is inverted relative to the object. Indeed, if you look at yourself in a concave mirror, your image will be upside down if your face is farther than the focal length from the mirror. But if the object (your face, for example) is between the mirror and the focal point, as in Figure 22-12, then the rays reflected from the mirror diverge (as shown); they appear to be coming from a point behind the mirror. Thus the image is behind the mirror and is right side up; and it is magnified. (It is also virtual since the rays don't actually pass through it.)

As an example, let us suppose you look at yourself in a concave mirror whose radius of curvature is 30 cm. If you put your face 10 cm from the mirror, where will the image be and what will be the magnification? Since $f = r/2 = 30\text{ cm}/2 = 15\text{ cm}$, and the object distance is $d_o = 20\text{ cm}$, then

$$\frac{1}{d_i} = \frac{1}{f} - \frac{1}{d_o} = \frac{1}{15\text{ cm}} - \frac{1}{10\text{ cm}} = \frac{2-3}{30\text{ cm}} = -\frac{1}{30\text{ cm}}$$

Since $\frac{1}{d_i} = -\frac{1}{30\text{ cm}}$ then $d_i = -30\text{ cm}$. Thus the image is 30 cm from the

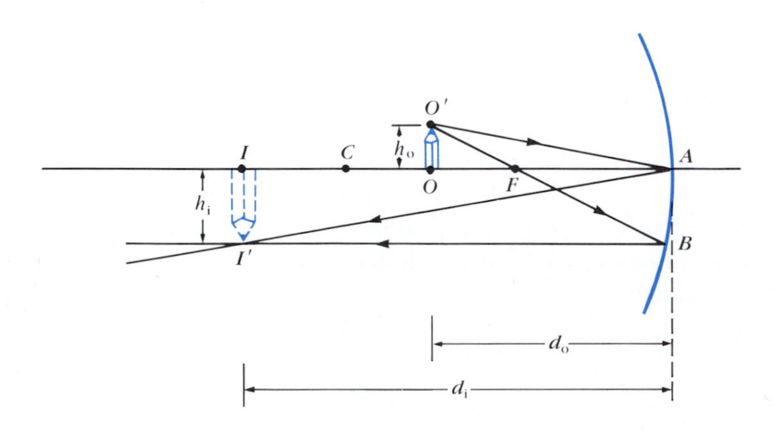

FIGURE 22-11
Diagram for deriving the mirror equation.

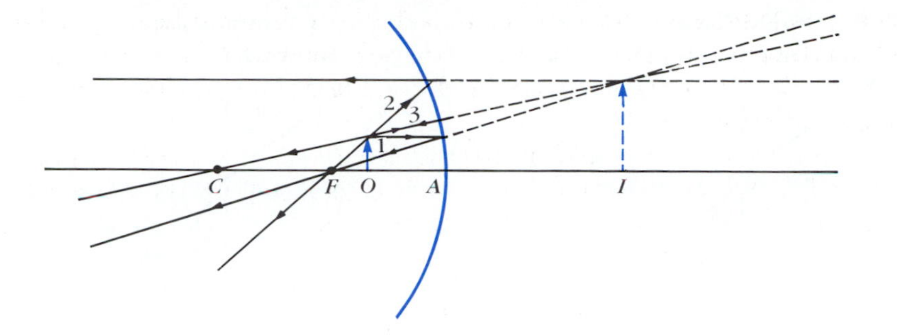

FIGURE 22-12
Object placed within the focal point F of a concave mirror. The image is *behind* the mirror and is *virtual*.

mirror (see Figure 22-12). The minus sign means the image is *behind* the mirror. [This illustrates a useful general rule: if the reflecting surface of the mirror faces to the left, as it has in all figures so far, distances are considered positive for points (object, image, and focal points) that are to the left; and they will be negative if they are to the right, behind the mirror, as the image distance was in this example.] To determine the magnification, $m = h_i/h_o$, we use the relation, $m = d_i/d_o = 30\text{ cm}/10\text{ cm} = 3\times$. So the image is 3 times larger than the object.

The analysis used for concave mirrors can be applied to *convex* mirrors. Figure 22-13(a) shows parallel rays falling on a convex mirror. The reflected rays diverge, but seem to come from point F behind the mirror. This is the *focal point*, and its distance from the center of the mirror is the focal length f. It is easy to show that again $f = r/2$. We see that an object at infinity produces a virtual image in a convex mirror. Indeed, no matter where the object is placed on the reflecting side of the mirror, the image will be right side up and smaller than the object, as shown in Figure 22-13(b). Thus convex mirrors are used to cover a wide field of view, as in a store (Figure 22-7), to aid motorists in seeing around a dangerous corner, and as a rearview or side mirror on a car or truck. The mirror equation holds for convex mirrors, but the focal length f must be considered negative.

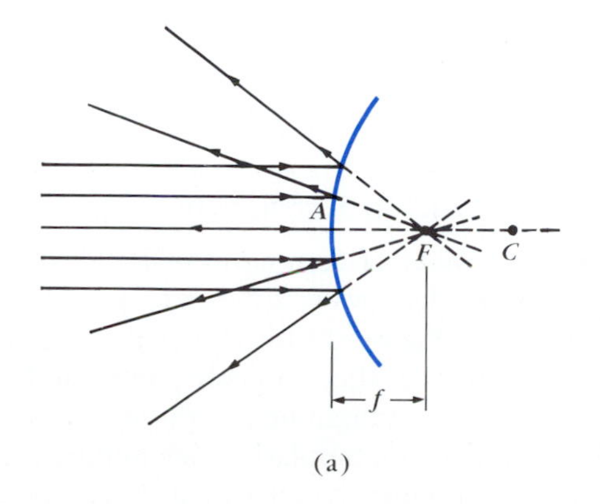

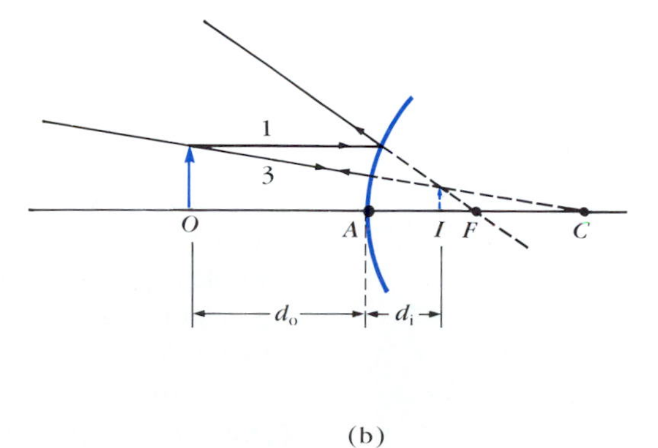

FIGURE 22-13
Convex mirror: (a) the focal point is at F, behind the mirror; (b) image I for the object at O is virtual, upright, and smaller than the object.

FIGURE 22-14
Refraction of light. (a) Light entering a medium in which its speed is *less* bends *toward* the perpendicular. (b) Light entering medium in which its speed is *faster* bends *away* from the perpendicular.

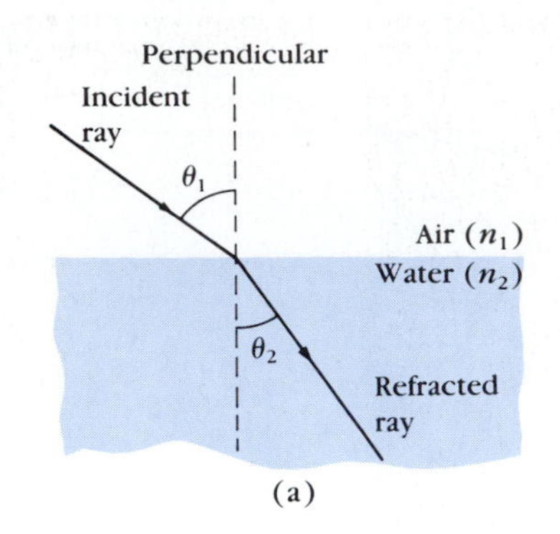

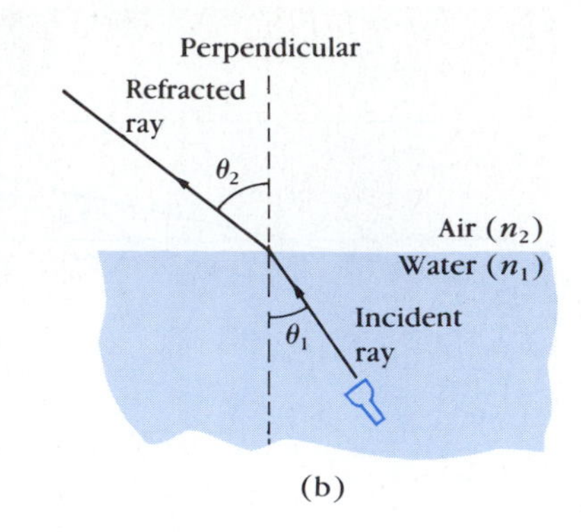

3. REFRACTION OF LIGHT

When light passes from one medium into another, part of the incident light is reflected at the boundary. The remainder passes into the new medium. If a ray of light is incident at an angle to the surface (other than perpendicular), the ray is bent as it enters the new medium. This bending is called **refraction**. Figure 22-14(a) shows a ray passing from air into water. The angle θ_1 is the *angle of incidence* and θ_2 is the *angle of refraction*. Notice that the ray bends toward the perpendicular when entering the water. This is always the case when the ray enters a medium where the speed of light is less. If light travels from one medium into a second where its speed is greater, the ray bends away from the perpendicular; this is shown in Figure 22-14(b) for a ray traveling from water to air. The only exceptions are when the light strikes the second medium perpendicular to its surface; the light then passes straight through without bending at all. Figure 22-14 illustrates another important fact: the reversibility of light rays. If a light ray were directed back along the path by which it had come, it would follow that same path even after refraction.

FIGURE 22-15
The penny *seems* to rise as water is poured into the cup.

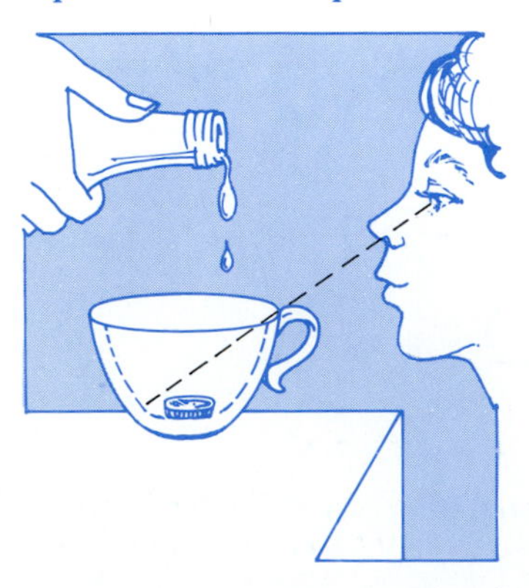

EXPERIMENT

Put a penny in the bottom of an empty cup. Position your head so the penny is just out of sight, as shown in Figure 22-15. Now have someone pour water slowly into the cup. Don't move your head. The penny becomes visible and seems to rise. Explain.

Some Optical Illusions Are Due to Refraction of Light

Refraction is responsible for a number of common optical illusions. For example, a person standing in waist-deep water appears to have shortened legs, Figure 22-16(a). As shown in Figure 22-16(b), the rays leaving the person's foot are bent at the surface; the observer's eye (and brain) assumes the rays to have traveled a straight-line path, and so the feet appear to be higher than they really are. Similarly, a swimming fish or a rock on the bottom of a lake seems higher than it actually is, and bodies of

(a)

Foot appears
to be here

(b)

FIGURE 22-16
Because of refraction, a person's legs appear shorter in water.

water do not appear to be as deep as they really are. The apparent bending of a stick placed in a glass of water (try it) is another example of an illusion created by refraction.

Mirages are produced by the bending of light rays in the atmosphere. On hot days, particularly in the desert, a layer of very hot air lies along the ground. Hot air is less dense than cool air, and so the speed of light in this layer is slightly greater than in the cooler air above. Because of refraction, the light rays are bent upward into a distant observer's eye and he "sees" the tree upside down, as if reflected in a lake (Figure 22-17). Motorists often see a mirage of water on the highway some distance ahead of them

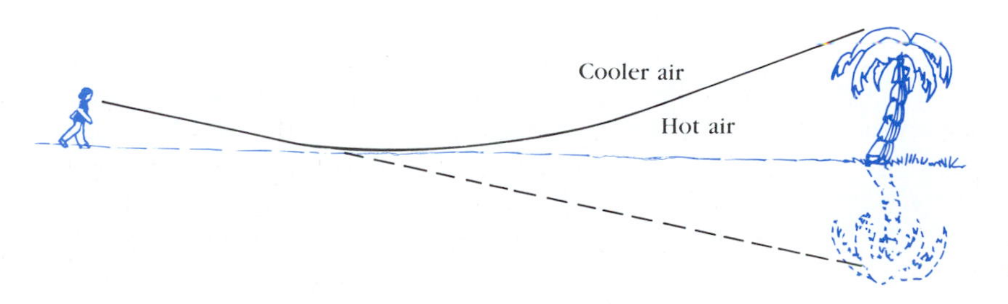

FIGURE 22-17
Mirages are caused by the bending of light in different densities of air.

FIGURE 22-18
A highway mirage. Vehicles appear to be reflected in water on the surface of the road. When you reach that point you find there actually is no water there.

(Figure 22-18). It looks as though the sky were reflected in this body of water; but in fact the motorist is seeing the sky directly, because of refraction.

4. TOTAL INTERNAL REFLECTION AND FIBER OPTICS

When light passes from one material into a second material where the index of refraction is less (say from water into air), the light bends away from the normal, as for ray A in Figure 22-19. At a particular incident angle, the angle of refraction will be 90°, and the refracted ray would skim the surface (ray B) in this case. The incident angle at which this occurs is called the *critical angle*. For any incident angle less than the critical angle, there will be a refracted ray, although part of the light will also be reflected at the boundary. However, for incident angles greater than the critical angle, there is no refracted ray at all, and *all of the light is reflected*, as for ray C in the diagram. This is called **total internal reflection**.

The critical angle depends on the speed of light in each of the two materials. For light traveling in water toward air, the critical angle is 49°. For light traveling in glass whose index of refraction is $n = 1.5$, the critical angle is 42°.

Total internal reflection can occur at a surface only if the speed of light in the material is less than the speed beyond the surface. For example, light passing from air to water does not undergo total internal reflection; but it will if the light is going from water toward air and the incident angle exceeds 49°. There are many interesting applications of this phenomenon. For example, when you look up from beneath the surface of the water in a lake, you see the outside world compressed into a circle whose edge makes a 49° angle with the vertical (Figure 22-20). And if the water is smooth, you will see reflections from the sides or bottom at angles greater than 49°.

FIGURE 22-19
Since $n_2 < n_1$, light rays are totally internally reflected if the incident angle is greater than the critical angle, as for ray C.

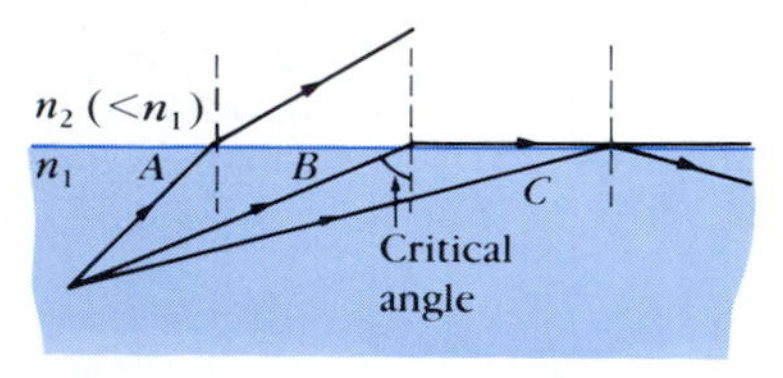

FIGURE 22-20
View looking upward from beneath the water; the surface of the water is smooth.

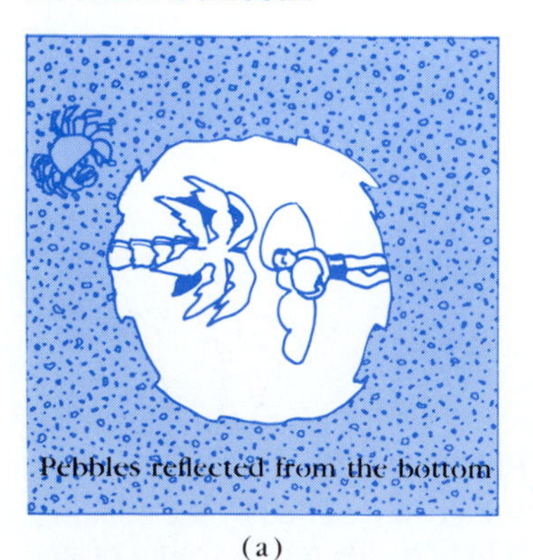

(a)

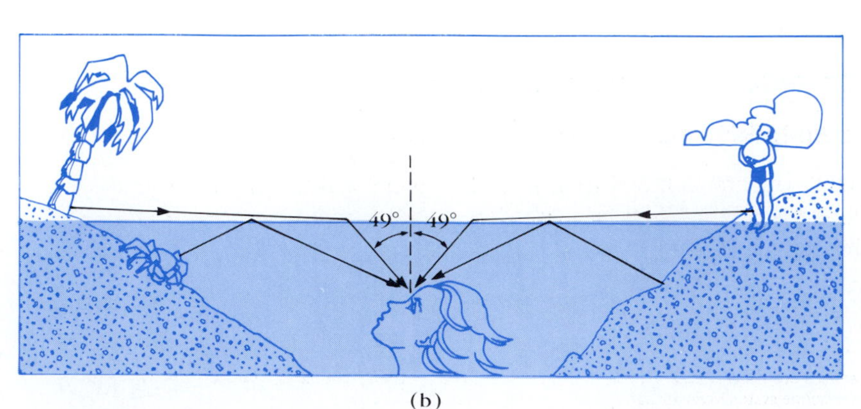

(b)

Many optical instruments, such as binoculars, use total internal reflection within a prism to reflect light. The advantage is that very nearly 100 percent of the light is reflected, whereas even the best mirrors reflect somewhat less than 100 percent. Thus the image is brighter. For glass with $n = 1.50$, the critical angle is 42°. Therefore, 45° prisms will reflect all the light internally as shown in the binoculars of Figure 22-21.

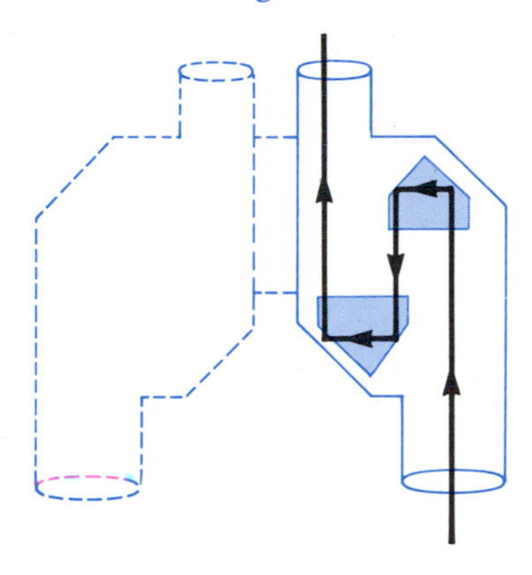

FIGURE 22-21
Prisms reflect light in binoculars.

Total internal reflection is the principle behind *fiber optics*. Very thin glass and plastic fibers can now be made as small as a few micrometers in diameter. A bundle of such tiny fibers is called a *light pipe* since light can be transmitted along it with almost no loss. Figure 22-22 shows how light traveling down a thin fiber makes only glancing collisions with the walls so that total internal reflection occurs. Even if the light pipe is bent into a complicated shape, the critical angle won't (usually) be exceeded so light is transmitted practically undiminished to the other end. This effect is used in decorative lamps and to illuminate water streams in fountains. Light pipes can be used to illuminate difficult places to reach, such as inside the human body. They can be used to transmit telephone calls and other communication signals; the signal is a modulated light beam (variable intensity of the light beam), and is transmitted with less loss than an electrical signal in a copper wire. One sophisticated use of fiber optics, particularly in medicine, is to transmit a clear picture (Figure 22-23). For example, a patient's stomach can be examined by inserting a light pipe through the mouth. Light is sent down one set of fibers to illuminate the stomach. The reflected light returns up another set of fibers. Light directly in front of each fiber travels up that fiber. At the opposite end, a viewer sees a series of bright and dark spots, much like a TV screen—that is, a picture of what lies at the opposite end.† The fibers must be optically insulated from one another, usually by a thin coating of material whose refractive index is less than that of the fiber. The fibers must be arranged precisely parallel to one another if the picture is to be clear. The more fibers there are, and the smaller they are, the more detailed the picture. Such an "endoscope" is useful for observing the stomach and other hard-to-reach places for surgery or searching for lesions without surgery.

FIGURE 22-22
Light reflected totally at the interior surface of a glass or transparent plastic fiber.

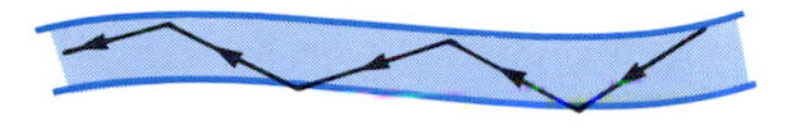

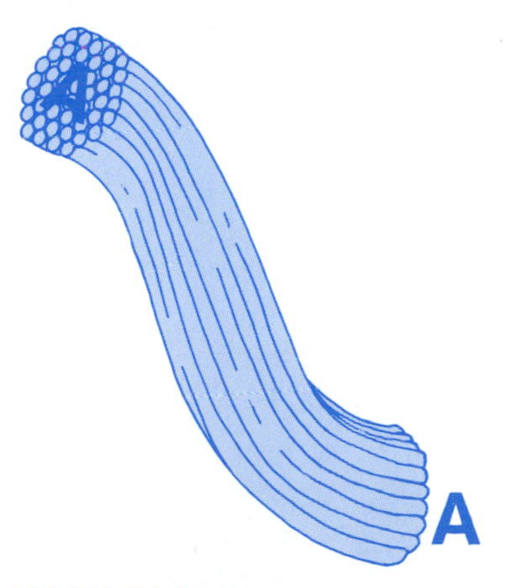

FIGURE 22-23
Fiber optic image.

SUMMARY

Light appears to travel in straight-line paths called *rays* at a speed v that depends on the *index of refraction*, n, of the material: $v = c/n$, where c is the speed of light in vacuum (300,000 km/s).

† Lenses are used at each end: at the object end to bring the rays in parallel and at the viewing end as a telescope system for viewing.

When light reflects from a flat surface, the angle of reflection equals the angle of incidence. This *law of reflection* explains why mirrors can form *images*. In a plane mirror, the image is virtual, upright, the same size as the object, and as far behind the mirror as the object is in front. Spherical mirrors can be concave or convex. A concave mirror can produce an image larger or smaller than the object, depending on where the object is placed. The *focal length* is that distance from the mirror where parallel rays (say, from an object at infinity) come to a focus. A convex mirror forms as image smaller than the object, and can cover a large field of view.

When light passes from one transparent medium into another, the rays bend or *refract*. If the rays do not strike the boundary perpendicularly, they bend toward the perpendicular if the speed of light is slower in the second medium; they bend away from the perpendicular if the speed of light is greater in the second medium.

If the speed of light is greater in the second medium, there is a *critical angle* beyond which *total internal reflection* occurs. In this case all the light is reflected from the surface and none passes through.

QUESTIONS

1. What is a shadow?
2. How does the size of your image in a flat mirror change with your increased distance from the mirror?
3. Provide arguments to show why the moon must have a rough surface rather than a polished mirrorlike surface.
4. When you look at yourself in a tall plane mirror, you see the same amount of your body whether you are close to the mirror or far away. (Try it and see.) Use ray diagrams to show why this is true.
5. When you look at the moon's reflection from a ripply sea it appears elongated. Explain.
6. Using the rules for the three rays discussed with reference to Figure 22-10, draw ray 2 for Figure 22-13(b).
7. What is the focal length of a plane mirror?
8. Does the mirror equation hold true for a plane mirror? Explain.
9. What is the magnification of a plane mirror?
10. When a wide beam of parallel light enters water at an angle, the beam broadens. Explain.
11. What is the angle of refraction when a light ray meets the boundary between two materials perpendicularly?
12. When you look down into a swimming pool or a lake, are you likely to underestimate or overestimate its depth? Explain. How does the apparent depth vary with the viewing angle? (Use ray diagrams.)

13. Draw a ray diagram to show why a stick appears bent when part of it is under water.
14. How are you able to "see" a round drop of water on a table even though the water is transparent and colorless?
15. Can a light ray traveling in air be totally reflected when it strikes a smooth water surface if the incident angle is right?
16. When light passes from glass into water, does the beam bend toward or away from the perpendicular?
17. Draw a ray diagram showing how the highway mirage of Figure 22-18 can occur. (*Hint*: See Figure 22-17.)
18. On certain occasions, sailors have reported seeing a ship or an island floating in the sky above the horizon. Explain how this could happen.
19. Stars seem to twinkle when seen from the earth, but not when seen by astronauts beyond the earth. Why?
20. When you look up at an object in the air from under water, does it seem to be the same size as when you look at it from out of the water?
21. The critical angle for total internal reflection in glass is smaller than in water because the speed of light is less in glass than in water. Explain.
22. Diamonds sparkle as a result of total internal reflection. Why do you see different colors? How does the very low speed of light in diamonds (Table 22-1) affect their "brilliance"?

EXERCISES

1. What is the speed of light in (a) ethyl alcohol (b) lucite?
2. The speed of light in ice is 2.29×10^8 m/s. What is the index of refraction of ice?
3. Light is emitted from an ordinary light bulb filament in wave-train bursts of about 10^{-8}s in duration. What is the length in space of such wave trains?
4. How long does it take light to reach us from the sun, 1.49×10^8 km away?
5. Our nearest star (other than the sun) is 4.2 light years away. That is, it takes 4.2 years for the light to reach earth. How far away is it in meters?
6. How far apart would Galileo and his assistant have had to be in order for the light to take one second to travel over and back between them?
7. What is the speed of light in air expressed in miles per hour?
8. How long did it take light to travel the 70 km in Michelson's experiment, Figure 22-2?

FIGURE 22-24

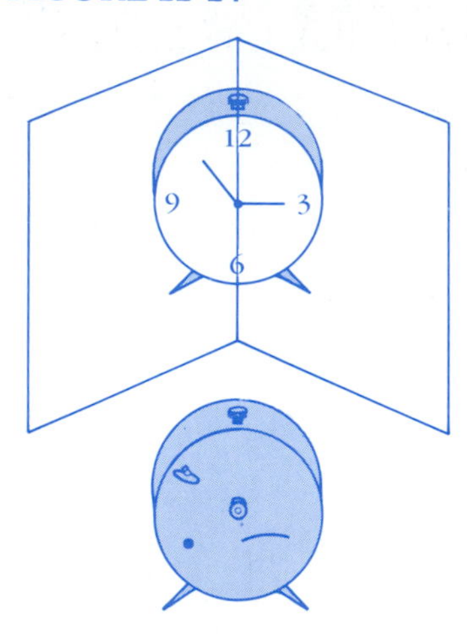

*9. Using your answer to exercise 8, calculate how many revolutions per second the mirror must rotate so that a new mirror face will just be in position to reflect the light into the observer's eye.

10. Suppose that you want to take a photograph of yourself as you look at your image in a mirror 2.0 m away. For what distance should the camera lens be focused?

11. What is the minimum-height plane mirror for which a person 1.84 m tall can see his or her whole self?

12. Stand up two plane mirrors so they form a right angle as in Figure 22-24. When you look into this double mirror you see yourself as others see you, instead of reversed as in a single mirror. Make a ray diagram to show why.

13. What is the radius of a concave reflecting surface that brings parallel light to a focus 13.8 cm in front of it?

14. What is the focal length of a 30-cm-diameter crystal ball whose surface reflects light?

15. How far from a concave mirror (radius 40 cm) must an object be placed if its image is to be at infinity?

16. Show with diagrams that the magnification of a concave mirror is less than 1 if the object is beyond the center of curvature C and is greater than 1 if it is within this point.

17. You try to look at yourself in a silvered ball of diameter 60 cm when you are 2.0 m away. Where is your image? Is it real or virtual? Can you see yourself clearly?

18. A dentist wants a small mirror that, when 2.0 cm from a tooth, will produce a 6.0 × upright image. What kind of mirror must be used and what must its radius of curvature be?

*19. A woman 1.60 m tall stands in front of a vertical plane mirror. What is the minimum height of the mirror and how high must its lower edge be above the floor if she is to be able to see her whole body? (Assume that her eyes are 10 cm below the top of her head.)

*20. A convex rearview car mirror has a radius of curvature of 40.0 cm. Determine the location of the image and its magnification for an object 10.0 m from the mirror.

*21. A 1.50-cm-high object is placed 20.0 cm from a concave mirror whose radius of curvature is 30.0 cm. Determine (a) the position of the image and (b) its size.

*22. A 2.0-cm-tall object is placed 20 cm from a spherical mirror. It produces a virtual image 3.0 cm high. (a) What type of mirror is being used? (b) Where is the image located? (c) What is the radius of curvature of the mirror?

LENSES AND OPTICAL INSTRUMENTS

CHAPTER

23

The most important simple optical device is no doubt the thin lens. The development of optical devices using lenses dates to the sixteenth and seventeenth centuries, although the earliest record of eyeglasses dates from the late thirteenth century. Today we find lenses in eyeglasses, cameras, magnifying glasses, telescopes, binoculars, microscopes, and medical instruments. A thin lens is usually circular and its two faces are portions of a sphere. (Although cylindrical surfaces are also possible, we will concentrate on spherical.) The two faces can be concave, convex, or plane (Figure 23-1). The importance of lenses is that they form images of objects.

We first look at the remarkable properties of lenses and how they can form images; and then we discuss how some basic optical devices work, including that magnificent one, the human eye.

FIGURE 23-1
Types of lenses.

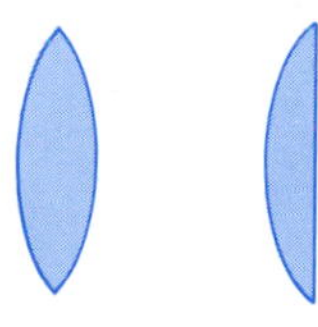

(a) Converging lenses

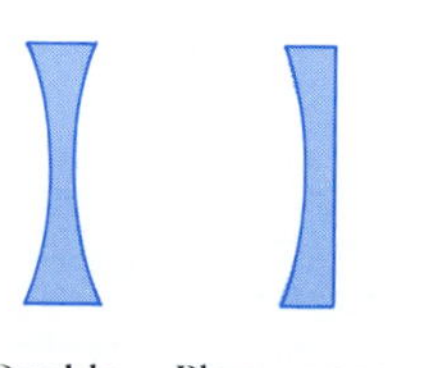

(b) Diverging lenses

1. LENSES AND IMAGES

Focal Point; Converging and Diverging Lenses

Consider the rays parallel to the axis of the double convex lens which is shown in cross section in Figure 23-2(a). We assume the lens is made of glass or transparent plastic so its index of refraction is greater than that of the outside air. From the law of refraction we can see that each ray is bent toward the axis at both lens surfaces (note the dashed lines indicating the perpendicular to each surface for the top ray). If parallel rays fall straight onto a thin lens as in Figure 23-2(a), they will be focused to a point called the **focal point**, F. (This will not be precisely true for a lens with spherical surfaces. But it will be very nearly true if the diameter of the lens is small compared to the radii of curvature of the two lens surfaces. This criterion

FIGURE 23-2
Parallel rays are brought to a focus by a converging lens.

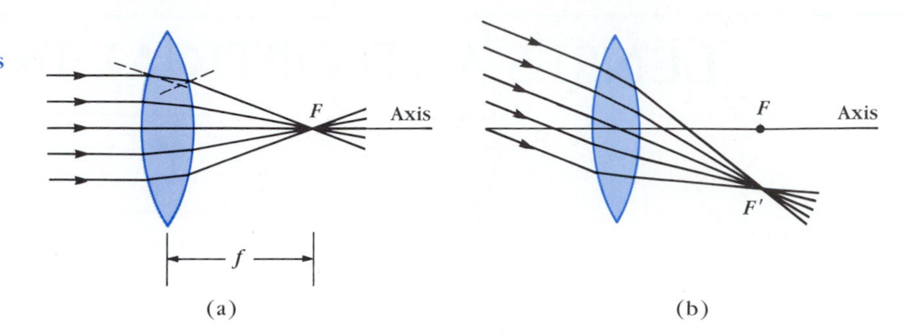

is satisfied by a *thin lens*, one that is very thin compared to its diameter, and it is only thin lenses that we consider here.)

Since the rays from a distant object are essentially parallel (see Figure 22-8), we can also say that *the focal point is the image point for an object at infinity*. Thus the focal point of a lens can be found by locating the point where the sun's rays (or those of some other distant object) are brought to a sharp image. The distance of the focal point from the center of the lens is called the **focal length**, f. A lens can be turned around so light can pass through it from the opposite side; the focal length is the same on both sides. If parallel rays fall on a lens at an angle, as in Figure 23-2(b), they focus at a point F'. The plane in which all points such as F and F' fall is called the *focal plane* of the lens.

Any lens which is thicker in the center than at the edges will make parallel rays converge to a point, and is called a *converging lens*, see Figure 23-1(a). Lenses which are thinner in the center than at the edges, Figure 23-1(b), are called *diverging lenses* because they make parallel light diverge, as shown in Figure 23-3. The *focal point F* of a diverging lens is defined as that point from which refracted rays, originating from parallel incident rays, seem to emerge as shown in the figure. And the distance from F to the lens is called the focal length, just as for a converging lens. The focal length is a measure of how "strong" a lens is; that is, how much it bends light rays. Thus the focal length is the most important specification of a lens. The focal length is determined by the curvature of the two surfaces and is the same on both sides of a lens; it is thus unchanged if the lens is turned around.

FIGURE 23-3
Diverging lens.

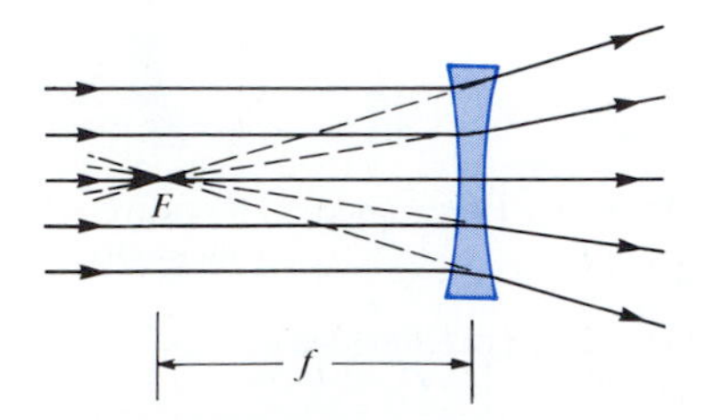

Ray Tracing to Find the Image

To find the image point by drawing rays would be difficult if we had to determine all the refractive angles. Instead, we can do it very simply by making use of certain facts we already know, such as that a ray falling straight on the lens passes (after refraction) through the focal point. In fact, we need consider only the three rays indicated in Figure 23-4. These three rays, emanating from a single point on the object, are drawn as if the lens were infinitely thin. Ray 1 is drawn parallel to the axis; therefore it is refracted by the lens so it passes through the focal point F behind the lens.

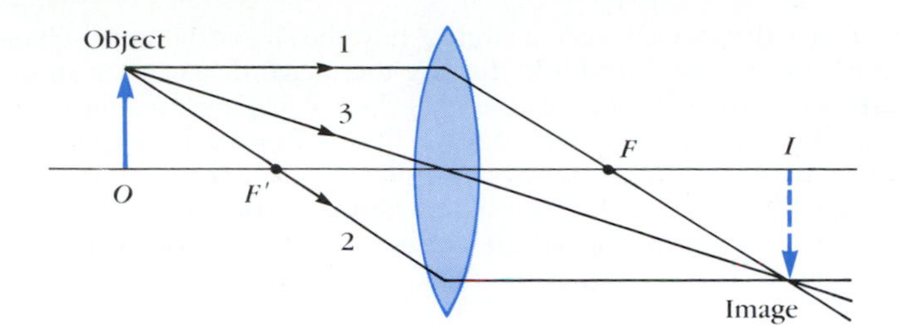

FIGURE 23-4
Finding image I by ray tracing—converging lens.

Ray 2 is drawn through the focal point F' on the same side of the lens as the object; it therefore emerges from the lens parallel to the axis. Ray 3 is directed toward the very center of the lens where the two surfaces are essentially parallel to each other; this ray therefore emerges from the lens at the same angle as it entered. Actually, any two of these rays will suffice to locate the image point, which is the point where they intersect. Drawing the third can serve as a check.

In this way we can find the image point for one point of the object (the top of the arrow in Figure 23-4). The image points for all other points on the object can be found similarly to determine the complete image of the object. Because the rays actually pass through the image for the case shown, it is a *real image*.

By drawing the same three rays we can determine the image position for a diverging lens, as shown in Figure 23-5. Note that ray 1 is drawn parallel to the axis, but does not pass through the focal point F behind the lens; instead it seems to come from the focal point F' in front of the lens (dashed line). Ray 2 is directed toward F and is refracted parallel by the lens. The three refracted rays seem to emerge from a point on the left of the lens. This is the image, I. Since the rays do not pass through the image, it is a *virtual image*.

EXPERIMENT—PROJECT

Obtain a simple lens—the lens from a pair of reading glasses, a magnifying glass, or a camera or projector lens. Use a light bulb as the object and project its image onto a piece of white paper. Be sure to position the paper so the image is sharp. Try various lens-to-object distances and notice how the

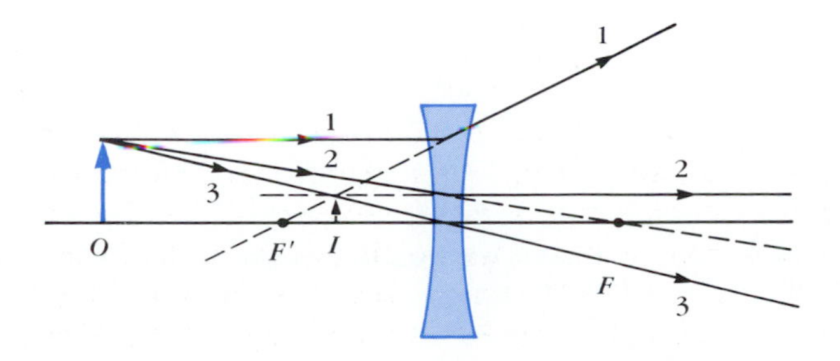

FIGURE 23-5
Finding image I by ray tracing—diverging lens.

lens-to-image distance changes, as well as how the size of the image changes. Also measure the focal length by finding the image distance for an object very far away. (Note: Do not put your eye where you expect the image to be; put the white paper there and observe the real image on paper.)

AND

Use the lens as a magnifier. Determine where an object must be placed so that the lens produces the largest image. (This time observe the image directly with your eye.) Is it at the focal point, beyond it, or in front of it? Also, determine the magnification of the lens by observing a piece of lined paper directly and through the lens. Compare the spacing between the lines.

The Lens Equation (optional)

The use of ray tracing to find the position of an image can be time consuming, especially if it is done accurately. The image position can be determined more quickly and accurately using the **lens equation**:

$$\frac{1}{d_o} + \frac{1}{d_i} = \frac{1}{f}.$$

This equation is easily derived using Figure 23-6.[†]

The lens equation relates the image distance d_i to the object distance d_o and the focal length f. It is the most useful equation in geometric optics. (Interestingly, it is exactly the same as the "mirror equation" of Chapter 22.) Note that if the object is at infinity, then $1/d_o = 0$, so $d_i = f$. Thus the focal length is the image distance for an object at infinity, as mentioned earlier.

The lens equation is valid for both converging and diverging lenses, and for *all* situations, but only if we use the following conventions:

1. The focal length is positive for converging lenses and negative for diverging lenses.

[†] For a converging lens, Figure 23-6(a), we consider the two rays shown, where h_o and h_i refer to the heights of the object and image, and d_o and d_i are their distances from the lens. The triangles $FI'I$ and FBA are similar, so

$$\frac{h_i}{h_o} = \frac{d_i - f}{f}$$

since length $AB = h_o$. Triangles OAO' and IAI' are similar. Therefore,

$$\frac{h_i}{h_o} = \frac{d_i}{d_o}.$$

We equate these two, divide by d_i, and rearrange to obtain

$$\frac{1}{d_o} + \frac{1}{d_i} = \frac{1}{f},$$

which is the lens equation. The derivation for a diverging lens is similar, Figure 23-6(b). (But consider the "conventions" discussed in a moment.)

2. The object distance is positive if it is on the side of the lens from which the light is coming (this is normally the case, although when lenses are used in combination, it might not be true); otherwise, it is negative.
3. The image distance is positive if it is on the opposite side of the lens from where the light is coming; if it is on the same side, d_i is negative.

The *magnification m* of a lens is defined as the ratio of the image height to object height, $m = h_i/h_o$. From Figure 23-6 we can see that

$$m = \frac{h_i}{h_o} = \frac{d_i}{d_o}.$$

As an example, let us determine the position and size of the image of a 20-cm-high giant insect placed 1.50 m from a +50-mm-focal-length camera

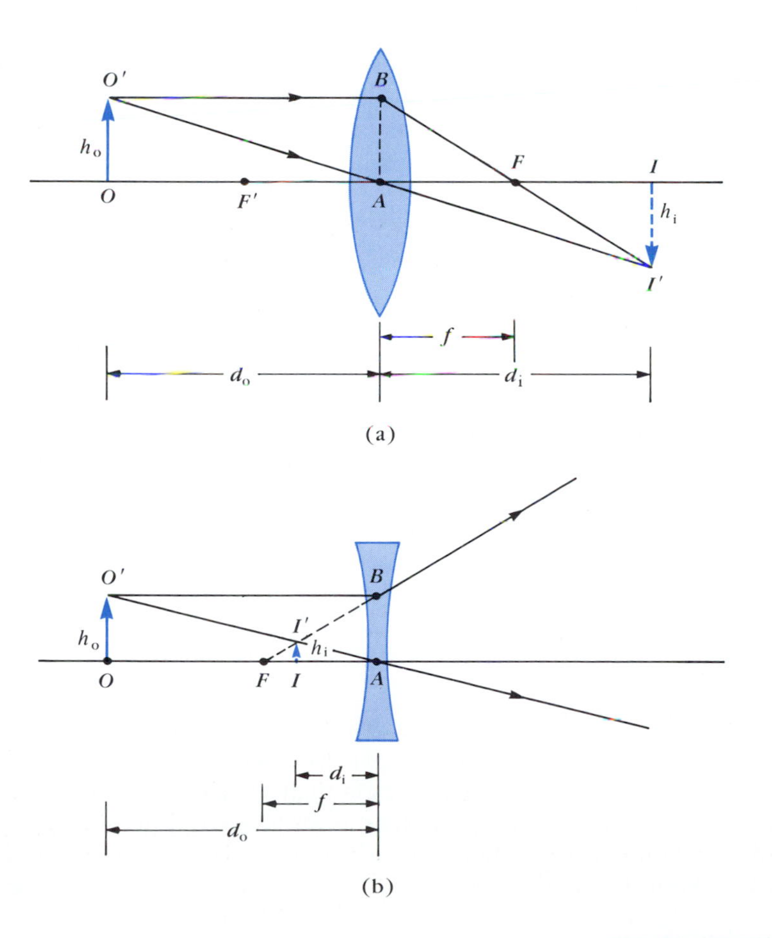

FIGURE 23-6
Deriving the lens equation for (a) converging and (b) diverging lenses.

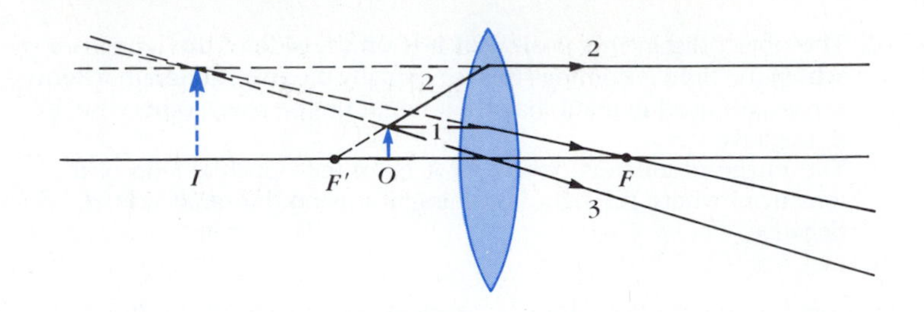

FIGURE 23-7
An object placed within the focal point of a converging lens produces an image on the same side of the lens as the object, and it is virtual.

lens. The camera lens is converging, with $f = +5.0$ cm, so the lens equation gives

$$\frac{1}{d_i} = \frac{1}{f} - \frac{1}{d_o} = \frac{1}{5.0\text{ cm}} - \frac{1}{150\text{ cm}} = \frac{30 - 1}{150\text{ cm}};$$

thus $d_i = 150\text{ cm}/29 = 51.7$ mm or 5.17 cm behind the lens. The magnification $m = d_i/d_o = 5.17\text{ cm}/150\text{ cm} = 0.0345$, so $h_i = (0.0345)(20\text{ cm}) = 0.69$ cm. Thus the image is 6.9 mm high. Notice that the image is 1.7 mm farther from the lens than would be the image for an object at infinity. This is an example of the fact that when focusing a camera lens, the closer the object is to the camera the farther the lens must be from the film (see following section).

As another example, suppose an object is placed 10 cm from a 15-cm-focal-length converging lens; let us determine the image position and size (a) analytically, (b) using a ray diagram. (a) Since $f = 15$ cm and $d_o = 10$ cm,

$$\frac{1}{d_i} = \frac{1}{15\text{ cm}} - \frac{1}{10\text{ cm}} = \frac{2}{30\text{ cm}} - \frac{3}{30\text{ cm}} = -\frac{1}{30\text{ cm}};$$

so $d_i = -30$ cm. Since d_i is negative, the image must be on the same side of the lens as the object. The magnification $m = (30\text{ cm})/(10\text{ cm}) = 3.0$; so the image is three times as large as the object. (b) The ray diagram is shown in Figure 23-7 and confirms the result in part (a).

FIGURE 23-8
A simple camera.

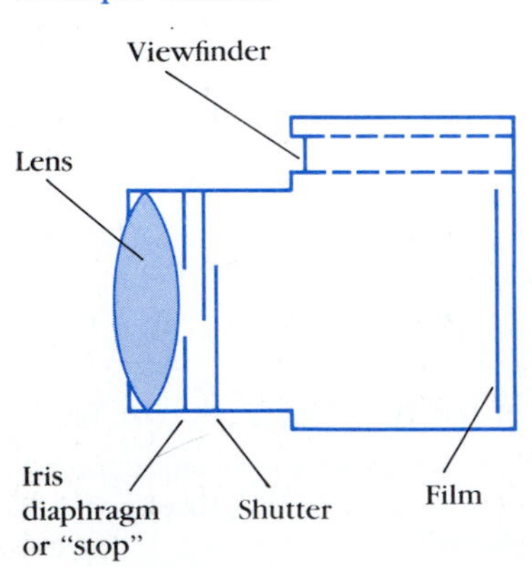

2. THE CAMERA

The basic elements of a camera are a lens, a lightproof box, a shutter to let light pass through the lens, and a sensitized plate or piece of film (Figure 23-8). When the shutter is opened briefly, light from external objects in the field of view are focused by the lens as an image on the film. The film contains light-sensitive chemicals which undergo change when light strikes them. In the development process chemical reactions cause the changed areas to turn black so that the image is recorded on the film.[†] You can see an image yourself by opening the camera back and viewing through a piece of tissue or wax paper at the position of the film with the shutter open.

There are three main adjustments on good-quality cameras: shutter speed, *f*-stop, and focusing; we now discuss them.

Shutter speed. This refers to how long the shutter is open and the film exposed. It may vary from a second or more ("time exposures") to 1/1000 s or less. To avoid blurring from camera movement, speeds faster than $\frac{1}{50}$ s are normally used; if the object is moving, faster shutter speeds are needed to "stop" the action.

f-stop. The amount of light reaching the film must be carefully controlled to avoid *underexposure* (too little light for any but the brightest objects to show up) or *overexposure* (too much light, so that all bright objects look the same, with a consequent lack of contrast and a "washed-out" appearance). To control the exposure, a "stop" or iris diaphragm of variable diameter is placed behind the lens (Figure 23-8). The size of the opening is varied to compensate for bright or dark days and for different shutter speeds. The opening is specified by the *f-stop*, defined as

$$f\text{-stop} = \frac{f}{D},$$

where f is the focal length of the lens and D the diameter of the opening. For example, a 50-mm-focal-length lens set at $f/2$ has an opening $D = 25$ mm; when set at $f/8$, the opening is only $6\frac{1}{4}$ mm. The faster the shutter speed (or the darker the day), the greater the opening must be to get a proper exposure. This corresponds to a smaller *f*-stop number. The smallest *f*-number of a lens is referred to as the "speed" of the lens. It is common to find $f/2.0$ lenses today, and even some as fast as $f/1.0$. Fast lenses are expensive to make and require many elements in order to reduce the defects present in simple thin lenses. The advantage of a fast lens is that it allows pictures to be taken under poorly lighted conditions.

Lenses normally stop down to $f/16$, $f/22$, or $f/32$. Although the lens opening can usually be varied continuously, there are nearly always markings for specific lens openings: the standard *f*-stop markings are 1.0, 1.4, 2.0, 2.8, 4.0, 5.6, 8, 11, 16, 22, and 32. Notice that each of these corresponds to a diameter of about $\sqrt{2} = 1.4$ times smaller than the previous one. Since the amount of light reaching the film is proportional to the area of the opening (and therefore proportional to the diameter squared), we see that each standard *f*-stop corresponds to a factor of 2 in light intensity reaching the film.

[†] This is called a *negative*, since the black areas correspond to bright objects and vice versa. The reverse process occurs during printing to produce a black-and-white ("positive") picture from the negative. Color film makes use of three dyes corresponding to the primary colors.

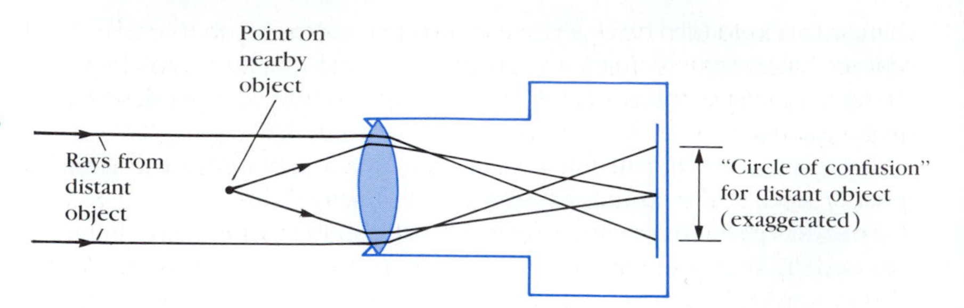

FIGURE 23-9
When the lens is positioned to focus on a nearby object, points on a distant object produce circles and are therefore blurred. (The effect is greatly exaggerated.)

Focusing is the operation of placing the film at the correct position for the sharpest image. The image distance is a minimum for objects at infinity (the symbol ∞ is used for infinity) and is equal to the focal length. For closer objects the image distance is greater than the focal length. This steeper angle with the lens, as shown in Figure 23-9, so that the angle of the rays leaving the lens will not be as great and the image is formed farther back. (This can also be seen from the lens equation, $1/f = 1/d_0 + 1/d_i$.) To focus on nearby objects, the lens must therefore be moved away from the film, and this is usually done by turning a ring on the lens.

If the lens is focused on a nearby object, a sharp image of it will be formed. But the rays from an object far away will be out of focus—they will form a circle on the film as shown (exaggerated) in Figure 23-9; the distant object will thus produce an image consisting of overlapping circles and will be blurred. These circles are called *circles of confusion*. If you want to have near and distant objects sharp at the same time, you can set the lens focus at an intermediate position. Neither near nor distant objects will then be perfectly sharp, but the circles of confusion may be small enough that the blurriness is not noticeable.

For a given distance setting, there is a range of distances over which the circles of confusion will be small enough that the images will be reasonably sharp. This is called the *depth of field*. For a particular circle of confusion diameter (typically taken to be 0.03 mm for 35-mm cameras), the depth of field depends on the lens openings. If the lens opening is smaller the circles of confusion will be smaller, since only rays through the central part of the lens are accepted, and these form a smaller circle of confusion (Figure 23-9). Hence at smaller lens openings, the depth of field is greater.

Other factors also affect the sharpness of the image, such as the graininess of the film, diffraction, and lens aberrations relating to the quality of the lens itself.

Lens Focal Length and Magnification

Camera lenses are categorized into normal, telephoto, and wide-angle, according to focal length and film size. A *normal lens* is one that covers the film with a field of view that corresponds approximately to that of normal vision. A normal lens for 35-mm film has a focal length in the vicinity of 50 mm. A *telephoto* lens, as its name implies, acts like a telescope to magnify images. They have longer focal lengths than a normal lens. As we saw in Section 1, the height of the image for a given object distance is proportional to the image distance; and the image distance will be greater for a lens with a longer focal length. For distant objects, the image height is very nearly proportional to the focal length (can you prove this?). Thus a 200-mm telephoto lens for use with a 35-mm camera gives a $4\times$ magnification over the normal 50-mm lens. A *wide-angle* lens has a shorter focal length than normal. A wider field of view is included and objects appear smaller. A *zoom lens* is one whose focal length can be changed so that you can zoom up to or away from the subject.

Wax paper

FIGURE 23-10
A pinhole camera.

EXPERIMENT 

Make a "pinhole camera" by removing one of the ends of a box and covering the opening with a piece of wax paper (or tissue paper). Now make a clean small hole in the opposite end (Figure 23-10). Stand in a darkened room and point your camera toward a bright scene outdoors. You will see the image of the scene on the wax paper. Draw a ray diagram to show how the image is formed. If you make the hole smaller, the image becomes clearer but dimmer. Why?

High-Quality Lenses Consist of Several Simple Lenses

It is difficult to design a lens that will have a large opening (small *f*-stop number) and that will still give a clear picture. For example, a simple thin lens does not focus all the rays from a single point on an object at one point, not even when the rays are parallel rays from a distant object. Rays that strike the edge of the lens are bent more and more radically, as shown in Figure 23-11(a). This is called "spherical aberration." Only the central part of a simple lens is usable to produce clear pictures. However, by combining several thin lenses (called "elements"), as in Figure 23-11(b), we can compensate for this and other aberrations. Today complex lenses that give very sharp images are normally designed with the aid of a computer that traces the rays through a hypothetical lens and adjusts the various elements to minimize aberrations.

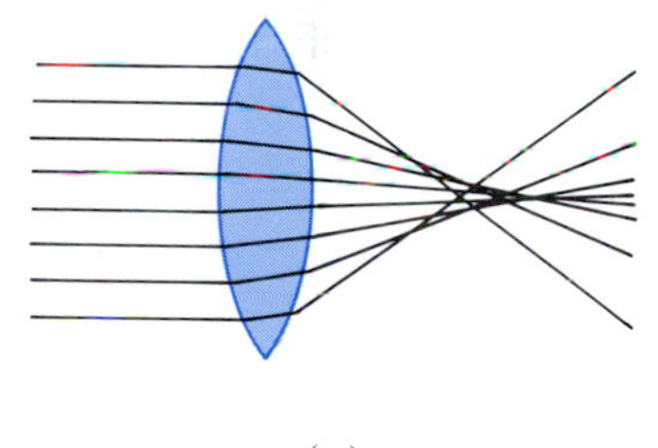

(a)

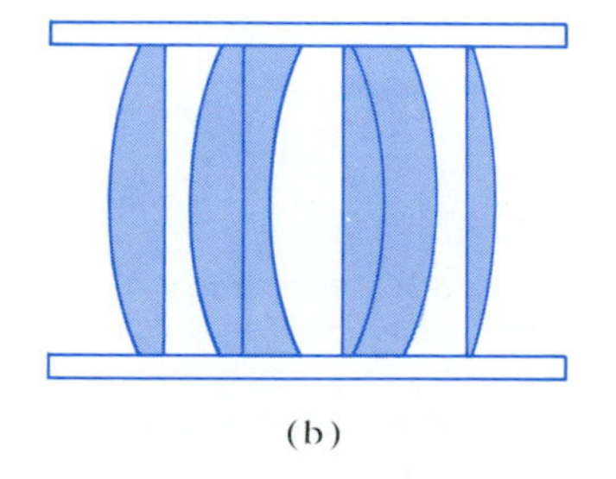

(b)

FIGURE 23-11
(a) Spherical aberration. Single lenses do not bring all the rays to a focus at one point, which leads to a blurry image. (b) A high-quality lens containing six elements.

3. THE HUMAN EYE; CORRECTIVE LENSES

The Eye

The human eye resembles a camera in its basic structure (Figure 23-12). The eye is an enclosed volume into which light passes through a lens. A diaphragm, called the *iris* (the colored part of your eye), adjusts automatically to control the amount of light entering the eye. The hole in the iris

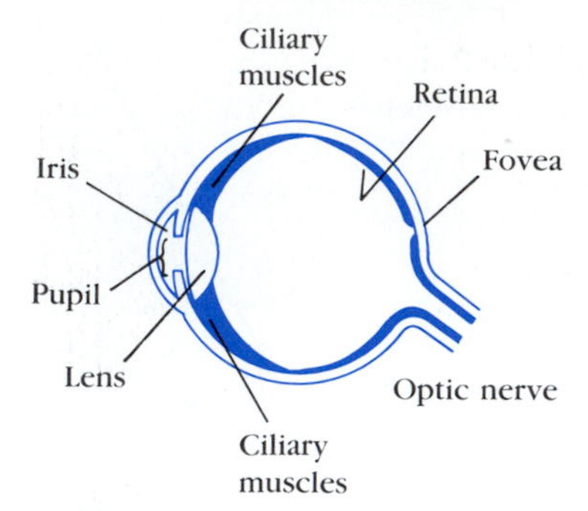

FIGURE 23-12
Diagram of a human eye.

through which light passes (the *pupil*) is black because no light is reflected from it (it's a hole) and very little light is reflected back out from the interior of the eye. The *retina*, which plays the role of the film in a camera, is on the curved rear surface. It consists of a complex array of nerves and receptors known as rods and cones that change light energy into electrical signals that travel along the nerves. The reconstruction of the image from all these tiny receptors is done mainly in the brain, although some analysis is apparently done in the complex interconnected nerve network at the retina itself. At the center of the retina is a small area called the *fovea*, about 0.25 mm in diameter, where the cones are very closely packed and the sharpest image and best color discrimination are found.

Unlike a camera, the eye contains no shutter. The equivalent operation is carried out by the nervous system, which analyzes the signals to form images at the rate of about 30 per second. This can be compared to motion picture or television cameras, which operate by taking a series of still pictures at a rate of 24 (movies) or 30 (television) per second. The rapid projection of these on a screen or television set gives the appearance of motion.

The lens of the eye does little of the bending of the light rays. Most of the refraction is done at the front surface of the *cornea*, which also acts as a protective covering. The lens acts as a fine adjustment for focusing at different distances. This is accomplished by the ciliary muscles (Figure 23-12) which change the curvature of the lens so that its focal length is changed. To focus on a distant object, the muscles are relaxed and the lens is thin—Figure 23-13(a). To focus on a nearby object, the muscles contract causing the center of the lens to be thicker, Figure 23-13(b), thus shortening the focal length. This adjustment is called *accommodation*. The closest distance at which the eye can focus clearly is called the *near point* of the eye. For young adults it is typically 25 cm, although younger children can often focus on objects as close as 10 cm; as people grow older, the ability to accommodate is reduced and the near point increases. A given person's *far point* is the farthest distance at which an object can be seen clearly. A so-called *normal eye* (a sort of average over the population) is defined as one having a near point of 25 cm and a far point of infinity.

The "normal" eye is more of an ideal than a commonplace. A large part of the population have eyes that do not accommodate within the normal range of 25 cm to infinity, or have some other defect. Two common defects are nearsightedness and farsightedness. Both can be corrected to a large extent with lenses—either eyeglasses or contact lenses.

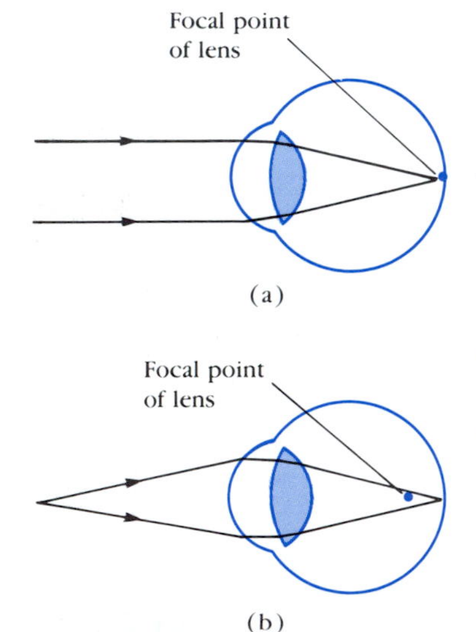

FIGURE 23-13
Accommodation by eye:
(a) the lens is relaxed, focused on infinity. (b) The lens is thickened, focused on nearby object.

Defects of the Eye and Corrective Lenses

A *nearsighted* or *myopic* eye is one that can focus only on nearby objects; that is, the far point is not infinity but some shorter distance. It is usually caused by an eyeball that is too long, although sometimes it is the curvature of the cornea that is too great. In either case, images of distant objects are focused in front of the retina. A diverging lens, because it

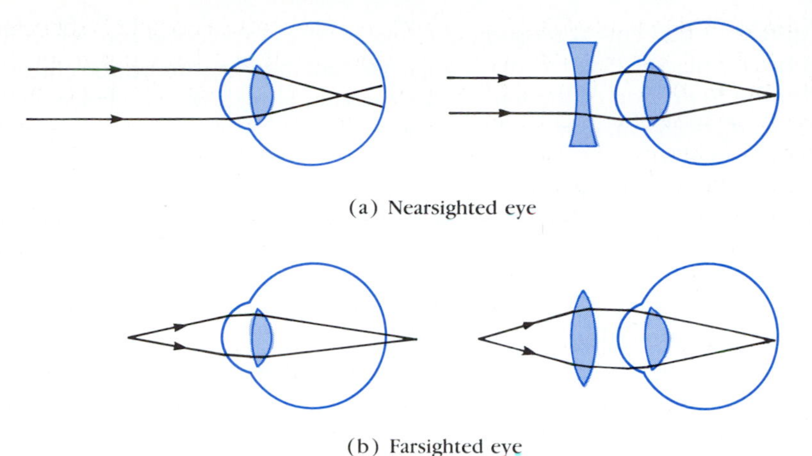

FIGURE 23-14
(a) A nearsighted eye, which cannot focus clearly on distant objects, can be corrected by use of a diverging lens. (b) A farsighted eye, which cannot focus clearly on nearby objects, can be corrected by use of a converging lens.

causes parallel rays to diverge, allows the rays to be focused at the retina—Figure 23-14(a)—and thus corrects this defect. *Farsightedness*, or *hyperopia*, refers to an eye which cannot focus on nearby objects. Although distant objects are usually seen clearly, the near point is somewhat greater than the "normal" 25 cm, which makes reading difficult. This defect is caused by an eyeball that is too short, or (less often) by a cornea that is not sufficiently curved. It is corrected by a converging lens—Figure 23-14(b). Similar to hyperopia is *presbyopia*, which refers to the lessening ability of the eye to accommodate as one ages and the near point moves out. Converging lenses also compensate for this.

Power of a Lens (optional)

Optometrists and ophthalmologists, instead of using the focal length, use the reciprocal of the focal length to specify the strength of lenses. This is called the *power*, P, of a lens:

$$P = \frac{1}{f}.$$

The unit for lens power is the diopter (D) which is an inverse meter: $1 \text{ D} = 1 \text{ m}^{-1}$. A 20-cm-focal-length lens, for example, has a power $P = 1/0.20 \text{ m} = 5.0 \text{ D}$. The power of a converging lens is taken as positive, whereas that for a diverging lens is negative (since f is negative—see convention 1 in Section 1).

Eyeglasses and Contact Lenses—Numerical Examples (optional)

A particular farsighted person has a near point of 100 cm. What lens power must reading glasses have so that this person can read a newspaper at a distance of 25 cm? We assume the lens is very close to the eye; when the object is placed 25 cm from the lens, we want the image to be 100 cm away on the *same* side of the lens, and so it will be as shown in Figure 23-15. Thus $d_o = 25$ cm, $d_i = -100$ cm and

$$\frac{1}{f} = \frac{1}{d_o} + \frac{1}{d_i} = \frac{1}{25\text{ cm}} + \frac{1}{-100\text{ cm}} = \frac{1}{33\text{ cm}}.$$

So $f = 33$ cm $= 0.33$ m. The power P of the lens is $P = 1/f = +3.0$ D. The plus sign indicates it is a converging lens.

As a second example, suppose a nearsighted eye has near and far points of 12 cm and 17 cm, respectively. What lens power is needed for this person to see distant objects clearly, and what then will be the near point? We assume that each lens is 2.0 cm from the eye (typical for eyeglasses); the lens must image distant objects ($d_o = \infty$) so they are 17 cm from the eye, or 15 cm in front of the lens ($d_i = -15$cm):

$$\frac{1}{f} = -\frac{1}{15\text{ cm}} + \frac{1}{\infty} = -\frac{1}{15\text{ cm}}.$$

So $f = -15$ cm $= -0.15$ m or $P = 1/f = -6.7$ D; the minus sign indicates it must be a diverging lens, as we expect—see Figure 23-14(a). For the near point, the image must be 12 cm from the eye or 10 cm from the lens, so $d_i = 0.10$ m and

$$\frac{1}{d_o} = \frac{1}{f} - \frac{1}{d_i} = -\frac{1}{0.15\text{ m}} + \frac{1}{0.10\text{ m}} = \frac{1}{0.30\text{ m}}.$$

So $d_o = 30$ cm, which means the near point when the person is wearing glasses is 30 cm in front of the lens.

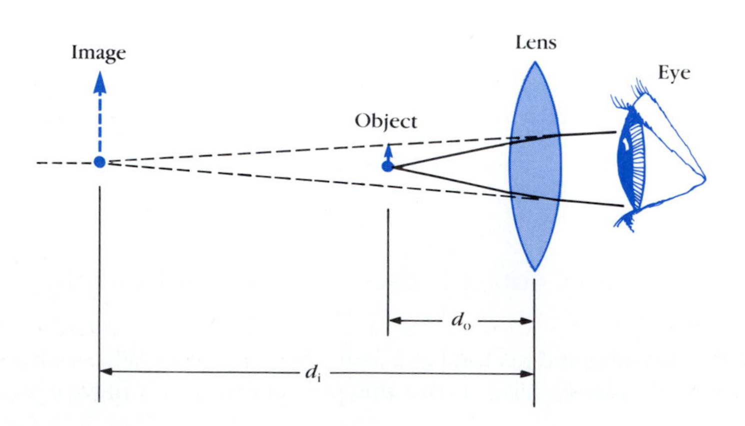

FIGURE 23-15
Lens of a reading glass.

Contact lenses could be used to correct the eye in this example. Since contacts are placed directly on the cornea, we would not subtract out the 2.0 cm. That is, for distant objects $d_i = -17$ cm, so $P = 1/f = -5.9$ diopters. Thus we see that a contact lens and an eyeglass lens will require slightly different focal lengths for the same eye because of their different placement relative to the eye.

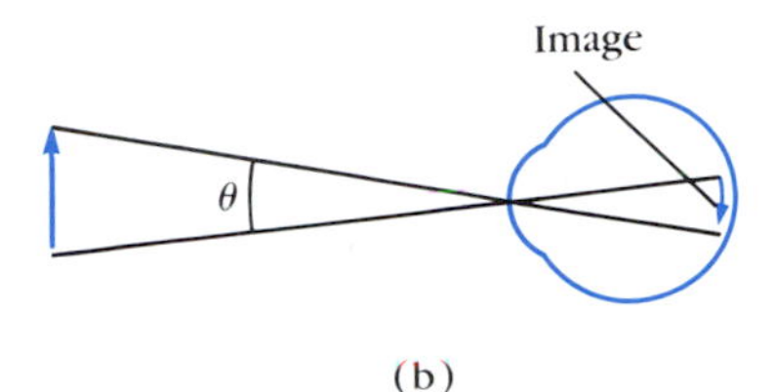

FIGURE 23-16
When the same object is viewed at a shorter distance, the image on the retina is greater; so the object appears larger and more detail can be seen. The angle θ the object subtends in (a) is greater than in (b).

4. THE MAGNIFYING GLASS

The remainder of this chapter will deal with optical devices that are used to produce magnified images of objects. We first discuss the *simple magnifier* or *magnifying glass*, which is simply a converging lens.

How large an object appears and how much detail we can see on it depends on the size of the image it makes on the retina. This, in turn, depends on the angle subtended by the object at the eye. For example, a penny held 30 cm from the eye looks twice as high as one held 60 cm away, because the angle it subtends is twice as great (Figure 23-16). When we want to examine detail on an object, we bring it up close to our eyes so that it subtends a greater angle. However, our eyes can accommodate only up to the near point, which we assume to be the standard distance of 25 cm.

A magnifying glass allows us to place the object closer to our eye so that it subtends a greater angle. As shown in Figure 23-17(a), the object is placed at the focal point or just within it. Then the converging lens produces a virtual image, which must be at least 25 cm from the eye if the eye is to focus on it. If the eye is relaxed, the image will be at infinity, and in this case the object is exactly at the focal point. (You make this slight adjustment yourself when you "focus" on the object by moving the lens.)

A comparison of part (a) of Figure 23-17 with part (b), in which the same object is viewed at the near point with the unaided eye, reveals that the angle the object subtends at the eye is much larger when the magnifier is used.

Magnifying Power (optional)

The **magnifying power**, M, of a lens is defined as the ratio of the angle subtended by the object when viewed through the lens to that subtended by the unaided eye without the lens at a distance of 25 cm:

$$M = \frac{\theta'}{\theta}$$

where θ and θ' are shown in Figure 23-17. This can be written in terms of the focal length f of the lens as follows. (We use θ in radians, as discussed in Chapter 9.) In part (a) of Figure 23-17, the angle θ' can be

FIGURE 23-17
Leaf viewed (a) through a magnifying glass and (b) with the unaided eye focused at its near point.

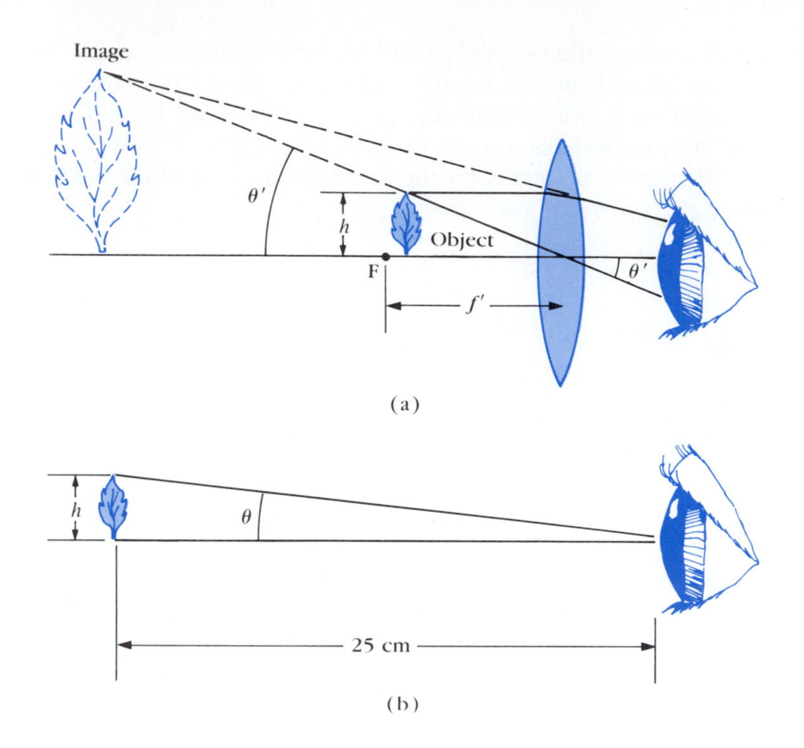

written approximately as $\theta' \approx h/f$ since the object is placed very near the focal point F (precisely at F if the eye is relaxed). In part (b) we see that the angle θ is given by $\theta \approx h/25\text{ cm}$. Thus

$$M = \frac{\theta'}{\theta} = \frac{\dfrac{h}{f}}{\dfrac{h}{25\text{ cm}}} = \frac{25\text{ cm}}{f},$$

where the focal length f of the magnifier must be given in cm. For example, a 10-cm-focal-length lens magnifies 25 cm/10 cm = 2.5 times.

5. THE MICROSCOPE AND THE TELESCOPE

Microscopes are used for much larger magnifications than can be attained with a magnifying glass. A microscope uses two lenses, as shown in Figure 23-18(a). The **objective** lens, the one closer to the object, has a very short focal length and magnifies the object by forming an enlarged real image. The second lens, the **eyepiece**, then magnifies the image made by the first lens, forming a virtual image that can be seen by the eye. The eyepiece thus acts like a simple magnifying glass.

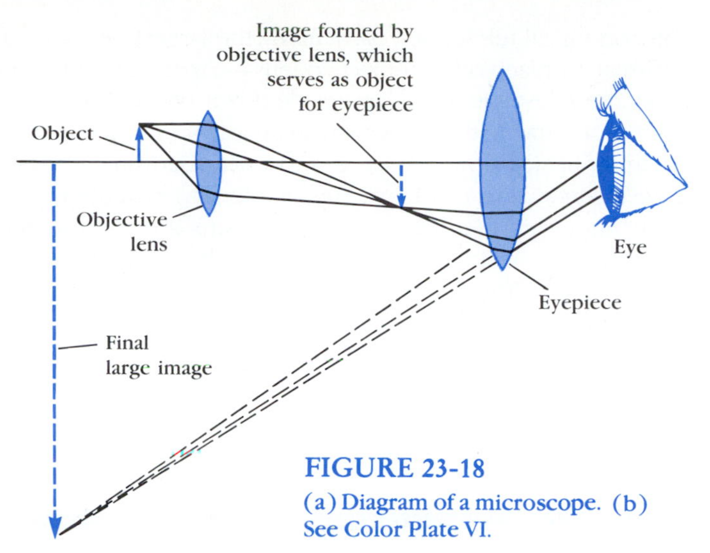

FIGURE 23-18
(a) Diagram of a microscope. (b) See Color Plate VI.

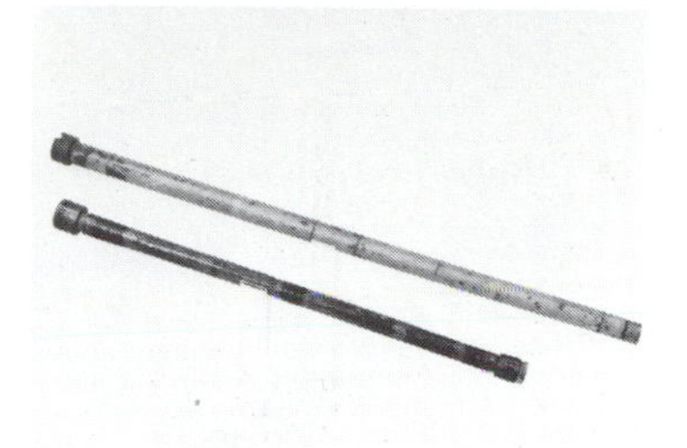

FIGURE 23-19
See Color Plate VI.

A **telescope** is used to magnify objects that are very far away; in most cases the object can be considered to be at infinity. Galileo, although he did not invent it, developed the telescope into a usable and important instrument; he was the first to train it on the heavens where he made a number of world-shaking discoveries (the moons of Jupiter, the phases of Venus, sunspots, the structure of the moon's surface, that the Milky Way is made up of a huge number of individual stars, among others). See Figure 23-19, Color Plate VI.

A *refracting telescope* has eyepiece and objective lenses—Figure 23-20(a)—and forms images much as the microscope does, except that the objective lens has a rather long focal length. This is necessary because a telescope is used to view very distant objects rather than very close objects as a microscope does. The telescope magnifies because, like the simple magnifier, the image subtends a greater angle than the distant object does.

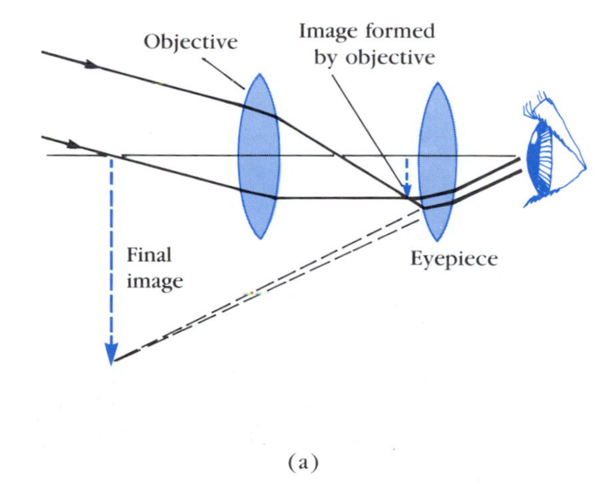

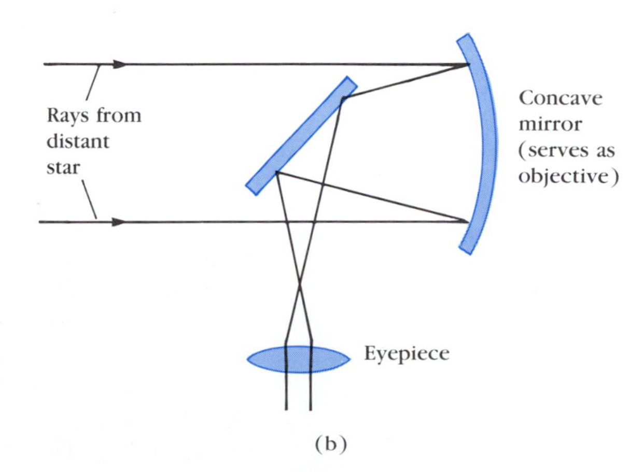

FIGURE 23-20
(a) Ordinary refracting telescope. (b) Astronomical telescope utilizing a mirror instead of a lens for the objective.

Many astronomical telescopes, particularly the largest ones, use a large concave mirror in place of the objective lens—Figure 23-20(b). Like the microscope, the telescope forms an upside-down image. This creates no real problem for viewing the heavens with an astronomical telescope, but on the earth an upside-down image is a distinct disadvantage. Therefore *terrestrial telescopes* use a third, intermediate lens to turn the image right side up. In binoculars (see Figure 22-21), prisms are used to turn the image upright; at the same time they reduce the length of the instrument to a manageable size. The prisms, as discussed in Chapter 22, reflect very nearly 100 percent of the light incident upon them because of total internal reflection.

SUMMARY

A lens uses refraction to produce a real or virtual image. Parallel rays of light are focused to a point, called the *focal point*, by a *converging lens*. The distance of the focal point from the lens is called the *focal length* f of the lens. After parallel rays pass through a *diverging lens*, they appear to diverge from a point, its focal point; and the corresponding focal length is considered negative. The position and size of the image formed by a lens of a given object can be found by ray tracing. Algebraically, the relation between image and object distances, d_i and d_o, and the focal length is given by the *lens equation*: $1/d_o + 1/d_i = 1/f$. The ratio of image height to object height, which equals the magnification, is given by d_i/d_o.

A *camera* lens forms an image on film by allowing light in through a shutter. The lens is focused by moving it relative to the film, and its f-stop (or lens opening) must be adjusted for the brightness of the scene and the chosen shutter speed. The f-stop is defined as the ratio of the focal length to the diameter of the lens opening.

The human *eye* also adjusts for the available light—by opening and closing the iris. It focuses by adjusting the shape of the lens to vary its focal length. The image is formed on the retina, which contains an array of receptors known as rods and cones.Diverging eyeglass or contact lenses are used to correct the defect of a nearsighted eye, which cannot focus well on distant objects; converging lenses are used to correct for defects in which the eye cannot focus on close objects.

A *simple magnifier* is a converging lens that forms a virtual image of an object placed at (or within) the focal point. The *magnification*, M, is equal to $M = 25\ \text{cm}/f$ where the focal length f of the lens is given in cm.

A *microscope* consists of an *objective* lens and an *eyepiece* lens; the latter magnifies the image formed by the objective lens into a very large final image. A *telescope* consists of an objective lens or mirror and an eyepiece that magnifies the real image formed by the objective lens. A terrestrial telescope uses an extra lens, or prisms (in a binocular), so that the final image is right side up.

QUESTIONS

1. State the difference between a real and a virtual image.
2. Can real images be projected on a screen? Can virtual images? Can either be photographed? Discuss carefully.
3. Why can't a prism be used to form an image?
4. A thin converging lens is moved closer to a nearby object. Does the real image formed change (a) in position (b) in size? If yes, describe how.
5. The thicker a double convex lens is in the center as compared to its edges, the shorter its focal length for a given lens diameter. Explain.
6. Light rays are reversible. Is this consistent with the lens equation?
7. In a close-up photograph of a person's face the background is usually out of focus. Explain.
8. Why is the depth of field greater and the image sharper when a camera lens is "stopped down" to a larger f number?
9. Why must a camera lens be moved farther from the film to focus on a closer object?
10. A "pinhole" camera uses a tiny pinhole instead of a lens. Show, using ray diagrams, how reasonably sharp images can be formed using such a pinhole camera. In particular, consider two point objects 2.0 cm apart that are 1.0 m from a 1-mm-diameter pinhole. Show that on a piece of film 5 cm behind the pinhole each object produces a tiny, easily resolvable spot.
11. The film in a camera plays a similar role to what part of the eye?
12. Bifocal glasses contain lenses that are split; the lower halves are a different focal length than the upper halves. Why is this, and what is the advantage?
13. As people grow older the ability of the eye's muscles to make the lens thicker and rounder is diminished, making it more difficult to focus on nearby objects. For this reason, reading glasses are commonly used by older people. Are reading glasses converging or diverging lenses?
14. Is the image formed on the retina of the human eye upright or inverted? Discuss the implications of this for our perception of objects.
15. Explain why swimmers with good eyes see distant objects as blurry when they are under water. Use a diagram, and also show why goggles correct this problem.

16. Reading glasses use converging lenses. A simple magnifier is also a converging lens. Are reading glasses therefore magnifiers? Discuss the similarities and differences between converging lenses as used for these two different purpose.

EXERCISES

1. A flower is 1.0 m from a 5.00-cm focal length camera lens. How far behind the lens will the image be?
2. If an object is placed 200 mm in front of a + 100-mm-focal-length lens, where will the image be?
3. An object is placed 50 mm in front of a + 50-mm-focal-length lens. Where will the image be? What if the object is placed 49 mm from the lens?
4. A sharp image is located 60 mm behind a 50-mm-focal-length converging lens. Calculate the object distance.
5. How far must a 50-mm-focal-length camera lens be moved from its infinity setting in order to sharply focus an object 3.0 m away?
6. A leaf is placed 70 cm in front of a −80-mm-focal-length lens. Where is the image? Is it real or virtual?
7. A certain lens focuses an object 20 cm away as an image 30 cm on the other side of the lens. What type of lens is it (converging or diverging) and what is its focal length? Is the image real or virtual?
8. How large is the image of the sun on the film used in a camera with a 50-mm-focal-length lens? The sun's diameter is 1.4×10^6 km and it is 1.5×10^8 km away.
9. Suppose you want to take a picture of yourself in a mirror 3 meters away. The camera should be focused for what distance? Can your image and the mirror frame both be in sharp focus?
10. What is the maximum diameter of a 50-mm, $f/1.4$ camera lens?
11. A 135-mm-focal-length lens has *f*-stops ranging from $f/3.5$ to $f/32$. What is the corresponding range of lens diameters?
12. A light meter reads that a lens setting of $f/8$ is correct for a shutter speed of $\frac{1}{250}$ s under certain conditions. What would be the correct lens opening for a shutter speed of $\frac{1}{500}$ s?
13. A light meter reports that a camera setting of $\frac{1}{100}$ s at $f/11$ will give a correct exposure. But the photographer wishes to use $f/16$ to increase the depth of field. What should the shutter speed be?
14. A 135-mm-focal-length lens can be adjusted so that it is 135 to 140 mm from the film. For what range of object distances can it be adjusted?

*15. A nature photographer wishes to photograph a 4.5-m-high elephant from a distance of 100 m. What focal-length lens should be used if the image is to fill the 24-mm height of the film?

16. What is the power of a 30.0-cm-focal-length lens?

17. What is the focal length of a −8.5-diopter lens? Is this lens converging or diverging?

18. A person's left eye is corrected by a −8.5-diopter lens, 2.0 cm from the eye. (a) Is this person near- or farsighted? (b) What is this person's far point without glasses?

19. Reading glasses of what power are needed for a person whose near point is 150 cm so that she can read at 25 cm? Assume a lens-eye distance of 2.0 cm.

20. A man's left eye can see objects clearly only if they are between 15 and 35 cm away. (a) What power of contact lens is required so that objects far away are sharp? (b) What, then, will be his near point?

21. What is the focal length of a magnifying glass of 3.5 × magnification for a relaxed normal eye?

22. A magnifier is rated at 3 × for a normal eye. What is its focal length?

23. What is the magnification of a 6.0-cm-focal-length converging lens for a normal eye?

24. What is the magnifying power of a +22-diopter lens? Assume a relaxed normal eye.

25. A child has a near point of 10 cm. What is the maximum magnification the child can obtain using an 8.0-cm-focal-length magnifier? Compare to that for a normal eye.

26. The magnification of a telescope is equal to the ratio f_o/f_e, where f_o is the focal length of the objective and f_e is the focal length of the eyepiece. If you have lenses with focal lengths of 20 cm and 40 cm, what two magnifications can you produce?

*27. How far from a 50-mm-focal-length lens must an object be placed if its image is to be magnified 3.0 × and be real?

*28. In a slide or movie projector, the film acts as the object whose image is projected on a screen. If a 100-mm-focal-length lens is to project an image on a screen 4.0 m away, how far from the lens should the slide be? If the slide is 35 mm wide, how wide will the picture be on the screen?

*29. A 35-mm slide (picture size is actually 24 by 36 mm) is to be projected on a screen 1.00 by 1.50 m placed 6.0 m from the projector. What focal-length lens should be used if the image is to cover the screen?

THE WAVE NATURE OF LIGHT

CHAPTER

24

That light carries energy is obvious to anyone who has used a magnifying glass to focus the sun's rays on a piece of paper and burned a hole in it. But how does light travel and in what form is this energy carried?

In our discussion of waves in Chapter 16, we noted that energy can be carried from place to place in basically two ways: by particles or by waves. In the first case, material bodies or particles can carry energy, such as a thrown baseball or rushing water. In the second case, water waves and sound waves, for example, can carry energy over long distances even though mass itself does not travel these distances.

In view of this, what can we say about the nature of light: does light travel as a stream of particles away from its source or does it travel in the form of waves that spread outward from the source? Historically, this question has turned out to be a difficult one. For one thing, light does not reveal itself in any obvious way as being made up of tiny particles nor do we see tiny light waves passing by, as we do water waves. Dutch scientist Christian Huygens (1629–1695), a contemporary of Newton, developed a wave theory of light. Newton himself, however, favored the idea that light consisted of tiny particles. The evidence seemed to favor first one side and then the other, until about 1830 when most physicists had accepted the wave theory. By the end of the nineteenth century, light was seen to be an *electromagnetic wave* (Chapter 21). Although modifications had to be made in the twentieth century, the wave theory of light has proved very successful. We now investigate the evidence for the wave theory and how it has explained a wide range of phenomena.

1. DIFFRACTION

In Chapter 16 we discussed waves and wave motion, and we found certain phenomena that waves exhibit but that particles don't. These were the phenomena of *interference* and *diffraction*.

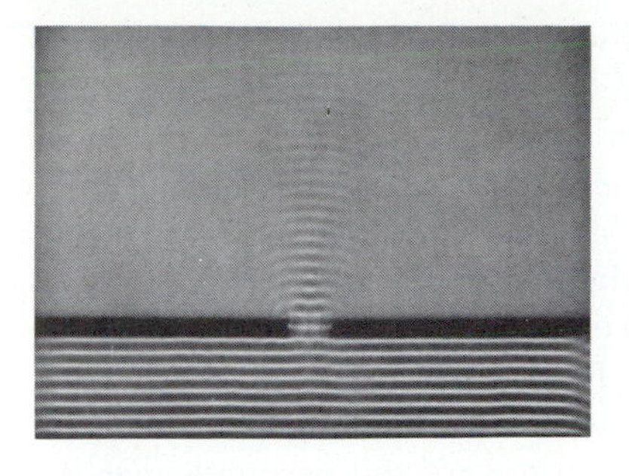

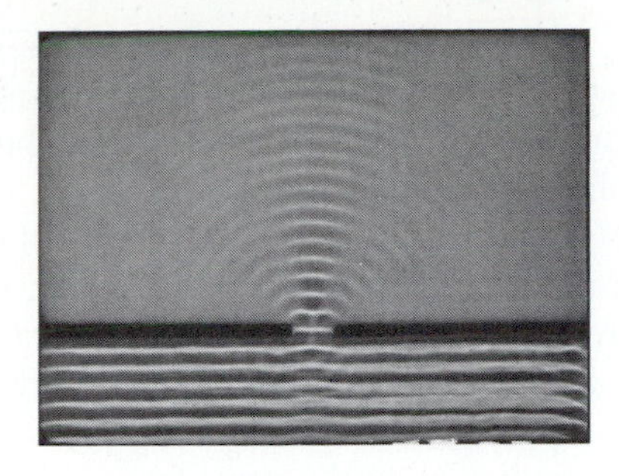

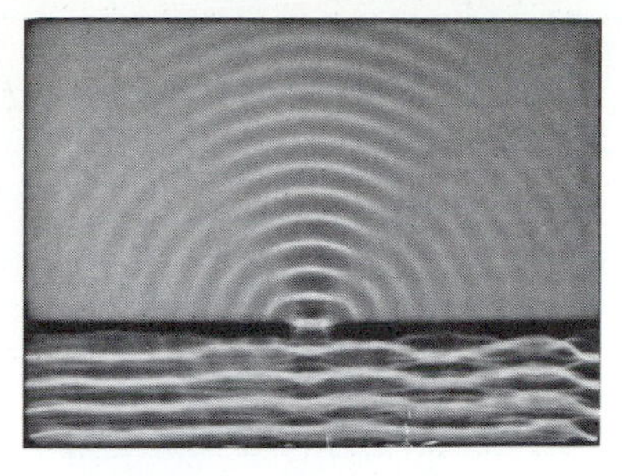

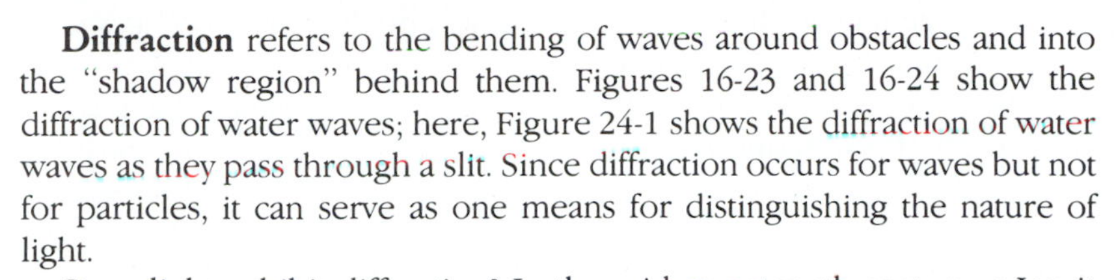

FIGURE 24-1
Diffraction of water waves. The waves come from below, pass through the slit, and spread into the shadow region on each side of the slit. Note the amount of diffraction is large only when the wavelength is not much smaller than the width of the slit.

Diffraction refers to the bending of waves around obstacles and into the "shadow region" behind them. Figures 16-23 and 16-24 show the diffraction of water waves; here, Figure 24-1 shows the diffraction of water waves as they pass through a slit. Since diffraction occurs for waves but not for particles, it can serve as one means for distinguishing the nature of light.

Does light exhibit diffraction? In the mid-seventeenth century, a Jesuit priest, Francesco Grimaldi (1618–1663) observed that when sunlight entered a darkened room through a tiny hole in a screen, the spot on the opposite wall was larger than would be expected from geometric rays (Figure 24-2). Grimaldi attributed this to the diffraction of light. Newton, who favored a particle theory, was aware of Grimaldi's result. He felt that Grimaldi's result was due to the interaction of light corpuscles ("little bodies") with the edges of the hole. If light were a wave, he said, the light waves should bend more than that observed. Newton's argument seems reasonable. Yet, as we can see in Figure 24-2 (see also Chapter 16), diffraction is large only when the size of the obstacle or the hole is on the order of the wavelength of the wave. Newton did not imagine that the wavelengths of visible light might be incredibly tiny, and thus diffraction effects would be very small. (Indeed this is why geometric optics using rays is so successful—normal openings and obstacles are much larger than the wavelength of the light, and so relatively little diffraction or bending occurs.)

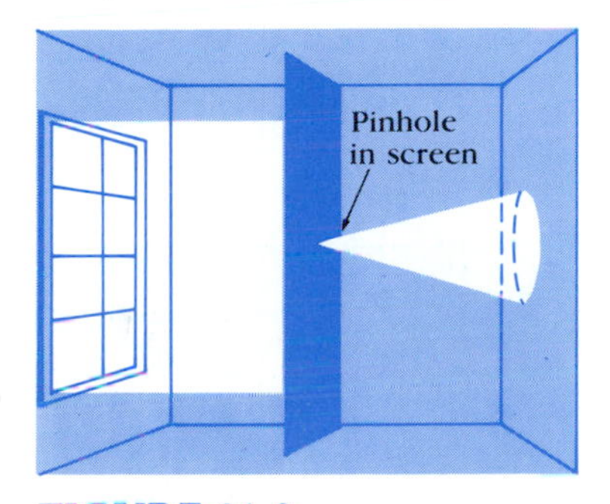

FIGURE 24-2
Grimaldi's experiment showing the diffraction of light.

2. WAVE THEORY EXPLAINS REFRACTION

The laws of reflection and refraction were well-known in Newton's time. The law of reflection could not distinguish the two theories. For when waves reflect from an obstacle, the angle of incidence equals the angle of reflection (Figure 16-18); the same is truc of particles—think of a tennis ball without spin striking a flat surface.

The law of refraction is another matter. The wave theory, but not the particle theory, can explain why light rays bend toward the perpendicular when they enter a region where the velocity of light is less, and away from the perpendicular when the velocity is greater. To see this, we make use of

FIGURE 24-3
(a) Soldiers in a line march from firm ground into mud where their marching speed is reduced. Those that reach the mud first are slowed down first. (b) Wave front acts similarly, explaining why a ray bends toward the perpendicular upon entering a medium in which the speed is less.

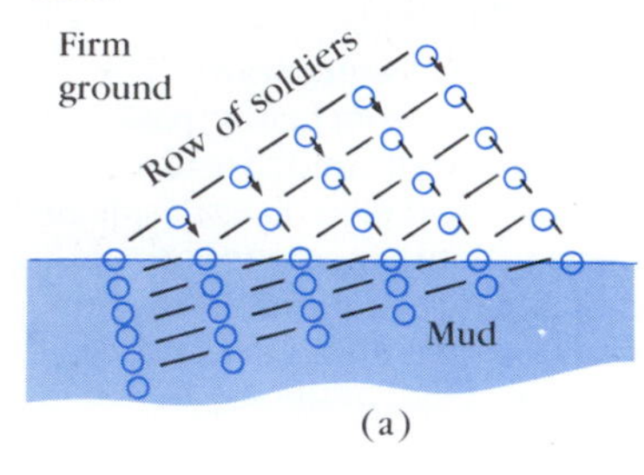

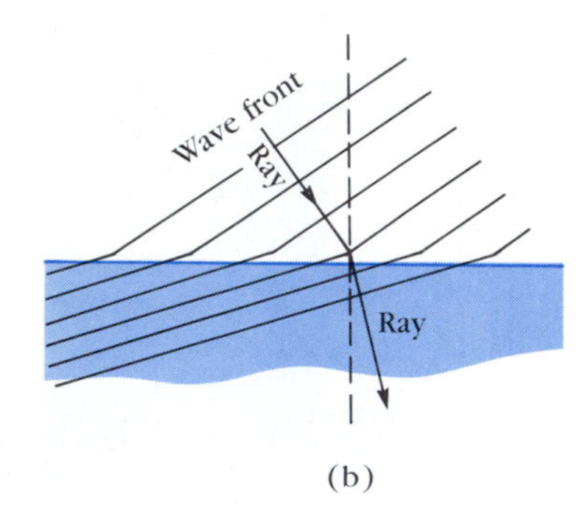

an analogy in which a line of soldiers walking arm in arm represents a wave front or wave crest—see Figure 24-3(a). At first the soldiers are marching on hard pavement, but then they approach a muddy region obliquely and they are slowed down. The soldiers who reach the mud first are slowed down first and the line of soldiers bends, Figure 24-3(a). The same thing happens to a wave front, whether it's a water wave or a light wave. The first part of the wave front to enter the slower medium starts to drag first and the wave front bends, Figure 24-3(b). The rays, which represent the direction in which the waves are moving and thus are perpendicular to the wave fronts, are bent toward the perpendicular, as found experimentally. If the light passes into a medium in which its speed is greater, the reverse effect occurs.

Newton's corpuscle theory predicted the opposite result: if the path of light corpuscles entering a new medium changed direction, it must be because the medium exerted a force on the corpuscles at the boundary. This force was assumed to act perpendicular to the boundary. When light entered a medium such as water where it was bent toward the perpendicular, the force must accelerate the corpuscles so the velocity is increased; only in this way will the refracted angle be less than the incident angle. In Newton's theory, then, the speed of light would be greater in the second medium. Thus the wave theory predicts that the speed of light in water (say) is less than in air and Newton's corpuscle theory predicts the reverse. An experiment to actually measure the speed of light in water and confirm the wave theory prediction was not conducted until 1850 (by the French physicist Jean Foucault), and by then the wave theory was fully accepted as we shall see in the next section.

3. INTERFERENCE OF LIGHT AND YOUNG'S DOUBLE-SLIT EXPERIMENT

In 1801, the Englishman Thomas Young (1773–1829) obtained convincing evidence for the wave nature of light and was even able to measure the wavelength. Figure 24-4(a) shows a diagram of Young's famous double-slit experiment. Light from a single source (Young used the sun) falls on a screen containing two closely spaced slits. If light consists of particles, we would expect to see two bright lines on a screen placed behind the slits as in Figure 24-4(b). But Young observed instead a series of bright lines

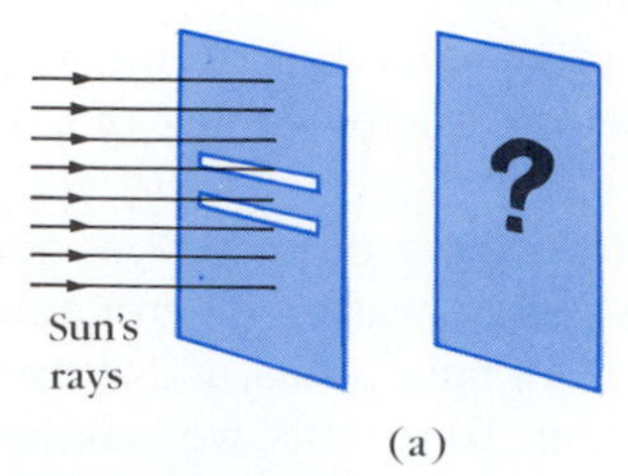

FIGURE 24-4
(a) Young's double-slit experiment. (b) If light consists of particles, we would expect to see two bright lines on the screen behind the slits. (c) Young observed many lines.

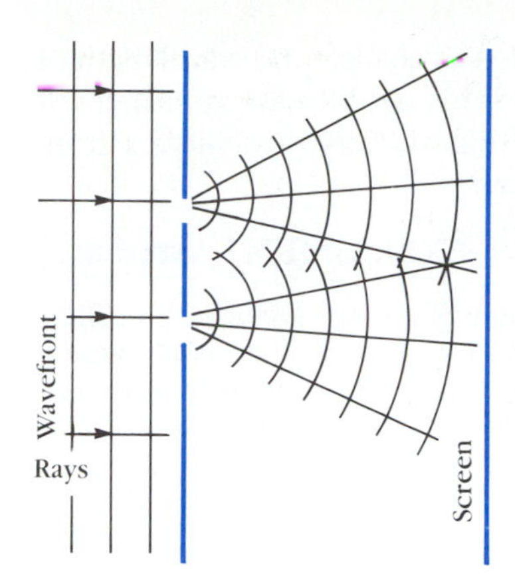

FIGURE 24-5

If light is a wave, light passing through one of two slits should interfere with light passing through the second slit.

as in Figure 24-4(c). Young was able to explain this result as a **wave-interference** phenomenon.

To see this, imagine plane waves of light of a single wavelength (called "monochromatic") falling on the two slits as shown in Figure 24-5. Because of diffraction, the waves leaving the two small slits spread out as shown. This is equivalent to the interference pattern produced when two rocks are thrown into a lake (Figure 16-22); in place of the two rocks acting as sources of waves, the two slits in the screen serve as two sources of light waves. In Figure 24-5, the light waves are shown spreading out from each slit just as water waves do. Note that the light *rays*, which we defined as the direction in which the light moves, are perpendicular to the wave fronts or crests. The interference pattern illustrated in Figure 24-5 is much like that for water waves. However, we cannot directly observe the interference pattern as it spreads through space. Rather, we observe only the illuminations produced by these wave patterns when they strike the screen. The series of bright and dark lines on the screen is also known as an **interference pattern**. To explain how the interference pattern is produced on the screen, we make use of the diagram of Figure 24-6. Here the waves that travel along each ray are shown individually, and the regions on the screen where constructive and destructive interference occur are indicated. We perceive areas of destructive interference—where there is no wave activity—as darkness, and areas of constructive interference as brightness. We see from Figure 24-6 that a bright area will occur when the paths of the rays from the two slits are the same—for example, at the center of the screen, Figure 24-6(a). The screen will also be bright where the paths of the two rays differ by exactly one wavelength (or two, or any whole number of wavelengths), since the waves will then interfere constructively, Figure 24-6(b). For those regions on the screen where the distances from the two slits differ by a half wavelength (or $1\frac{1}{2}$ wavelengths, $2\frac{1}{2}$ wavelengths, and so on), the two waves will destructively interfere, Figure 24-6(c). Thus the wave theory predicts a pattern of bright and dark lines called *fringes* that is precisely in accord with experimental results, Figure 24-4(c).

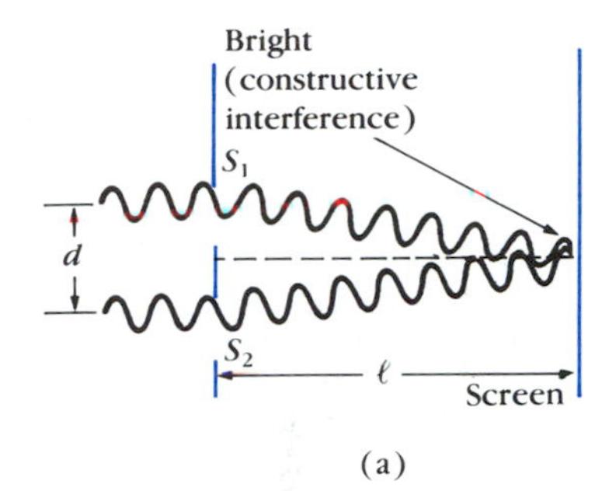

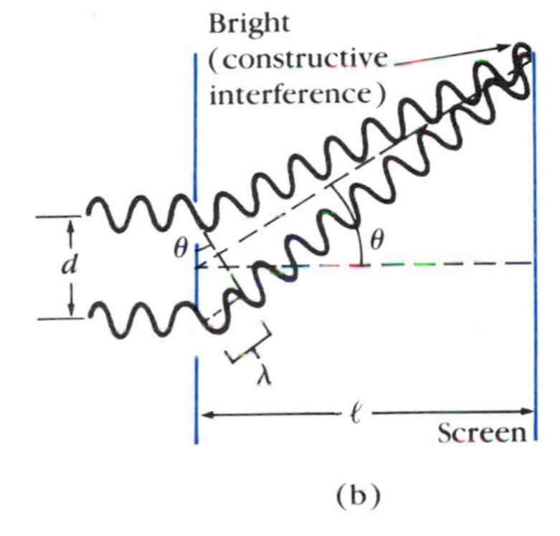

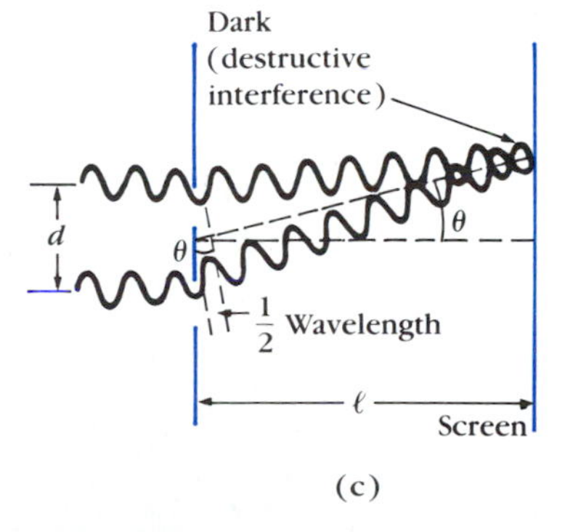

FIGURE 24-6

How the wave theory explains the pattern of lines seen in the double-slit experiment.

Using simple geometry, Young was able to calculate the wavelength of visible light with the data he obtained from his double-slit experiment. It was found that the wavelength of visible light is very tiny, varying from about 4×10^{-7} meters to 7.5×10^{-7} meters.

EXPERIMENT

Hold a handkerchief at arm's length in front of your eyes and view a small source of light such as a distant street lamp. You will see an array of spots that is larger than the source seen directly. This is an interference pattern produced by the light traveling through the many tiny holes between the threads of the cloth. Observe the interference patterns produced by fabrics of fine weave and of coarse weave. How does the fineness of the weave affect the pattern?

Young's experiment was a crucial one, for it had at last settled the argument over whether light has a wave or a particle nature. There was no way to explain Young's results using a particle view of light. Continued experimentation during the nineteenth century further confirmed the idea that light is made up of waves, and by the end of the century the wave theory of light stood on firm experimental and theoretical bases.

Calculation of Wavelengths (optional)

To determine exactly where the bright lines fall in the double-slit experiment, first note that Figure 24-6 is somewhat exaggerated; in real situations, the distance d between the slits is very small compared to the distance ℓ to the screen. The rays from each slit for each case will therefore be essentially parallel, and θ is the angle they make with the horizontal. The angle at which constructive interference occurs can be written in terms of the wavelength of the light, λ, if we write the angle θ in radians. (As discussed in Section 1 of Chapter 9, any angle is equal to the arc length it subtends divided by the distance to the arc.) Thus, in the shaded triangle of Figure 24-6(b), the angle θ can be written (in radians) as

$$\theta = \frac{\lambda}{d}$$

(as long as θ is small) where d is the distance between the two slits; λ is the wavelength of the light, and in Figure 24-6(b) it represents the extra distance the lower ray must travel for constructive interference to occur. This angle θ represents the angle at which the first bright fringe occurs above the central one [at $\theta = 0$ as in Figure 24-6(a)]. Other bright fringes occur when the extra path length is 2λ, 3λ, and so on, corresponding to angles

$$\theta = 2\frac{\lambda}{d}, 3\frac{\lambda}{d}, \text{ and so on.}$$

Similarly, using Figure 24-6(c), we can show that completely destructive interference occurs at angles θ given by $\theta = \frac{1}{2}\lambda/d$, $\frac{3}{2}\lambda/d$, and so on.

To see how Young measured the wavelengths of visible light, suppose yellow light from a distance source falls on the two slits which are a distance $d = 0.10$ mm apart. The interference pattern is observed on a screen 1.0 m away (ℓ in Figure 24-6), and the distance between the fringes on the screen is observed to be 5.7 mm. What is the wavelength of this light? From Figure 24-6(b), the angle θ is not only equal to λ/d, but it is also equal to the distance between the central fringe and the first one above it (call it x, which in this example is 5.7 mm) divided by ℓ:

$$\theta = \frac{x}{\ell} = \frac{\lambda}{d}.$$

Thus

$$\lambda = \frac{x}{\ell}d = \frac{(5.7\ \text{mm})}{(1.0\ \text{m})}(0.10\ \text{mm})$$

$$= \frac{(5.7 \times 10^{-3}\ \text{m})}{(1.0\ \text{m})}(1.0 \times 10^{-4}\ \text{m})$$

$$= 5.7 \times 10^{-7}\ \text{m}$$

or 570 nm (570 nanometers).

Making similar measurements, it was found that visible light has wavelengths in the range of about 400 nm (violet) to 750 nm (red), as mentioned earlier.

4. DIFFRACTION AROUND OBJECTS AND LIMIT ON MAGNIFICATION

In Figure 24-1 we saw that water waves passing through a small hole are diffracted significantly only when the size of the hole is not much larger than the wavelength of the waves. If instead of waves passing through a hole we consider waves diffracting around an object, as was discussed in Chapter 16 (see Figure 16-24), we have a similar result: if the object is as small as or smaller than the wavelength of the waves, there is considerable diffraction around the object; but the larger the object is in comparison to the wavelength of the waves, the smaller will be the effect of diffraction. This can be said another way: *the larger the wavelength compared to the size of the object, the more diffraction will occur.* See especially Figure 16-24(c) and (d). This is why light waves, whose wavelengths are very small, do not bend around corners in everyday situations; but soundwaves, whose wavelengths are large, do bend around corners. It is because of the very tiny wavelengths that light seems to travel in straight lines—there is very little diffraction under normal circumstances. It is also for this reason that the ray model we discussed earlier was so successful.

FIGURE 24-7
Illumination of an object in a microscope.

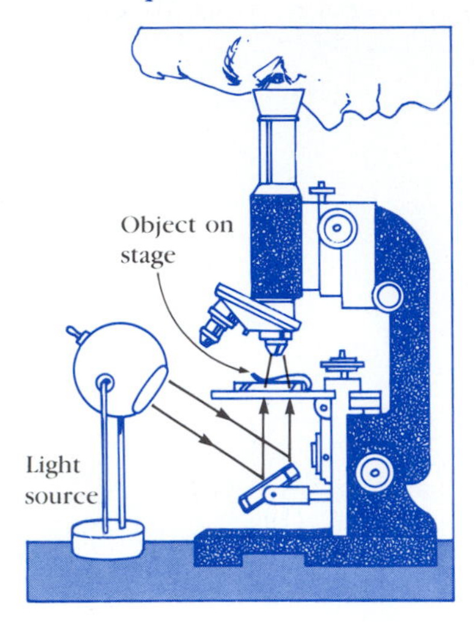

Diffraction results in a very important principle when applied to light waves. Objects whose dimensions are about the size of or smaller than the wavelength of the light that illuminates them cannot be clearly seen. For example, to view very small objects we usually use a microscope; the light used to illuminate the object is either reflected off the surface of the object or, more commonly, is transmitted through the object from below (Figure 24-7). In the latter case, we "see" the object because it stops the light from passing through—we see its shadow.

If an object viewed in a microscope is not much larger than the wavelength of the light used, diffraction effects will be very important, giving rise to a distorted image or, if the object is really small, to no image at all—see the corresponding effects for water waves in Figure 16-24(a) and (b). Of course, if the object is much larger than the wavelength of light used, diffraction is not important and a clear image can be formed. The wave nature of light thus puts a limit on the size of the tiniest object, or parts of an object, that can be clearly seen. This is summarized in the following general rule:

It is not possible to observe objects clearly whose dimensions are smaller than the wavelength of the light used to illuminate them.

The useful magnification of microscopes using visible light, even if the lenses were perfect, is limited to about 1000 times because of diffraction. Further magnification would merely result in magnifying the distorted patterns caused by diffraction around the edges of tiny objects. However, electron microscopes can have much greater magnification because, as we shall see in Chapter 26, electrons have very small wavelengths.

5. THE VISIBLE SPECTRUM AND DISPERSION; RAINBOWS AND DIAMONDS

The two most obvious properties of light are readily describable in terms of the wave theory of light: intensity (or brightness) and color. The *intensity* of light is related to the amplitude of the wave, just as for any wave (Chapters 16 and 17). The *color* of the light is related to the wavelength or frequency of the light. Visible light—that to which our eyes are sensitive—falls in the wavelength range of about 400 nm to 750 nm.† This is known as the *visible spectrum*, and within it lie the different colors from violet to red as shown in Figure 24-8 and Color Plate VII. Light with wavelengths shorter than 400 nm is called ultraviolet (UV) and that with wavelengths greater than 750 nm is called infrared (IR).* Although human

† Sometimes the Angstrom (Å) unit is used when referring to light: $1 Å = 1 \times 10^{-10}$ m; then visible light falls in the wavelength range of 4000 Å to 7500 Å.

* The complete electromagnetic spectrum is illustrated in Figure 21-10.

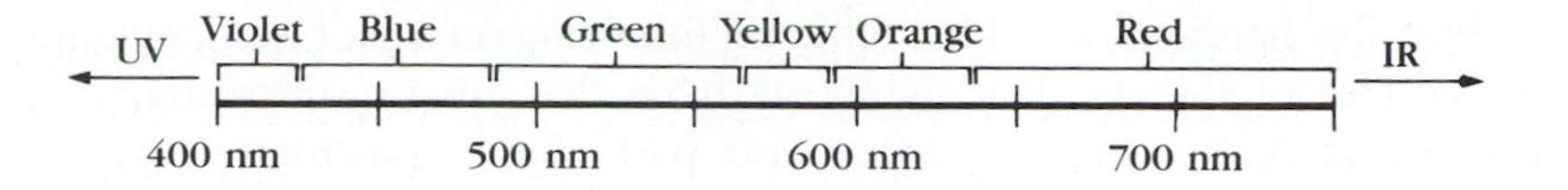

FIGURE 24-8
The spectrum of visible light, showing the range of wave lengths for the various colors. See also Color Plate VII.

eyes are not sensitive to UV or IR, some types of photographic film do respond to them.

It is a familiar fact that a prism separates white light into a rainbow of colors. This is due to the fact that the index of refraction of a material depends on the wavelength. That is, the speed of light in a material (but not in a vacuum) is slightly different for different wavelengths. White light is a mixture of all visible wavelengths; and when incident on a prism as in Figure 24-9, the different wavelengths are bent to varying degrees. The index of refraction is greater for the shorter wavelengths, so violet light is bent the most and red the least as indicated. This spreading of white light into the full spectrum is called **dispersion**.

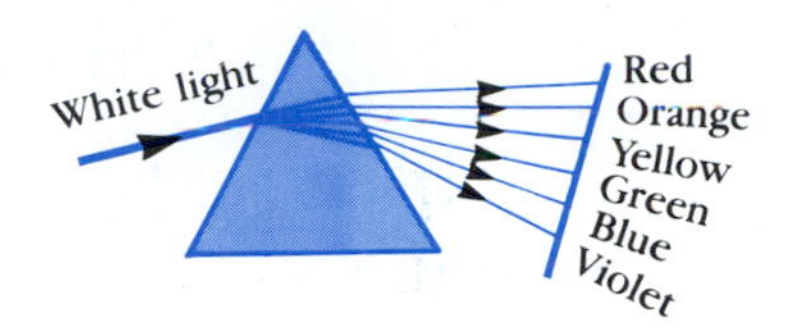

FIGURE 24-9
White light dispersed by a prism into the visible spectrum.

Rainbows are a spectacular example of dispersion (by water in this case). You can see rainbows when you look at falling water with the sun at your back. Figure 24-10 shows how red and violet rays are bent by spherical water droplets and are reflected off the back surface. Red is bent the least and so reaches the observer's eyes from droplets higher in the sky. Thus the top of the rainbow is red.

Diamonds achieve their brilliance from a combination of dispersion and total internal reflection. Since diamonds have the incredibly high index of refraction of about 2.4, the critical angle for total internal reflection is only 25°. Incident light therefore strikes many of the internal surfaces before it strikes one at less than 25° and emerges. After many such reflections the light has traveled far enough that the colors have become sufficiently separated to be seen individually and brilliantly by the eye after leaving the crystal.

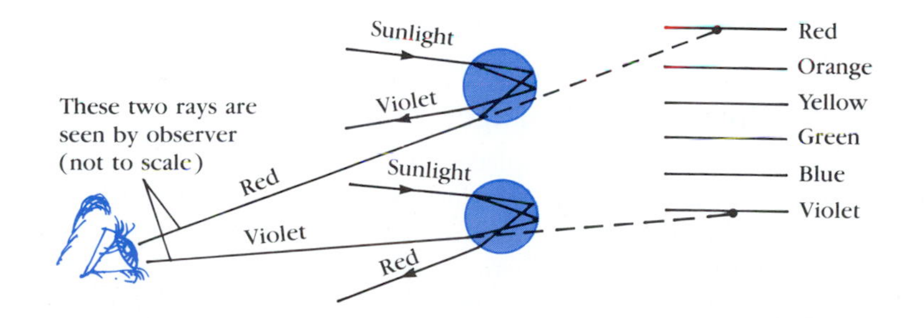

FIGURE 24-10
Formation of a rainbow.

6. INTERFERENCE BY THIN FILMS

Interference of light gives rise to many everyday phenomena such as the bright colors reflected from soap bubbles and from thin oil films on water. In these and other cases, the colors are a result of constructive interference between light reflected from the two surfaces of the thin film. To

FIGURE 24-11
Light reflected from upper and lower surfaces of a thin film of oil lying on water.

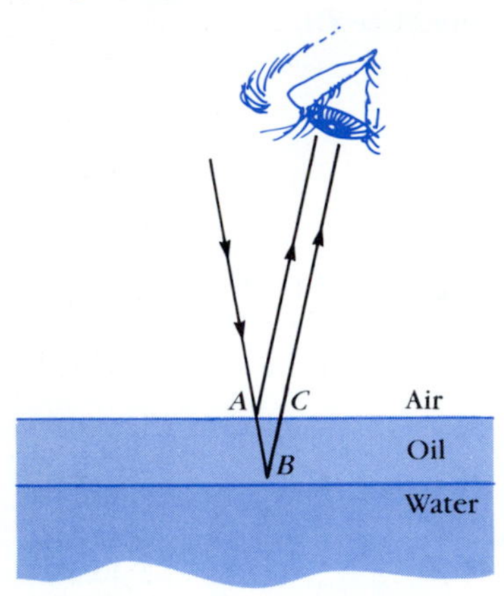

see how this happens, consider a thin oil film lying on top of water (Figure 24-11). Part of the incident light (say from the sun or street lights) is reflected at A on the top surface and part of the transmitted light is reflected at B on the lower surface. The part reflected at the lower surface must travel the extra distance ABC. If the distance ABC is equal to one or a whole number of wavelengths, the two waves will interfere constructively and the light will be bright. But if ABC equals $\frac{1}{2}\lambda$, $\frac{3}{2}\lambda$, and so on, the two waves will be out of phase and destructive interference occurs. When white light falls on such a film, the path ABC will equal λ (or a whole number multiple of λ) for only one wavelength at a given viewing angle. This color will be seen as very bright. For light viewed at a slightly different angle, the path ABC will be longer or shorter and a different color will undergo constructive interference. Thus, for an extended source emitting white light, a series of bright colors will be seen next to one another. Variations in thickness of the film will also alter the length ABC and therefore affect the color of light that is most strongly reflected. This *thin film interference* is the origin of the bright colors seen on soap bubbles and oil films.

An important application of thin-film interference is in the coating of glass to make it "nonreflecting," particularly for lenses. Glass reflects about 4 percent of the light incident upon it. Good-quality cameras, microscopes, and other optical devices may contain six to ten thin lenses. Reflection from all these surfaces can reduce the light level considerably and multiple reflections produce a background haze that reduces the quality of the image. Reflections can be reduced considerably by applying a very thin coating on the lens surfaces so that destructive interference will occur nearly completely for wavelengths in the center of the spectrum (around 550 nm). The extremes of the spectrum—red and violet—will not be reduced as much; since a mixture of red and violet produces purple, the light seen reflected from such coated lenses is purple. Lenses containing two or three separate coatings can more effectively reduce a wider range of reflecting wavelengths.

Newton's Rings; Phase Change (optional)

When a curved glass surface is placed in contact with a flat glass surface, Figure 24-12, a series of concentric rings is seen when illuminated from above by monochromatic light. These are called *Newton's rings*, and they are due to interference between rays reflected by the top and bottom surfaces of the *air gap* between the two pieces of glass. Because this gap (which is equivalent to a thin film) increases in width from the central contact point out to the edges, the extra path length for the lower ray (equal to BCD) varies; where it equals 0, $\frac{1}{2}\lambda$, λ, $\frac{3}{2}\lambda$, 2λ, and so on, it corresponds to constructive and destructive interference; and this gives rise to the series of bright and dark lines seen in Figure 24-12(b).

Note that the point of contact of the two glass surfaces (A in Figure 24-12(a)) is dark in Figure 24-12(b). Since the path difference is zero here, we expect the rays reflected from each surface to be in phase and this point to be bright. But it is dark, which tells us the two rays must be interfering destructively; this can happen only because one of the waves flips over upon reflection, so it is $\frac{1}{2}$ cycle, or $\frac{1}{2}\lambda$, out of phase with the other wave. Indeed, this and other experiments reveal that *a beam of light reflected by a material whose index of refraction is greater than that in which it is traveling changes phase by $\frac{1}{2}$ cycle*. If the index is less than that of the material in which it is traveling, no phase change occurs. (This corresponds to the reflection of a wave traveling along a rope when it reaches the end; as we saw in Figure 16-17, if the end is tied down the rope changes phase and the pulse flips over; but if the end is free, no phase change occurs.) Thus the ray reflected by the curved surface above the air gap in Figure 24-12(a) undergoes no change in phase. That reflected at the lower surface, where the beam in air strikes the glass, undergoes a $\frac{1}{2}\lambda$ phase change. Thus the two rays reflected at the point of contact A of the two glass surfaces (where the air gap approaches zero thickness) will be $\frac{1}{2}\lambda$ out of phase, and a dark spot occurs. Other dark bands will occur when the path difference BCD in Figure 24-12(a) is equal to an integral number of wavelengths. Bright bands will occur when the path difference is $\frac{1}{2}\lambda$, $\frac{3}{2}\lambda$, and so on, since the phase change at one surface effectively adds another $\frac{1}{2}\lambda$.

FIGURE 24-12
Newton's rings.
The circular patches at various places on the photograph of the rings are an artifact (moiré pattern) of the printing process.

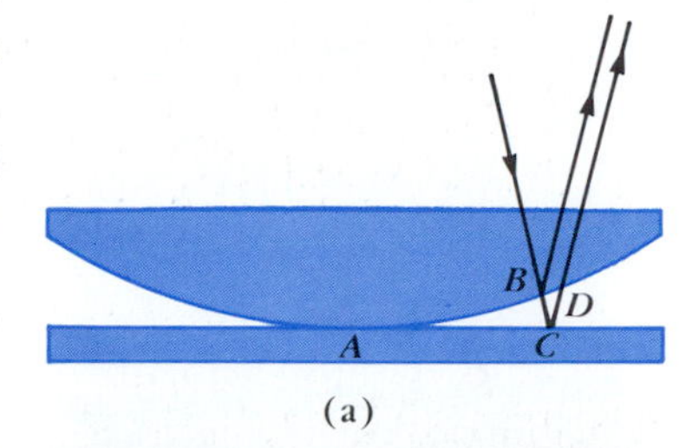

(a)

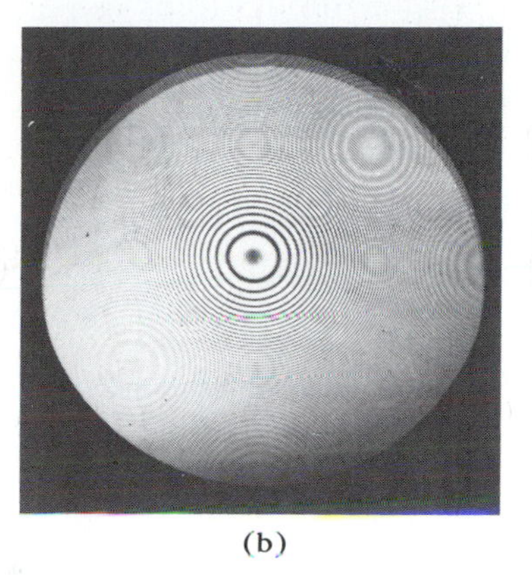
(b)

An Example (optional)

Suppose a soap bubble appears green ($\lambda = 540$ nm) at its point nearest the viewer. What is its minimum thickness? (The bubble is filled with air and we assume the index of refraction $n = 1.35$ inside the soap film.) The light is reflected perpendicularly from the point on a spherical surface nearest the viewer. Therefore the path difference is $2t$ where t is the thickness of the soap film. Light reflected from the outer surface undergoes a $\frac{1}{2}\lambda$ phase change, whereas that reflected on the inner surface does not. Therefore green light is bright when the minimum path difference equals $\frac{1}{2}\lambda$. Although $\lambda = 540$ nm in the air, inside the film it will be less because the speed of light is less, $v = c/n$. Since $v = \lambda f$ and the frequency f doesn't change when the light enters the film, then the wavelength inside soap film is $\lambda/n = 540\text{ nm}/1.35 = 400$ nm. Thus $2t = (400\text{ nm})$ or $t = 100$ nm is the minimum thickness of the soap bubble.

7. POLARIZATION

Once the evidence had revealed that light travels as a wave, a new question arose: Are light waves longitudinal or transverse? The properties of light waves we have studied so far would occur for either.

In the nineteenth century it was recognized that light is transverse because it displays the phenomenon of **polarization**, which occurs only

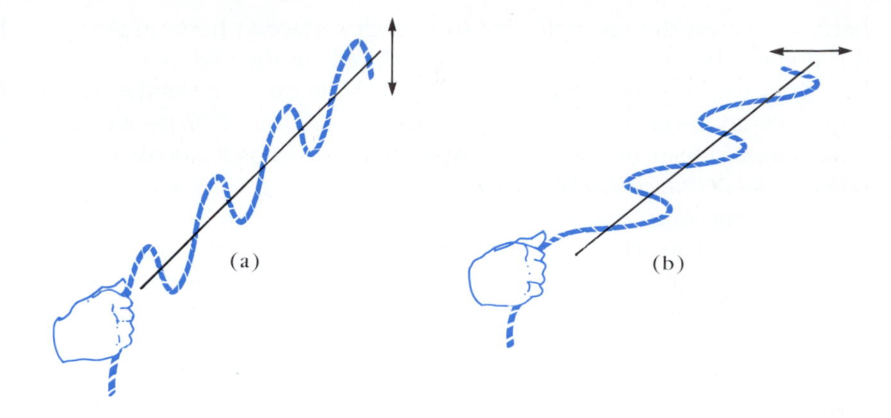

FIGURE 24-13
Transverse waves on a rope polarized (a) in a vertical plane and (b) in a horizontal plane.

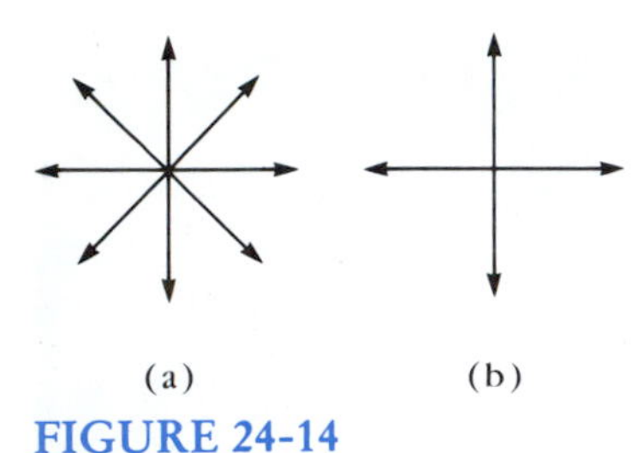

FIGURE 24-14
(a) An unpolarized wave has vibrations in all directions. (b) These vibrations can be resolved into two mutually perpendicular components.

for transverse waves and not for longitudinal waves. To see what is meant by polarization, let us first examine a transverse wave, such as a wave on a rope. A rope may be set into vibration in a vertical plane as shown in Figure 24-13(a) or in a horizontal plane, Figure 24-13(b). In either case, the wave is said to be **plane polarized**. A transverse wave can also be **unpolarized**, which means that the source has vibrations in many planes at once, as in Figure 24-14(a). These various planes of vibration can be resolved into components (like vectors; see Chapter 5) along the horizontal and vertical directions, as shown in Figure 24-14(b). Thus, an unpolarized wave can be considered as the sum of polarized beams in the horizontal and vertical directions.

Now, if we place a barrier with a slit cut in it in the path of a transverse wave, Figure 24-15, a plane-polarized wave will pass through if it is polarized parallel to the slit (a) but will be blocked if it is perpendicular to the slit (b). Thus if two such barriers with slits are placed one behind the other, one with its slit vertical and the other with its slit horizontal, a plane-polarized wave will be stopped by one or the other of them. Even if the waves are unpolarized, they cannot pass through. For, as shown in (c), the first (vertical) slit eliminates all but the vertical component so that the unpolarized wave is turned into a plane-polarized wave; and this in turn is eliminated by the second slit. It is important to note that a longitudinal wave would not be stopped by such a pair of slits.

Now you can show for yourself that light is a transverse wave.

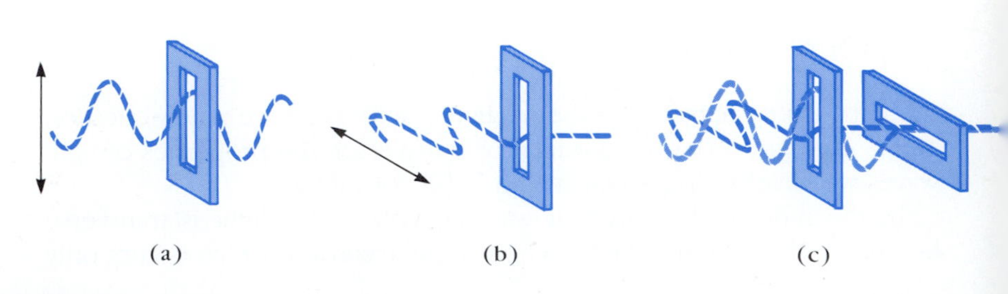

FIGURE 24-15
Vertically polarized wave (a) passes through a vertical slit, but a horizontally polarized wave (b) will not. In (c) crossed polarizers stop both components.

EXPERIMENT

Obtain two pairs of polarized sunglasses. Put on one pair and, keeping only one eye open, place one lens of the other pair in front of your open eye, as shown in Figure 24-16. Now rotate this lens. You should see that the light coming through the two lenses nearly vanishes at one orientation and is brightest at an orientation perpendicular to this.

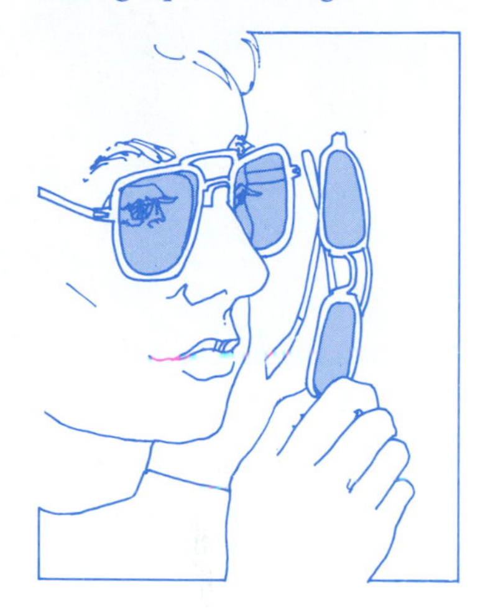

FIGURE 24-16
Looking through two polarizers. When they are "crossed" very little light passes through.

The barriers with slits in Figure 24-15 are called **polarizers**. They are so called because if an unpolarized wave encounters one such barrier, the transmitted wave is plane-polarized. Each of the lenses of polarized sunglasses is a polarizer for light waves. Such polarizing material is generally made of certain long molecules that are all aligned parallel to one another; consequently, they create the equivalent of the slits shown in Figure 24-15 and permit only one component of polarization to pass through. The light is dimmest when the polarizers are "crossed" (the "slits" crossed) and brightest when their "slits" are parallel.

The light from most natural sources, such as the sun and ordinary light bulbs, is unpolarized. When such light passes through a polarizing sheet, it comes out plane-polarized. In this process, 50 percent of the unpolarized incident light is blocked. Thus, polarized sunglasses cut out much more light than ordinary sunglasses with the same tint.

The light reflected from most nonmetallic surfaces is partially polarized. That is, one component is absorbed more than the component perpendicular to it. You can check this by rotating polarized sunglasses while looking at a lake, a window, or a highway on a bright day. Generally, the light reflected from a surface has less vibration perpendicular to the surface than parallel to it (Figure 24-17). Since most reflecting surfaces outdoors are horizontal—including highways—the lenses of polarized sunglasses are usually oriented vertically so as to eliminate the stronger horizontal component. That's why they are so effective in reducing glare. This is well known by fishermen who wear Polaroids to eliminate reflected

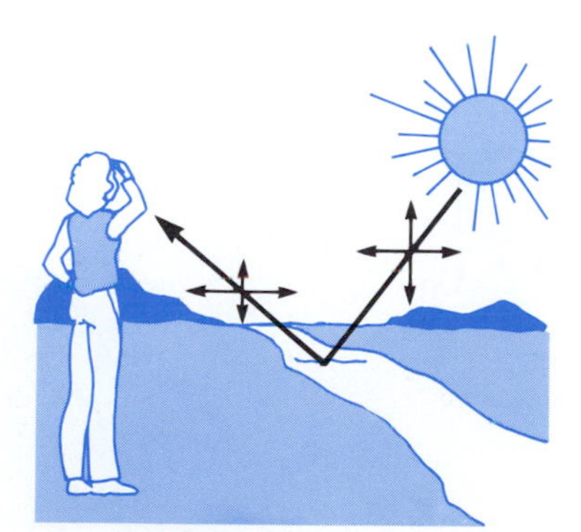

FIGURE 24-17
Light reflected from most surfaces is partially polarized. The vertical component is absorbed more by the surface.

FIGURE 24-18
Photograph of the surface of water (a) allowing all light into the lens and (b) using a polarizer that absorbs most of the light reflected from the surface; therefore, the river bottom can be seen more clearly.

glare from the surface of a lake or stream, and thus see beneath the water more clearly (Figure 24-18).

8. LIGHT SCATTERING; BLUE SKIES AND RED SUNSETS

In a gas, the molecules are far apart and most of the light incident on the gas passes through without being absorbed. However, the oxygen and nitrogen molecules of the air do "scatter" the light (absorb and reemit it) in all directions, and it is the violet and blue colors that are scattered the most. This is what gives air its bluish color and is why if you look at the sky in any direction, other than directly at the sun, you see blue light, Figure 24-19. Of course not all the blue is scattered in this way, nor does all the

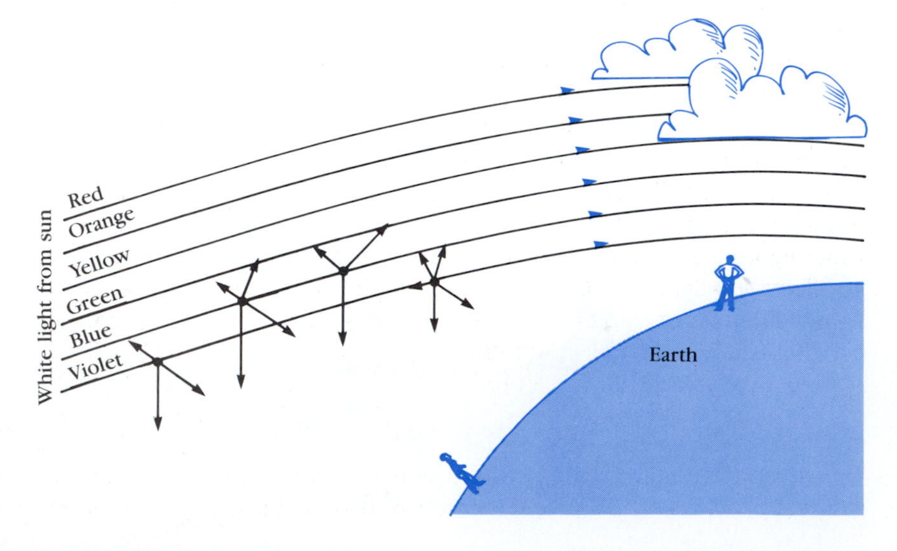

FIGURE 24-19
Blue light is preferentially scattered by air molecules, which can thereby enter an observer's eye. At sunset (upper observer) the light reaching the earth has had much of the blue removed and hence appears reddish when reflected off of objects such as clouds.

red and orange light pass through unscathed. But much less of the red is scattered than of the blue. Thus the sky looks blue.

The same mechanism gives rise to red sunsets. Late in the day the sun passes through a greater thickness of atmosphere than at midday, say. A good part of the violet and blue light, and even some of the green, has been scattered by the atmosphere. The light that remains and which passes through to the surface of the earth is thus primarily the red, orange, and yellow wavelengths. Thus the light that reaches the earth is reddish orange, and this is the color we see reflected from clouds and dust near the surface of the earth at sunset.

9. COLOR AND COLOR MIXING

Where Do Objects Get Their Color?

Some objects have a certain color because they emit light of particular wavelengths; these emitters or sources of light include light bulbs, neon signs, "red-hot" irons, the sun, and the stars.

But most objects in our environment appear colored not because they are sources of light but because they reflect part of the light that strikes them. Normally we see objects illuminated by white light—from the sun or from electric light bulbs—so all visible wavelengths are incident on the object. Depending on the materials from which the object is made, certain of these wavelengths are absorbed and others are reflected, and it is the reflected light that reaches our eyes. A rose is red because the wavelengths corresponding to red are reflected from its petals and other wavelengths are absorbed. Leaves are green because only green light is strongly reflected. Some clothes are yellow, others red or blue, depending on the wavelengths of light that the material reflects. An object that appears white reflects all wavelengths. A black object reflects very little and absorbs essentially all the light that strikes it.

What happens to the light energy that an object absorbs? It is transformed into heat. Thus a black object will be warmer than a white object when subjected to the same illumination. This is why in hot weather people feel more comfortable in light-colored clothing.

The color of the light illuminating an object affects the color of the object. For example, if an object that appears red when illuminated by white light is instead illuminated by red light, it will still look red because it reflects red light. But an object that is green when illuminated by white light will look black when illuminated by red light, because it absorbs red light. Similarly, a red object illuminated by green light will look black, but a green object will still look green.

Mixing Color Pigments—Subtractive Mixing

The mixing of different-colored pigments (or paints) to obtain different colors is part of the painter's art. Artists have long known that nearly

FIGURE 24-20
The subtractive primaries (for pigments) showing approximately the wavelengths reflected by each.

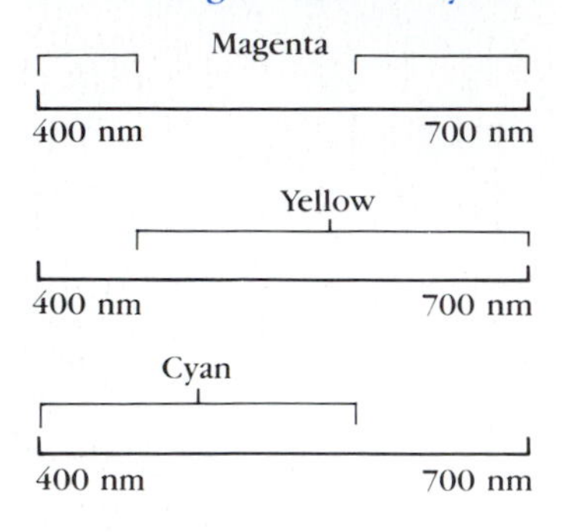

all colors can be obtained from mixing only the three so-called *primary colors*, which may be taken as red, yellow, and blue. Actually, there are other three-color sets that can also produce a wide variety of colors when mixed. But the set that seems to produce the widest range, and that is considered today as the standard set of primary colors for mixing pigments, consists of the colors *magenta* (which is reddish-purple), *yellow*, and *cyan* (turquoise).

Each of these colors absorbs part of the white light that falls on it and reflects the rest. The range of frequencies that each reflects is indicated in Figure 24-20. Let's see how we can produce green by mixing cyan and yellow. When white light falls on a mixture of cyan and yellow, the yellow pigment absorbs light at the blue end of the spectrum, as shown in Figure 24-20, but reflects other frequencies. The cyan, on the other hand, absorbs all the wavelengths in the red region of the spectrum, whether they are from the original light or reflected by the molecules of the yellow pigment. The only light that won't be absorbed is that which both pigments reflect; this will be the wavelengths in the central part of the spectrum, which are predominantly green. Hence the mixture looks green. Because the result of this mixing depends on what colors are absorbed or subtracted out, it is called **subtractive mixing**.

FIGURE 24-21
See Color Plate VII.

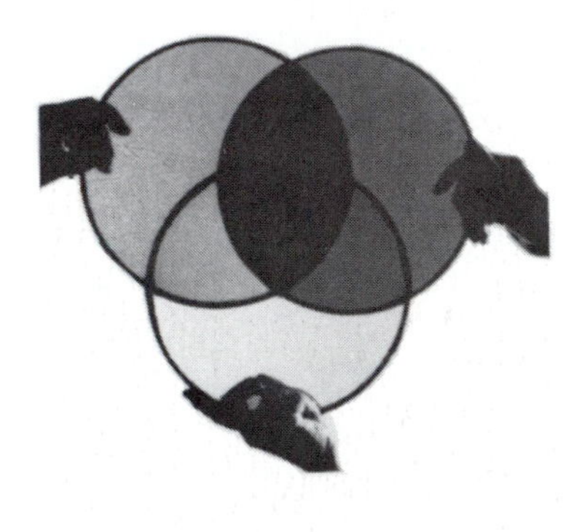

The result of mixing any two of the subtractive primary colors in equal amounts is shown in Figure 24-21 (Color Plate VII). We saw how green was obtained from a mixture of cyan and yellow. You can make a similar analysis for the mixtures of magenta and yellow, and of cyan and magenta. Other colors are obtained by mixing two or three primary colors in *un*equal amounts. A mixture of all three in equal amounts produces black (center of Figure 24-21), since light from all parts of the spectrum will be absorbed. Actually, gray or brown is sometimes obtained if the mixture is not exactly equal or if the pigments are not precisely those indicated in Figure 24-20.

Complementary colors are any two colors that, when mixed, produce black (or perhaps gray or muddy brown). For example, blue and yellow produce black, as in Figure 24-21. Since the pigments of complementary colors have no common region of the spectrum in which they overlap, little or no light is reflected.

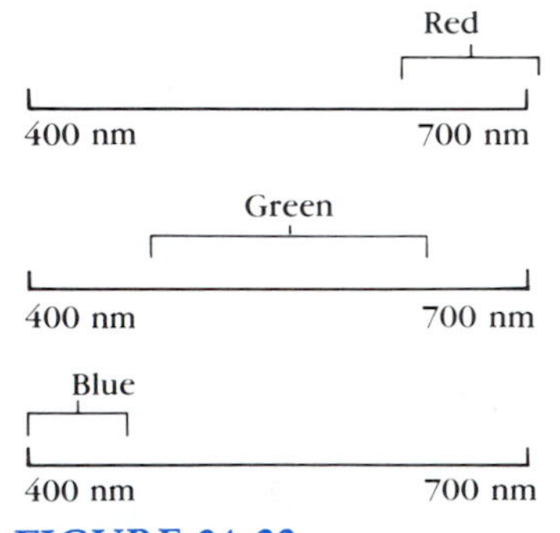

FIGURE 24-22
The additive primaries, showing approximately the spectrum of light each contains.

Mixing Colored Light—Additive Mixing

If you shine colored spotlights on a screen, new colors will be produced where they overlap. This kind of color mixing is called **additive mixing**, because the color you see is the *sum* of the spectrum of colors contained in each light. Again, nearly all colors can be obtained from three colors, but they are not the same as the subtractive primaries. Rather, the *additive primaries* are usually taken to be red, green, and blue, and the spectrum of light each contains is shown in Figure 24-22. The superposition of a red light and a green light on a screen means that all the wavelengths each contains will reach the screen, and so the screen will appear yellow (note the spectrum of yellow in Figure 24-20). Other combinations of the additive primaries are shown in Figure 24-23 (Color Plate VII). Mixing all three primaries together in equal proportions produces white, since all frequencies are present. And the mixing of two complementary colors also produces white (they are truly complementary only if one contains all the frequencies the other does not). For example, Figure 24-23 shows that yellow and blue produce white.

Color television works on the principle of color addition. If you look closely at the picture on a color set, you will see that there are tiny dots or lines of the three primary colors: red, green, and blue. When you move away from the set, these tiny spots of color merge in your perception. If all three colors are equally bright, white results. A wide range of colors is achieved by varying the intensity of each of the colored dots, Figure 24-24 (Color Plate VII). (A description of a TV tube is given in Chapter 21.) To prove to yourself that the eye alone can mix these colors, try the following experiment.

FIGURE 24-23
See Color Plate VII.

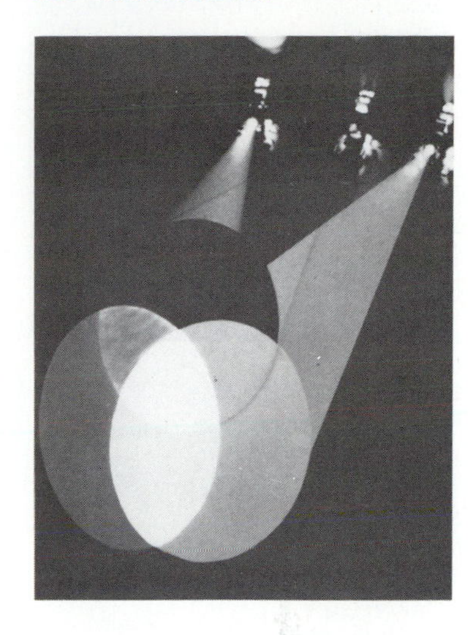

FIGURE 24-24
See Color Plate VII.

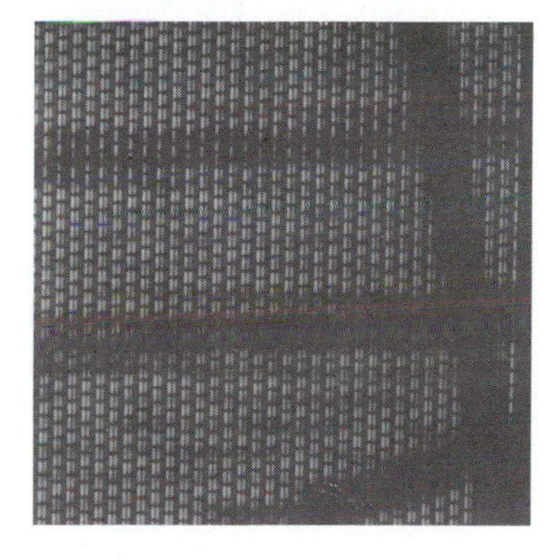

EXPERIMENT

Cut a filing card (or a piece of thin white cardboard) into a circle and paint one side with the three primary additives, as shown in Figure 24-25(a), and the other side with red and green, as in (b). Make two small holes on either side of center and run a string through them. Hold the string in both hands and rotate the card until the string is very twisted, Figure 24-25(c). When you release the card and pull gently on the string, the card will spin rapidly. One side will appear white and the other yellow. Explain.

If you look closely at a four-color illustration in a book, or a poster, you will see that it too is made up of many tiny colored dots much like those on a TV screen. Here, however, some colors overlap, so that the subtractive process is involved as well as the additive. Again a wide variety of colors can be obtained by using three primary colors, plus black (for emphasis and contrast).

The Impressionist Painters

In general, the process of mixing colors additively tends to produce a brighter color than mixing subtractively with pigments. This fact was used by many painters in the latter part of the nineteenth century, particularly those known as Impressionists and Postimpressionists. An example is

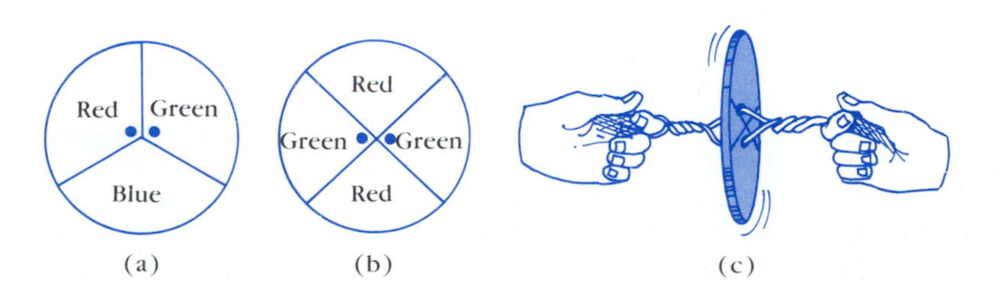

FIGURE 24-25
Experiment to show how the eye mixes color additively.

FIGURE 24-26
See Color Plate VIII.

Claude Monet's *La Grenouillière* (The Frog Pond), Figure 24-26 (Color Plate VIII). These artists tended to use only a small number of paints. They applied the colors with adjacent brush strokes (being sure any underlying paint was dry, so that no mixture of pigments occurred) and left it to the eye of the observer to do the mixing, much as color TV does today. This technique enabled them to produce intense and vibrant colors that seemed to capture the scintillating quality of sunlight and the texture of real objects. The fact that the appearance of the picture changes somewhat as the observer moves away gives the picture a kind of life of its own. Close up, the individual colors from each brushstroke are clear, whereas at a distance they fuse into a single color by additive mixture. When the observer stands at some intermediate point, the eye apparently sees the colors alternately fused and separated, and the painting seems to sparkle or scintillate. Try this with the Monet reproduction, Figure 24-26 (Color Plate VIII).

10. COLOR VISION

The retina of our eyes contains two kinds of light receptors, **rods** and **cones**. The rods are very sensitive to light (they respond even to small amounts of light), but they do not distinguish colors. The cones, on the other hand, distinguish the different colors but require brighter light to function. The eyes of most mammals other than humans and primates have only rods and thus do not perceive color.

Near the center of the retina is an area of very closely packed cones called the **fovea**. The sharpest image and the best color discrimination are therefore found in this area. When we view a scene, the eye moves so that the area of greatest interest is focused on the fovea. There are no rods in the fovea, but they are present in the regions surrounding it; in the peripheral regions of the eye, the rods predominate. You can prove this for yourself by having someone bring a brightly colored object from behind you so that you see it only peripherally. You won't be able to tell what color it is! When we see things directly, however, our brain usually remembers their color even when they are off to the side.

In very dim light, objects off to one side seem brighter than those focused at the fovea because the rods are more sensitive to light. However, we can't perceive colors well in dim light because the rods don't distinguish colors. Thus, although most stars appear plain white at night, long photographic time exposures reveal that some stars are blue and that others are red or orange.

EXPERIMENT

Sit in a dark room and let your eyes become adapted to the dark. Open a magazine to a colored picture. Try to tell what colors are present and where. Then turn on a bright light and see if you were right.

A number of theories have been proposed to explain how the cones distinguish different colors. The most likely theory assumes the existence of three kinds of cone corresponding to the three additive primary colors. Although the evidence is not fully conclusive, the fact that only three colors can produce all the colors we perceive lends support to this theory.

The rods and cones in the eye are connected by way of specialized cells to nerves that carry the information to the brain, which produces the sensation of color and arranges the information from the different receptors into a coherent whole, the "picture" we see. Actually, some of the receptors (rods and cones) are interconnected within the eye itself, so that some of the "thinking" or organizing is actually performed there.

SUMMARY

The wave theory of light is strongly supported by the fact that light exhibits interference and diffraction; this theory also explains the refraction of light and the fact that light travels more slowly in transparent solids and liquids than it does in air.

Diffraction refers to the fact that light, like other waves, bends around objects it passes and spreads out after passing through narrow slits.

Young's double-slit experiment clearly demonstrated the *interference* of light; the observed bright spots of the interference pattern were explained as constructive interference between the beams coming through the two slits, where they differ in path length by a whole number of wavelengths. The dark areas in between are due to destructive interference when the path lengths differ by $\frac{1}{2}\lambda$, $\frac{3}{2}\lambda$, and so on.

The wave nature of light limits the sharpness of images. Because of diffraction, it is not possible to discern details smaller than the wavelength of the radiation being used. This limits the useful magnification of a light microscope to about $1000\times$.

The wavelength of light determines its color; the *visible spectrum* extends from 400 nm (violet) to about 750 nm (red). Glass prisms break white light down into its constituent colors because the index of refraction varies with wavelength, a phenomenon known as *dispersion*.

Light reflected from the front and rear surfaces of a thin film of transparent material can interfere. Such thin-film interference has many practical applications, such as lens coatings and Newton's rings.

In *unpolarized light* the waves vibrate at all angles. If the wave vibrations are restricted to one plane, the light is said to be *plane-polarized*. When an unpolarized light beam passes through a Polaroid sheet, the emerging beam is plane-polarized. Light can also be partially or fully polarized by reflection. That light can be polarized shows that it must be a transverse wave.

Blue skies and red sunsets result from the scattering of light by molecules of air, which is stronger for shorter (blue) wavelengths.

Since most objects do not emit light, their color is due to the wavelengths of light they reflect. Colors occurring because of reflection, such as in pigments, combine *subtractively* to produce other colors. The resulting color of the mixture is due only to those wavelengths that are not absorbed (and are therefore reflected) by the pigments in the mixture. On the other hand, colors occurring because of the emission of light, as in colored lights, combine *additively*: the resulting color is due to the sum of all the wavelengths emitted by the sources.

The eye contains two kinds of light receptors, rods and cones. The *rods* are more sensitive to light but the *cones* distinguish colors.

QUESTIONS

1. Why is light sometimes described as rays and sometimes as waves?
2. We can hear sounds around corners, but we cannot see around corners; yet both sound and light are waves. Explain the difference.
3. Two rays of light from the same source destructively interfere if their path lengths differ by how much?
4. Why was the observation of the double-slit interference pattern more convincing evidence for the wave theory of light than the observation of diffraction?
5. Why can't a light microscope be used to observe molecules in a cell?
6. If magnifications greater than 1000 × were used in a light microscope, the interference patterns formed by diffraction around objects in the field of view would be evident. Do you think therefore that greater magnifications, say 5000 ×, would give a misleading picture of what the object actually looks like?
7. Is it better to use red light or blue light while photographing a tiny object through a microscope when you want to obtain maximum definition?
8. When white light passes through a flat piece of window glass, it is not broken down into its spectrum. Why not?
9. Why are Newton's rings (Figure 24-12) closer together farther from the center?
10. Sunlight will not pass through two Polaroids whose axes are at right angles. What happens if a third Polaroid, with an axis at 45° to each of the other two, is placed between them?
11. How could you use polarized sunglasses to tell if light from the sky (not directly from the sun) is polarized? Try it and see.
12. How would you go about determining whether a given pair of sunglasses were polarized or not?

13. What would be the color of the sky if the earth had no atmosphere?
14. If the earth's atmosphere were 50 times denser than it is, would sunlight still be white or would it be some other color?
15. Give an example of a nonspectral color. What is meant by a nonspectral color?
16. Why are you more comfortable in light-colored clothes on a hot day than in dark clothes?
17. What is the physical difference between red and orange light?
18. Describe the physical difference between the light from an object that appears yellow and one that appears green.
19. A performer wears blue clothes. How could you use spotlights to make them appear black?
20. A red spotlight and a blue spotlight overlap. What color will they produce?
21. An object appears magenta when illuminated by white light (see Figure 24-20). What color will it appear if it is illuminated by (a) blue light (b) yellow light (c) red light?
22. An object appears green in white light. What color will it appear to be if illuminated by (a) magenta light (b) cyan light (c) pure blue light?
23. Which colors are complementary to (a) magenta (b) red (c) green?
24. Impressionist (and other) painters found that shadows are more accurately portrayed if the objects in the shadows are tinted with their complementary color. Explain.
25. Cats do not see in color. How do their eyes differ from ours?
26. Why do objects seen in the moonlight appear so colorless?

EXERCISES

1. A screen containing two slits 0.100 mm apart is 1.20 m from the viewing screen. Light of wavelength $\lambda = 500$ nm falls on the slits from a distant source. Approximately how far apart will the bright interference fringes be on the screen?
2. Use an argument, similar to that used for Figure 24-3, to show why light entering a medium where its speed is faster will bend away from the perpendicular.
3. Monochromatic light falls on two slits 0.020 mm apart. The fringes on a screen 3.00 m away are 9.2 cm apart. What is the wavelength of the light?
4. A parallel beam of 600 nm light falls on two small slits 5.0×10^{-2} mm apart. How far apart are the fringes on a screen 5.0 m away?

5. Light of wavelength 680 nm falls on two slits and produces an interference pattern in which the fringes are 7.0 mm apart on a screen 1.0 m away. What is the separation of the two slits?
6. A red light has a wavelength of 620 nm. What is the frequency?
7. A light wave has a frequency of 5.8×10^{14} cycles per second. What is the wavelength of the light?
8. If the wavelength of green light is 520 nm in air, what will its wavelength be when it passes through glass with an index of refraction of 1.6? (*Hint*: The frequency does not change.)
9. A lens appears greenish yellow ($\lambda = 570$ nm is strongest) when white light reflects from it. What minimum thickness of coating ($n = 1.25$) do you think is used on such a (glass) lens, and why?

*10. A total of 33 bright and 33 dark Newton's rings are observed when 450-nm light falls normally on a planoconvex lens resting on a flat glass surface (Figure 24-12). How much thicker is the center than the edges?

*11. If a soap bubble is 120 nm thick, what color will appear at the center when illuminated normally by white light? Assume that $n = 1.34$.

12. A single polarizer ideally transmits 50 percent of the unpolarized light falling on it. How much will be transmitted if two screens are used (a) with their axes parallel (b) with their axes crossed at 90°?

*13. Television and radio waves can reflect from nearby mountains or from airplanes, and the reflections can interfere with the direct signal from the station. Determine what kind of interference will occur when 75-MHz television signals arrive at a receiver directly from a distant station and reflected from an airplane 118 m directly above the receiver. (Assume no change in phase of the signal upon reflection.)

SPECIAL THEORY OF RELATIVITY

CHAPTER

25

Physics at the end of the nineteenth century looked back on a period of great progress. The theories developed over the preceding three centuries had been very successful in explaining a wide range of natural phenomena. Newtonian mechanics beautifully explained the motion of objects on earth and in the heavens; furthermore, it formed the basis for successful treatments of fluids, wave motion, and sound. Kinetic theory, on the other hand, explained the behavior of gases and other materials. And Maxwell's theory of electromagnetism not only brought together and explained electric and magnetic phenomena, but it predicted the existence of electromagnetic waves that would behave in every way just like light—so light came to be thought of as an electromagnetic wave. Indeed, it seemed that the natural world, as seen through the eyes of physicists, was very well explained; a few puzzles remained, and it was felt that these would soon be explained using already known principles.

But it did not turn out to be so simple. Instead, those puzzles were only to be solved by the introduction, in the early part of the twentieth century, of two revolutionary new theories that changed our whole conception of nature: the *theory of relativity* and *quantum theory*.

Physics as it was known at the end of the nineteenth century (what we've covered thus far in this book) is referred to as **classical physics**. The new physics that grew out of the great revolution at the beginning of the twentieth century is now called **modern physics**. In this chapter we present the special theory of relativity, which was first proposed by Albert Einstein in 1905 (Figure 25-1). In the following chapters we discuss the equally momentous quantum theory, nuclear and elementary particle physics, and finally Einstein's general theory of relativity and cosmology.

FIGURE 25-1
Albert Einstein.

1. REFERENCE FRAMES AND THE ADDITION OF VELOCITIES

The theory of relativity carefully examines how we observe physical *events*. It thus deals with the very foundation of scientific investigation. The first question we must answer is "How do we specify an event?" The answer is that we designate both the time of the event and the place at which it occurred. For example, the statement "The first shot of the American Revolution was fired in Concord on April 19, 1775" specifies both the place and the time of this event. If we don't know where Concord is, we can find its coordinates on a map, namely, its latitude and longitude. The theory of relativity is concerned with the times and places at which events take place as observed from different points of view, or different "reference frames."

Measurements Are Made Relative to a Frame of Reference

Whenever we describe an event we have observed, we describe it from a particular frame of reference. Let's take an example. You are traveling on a train and observe a bird flying above you in the sky. You measure its speed and say, "That bird is flying at 35 kilometers per hour." A friend who is traveling with you replies, "Your statement is not really complete: the bird's speed is 35 km/hr relative to what? Is it 35 km/hr with respect to the earth or is it 35 km/hr with respect to the train? What's your frame of reference?" You ignore your friend and take a seat in the club car. You note that the train is traveling 100 km/hr. A waiter walks past swiftly—according to your estimate, about 5 km with respect to the train. To an observer standing on the ground, however, the waiter's speed is considerably greater. If the waiter is walking toward the front of the train, his speed is 100 km/hr plus 5 km/hr, or 105 km/hr *with respect to the earth* (Figure 25-2). If he is walking toward the back of the train, his speed is 100 km/hr minus 5 km/hr, or 95 km/hr with respect to the earth. The waiter's speed with respect to the earth is certainly much different from what it is with respect to the train.

Clearly, then, whenever we talk about a speed or a velocity, we must specify the reference point or **reference frame** from which it is measured. In the examples just described, we mentioned two reference frames: the

FIGURE 25-2
A train moving at 100 km/hr. The waiter walking up the aisle has a different speed according to the two observers. The man seated on the train observes the waiter passing him with a speed of 5 km/hr. The man on the ground observes the waiter passing him with a speed of 105 km/hr.

earth and the train. The velocities we spoke of were measured with respect to one or the other of these reference frames. In everyday speech we seldom specify the reference frame, since it is usually understood from the context. Often the reference frame we have in mind is the earth, but not always.

Addition of Velocities

Let's follow that waiter walking on the train. Suppose we know his velocity with respect to the train, call it v_1, and the velocity of the train with respect to the earth, call it v_2. To find the velocity, v, of the waiter with respect to the earth, we simply *add* the two velocities if they are both in the same direction: $v = v_2 + v_1$. If the two velocities are in opposite directions, we subtract them: $v = v_2 - v_1$. This simple procedure is known as the *addition of velocities*.

In a similar way, for a boat traveling up or down a river we can calculate the velocity with respect to the river bank by adding the velocity of the river current to the velocity of the boat with respect to the water (Figure 25-3). Suppose the boat can go 10 km/hr in still water. If the speed of the river is 3 km/hr, how fast will the boat be going with respect to the earth? If the boat is going downstream, its velocity with respect to the earth will be: $v = 10\ \text{km/hr} + 3\ \text{km/hr} = 13\ \text{km/hr}$. If the boat is going upstream, it will be fighting the current and its velocity will be less: $v = 10\ \text{km/hr} - 3\ \text{km/hr} = 7\ \text{km/hr}$. These results make sense, for obviously it takes longer to travel the same distance going upstream than going downstream.

Here's another example. A ballplayer who is standing still throws a baseball with a velocity of 20 m/s. A second player runs toward the ball at a speed of 5 m/s with respect to the ground and catches it. The ball strikes the second player's glove harder than it would had he been at rest, because the velocity of the ball with respect to the second ballplayer is

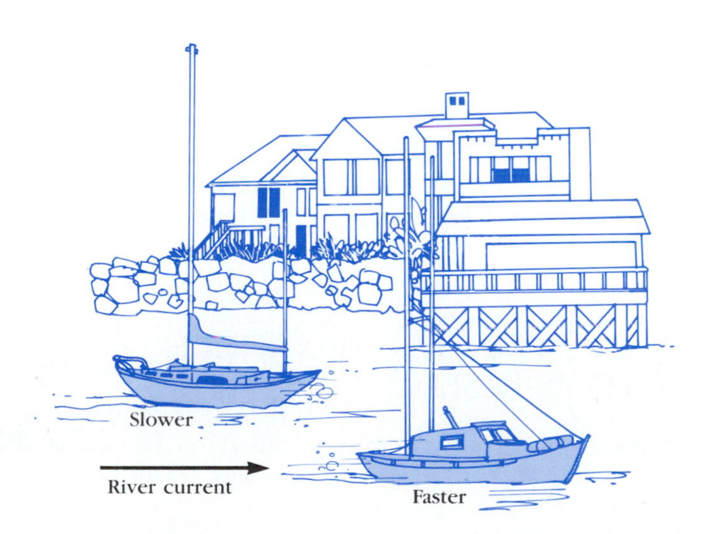

FIGURE 25-3
A boat moving upstream moves slower with respect to the earth than a boat moving downstream.

$v = 20$ m/s + 5 m/s = 25 m/s. If the second ballplayer were running *away* from the ball at 5 m/s, the velocity of the ball with respect to the player would be $v = 20$ m/s − 5 m/s = 15 m/s, and it would be a "softer" catch.

What happens if the player who throws the ball is moving, say, at 5 m/s, and the player who catches the ball is at rest? The thrower still throws the ball at 20 m/s but the velocity of the ball with respect to the second player will be either 25 m/s or 15 m/s, depending on whether the thrower is moving toward or away from the second player. We might say that the ball acquires the additional velocity of the thrower's frame of reference.

Clearly, then, the velocity of an object is different when viewed from different reference frames. In other words, velocity is a *relative* concept. And we say that "motion is relative." This, of course, is not new. It was understood centuries ago by Galileo and Newton.

Position and Time Are Relative

The position of an object is also a relative concept. San Francisco is 600 kilometers from Los Angeles but 4000 kilometers from New York City. It all depends on your point of reference. If someone were to say that Chicago is 1000 kilometers away, you would need to know the speaker's point of reference: 1000 kilometers away from where?

What about time? Is time a relative or an absolute concept? Consider this hypothetical statement: "A football game was played between Michigan and California on January 1, 1989, beginning at 1:30 P.M., PST; exactly 53 seconds after the game began, California scored a touchdown." Presumably, you will accept this as a complete statement. Whether the observer is in the stands or in an airplane flying overhead, experience leads us to believe that he would observe that the touchdown occurred 53 seconds after the start of the game. In other words, we think of time as an absolute concept, not a relative one. Until the twentieth century, physicists were convinced of the absolute nature of time. That conviction is actually built into Newtonian mechanics. In the early part of this century, however, Einstein argued in his special theory of relativity that the absolute concept of time is not correct. For example, the observer in the airplane would not agree that the touchdown occurred 53 seconds after the game began. He would observe that it took somewhat longer, perhaps 54 seconds or maybe 67 seconds—the exact value depends on the velocity of his reference frame, the airplane. Strange, you say? We will soon see how this comes about.

2. THE TWO PRINCIPLES ON WHICH RELATIVITY IS BASED

Einstein's special theory of relativity is based on two principles: the so-called "principle of relativity" and the "principle of the constancy of the speed of light."

The Relativity Principle

The first principle, which had in essence been stated earlier by Galileo and others before him, is the **relativity principle**:

> **The laws of nature are the same in all reference frames that move uniformly (i.e. at constant velocity) with respect to each other.**

In other words, the laws of physics have the same form on land as they do, say, on a train moving in a straight line at 100 km/hr. For example, playing table tennis or billiards on a train or a ship is no different than in a building, unless the train or ship lurches, in which case it is not (for that moment) moving uniformly but is accelerating. You can feed yourself normally in a uniformly moving Boeing 757 aircraft; if you drop a fork, it falls just as it does on the earth. Moving objects follow the same laws on a uniformly moving train, ship, or plane as they do on land. One way to convince yourself is to do the following experiment:

EXPERIMENT

Take a ride in an automobile (not a convertible) with someone else driving. Close the windows to avoid any wind inside the car. With the car moving at a constant speed, hold an object (such as a ball or a coin) above you and let it drop. Notice its path. Is it any different from what it would be if you were at rest on the ground when you dropped the object?

Whether you drop an object from a fixed position on the earth or you drop an object from a fixed position inside a vehicle moving at a uniform speed, the object falls in a straight line. (However, if you drop an object out the window of a moving vehicle, its path is curved because the air—wind—exerts an extra force on the object.)

We said that if you dropped the object inside the car it would fall straight down. And it does—with respect to the car. But to an observer standing along the road, the object would follow a curved path (Figure 25-4). The path of the object is different for the two observers. But this does not violate the relativity principle, because it is only the *laws* of physics that are the same in these two reference frames. It is the *paths* that look different. The curved path seen by an observer on land is in accord

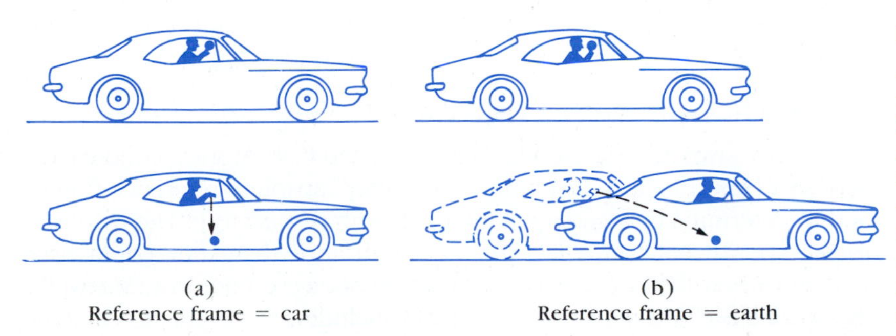

FIGURE 25-4
A coin is dropped by a person in a moving car. (a) In the reference frame of the car, the coin falls straight down. (b) In the reference frame fixed on the earth, the coin follows a curved (parabolic) path. The upper views show the moment of the coin's release, and the lower views a short time later.

with the same laws of physics as the straight path seen by the observer in the car. To the observer in the car the coin was dropped from a fixed position, so fell straight down. To the observer on land, the object had an initial velocity (equal to the velocity of the car) and therefore followed a curved path like a projectile.

The Relativity Principle Asserts That There Is No Absolute Reference Frame at Rest

The relativity principle tells us that one reference frame is as valid as any other that moves uniformly with respect to the first for specifying the laws of physics. Furthermore, an observer on a train moving uniformly with respect to the earth is perfectly justified in saying that he is at rest and the earth is moving beneath him. What we really mean by "relativity" is that one reference frame is as valid as another. *There is no preferred reference frame*. There is no reference frame (or object) that we can claim as being at absolute rest, and which therefore would be a preferred one from which to make measurements.

Scientists before Einstein assumed that there must be a "preferred" reference frame, one that is truly at rest. Early in history it was thought that the earth was this absolute reference frame at rest and that the planets, the sun, and the stars revolved around the earth. Later, Copernicus and Galileo found that the motion of the heavenly bodies was most easily explained by assuming that the earth moves about the sun. So perhaps we could claim that the sun is absolutely at rest? Alas, that assumption is also erroneous, since the sun moves with respect to the center of our galaxy. And the galaxy moves with respect to . . . well, let's say the distant stars (or do the distant stars move with respect to our galaxy?) As Galileo said, "Everything moves." Everything moves with respect to everything else. We cannot tell by experiment whether our galaxy moves with respect to others or whether others move with respect to ours. That is what relativity tells us: The laws of physics are the same in every uniformly moving reference frame, and so we cannot tell which is the moving one and which is the one at rest. All we can say is that two reference frames are moving *with respect to each other*.

Even in the nineteenth century most physicists believed that there was an absolute reference frame that is at rest. In searching for it they discovered something startling, whose explanation came only in 1905 with Einstein's theory of relativity. Here is what they found.

The Michelson–Morley Experiment

In the 1880s two Americans, A. A. Michelson and E. W. Morley, conducted a series of experiments to find the sought-after "absolute reference frame" and to determine the velocity of the earth with respect to it. The results of their experiments had a far greater significance than they could have realized when they began. Their experiments were based on Maxwell's theory of electromagnetism (Chapter 21), including the prediction that the velocity of light is 300,000 km/s. Scientists expected that this must be its velocity with respect to the "absolute reference frame." The velocity of light in any other reference frame would presumably be different, since the velocity of that reference frame would have to be added on.

By measuring the speed of light and determining in what reference frame it has the velocity predicted by Maxwell, Michelson and Morley sought to determine whether this absolute frame was fixed to the earth, to the sun, or what. Similar experiments were subsequently performed by others. We now discuss one set of experiments in a very simplified way. As shown in Figure 25-5, the light coming from a distant star can be observed at times of the year six months apart, say in January and July. For the purpose of argument assume that the star is at rest with respect to the sun and the velocity of light as it leaves the star is $c = 300{,}000$ km/s. Now the earth moves in its orbit with a velocity, call it u, equal to 30 km/s with respect to the sun. This is very fast indeed, but it is small compared to the velocity of light. Nonetheless, Michelson and Morley had excellent equipment, namely the "interferometer" invented by Michelson, which could detect even small differences in the velocity of light. If the light leaves the star with velocity c and the earth is moving toward the star with velocity u (January in Figure 25-5), then an observer on earth would expect to measure the velocity of the star's light to be $c + u$. Six months later, in July, the earth is going away from the star, and hence the star's light should have a velocity with respect to the earth of $c - u$. The velocity of the light from the star should be less in July when the earth is moving away from the star than it is in January when the earth is approaching the star.

To the astonishment of scientists, Michelson and Morley found that the velocity of the light from the star with respect to the earth was exactly the same in July as it was in January! They tested all sorts of different cases using different sources of light. In every case the measured velocity of the light was the same. Michelson and Morley had set out to find a fairly simple entity, the "reference frame at absolute rest," but they discovered instead a far more intriguing and complex phenomenon.

Many attempts were made in the ensuing years to explain the results of the Michelson–Morley experiments. Most were complicated explanations

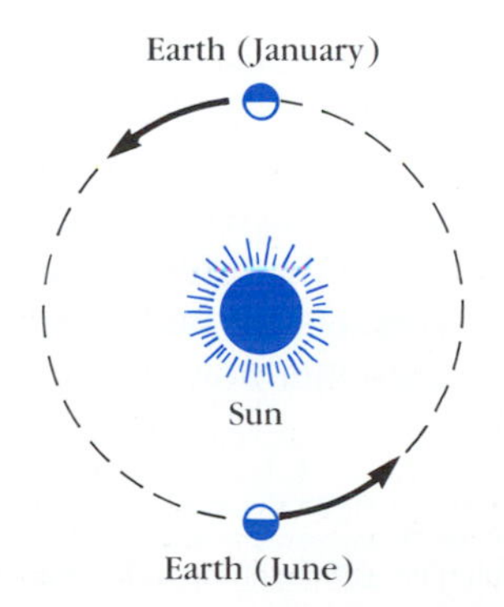

FIGURE 25-5
The orientation of the Sun, Earth, and a distant star at two different times; in January the Earth is moving toward the star, in June away from it.

and were eventually discredited by internal contradictions or by further experimental results. Einstein finally clarified the situation.

The Second Principle: The Constancy of the Speed of Light

Einstein, convinced that there was no reference frame at absolute rest, proposed a second principle called the **principle of the constancy of the speed of light**:

> **The speed of light in empty space is the same for all observers regardless of their velocity or the velocity of the source of the light.**

This second principle of the special theory of relativity is consistent, in a simple way, with the result of the Michelson–Morley experiment.† It says that an observer traveling toward a source of light would measure exactly the same velocity for that light as would an observer at rest with respect to the light. If a ballplayer runs toward a baseball coming toward him, the velocity of the baseball is the sum of the player's velocity with respect to the ground plus the velocity of the ball with respect to the ground, as we saw earlier. But for light, the velocity with respect to the moving observer is the *same* as its velocity with respect to the ground. We do not add the velocity of the observer! Now this violates common sense. How can this bizarre fact be reconciled with our everyday experience? Part of the problem is that in our everyday experience we do not measure velocities anywhere near the velocity of light, and so we cannot expect our everyday experience to be helpful in situations at very high speeds.

The special theory of relativity is based entirely on the two principles we have discussed: (1) the relativity principle and (2) the constancy of the speed of light. From these two principles Einstein derived a number of extremely interesting results, which we now discuss.

3. SIMULTANEITY AND TIME DILATION

Simultaneity Is Not Absolute

Perhaps the most striking aspect of the theory of relativity, at least philosophically, is that we can no longer regard time as an absolute quantity. No one doubts that time flows on and never turns back. Yet the time between two events, and even whether two events are simultaneous, depends on the reference frame in which the events are being observed.

Two events are said to occur simultaneously if they occur at exactly the same time. But how do we tell if two events occur precisely at the same

† The Michelson–Morley experiment can also be considered as evidence for the first postulate, for it was intended to measure the motion of the earth relative to an absolute reference frame; its failure to do so implies the absence of any such preferred frame.

time? If they occur at the same point in space—such as two apples falling on your head at the same time—it is easy. But if the two events occur at widely separated places, it is more difficult to know since we have to take into account the time it takes for the light to reach us. Because light travels at finite speed, a person who sees two events must calculate back to find out when they actually occurred. For example, if two events are *observed* to occur at the same time but one actually took place farther from the observer than the other, then the former must have occurred earlier, and the two events were not simultaneous.

To avoid making calculations, we will now make use of a simple thought experiment. Einstein himself thought up similar simple experiments that, though not always practical to carry out, do illustrate the important principles; he called these "thought experiments"—in German, "gedanken" experiments. We assume that an observer, called O, is located exactly halfway between points A and B where two events occur (Figure 25-6). The two events may be lightning that strikes at both points A and B, as shown, or any other type of event. For brief events like lightning strikes, only short pulses of light will travel outward from A and B and reach O. O "sees" the events when the pulses of light reach point O. If the two pulses reach O at the same time, then the two events had to be simultaneous. This is because the two light pulses travel at the same speed; and since the distance OA equals OB, the time for the light to travel from A to O and from B to O must be the same. Observer O can then definitely state the two events occurred simultaneously. On the other hand, if O sees the light from one event before that from the other, then it is certain the former event occurred first.

FIGURE 25-6
A moment after lightning strikes points A and B, the pulses of light are traveling toward O, but O "sees" the lightning only when the light reaches him or her.

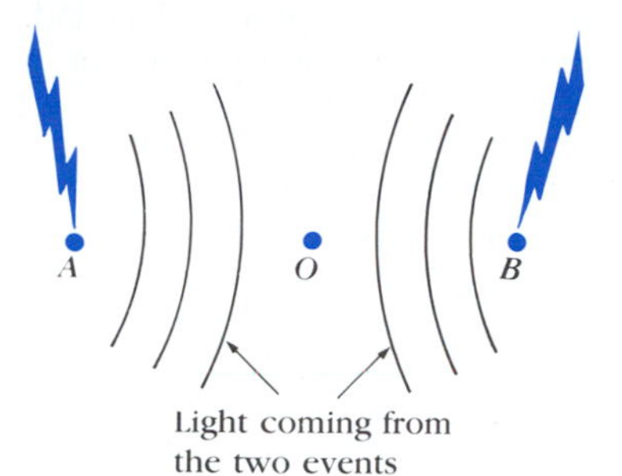

The question we really want to examine is this: If two events are simultaneous to an observer in one reference frame, are they also simultaneous to another observer moving with respect to the first? Let us call the observers O_1 and O_2 and assume that they are fixed in reference frames 1 and 2 that move with speed v relative to one another. These two reference frames can be thought of as railroad cars (Figure 25-7). O_2 says that O_1 is moving to the right with speed v, as in (a); and O_1 says O_2 is moving to the left, as in (b). Both viewpoints are legitimate according to the relativity principle. (There is, of course, no third point of view that will tell us which one is "really" moving.)

Now suppose two events occur that are seen by both observers. Let us assume again that the two events are the striking of lightning and that the lightning marks both trains where it struck: at A_1 and B_1 on O_1's train, and

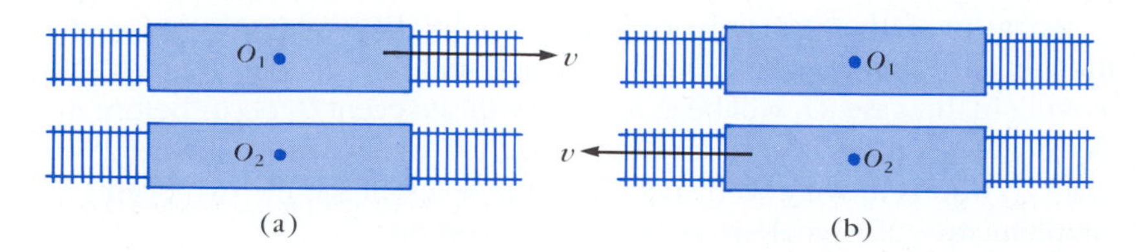

FIGURE 25-7
Observers O_1 and O_2 on two different trains (two different reference frames), are moving with relative velocity v. O_2 says that O_1 is moving to the right (a); O_1 says that O_2 is moving to the left (b). Both viewpoints are legitimate—it all depends on your reference frame.

FIGURE 25-8
Two reference frames as viewed by an observer O_2. To O_2, the other reference frame (O_1) is moving to the right. In (a) one lightning bolt strikes in both reference frames, at A_1 and A_2, and a second lightning bolt strikes at B_1 and B_2. According to the observer O_2,the two bolts of lightning strike simultaneously. (b) A moment later the light from the two events reaches O_2 at the same time (simultaneously). But in the other (O_1) reference frame, the light from B_1 has already reached O_1 whereas the light from A_1 has not yet reached O_1. So in O_1's reference frame, the event at B_1 must have preceded the event at A_1. Time is not absolute!

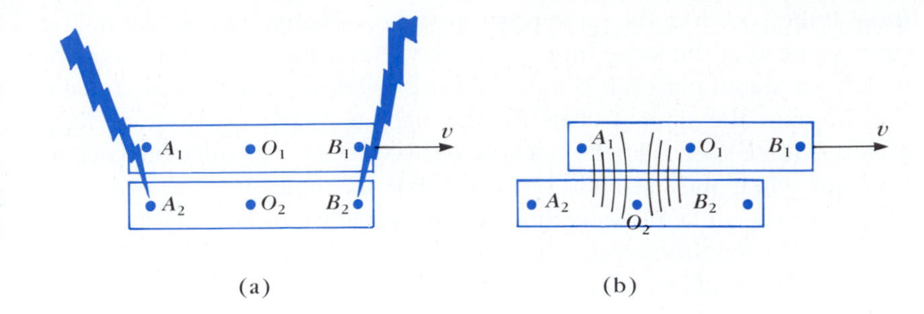

at A_2 and B_2 on O_2's train. For simplicity, we assume O_1 happens to be exactly halfway between A_1 and B_1, and that O_2 is halfway between A_2 and B_2. We must now put ourselves in the one reference frame or the other. Let us put ourselves in O_2's reference frame, so we see O_1 moving to the right with speed v. Let us also assume that the two events occur *simultaneously* in O_2's frame and just at the time when O_1 and O_2 are opposite each other, Figure 25-8(a). A short time later, Figure 25-8(b), the light reaches O_2 from A_2 and B_2 at the same time. Since O_2 knows (or measures) the distances O_2A_2 and O_2B_2 as equal, he knows the two events are simultaneous.

But what does observer O_1 see? From our (O_2) reference frame, we can see that O_1 moves to the right during the time the light is traveling to O_1 from A_1 and B_1. As can be seen in Figure 25-8(b), the light from B_1 has already passed O_1, whereas that from A_1 has not yet reached O_1. Therefore, it is clear that O_1 sees the light from B_1 before that from A_1. Now O_1's frame is as good as O_2's. Light travels at the same speed for O_1 as for O_2, and is the same from A_1 to O_1 as from B_1 to O_1. Since the distance O_1A_1 equals O_1B_1, observer O_1 must conclude that the event at B_1 occurred before the event at A_1.

We thus find that *two events which are simultaneous to one observer are not necessarily simultaneous to a second observer*.

It may be tempting to ask: "Which observer is right, O_1 or O_2?" The answer, according to relativity, is that they are *both* right. There is no "best" reference frame we can choose to determine which observer is right. Both frames are equally good. We can only conclude that simultaneity is not an absolute concept, but is relative. We are not aware of it in everyday life, however, since the effect is noticeable only when the relative speed of the two reference frames is very large (near the speed of light) or the distances involved are very large.

Because of the principle of relativity, the argument given for the thought experiment of Figure 25-8 can be done from O_1's reference frame as well. In this case, O_1 will be at rest and will see event B_1 occur before A_1. But O_1 will recognize (by drawing a diagram equivalent to Figure 25-8) that O_2, who is moving with speed v to the left, will see the two events as simultaneous. The analysis is left as an exercise.

Time Dilation: Moving Clocks Run Slowly

The fact that two events simultaneous to one observer may not be simultaneous to a second observer suggests that time itself is not absolute; could it be that time passes differently in one reference frame than another? This is, indeed, just what Einstein's theory of relativity predicts, as the following thought experiment shows.

Figure 25-9 shows a spaceship traveling past earth at high speed. The point of view of an observer on the spaceship is shown in part (a) and that of an observer on earth in part (b). Both observers have accurate clocks. The person on the spaceship (a) flashes a light and measures the time it takes the light to travel across the spaceship and return after reflecting from a mirror. The light travels a distance $2D$ at speed c so the time required, which we call t_0, is

$$t_0 = \frac{2D}{c}$$

since velocity = distance/time or $c = 2D/t_0$.

The observer on earth—Figure 25-9(b)—observes the same process. But to this observer, the spaceship is moving; so the light travels the diagonal path shown in going across the spaceship, reflecting off the mirror, and returning to the sender. Although the light travels at the same speed to this observer (the second postulate), it travels a greater distance. Hence the time required, as measured by the earth observer, will be greater than that measured by the observer on the spaceship. The time, t,

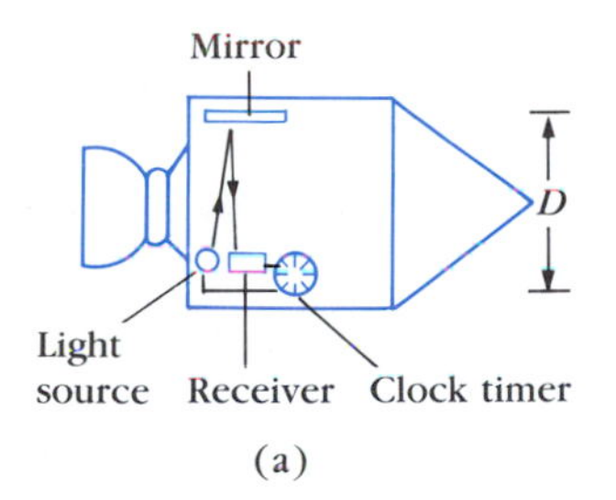

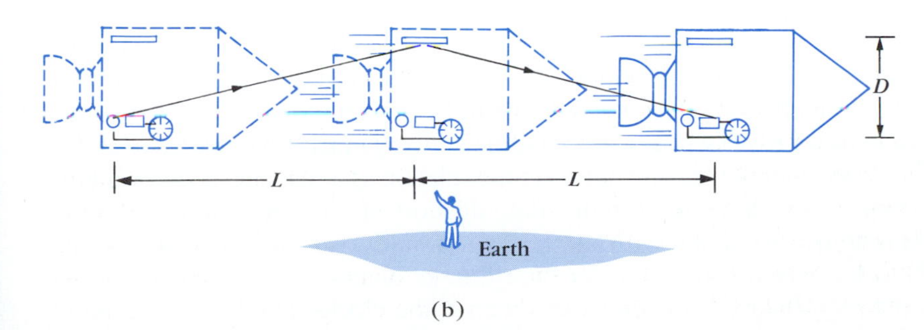

FIGURE 25-9
Time dilation can be shown by a thought experiment: the time it takes for the light to travel over and back on a spaceship is longer for the earth observer (b) than for the observer on the spaceship (a).

as observed by the earth observer can be calculated as follows. In the time t, the spaceship travels a distance $2L = vt$ where v is the speed of the spaceship—Figure 25-9(b). Thus, the light travels a total distance on its diagonal path of $2\sqrt{D^2 + L^2}$ (we used the theorem of Pythagoras for right triangles) and therefore

$$c = \frac{2\sqrt{D^2 + L^2}}{t} = \frac{2\sqrt{D^2 + v^2t^2/4}}{t}.$$

We square both sides and solve for t to find

$$c^2 = \frac{4D^2}{t^2} + v^2,$$

$$t = \frac{2D}{c^2 - v^2} = \frac{2D}{c\sqrt{1 - v^2/c^2}}.$$

We combine this with the formula above for t_0 and find:

$$t = \frac{t_0}{\sqrt{1 - v^2/c^2}}.$$

Since $\sqrt{1 - v^2/c^2}$ is always less than 1, we see that t is always greater than t_0. That is, the time between the two events (the sending of the light and its reception at the receiver) is *greater* for the earth observer than for the traveling observer. This is a general result of the theory of relativity, and is known as **time dilation**. Stated simply, the time dilation effect says that

moving clocks are measured to run slowly.

However, we should not think that the clocks are somehow at fault. To the contrary; we assume the clocks are good ones. Time is actually measured to pass more slowly in any moving reference frame as compared to your own. This remarkable result is an inevitable outcome of the two postulates of the theory of relativity.

The concept of time dilation may be hard to accept, for it violates our commonsense understanding. We can see from the above equation that the time-dilation effect is negligible unless v is reasonably close to c. The speeds we experience in everyday life are very much smaller than c, so it is little wonder we haven't noticed time dilation as a real effect. Experiments have been done to test the time-dilation effect, and have confirmed Einstein's predictions. In 1971, for example, extremely precise atomic clocks were flown around the world in jet planes. Since the speed of the planes (10^3 km/hr) is much less than c, the clocks had to be accurate to nanoseconds (10^{-9} s) in order to detect the time-dilation effect. They were this accurate and confirmed the time-dilation equation to within experimental error. Time dilation had been confirmed decades earlier, however, by observation of "elementary particles" (see Chapter 31) which have very small masses (typically 10^{-30} to 10^{-27} kg) and so require little energy to be accelerated to speeds close to c. Many of these elementary particles are not stable and decay after a time into simpler particles. One example is the muon whose mean lifetime is $2.2 \mu s$ (microseconds) when at rest. Careful experiments showed that when a muon is traveling at high speeds, its lifetime increases just as predicted by the time-dilation formula. For example, let us calculate the mean lifetime of a muon as measured in the laboratory if it is traveling $0.60\,c = 1.8 \times 10^8$ m/s. (Its mean life at rest is 2.2×10^{-6} s.) If an observer were to move along with the muon (the muon would be at rest to this observer), the muon would have a mean life of 2.2×10^{-6} s. To an observer in the lab, the muon lives longer because of time dilation. From the time-dilation equation, with $v = 0.60\,c$, we get

$$t = \frac{t_0}{\sqrt{1 - v^2/c^2}} = \frac{2.2 \times 10^{-6}\text{ s}}{\sqrt{1 - 0.36\,c^2/c^2}} = \frac{2.2 \times 10^{-6}\text{ s}}{\sqrt{0.64}} = 2.8 \times 10^{-6}\text{ s}.$$

We must comment about the time-dilation formula and the meaning of t and t_0. The equation is true only when t_0 represents the time interval between the two events in a reference frame where the two events occur *at the same point in space*—as in Figure 25-9(a). This time interval, t_0, is often called the *proper time*. Then t represents the time interval between the two events as measured in a reference frame moving with speed v with respect to the first. In the preceding example, t_0 (and not t) was set equal to 2.2×10^{-6} s because it is only in the rest frame of the muon that the two events ("birth" and "decay") occur at the same point in space.

Space Travel

Time dilation has aroused interesting speculation about space travel. Under the old time regime, to reach a star 100 light-years away would not be possible by ordinary mortals (1 light-year is the distance light can travel in 1 year). Even if a rocket ship could travel at close to the speed of light, it would take over 100 years to reach such a star. But time dilation tells us that the time involved would be less for an astronaut. In a spaceship traveling at $v = 0.999\,c$, the time for such a trip would be only about $t_0 = t\sqrt{1 - v^2/c^2} = (100\text{ yr})(\sqrt{1 - (0.999)^2}) = 4.5\text{ yr}$. Thus a person could make such a trip. Time dilation allows such a trip, but the enormous practical problems of achieving such speeds will not be overcome in the foreseeable future.

Notice, in this example, that where 100 years pass on earth only 4.5 years pass for the astronaut on the trip. Is it just the clocks that slow down for the astronaut? The answer is no. All processes, including life processes, are seen to run more slowly for the astronaut according to the earth observer. But to the astronaut, time passes in the normal way. The astronaut experiences 4.5 years of normal sleeping, eating, reading, and so on; and people on earth experience 100 years of ordinary activity.

4. LENGTH CONTRACTION

Because time intervals are different in different reference frames, we might expect space intervals—lengths and distances—to be different as well. This is indeed the case, and we illustrate with a thought experiment.

An observer on earth watches a spacecraft traveling at speed v from earth to, say, Neptune, Figure 25-10(a). The distance between the planets, as measured by an earth observer, is L_0. The time required for the trip, measured from earth, is

$$t = L_0/v.$$

In Figure 25-10(b) we see the point of view of an observer on the spacecraft. In this frame of reference the spaceship is at rest; the earth and Neptune move with speed v. (We assume that v is much greater than the relative speed of Neptune and Earth, so the latter can be ignored.) But the time between the departure of Earth and the arrival of Neptune is less for the spacecraft observer than for the earth observer because of time dilation: the time for the trip as viewed by the spacecraft is $t_0 = t\sqrt{1 - v^2/c^2}$. Since the spacecraft observer measures the same speed but less time between these two events, he must also measure the distance as less. If we let L be the distance between the planets as viewed by the spacecraft observer, then $L = vt_0$. We have already seen that $t_0 = t\sqrt{1 - v^2/c^2}$ and $t = L_0/v$, so we have $L = vt_0 = vt\sqrt{1 - v^2/c^2} = L_0\sqrt{1 - v^2/c^2}$. That is,

$$L = L_0\sqrt{1 - v^2/c^2}.$$

This is a general result of the special theory of relativity and applies to

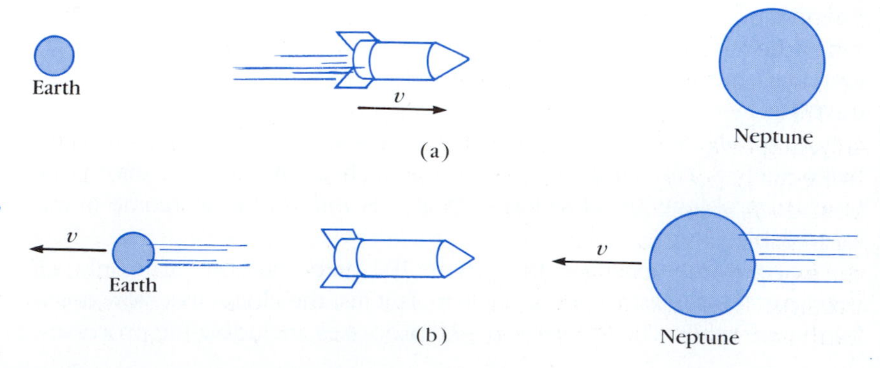

FIGURE 25-10
(a) A spaceship traveling at very high speed from Earth to Neptune, as seen from Earth's frame of reference. (b) As viewed by an observer on the spaceship, Earth and Neptune are moving at the very high velocity v.

lengths of objects as well as to distance. The result can be stated most simply in words as:

The length of an object is measured to be shorter when it is moving than when it is at rest.

This is called **length contraction**. The length L_0 is called the *proper length*; it is the length of the object (or distance between two objects) as measured by an observer at rest with respect to it. L is the length that will be measured when the object travels by an observer at speed v. It is important to note that length contraction occurs *only along the direction of motion*. For example, the moving spaceship in Figure 25-10(a) is shortened in length, but its height is the same as when it is at rest.

Length contraction, like time dilation, is not noticeable in everyday life because the factor $\sqrt{1 - v^2/c^2}$ differs from 1.00 significantly only when v is very large.

As an example, suppose a spaceship passes the earth at speed $v = 0.80\ c$. What will be the changes in length of a meter stick as it is slowly rotated from vertical to horizontal by a person inside as viewed (a) by another person in the spaceship (b) by an observer on earth? (a) The meter stick looks 1.0 m long in all orientations since it is at rest. (b) The meter stick varies in length from 1.0 m (vertical) to $L = (1.0\text{ m}) \times \sqrt{1 - (0.80)^2} = 0.60\text{ m}$ long in the horizontal direction (assuming that is the direction of motion).

5. THE TWIN PARADOX

Not long after Einstein proposed the special theory of relativity, an apparent paradox was pointed out. According to this "twin paradox," suppose one of a pair of 20-year-old twins takes off in a spaceship traveling at very high speed to a distant star and back again while the other twin remains on earth. According to the earth twin, the traveling twin will age less. Whereas 20 years might pass for the earth twin, perhaps only 1 year (depending on the spacecraft's speed) would pass for the traveler. Thus, when the traveler returns, the earthbound twin could expect to be 40 years old whereas his twin would be only 21.

This is the viewpoint of the twin on the earth. But what about the traveling twin? Since "everything is relative," all inertial reference frames are equally good. Won't the traveling twin make all the claims the earth twin does, only in reverse? Can't the astronaut twin claim that since the earth is moving away at high speed, time passes more slowly on earth and the twin on earth will age less? This is the opposite of what the earth twin predicts. They cannot both be right, for after all the spacecraft returns to earth and a direct comparison of ages and clocks can be made.

The resolution lies in the fact that the starting and ending points are fixed relative to the earth twin. The traveling twin must accelerate at the beginning and end of the trip and also when turning around far out in space. We assume these acceleration periods occupy only a tiny portion of the total time. But while traveling at constant velocity on both outward and homeward journeys, the traveling twin measures the distance he must travel as shorter than the distance his earthbound twin measures. Why? Because of length contraction, and it is the earth observer who measures the proper length. Since the traveling twin measures a shorter travel distance, and both twins measure the same relative speed, this confirms that the traveling twin measures a shorter time and thus returns to earth having aged less than the earthbound twin. Even when the acceleration periods are considered, Einstein's general theory of relativity, which deals with accelerating reference frames, confirms this result.

6. FOUR-DIMENSIONAL SPACE-TIME

Let us suppose a person is on a train moving at a very high speed, say $0.65c$. This person begins a meal at 7:00 and finishes at 7:15, according to a clock on the train. The two events, beginning and ending the meal, take place at the same point on the train. So the proper time between these two events is 15 minutes. To observers on earth, the meal will take longer—20 minutes according to time dilation. Let us assume that the meal was served on a 20-cm-diameter plate. To observers on the earth, the plate is only 15 cm wide (length contraction). Thus to observers on the earth the food looks smaller but lasts longer.

In a sense these two effects, time dilation and length contraction, balance each other. When viewed from the earth, what the food seems to lose in size it gains in length of time it lasts. Space, or length, is exchanged for time.

Considerations like this led to the idea of *four-dimensional space-time*: space takes up three dimensions and time is a fourth dimension. Space and time are intimately connected. Just as when we squeeze a balloon we make one dimension larger and another smaller, so when we examine objects and events from different reference frames a certain amount of space is exchanged for time, or vice versa.

Although the idea of four dimensions may seem strange, it refers to the fact that any object or event is specified by four quantities—three to describe where in space and one to describe when in time. The really unusual aspect of four-dimensional space-time is that space and time can intermix: a little of one can be exchanged for a little of the other when the reference frame is changed.

It is difficult for most of us to understand the idea of four-dimensional space-time. Somehow we feel, just as physicists did before the advent of relativity, that space and time are completely separate entities. Yet we have found in our thought experiments that they are not completely separate. Our difficulty in accepting this is reminiscent of the situation in the seventeenth century at the time of Galileo and Newton. Before Galileo, the vertical direction—that in which objects fall—was considered to be distinctly different from the two horizontal dimensions. Galileo showed that the vertical dimension differs only in that it happens to be the direction in which gravity acts. Otherwise, all three dimensions are equivalent, a fact that we all accept today. Now we are asked to accept one more dimension, time, which we had previously thought of as being somehow different. This is not to say that there is no distinction whatsoever between space and time. What relativity has shown is that space and time determinations are not independent of one another.

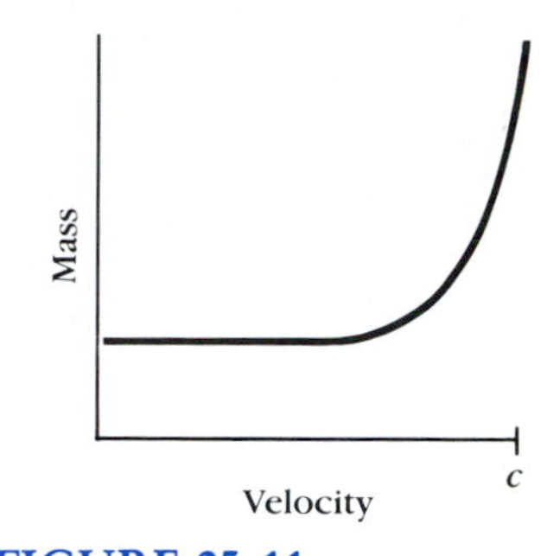

FIGURE 25-11
Increase in mass of an object as its velocity increases.

7. MASS INCREASE

The three basic quantities we can measure are length, time, and mass. As we have seen, the first two are relative; measurements of length and time intervals are different in different moving reference frames. And the same is true of mass. According to the theory of relativity:

Moving objects have increased mass.

Quantitatively (we will not go into details), the theory predicts that if an object has a mass m_0 when it is at rest, then when moving with a velocity v past an observer, the object's mass will be measured to be

$$m = \frac{m_0}{\sqrt{1 - v^2/c^2}}.$$

Since the denominator is always less than 1 (unless $v = 0$), the mass m will always be more than the rest mass m_0.

Relativistic mass increase has been tested countless times on tiny elementary particles (such as muons), and the mass has been found to increase in accord with this formula (see Figure 25-11).

FIGURE 25-12
Rocket 2 is fired from rocket 1 with speed $u = 0.60\ c$. What is the speed of rocket 2 with respect to the earth?

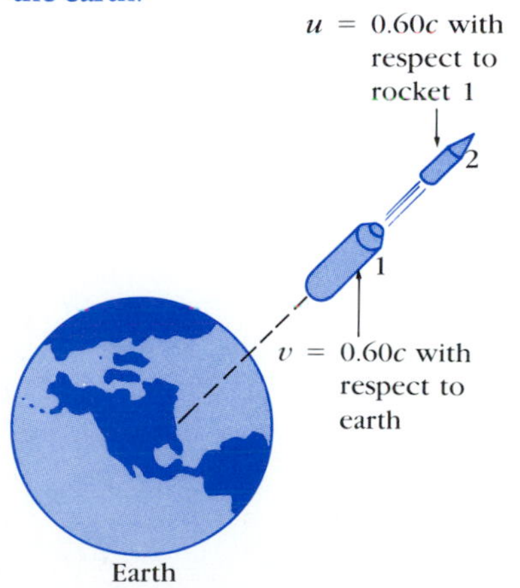

8. RELATIVISTIC ADDITION OF VELOCITIES AND THE ULTIMATE SPEED

Suppose a rocket ship traveling away from the earth with velocity v sends off a second rocket, as shown in Figure 25-12. If the velocity of the second rocket as observed by people on the first rocket is u, then we might expect that the velocity of rocket 2 as seen by observers on earth would be $v' = u + v$. (This is like the waiter on the train in Figure 25-2 who is going 105 km/hr with respect to the ground.) However, Einstein showed that since both length and time are different in different moving frames of reference, velocity cannot be added up so simply. Instead, Einstein showed that we must use the formula

$$v' = \frac{v + u}{1 + \frac{vu}{c^2}},$$

where c is the velocity of light.

If v and u are each small compared to the velocity of light c, as is usually the case in everyday experience, then $\frac{uv}{c^2}$ is practically zero and $v' \approx \frac{v+u}{1} = v + u$, which is just our commonsense formula. Thus Einstein's formula agrees with everyday experience for ordinary velocities that are much smaller than the velocity of light.

If the velocity of an object is close to the speed of light, rather bizarre effects occur. Suppose for example that rocket 1 in Figure 25-12 is launched from the earth at a velocity of 180,000 km/s, which is 6/10 the speed of light, and that rocket 2 is shot from the nose of the first at a velocity of $0.6\,c$ with respect to the first rocket. Common sense tells us that the second rocket should have a velocity of $0.6\,c + 0.6\,c = 1.2\,c$ with respect to the earth; but our "common sense" arises from everyday experience, which does not include objects moving at such high velocities. Instead, we must use Einstein's formula. Substituting $v = 0.6\,c$ and $u = 0.6\,c$ in Einstein's addition of velocities formula, we find

$$v' = \frac{v + u}{1 + vu/c^2} = \frac{0.6\,c + 0.6c}{1 + 0.36\,c^2/c^2} = \frac{1.2\,c}{1.36} = 0.88\,c,$$

or 264,000 km/s, and not $1.2\,c$. In general, no matter how close the two velocities are to the speed of light, their sum will not exceed c. Try various velocities (less than c) in the formula and prove it to yourself. Because it is not possible to exceed the speed of light by adding together two velocities that are each less than the speed of light, we have here the first inkling that the speed of light is somehow a speed limit. And Einstein's theory of relativity does, indeed, predict that no ordinary object can move faster than the speed of light.

The Speed of Light: The Ultimate Speed

That the speed of light is the maximum velocity for objects in the universe can be seen from the time dilation, length contraction, and mass increase formulas. Perhaps it is easiest to use the mass increase formula, $m = \frac{m_0}{\sqrt{1 - (v^2/c^2)}}$. As a body is accelerated to greater and greater speeds, its mass becomes larger and larger. Indeed, if v were to equal c, the denominator in this equation would be zero and the mass m would become infinite. To accelerate an object to $v = c$ would thus require infinite energy, and so is not possible. The fact that light itself travels at the speed c implies that its rest mass (if such a thing makes sense) must be zero; of course, light is never at rest.

If v were to exceed c, then the factor $\sqrt{1 - v^2/c^2}$ would be the square root of a negative number, which is imaginary: so lengths, time intervals, and mass would not be real. We conclude, then, that ordinary objects cannot equal or exceed the speed of light. However, as was pointed out in the late 1960s, Einstein's equations do not rule out the possibility that objects exist whose speed is *always* greater than c. If such particles exist (the name "tachyon"—meaning "fast"—was proposed) the rest mass m_0 would have to be imaginary; in this way the mass m would be the ratio of two imaginary numbers for $v > c$, which is real. For such hypothetical particles, c would be a *lower* limit on their speed. In spite of extensive searches for tachyons, none has been found. It seems that the speed of light *is* the ultimate speed in the universe.

9. $E = mc^2$; MASS AND ENERGY

The equation $E = mc^2$ is perhaps the most famous equation in all of physics. What does it mean? When a steady net force is applied to an object of rest mass m_0, the object increases in speed. Since the force is acting through a distance, work is done on the object and its energy increases. As the speed of the object approaches c, the speed cannot increase indefinitely since it cannot exceed c. On the other hand, the mass of the object increases with increasing speed. That is, the work done on an object not only increases its speed but also contributes to increasing its *mass*. Normally the work done on an object increases its energy. This new twist from the theory of relativity leads to the idea that mass is a form of energy, a crucial part of Einstein's theory of relativity.

Einstein showed that the kinetic energy of an object at high speeds is not given by $\text{KE} = \frac{1}{2}mv^2$, not even if m in this relation is the relativistic mass, $m = m_o/\sqrt{1 - v^2/c^2}$. Rather, the KE is given by

$$\text{KE} = mc^2 - m_0c^2$$

where m is the relativistic mass at speed v and m_0 is the object's mass when at rest ($v = 0$). Thus even when an object is at rest it has an energy, called its **rest energy**, $E_0 = m_0c^2$. If the object has speed v, its *total energy* (assuming no potential energy) is $E = mc^2$. From the preceding equation we have

$$E = mc^2$$

or

$$E = \text{KE} + m_0c^2.$$

Here we have Einstein's famous formula $E = mc^2$. And we see that the total energy of an object is its kinetic energy plus its rest energy.

The relation $E = mc^2$ is now believed to apply to all processes, although the changes are often too small to measure. For example, when you heat a pot of water on the stove, you are putting heat energy into the water; and the mass of the water increases a tiny amount.[†] When electric energy flows into a light bulb or an electric frying pan, the mass of the bulb or pan increases. In a chemical reaction in which heat energy is gained or lost, the masses of the reactants and of the products will be different. Ordinarily, however, the change in mass is too small to be detected.

The law of conservation of energy remains valid in the theory of relativity but only as long as mass is considered to be a form of energy. Indeed, Einstein predicted that mass can be changed into energy and energy can be changed into mass according to $E_0 = m_0c^2$. This prediction has been amply verified by experiment. Light energy has been transformed into tiny elementary particles that have mass, such as muons or electrons. The reverse has been observed as well. Perhaps the most graphic illustration occurs in an atomic bomb or a nuclear reactor. In both cases, mass is transformed into energy on a large scale. Indeed, because of the c^2 in the mass-energy equation, we see that even a small mass "contains" an incredible amount of energy. For example, one kilogram of a substance is equivalent to an amount of energy: $E = mc^2 = 1\ \text{kg} \times (300{,}000{,}000\ \text{m/s})^2 = 90{,}000{,}000{,}000{,}000{,}000$ joules $= 9 \times 10^{16}$ J. This amount of energy would supply the needs of a city of a million people for several years. It is not easy, however, to obtain this energy.

The conversion of mass into energy does not occur spontaneously except in certain specific situations. In Chapter 30 we will discuss how it is done in the bomb and in nuclear reactors. Note that the equation $E = mc^2$ does *not* say that an object must travel at speed c in order to have this much energy. It has this energy even when at rest. The c^2 is merely a sort of "proportionality constant." (See Chapter 1.) By multiplying the mass m by c^2, we determine how much energy is released if the mass m is converted into energy.

10. MR. TOMPKINS

A clever book by George Gamow, *Mr. Tompkins in Wonderland*, delightfully illustrates the results of relativity and what it would be like if ordinary speeds were close to the speed of light. In the excerpt* that follows, a

[†] This example is easy to see because as heat is added, the temperature and therefore the average speed of the molecules increases; and the relativistic mass formula tells us that the mass also increases.

* George Gamow, *Mr. Thompkins in Wonderland* (Cambridge, England: Cambridge University Press 1940). Reprinted by permission of the publisher. This work has been published along with *Mr. Tompkins Meets the Atom* in paperback form as *Mr. Tompkins in Paperback* (Cambridge University Press, 1967).

Mr. Tompkins has fallen asleep while listening to a lecture on the theory of relativity. He finds himself dreaming of a world in which the natural speed limit, the speed of light, is only 20 mi/hr:

> . . . to his surprise there was nothing unusual happening around him; even a policeman standing on the opposite corner looked as policemen usually do. The hands of the big clock on the tower down the street were pointing to five o'clock and the streets were nearly empty. A single cyclist was coming slowly down the street and, as he approached, Mr. Tompkins' eyes opened wide with astonishment. For the bicycle and the young man on it were unbelievably shortened in the direction of the motion, as if seen through a cylindrical lens. The clock on the tower struck five, and the cyclist, evidently in a hurry, stepped harder on the pedals. Mr. Tompkins did not notice that he gained much in speed, but, as the result of his effort, he shortened still more and went down the street looking exactly like a picture cut out of cardboard. Then Mr. Tompkins felt very proud because he could understand what was happening to the cyclist—it was simply the contraction of moving bodies, about which he had just heard. "Evidently nature's speed limit is lower here," he concluded, "that is why the bobby on the corner looks so lazy, he need not watch for speeders." In fact, a taxi moving along the street at the moment and making all the noise in the world could not do much better than the cyclist, and was just crawling along. Mr. Tompkins decided to overtake the cyclist who looked a good sort of fellow, and ask him all about it. Making sure that the policeman was looking the other way, he borrowed somebody's bicycle standing near the kerb and sped down the street. He expected that he would be immediately shortened, and was very happy about it as his increasing figure had lately caused him some anxiety. To his great surprise, however, nothing happened to him or to his cycle. On the other hand, the picture around him completely changed. The street grew shorter, the windows of the shops began to look like narrow slits, and the policeman on the corner became the thinnest man he had ever seen.
>
> "By Jove!" exclaimed Mr. Tompkins excitedly, "I see the trick now. This is where the word *relativity* comes in. Everything that moves relative to me looks shorter for me, whoever works the pedals!" He was a good cyclist and was doing his best to overtake the young man. But he found that it was not at all easy to get up speed on this bicycle. Although he was working on the pedals as hard as he possibly could, the increase in speed was almost negligible. His legs already began to ache, but still he could not manage to pass a lamppost on the corner much faster than when he had just started. It looked as if all his efforts to move faster were leading to no result. He understood now very well why the cyclist and the cab he had just met could not do any better, and he remembered the words of the professor about the impossibility of surpassing the limiting velocity of light. He noticed, however, that the city blocks became still shorter and the cyclist riding ahead of him did not now look so far away. He overtook the cyclist at

the second turning, and when they had been riding side by side for a moment, was surprised to see the cyclist was actually quite a normal, sporting-looking young man. "Oh, that must be because we do not move relative to each other," he concluded; and he addressed the young man.

"Excuse me, sir!" he said, "Don't you find it inconvenient to live in a city with such a slow speed limit?"

"Speed limit?" returned the other in surprise, "we don't have any speed limit here. I can get anywhere as fast as I wish, or at least I could if I had a motor-cycle instead of this nothing-to-be-done-with old bike!"

"But you were moving very slowly when you passed me a moment ago," said Mr. Tompkins. "I noticed you particularly."

"Oh you did, did you?" said the young man, evidently offended. "I suppose you haven't noticed that since you first addressed me we have passed five blocks. Isn't that fast enough for you?"

"But the streets became so short," argued Mr. Tompkins.

"What difference does it make anyway, whether we move faster or whether the street becomes shorter? I have to go ten blocks to get to the post office, and if I step harder on the pedals the blocks become shorter and I get there quicker. In fact, here we are," said the young man getting off his bike.

Mr. Tompkins looked at the post office clock, which showed half-past five. "Well!" he remarked triumphantly, "it took you half an hour to go this ten blocks, anyhow—when I saw you first it was exactly five!"

"And did you *notice* this half hour?" asked his companion. Mr. Tompkins had to agree that it really seemed to him only a few minutes. Moreover, looking at his wrist watch he saw it was showing only five minutes past five. "Oh!" he said, "is the post office clock fast?" "Of course it is, or your watch is too slow, just because you have been going too fast. What's the matter with you, anyway? Did you fall down from the moon?" and the young man went into the post office.

After this conversation, Mr. Tompkins realized how unfortunate it was that the old professor was not at hand to explain all these strange events to him. The young man was evidently a native, and had been accustomed to this state of things even before he had learned to walk. So Mr. Tompkins was forced to explore this strange world by himself. He put his watch right by the post office clock, and to make sure that it went all right waited for ten minutes. His watch did not lose. Continuing his journey down the street he finally saw the railway station and decided to check his watch again. To his surprise it was again quite a bit slow. "Well, this must be some relativity effect, too," concluded Mr. Tompkins; and decided to ask about it from somebody more intelligent than the young cyclist.

The opportunity came very soon. A gentleman obviously in his forties got out of the train and began to move towards the exit. He was met by a very old lady, who, to Mr. Tompkins's great surprise, addressed him as "dear Grandfather." This was too much for Mr. Tompkins. Under the excuse of helping with the luggage, he started a conversation.

"Excuse me, if I am intruding into your family affairs," said he, "but are you really the grandfather of this nice old lady? You see, I am a stranger here, and never . . . " "Oh, I see," said the gentleman, smiling with his moustache. "I suppose you are taking me for the Wandering Jew or something. But the thing is really quite simple. My business requires me to travel quite a lot, and, as I spend most of my life in the train, I naturally grow old much more slowly than my relatives living in the city. I am so glad that I came back in time to see my dear little granddaughter still alive! But excuse me, please, I have to attend to her in the taxi," and he hurried away leaving Mr. Tompkins alone again with his problems.

SUMMARY

The *theory of relativity* examines how we observe physical events. All observations and measurements are made in some *frame of reference*. The velocity of an object depends on the frame of reference in which it is measured, so we say that velocity is a relative concept.

The theory of relativity is based on two principles: the *relativity principle*, which states that the laws of physics are the same in all uniformly moving reference frames—i.e., there is no preferred reference frame that can be considered at absolute rest; and the principle of the *constancy of the speed of light*, which states that the speed of light in empty space has the same value in all inertial reference frames.

One intriguing aspect of relativity theory is that two events that are simultaneous in one reference frame may not be simultaneous in another. Other effects are *time dilation*: moving clocks are measured to run slowly; *length contraction*: the length of an object is measured to be shorter when it is moving than when it is at rest; *mass increase*: the mass of a body increases with speed; and velocity addition must be done in a special way. All these effects are significant only at high speeds, close to the speed of light, which itself is the ultimate speed in the universe.

The theory of relativity has changed our notions of space and time, and of mass and energy. Space and time are seen to be intimately connected, with time being the fourth dimension in addition to space's three dimensions. Mass and energy are interconvertible; the equation

$$E = mc^2$$

tells how much energy E is needed to create a mass m, or vice versa. The law of conservation of energy must include mass as a form of energy.

QUESTIONS

1. A man is standing on the top of a moving railroad car. He throws a heavy ball straight up (it seems to him) in the air. Ignoring air resistence, will the ball land *on* the car or *behind* it?
2. According to the principle of relativity, it is just as legitimate for a person riding in a uniformly moving automobile to say that the automobile is at rest and the earth is moving as it is for a person on the ground to say that the car is moving and the earth is at rest. Do you agree, or are you reluctant to accept this? Discuss the reasons for your response.
3. Does the earth *really* go around the sun? Or is it also valid to say that the sun goes around the earth? Discuss in view of the first principle of relativity (that there is no preferred frame).
4. If you were on a space vehicle traveling at half the speed of light away from a bright star, with what speed would the star's light go past you?
5. Suppose you were placed in a laboratory that could "float" on a layer of air just above the ground and there was one small window you could look out. If you saw a row of trees going past you outside the window, could you perform an experiment to prove (a) that the trees were at rest and you were moving uniformly or (b) that you were at rest and the trees were moving uniformly? Explain.
6. Does the theory of relativity show that Newtonian mechanics is wrong?
7. Will two events that occur at the same place and same time for one observer be simultaneous to a second observer moving with respect to the first?
8. The time-dilation effect says that "moving clocks run slowly." Actually, this effect has nothing to do with motion affecting the functioning of clocks. What then does it deal with?
9. Does time dilation mean that time actually passes more slowly in moving reference frames or that it only *seems* to pass more slowly?
10. Today's subways are said to age people prematurely. Suppose that in the future, subway trains that traveled very close to the speed of light could be designed. How do you think this would affect the aging process?
11. A young-looking woman astronaut has just arrived home from a long trip. She rushes up to an old gray-haired man and in the ensuing conversation refers to him as her son. How is this possible?
12. If you were traveling away from Earth at a speed of $0.5c$, would you notice a change in your heartbeat? Would your mass, height, or waistline change? What would observers on earth using telescopes say about these things?

13. Is it possible for a person to live long enough to travel to a star 10,000 light-years away? (A light-year is the distance that light travels in one year.)
14. Suppose you were to journey to a distant star 100 light-years away. If you traveled at a very high speed, would the distance you had to travel still be 100 light-years? Explain.
15. It takes light four years to reach us from the nearest star. Is it possible for an astronaut to travel to that star in one year?
16. What is the fourth dimension?
17. Do length contraction, time dilation, and mass increase occur at ordinary speeds, say, 90 km/hr?
18. Consider an object of mass m to which is applied a constant force for an indefinite period of time. Discuss how its velocity and mass change with time.
19. Suppose the speed of light were infinite. What would happen to the relativistic predictions of length contraction, time dilation, and mass increase?
20. Why don't we notice the special effects of the theory of relativity in everyday life? Be specific.
21. The predictions of the theory of relativity seem to contradict certain of our everyday intuitive notions and therefore don't seem to "make sense." What are those notions? Examine them in detail and determine if any physically measurable contradiction exists.
22. Explain how the length-contraction and time-dilation formulas might be used to indicate that c is the limiting speed in the universe.
23. Does the equation $E = mc^2$ conflict with the conservation of energy principle? Explain.
24. Does $E = mc^2$ apply to particles that travel at the speed of light? Does it apply only to them?
25. If mass is a form of energy, does this mean that a spring has more mass when compressed than when relaxed?
26. It is not correct to say that "matter can neither be created nor destroyed." What must we say instead?
27. A neutrino is an elementary particle with zero rest mass that travels at the speed of light. Could you ever catch up to a neutrino that passed you?
28. A red-hot iron bar is cooled to room temperature. Does its mass change?
29. Mr. Tompkins finds that the clock in the post office advanced 30 minutes while his own watch advanced only 5 minutes. Since the post office was moving with respect to him, shouldn't it have been the reverse? Explain.

30. As Mr. Tompkins pedals along the street in Wonderland, does he notice his bicycle getting shorter as he pedals faster? Do people standing on the street notice his bicycle getting shorter?

EXERCISES

1. A boat whose speed in still water is 10 km/hr is traveling along a river whose current is 4 km/hr. What is the net speed of the boat when going upstream? When going downstream?
2. How long will it take the boat in the preceding problem to make a round trip of 60 km (30 km upstream and 30 km back again)?
3. An airplane with a maximum speed of 1000 km/hr flies into a headwind of 200 km/hr (with respect to the earth). What is the speed of the airplane with respect to the earth?
4. Analyze the thought experiment of Section 3 from O_1's point of view. (Make a diagram analogous to Figure 25-8.)
5. A beam of a certain type of elementary particle travels at a speed of 2.8×10^8 m/s. At this speed, the average lifetime is measured to be 2.5×10^{-8} s. What is the particle's lifetime at rest?

*6. What is the speed of a beam of pions if their average lifetime is measured to be 3.5×10^{-8} s? At rest, their lifetime is 2.6×10^{-8} s.

7. A spaceship passes you at a speed of 0.80 c. You measure its length to be 90 m. How long would it be when at rest?
8. You are sitting in your sports car when a very fast sports car passes you at a speed of 0.18 c. A person in the car says his car is 6.00 m long and yours is 6.15 m long. What do you measure for these two lengths?

*9. Suppose you decide to travel to a star 60 light-years away. How fast would you have to travel so the distance would be only 20 light-years?

10. If you were to travel to a star 50 light-years from earth at a speed of 2.0×10^8 m/s, what would you measure this distance to be?
11. A certain star is 36 light-years away. How long would it take a spacecraft traveling 0.98 c to reach that star from Earth as measured by observers (a) on Earth (b) on the spacecraft? (c) What is the distance traveled according to observers on the spacecraft? (d) What will the spacecraft occupants compute their speed to be from the results of (b) and (c)?
12. A friend of yours travels by you in her fast sports car at a speed of 0.76 c. It appears to be 5.80 m long and 1.45 m high. (a) What will be its length and height at rest? (b) How many seconds did you see elapse on your friend's watch when 20 s passed on yours? (c) How fast did you appear to be traveling to your friend? (d) How many seconds did she see elapse on your watch when she saw 20 s pass on hers?

13. A person on a rocket traveling at half the speed of light (with respect to the earth) observes a meteor come from behind and pass him at a speed he measures as half the speed of light. How fast is the meteor moving with respect to the earth?

14. Two spaceships leave the earth in opposite directions, each with a speed of one-half the speed of light with respect to the earth. (a) What is the velocity of spaceship 1 relative to spaceship 2? (b) What is the velocity of spaceship 2 relative to spaceship 1?

15. Show that the relativistic-addition-of-velocities formula is consistent with the second principle of relativity; that is, if the object is light so that $v = c$, then show that $v' = c$ also.

16. What is the mass of an electron traveling at $v = 0.85\,c$?

*17. At what speed will an object's mass be twice its rest mass?

*18. At what speed v will the mass of an object be 1 percent greater than its rest mass?

*19. Derive a formula showing how the density of an object changes with speed v.

EARLY QUANTUM THEORY

CHAPTER

26

The second aspect of the revolution that shook the world of physics in the early part of the twentieth century (the first half was Einstein's theory of relativity) is the quantum theory. Unlike the special theory of relativity —whose basic tenets were put forth mainly by one person in a single year—the revolution of quantum theory required almost three decades to unfold, and many scientists contributed to its development. It began in 1900 with Planck's quantum hypothesis and culminated in the mid-1920s with the theory of quantum mechanics of Schrödinger and Heisenberg, which has been very effective in explaining the structure of matter.

1. PLANCK'S QUANTUM HYPOTHESIS

One of the observations that was unexplained at the end of the nineteenth century was the spectrum of light emitted by solids at high temperatures. When a solid is heated to higher temperatures, the spectrum of light it emits includes radiation of higher and higher frequencies. If the temperature is high enough, the radiation will contain frequencies in the visible region, and the object will glow. For example, when iron is heated to 600 or 700°C it glows with a red color—we say it is "red hot"; most of the spectrum of light it emits is not visible but a small amount is in the visible region at the low frequency end, the red. As the iron is heated to higher temperatures it becomes orange and then white, when all visible light frequencies are represented. Similarly, the whitish-yellow light that we get from an ordinary incandescent light bulb is emitted from the white-hot tungsten filament heated to over 2000°C by the electric current passing through it. At any particular temperature, the brightness of the emitted light at each frequency can be measured.

The spectrum of light emitted by a hot dense object is shown in Figure 26-1. As can be seen, the spectrum contains a continuous range of frequencies, and at higher temperatures there is more radiation and it is peaked at higher frequencies (shorter wavelengths). Such a continuous spectrum is emitted by any heated solid or liquid, and even by dense gases. (The 6000-K curve in Figure 26-1 corresponds to the temperature of the sun, and the visible part of the spectrum is seen to be around the peak.)

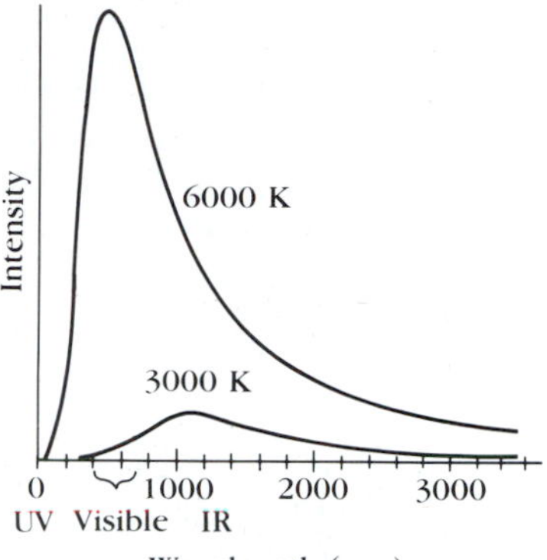

FIGURE 26-1
Spectrum of frequencies emitted by a blackbody at two different temperatures.

The experimental curves of Figure 26-1 are for a body that approximates an idealized nonreflecting object. Such a body would appear black (when not too hot) and the radiation it emits is the simplest to deal with. The problem that faced scientists in the 1890s was to explain this *blackbody radiation*. Maxwell's electromagnetic theory had shown that oscillating electric charges produce electromagnetic waves. The radiation emitted by a hot body was assumed to be due to the oscillations of electric charges in the molecules of the material. Although this explains where the radiation comes from, it did not correctly predict the observed spectrum of emitted light. Indeed, one theory based on Maxwell's electromagnetic theory predicted that the intensity increases with frequency as shown by the dashed curve in Figure 26-2. Thus all objects would glow with a blue color, and in fact the curve goes to infinity in the UV region. Clearly, classical theory was inconsistent with the experimental results.

The break came in 1900. In that year the German physicist Max Planck (1858—1947) proposed a new theory that was fully in accord with the experimental data (solid curve in Figure 26-2). But to achieve this accord Planck had to make a radical assumption. He hypothesized that the vibrating molecules in a heated material can vibrate only with certain discrete amounts of energy. In particular, Planck assumed that the minimum energy of vibration E_{min} is proportional to the natural frequency of oscillation f:

$$E_{\text{min}} = hf;$$

here h is a constant, now called *Planck's constant*, whose value was estimated by Planck by fitting his formula for the blackbody radiation curve (h appears in this formula) to experiment. The value accepted today is

$$h = 6.626 \times 10^{-34}\ \text{J} \cdot \text{s}.$$

Furthermore, Planck said, the energy of any molecular vibration could only be some whole-number multiple of this minimum energy:

$$E = nhf, \qquad n = 1,2,3,\ldots$$

This is often called **Planck's quantum hypothesis**. (*Quantum* means

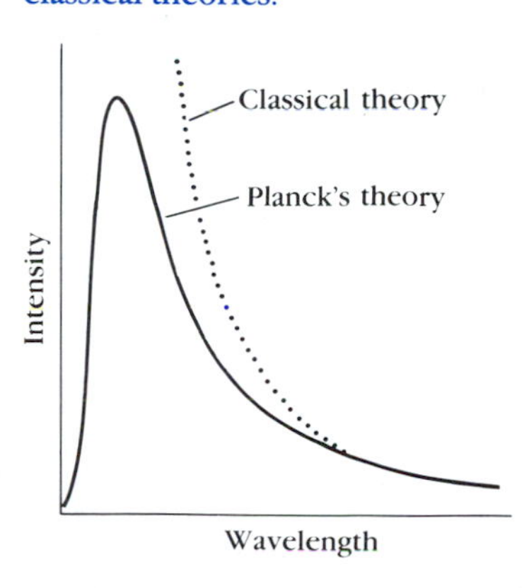

FIGURE 26-2
Comparison of the Planck and classical theories.

"a fixed amount.") The quantum hypothesis says that the energy can be $E = hf$, or $2hf$, or $3hf$, and so on, but there cannot be vibrations whose energy lies between these values. That is, energy is not a continuous quantity as had been believed for centuries; rather it is *quantized*—it exists only in discrete amounts. The smallest amount of energy possible (hf) is called the *quantum of energy*. Another way of expressing the quantum hypothesis is that not just any amplitude of vibration is possible. The possible values for the amplitude are related to the frequency f.

The concept of quantization is not entirely new. The mass of a block of copper, for example, is a whole-number multiple of the mass of one copper atom. Nowhere will you find a copper block with a fractional number of atoms, say $936\frac{1}{2}$. Similarly, electric charge is quantized. The smallest charge found in nature is e, the charge on the electron. And the net charge on any object is a whole-number multiple of this quantum of charge. Thus a body may have a charge of $1e$, $2e$, $3e$, or $317e$, but never $317.2e$. Nonetheless, the idea that energy is quantized was not easy to accept. But in view of Einstein's epochal idea that mass and energy are equivalent ($E = mc^2$), since mass is quantized, why not energy?

2. PHOTON THEORY OF LIGHT AND THE PHOTOELECTRIC EFFECT

Einstein's Photon Theory

In 1905, the same year he introduced the special theory of relativity, Einstein made a bold extension of the quantum idea by proposing a new theory of light. Planck had assumed that the vibrational energy of molecules in a radiating object is quantized with energy $E = nhf$. Einstein reasoned that if the energy of the molecular oscillators is thus quantized, then to conserve energy the light must be emitted in packets or quanta, each with an energy

$$E = hf.$$

Again h is Planck's constant. And since all light ultimately comes from a radiating source, Einstein proposed that *light is transmitted as tiny particles*, or **photons**, as they are now called, rather than as waves. This, too, was a radical departure from classical ideas. And Einstein proposed a simple test of his photon theory: careful experiments on the photoelectric effect.

The Photoelectric Effect

The **photoelectric effect** refers to the fact that when light shines on a metal surface, electrons are emitted from the surface. (The photoelectric effect occurs in other materials but is most easily observed with metals.) This effect can be observed using the apparatus shown in Figure 26-3. A metal plate P along with a smaller electrode C are placed inside an evacuated glass tube, called a *photocell*. The two electrodes are connected to an ammeter and a battery as shown. When the photocell is in the dark, the ammeter reads zero. But when light of sufficiently high frequency is shone on the plate, the ammeter indicates a current flowing in the circuit. To explain completion of the circuit we can imagine electrons flowing across the tube from the plate to the "collector" C as shown in the diagram.

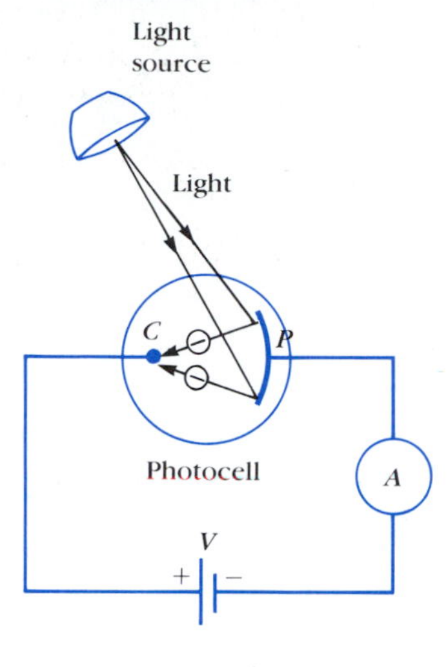

FIGURE 26-3
Photoelectric effect.

That electrons should be emitted when light shines on a metal is consistent with the electromagnetic (EM) wave theory of light, since the electric field of an EM wave could exert a force on electrons in the metal and thrust some of them out. Einstein pointed out, however, that the wave theory and the photon theory of light give very different predictions on the details of this effect.

Recall that the two main properties of a wave are its amplitude and its frequency (or wavelength); amplitude refers to the height of the crests of the wave and frequency to the number of crests passing a given point per second. The amplitude of a light wave corresponds to the brightness of the light and the frequency corresponds to the color. A brighter light would thus mean that the electric and magnetic fields are larger. So, according to the wave theory, we would expect that when a brighter light is shined on a metal the electrons should feel a greater force due to the greater electric field and therefore should be ejected with greater kinetic energy.

According to Einstein's theory, when a beam of light strikes a metal, each individual photon can collide with a single electron, causing it to be ejected from the metal. In the process the electron absorbs all the energy of the photon, and the photon ceases to exist. Since electrons are held in the metal by attractive forces, some minimum energy is required just to get an electron out of the metal; this is called the *work function*, W. If the frequency f is so low that the photon energy hf is less than W, then the photons will not have sufficient energy to knock electrons out of the metal. This minimum frequency is called the "cutoff frequency," and we designate it by f_0. If the energy of the photons is greater than W, then a photon will have sufficient energy to knock an electron completely out of the metal. During the collision, energy is conserved; so part of the photon's energy goes into overcoming the forces holding the electron in the metal and what is left over goes into the kinetic energy of the freed electron. Thus, according to the photon theory, if the frequency of the light is increased the kinetic energy of the emitted electrons is increased. This is shown in the graph of Figure 26-4. Finally, according to Einstein's theory, an increase in light intensity with no change in its frequency f means only that the light beam contains a greater number of photons, each still of energy hf. With the increase in number of photons, more collisions with electrons can occur and more electrons will be ejected; but since the

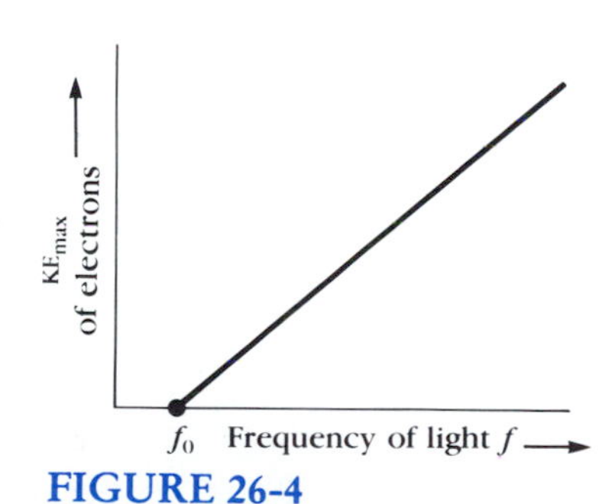

FIGURE 26-4
Photoelectric effect: maximum kinetic energy of ejected electrons increases linearly with frequency of incident light. No electrons are emitted if $f < f_0$.

energy of each photon has not been changed, the kinetic energy of the electrons is not changed.

In summary, Einstein's photon theory predicts that:

1. Increasing the intensity of the light increases the number of electrons ejected, but their energies remain the same.
2. Light of higher frequency causes electrons to be ejected with greater kinetic energy.
3. No electrons are emitted if the frequency of the light is below a certain value.

These three predictions are very different from those of the wave theory, which predicts that (1) the energy of the ejected electrons should be greater for a greater intensity of light, since the greater amplitude of the wave could thrust out electrons with greater speed; (2) frequency would have no effect on the electron energy; (3) there would be no "cutoff" frequency—electrons would be ejected at any frequency.

When careful experiments were carried out in 1913–14, the results were fully in accord with Einstein's photon theory! It was found, for example, that violet light readily ejects electrons from sodium; when yellow light is used instead, the electrons have less energy; and if red light is used, no electrons are ejected at all. And the intensity of the light was found to have no effect on the energies of the ejected electrons.

Einstein's brilliant theory was thus affirmed, and it was a crucial factor in the development of quantum theory.

The Electron Volt, a Unit of Energy

The joule is a very large unit for dealing with energies of electrons, atoms, or molecules. Therefore, in atomic and nuclear physics (and in some other fields as well), we often use another unit called the **electron volt** (abbreviated eV). One electron volt is defined as the energy acquired by an electron when moving through a voltage drop of 1 volt. Since the charge on an electron is 1.60×10^{-19} coulomb and, as we saw in Section 6 of Chapter 18, the change in potential energy equals charge times voltage (qV), then one eV is equal to $(1.60 \times 10^{-19}\ \mathrm{C})(1\ \mathrm{V}) = 1.60 \times 10^{-19}$ joules. That is

$$1\ \mathrm{eV} = 1.60 \times 10^{-19}\ \mathrm{J}.$$

This is simply a conversion factor between two energy units. If an electron falls through a potential difference $V = 5$ volts, it will lose 5 eV of potential energy and will gain 5 eV of kinetic energy. In joules, this corresponds to $(5\ \mathrm{eV})(1.6 \times 10^{-19}\ \mathrm{J/eV}) = 8 \times 10^{-19}\ \mathrm{J}$.

Now let us take a numerical example of the photoelectric effect. Suppose the cut-off wavelength for the metal sodium was found to be $\lambda_0 = 545$ nm; then the cut-off frequency is $f_0 = c/\lambda_0 = (3.00 \times 10^8\ \mathrm{m/s})/(545 \times 10^{-9}\ \mathrm{m}) = 5.50 \times 10^{14}$ Hz. The energy of such a photon is $hf_0 = (6.63 \times 10^{-34}\ \mathrm{J \cdot s})(5.50 \times 10^{14}\ \mathrm{s}) = 3.65 \times 10^{-19}\ \mathrm{J}$, and this then is equal to the minimum energy W required to knock an electron out of the metal. We can state this energy in terms of electron volts using the conversion factor: $(3.65 \times 10^{-19}\ \text{J})/(1.60 \times 10^{-19}\ \text{J/eV}) = 2.28\ \text{eV}$.

Although the electron volt is handy for *stating* the energies of electrons and atoms, it is not a proper SI unit; for calculations it should be converted to joules.

Applications of the Photoelectric Effect

The photoelectric effect, besides playing an important historical role in confirming the photon theory of light, also has many practical applications. Burglar alarms and automatic door openers often make use of the photocell circuit of Figure 26-3. When a person interrupts the beam of light, the sudden drop in current in the circuit activates a switch—often a solenoid—to operate a bell or open the door. UV or IR light is sometimes used in burglar alarms because of its invisibility. Many smoke detectors use the photoelectric effect to detect tiny amounts of smoke that interrupt the flow of light and so alter the electric current. Photographic lightmeters use this circuit as well. The brighter the light the more current that flows in the circuit, as indicated by the ammeter needle. Similar but smaller devices are used in medical diagnoses, such as to measure blood flow. One type of film soundtrack is a variably shaded narrow section at the side of the film; light passing through the film is thus "modulated" and the output electrical signal of the photocell detector follows the frequencies on the sound track.

For many applications today, the vacuum-tube photocell of Figure 26-3 has been replaced by a semiconductor device known as a *photodiode*. In these semiconductors the absorption of a photon liberates a bound electron, which changes the conductivity of the material. Thus the current through a photodiode changes when light shines on it, and so its operation is similar to a photocell.

Photon Theory in Biology

The photon theory of light is often used in biology and medicine, particularly where low light levels occur. For example, the sensitivity of the eyes is specified in terms of photons. Research indicates that receptors (rods and cones) in the retina actually respond to single photons. However, the brain does not register "seeing" for a single photon. Instead several photons (at least 5 to 10 and usually more) are needed, either simultaneously at one receptor or at several adjacent receptors. This coincidence requirement is a valuable mechanism, for it means that our minds are not bothered by random signals that are essentially meaningless. The photon concept is also used in the analysis of *photosynthesis*. Photosynthesis is the complex biochemical process by which pigments such as chlorophyll in plants use the energy of light from the sun to change CO_2 to useful carbohydrate with the release of molecular oxygen.

FIGURE 26-5
Compton effect.

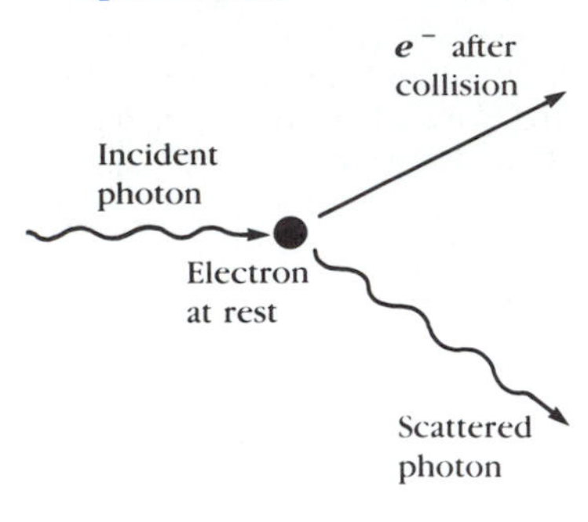

Photon Interactions and Pair Production (optional)

The photon is truly a relativistic particle, for it travels at the speed of light. (It *is* light!) Thus we must use relativistic formulas for dealing with its mass, energy, and momentum. The mass m of any particle is given by $m = m_0/\sqrt{1 - v^2/c^2}$. Since $v = c$ for a photon, the denominator is zero. Thus the rest mass, m_0, of a photon must also be zero, for otherwise its energy, $E = mc^2$, would be infinite. Of course, a photon never is at rest.

When a photon passes through matter, it interacts with the atoms and electrons. There are four important types of interactions that a photon can undergo. First, the photon can be scattered off an electron (or a nucleus) and in the process loose some energy; this is called the Compton effect (Figure 26-5). But notice that the photon is not slowed down. It still travels with speed c, but its frequency will be less. A second type of interaction is the photoelectric effect; a photon may knock an electron out of an atom and in the process itself disappear. The third process is similar. The photon may knock an atomic electron to a higher-energy state in the atom if its energy is not sufficient to knock it out altogether. In this process the photon also disappears, and all its energy is given to the atom; such an atom is then said to be in an excited state, and we shall discuss this more later. Finally, a photon can actually create matter. The most common process is the production of an electron and a positron, Figure 26-6. (A positron has the same mass as an electron, but the opposite charge, $+e$.)† This is called **pair production** and the photon disappears in the process. This is an example of rest mass being created from pure energy, and it occurs in accord with Einstein's equation $E = mc^2$. Notice that a photon cannot create a single electron since electric charge would not then be conserved.

What is the minumum energy of a photon, and its wavelength, that can produce an electron-positron pair? The mass of either an electron or a positron is 9.1×10^{-31} kg. Because $E = mc^2$, the photon must have energy $E = 2(9.1 \times 10^{-31}\,\text{kg})(3.0 \times 10^8\,\text{m/s})^2 = 1.64 \times 10^{-13}\,\text{J}$; in electron volts, this is $(1.64 \times 10^{-13}\,\text{J})/(1.6 \times 10^{-19}\,\text{J/eV}) = 1.02 \times 10^6\,\text{eV} = 1.02\,\text{MeV}$ (million electron volts). A photon with energy less than 1.02 MeV cannot undergo pair production. Since $E = hf = hc/\lambda$, the wavelength of a 1.02-MeV photon is

$$\lambda = \frac{hc}{E} = \frac{(6.6 \times 10^{-34}\,\text{J}\cdot\text{s})(3.0 \times 10^8\,\text{m/s})}{(1.64 \times 10^{-13}\,\text{J})} = 1.2 \times 10^{-12}\,\text{m},$$

which is 0.0012 nm. Thus the wavelength must be very short. Such photons

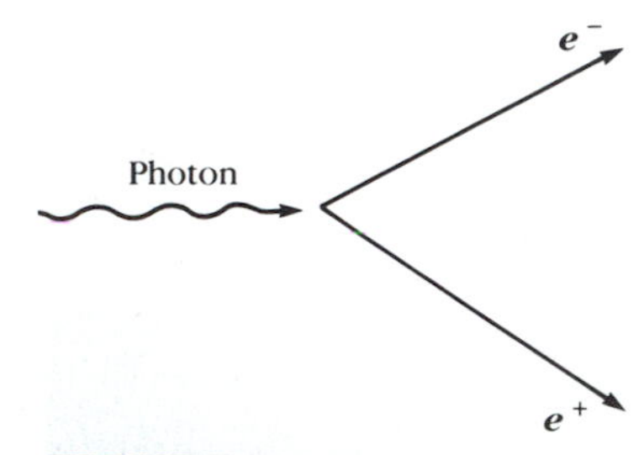

FIGURE 26-6
Pair production: a photon disappears and produces an electron and a positron.

† Positrons do not last long in nature because, when they collide with an electron, they annihilate each other and produce two or more photons. This pair annihilation process is also an example of $E = mc^2$.

are in the gamma-ray (or very short X-ray) region of the electromagnetic spectrum.

3. X RAYS

In 1895, W. C. Roentgen (1845–1923) discovered that when electrons (then called cathode rays) were accelerated by a high voltage in a vacuum tube and allowed to strike a glass or metal surface, certain minerals placed nearby would fluoresce and film placed nearby in a closed wrapper would become exposed. Roentgen attributed these effects to a new type of radiation, which he termed **X rays** after the algebraic symbol x, meaning an unknown quantity. He showed that X rays could pass readily through cardboard and other opaque materials. He soon found that X rays penetrated through some materials better than through others, and within a few weeks he presented the first X-ray photograph (of his wife's hand). The production of X rays today is done in a tube (Figure 26-7) similar to Roentgen's, typically using voltages of 30 to 150 kV.

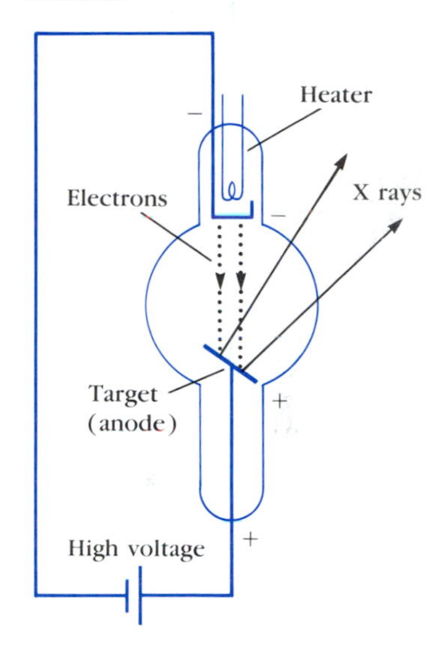

FIGURE 26-7
X-ray tube.
Electrons emitted by a heated filament in a vacuum tube are accelerated by high voltage. When they strike the surface of the anode, the "target," X rays are emitted.

The question naturally arose as to the nature of these X rays. Since they could not be deflected by electric or magnetic fields, it was clear that X rays were not charged particles such as electrons. It was thought, therefore, that they might be a form of invisible light. However, it was nearly 20 years before this hypothesis was tested. From Einstein's photon theory it appeared that at the high voltages used, if X rays were indeed light, their wavelengths would be on the order of 0.1 nm. To detect light waves of such short wavelength would be very difficult; for as we saw in Chapter 24, properties of light such as interference and diffraction are noticeable only when the size of diffracting objects or slits is on the order of the wavelength of the light. Around 1912 it was suggested by Max von Laue (1879–1960) that if the atoms in a crystal were arranged in a regular array, see Figure 11-5(a)—a theory generally held by scientists though not then fully tested—the spaces between atoms might serve as slits to demonstrate wave interference for very short wavelengths on the order of the spacing between atoms, estimated to be about 10^{-10} m (0.1 nm). Experiments soon showed that X rays scattered from a crystal did indeed show the peaks and valleys of an interference pattern. Thus it was shown, in a single blow, that X rays have a wave nature and that atoms are arranged in a regular way in crystals. Today X rays are recognized as electromagnetic radiation with wavelengths in the range of about 10^{-2} nm to 10 nm, the range readily produced in an X-ray tube.

The mechanism for the production of X rays can be explained using the photon theory of light. When the electrons accelerated by the high voltage of the tube in Figure 26-7 strike a target plate, they collide with atoms of the material and are rapidly decelerated. Some of the energy that is lost becomes heat. But some of the energy is transformed into photons of light:

recall from Chapter 21 that an accelerating electric charge can emit radiation. In the present case, the loss in KE of a decelerated electron will be equal to the energy of the emitted photon, $E = hf$.

X-ray machines used in hospitals throughout the world today make use of this X-ray quantum phenomenon.

4. WAVE-PARTICLE DUALITY

The photoelectric effect, X rays, and other experiments have placed the particle theory of light on a firm experimental basis. But what about the classic experiments of Young and others (Chapter 24) on interference and diffraction which showed that the wave theory of light also rests on a firm experimental basis?

We seem to be in a dilemma. Some experiments indicate that light behaves like a wave; and others indicate that it behaves like a stream of particles. These two theories seem to be incompatible but both have been shown to have validity. This is a rather unusual situation. Often in the history of science a physical theory has had to be discarded because it didn't agree with experimental results or a theory has had to be modified to account for new experimental findings. But in the case of light, two incompatible theories have both been shown to have validity. Which theory is correct, the wave theory or the particle theory, seems to depend on the experiment being performed. Some experiments with light can only be explained by using a wave theory, whereas others can only be explained by using a particle, or "quantum," theory. Eventually physicists came to realize that this duality of light had to be accepted as a fact of life. It is referred to as the **wave-particle duality**. Apparently, light is a more complex phenomenon than just a simple wave or a simple beam of particles.

To clarify the situation, the great Danish physicist Niels Bohr (1885–1962) proposed his famous **principle of complementarity**. It states that to understand any given experiment we must use either the wave or the photon theory, but not both. Yet we must be aware of both the wave and particle aspects of light if we are to have a full understanding of light. Therefore these two aspects of light complement one another.

It is not possible to "visualize" this duality. We cannot picture a combination of wave and particle. Instead, we can only recognize that the two aspects of light are different "faces" that light shows to experimenters.

Part of the difficulty stems from how we think. Visual pictures in our minds are based on what we see in the everyday world. We apply the concepts of waves and particles to light because in the macroscopic world we see that energy is transferred from place to place by these two methods. We cannot see directly whether light is a wave or particle—so we do indirect experiments. To explain the experiments we apply the models of waves or of particles to the nature of light. But these are abstractions of the human mind. When we try to conceive of what light really "is," we insist on a visual picture. Yet there is no reason why light should conform to these visual images taken from the macroscopic world. The "true" nature of light—if that means anything—is not possible to visualize. The best we can do is recognize that our knowledge is limited to indirect experiments and that in terms of everyday language and images, light reveals both wave and particle properties.

It is worth noting that Einstein's equation $E = hf$ itself links the particle and wave properties of a light beam. In this equation E refers to the energy of a particle; and on the other side of the equation we have the frequency f of the corresponding wave.

5. WAVE NATURE OF MATTER

In 1923 Louis de Broglie (1892–) extended the idea of the wave-particle duality. He sensed deeply the symmetry in nature and argued that if light sometimes behaves like a wave and sometimes like a particle, then perhaps those things in nature thought to be particles—such as electrons and other material objects—might also have wave properties. De Broglie proposed that the wavelength λ of a particle of mass m moving with a velocity v would be

$$\lambda = \frac{h}{mv},$$

where h is again Planck's constant. This is sometimes called the **de Broglie wavelength** of a particle. Note that this equation relates a wave property, the wavelength λ, and particle properties, the mass m and velocity v.

Let us calculate the de Broglie wavelength of, say, a 0.20-kg ball moving with a speed of 15 m/s. From the preceding formula, $\lambda = h/mv = (6.6 \times 10^{-34}\ \text{J}\cdot\text{s})/(0.20\ \text{kg})(15\ \text{m/s}) = 2.2 \times 10^{-34}\ \text{m}$. This is an incredibly small wavelength. Even if the speed were imperceptible, say 0.0001 m/s, the wavelength would be about 10^{-29} m. Indeed, the wavelength of any ordinary object is much too small to be measured and detected. The problem is that the properties of waves, such as interference and diffraction, are significant only when the size of objects or slits is not much larger than the wavelength. And there are no known objects or slits to diffract waves only 10^{-30} m long; so the wave properties of ordinary objects go undetected.

But tiny elementary particles such as electrons are another matter. Since the mass m appears in the denominator of the de Broglie wavelength formula, a very small mass should give a much larger wavelength. Indeed, the wavelength of an electron ($m = 9.1 \times 10^{-31}$ kg) traveling with a speed of, say, 6.0×10^{6} m/s (a reasonable speed for an electron) will be $\lambda = h/mv = (6.6 \times 10^{-34}\ \text{J}\cdot\text{s})/(9.1 \times 10^{-31}\ \text{kg})(6.0 \times 10^{6}\ \text{m/s}) = 1.2 \times 10^{-10}$ m, or 0.12 nm. Thus we see that electrons can have wavelengths on

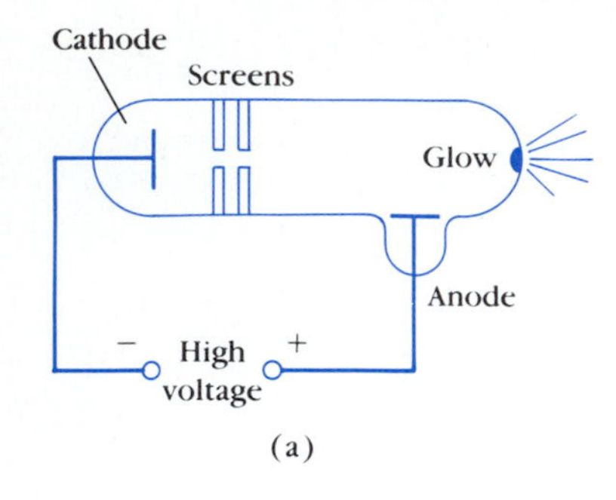

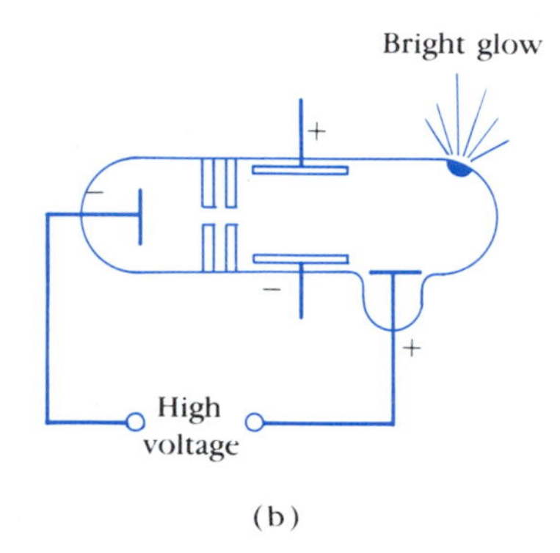

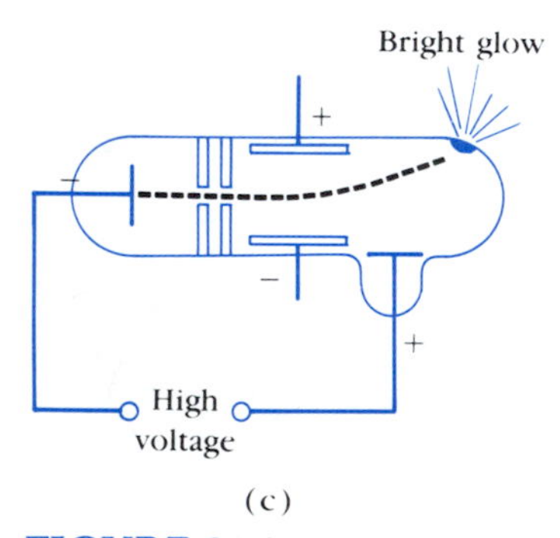

FIGURE 26-8

A gas is maintained at a low pressure in a glass tube into which two terminals intrude. (a) A high voltage causes the tube to glow at the end opposite the negative terminal. (b) J. J. Thomson showed that the electric field produced by two oppositely charged plates caused the glow to shift in position. (c) Thomson interpreted this phenomenon to be negatively charged particles (dotted line) that were moving away from the negative terminal.

the order of 10^{-10} m. Although small, this wavelength can be dealt with. The spacing of atoms in a crystal is on the order of 10^{-10} m, and the orderly array of atoms in a crystal could be used to demonstrate interference, as had already been done for X rays (see Section 3). In 1927, C. J. Davisson and L. H. Germer performed the crucial experiment. They scattered electrons from the surface of a metal crystal and observed that the electrons came off in regular peaks. When they interpreted these peaks as the result of interference, the wavelength of the electrons was found to be just that predicted by de Broglie. Later experiments showed that protons, neutrons, and other particles also have wave properties.

Just as in the case of light, the wave and particle aspects of matter are complementary. We must be aware of both the particle and the wave aspects in order to have a full understanding of matter. But again we must recognize that a "wave–particle" is impossible to visualize.

6. THE ELECTRON

At this point we might ask: What is the evidence that an electron is a *particle*? To answer this question we must go back to the early experiments that indicated that something, which came to be known as an electron, actually existed.

Toward the end of the nineteenth century, studies were being done on the discharge of electricity through rarefied gases. One apparatus, diagrammed in Figure 26-8(a) was a glass tube fitted with electrodes and evacuated so only a small amount of gas remained inside. When a very high voltage was applied to the end of the tube opposite the electrodes, the negative electrode (called the *cathode*) would glow. If one or more screens containing a small hole were inserted as shown, the glow was restricted to a tiny spot on the end of the tube. It seemed as though something being emitted by the cathode traveled to the opposite end of the tube; these were named *cathode rays*.

There was much discussion at the time about what these rays might be. Some scientists thought they might resemble light. But the observation that the bright spot at the end of the tube could be deflected to one side by an electric or magnetic field suggested that cathode rays could be charged particles—Figure 26-8(b) and (c)—and the direction of the deflection was consistent with a negative charge. Soon these cathode rays came to be called *electrons*.

Notice that in these and in subsequent experiments the existence of a negatively charged particle, called an electron, was inferred from the experimental data. No one, however, has actually seen an electron directly. The drawings we sometimes make of electrons as tiny spheres with a negative charge on them are merely convenient pictures (now recognized to be inaccurate). Again, we must rely on experimental results, some of which are best interpreted using a particle picture and others using a wave picture. And again, these are mere pictures that we use to extrapolate from the macroscopic world to the tiny microscopic world of the atom. And there is no reason to expect that these pictures somehow reflect the reality of an electron. We thus use a wave or a particle model (whichever works best in a given situation) so that we can talk about what is happening. But we should not be led to believe that an electron *is* a wave or a particle. Perhaps it is easiest to think of an electron as being something between a wave and a particle. But it may be more accurate to say that an electron is the set of its properties that we can measure. Bertrand Russell said it well when he remarked that an electron is "a logical construction."

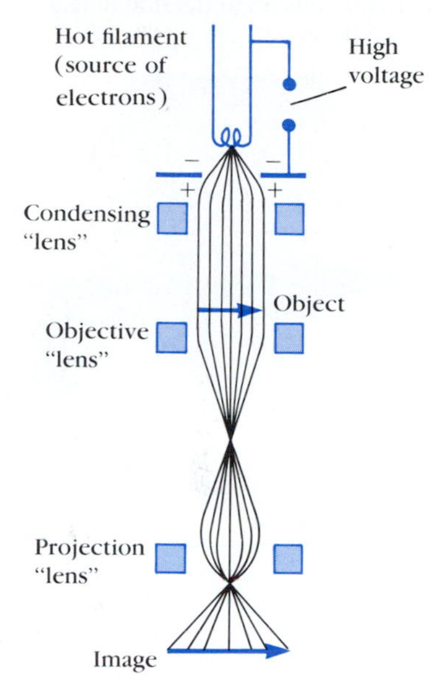

FIGURE 26-9
Electron microscope (transmission type). The squares represent magnetic field coils for the "magnetic lenses."

7. THE ELECTRON MICROSCOPE

Soon after the experimental confirmation that electrons have wave properties, it was suggested that electrons could be used to magnify tiny objects with much greater detail than was attainable with an ordinary light microscope. The resolution of a microscope—that is, the tiniest object that it can display clearly—is no smaller than the wavelength that is used. So with visible light, a microscope cannot be used to observe anything smaller than about 400 nm (1 nm = 10^{-9} meter). A beam of electrons, however, can easily be produced with a wavelength of less than 1 nm. Thus, the electron microscope can increase our magnifying ability by a factor of a thousand or more. The first electron microscopes were built in the 1930s. Figures 26-9 and 26-10 are diagrams of two types: the

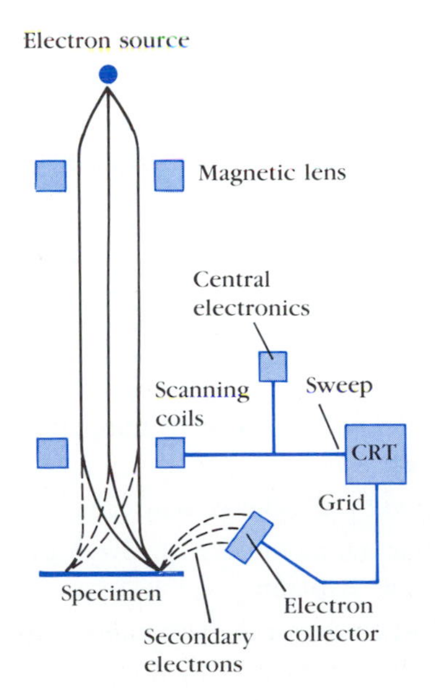

FIGURE 26-10
Scanning electron microscope. Scanning coils move an electron beam back and forth across the specimen. Secondary electrons produced when the beam strikes the specimen are collected and modulate intensity of the beam in the CRT to produce a picture.

FIGURE 26-11
(a) Transmission electron microscope photograph of plant cell wall ($\times$51,000). (b) Scanning electron microscope photograph of plant cells on underside of leaf ($\times$12,000).

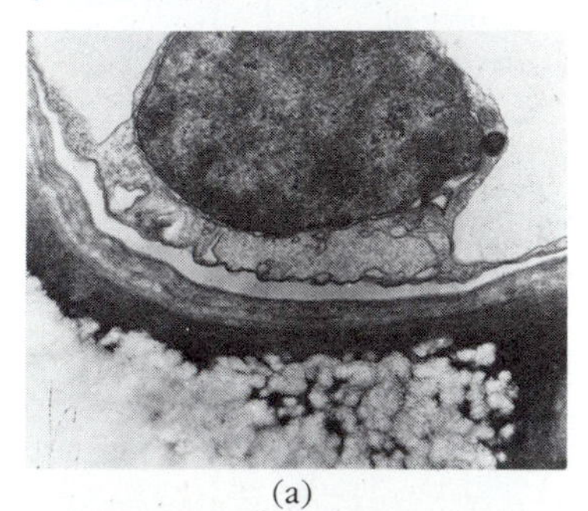

(a)

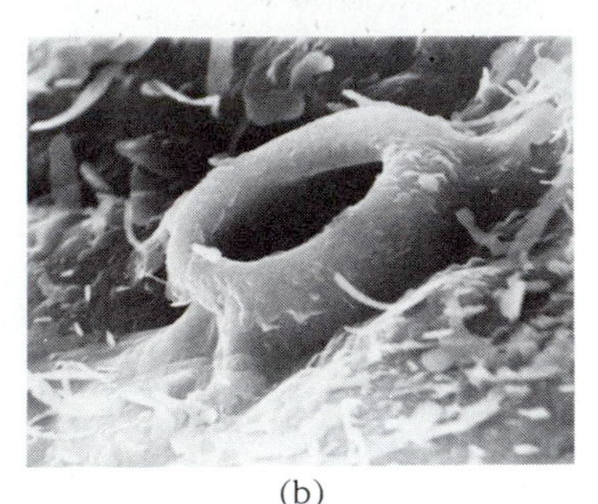

(b)

transmission electron microscope, which produces a two-dimensional image of a thin slice of material, and the more recently developed *scanning electron microscope*, which produces "volume" images with a three-dimensional quality. In each design the objective and eyepiece lenses are actually magnetic fields that exert forces on the electrons to bring them to a focus; the fields are produced by carefully designed current-carrying coils of wire. Photographs using each type are shown in Figure 26-11.

SUMMARY

Quantum theory has its origins in *Planck's quantum hypothesis* that molecular oscillations are *quantized*. Their energy E can only be integer (n) multiples of hf, where h is Planck's constant and f is the natural frequency of oscillation: $E = nhf$. This hypothesis explained the spectrum of radiation emitted by (black) bodies at high temperature.

Einstein proposed that light must be emitted as quanta (or particles) we now called *photons*, each with energy $E = hf$. He proposed the photoelectric effect as a test for the photon theory of light. In this effect, the photon theory says that each incident photon can strike an electron in a material and eject it if it has sufficient energy; the maximum energy of ejected electrons is then linearly related to the incident-light frequency. The photon theory is also supported by the Compton effect and the observation of electron-positron *pair production*.

X rays are high-energy photons. They are produced when high-speed electrons are decelerated upon striking a metal plate.

The *wave–particle duality* refers to the idea that light and matter (such as electrons) have both wave and particle properties. The wavelength of a material object is $\lambda = h/mv$, where m is the mass and v is the speed of the object. The *principle of complementarity* states that we must be aware of both the particle and wave properties of light and matter for a complete understanding of them.

QUESTIONS

1. What can be said about the relative temperature of whitish-yellow, reddish, and bluish stars?
2. If energy is radiated by all objects, why can't we see them in the dark?
3. Does a light bulb at a temperature of 2500 K produce as white a light as the sun at 6000 K? Explain.
4. Why do jewelers often examine diamonds in daylight rather than with indoor light?

5. "Orthochromatic" film is not sensitive to red light. Give an explanation based on the photon theory of light.
6. If the threshold wavelength in the photoelectric effect increases when the emitting metal is changed, what can you say about the work functions of the two metals?
7. Explain why the existence of a cutoff frequency in the photoelectric effect more strongly favors a particle theory and not a wave theory of light.
8. UV light causes sunburn, whereas visible light does not. Explain.
9. If an X-ray photon collides with an electron, does its wavelength change? If so, does it increase or decrease?
10. Explain how the photoelectric circuit of Figure 26-3 could be used in (a) a burglar alarm, (b) a smoke detector, (c) a photographic light meter, (d) a film sound track.
11. Describe what we mean by a "particle" and by a "wave."
12. Why do you suppose the particle nature of the electron was "discovered" before its wave nature?
13. An electron has a wavelength of 5.3×10^{-7} m. This is the wavelength of green light. Will the electron appear green? Explain why or why not.
14. Why do we say that light has wave properties? Why do we say that light has particle properties?
15. Why do we say that electrons have wave properties? Why do we say that electrons have particle properties?
16. What is the difference between a photon and an electron? Be specific.
17. If an electron and a proton travel at the same speed, which has the shorter wavelength?
18. How can you distinguish between an electron and an X ray that have the same wavelength?
19. Photographs have been made with electron microscopes that purport to show the outline of large molecules and even larger atoms like gold. Does this prove that atoms and molecules exist? Discuss in view of the fact that our understanding of the operation of an electron microscope and the photographs it produces is based on the assumption that atoms and electrons exist. You might consider if circular reasoning is involved; or if the atomic theory is self-consistent.

EXERCISES

1. An HCl molecule vibrates with a natural frequency of 8.1×10^{13} Hz. What is the difference in energy (in joules and electron volts) between possible values of the oscillation energy?

2. A child's swing has a natural frequency of 0.40 Hz. What is the separation between possible energy values (in joules)? If the swing reaches a height of 30 cm above its lowest point and has a mass of 20 kg (including the child), what is the value of the quantum number n?
3. What is the frequency of a photon whose energy is 1.0×10^{-20} J?
4. What is the wavelength of a photon whose energy is 2.6×10^{-19} J?
5. What is the energy of ultraviolet photons of frequency 1.0×10^{15} Hz?
6. What is the energy of a photon of wavelength (a) 400 nm and (b) 700 nm?
7. Calculate the wavelength of a 4.6-eV photon in nanometers.
8. What is the energy of photons (in eV) emitted by a 100-MHz FM radio station?
9. If a certain metal has a work function of 4.2×10^{-19} joules, will blue light ($\lambda = 4.5 \times 10^{-7}$ m) eject electrons? Will ultraviolet light of wavelength $\lambda = 3.6 \times 10^{-7}$ m?
10. What is the cutoff frequency for the metal in the above problem?
11. What is the longest wavelength of light that will eject electrons from a metal whose work function is 3.8×10^{-19} J?
12. What is the lowest frequency of light that will bring about the emission of electrons from a metal if the minimum energy W required is 2.5 eV? What wavelength and color is this?
13. The work function W of a certain metal is 2.1 eV. Will blue light ($\lambda = 4.5 \times 10^{-7}$ m) eject electrons? If so, what maximum kinetic energy will they have?
14. How much total kinetic energy will an electron-positron pair have if produced by a 3.6-MeV photon?
15. What is the minimum photon energy needed to produce a $\mu^{+} - \mu^{-}$ pair? The mass of each μ is 207 times the mass of the electron. What is the wavelength of such a photon?
16. Calculate the wavelength of an electron traveling 7.8×10^{5} m/s.
17. What is the wavelength of a proton ($m = 1.67 \times 10^{-27}$ kg) traveling 1.2×10^{6} m/s?
18. Calculate the wavelength of a tiny rock whose mass is 2.0 grams and moves (a) 20 m/s (b) 0.00001 m/s. How do these compare with the wavelengths of visible light?
19. How fast must an electron be moving to have a wavelength of 7.0×10^{-7} m? This is the wavelength of red light; will the electrons appear red?
20. What must be the speed of a 100-gram ball if its wavelength is to be 5.1×10^{-7} m, equivalent to that of green light?

*21. Sunlight reaching the earth has an intensity of 1300 W/m^2. How many photons per square meter per second does this represent? Take the average wavelength to be 550 nm.

*22. What is the maximum speed electrons can have that are ejected from a metal whose work function is 2.2 eV when illuminated by light whose frequency is 7.5×10^{16} Hz?

QUANTUM THEORY OF THE ATOM

CHAPTER

27

The idea that matter is made up of atoms was accepted by most scientists by 1900. With the discovery of the electron in the 1890s, it became clear that the atom itself must have a structure and that electrons were part of that structure. However, as we shall see in this chapter, the structure of the atom could not be understood until the quantum theory and the wave–particle duality were applied to it. Indeed, the search for an understanding of atomic structure led to the further development of the quantum theory and the introduction of the far-reaching and powerful theory of *quantum mechanics*. We now trace the development of our modern understanding of the atom and of the quantum theory with which it is intertwined.[†]

1. EARLY MODELS OF THE ATOM

A typical model of the atom in the 1890s visualized the atom as a homogeneous sphere of positive charge inside of which there were the negatively charged electrons, a little like plums in a pudding (Figure 27-1). J.J. Thomson (1856–1940), soon after the discovery of the electron in 1897, argued that the electrons in this model should be moving.

Around 1911, Ernest Rutherford (1871–1937) and his colleagues performed experiments which indicated that the "plum-pudding model" of

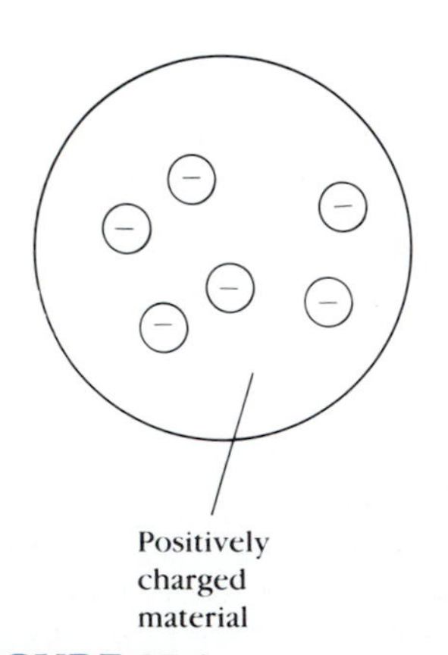

FIGURE 27-1
Plum-pudding model of the atom.

[†] Some readers may prefer that we tell them the facts as we know them today, and not bother with the historical background and its "outmoded" theories. Yet, such an approach would not only ignore the creative aspect of science and thus give a false impression of how science develops, but it would not really be possible to understand today's view of the atom and the quantum without discussing the concepts that led to it.

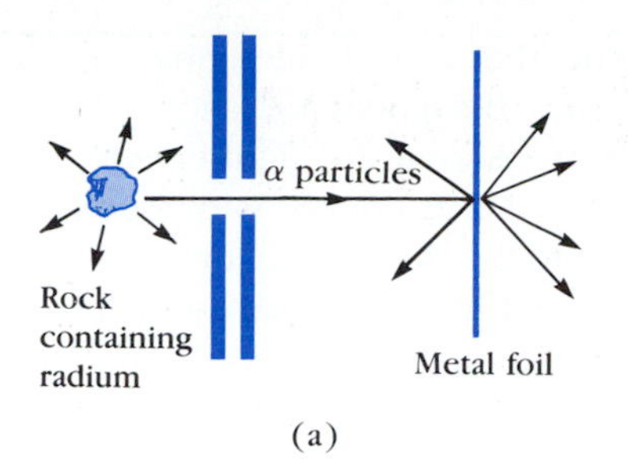

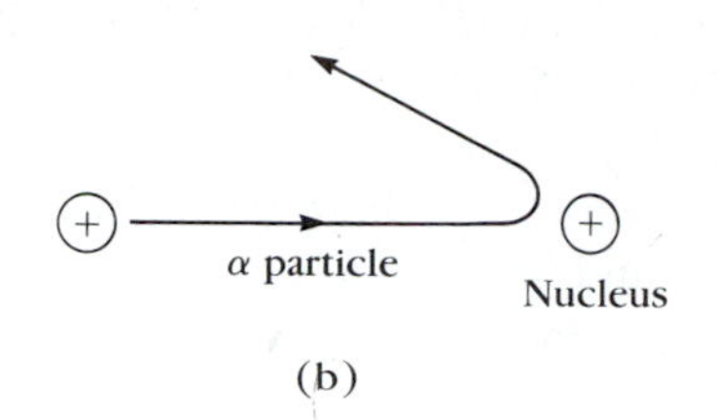

FIGURE 27-2
(a) Experimental setup for Rutherford's experiment: α particles emitted by radium strike metallic foil and some rebound backward. (b) Backward rebound of α particles explained as repulsion from heavy positively charged nucleus.

the atom could not be correct. In these experiments a beam of positively charged "alpha (α) particles" was directed at a thin sheet of metal foil, such as gold—Figure 27-2(a). (These newly discovered α particles were emitted by certain radioactive materials and were soon shown to be ionized helium atoms—see Chapter 29). It was expected from the plum-pudding model that the alphas would not be deflected significantly to one side since they never approached any massive concentration of positive charge to strongly repel them, nor would collision with an electron cause significant deflection since the mass of the alphas was many thousands of times that of an electron.

The experimental results completely contradicted these predictions. It was found that most of the α particles passed through the foil unaffected, as if the foil were mostly empty space. Of those deflected, it was found that a few of them were deflected at very large angles; some even rebounded back in nearly the direction from which they had come. This could happen, Rutherford reasoned, only if the positively charged particles were being repelled by a massive positive charge concentrated in a very small region of space—Figure 27-2(b). He concluded that the atom must consist of a tiny but massive positively charged nucleus containing over 99.9 percent of the mass of the atom and surrounded by electrons some distance away. The electrons must be moving in orbits about the nucleus—much like the planets moving around the sun—since if they were at rest, they would fall into the nucleus because of electrical attraction (Figure 27-3). From his experiments Rutherford concluded that the nucleus must have a radius of about 10^{-15} to 10^{-14} m. From kinetic theory, and especially Einstein's analysis of Brownian movement, the radius of atoms was estimated to be about 10^{-10} m. Thus the electrons must orbit the nucleus at a distance 10,000 to 100,000 times the radius of the nucleus itself, so an atom would be mostly empty space.

Rutherford's "planetary" model of the atom was a major step toward how we view the atom today. It was, however, not a complete model and presented some rather significant problems. For example, it was at once recognized that there was no reason for the hypothesized electron orbits to remain stable. Since the electrons are moving in orbits in which

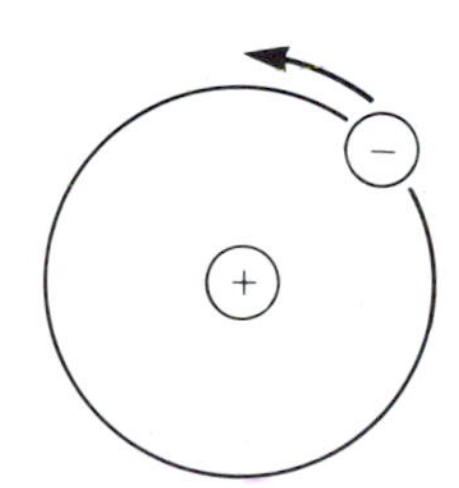

FIGURE 27-3
Rutherford's model of the atom, in which electrons orbit a tiny positive nucleus (not to scale). The atom is visualized as mostly empty space.

the direction of their velocity is changing, they are "accelerating"; and an accelerating electric charge must emit electromagnetic radiation, or light, as we saw in Chapter 21. When the electrons emit this light energy, they must lose kinetic energy to compensate—since energy is conserved—and so we would expect them to spiral into the nucleus; that would mean the end of the atom. This obviously does not happen, since the matter all around us is made up of atoms and is quite stable. Rutherford's model of the atom clearly required some modification. By 1910 the necessary clues were at hand.

2. ATOMIC SPECTRA: KEY TO THE STRUCTURE OF THE ATOM

At the beginning of Chapter 26 we saw that heated solids (as well as liquids and dense gases) emit light with a continuous range of frequencies. The dominant frequency range depends on the temperature of the body. The radiation from solids is assumed to be due to the vibration of the atoms and molecules in the solid, and this motion is governed by the interaction of each atom or molecule with its neighbors. In a rarefied gas, on the other hand, the atoms or molecules are presumed to be so far apart that they do not interact significantly with each other, except during infrequent collisions. Thus the light emitted from rarefied gases must come from the atoms themselves. Hence it is to the spectra of light emitted by gases that we must turn in order to gain knowledge of the internal structure of atoms.

For rarefied gases to emit light they must be excited by intense heating or, more commonly, by applying a high voltage to a "discharge tube" containing the gas at low pressure (Figure 27-4). The light emitted by

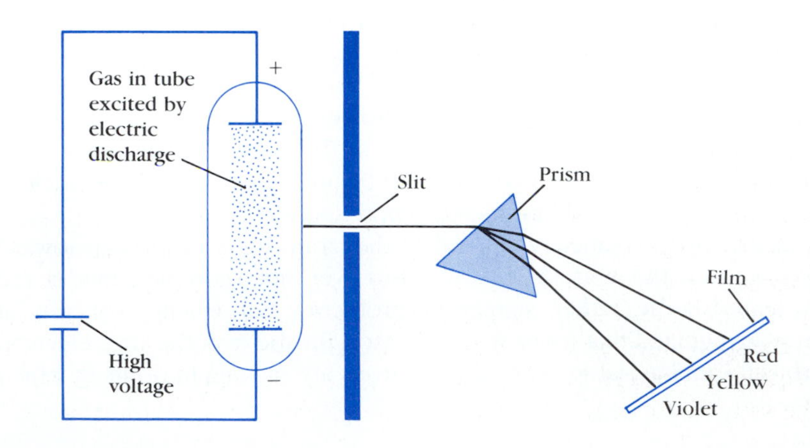

FIGURE 27-4
Rarefied gas in a "discharge tube" excited by high voltage; emitted light passes through the prism of the spectroscope which separates colors.

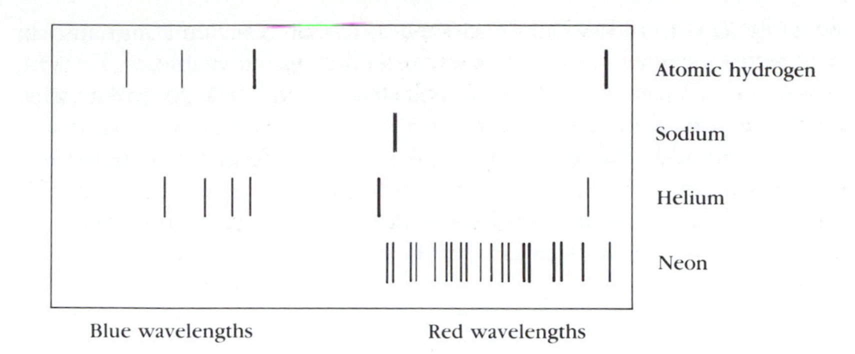

FIGURE 27-5
Spectra of various elements in the gas state.

heated or electrically excited gases had been observed early in the nineteenth century. This light was strange: Instead of a continuous spectrum of wavelengths, like that emitted by heated solids, the light emitted by an excited gas consisted only of certain discrete wavelengths or frequencies; this is called a discontinuous, or "line," spectrum (Figure 27-5).

The essential tool for observing and measuring the spectrum of light from a source is a spectroscope. And the essential part of a spectroscope is a prism, which, as we saw in Chapter 24, separates light into its component colors. The prism bends light of different wavelengths by different amounts, the shorter wavelengths of light being bent more than the longer wavelengths. For example, when white light passes through a prism each of the component colors or wavelengths is bent by a different amount, and all the colors of the rainbow from red to violet are seen. In the spectroscope, a viewing screen or photographic film is placed behind the prism, as shown in Figure 27-4. Each component wavelength arrives at a different point, so the position on the screen or film is a measure of the wavelength. Film sensitive to ultraviolet or infrared light can be used to detect radiation in these nonvisible parts of the spectrum. Figure 27-5 shows the **emission spectra** of various elements in the gaseous state. Each is a **line spectrum**; that is, only certain distinct wavelengths are present and each appears as a line on the film. These discrete wavelengths are different for each element and are therefore characteristic of the element. In a sense, the spectrum of each kind of atom or molecule is a "fingerprint"—it is unique and can be used for identification. By examining the spectral lines emitted by a gaseous object—for example, a star, the atmosphere of a planet, or the flame of a burning object—scientists can determine what elements and compounds are present. Chemists have made great use of the spectroscope as a tool for chemical analysis, a field of research known as spectroscopy.

If a continuous spectrum passes through a gas, dark lines are observed in the spectrum that correspond to lines normally emitted by the gas. This is called an *absorption spectrum*. It became clear that gases absorb light at the same frequencies at which they emit it. With the use of film sensitive to

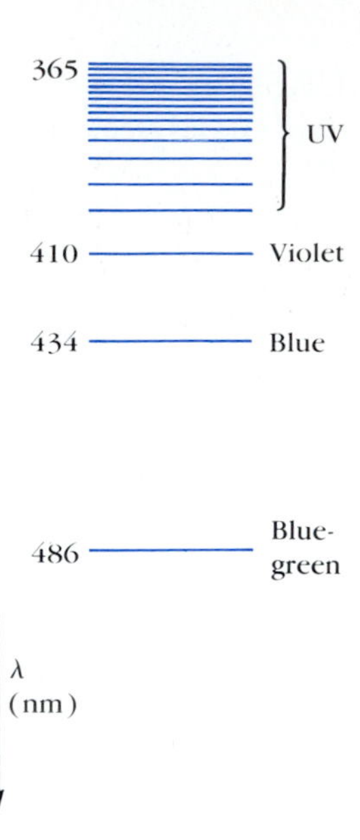

FIGURE 27-6
Balmer series of lines for hydrogen.

ultraviolet and to infrared light, it was found that gases emit and absorb discrete frequencies in these regions as well as in the visible.

Now any reasonable theory of atomic structure will have to explain why atoms emit and absorb light of only certain discrete frequencies. Indeed, the theory should be able to predict just what these frequencies are for any particular atom.

Hydrogen is the simplest atom—it has only one electron orbiting its nucleus. And it also happens to have the simplest spectrum. The spectrum of most atoms shows little regularity; but for hydrogen, as seen in Figure 27-5, the spacing between lines decreases in a regular way. Indeed, in 1885, J. J. Balmer (1825–1898) showed that the four visible lines in the hydrogen spectrum (measured to be 656, 486, 434, and 410 nm) would fit a simple formula, which we write as:

$$\frac{1}{\lambda} = R\left(\frac{1}{2^2} - \frac{1}{n^2}\right),$$

where n takes on the values 3, 4, 5, 6 for the four lines and R is a constant that has the value $R = 1.097 \times 10^7\ \mathrm{m}^{-1}$. Later it was found that this *Balmer series* of lines extended into the UV region, ending at $\lambda = 365$ nm, as shown in Figure 27-6. Balmer's formula worked for these lines as well as for higher integer values of n. The lines near 365 nm became too close together to distinguish, but the limit of the series at 365 nm corresponds to $n = \infty$, so $1/n^2 = 0$ in the formula.

Later experiments on hydrogen showed that there were other series of lines in the UV and IR, and these additional series had a pattern just like the Balmer series but at different wavelengths.

The Rutherford model, as it stood, was unable to explain why atoms emit line spectra. It had other difficulties as well. According to the Rutherford model, electrons orbit the nucleus and since their paths are curved the electrons are accelerating. Hence they should give off light like any other accelerating electric charge (Chapter 21). Then, since energy is conserved, the electron's own energy must decrease to compensate and electrons would be expected to spiral into the nucleus. As they spiraled inward, their frequency would increase gradually and so too would the frequency of the light emitted. Thus the two main difficulties of the Rutherford model are (1) it predicts that light of a continuous range of frequencies will be emitted, whereas experiment shows line spectra, and (2) it predicts that atoms are unstable—electrons quickly spiral into the nucleus—but we know that atoms in general are stable, since the matter around us is stable.

Clearly Rutherford's model was not sufficient. Some sort of modification was needed, and it was Niels Bohr who provided it by adding an essential idea—the quantum hypothesis.

3. THE BOHR MODEL

Bohr had studied in Rutherford's laboratory for several months in 1912 and was convinced that Rutherford's planetary model of the atom had validity. But to make it work he felt that the newly developing quantum theory would somehow have to be incorporated in it. The work of Planck and Einstein had shown that in heated solids the energy of oscillating electric charges must change discontinuously—from one discrete energy state to another with the emission of a quantum of light. Perhaps, Bohr argued, the electrons in an atom also cannot lose energy continuously but must do so in quantum "jumps." In working out his theory during the next year, Bohr postulated that electrons move about the nucleus in circular orbits, but that only certain orbits are allowed (Figure 27-7). He further postulated that an electron in each orbit would have a definite energy and would move in the orbit *without radiating energy* (even though this violated classical ideas). He thus called the possible orbits **stationary states**. Light is emitted, he hypothesized, only when an electron jumps from one stationary state to another of lower energy. When such a jump occurs, a single photon of light would be emitted. Since energy must be conserved, the photon energy, hf, is given by

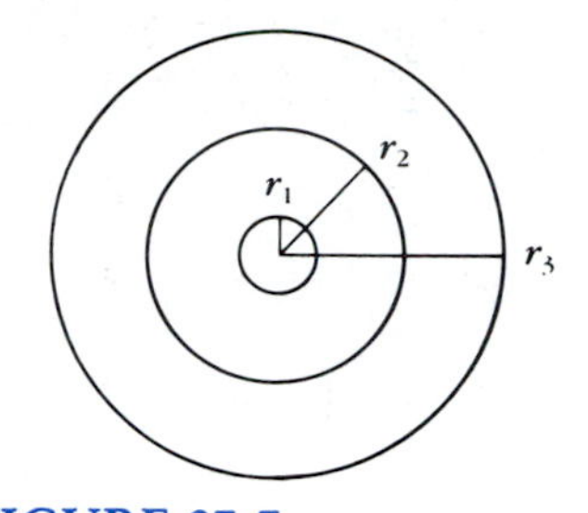

FIGURE 27-7
Bohr model of the hydrogen atom. Its one electron can revolve in any one of many circular orbits that can have only certain radii. The three smallest orbits are shown.

$$hf = E_u - E_\ell$$

where E_u is the energy of the upper state and E_ℓ the energy of the lower state.

Bohr next set out to determine what energies these orbits have, since then the spectrum of light emitted can be predicted from the preceding equation. When he became aware of the Balmer formula in early 1913, he had the key he was looking for. Bohr quickly found that his theory would be in accord with the Balmer formula if he assumed the electron's angular momentum L was equal to an integer n times $h/2\pi$. We saw in Chapter 9 that the angular momentum of a particle of mass m moving in a circle of radius r with speed v is $L = m\omega r^2$, or (since $\omega = v/r$), $L = mvr$. **Bohr's quantum condition** then is

$$L = mvr_n = n\frac{h}{2\pi}. \qquad [n = 1,\ 2,\ 3, \ldots .]$$

Here n is an integer and r_n is the radius of the nth possible orbit. The allowed orbits are numbered 1, 2, 3 ... according to the value of n, which is called the **quantum number** of the orbit. This quantum condition did not have a firm theoretical foundation. Bohr had searched for some "quantum condition"; and such tries as $E = hf$ (where E represents the energy of the electron in an orbit) did not give results in accord with experiments. Bohr's reason for using it was simply that it worked.

Starting with this quantum condition, and using the laws of classical physics, Bohr derived[†] a formula for the radii of electron orbits,

$$r_n = n^2 r_1 \qquad [n = 1, 2, 3 \ldots]$$

where

$$r_1 = 0.53 \times 10^{-10}\ \text{m}$$

is the radius of the smallest (innermost) orbit, and is often called the **Bohr radius**. The radii of the larger orbits increase as n^2. The next possible orbit, $n = 2$, has a radius $(2)^2 = 4$ times as large:

$$r_2 = 4r_1 = 2.1 \times 10^{-10}\ \text{m}.$$

The third orbit, $n = 3$, has a radius $(3)^2 = 9$ times as large as the smallest one,

$$r_3 = 9r_1 = 4.8 \times 10^{-10}\ \text{m}$$

and so on. The first three are shown in Figure 27-7. Notice that according to Bohr's model, an electron can exist only in orbits whose radii are $n^2 \times r_1$; it cannot exist in between. (Be careful not to believe that these well-defined orbits actually exist. The Bohr model is only a model, not reality; and, as we shall see, the idea of electron orbits has been rejected. Instead, today electrons are better thought of as forming "clouds," as discussed later in this chapter.)

[†] The derivation is not difficult. An electron in a circular orbit of radius r_n would have a centripetal acceleration v^2/r_n produced by the electrical force of attraction between the negative electron and the positive nucleus according to Coulomb's law, $F = kq_1q_2/r^2$ (see Chapter 18), where the charge on the electron is $q_1 = -e$, and that on the nucleus is $q_2 = +e$. From Newton's second law, $F = ma$, we substitute $a = v^2/r_n$ and Coulomb's law for F, and obtain

$$k\frac{e^2}{r_n^2} = \frac{mv^2}{r_n}.$$

We solve this for r_n and substitute for v from the quantum condition ($L = mvr_n = nh/2\pi$, so $v = nh/2\pi mr_n$):

$$r_n = \frac{ke^2}{mv^2} = \frac{ke^2 4\pi^2 m r_n^2}{n^2h^2}.$$

We solve for r_n (it appears on both sides, so we cancel one of them) and find

$$r_n = \frac{n^2h^2}{4\pi^2 mke^2}.$$

This equation gives the radii of the possible orbits. When we insert numbers for all the values (except for n, which is any whole number), we get

$$r_n = \frac{(6.626 \times 10^{-34}\ \text{J}\cdot\text{s})^2}{4(3.14)^2(9.11 \times 10^{-31}\ \text{kg})(9.00 \times 10^9\ \text{N}\cdot\text{m}^2/\text{C}^2)(1.602 \times 10^{-19}\ \text{C})^2}n^2$$

$$r_n = (0.53 \times 10^{-10}\ \text{m}) \times n^2.$$

In each of these possible orbits, or stationary states, the electron would have a definite energy. Bohr showed that the energy of each of these stationary states is given by

$$E_n = E_1 \frac{1}{n^2} \qquad [n = 1, 2, 3 \ldots.]$$

where the energy of the lowest state, E_1, is

$$E_1 = -2.17 \times 10^{-18}\ \text{J}.$$

In terms of the commonly used electron-volt unit (Chapter 26, Section 2), we get $(2.17 \times 10^{-18}\ \text{J})/(1.60 \times 10^{-19}\ \text{J/eV}) = 13.6\ \text{eV}$, so

$$E_1 = -13.6\ \text{eV}.$$

For the larger orbits

$$E_n = (-13.6\ \text{eV})/n^2,$$

so for $n = 2$ and $n = 3$ we have

$$E_2 = E_1/4 = -3.40\ \text{eV}$$
$$E_3 = E_1/9 = -1.51\ \text{eV}.$$

Thus we see that not only are the orbit radii quantized but so is the energy.

Notice that although the energy for the larger orbits has a smaller numerical value, all the energies are less than zero. Thus -3.4 eV is a greater energy than -13.6 eV. Hence the orbit closest to the nucleus (r_1) has the lowest energy. The reason the energies have negative values is because we consider an electron which is completely free of an atom as having energy equal to zero if it is not moving. (If a free electron is moving, its energy is greater than zero.) An electron free of an atom would be far from the atom, corresponding to $n \approx \infty$ (infinity); and from the above energy equation, for $n = \infty$, $E = E_1/n^2 = 0$. When an electron is bound to an atom, there must be an input of energy to remove it from its atom. Thus, when bound to an atom an electron must have less energy than a free electron—that is, energy less than zero. Incidentally, the energy needed to remove an electron from an atom is called the **binding energy** or **ionization energy**. The ionization energy for hydrogen has been measured to be 13.6 eV; and this corresponds precisely to removing an electron from the lowest state, $E_1 = -13.6$ eV, up to $E = 0$ where it will be free.

It is useful to show the various possible energy values as horizontal lines on an energy-level diagram. This is shown for hydrogen in Figure 27-8. The quantum number n that labels the orbit radii also labels the energy levels. The lowest *energy level* or *energy state* has energy E_1, and is

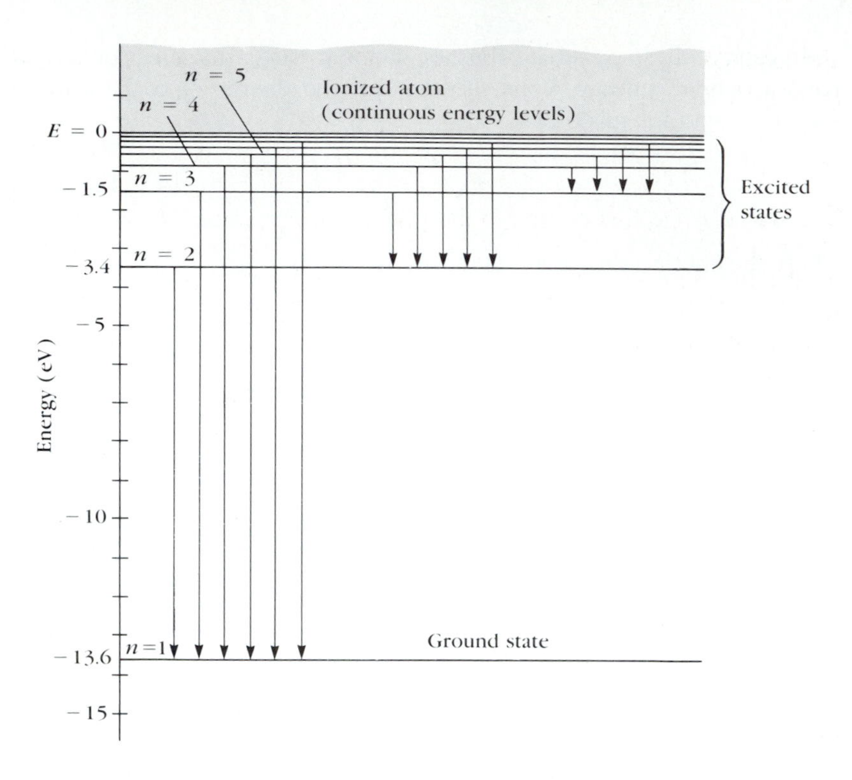

FIGURE 27-8
Energy-level diagram for the hydrogen atom.

called the **ground state**. The higher states, E_2, E_3, and so on, are called **excited states**. The electron in a hydrogen atom can be in any one of these levels according to Bohr theory. But it could never be in between, say at -9.0 eV. At room temperature, nearly all H atoms will be in the ground state. At higher temperatures, or during an electric discharge when there are numerous collisions between free electrons and atoms, many electrons can be in excited states. Once in an excited state, an electron can jump down to a lower state and give off a photon in the process. This is, according to the Bohr model, the origin of the line spectra emitted by excited gases. The vertical arrows in Figure 27-8 represent the transitions or jumps that correspond to the various observed spectral lines. For example, an electron jumping from the level $n = 3$ to $n = 2$ gives rise to the 656-nm line in the Balmer series, and the jump from $n = 4$ to $n = 2$ gives rise to the 486-nm line (see Figure 27-6).

Calculating Spectral Lines (optional)

We can predict wavelengths of the spectral lines using the Bohr theory. For the $n = 5$ to $n = 2$ transition, the initial (upper) state has $n = 5$ and its energy is $E_5 = -13.6\ \text{eV}/(5)^2 = -13.6\ \text{eV}/25 = -0.54$ eV. The energy of the lower state ($n = 2$) is $E_2 = -13.6\ \text{eV}/(2)^2 = -3.40$ eV. The energy of the emitted photon is equal to the difference of these two energies (energy lost by atom = energy given to photon):

$$hf = E_u - E_\ell = -0.54\text{ eV} - (-3.40\text{ eV}) = 2.86\text{ eV}.$$

We must change this energy to the proper SI unit, joules: $hf = (2.86\text{ eV})(1.6 \times 10^{-19}\text{ J/eV}) = 4.58 \times 10^{-19}\text{ J}$. Then $f = (4.58 \times 10^{-19}\text{ J})/(h) = 4.58 \times 10^{-19}\text{ J}/6.63 \times 10^{-34}\text{ J}\cdot\text{s} = 6.91 \times 10^{14}\text{ Hz}$. We want to determine the wavelength, which is $\lambda = c/f = (3.00 \times 10^8\text{ m/s})/(6.91 \times 10^{14}\text{ Hz}) = 4.34 \times 10^{-7}\text{ m} = 434\text{ nm}$. This is the third line of the Balmer series (see Figure 27-6).

If we combine the equation $hf = E_u - E_\ell$ with Bohr's formula for the energy, $E_n = E_1/n^2$, we can write $hf = E_1\,(1/n'^2 - 1/n^2)$ where n' and n are the quantum numbers of the upper and lower states respectively. Then, since $f = c/\lambda$, or $1/\lambda = f/c$, we see that

$$\frac{1}{\lambda} = \frac{f}{c} = \frac{E_1}{hc}\left(\frac{1}{n'^2} - \frac{1}{n^2}\right).$$

This equation has the same form as Balmer's formula (Section 2) with $n' = 2$. Even the constant, E_1/hc, has the value equal to Balmer's R. Thus we see that the Balmer series of lines corresponds to transitions or "jumps" that bring the electron from upper states down to the second ($n = 2$) energy state (see Figure 27-8). Similarly, $n' = 1$, $n' = 3$, and so on, correspond to other series of lines (Figure 27-8).

Significance of the Bohr Model; the Correspondence Principle

The great success of Bohr's theory is that it explains why atoms emit line spectra and accurately predicts, for hydrogen, the wavelengths of emitted light. The Bohr theory also expalins absorption spectra: photons of just the right wavelength can knock an electron from one energy level to a higher one. To conserve energy, the photon must have just the right energy. This explains why a continuous spectrum passing through a gas will have dark (absorption) lines at the same frequencies as the emission lines. The Bohr theory also ensures the stability of atoms. It establishes stability by fiat: The ground state is the lowest state for an electron and there is no lower energy level to which it can go and emit more energy. Finally, as we saw above, the Bohr theory correctly predicts how much energy is required to remove an electron from a hydrogen atom. This, the ionization energy, is measured for hydrogen to be 13.6 eV and is fully in accord with predictions of the Bohr theory for an upward jump of an electron from the ground state ($E_1 = -13.6\text{ eV}$) to where it is free of the atom ($E = 0$) (Figure 27-8). (Note that an electron can have any energy above $E = 0$, for here it is free.)

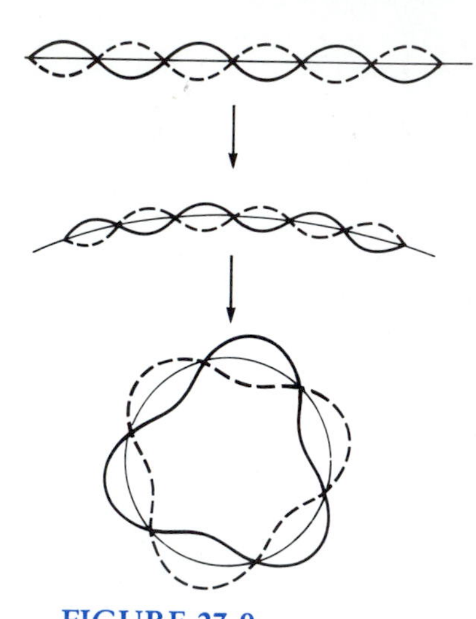

FIGURE 27-9
An ordinary standing wave compared to a circular standing wave.

We should note that Bohr made some radical assumptions that were at variance with classical ideas. He assumed that electrons in fixed orbits do not radiate light even though they are accelerating, and he assumed that the angular momentum is quantized. Furthermore, he was not able to say how an electron moved when it made a transition from one energy level to another. On the other hand, there is no real reason to expect that in the tiny world of the atom electrons would behave as ordinary-sized objects do. Nonetheless, he felt that where quantum theory overlaps with the macroscopic world it should predict classical results. This he called the **principle of correspondence**. This principle does seem to work for his theory of the hydrogen atom. The orbit sizes and energies are quite different for $n = 1$ and $n = 2$, say. But orbits number 100,000,000 and 100,000,001 would be very close in size and energy (see Figure 27-8). Indeed, jumps between such large orbits, which would approach everyday sizes, would be imperceptible; such orbits would thus appear to be continuous, which is what we expect in the everyday world.

Bohr's theory was largely of an ad hoc nature. Assumptions were made so that theory would agree with experiment. But Bohr could give no reason why the orbits were quantized. Ten years later, a reason was found by de Broglie.

4. DE BROGLIE'S HYPOTHESIS

We saw in Chapter 26 that in 1923, Louis de Broglie proposed that material particles such as electrons have a wave nature. And this hypothesis was confirmed by experiment several years later.

One of de Broglie's original arguments in favor of the wave nature of electrons was that it provided an explanation for Bohr's theory of the hydrogen atom.

As discussed earlier, de Broglie hypothesized that a particle of mass m moving with speed v would have a wavelength

$$\lambda = \frac{h}{mv}.$$

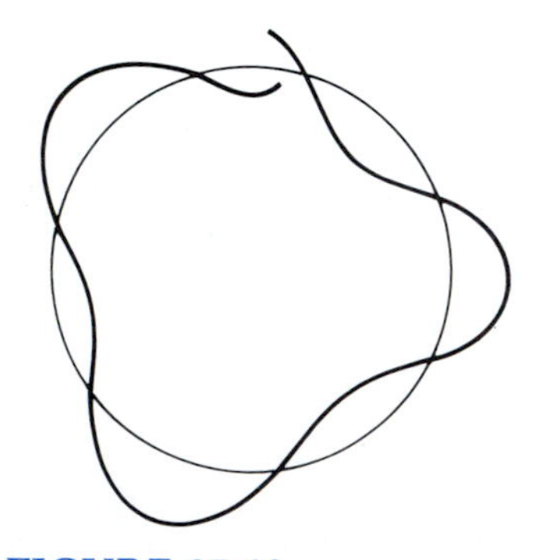

FIGURE 27-10
When a wave does not close (and hence interferes with itself), it rapidly dies out.

Each electron orbit in an atom, he proposed, is actually a standing wave As we saw in Chapter 16, when a violin or guitar string is plucked a vast number of wavelengths are excited. But only certain ones—those that have nodes at the ends—are sustained. These are the *resonant* modes of the string. All other wavelengths interfere with themselves upon reflection and their amplitudes quickly drop to zero. Since electrons move in circles, according to Bohr's theory, de Broglie argued that the electron wave must be a *circular* standing wave that closes on itself (Figure 27-9). If the wavelength of a wave does not close on itself, as in Figure 27-10, destructive interference takes place as the wave travels around the loop and it quickly dies out. Thus, the only waves that persist are those for which the circumference of the circular orbit contains a whole number of wavelengths (Figure 27-11). The circumference of a Bohr orbit of radius r is $2\pi r$ so we must have

$$2\pi r = n\lambda, \qquad n = 1, 2, 3 \ldots.$$

When we substitute $\lambda = h/mv$, we get

$$2\pi r = \frac{nh}{mv},$$

or

$$mvr = \frac{nh}{2\pi}.$$

This is just the *quantum condition* proposed by Bohr on an *ad hoc* basis (Section 3). And it is from this equation that the discrete orbits and energy levels were derived. Thus we have an explanation for the quantized orbits and energy states in the Bohr model: they are due to the wave nature of the electron and the fact that only resonant "standing" waves can persist. This implies that the *wave–particle duality* is at the root of atomic structure.

It should be noted in viewing the circular electron waves of Figure 27-11 that the electron is not to be thought of as following the oscillating wave pattern. In the Bohr model of hydrogen, the electron, considered as a particle, moves in a circle. The circular wave, on the other hand, represents the *amplitude* of the electron "matter wave," and in Figure 27-11 the wave amplitude is shown superimposed on the circular path of the particle orbit for convenience.

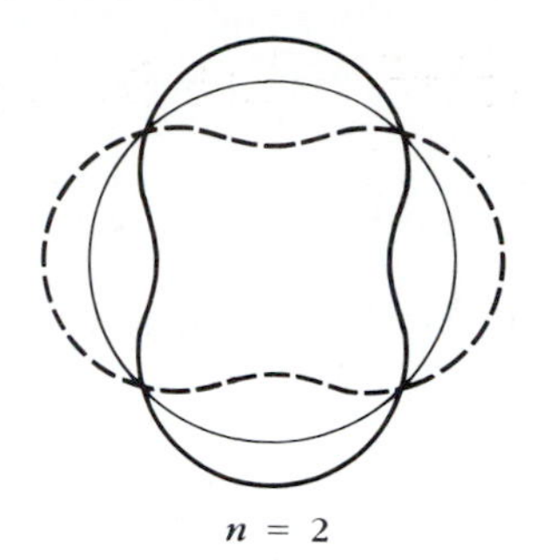

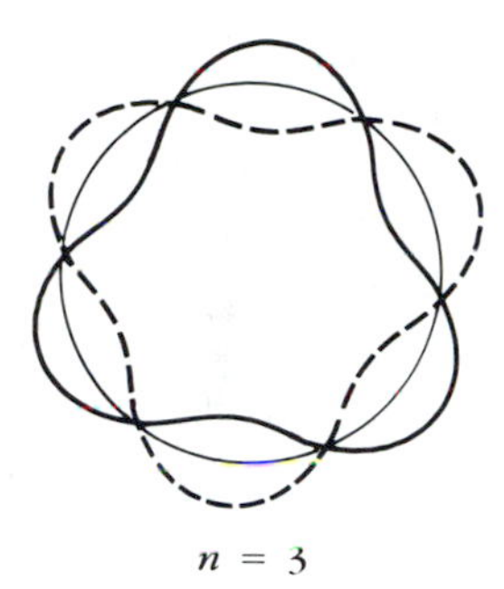

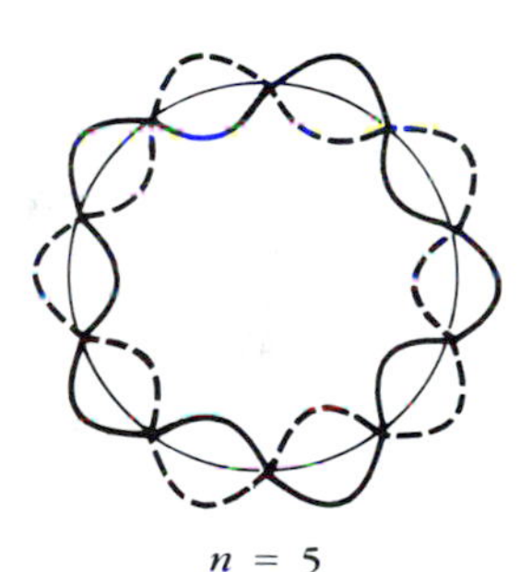

FIGURE 27-11
Standing circular waves for two, three, and five wavelengths on the circumference; n, the number of wavelengths, is also the quantum number.

5. QUANTUM MECHANICS

A New Theory

The Bohr theory of the atom gave us a first (though rough) picture of what an atom is like. But even with the theoretical basis provided by de Broglie's matter-wave hypothesis, there still were limitations. The Bohr theory successfully accounted for the spectrum of lines emitted by the hydrogen atom; but it was incapable of explaining the spectra of more complex atoms—atoms containing two or more electrons. Furthermore, the problem of the wave–particle duality was far from resolved. A more comprehensive theory was clearly needed. It was not long in coming. Less than two years after de Broglie presented his matter-wave hypothesis, two physicists independently developed the desired theory: Erwin Schrödinger (1887–1961) and Werner Heisenberg (1901–1976).

The new theory, known as **quantum mechanics**, unified the wave–particle duality into a single consistent but highly mathematical theory. The wave and particle aspects no longer had to be applied on the basis of which worked best in a given situation. Instead, the wave and particle aspects were integrated unequivocally into one single formalism.

Quantum mechanics as a theory is accepted today by nearly all physicists as the fundamental theory underlying physical processes. A theory is as good as the experimental facts that it can confirm and predict, and quantum mechanics is very good indeed.

Although quantum mechanics deals mainly with the microscopic world of atoms and light, in the macroscopic world of everyday life we do perceive light and we take for granted that ordinary objects are made up of atoms. Therefore if this new theory is really good, it must also account for the verified results of classical physics when it is applied to macroscopic phenomena; that is, when applied to macroscopic phenomena, the laws of quantum mechanics must yield the old classical laws. This is the *principle of correspondence* enunciated by Bohr, discussed in Section 3. Quantum mechanics fulfills this requirement as well. This doesn't mean we throw away classical theories such as Newton's laws. In the everyday world the latter are far easier to apply and give an accurate description. But when we deal with high speeds, close to the speed of light, we must use the theory of relativity; and when we deal with the tiny world of the atom, we use quantum mechanics.

Although we won't go into the detailed mathematics of quantum mechanics, we will discuss the main ideas and how they involve the wave and particle properties of matter to explain atomic structure and other applications.

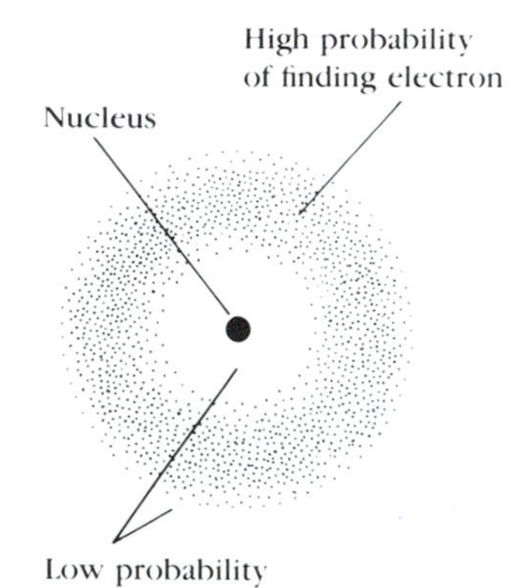

FIGURE 27-12

Electron cloud or "probability distribution" for the ground state of the hydrogen atom. The cloud is densest—indicating the highest probability—at a distance from the nucleus of 0.53×10^{-10} m, which is just what the Bohr model predicts for the radius of the first orbit. But unlike the Bohr model, quantum mechanics tells us the electron can be within or beyond that distance at any given time.

Atoms: Energy Is Quantized but Orbits Are Diffuse

Quantum mechanics applied to atoms is far more complete than the old Bohr theory. It retains certain aspects of the older theory, such as that electrons in an atom exist only in discrete states of definite energy and that a photon of light is emitted (or absorbed) when an electron makes a transition from one state to another. In fact, quantum mechanics predicts exactly the same energy levels for hydrogen as the older Bohr theory did (Figure 27-8). But quantum mechanics is able to predict the energy levels for the more complex atoms as well, something the Bohr theory could not do.

But quantum mechanics is not simply an extension of the Bohr theory. It is a far deeper theory, and has provided us with a very different view of the atom. According to quantum mechanics, electrons do not exist in the well-defined circular orbits of the Bohr theory. Rather, the electron (because of its wave nature) is spread out in space, as a "cloud" of negative charge. The shape and size of the electron cloud can be calculated for a given state of an atom. The electron cloud for the ground state of the hydrogen atom is spherically symmetric as shown in Figure 27-12. The electron cloud roughly indicates the "size" of an atom; but just as a cloud does not have a distinct border, atoms do not have a precise boundary or a well-defined size. Not all electron clouds have a spherical shape: some are "dumbbell" shaped as in Figure 27-13, and these latter are especially important in making chemical bonds between atoms, as we shall see in the next chapter.

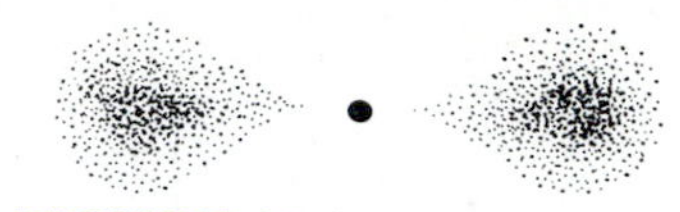

FIGURE 27-13
Dumbbell-shaped electron probability distribution in an atom.

The electron cloud can be interpreted using either the particle or the wave viewpoint. Remember that by a particle we mean something that is *localized* in space—it has a definite position at any moment. But a wave is spread out in space. Thus the electron cloud, spread out in space as in Figure 27-12, is a result of the wave nature of electrons. Electron clouds can also be interpreted as **probability distributions** for a particle. If you were to make 500 different measurements of the position of an electron (thinking of it as a particle), the majority of the results would show the electron at points where the probability is high (dark area in Figure 27-12); only occasionally would the electron be found where the probability is low. We cannot predict the path an electron will follow. After one measurement of its position we cannot predict exactly where it will be at a later time. We can only calculate the probability that it will be found at different points. This is clearly different from classical Newtonian physics. Indeed, as Bohr later pointed out, since an electron is not simply a particle, it is meaningless to even ask how it gets from one state to another when the atom emits a photon of light.

The fact that we cannot predict the exact position of an electron in an atom but can only give the probability distribution is essentially a result of its wave properties. This fact has deep philosophic consequences, and we discuss it more fully in the next chapter.

Also in Chapter 28 we shall look in more detail at what quantum mechanics tells us about atomic structure, and the structure of molecules and matter in general.

SUMMARY

Early models of the atom include Thomson's modified plum-pudding model and Rutherford's planetary (or nuclear) model. Rutherford's model, which was created to explain the back-scattering of alpha particles from thin metal foils, assumes that an atom consists of a massive positively charged nucleus surrounded (at a relatively great distance) by electrons.

To explain the *line spectra* emitted by atoms, as well as their stability, Bohr proposed a theory which postulated (1) that electrons bound in an atom can only occupy orbits for which the angular momentum is quantized, which results in discrete values for the radius and energy; (2) that an electron in such a *stationary state* emits no radiation; (3) that if an electron jumps to a lower state it emits a photon whose energy equals the difference in energy between the two states; (4) that the angular momentum L of atomic electrons is quantized by the rule $L = nh/2\pi$, where n is an integer called a *quantum number*. The $n = 1$ state in hydrogen is the *ground state*, which has an energy $E_1 = -13.6$ eV; higher values of n correspond to *excited states* and their energies are $E_n = -13.6\ \text{eV}/n^2$. Atoms are excited to these higher states by collisions with other atoms or electrons or by absorption of a photon of just the right frequency.

De Broglie's hypothesis that electrons (and other matter) have wavelength $\lambda = h/mv$ gave an explanation for Bohr's quantized orbitals by bringing in the wave-particle duality: the orbits correspond to circular standing waves in which the circumference of the orbit equals a whole number of wavelengths.

In 1925, Schrödinger and Heisenberg worked out a new theory, *quantum mechanics*, which integrated the wave and particle aspects into a unified whole. Quantum mechanics is now considered to be the basic theory at the atomic level. It is a statistical theory rather than a deterministic one. According to quantum mechanics, the electrons in an atom do not have well-defined orbits but instead exist as a "cloud." Electron clouds can be interpreted as an electron wave spread out in space, or as a probability distribution for electrons as particles.

QUESTIONS

1. In Rutherford's planetary model of the atom, what prevents the electrons from flying off into space?
2. Which of the following can emit a line spectrum: (a) gases, (b) liquids, (c) solids. Which can emit a continuous spectrum?
3. Why doesn't the O_2 gas in the air around us give off light?
4. How can you tell if there is oxygen on the sun?
5. Some materials produce colors other than yellow when thrown into a fire. Why?
6. Explain how the closely spaced energy levels for hydrogen near the top of Figure 27-8 correspond to the closely spaced spectral lines at the top of Figure 27-6.
7. Discuss the differences between Rutherford's and Bohr's theory of the atom. Be specific.
8. In a helium atom, which contains two electrons, do you think that on the average the electrons are closer to the nucleus or farther away than in a hydrogen atom? Why?
9. How can the spectrum of hydrogen contain so many lines when hydrogen contains only one electron?
10. What happens to an atom when it emits light?

11. Is energy conserved when an atom emits a photon of light? Explain.
12. What do we mean when we say an atom is "excited"?
13. Does an atom have more energy when it is excited? If so, what kind of energy is it?
14. In an atom, why can't an electron orbit exist whose circumference is equal to $2\frac{1}{2}$ wavelengths?
15. Discuss the differences between Bohr's view of the atom and the quantum mechanical view.
16. If Planck's constant were much larger than it is, how would this affect our everyday life?

EXERCISES

1. When an atom jumps from the first excited state in hydrogen to the ground state, light of frequency 2.46×10^{15} Hz is emitted. What is the difference in energy between these two states?
2. An electron jumps within an atom from an energy level $E_u = -2.2 \times 10^{-19}$ J to $E_\ell = -8.0 \times 10^{-19}$ J. What is the frequency of light emitted?
3. An electron jumps from a level $E_u = -3.5 \times 10^{-19}$ J to $E_\ell = -1.20 \times 10^{-18}$ J. What is the wavelength of emitted light?
4. An atom emits light of frequency $f = 4.0 \times 10^{14}$ Hz when an electron jumps to the ground state of $E_\ell = -9.4 \times 10^{-19}$ J. What was the energy of the excited state?
5. Sodium emits bright light of wavelength 589 nm. Calculate the energy difference in the atomic energy levels when such a transition occurs.
6. Determine the wavelength of the second Balmer line ($n = 4$ to $n = 2$ transition) using Figure 27-8.
7. Determine the wavelength for the $n = 2$ to $n = 1$ transition in hydrogen using Figure 27-8.
8. Calculate the wavelength of the fourth Balmer line ($n = 6$ to $n = 2$).
9. What is the wavelength of light emitted when an $n = 50$ to $n = 30$ transition occurs in hydrogen? In what region of the spectrum does such a photon lie?
10. Light of wavelength 388 nm is emitted by a hydrogen discharge tube. Between what energy levels is the atom jumping?
11. How much energy is needed to ionize a hydrogen atom in the $n = 3$ state?

*12. An excited hydrogen atom could, in principle, have a radius of 1.0 mm. What would be the value of n for a Bohr orbit of this size? What would its energy be?

*13. Calculate the ratio of the gravitational to electrical force for the electron in a hydrogen atom. Can the gravitational force be safely ignored?

*14. Use Coulomb's law and the fact that a particle moving in a circle must have a force on it equal to mv^2/r to determine the speed of an electron in the ground state of the hydrogen atom.

*15. By what fraction does the mass of an H atom decrease when it makes an $n = 2$ to $n = 1$ transition?

*16. What is the longest wavelength of light capable of ionizing a hydrogen atom in the ground state?

*17. At low temperatures, nearly all the atoms in hydrogen gas will be in the ground state. What minimum frequency photon is needed if the photoelectric effect is to be observed?

QUANTUM-MECHANICAL VIEW OF THE WORLD

CHAPTER

28

Since its development in the 1920s, quantum mechanics has had a profound influence on our lives, both intellectually and technologically. Even the way we view the world has changed. In this chapter we will discuss how quantum mechanics has given us an understanding of the structure of atoms, molecules, and matter in bulk. We will also discuss a number of applications including semiconductor electronics and lasers. We begin first with a discussion of one of the most famous outcomes of quantum mechanics, the uncertainty principle, followed by some of the philosophic implications of quantum mechanics.

1. THE HEISENBERG UNCERTAINTY PRINCIPLE

Whenever a measurement is made, some uncertainty or error is always involved. For example, you cannot make an absolutely exact measurement of the length of a table. Even with a measuring stick that has markings 1 mm apart there will be an inaccuracy of about $\frac{1}{2}$ mm or so. More precise instruments will produce more precise measurements; but there is always some uncertainty involved in a measurement no matter how accurate the measuring device. We expect that by using more precise instruments the uncertainty in a measurement can be made indefinitely small.

But according to quantum mechanics there is actually a limit to the accuracy of certain measurements. This limit is not a restriction on how well instruments can be made; rather, it is inherent in nature. It is the result of two factors: the wave–particle duality and the unavoidable interaction between the thing observed and the observing instrument. Let us look at this in more detail.

To make a measurement of an object without somehow disturbing it, at least a little, is not possible. Consider trying to locate a Ping-Pong ball in a completely dark room. You grope about trying to find its position; and just when you touch it with your finger, it bounces away. Whenever we measure the position of an object, whether it's a Ping-Pong ball or an electron, we always touch it with something else which gives us the information about its position. To locate a lost Ping-Pong ball in a dark room you could probe about with your hand or a stick, or you could shine a light and detect the light reflecting off the ball. When you search with your hand or a stick, you find the ball's position when you touch it. But when you touch the ball you unavoidably bump it and give it some momentum; thus you won't know its *future* position. The same would be true, but to a much lesser extent, if you observe the Ping-Pong ball using light: in order to "see" the ball at least one photon[†] must scatter from it, and the reflected photon must enter your eye or some other detector. When a photon strikes an ordinary-sized object it does not appreciably alter the motion or position of the object. But when a photon strikes a very tiny object like an electron, it can transfer much of its momentum to the object and thus greatly change the object's motion and position in an unpredictable way. The mere act of measuring the position of an object at one time makes our knowledge of its future position inaccurate.

Now let us see how the wave–particle duality applies. Imagine a thought experiment in which we are trying to measure the position of an object (say an electron) with photons, although the arguments would be similar if we were using an electron microscope. As we saw in Chapter 24, objects can be seen to an accuracy no greater than the wavelength of the radiation used. If we want an accurate position measurement, we must use a short wavelength. But a short wavelength corresponds to high frequency and high energy (since $E = hf$); and the more energy the photons have, the more momentum they can give the object when they strike it. If photons of longer wavelength and correspondingly lower energy are used, the object's motion when struck by the photons will not be affected as much; but its position will be less accurately known. Thus the act of observing produces a significant uncertainty in either the *position* or the *momentum* of the electron. This is the essence of the *uncertainty principle* first enunciated by Heisenberg in 1927.

Quantitatively, we can make an approximate calculation of the magnitude of this effect. If we use light of wavelength λ, the position can be measured at best to an accuracy of about λ. That is, the uncertainty in the position measurement, Δx, is approximately

$$\Delta x \approx \lambda .$$

Here, the symbol Δ means "uncertainty in"; so Δx stands for the uncertainty in position, x. Now, suppose that the object can be detected by

[†] Instruments have been developed that can detect single photons. Our eyes, on the other hand, are not sensitive to individual photons; so many photons will be required if we are to "see" the object with our eyes.

a single photon. The photon has a momentum $mv = h/\lambda$ (de Broglie's wavelength formula), and when it strikes our object it will give some or all of this momentum to the object. Therefore, the final momentum of our object will be uncertain in the amount

$$\Delta mv \approx \frac{h}{\lambda}$$

since we can't tell beforehand how much momentum will be transferred. The product of these uncertainties is

$$(\Delta x)(\Delta mv) \approx h.$$

Of course, the uncertainties could be worse than this, depending on the apparatus and the number of photons needed for detection. In Heisenberg's more careful calculation, he found that at the very best

$$(\Delta x)(\Delta mv) \gtrsim \frac{h}{2\pi}.$$

This is a mathematical statement of **Heisenberg's uncertainty principle**. It tells us that we cannot measure both the position and momentum of an object precisely at the same time. The more accurately we try to measure the position, so that Δx is small, the greater will be the uncertainty in momentum, Δmv. If we try to measure the momentum very precisely, then the uncertainty in the position becomes large. The uncertainty principle does not forbid single exact measurements, however. For example, in principle we could measure the position of an object exactly (then $\Delta x = 0$). But its momentum, and hence its velocity, would be completely unknown. Thus, although we might know the position of the object exactly at one instant, we could have no idea at all where it would be a moment later.

Another useful form of the uncertainty principle relates energy and time. Heisenberg's calculation showed that, at best,

$$(\Delta E)(\Delta t) \gtrsim \frac{h}{2\pi}.$$

This form of the uncertainty principle tells us that the energy of an object can be uncertain, or may even be nonconserved, by an amount ΔE for a time $\Delta t \approx h/(2\pi\Delta E)$.

We have been discussing the position and velocity of an electron as if it were a particle. But it isn't a particle. Indeed, we have the uncertainty principle because an electron—and matter in general—is not purely particulate. What the uncertainty principle really tells us is that if we insist on thinking of the electron as a particle, then there are certain limitations on this simplified view—namely that the position and velocity cannot both be known precisely at the same time and that the energy can be uncertain (or nonconserved) in the amount ΔE for a time $\Delta t \approx h/(2\pi\Delta E)$.

Because Planck's constant, h, is so small, the uncertainties expressed in the uncertainty principle are usually negligible on the macroscopic level. But at the level of the atom, the uncertainties are significant. Because ordinary sized objects are made up of atoms containing nuclei and electrons, the uncertainty principle is relevant to our understanding of all of nature. The uncertainty principle expresses, perhaps most clearly, the probabilistic nature of quantum mechanics; thus it is often used as a basis for philosophic discussion.

2. PHILOSOPHIC IMPLICATIONS OF QUANTUM MECHANICS

Distinction Between Nature and Our Our Description of Nature

The model of an atom that shows electrons moving in orbits around a nucleus treats electrons as if they were actually particles—that is, as if an electron had a definite position at each point in time. Since electrons are not simply particles, they cannot be represented as following particular paths in space and time. This suggests that a description of matter in space and time may not be completely correct. This deep and far-reaching conclusion has been a lively topic of discussion among philosophers. Perhaps the most important and influential philosopher of quantum mechanics was Niels Bohr. He argued that a space–time description of actual atoms and electrons is not possible. But, he pointed out, a description of experiments on atoms or electrons must be given in terms of space and time and other concepts familiar to ordinary experience, such as waves and particles. Yet we must not let our *descriptions* of experiments lead us into believing that atoms or electrons themselves actually exist in space and time as particles. This distinction between our interpretation of experiments and what is "really" happening in nature is crucial.

This brings us back to our discussion in Chapter 1 of the similarities between the arts and sciences. The scientist is not always a logician—he is often much like a poet. This is no better illustrated than when a physicist talks about the ultimate constituents of matter. When a physicists says "an electron is like a particle," he is making a metaphorical comparison like the poet who says "love is like a rose." In both images a concrete object, a rose or a particle, is used to illuminate an abstract idea, love or electron.

Probability Versus Determinism

The classical Newtonian view of the world is a deterministic one (see Section 6 of Chapter 6). One of its basic ideas is that once the position and velocity of an object are known at a particular time, its future position can be predicted if the forces on it are known. For example, if a stone is thrown a number of times with the same initial velocity and angle, and the forces on it remain the same, the path of the projectile will always be the same. If the forces are known (gravity and air resistance, if any) the stone's path, and where it will travel, can be precisely predicted. This mechanistic view implies that the future unfolding of the universe, assumed to be made up of particulate bodies, is completely determined.

But in the twentieth century we find that the basic entities, such as electrons, cannot even be considered to be particles since they have wave properties as well. And we therefore cannot know both the position and velocity of an object accurately. Indeed, the classical deterministic view of the physical world has been radically altered by quantum mechanics. Quantum mechanics only allows us to calculate the probability that the electron (when thought of as a particle) will be observed at various different places. Only approximate predictions are possible. There seems to be an inherent unpredictability in nature.

Although a few physicists have not given up the deterministic view of nature and have refused to accept quantum mechanics as a complete theory—one was Einstein—nonetheless, the vast majority of physicists do accept quantum mechanics and the probabilistic view of nature. This view, which as presented here is the generally accepted one, is called the *Copenhagen interpretation* of quantum mechanics in honor of Niels Bohr's home, as it was largely developed there through discussions between Bohr and other prominent physicists.

Let us now investigate what quantum mechanics says about causality. If you drop a rock, it generally falls to the ground. But have you ever seen a violation of this phenomenon—for example, a stone suddenly jump upward? Probably not. Similarly, when it is cold enough, we generally observe that water freezes. Our experience tells us that it always does. This repeatability of similar events is what leads to the idea of causality. But causality does not hold at the atomic level. Electrons that are treated the same in a given experiment will not all end up in the same place; certain probabilities exist that an electron will arrive at different points (Figure 28-1).

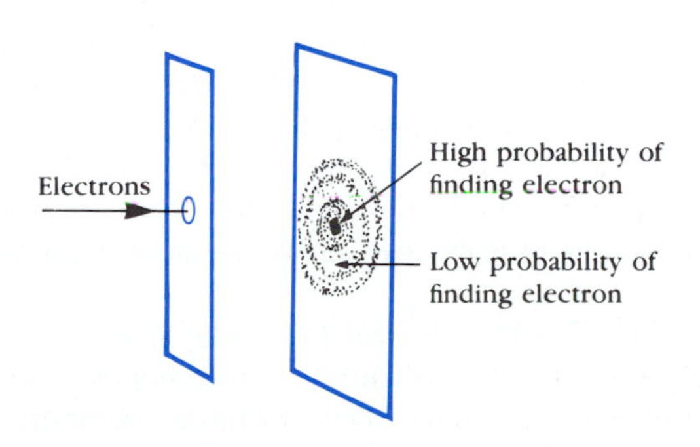

FIGURE 28-1

Electrons moving horizontally approach a slit. We cannot predict where they will go after passing through the slit. We can only predict the probability of finding them at various points. After many electrons have passed through the slit, the position of their arrival points on the screen will look like a diffraction pattern of waves passing through the slit. Why? Because electrons have wave properties, too.

Since matter is considered to be made up of atoms, even ordinary-sized objects are expected to be governed by chance and probability rather than by strict determinism. For example, there is a finite (but very small) probability that when you throw a stone its path will suddenly curve upward instead of following the downward-curved parabola of normal projectile motion, or that a lake will freeze on a hot summer's day. Quantum mechanics predicts with very high probability that ordinary objects will behave just as the classical laws of physics predict; but these predictions are probabilities, not certainties. The reason macroscopic objects behave in accordance with classical laws with very high probability is that there are large numbers of molecules involved: when large numbers of objects are present in a statistical situation, deviations from the average are negligible. It is the average configuration of vast numbers of molecules that follows the so-called fixed laws of classical physics with such high probability and gives rise to an apparent "determinism." Deviations from classical laws are readily observed when small numbers of molecules are dealt with. We can say, then, that although there are no precise deterministic laws in quantum mechanics, there are statistical laws based on probability.

3. COMPLEX ATOMS; PAULI EXCLUSION PRINCIPLE

The Arrangement of Electrons in Atoms

We have discussed the hydrogen atom in detail, and the fact that its single electron can be in the ground state or in an excited state. What about more complex atoms, such as helium with 2 electrons, oxygen with 8, or gold with 79 electrons? First let us recall that the number of electrons in a neutral atom, called its *atomic number*, Z, is also the number of positive charges (protons) in the nucleus, and it is this that determines what kind of atom it is. That is, Z determines most of the properties that distinguish one kind of atom from another.

As in the case of hydrogen, each atom has its own characteristic set of possible energy levels. When an atom is in its ground state, we might expect that the electrons would all be in the lowest energy state. But this is not the case. Instead, as Wolfgang Pauli (1900–1958) pointed out in 1925, the arrangement of electrons in an atom can be explained if we assume that

no two electrons in an atom can occupy the same quantum state.

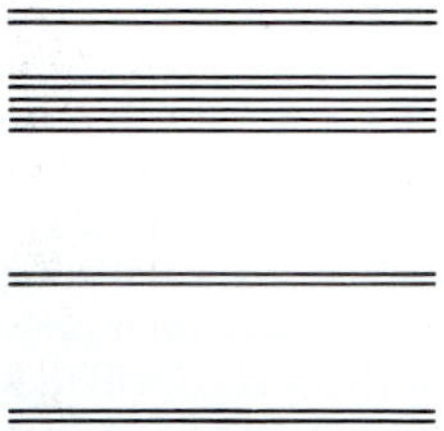

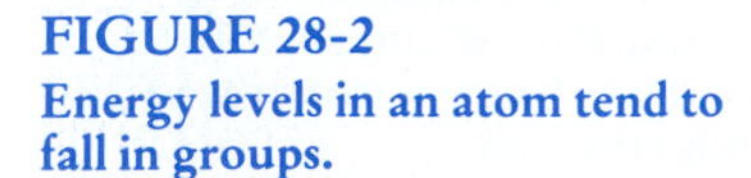

FIGURE 28-2
Energy levels in an atom tend to fall in groups.

This is known as the **Pauli exclusion principle**. Thus only one electron can be in each state. However, the states tend to group in sets (Figure 28-2) known as shells and subshells. The quantum number, n, that we

SHELL	MAXIMUM NO. OF ELECTRONS IN SHELL	MAXIMUM NO. OF ELECTRONS ARRANGED ACCORDING TO SUBSHELLS
$n = 1$	2	2
$n = 2$	8	2 + 6
$n = 3$	18	2 + 6 + 10
$n = 4$	32	2 + 6 + 10 + 14
$n = 5$	50	2 + 6 + 10 + 14 + 18
$n = 6$	72	2 + 6 + 10 + 14 + 18 + 22

TABLE 28-1 Numbers of Electrons in Atomic Shells and Subshells

discussed for Bohr's treatment of hydrogen is also useful for other atoms, and it labels the shells. In the lowest shell, for which $n = 1$, there are actually two very close states, so there can be two electrons. For the $n = 2$ shell there are two groups of energy levels, or subshells: one can contain two electrons and the other can hold six electrons. For the $n = 3$ shell there are three subshells that can contain 2, 6, and 10 electrons, respectively. See Table 28-1. Normally an atom is in the ground state, and the electrons fill the lowest energy states possible.

Thus for the ground state of helium (He), which is the atom that contains two electrons, both electrons are in the $n = 1$ shell. The next largest atom, lithium (Li), with three electrons, has two of them in the $n = 1$ shell and one in the $n = 2$ shell. The ground state of Berylium (Be), with 4 electrons, has two electrons with $n = 1$ and two with $n = 2$. When we get to Neon, with ten electrons, both the $n = 1$ and $n = 2$ shells are filled. Then sodium, with 11 electrons, has two with $n = 1$, eight with $n = 2$, and one with $n = 3$. And so on for atoms with more and more electrons.

We have been describing the ground states of complex atoms. If one or more of the electrons is in a higher energy level so that a "vacancy" exists at a lower level, the atom is said to be in an excited state. When an electron drops down to fill the vacancy, it emits a photon of light. The spacing of the energy levels is different in every atom and gives rise to the atom's characteristic spectrum (Figure 27-5). Under normal conditions, an atom spends most of its time in the ground state.

The electrons that occupy the lowest energy levels in a complex atom are, on the average, closest to the nucleus. The electron clouds or probability distributions take on specific shapes. For example, in oxygen the two electrons in the lowest energy level have spherically symmetric probability distributions. The next two electrons also have spherically

† There are several other quantum numbers that relate to the atom's angular momentum, but we won't need to discus them here.

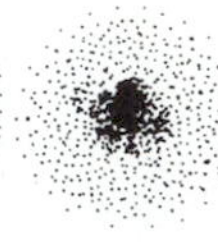

FIGURE 28-3
Electron probability distribution in oxygen.

shaped probability distributions, but on the average they are farther from the nucleus (Figure 28-3). The next two electrons have dumbbell-shaped orbits. The final two electrons also have dumbbell-shaped orbits but around a different axis—that is, they are moving in and out of the paper in Figure 28-3; one end of the dumbbell is above the paper and the other is below, so they are not shown in the diagram.

Although the spatial distribution of the electron can be calculated for the various states, it is difficult to measure them experimentally. Indeed, most of the experimental information about the atom has come from careful examination of the emission spectra under various conditions.

The Periodic Table

A century ago, Dmitri Mendeleev (1834–1907) arranged the then known elements into what we now call the *periodic table* of the elements. The atoms were arranged according to increasing mass, but also so that elements with similar chemical properties would fall in the same column, as we saw in Chapter 11 (see Table 11-2). Now we can explain why the elements show this repeating regularity in properties. All the noble gases (in the last column of the table) have completely filled shells or subshells. That is, their outermost subshell is completely full, the electron distribution is spherically symmetric, and there is no opportunity for additional electrons. This is the reason they are nonreactive (more on this when we discuss molecules and bonding). Column seven contains the "halogens," which lack one electron from a filled shell. Because of the shapes of the orbits, an additional electron can be accepted from another atom, and hence these are quite reactive with a valence of -1 (meaning that when such an electron is accepted, the resulting ion has a net charge of $-1e$). At the left of the table, the first column contains the alkali metals, all of which have a single outer electron. This electron spends most of its time outside the inner closed shells and subshells which shield it from most of the nuclear charge; indeed, it is relatively far from the nucleus and is attracted to it by a net charge of only $+1e$ because of the shielding effect of the other electrons. Hence this outer electron is easily removed and can spend much of its time around another atom, forming a molecule; this is why the alkali metals have a valence of $+1$. The other columns of the table can be treated similarly.

4. MOLECULAR BONDING

The study of molecules and their formation is considered both physics and chemistry. One of the great successes of quantum mechanics was to give scientists, at last, an understanding of the nature of chemical bonds.

By a molecule we mean a group of two or more atoms that are strongly held together so that it functions like a single unit. When atoms make such an attachment, we say that a chemical **bond** has been formed. This bond usually comes about when one or more electrons are shared by both atoms, and it is the electrical attraction of the nuclei of both atoms to these shared electrons that holds the two atoms together. Let's take a concrete example: the formation of magnesium oxide from atoms of magnesium and oxygen. A magnesium atom contains twelve electrons, but only the two outermost electrons—the so-called "valence" electrons—play a role in making a chemical bond. The other ten electrons make up a spherically symmetric probability distribution that shields the two outer electrons from the nucleus. Therefore the two valence electrons, instead of feeling the attraction of all twelve positive charges in the nucleus, feel a net attraction of only two plus charges. Since the two valence electrons are rather far from the nucleus, the electric force holding them on the magnesium atom is not very strong. Consequently, when a magnesium atom approaches an oxygen atom, Figure 28-4(a), the valence electrons of magnesium can be attracted to the oxygen atom if they happen to be in the correct orientation. The attraction occurs because the oxygen atom has two dumbbell-shaped orbits along two axes, but not along the third; so an electron that happens to move into this "opening"—see Figure 28-4(a)—will feel the net attraction of four positive charges. This can occur because in the particular region shown, the oxygen nucleus with its eight protons is shielded by only four electrons, the other four electrons being in dumbbell-shaped orbits off to the side.

Thus, the two valence electrons of magnesium are attracted to both their own nucleus and to the nucleus of the oxygen atom; they then can orbit both atoms but tend to spend most of their time between the two atoms, Figure 28-4(b). The positively charged nuclei of the two atoms are attracted to this concentration of negative charge between them, and this electric attraction is the "bond" that holds the atoms together. This is known as a **covalent bond**.

Another common type of bond is the **ionic bond**, which in a sense is an extreme case of the covalent bond. Instead of the electrons being shared

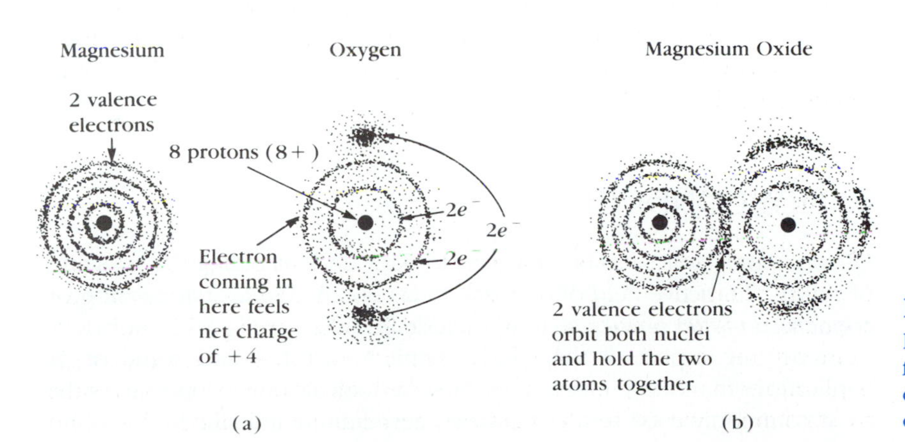

FIGURE 28-4
Electron probability distributions for (a) a magnesium atom and an oxygen atom, (b) a magnesium oxide molecule.

equally, they are shared unequally. For example, in sodium chloride (NaCl) the outer electron of the sodium spends nearly all its time around the chlorine. The chlorine atom has a net negative charge as a result of the extra electron, whereas the sodium is positive. The electrostatic attraction between these two charged atoms is what holds them together and forms the bond. This is called an **ionic bond** because it is the attraction between the two ions (Na^+ and Cl^-) that holds them together. For all types of bonds, we again see the great importance of the electric force.

5. BONDING IN SOLIDS

The molecules of a solid are held together in a number of ways, most commonly by covalent bonding (such as between the carbon atoms of diamond) or ionic bonding (as in a NaCl crystal). Often the bonds are partially covalent and partially ionic.

Another type of bond occurs in metals. Metals have relatively free outer electrons, and present theories indicate that in a metallic solid these outer electrons roam rather freely among all the metal atoms which, without their outer electrons, act like positive ions. The electrostatic attraction between the metal ions and the negative electron "gas" (as it is called) is what is believed to hold the solid together. This theory nicely accounts for the shininess of smooth metal surfaces: the free electrons can vibrate at any frequency; so when light of almost any frequency falls on a metal, the electrons can vibrate in response and re-emit light of that same frequency. Hence the reflected light has the same frequency as the incident light. Compare this to ordinary materials that have a distinct color—the electrons exist only in certain energy states and thus can resonate only at certain frequencies.

The atoms or molecules of some materials, such as the inert gases, can only form *weak bonds* with each other. These bonds, called *van der Waals bonds*, are due to simple electrostatic attraction between molecules (no electron sharing) and are not strong enough to hold the atoms or molecules together as a liquid or solid at room temperature. Such materials condense only at very low temperatures, where the atomic kinetic energy is small and the bonds can then hold the atoms together.

6. SEMICONDUCTORS

Quantum mechanics has been a great tool for understanding the structure of solids. This active field of research today is called *solid-state physics* (*or condensed-matter physics*, so as to include liquids as well). We now briefly examine one aspect of this field—semiconductors—and some of its applications in modern electronics. First we look at some properties of the most commonly used semiconductors, germanium and silicon.

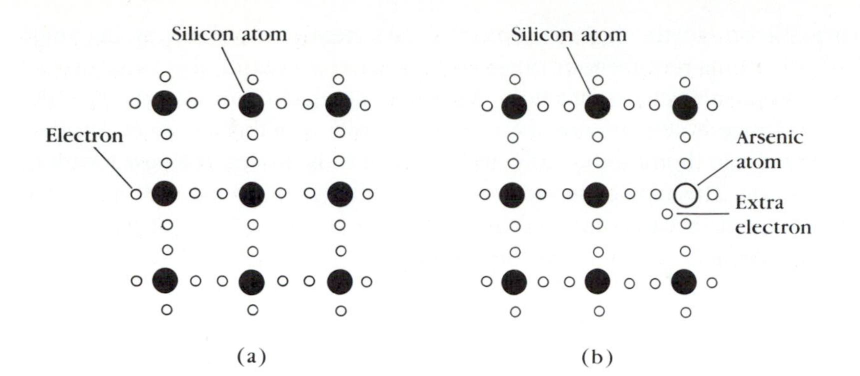

FIGURE 28-5
A silicon crystal.
(a) Four (outer) electrons surround each silicon atom.
(b) Silicon crystal doped with a few arsenic atoms; the extra electron doesn't fit into the crystal lattice and so is free to move about. This is an *n*-type semiconductor.

An atom of silicon or germanium has four outer electrons which act to hold the atoms in the regular lattice structure of the crystal—Figure 28-5(a). Germanium and silicon acquire useful properties for use in electronics only when a tiny amount of impurity is introduced into the crystal structure (about 1 part in a million). This is called "doping" the semiconductor. Two kinds of semiconductors can be made, depending on the type of impurity used. If the impurity is a material whose atoms have five outer electrons, such as arsenic, we have the situation shown in Figure 28-5(b). Only four of arsenic's electrons fit into the crystal structure. The fifth does not fit in and can move relatively freely, much like the electrons in a conductor. Because of this small number of extra electrons, a doped semiconductor becomes slightly conducting. An arsenic-doped silicon crystal is called an *n-type semiconductor* because electrons (*n*egative charge) carry the electric current.

In a *p-type semiconductor*, a small amount of impurity with three outer electrons—such as gallium—is added to the semiconductor. As shown in Figure 28-6(a), there is a "hole" in the lattice structure next to a gallium atom since it has only three outer electrons. Electrons from nearby silicon atoms can jump into this hole and fill it. But this leaves a "hole"' where that electron had previously been—Figure 28-6(b). The vast majority of atoms

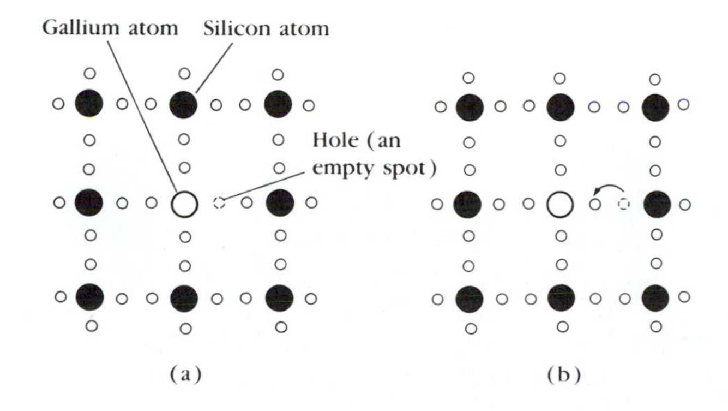

FIGURE 28-6
A *p*-type semiconductor, gallium-doped silicon.
(a) Gallium has only three outer electrons, so there is an empty spot or *hole* in the structure.
(b) Electrons from silicon atoms can jump into the hole and fill it. As a result, the hole moves to a new location (to the right in this figure).

are silicon, so the hole is almost always next to a silicon atom. Since silicon atoms require four outer electrons to be neutral, this means there is a net positive charge at a hole. Whenever an electron moves to fill a hole, the positive hole is then at the previous position of that electron. Another electron can then fill this hole and the hole thus moves to a new location, and so on. This type of semiconductor is called *p-type* because the *p*ositive holes seem to carry the electric current. Note, however, that *p*-type and *n*-type semiconductors have *no net charge* on them.

7. SEMICONDUCTOR ELECTRONICS

The Diode

Semiconductor diodes and transistors are essential components of modern electronic devices. The miniaturization achieved today allows many thousands of diodes, transistors, resistors, and so on, to be placed on a single *chip* only a centimeter on a side. We now briefly discuss the operation of diodes and transistors.

When an *n*-type semiconductor is joined to a *p*-type semiconductor, a "*p-n* junction diode" is formed. When a voltage is applied across such a diode, current will flow only if it is connected as shown in Figure 28-7(a). The positive holes in the *p*-type semiconductor are repelled by the positive terminal of the battery and the electrons in the *n*-type are repelled by the negative terminal of the battery. They meet in the middle, at the "junction," and the electrons cross over and fill the holes. In the meantime, the positive terminal of the battery is continually pulling electrons off the *p* end, forming new holes, and electrons are being supplied by the negative terminal at the *n* end. Consequently a large current flows through the diode. But when the voltage is reversed, as in Figure 28-7(b), the holes in the *p* end are attracted to the battery's negative terminal and the electrons in the *n* end are attracted to the positive terminal. The current carriers do not meet near the junction, and no current flows. Thus a semiconductor diode allows current flow in one direction only. It can thus be used to change ac into dc, which is called *rectification*. The diode is then said to be

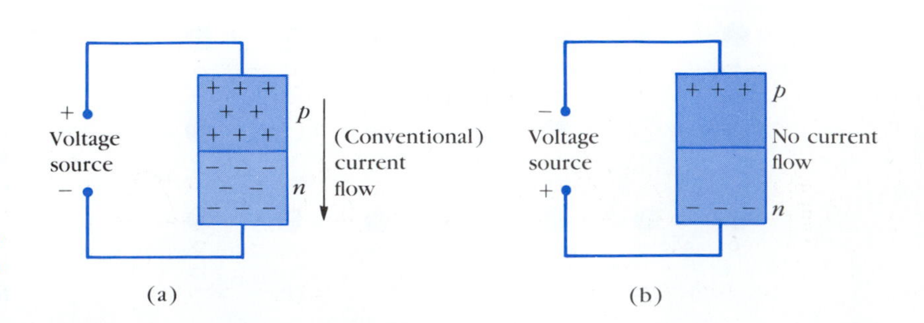

FIGURE 28-7
Schematic diagram showing how a semiconductor diode operates. Current flows when the voltage is connected as in (a), but not when connected as in (b).

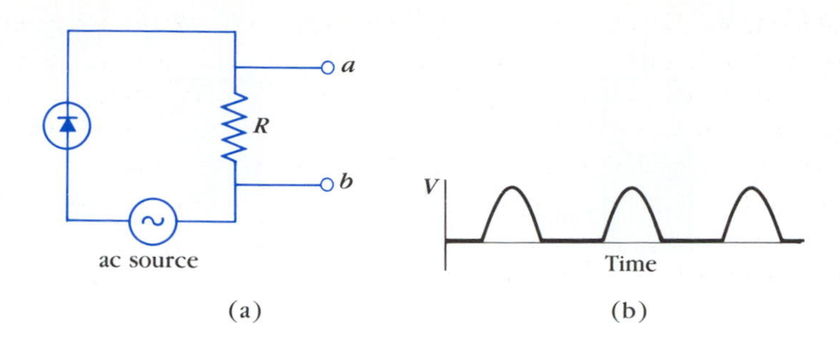

FIGURE 28-8
(a) Simple rectifier circuit using a semiconductor diode. (b) Output voltage across *R* as a function of time.

a *rectifier*. A simple rectifier circuit is shown in Figure 28-8(a) where the arrow inside the symbol for a diode indicates the direction in which a diode conducts conventional (+) current. The ac source applies a voltage across the diode alternately positive and negative. Only during half of each cycle will a current pass through the diode; so only then is there a current through the resistor *R*. Hence the voltage *V* across *R* as a function of time looks like that shown in the graph in Figure 28-8(b). This *half-wave rectification* is not exactly dc, but it is unidirectional. More useful is a *full-wave rectifier* circuit which uses two diodes (or sometimes four) as shown in Figure 28-9(a). At any given instant, either one diode or the other will conduct current to the right. Therefore the output across the load resistor *R* will be as shown in Figure 28-9(b); actually this is the voltage if the capacitor *C* were not in the circuit. The capacitor tends to store charge, and thus helps to smooth out the current as shown in Figure 28-9(c).

Rectifier circuits are important because most line voltage is ac and most electronic devices require a dc voltage for their operation. Hence, diodes are found in nearly all electronic devices including radio and TV sets, calculators, and computers.

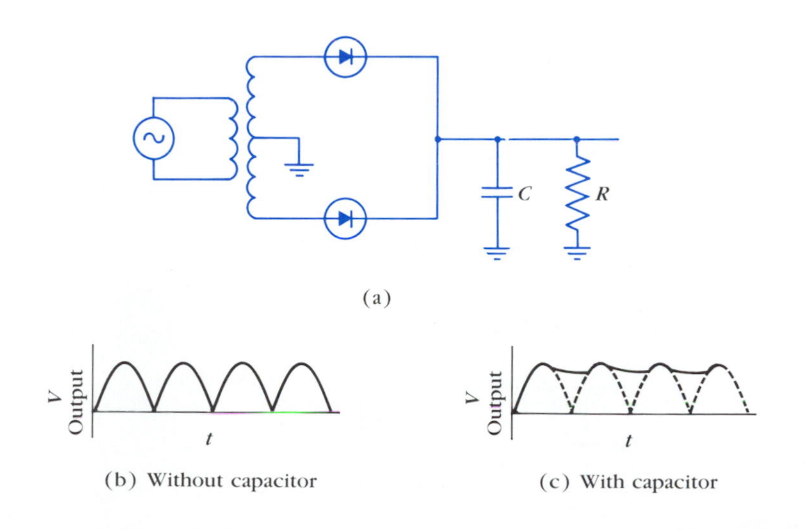

FIGURE 28-9
(a) Full-wave rectifier circuit, including a transformer so the magnitude of the voltage can be changed. (b) Output voltage in the absence of capacitor *C*.
(c) Output voltage with capacitor in the circuit.

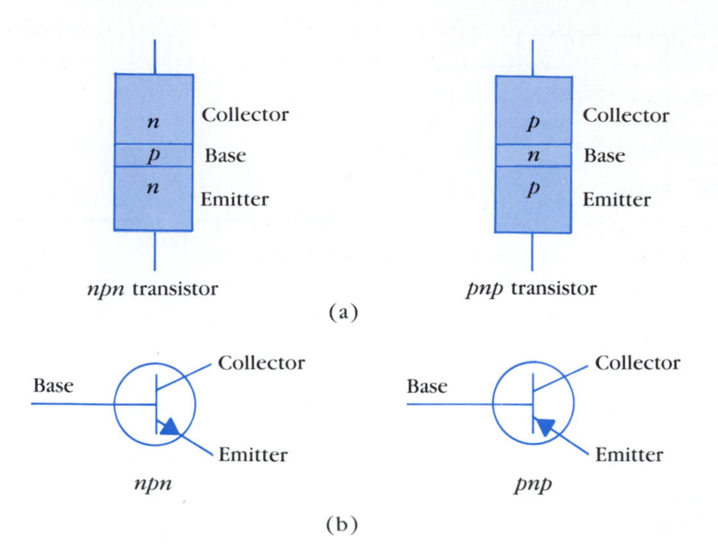

FIGURE 28-10
(a) Schematic diagram of *npn* and *pnp* transistors. (b) Symbols for *npn* and *pnp* transistors.

The Transistor

A simple *junction transistor* consists of a crystal of one type of semiconductor sandwiched between two crystals of the opposite type. Both *pnp* and *npn* transistors are made and they are shown schematically in Figure 28-10(a). The three semiconductors are given the names collector, base, and emitter. The symbols for *npn* and *pnp* transistors are shown in Figure 28-10(b). The arrow is always placed on the emitter and indicates the direction of (conventional) current flow in normal operation.

An *npn* transistor used as an amplifier is shown in Figure 28-11. If the input signal voltage applied to the base is positive, electrons in the emitter are attracted into the base. Since the base region is very thin (perhaps 1 μm), most of the electrons flow right across into the collector, which is maintained at a positive voltage by the battery. A large current then flows between collector and emitter. A small variation in the base voltage due to an input signal causes a large change in the collector current and therefore a large change in the voltage drop across the output resistor R. Hence a transistor can amplify a small signal into a larger one. In fact transistors are the basic elements in modern electronic amplifiers of all sorts.

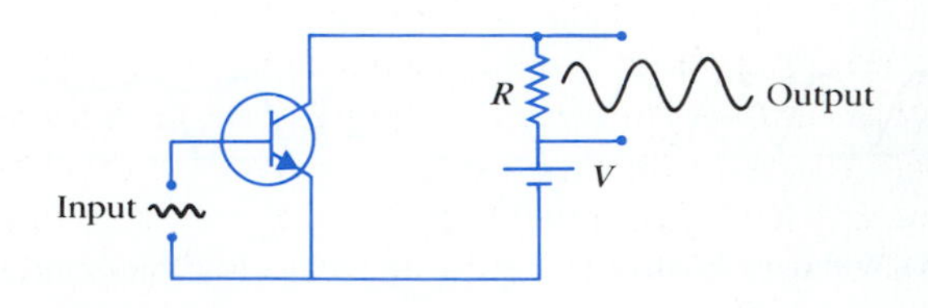

FIGURE 28-11
An *npn* transistor used as an amplifier.

A *pnp* transistor operates in the same fashion, except that holes move instead of electrons. The collector voltage is negative, and so is the base voltage in normal operation.

Integrated Circuits, or Chips

Transistors were a great advance in miniaturization of electronic circuits. Although individual transistors are very small, they are huge compared to *integrated circuits*, or "chips." Tiny amounts of impurities can be placed at particular locations within a single silicon crystal. These can be arranged to form diodes, transistors, and resistors (which are simply undoped semiconductors). Capacitors and other devices can also be formed, although they are often connected separately. A tiny chip, only 1 cm on a side, may contain thousands of transistors and other circuit elements.

8. FLUORESCENCE AND PHOSPHORESCENCE

The materials used to coat the inside of fluorescent light tubes and the materials used on watch dials that glow in the dark have unique properties known as **fluorescence** and **phosphorescence**.

Materials that emit visible light when ultraviolet light is shined upon them are said to be **fluorescent**. Fluorescence can be readily explained as a quantum phenomenon at the atomic level. When a photon of ultraviolet light is absorbed by an atom of the fluorescent material, an electron is raised to a higher energy state. The electron, instead of jumping back to its original energy state in a single jump, may instead return in two or more jumps as shown in Figure 28-12. In this case, each of the emitted photons will have less energy than the energy of the original incoming ultraviolet photon. Therefore the emitted photons will have lower frequency and thus may be photons of visible light.

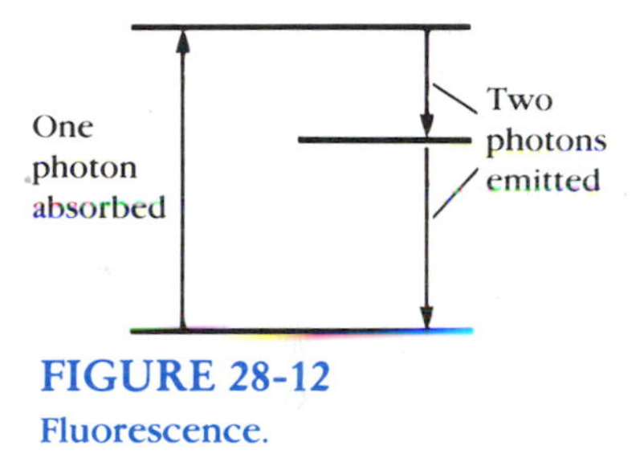

FIGURE 28-12
Fluorescence.

Fluorescent rocks are often seen in museums. When ultraviolet is shined on them, beautiful reds, greens, and violets are seen. The color emitted depends on the energy levels of the particular atoms that are excited. Fluorescent paints work in a similar way.

Fluorescent light bulbs work in a two-step process. The applied voltage accelerates electrons that strike atoms of the gas in the tube and cause them to be excited. When the excited atoms jump down to their normal levels, they emit UV photons which strike a fluorescent coating on the inside of the tube. The light we see is a result of this material fluorescing in response to the UV light striking it.

Materials such as those used for luminous watch dials are said to be **phosphorescent**. In a phosphorescent substance, atoms can be excited by absorption of a photon to an energy level said to be "metastable." When an atom is raised to a normal excited state, it drops back down within about 10^{-8}s. *Metastable* states last much longer—even a few seconds or more. In a collection of such atoms, many of the atoms will descend to the lower state fairly soon but many will remain in the excited state for over an hour. Hence light will be emitted even after long periods. When you put your watch dial close to a bright lamp, it excites many atoms to metastable states and you can see the glow a long time after.

9. THE LASER AND HOLOGRAPHY

The laser is a device that can produce a very narrow, concentrated beam of coherent light. (By coherent, we mean that the light waves across the width of the beam are in phase with each other.) The name **laser** is an acronym for "**L**ight **A**mplification by **S**timulated **E**mission of **R**adiation," which is a concise description of how a laser works. There are many types of laser that use either solid or gaseous materials, but they all work in basically the same way (Figure 28-13). The material to be used is placed between two mirrors, one of which is partially transparent (perhaps 1 or 2 percent). The atoms of the "lasing" material are periodically excited to higher energy states by putting in a large amount of energy, often light energy from a photo flash tube. Just as in the case of phosphorescence, many of the atoms remain in the excited state, although some drop almost immediately to the ground state, emitting a photon. An atom that has just dropped to the ground state is shown on the left in Figure 28-13. If the emitted photon strikes another atom that is still in the same excited state, it *stimulates* the second atom to drop to the lower energy level with the emission of a photon that moves in the same direction as the first photon. This is called **stimulated emission**, and so there are now two photons moving in the same direction. These two photons may strike two more excited atoms, and two new photons will be emitted. Then there will be four photons moving in the same direction. As the process continues, the number of photons multiplies. When the photons strike the end mirrors, most of them are reflected; as they move in the opposite direction, they continue to stimulate other excited atoms to emit photons. As the photons move back and forth between the reflecting mirrors, a small percentage of them pass through the partially transparent mirror at one end. It is these photons that make up the usable laser beam.

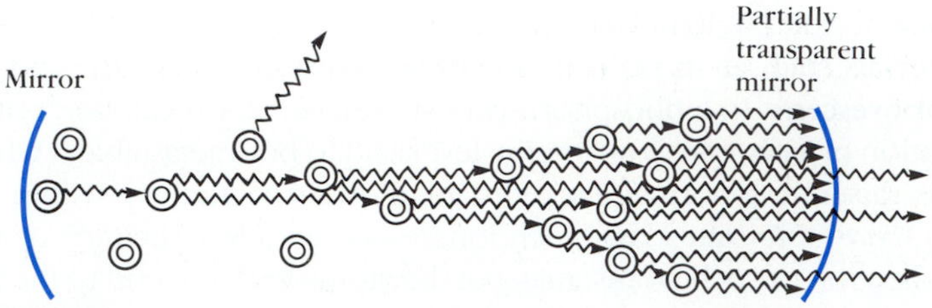

FIGURE 28-13
Laser diagram showing excited atoms stimulated to emit light.

Inside the tube some spontaneously emitted photons will not be emitted parallel to the axis, and these will merely go out the side of the tube and not contribute to the main beam. Thus the beam can be very narrow.

The light emitted from a laser is unique in several ways. Since it is produced by one or more atomic transitions, it consists of a single frequency or a series of distinct frequencies. Furthermore, unlike most light sources in which the light is emitted in all directions, laser light comes out parallel in a straight, narrow beam.

It is the high concentration of light energy over a very small area that makes the laser so useful. A laser is *not a source of energy*—energy must be put in, and the laser then *converts* this energy into an intense narrow beam of light energy. Lasers have many practical uses. They have become a useful surgical tool; the narrow intense beam can be used to destroy tissue in a localized area or, because of the heat produced, a laser beam can be used to "weld" broken tissue such as a detached retina. Tiny organelles within a living cell have been destroyed by researchers using lasers to study how the absence of that organelle affects the behavior of the cell. The finely focused beam of a laser has been used to destroy cancerous and precancerous cells; at the same time the heat seals off capillaries and lymph vessels, thus "cauterizing" the wound in the process to prevent spread of the disease. The intense heat produced in a small area by a laser beam is also used for welding and machining metals and for drilling tiny holes in hard materials.† The precise straightness of a laser beam is also useful to surveyors for lining up equipment accurately, especially in inaccessible locations.

One of the most interesting applications of laser light is the production of three-dimensional images called **holograms**. In an ordinary photograph, the film simply records the intensity of light reaching it at each point; when the photograph or transparency is viewed, light reflecting from it or passing through it gives us a two-dimensional picture. In holography, the images are formed by interference without lenses. When a laser hologram is made on film, a broadened laser beam is split into two parts by a half-silvered mirror (Figure 28-14). One part goes directly to the film; the rest passes to the object to be photographed, from which it is reflected to the film. Light from every point on the object reaches each point on the film, and the interference of the two beams allows the film to record both the intensity and relative phase of the light waves at each point. Thus a hologram is a complex interference pattern. After the film is developed, it is placed again in a laser beam and a three-dimensional image of the object is seen. You can walk around such an image and see it

† Science fiction stories and movies lead us to believe that hand-held lasers may someday be used as weapons to destroy ordinary-sized objects. This is highly unrealistic. Only a giant-sized laser could possibly transform enough energy to vaporize an ordinary-sized object.

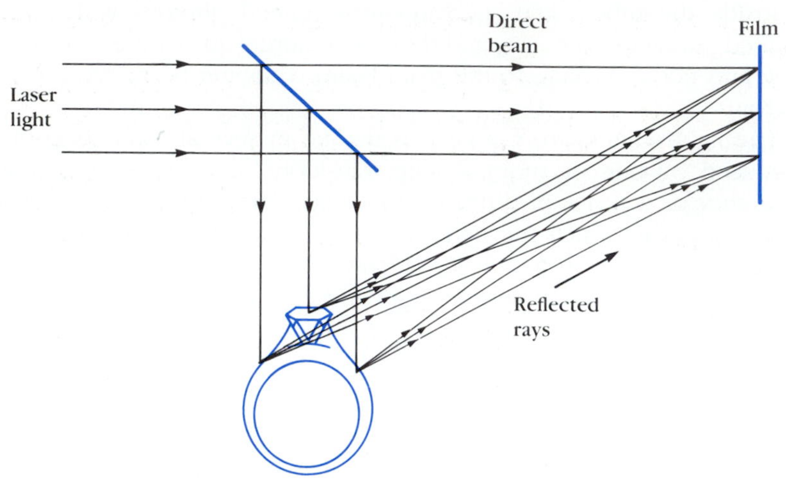

FIGURE 28-14
Making a hologram.

from different sides as if it were the original object. Yet, if you try to touch it with your hand, there will be nothing material there.

The principles of holography are currently being applied to moving pictures to produce truly three-dimensional movies.

SUMMARY

An important aspect of quantum mechanics is the *Heisenberg uncertainty principle*. It results from the wave–particle duality and the unavoidable interaction between the observed object and the observer. One form of the uncertainty principle states that the position and momentum of an object cannot both be measured precisely at the same time; the products of the uncertainties, $(\Delta x)(\Delta mv)$, can be no less than $h/2\pi$. Another form states that the energy can be uncertain, or nonconserved, by an amount ΔE for a time $\Delta t \approx h/(2\pi\Delta E)$.

According to quantum mechanics, the basic processes in nature are not described deterministically. Instead we describe nature using statistical laws.

The arrangement of electrons in multielectron atoms is governed by the *Pauli exclusion principle*, which states that no two electrons can occupy the same quantum state. Electrons, as a result, are grouped into shells (according to the value of n) and subshells. This shell structure gives rise to a periodicity in the properties of the elements.

Quantum mechanics explains the bonding together of atoms to form molecules. In a *covalent bond*, the electron clouds of two or more atoms overlap and the positive nuclei are attracted to this concentration of negative charge between them, which forms the bond. An *ionic bond* is an extreme case of a covalent bond in which one or more electrons from one atom spend much more time around the other atom than around their own; the atoms then act as oppositely charged ions which attract each other—which is the bond.

In metals, the electrostatic force between the free electrons and the positive ions is accepted as the mechanism for the metallic bond.

Semiconductor devices make use of two types of doped semiconductors: *n*-type, which contain impurities with extra electrons that can move rather freely and *p*-type, which contain impurities with fewer than the normal number of electrons so that positively charged "holes" can move about.

A semiconductor diode consists of a *pn*-junction and allows current to flow in one direction only; it can be used as a rectifier to change ac to dc. Common transistors consist of three semiconductor sections, either as *pnp* or *npn*. Transistors can amplify electrical signals and find many other uses. An integrated circuit consists of a tiny semiconductor crystal or "chip" on which many transistors, diodes, resistors, and other circuit elements have been constructed using careful placement of impurities.

In *fluorescence*, atoms are excited from their lowest, or ground, energy state to higher excited states by ultraviolet light. The atoms then emit visible light of a lower frequency when they return in several steps to the ground state. In *phosphorescence*, the atoms continue emitting visible light over a period of time after the exciting source is removed.

A *laser* is a device that produces an intense, narrow beam of light based on stimulated emission of photons from atoms. *Holograms* are three-dimensional images produced with laser light.

QUESTIONS

1. Explain why Bohr's theory of the atom is not compatible with quantum mechanics, particularly the uncertainty principle.
2. What does the uncertainty principle make us uncertain about?
3. Explain why the more massive an object is the easier it becomes to predict its future position.
4. In view of the uncertainty principle, why does a baseball seem to have a well-defined position and speed whereas an electron does not?
5. Discuss whether something analogous to the uncertainty principle operates in a public opinion survey. That is, do we alter what we are trying to measure when we take such a survey?
6. A cold thermometer is placed in a hot bowl of soup. Will the temperature reading of the thermometer be the same as the temperature of the hot soup before the measurement was made?
7. The sun rose this morning. It rose yesterday morning, too. What evidence do we have that it will rise tomorrow? Do you actually *know* it will rise tomorrow?

8. On numerous occasions it has been observed that a stone thrown up in the air comes back down again. Can you therefore predict with certainty that this will happen again?
9. Compare the Newtonian and modern quantum views of physical reality.
10. In what ways is Newtonian mechanics contradicted by quantum mechanics?
11. Why do chlorine and iodine exhibit similar properties?
12. Explain why potassium and sodium exhibit similar properties.
13. If conduction electrons are free to roam about in a metal, why don't they leave the metal entirely?
14. A silicon semiconductor is doped with phosphorus. What type of semiconductor will this be?
15. Explain how a transistor could be used as a switch.
16. Describe how a *pnp* transistor can operate as an amplifier.
17. Can a diode be used to amplify a signal?
18. What purpose does the capacitor in Figure 28-9 serve?
19. If the battery, V, were reversed in Figure 28-11, how would the amplification be altered?
20. Certain dyes and other materials fluoresce by emitting visible light when UV light falls on them. Can infrared light produce fluorescence?
21. Why do fluorescent colors seem so bright?
22. Explain how a 0.0005-W laser beam, photographed at a distance, can seem much stronger than a 1000-W street lamp.

EXERCISES

1. If an electron is known to have a speed of 2.0×10^5 m/s give or take 50 percent, what is the uncertainty in its position?
2. An electron's position is known to be within a region that is 2.0×10^{-10} m in diameter. What is the minimum speed it is likely to have?
3. The speed of a 100-gram baseball is measured to be 36.405 m/s to an accuracy of 0.001 percent. What is the uncertainty in its position?
4. The position of a 1-gram stone is measured to within 0.1 mm. What is the uncertainty in its speed?
5. A proton is traveling with a speed of $(8.880 \pm 0.012) \times 10^5$ m/s. With what maximum accuracy can its position be ascertained?
6. If an electron's position can be measured to an accuracy of 1.6×10^{-8} m, how accurately can its velocity be known?

7. How many different states are possible for an electron whose principal quantum number is $n = 4$?
8. How many electrons can be in the $n = 5$ shell?
9. What is the maximum number of electrons that can be in an atom if the ground state is not to have $n = 4$ or higher levels?

*10. A 12-g bullet leaves a rifle at a speed of 480 m/s. (a) What is the wavelength of this bullet? (b) If the position of the bullet is known to an accuracy of 0.65 cm (radius of the barrel), what is the minimum uncertainty in its momentum? (c) If the accuracy of the bullet were determined only by the uncertainty principle (an unreasonable assumption), by how much might the bullet miss a pinpoint target 300 m away?

*11. A laser used to weld detached retinas emits 25-ms-long pulses of 640-nm light which average 0.50-W output during a pulse. How much energy can be deposited per pulse and how many photons does each pulse contain?

NUCLEAR PHYSICS AND RADIOACTIVITY

CHAPTER

29

In the early part of the twentieth century Rutherford's experiments led to the idea that at the center of an atom there is a tiny but massive nucleus. At the same time that the quantum theory was being developed and scientists were attempting to understand the structure of the atom and its electrons, investigations into the nucleus itself had also begun. In this chapter and the next, we study this tiny and mysterious world of **nuclear physics**.

1. STRUCTURE OF THE NUCLEUS

An important question to physicists in the early part of this century was whether the nucleus had a structure and what this structure might be. It turns out that the nucleus is a complicated entity not completely understood even today. However, by the early 1930s a model of the nucleus had been developed that still has validity. According to this model, a nucleus is considered as an aggregate of two types of particles: protons and neutrons. (Of course, we must remember that these "particles" also have wave properties; but for ease of visualization and language we simply refer to them as "particles.") A **proton** is the same thing as the nucleus of the simplest atom, hydrogen. It has a positive charge ($= +e = +1.6 \times 10^{-19}$ C) and a mass $m_p = 1.672 \times 10^{-27}$ kg. The **neutron**, whose existence was ascertained only in 1932 by the Englishman James Chadwick (1891–1974), is electrically neutral ($q = 0$), as its name implies; its mass, which is almost identical to that of the proton, is 1.675×10^{-27} kg. These two constituents of a nucleus, neutrons and protons, are referred to collectively as **nucleons**.

Although the hydrogen nucleus consists of a single proton alone, the nuclei of other elements consist of both neutrons and protons (Figure 29-1). The different types of nuclei are often referred to as *nuclides*. The number of protons in a nucleus (or nuclide) is called the **atomic number** and is designated by the symbol Z. The total number of nucleons, neutrons plus protons, is designated by the symbol A and is called the **atomic mass number**. This name is used since the mass of a nucleus is very close to A times the mass of one nucleon. A nuclide with 7 protons and 8 neutrons thus has $Z = 7$ and $A = 15$. The *neutron number* N equals $A - Z$.

The number of electrons in a neutral atom determines the properties of that atom and, in fact, determines the kind of atom it is: carbon, oxygen, iron, or whatever. Since the charge on the proton is the same as the charge on the electron, but of opposite sign, the number of protons in the nucleus is the same as the number of "orbiting" electrons. Therefore, the atomic number Z is equal to the number of electrons in a neutral atom and thus determines the kind of atom it is (see Table 11-2). Because neutrons are neutral, they do not affect the number of electrons surrounding the nucleus; accordingly, the number of neutrons has little effect on the properties and behavior of the atom. However, neutrons do have an effect on the stability of the nucleus as a whole, as we shall see shortly.

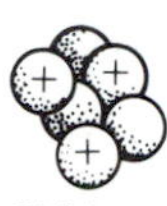

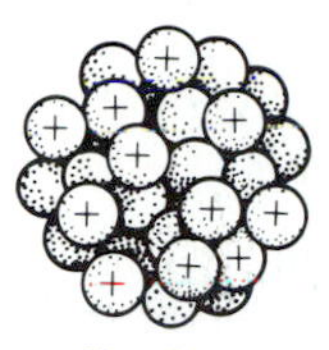

FIGURE 29-1
Different nuclei are made up of different numbers of protons and neutrons.

Isotopes

To specify a given nucleus, we need give only Z and A. It is common practice to use a special symbol to represent this information:

$$^{A}_{Z}X,$$

where X is the chemical symbol for the element (e.g., C for carbon, H for hydrogen, etc.; see Tables 11-1 and 11-2), Z is the atomic number (the number of protons), and A is the atomic mass number (the total number of nucleons). For example, $^{14}_{6}C$ means a carbon nucleus that has six protons and eight neutrons for a total of fourteen nucleons. Since the atomic number defines the type of atom, it is redundant to give both the symbol for the element and its atomic number. If the nucleus is nitrogen, for example, we know immediately that $Z = 7$. The subscript Z is thus sometimes dropped and the $^{15}_{7}N$ nucleus is then written simply ^{15}N; in words, we say "nitrogen fifteen." The subscript Z is often kept, nonetheless, for convenience.

For a particular type of atom (say carbon), nuclei are found that contain different numbers of neutrons, although they all have the same number of protons. For example, carbon nuclei have 6 protons but may have 5,6,7,8,9, or 10 neutrons. Nuclei that contain the same number of protons but different numbers of neutrons are called **isotopes**. Thus $^{11}_{6}C$, $^{12}_{6}C$, $^{13}_{6}C$, $^{14}_{6}C$, $^{15}_{6}C$, and $^{16}_{6}C$ are all isotopes of carbon. Similarly, the known isotopes of hydrogen are $^{1}_{1}H$, $^{2}_{1}H$, $^{3}_{1}H$ (Figure 29-2). Of course, the isotopes of a given element are not all equally common. For example, 98.9 percent of naturally occurring carbon (on earth) is the isotope $^{12}_{6}C$ and 1.1 percent is

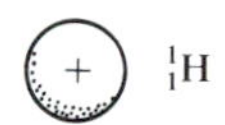

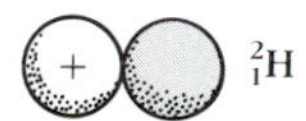

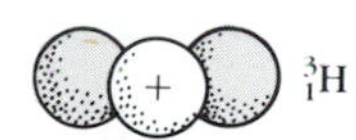

FIGURE 29-2
The three isotopes of hydrogen.

$^{13}_{6}C$. These percentages are referred to as the *natural abundances*.† Many isotopes that do not occur naturally can be produced in the laboratory by means of nuclear reactions (more on this later). Indeed, all elements beyond uranium ($Z > 92$) do not occur naturally and are only produced artificially.

Nuclear Size (optional)

The approximate size of nuclei was determined originally by Rutherford by the scattering of charged particles (See Figure 27-2). The most accurate recent measurements have been done by the scattering of high-speed electrons off nuclei. It is found that nuclei have a roughly spherical shape with a radius that increases with A according to the approximate formula

$$r \approx (1.2 \times 10^{-15}\ \text{m})(A^{\frac{1}{3}}).$$

Since the volume of a sphere is $V = \frac{4}{3}\pi r^3$, we see that the volume is proportional to the number of nucleons, $V \propto A$. This is what we would expect if nucleons were like impenetrable billiard balls: if you double the number of balls, you double the total volume. This result indicates that to form a nucleus, nucleons combine as if they were each impenetrable spheres, and all nuclei have nearly the same density.

Nuclear Masses (optional)

Nuclear masses are specified in *unified atomic mass units* (u). On this scale, a neutral $^{12}_{6}C$ atom is given the precise value 12.0000 u. A neutron then has a mass of 1.0087 u, a proton 1.0073 u, and a neutral hydrogen atom, $^{1}_{1}H$ (proton plus electron), 1.0078 u.

Masses are often specified using the electron-volt energy unit (see Section 2 of Chapter 26). This can be done because mass and energy are related, and the precise relationship is given by Einstein's famous equation $E = mc^2$ (Chapter 25). Since the mass of a neutral $^{1}_{1}H$ atom is 1.673×10^{-27} kg or 1.0078 u, then $1.0000\ \text{u} = (1.0000/1.0078) \times (1.673 \times 10^{-27}\ \text{kg}) = 1.660 \times 10^{-27}\ \text{kg}$. This is equivalent to an energy $E = mc^2 = (1.66 \times 10^{-27}\ \text{kg})(3.00 \times 10^8\ \text{m/s})^2 = 1.494 \times 10^{-10}\ \text{J}$. Since $1\ \text{eV} = 1.60 \times 10^{-19}\ \text{J}$, this is equal to 931×10^6 eV or 931 MeV (million electron volts). Thus

$$1\ \text{u} = 1.66 \times 10^{-27}\ \text{kg} = 931\ \text{MeV}/c^2.$$

(Note that, since $E = mc^2$, we state masses in MeV/c^2, although sometimes

† The mass values for the elements as given in the periodic table (Table 11-2) are averaged over the natural abundances.

TABLE 29-1 Rest masses in kilograms, atomic mass units, and MeV/c^2

OBJECT	MASS		
	kg	u	MeV/c^2
Electron	9.11×10^{-31}	0.00055	0.51
Proton	1.672×10^{-27}	1.00728	938.3
$^1_1$H atom	1.673×10^{-27}	1.00783	938.8
Neutron	1.675×10^{-27}	1.00867	939.6

this is shortened to MeV.) The rest masses of some of the basic particles are given in Table 29-1.

Binding Energy of Nuclei (optional)

A nucleus can be broken into its constituent protons and neutrons only by putting in a certain amount of energy, which is called the **binding energy** of the nucleus. The total mass of a given nucleus is always less than the sum of the masses of the free protons and free neutrons that make up the nucleus. The binding energy represents the difference in mass between the intact nucleus and the free protons and neutrons. That is, energy is changed into mass as Einstein predicted with his famous equation $E = mc^2$. For example, the mass of a neutral $^4_2$He atom is 4.0026 u. The mass of two neutrons and two protons (including the two electrons) is

$$2m_{\text{n}} = 2.0174\ \text{u}$$

$$2m_{^1_1\text{H}} = \underline{2.0156\ \text{u}}$$

$$4.0332\ \text{u}$$

Thus the free nucleons have a total mass of 4.0332 u. The difference between this and the mass of the "bound" $^4_2$He is 4.0332 u − 4.0026 u = 0.0306 u. In MeV, this is (0.0306 u)(931 MeV/u) = 28 MeV. Thus the mass of a $^4_2$He nucleus is 0.0306 u, or 28 MeV/c^2, *less* than the sum of the masses of its constituent nucleons. It would take an input of 28 MeV of energy to break the $^4_2$He nucleus apart.

If the mass of, say, a $^4_2$He nucleus were exactly equal to the mass of two neutrons plus two protons, the nucleus would immediately fall apart without our having to put in any energy. To be stable, then, the mass *must* be less than that of its constituents.

This situation can be compared to the binding energy of electrons in an atom. We saw in Chapter 27 that the binding energy of the one electron in the hydrogen atom, for example, is 13.6 eV. Indeed, the mass of a $^1_1$H atom is less than that of a single proton plus a single electron by 13.6 eV.

Compared to the total mass (939 MeV), this is incredibly small (1 part in 10^8), and for practical purposes the mass difference can be ignored. The binding energies of nuclei are on the order of 10^6 times greater than the binding energies of electrons and are therefore far more important.

2. NUCLEAR FORCES

An interesting question arises when we consider that nuclei contain protons—often in large numbers—and these protons repel each other because they are positively charged. Since the average distance between protons is approximately 10^{-15} m, Coulomb's law tells us that the repulsive forces must be extremely large. What, then, holds the nucleus together? Why doesn't it fly apart?

Since stable nuclei *do* stay together, it is clear that another force must be acting. Because this new force is stronger than the electric force (which in turn is much stronger than gravity at the nuclear level) it is called the **strong nuclear force**. The strong nuclear force is an attractive force that acts between all nucleons—protons and neutrons alike. Thus protons attract each other via the nuclear force at the same time they repel each other via the electric force. Neutrons, because they are electrically neutral, only attract other neutrons or protons via the nuclear force.

The nuclear force turns out to be far more complicated than the gravitational and electromagnetic forces. A precise mathematical description is not yet possible. Nonetheless, a great deal of work has been done to try to understand the nuclear force. One important aspect of the strong nuclear force is that it is a *short-range* force: it acts only over a very short distance. It is very strong between two nucleons if they are less than about 10^{-15} m apart; but it drops to zero if they are separated by a distance greater than this. Compare this to electric and gravitational forces, which can act over great distances and are therefore called *long-range* forces. The strong nuclear force has some strange quirks. For example, if a nuclide contains too many or too few neutrons relative to the number of protons, the nuclear force is weakened; nuclides that are too unbalanced in this regard are unstable. Also, for nuclei with many protons, as Z increases, the electrical repulsion increases; so a greater number of neutrons—which exert only the attractive nuclear force—are required to maintain stability. For very large Z, no number of neutrons can overcome the greatly increased electric repulsion; indeed, there are no completely stable nuclides above $Z = 82$.

What we mean by a stable nucleus is one that stays together indefinitely. What then is an unstable nucleus? It is one that comes apart; and this results in radioactive decay. Before we discuss the important subject of radioactivity (the next section), we note that there is a second type of nuclear force that is much weaker than the strong nuclear force. It is called the **weak nuclear force**, and we are aware of its existence only because it shows itself in certain types of radioactive decay. These two nuclear forces, the strong and the weak, together with the gravitational and electromagnetic forces, comprise the four known types of force in nature (more on this in Chapter 31).

3. DISCOVERY OF RADIOACTIVITY

Nuclear physics had its beginnings in 1896. In that year, Henri Becquerel (1852–1908) made an important discovery: in his studies of phosphorescence, he found that a certain mineral (which happened to contain uranium) would darken a photographic plate even when the plate was wrapped to exclude light. It was clear that the mineral emitted some new kind of radiation which, unlike X rays, occurred without any external stimulus. This new phenomenon eventually came to be called **radioactivity**.

Soon after Becquerel's discovery, Marie Curie (1867–1934) and her husband, Pierre Curie (1859–1906), isolated two previously unknown elements which were very highly radioactive. These were named polonium and radium. Other radioactive elements were soon discovered as well. The radioactivity was found in every case to be unaffected by the strongest physical and chemical treatments, including strong heating or cooling and the action of strong chemical reagents. It soon became clear that the source of radioactivity must be deep within the atom, that it must emanate from the nucleus. And it became apparent that radioactivity is the result of the *disintegration* or *decay* of an unstable nucleus. Certain isotopes are not stable under the action of the nuclear force, and they decay with the emission of some type of radiation or "rays."

Many unstable isotopes occur in nature, and such radioactivity is called "natural radioactivity." Other unstable isotopes can be produced in the laboratory by nuclear reactions; these are said to be produced "artificially" and they are said to have "artificial radioactivity."

4. α, β, AND γ DECAY

Rutherford and others began studying the nature of the rays emitted in radioactivity about 1898. They found that the rays could be classified into three distinct types according to their penetrating power. One type of radiation could barely penetrate a piece of paper. The second type could pass through as much as 3 mm of aluminum. The third was extremely penetrating: it could pass through several centimeters of lead and still be detected on the other side. They named these three types of radiation alpha (α), beta (β), and gamma (γ), respectively, after the first three letters of the Greek alphabet.

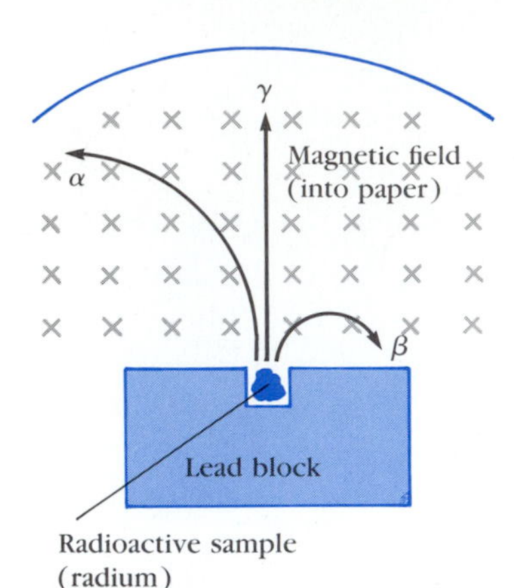

FIGURE 29-3

Alpha and beta rays are bent in opposite directions by a magnetic field, whereas gamma rays are not bent at all.

Each type of ray was found to have a different charge, and hence is bent differently in a magnetic field (Figure 29-3); α rays are positively charged, β rays are negatively charged, and γ rays are neutral. It was soon found that all three types of radiation consisted of familiar kinds of particles. Gamma rays are very high energy photons whose energy is even higher than that of X rays. Beta rays are electrons, identical to those that orbit the nucleus (but they are created within the nucleus itself). Alpha rays (or α particles) are simply the nuclei of helium atoms, ${}^4_2\text{He}$; that is, an α ray consists of two protons and two neutrons bound together.

We now discuss each of these three types of radioactivity, or decay, in more detail.

Alpha Decay

When a nucleus emits an α particle (${}^4_2\text{He}$), it is clear that the remaining nucleus will be different from the original: for it has lost two protons and two neutrons. Radium 226 (${}^{226}_{88}\text{Ra}$), for example, is an α emitter. It decays to a nucleus with $Z = 88 - 2 = 86$ and $A = 226 - 4 = 222$. The nucleus with $Z = 86$ is radon (Rn). Thus the radium decays to radon with the emission of an α particle (Figure 29-4). This is written

$${}^{226}_{88}\text{Ra} \rightarrow {}^{222}_{86}\text{Rn} + {}^4_2\text{He}.$$

It is clear that when α decay occurs, a new element is formed. The *daughter* nucleus (${}^{222}_{86}\text{Rn}$ in this case) is different from the *parent* nucleus (${}^{226}_{88}\text{Ra}$ in this case). This changing of one element into another is called **transmutation**.

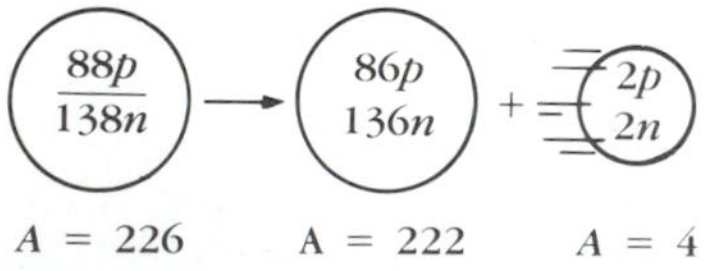

88p 138n → 86p 136n + 2p 2n

$A = 226$ $A = 222$ $A = 4$

FIGURE 29-4

Radium-226 decays with emission of an alpha particle (${}^4_2\text{He}$ nucleus) to Radon-222.

Alpha decay occurs because the strong nuclear force is unable to hold very large nuclei together. Because the nuclear force is a short-range force, it acts only betweeen neighboring nucleons. But the electric force can act clear across the nucleus. For very large nuclei, the large Z means the repulsive electric force is very large (Coulomb's law) and acts between all protons; the strong nuclear force, since it acts only between neighboring nucleons, is overpowered and is unable to hold the nucleus together.

Beta Decay and the Neutrino

Transmutation of elements also occurs when a nucleus decays by β decay—that is, with the emission of an electron or β particle. The nucleus ${}^{14}_6\text{C}$, for example, decays as follows:

$${}^{14}_6\text{C} \rightarrow {}^{14}_7\text{N} + e^-.$$

No nucleons are lost when an electron is emitted, and the total number of nucleons, A, is the same in the daughter as in the parent. But because an electron has been emitted, the charge on the daughter is different from the parent. The parent had $Z = +6$. In the decay, the escaping electron carries off a charge of -1, so the nucleus remaining behind (from charge conservation) must have an extra + charge for a total of 7. So the daughter has $Z = 7$, which is a nitrogen nucleus.

It must be carefully noted that the electron emitted in β decay is *not* an orbital electron. Instead, the electron is created *within the nucleus itself*. It is as if one of the neutrons changes to a proton and in the process (to conserve charge) throws off an electron. Indeed, free neutrons actually do decay in this fashion: $n \rightarrow p + e^-$. Because of their origin in the nucleus, the electrons emitted in β decay are often referred to as "β particles," rather than as electrons, to remind us of their origin; they are, nonetheless, indistinguishable from orbital electrons.

In any decay, the mass of the daughter nucleus plus that of the β particle is less than the mass of the parent nucleus. The difference in mass appears as kinetic energy. For example in the decay ${}^{14}_{6}C \rightarrow {}^{14}_{7}N + e^-$, the calculated release of energy is 160 keV (160,000 eV). Thus, we would expect the emitted electron to have a kinetic energy of 156 keV. (The daughter nucleus, because its mass is very much larger than that of the electron, recoils with very low velocity and hence gets very little of the kinetic energy.) Indeed, very careful measurements indicate that a few emitted β articles do have kinetic energy close to this calculated value; but the vast majority of emitted electrons have somewhat less energy. In fact, the energy of the emitted electron can be anywhere from zero up to the maximum value as calculated above. This was found to be true for any β decay. It was as if the law of conservation of energy was being violated! Careful experiments indicated that linear momentum and angular momentum also did not seem to be conserved. Physicists were troubled at the prospect of having to give up these laws, which had worked so well in all previous situations. In 1930, Wolfgang Pauli proposed an alternate solution: Perhaps a new particle, which was extremely difficult to detect, was emitted during β decay in addition to the electron. This hypothesized particle could be carrying off the energy, momentum, and angular momentum required to maintain the conservation laws. This new particle was named the **neutrino**—meaning "little neutral one"—by the great Italian physicist Enrico Fermi (1901–1954) who in 1933 worked out a detailed theory of β decay (Figure 29-5). (It was Fermi, in this theory, who postulated the existence of the fourth force in nature, which we call the weak nuclear force.) The neutrino has zero charge and seems to have zero rest mass, although there are recent suggestions that it may have a very tiny rest mass. If its rest mass is zero, it is much like a photon in that it is neutral and travels at the speed of light; but it is far more difficult to detect. In 1956, complex experiments produced further evidence for the existence of the neutrino; but by then most physicists had already accepted its existence.

FIGURE 29-5
Enrico Fermi (1901–1954) made significant contributions to experimental and theoretical physics.

The symbol for the neutrino is the Greek letter nu (ν). The correct way of writing the decay of $^{14}_{6}C$ is then

$$^{14}_{6}C \rightarrow\, ^{14}_{7}N + e^- + \bar{\nu}.$$

The bar ($^{-}$) over the neutrino symbol is to indicate that it is an "antineutrino." (More on antiparticles in Chapter 31).

Many isotopes decay by electron emission; they are always isotopes that have too many neutrons compared to the number of protons. But what about unstable isotopes that have too few neutrons compared to their number of protons? These, it turns out, decay by emitting a **positron** instead of an electron. A positron (sometimes called an e^+ or β^+ particle) has the same mass as the electron, but it has a positive charge of $+1e$. Because it is so like an electron, except for its charge, it is called the *antiparticle*† to the electron. An example of a β^+ decay is that of $^{19}_{10}Ne$:

$$^{19}_{10}Ne \rightarrow\, ^{19}_{9}F + e^+ + \nu.$$

In β decay it is the weak nuclear force that plays the crucial role. The neutrino is unique in that it interacts with matter only via the weak force, which is why it is so hard to detect.

Gamma Decay

Gamma rays are photons having very high energy. And the decay of a nucleus by emission of a γ ray is much like emission of photons by excited atoms. Like an atom, a nucleus itself can be in an excited state. When it jumps down to a lower energy state, or to the ground state, it emits a photon. The possible energy levels of a nucleus are much farther apart in energy than those of an atom: on the order of keV or MeV, as compared to a few eV for electrons in an atom. Hence the emitted photons have energies that are thousands (and even millions) of times greater than those of photons emitted when electrons jump to lower levels in an atom. Since a γ ray carries no charge, there is no change in the element as a result of a γ decay.

How does a nucleus get into an excited state? It may occur because of a violent collision with another particle; or more commonly the nucleus remaining after a previous radioactive decay may be in an excited state. A typical example is shown in the energy-level diagram of Figure 29-6. $^{12}_{5}B$ can decay by β decay directly to the ground state of $^{12}_{6}C$; or it can go by β decay to an excited state of $^{12}_{6}C$ which then decays by emission of a 4.4 MeV γ ray to the ground state.

FIGURE 29-6

Energy-level diagram showing how $^{12}_{5}B$ can decay to the ground state of $^{12}_{6}C$ by β decay (total energy released = 13.4 MeV) or can β-decay to an excited state of $^{12}_{6}C$ (indicated by *), which subsequently decays to its ground state by emitting a 4.4-MeV γ ray.

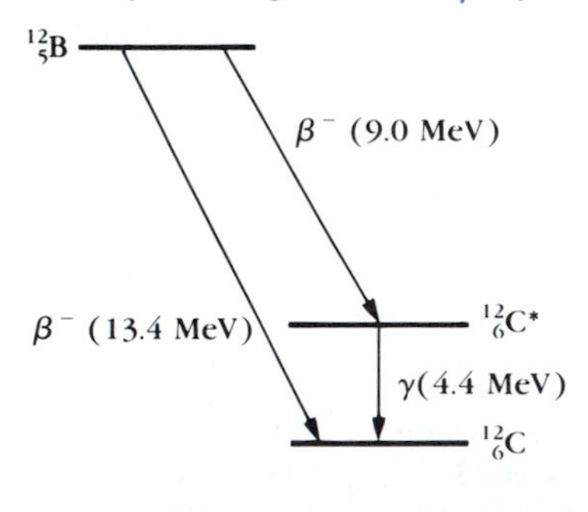

† Discussed in Chapter 31. Briefly, an antiparticle has the same mass as its corresponding particle, but opposite charge.

Conservation Laws

In all three types of radioactive decay the classical conservation laws hold. Energy, linear momentum, and angular momentum, are all conserved; these quantities are the same before the decay as after. But notice also that in all three types of decay the *total* atomic number Z, which represents the electric charge, does not change; nor does the total number of nucleons, A. This is a result of the conservation of electric charge and of a new conservation law, the conservation of nucleons:

> **the total number of nucleons always remains constant, although one type can be changed into the other type.**

5. HALF-LIFE AND RADIOACTIVE DATING

Half-life

A macroscopic sample of any radioactive isotope consists of a vast number of radioactive nuclei. These nuclei do not all decay at one time. Rather, they decay one by one over a period of time. This is a random process: we can't predict exactly when a given nucleus will decay. But we can determine, on a probabilistic basis, approximately how many nuclei in a sample will decay over a given time period.

Some radioactive isotopes are much more unstable than others. The more unstable an isotope is, the more quickly a sample of many such nuclei will decay. A highly unstable isotope is said to be very radioactive because a macroscopic sample of many such nuclei will emit alpha, beta, or gamma rays at a relatively high rate.

Radioactive decay is a "one-shot" process. That is, once a particular parent nucleus has decayed into the daughter nucleus, it cannot do it again; there is one less parent nucleus to decay (Figure 29-7). The number of parent nuclei is thus continually decreasing.

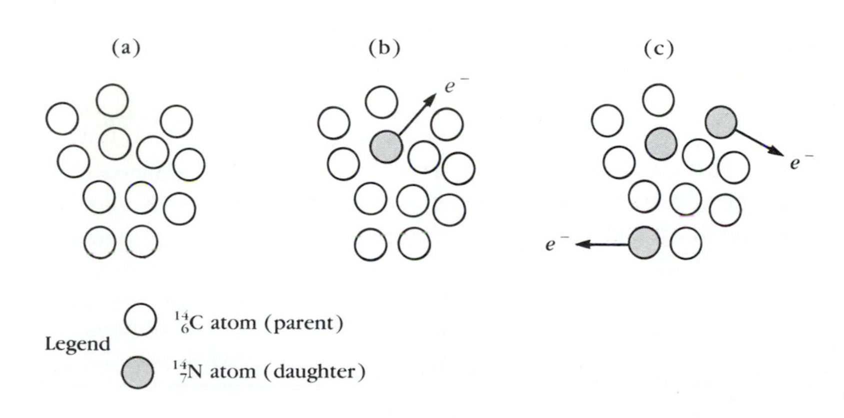

FIGURE 29-7
In a sample of radioactive nuclei, the nuclei decay one by one. Thus the number of parent nuclei is continually decreasing. As soon as a ^{14}C nucleus emits the electron, the nucleus has become a ^{14}N nucleus.

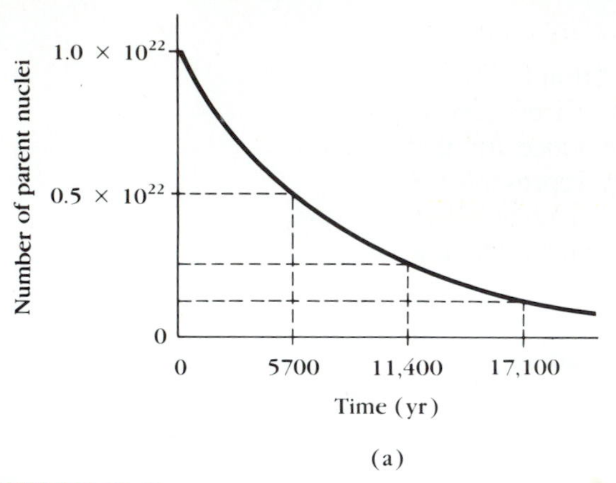

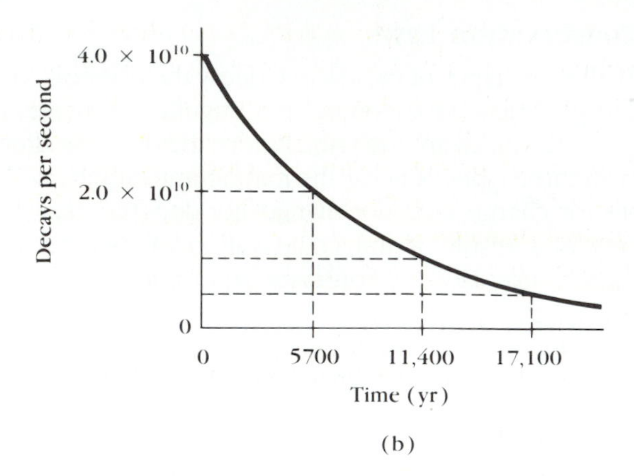

FIGURE 29-8
(a) The number of parent nuclei in a given sample of $^{14}_{6}C$ decreases exponentially. (b) The number of decays per second also decreases exponentially. The half-life of $^{14}_{6}C$ is about 5700 yr, which means that the number of parent nuclei and the rate of decay decrease by half every 5700 yr.

A quantitative measure of the longevity of a particular isotope is its **half-life**. The half-life of an isotope is the time it takes for half the original amount of the isotope in a given sample to decay. For example, the half-life of $^{14}_{6}C$ is about 5700 years. If at some time a piece of petrified wood contains, say, 1.0 g of $^{14}_{6}C$, then after 5700 years it will contain only 0.5 g. After 5700 more years (a total of 11,400 years) half of the remaining 0.5 g will have decayed, so that only 0.25 g remain; and so on. Figure 29-8 (a) shows a plot of remaining parent nuclei versus time; this is called an "exponential" curve.

The half-lives of known radioactive isotopes vary from as short as 10^{-22} seconds to as long as many billions of years (Table 29-2). The half-life of very long-lived isotopes, e.g. ^{238}U with a half-life of $4\frac{1}{2}$ billion years, is determined not by waiting for $4\frac{1}{2}$ billion years but by measuring only the beginning of a curve such as that of Figure 29-8(a).

The longer the half-life of an isotope, the more slowly it decays and the lower its *activity* or rate of emission of alpha, beta, or gamma rays.

TABLE 29-2 Some Commonly Occurring Radioactive Isotopes

ISOTOPE	PARTICLE EMITTED	HALF-LIFE
Carbon 14 ($^{14}_{6}C$)	β^-	5730 years
Cobalt 60 ($^{60}_{27}Co$)	β^-, γ	5.3 years
Strontium 90 ($^{90}_{38}Sr$)	β^-	29 years
Iodine 131 ($^{131}_{53}I$)	β^-, γ	8 days
Polonium 214 ($^{214}_{84}Po$)	α, γ	1.6×10^{-4} seconds
Radium 226 ($^{226}_{88}Ra$)	α, γ	1600 years
Uranium 238 ($^{238}_{92}U$)	α, γ	4.5×10^9 years

Conversely, very active isotopes have short half-lives. In any radioactive sample the activity depends not only on the half-life but on the number of nuclei present. For a given radioactive isotope the more nuclei that are present, the more nuclei that can decay. As decay proceeds and the number of parent nuclei decreases, the rate at which alpha, beta, or gamma rays are emitted decreases as well. Thus the number of particles emitted by a radioactive sample (its "activity") decreases in time and also follows an exponential curve, as shown in Figure 29-8 (b).

Radioactive Dating

Radioactive decay has many interesting applications. One is the technique of *radioactive dating* in which the age of ancient materials can be determined.

The age of any object made from once living matter, such as wood, can be determined using the natural radioactivity of $^{14}_{6}C$. All living plants absorb carbon dioxide (CO_2) from the air, utilizing the carbon and expelling oxygen. The vast majority of these carbon atoms are $^{12}_{6}C$, but a small fraction, about 1.3 in 10^{12}, is the radioactive isotope $^{14}_{6}C$. The ratio of $^{14}_{6}C$ to $^{12}_{6}C$ in the atmosphere has remained roughly constant over many thousands of years, in spite of the fact that $^{14}_{6}C$ decays with a half-life of about 5700 yr. This is because neutrons in the cosmic radiation that impinges on the earth from outer space collide with atoms of the atmosphere. In particular, collisions with nitrogen nuclei produce the following nuclear transformation: $n + {}^{14}_{7}N \rightarrow {}^{14}_{6}C + p$. That is, a neutron strikes and is absorbed by a $^{14}_{7}N$ nucleus, and a proton is knocked out in the process. The remaining nucleus is $^{14}_{6}C$. This continual production of $^{14}_{6}C$ in the atmosphere roughly balances the loss of $^{14}_{6}C$ by radioactive decay.

As long as a plant or tree is alive, it continually uses the carbon from carbon dioxide in the air to build new tissue and to replace old. Animals eat plants, so they too are continually receiving a fresh supply of carbon for their tissues. Organisms cannot distinguish† $^{14}_{6}C$ from $^{12}_{6}C$; and since the ratio of $^{14}_{6}C$ to $^{12}_{6}C$ in the atmosphere remains nearly constant, the ratio of the two isotopes within the living organism remains nearly constant as well. But when an organism dies, carbon dioxide is no longer absorbed and utilized; and because the $^{14}_{6}C$ decays radioactively, the ratio of $^{14}_{6}C$ to $^{12}_{6}C$ in a dead organism decreases in time. Since the half-life of $^{14}_{6}C$ is about 5700 yr, the $^{14}_{6}C/^{12}_{6}C$ ratio decreases by half every 5700 yr. If, for example, the $^{14}_{6}C/^{12}_{6}C$ ratio of an ancient wooden tool is half of what it is in living trees, then the object must have been made from a tree that was felled about 5700 year ago.

Actually, corrections must be made for the fact that the $^{14}_{6}C/^{12}_{6}C$ ratio in the atmosphere has not remained precisely constant over time. The

† Organisms operate almost exclusively via chemcial reactions—which involve only the electrons of the atom; extra neutrons have almost no effect.

determination of what this ratio has been over the centuries has required using techniques such as comparing the expected ratio to actual ratio for objects whose age is known, such as very old trees whose annual rings can be counted.

Carbon dating is useful only for determining the age of objects less than about 40,000 yr old. The amount of $^{14}_{6}C$ remaining in older objects is usually too small to measure accurately. However, radioactive isotopes with longer half-lives can be used in certain circumstances to obtain the age of older objects. For example, the decay of $^{238}_{92}U$, because of its long half-life of 4.5×10^9 years, is useful to determine the age of rocks on a geologic time scale. When molten material solidifies into rock, the uranium present in the material becomes fixed in its position and the daughter nuclei that result from the decay of uranium will also be fixed in that position. Thus by measuring the amount of $^{238}_{92}U$ remaining in the material relative to the amount of daughter nuclei, a scientist can determine the time when the rock solidified.

Radioactive dating methods using $^{238}_{92}U$ and other isotopes have shown the age of the oldest earth rocks to be about 4×10^9 yr. The age of rocks in which the oldest fossilized organisms are embedded indicates that life appeared at least 3 billion years ago. The earliest fossilized remains of mammals are found in rocks 200 million years old, and the first human-like creatures seem to have appeared about 2 million years ago. Radioactive dating has been indispensable for the reconstruction of earth's history and the evolution of its biological organisms.

6. DETECTION OF RADIATION

Individual particles such as electrons, protons, α particles, neutrons, and γ rays are not detected directly by our senses. Consequently, a variety of instruments have been developed to detect them.

FIGURE 29-9
Diagram of a Geiger counter.

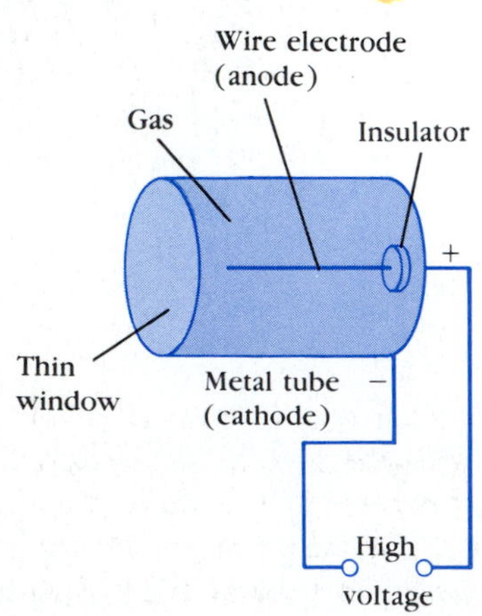

One of the most common is the *Geiger counter*. As shown in Figure 29-9, it consists of a cylindrical metal tube filled with a certain type of gas. A long wire runs down the center and is kept at a high positive voltage with respect to the outer cylinder. The voltage is just slightly less than that required to ionize the gas atoms (i.e., pull the electrons off their atoms). When a charged particle enters through a thin "window" at one end of the tube, it ionizes a few atoms of the gas. The freed electrons are attracted toward the positive wire and as they are accelerated they strike and ionize additional atoms. An "avalanche" of electrons is quickly produced, and when it reaches the wire "anode" it produces a voltage pulse. The pulse, after being amplified, can be sent to an electronic counter, which keeps track of how many particles have been detected. Or the pulses can be sent to a loudspeaker and each detection of a particle is heard as a "click."

A *scintillation counter* makes use of a solid, liquid, or gas known as a "scintillator." The atoms of the scintillator are easily excited when they are struck by an incoming particle and emit light when they drop down to their ground states. The emitted photons strike a metal that emits electrons (the photoelectric effect), which are multiplied by a special "photomultiplier tube" into detectable electric signals. Scintillation counters can be used not only to detect charged particles—to which Geiger counters are mainly limited—but also to detect neutral particles such as gamma rays and neutrons.

A *semiconductor detector* consists of a reversed biased *pn* junction diode. A particle passing through the junction causes ionization, and the released charges produce a short electrical pulse that can be counted just as for Geiger and scintillation counters.

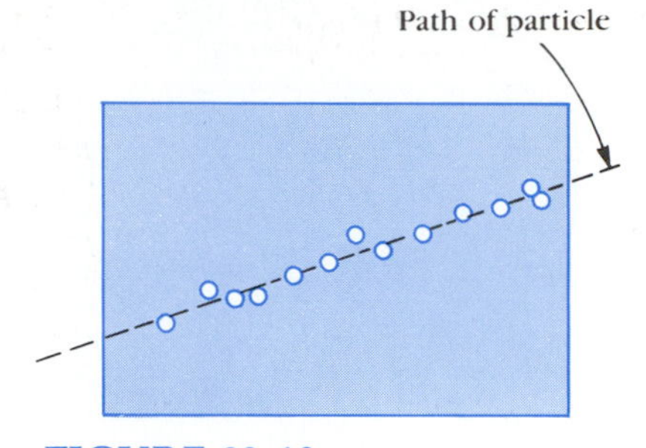

FIGURE 29-10
Bubbles formed by the passage of a charged particle in a cloud or bubble chamber.

With a *cloud chamber* and the more recent and more effective *bubble chamber*, the *paths* of charged particles can be seen. In a cloud chamber, a gas is cooled to a temperature slightly below its usual condensation point. The gas molecules do not condense immediately unless dust or an ionized molecule is present for them to condense on. However, when a charged particle passes through such a "supercooled" gas it ionizes the molecules along its path, and tiny bubbles form on the resulting ions; the row of bubbles indicates the path of the charged particle (Figure 29-10). In the bubble chamber, invented in 1952 by D. A. Glaser, the path of a particle is seen as a series of gaseous bubbles within a "superheated" liquid: the temperature of the liquid is slightly above its boiling point, and the bubbles characteristic of boiling will form around the ions produced by the passage of a charged particle. A photograph of the chamber thus reveals the path of particles that recently passed through. Because the bubble chamber uses a liquid, the density of atoms is much greater than in a cloud chamber and hence it is a much more efficient device for observing particles and their interactions with the nuclei of the liquid (Figure 29-11).

FIGURE 29-11
Bubble chamber photo of particle tracks.

A *spark chamber* or *wire ionization chamber* consists of a series of closely spaced long wires in planes at right angles to one another, immersed in a gas, and kept at high potential difference. When a charged particle passes through, the ions produced in the gas avalanche and produce a spark. When the sparks strike each wire, electric pulses travel along the wires; and the positions and path of the particle can be determined electronically.

SUMMARY

Nuclear physics is the study of atomic nuclei. Nuclei contain *protons* and *neutrons*, which are collectively known as *nucleons*. The total number of nucleons, A, is the *atomic mass number*. The number of protons, Z, is the *atomic number*. The number of neutrons equals $A - Z$. *Isotopes* are nuclei with the same Z, but with different numbers of neutrons. For an element X, an isotope of given Z and A is represented by ${}^{A}_{Z}X$. The nuclear radius is proportional to $A^{\frac{1}{3}}$, indicating that all nuclei have about the same density. Nuclear masses are specified in atomic mass units (u), where the mass of ${}^{12}_{6}C$ is defined as exactly 12.000 u, or in terms of their energy equivalent (because $E = mc^2$), where 1 u = 931 MeV/c^2. The mass of a nucleus is less than the sum of the masses of its constituent nucleons. The difference in mass (times c^2) is the *total binding energy*; it represents the energy needed to break the nucleus into its constituent nucleons.

Nuclei are held together by the *strong nuclear force*; the *weak nuclear force* makes itself apparent in β decay; these two forces, plus the gravitational and electromagnetic forces, are the four known types of force.

Unstable nuclei undergo *radioactive decay*; they change into other nuclei with the emission of an α, β, or γ particle. An α particle is a ${}^{4}_{2}He$ nucleus; a β particle is an electron or positron; and a γ ray is a high-energy photon. In β decay, a *neutrino* is also emitted. The transformation of the parent into the daughter nucleus is called *transmutation* of the elements.

Radioactive decay is a statistical process. The *half-life*, $T_{\frac{1}{2}}$, is the time required for half the nuclei of a radioactive sample to decay.

Electric charge, linear and angular momentum, mass-energy, and *nucleon number* are *conserved* in all decays. Radioactive decay occurs spontaneously only when the rest mass of the products is less than the mass of the parent nucleus; the loss in mass appears as kinetic energy of the products. Radioactivity can be used for dating the age of many materials.

Detectors of radiation include Geiger counters, scintillation counters, and semiconductor detectors. The paths of elementary particles can be seen in bubble chambers, cloud chambers, and reconstructed in spark or ionization chambers.

QUESTIONS

1. How do we know there is such a thing as the strong nuclear force?
2. What does atomic number refer to? What does atomic mass number refer to?
3. What do isotopes have in common?
4. Explain why we believe different nuclei have nearly the same density.
5. What are the elements represented by the X in the following: (a) ${}^{232}_{92}X$, (b) ${}^{18}_{7}X$, (c) ${}^{1}_{1}X$, (d) ${}^{82}_{38}X$, (e) ${}^{247}_{97}X$?
6. How many protons and how many neutrons do each of the isotopes in question 5 have?
7. Why are the atomic masses of many elements not close to whole numbers?

8. What is the experimental evidence in favor of radioactivity being a nuclear process?
9. The $^{3}_{1}H$ nucleus decays by beta emission into what nucleus?
10. The isotope $^{64}_{29}Cu$ is unusual in that it can decay by γ, β^-, and β^+ emission. What is the resulting nuclide for each case?
11. A $^{238}_{92}U$ nucleus α decays to a nucleus containing how many neutrons?
12. Describe, in as many ways as possible, the differences between α, β, and γ rays.
13. Which is most like an X ray: an alpha, beta, or gamma ray?
14. What element is formed by the radioactive decay of (a) $^{24}_{11}Na(\beta^-)$; (b) $^{22}_{11}Na(\beta^+)$; (c) $^{210}_{84}Po(\alpha)$?
15. What element is formed by the decay of (a) $^{32}_{15}P(\beta^-)$; (b) $^{35}_{16}S(\beta^-)$; (c) $^{211}_{83}Bi(\alpha)$?
16. Fill in the missing particle or nucleus:
    (a) $^{45}_{20}Ca \rightarrow ? + e^- + \bar{\nu}$
    (b) $^{58}_{29}Cu \rightarrow ? + \gamma$
    (c) $^{46}_{24}Cr \rightarrow ^{46}_{23}V + ?$
    (d) $^{234}_{94}Pu \rightarrow ? + \alpha$
    (e) $^{239}_{93}Np \rightarrow ^{239}_{92}U + ?$
17. Immediately after a $^{238}_{92}U$ nucleus decays to $^{234}_{90}Th + ^{4}_{2}He$, the daughter thorium nucleus still has 92 electrons circling it. Since thorium normally holds only 90 electrons, what do you suppose happens to the two extra ones?
18. Why are many artificially produced radioactive isotopes rare in nature?
19. Can hydrogen or deuterium emit an α particle?
20. Can $^{14}_{6}C$ dating be used to measure the age of stone walls and tablets of ancient civilizations?

EXERCISES

1. What is the approximate radius of a $^{64}_{29}Cu$ nucleus?
2. Approximately what is the value of A for a nucleus whose radius is 3.6×10^{-15} m?
3. What is the rest energy of an α particle in MeV?
4. A pi meson has a mass of 139 MeV/c^2. What is this in atomic mass units?
5. Given the masses of the particles in Table 29-1 in atomic mass units, show that the masses in MeV/c^2 are correct as given there.
6. What fraction of a sample of $^{68}_{32}Ge$, whose half-life is about 9 months, will remain after 4.5 yr?

7. In a series of decays, the nuclide ${}^{235}_{92}U$ becomes ${}^{207}_{82}Pb$. How many α and β^- particles are emitted in this series?
8. Some ancient rocks are found to contain equal amounts of ${}^{238}_{92}U$ and its daughter nuclides. How old are the rocks?
9. The half-life of a radioactive isotope is one day. How much of the original material will be left after two days?
10. If only $\frac{1}{16}$ of a radioactive sample remains after one year, what is the half-life of the material?
11. A radioactive material registers 1280 counts per minute on a Geiger counter at one time, and 6 hr later registers 320 counts per minute. What is its half-life?
12. The ${}^{3}_{1}H$ isotope of hydrogen, which is called *tritium* (because it contains three nucleons), has a half-life of 12.33 yr. It can be used to measure the age of objects up to about 100 yr. It is produced in the upper atmosphere by cosmic rays and is brought to earth by rain. As an application, determine approximately the age of a bottle of wine whose ${}^{3}_{1}H$ radiation is about $\frac{1}{10}$ that present in new wine.

*13. (a) Determine the density of nuclear matter in kg/m^3. (b) What would be the radius of the earth if it had its actual mass but had the density of nuclei? (c) What would be the radius of a ${}^{238}_{92}U$ nucleus if it had the density of the earth?

*14. Calculate the total binding energy of a ${}^{238}_{92}U$ nucleus whose mass is 236.32 times the mass of a proton.

*15. What is the binding energy of a ${}^{11}_{5}B$ nucleus whose mass is 18.276×10^{-27} kg.

*16. Calculate the binding energy of a deuterium nucleus, ${}^{2}_{1}H$ (mass = 2.014102 u).

*17. When ${}^{23}_{10}Ne$ (mass = 22.9945 u) decays to ${}^{23}_{11}Na$ (mass = 22.9898 u), what is the maximum kinetic energy of the emitted electron? What is its minimum energy? What is the energy of the neutrino in each case?

*18. A ${}^{232}_{92}U$ nucleus emits an α particle with KE = 5.32 MeV. What is the final nucleus and what is the approximate atomic mass (in u) of the final atom?

*19. ${}^{124}_{55}Cs$ has a half-life of 30.8 s. (a) If we have 7.5 μg initially, how many nuclei are present? (b) How many are present 2.0 min later?

NUCLEAR ENERGY AND THE EFFECTS AND USES OF RADIATION

CHAPTER

30

We continue our study of nuclear physics in this chapter. We begin with a discussion of nuclear reactions, after which we examine the enormous energy-releasing processes of fission and fusion. The remainder of the chapter deals with the effects of nuclear radiation when it passes through matter, particularly biological matter, and how radiation is used medically for therapy and diagnosis, including recently developed imaging techniques.

1. NUCLEAR REACTIONS AND THE ARTIFICIAL TRANSMUTATION OF ELEMENTS

When a nucleus undergoes α or β decay, the daughter nucleus is that of a different element from the parent. The transformation of one element into another, called *transmutation*, also occurs by means of nuclear reactions. A **nuclear reaction** is said to occur when a given nucleus is struck by another nucleus, or by a simpler particle such as a γ ray or neutron, so that an interaction takes place. Ernest Rutherford was the first to report seeing a nuclear reaction. In 1919 he observed that some of the α particles passing through nitrogen gas were absorbed and protons emitted. He concluded that nitrogen nuclei had been transformed into oxygen nuclei via the reaction

$$^{4}_{2}\text{He} + {}^{14}_{7}\text{N} \rightarrow {}^{17}_{8}\text{O} + p$$

where $^{4}_{2}\text{He}$ is an α particle and p is a proton.

Since then, a great many nuclear reactions have been observed. Indeed, many of the radioactive isotopes used in the laboratory are made by means of nuclear reactions. Nuclear reactions can be made to occur in the laboratory, but they also occur regularly in nature. In Chapter 29 we saw an example of this: ${}^{14}_{6}C$ is continually being made in the atmosphere via the reaction $n + {}^{14}_{7}N \rightarrow {}^{14}_{6}C + p$.

In any nuclear reaction, the total electric charge is conserved and the total number of nucleons is conserved. These conservation laws are often useful. For example, suppose a neutron is observed to strike an ${}^{16}_{8}O$ nucleus and a deuteron is given off. (A *deuteron*, or *deuterium*, is the isotope of hydrogen containing one proton and one neutron, ${}^{2}_{1}H$.) We can determine the resulting nucleus as follows. We have the reaction $n + {}^{16}_{8}O \rightarrow ? + {}^{2}_{1}H$. The total number of nucleons initially is $16 + 1 = 17$ and the total charge is $8 + 0 = 8$; the same totals apply to the right side of the reaction. Hence the product nucleus must have $Z = 7$ and $A = 15$. From the periodic table, we find that it is nitrogen that has $Z = 7$, so the nucleus produced is ${}^{15}_{7}N$.

The artificial transmutation of elements took a great leap forward in the 1930s when Enrico Fermi realized that neutrons would be the most effective projectiles for causing nuclear reactions and in particular for producing new elements. Because neutrons have no net electric charge, they are not repelled by positively charged nuclei as protons or alpha particles are. Hence the probability of a neutron reaching the nucleus and causing a reaction is much greater than for charged projectiles,[†] particularly at low energies. Between 1934 and 1936, Fermi and his co-workers in Rome produced many previously unknown isotopes by bombarding different elements with neutrons. Fermi realized that if the heaviest known element, uranium, were bombarded with neutrons, it might be possible to produce new elements whose atomic numbers were greater than that of uranium. After several years of hard work it was suspected that two new elements had been produced, neptunium ($Z = 93$) and plutonium ($Z = 94$). The full confirmation that such "transuranic" elements could be produced came several years later at the University of California, Berkeley. The reactions are shown in Figure 30-1.

(a) $n + {}^{238}_{92}U \rightarrow {}^{239}_{92}U$

Neutron captured by ${}^{238}_{92}U$

(b) ${}^{239}_{92}U \rightarrow {}^{239}_{93}Np + e^- + \bar{\nu}$

${}^{239}_{92}U$ decays by beta decay to neptunium 239

(c) ${}^{239}_{93}Np \rightarrow {}^{239}_{94}Pu + e^- + \bar{\nu}$

${}^{239}_{93}Np$ itself decays by beta decay to produce plutonium 239

FIGURE 30-1
Series of reactions in which neptunium and plutonium are produced after bombardment of uranium 238 by neutrons.

It was soon shown that what Fermi actually had observed when he bombarded uranium was an even stranger process—one that was destined to play an extraordinary role in the world at large.

2. NUCLEAR FISSION AND NUCLEAR REACTORS

The Discovery of Fission

In 1938, the German scientists Otto Hahn and Fritz Strassmann made an amazing discovery. Following up on Fermi's work, they found that uranium bombarded by neutrons sometimes produced smaller nuclei

[†] That is, positively charged particles. Electrons rarely cause nuclear reactions because they do not partake of the strong nuclear force.

which were roughly half the size of the original uranium nucleus. Lise Meitner and Otto Frisch, two refugees from Nazi Germany working in Scandinavia, quickly realized what had happened: the uranium nucleus, after absorbing a neutron, actually had split into two roughly equal pieces. This was startling, for until then the known nuclear reactions involved knocking out only a tiny fragment (for example, n, p, or α) from a nucleus.

This new phenomenon was named **nuclear fission** because of its resemblance to biological fission (cell division). It occurs much more readily for $^{235}_{92}U$ than for the more common $^{238}_{92}U$. The process can be visualized by imagining the uranium nucleus to be like a liquid drop. According to this *liquid-drop model*, the neutron absorbed by the $^{235}_{92}U$ nucleus gives the nucleus extra internal energy (like heating a drop of water). This intermediate state, or *compound nucleus*, is $^{236}_{92}U$ (because of the absorbed neutron). The extra energy of this nucleus (it is in an excited state) appears as increased motion of the individual nucleons, which causes the nucleus to take on abnormal elongated shapes (Figure 30-2). When the nucleus elongates into the shape shown in Figure 30-2(c), the attraction of the two ends via the short-range nuclear force is greatly weakened by the excess separation distance, but the electric repulsive force is weakened only slightly and becomes predominant; so the nucleus splits in two. The two resulting nuclei, N_1 and N_2, are called *fission fragments*, and in the process a number of neutrons (typically two or three) are also given off. The reaction can be written

$$n + {}^{235}_{92}U \rightarrow {}^{236}_{92}U \rightarrow N_1 + N_2 + \text{neutrons}.$$

The compound nucleus, $^{236}_{92}U$, exists for less than 10^{-12} s, so the process occurs very quickly. The two fission fragments have roughly half the mass of the uranium, although rarely are they exactly equal in mass. A typical fission reaction is

$$n + {}^{235}_{92}U \rightarrow {}^{141}_{56}Ba + {}^{92}_{36}Kr + 3n,$$

although many others also occur.

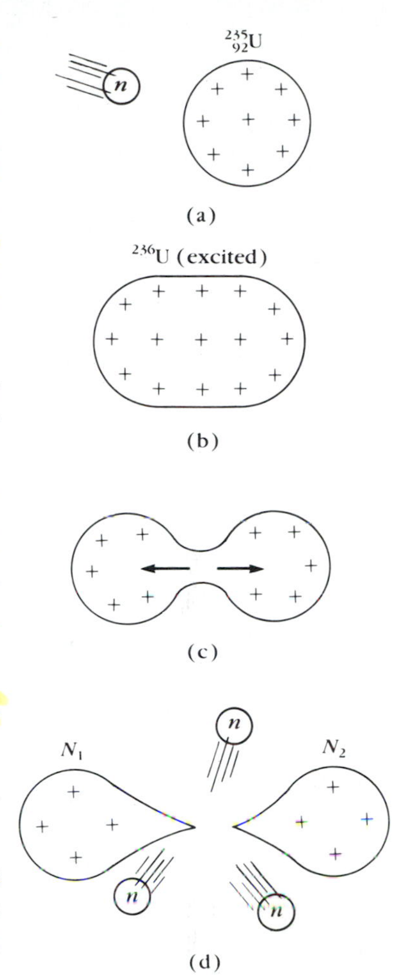

FIGURE 30-2
Fission of a $^{235}_{92}U$ nucleus after capture of a neutron.

Energy Release and Chain Reactions

A tremendous amount of energy is released in a fission reaction. This is because the mass of $^{235}_{92}U$ is considerably greater than that of the fission fragments. When we calculate the sum of the masses of the fission fragments and free neutrons, we find that their total mass is about 200 MeV/c^2 less than the mass of the original $^{235}_{92}U$ and incident neutron. This loss of mass appears as kinetic energy of the fission fragment nuclei (N_1 and N_2) and released neutrons. Thus some 200 MeV of energy is released in the fission of each uranium nucleus. This is an enormous amount of energy on the nuclear scale.

At a practical level, the energy from one fission is, of course, tiny. It was soon recognized, however, that a great deal of energy at the macroscopic level would be available if many such fissions could occur at once. A number of physicists, including Fermi, recognized that the neutrons released in each fission could be used to create a **chain reaction**: one neutron initially causes one fission of a uranium nucleus; the two or three neutrons released can go on to cause additional fissions, so the process multiplies as shown schematically in Figure 30-3.

Thus the possibility of obtaining the enormous energy available in nuclear fission seemed at hand. The problem was to find out if a *self-sustaining chain reaction* was actually possible in practice. Fermi and his co-workers at the University of Chicago set out to build the first **nuclear reactor**.

Several problems had to be overcome. First, the probability that a $^{235}_{92}U$ nucleus will absorb a neutron is high only for slow neutrons; but the neutrons emitted during a fission, and which are needed to sustain a chain reaction, are moving very fast. A substance known as a **moderator** must be used to slow down the neutrons. The most effective moderator consists of atoms whose mass is as close as possible to that of the neutrons. (To see why this is true, recall that a billiard ball striking an equal mass at rest can itself be stopped in one collision; yet a billiard ball striking a much heavier object bounces off with nearly the same speed with which it hit.) The best moderator would thus contain $^{1}_{1}H$ atoms; unfortunately, $^{1}_{1}H$ tends to absorb neutrons. But *deuterium*, $^{2}_{1}H$, does not absorb many neutrons and is thus

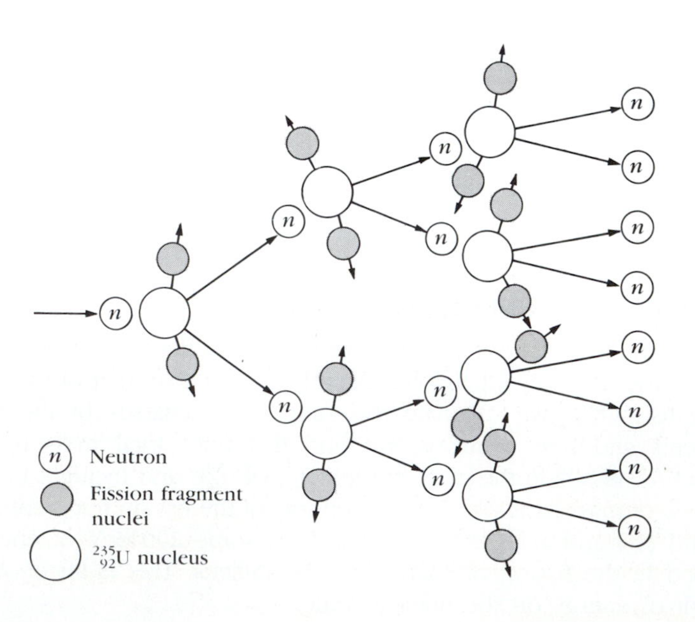

FIGURE 30-3
Chain reaction.

an ideal moderator. Either ^1_1H or ^2_1H can be used in the form of water; in the latter case, it is *heavy water*, where the hydrogen atoms have been replaced by deuterium. Another common moderator is *graphite*, which consists of $^{12}_6C$ atoms.

A second problem is that the neutrons produced in one fission may be absorbed and produce other nuclear reactions with other nuclei in the reactor, rather than produce further fissions. For example, $^{238}_{92}U$ absorbs neutrons, as do the ^1_1H nuclei used as the moderator in "light-water" reactors. Naturally occurring uranium contains 99.3 percent $^{238}_{92}U$ and only 0.7 percent fissionable $^{235}_{92}U$. To increase the probability of fission of $^{235}_{92}U$ nuclei, natural uranium is often **enriched** to increase the percentage of $^{235}_{92}U$ through processes such as diffusion or centrifugation.

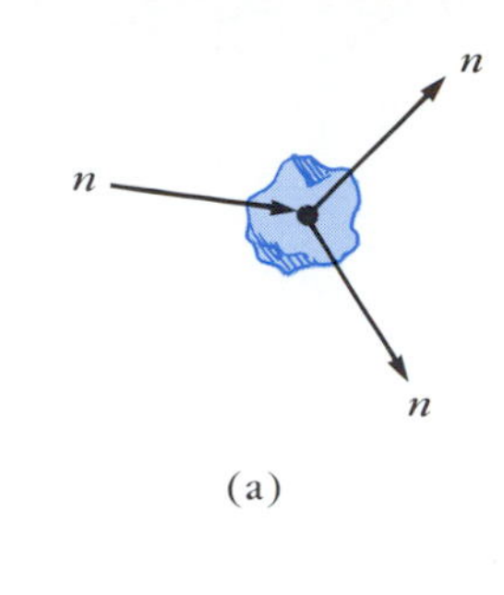

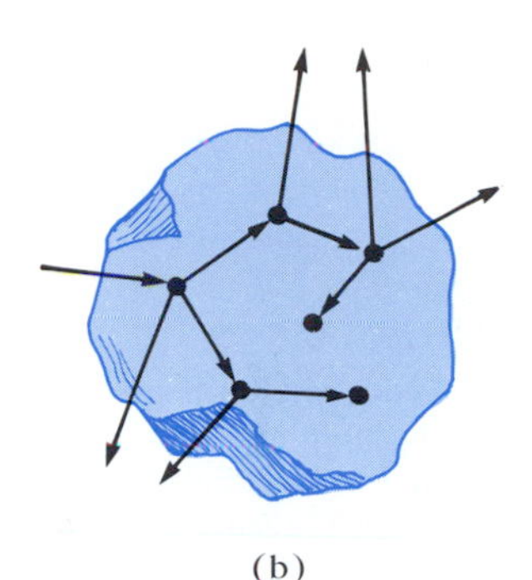

FIGURE 30-4
If the amount of uranium exceeds the critical mass, as in (b), a sustained chain reaction is possible. If the mass is less than critical, as in (a), most neutrons escape before additional fissions occur, and the chain reaction is not sustained.

The third problem is that some neutrons will escape through the surface of the reactor fuel before they cause further fissions (Figure 30-4). Thus the mass of fuel must be sufficiently large for a self-sustaining chain reaction to take place. The minimum mass of uranium needed is called the **critical mass**. The value of the critical mass depends on the moderator, the fuel ($^{239}_{94}Pu$ may be used instead of $^{235}_{92}U$), and how much the fuel is enriched, if at all. Typical values are on the order of a few kilograms (that is, not grams nor thousands of kilograms).

To have a self-sustaining chain reaction, it is clear that on the average at least one neutron produced in each fission must go on to produce another fission. The average number of neutrons per each fission that go on to produce further fissions is called the *multiplication factor*, f. For a self-sustaining chain reaction we must have $f > 1$. If $f < 1$, the reactor is "subcritical"; if $f > 1$, it is "supercritical." Reactors are equipped with movable *control rods* (usually of cadmium or boron), whose function is to absorb neutrons and maintain the reactor at just barely "critical," $f = 1$.

The First Nuclear Reactor

The first nuclear reactor was built by Fermi in 1942 (Figure 30-5) and was ready for testing in December of that year. Natural uranium was used with graphite as the moderator. The reactor was equipped with cadmium control rods that absorb neutrons and prevent the reactor from producing too much energy in a short time, which might lead to its destruction. With the rods in place, the cadmium—which strongly absorbs neutrons—keeps the reactor "subcritical" ($f < 1$ and no chain reaction). When the control rods are pulled out fewer neutrons are absorbed, so more are available to cause fission.

If all went well, it was hoped that a sustained chain reaction would occur. On December 2, 1942, the critical test was made: Fermi slowly withdrew the control rods—and the reactor went "critical." The first self-sustaining chain reaction had been produced. Arthur Compton, who had witnessed the event, announced it as follows: "The Italian navigator

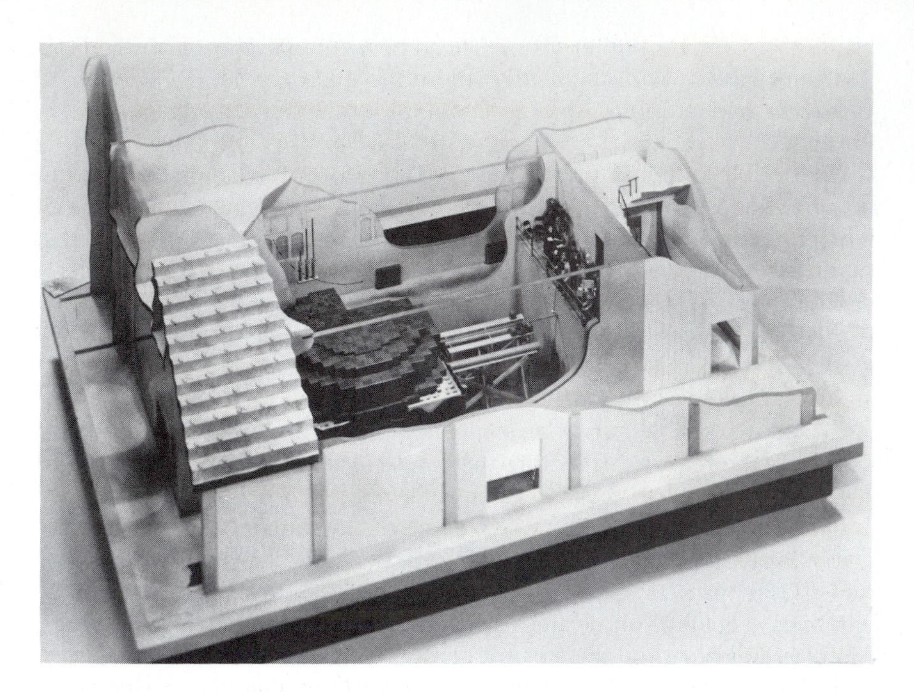

FIGURE 30-5
Sketch of the first nuclear reactor, built by Fermi under the grandstand of Stagg Field at the University of Chicago.

has just landed in the new world." For better or worse, the nuclear age had begun. The possibilities for the future seemed immense. It was clear that nuclear energy could be used to help mankind in peaceful endeavors; but it could also be used for destructive purposes, and—to the great misfortune of mankind—that was how it was first used.

The Bomb

The building of the first reactor in 1942 was part of a plan to see if a bomb based on nuclear energy was feasible. In early 1940, with Europe already at war, Hitler banned the sale of uranium from the Czech mines he had recently expropriated. Research into the fission process suddenly was enshrouded in secrecy by Germany and by the western powers. Physicists in the United States were alarmed. A group of them approached Einstein—a man whose name was a household word—to sign a letter to President Roosevelt about the possibilities of using nuclear fission for a bomb far more powerful than any previously known, and inform him that Germany might already have begun development of such a bomb.

Roosevelt responded by authorizing the program known as the Manhattan Project, to see if a bomb could be built. Work began in earnest after Fermi's demonstration that a sustained chain reaction was possible. A new secret laboratory was developed on an isolated mesa in New Mexico known as Los Alamos. Under the direction of J. Robert Oppenheimer (1904–1967), it became the home of famous scientists from all over

Europe and the United States; the roster of names of those who worked there during the next couple of years is really a who's who of science.

One of the challenges was how to build a bomb that was subcritical during transport but which could be made critical (to produce a chain reaction) at just the right moment. This was achieved by using two pieces of uranium each of which was less than the critical mass but when placed together became greater than the critical mass. The two masses would be kept separate until the moment of detonation arrived; then a special kind of gun would force the two pieces together very quickly, a chain reaction of explosive proportions would occur, and a tremendous amount of energy would be released very suddenly. The first fission bomb (popularly known as an "atomic bomb") was tested in the New Mexico desert in July 1945. It was successful. In early August, a fission bomb using uranium was dropped on Hiroshima and a second, using plutonium, was dropped on Nagasaki. World War II ended shortly thereafter, but the dropping of these bombs aroused lasting controversy.

Scientists were later criticized for working on the bomb, and many regretted having done so. Critical judgment, however, should perhaps be placed in the context of the times. In the early 1940s, while Hitler's armies were overrunning Europe, it was widely believed that German scientists were trying to build a bomb (they never did). Japan had entered the war, and the world seemed in danger of being overwhelmed by oppressive powerful regimes. The scientists at Los Alamos were aware they were developing a dangerous weapon. But they had to make a choice. Stopping Hitler and his allies was uppermost in their minds.

Nuclear Fallout

Besides its great destructive power, a fission bomb produces highly radioactive fission fragment nuclei. Fission fragments are radioactive because they have too many neutrons to be stable, and so decay by β^- emission to nuclei with more nearly equal numbers of protons and neutrons. (The fission fragments have an excess of neutrons because the original uranium nuclei have nearly $1\frac{1}{2}$ times as many neutrons as protons. The uranium nuclei *need* extra neutrons, as discussed in Section 2 of Chapter 29, for added nuclear attraction to offset the large electrical repulsion among its 92 protons.) When a fission bomb explodes, these radioactive isotopes are released into the atmosphere and are known as *radioactive fallout*.

The testing of nuclear bombs in the atmosphere following World War II was also a cause of concern, for the movement of air masses spread the fallout all over the globe. Radioactive fallout eventually settles to the earth—particularly in rainfall—and is absorbed by plants and grasses and enters the food chain. This is a far more serious problem than the same radioactivity on the exterior of our bodies, since α and β particles are largely absorbed by clothing and the outer (dead) layer of skin. But once inside our bodies via food, the isotopes are in direct contact with living cells. One particularly dangerous radioactive isotope is $^{90}_{38}Sr$, which is chemically much like calcium and becomes concentrated in bone, where it causes bone cancer and destruction of bone marrow. The 1963 treaty, signed by over 100 nations, that bans nuclear-weapons testing in the atmosphere was chiefly motivated because of the hazards of fallout.

Nuclear Reactors and Power Plants

Soon after World War II, peaceful uses of nuclear energy were developed. Nuclear reactors were built for use in research and to produce electric power. Fission produces many neutrons and a "research reactor" is basically an intense source of neutrons. These neutrons can be used as projectiles in nuclear reactions to produce nuclides not found in nature, including isotopes used as tracers and for medical therapy.

A "power reactor" is used to produce electricity for homes and industry. The energy released in the fission process appears as heat, which is used to boil water and produce steam to drive a turbine connected to an electric generator. Figure 30-6 shows a typical design.

The *core* of a nuclear reactor consists of the fuel and a moderator (water in most commercial United States reactors). The fuel is usually uranium enriched so that it contains 2 to 4 percent $^{235}_{92}U$. The difference between a nuclear power reactor and a nuclear bomb is the rate of release of the energy: in a bomb, the neutron multiplication factor is somewhat greater than 1, so the chain reaction occurs swiftly and an explosion occurs; in a nuclear reactor the multiplication of neutrons is held very

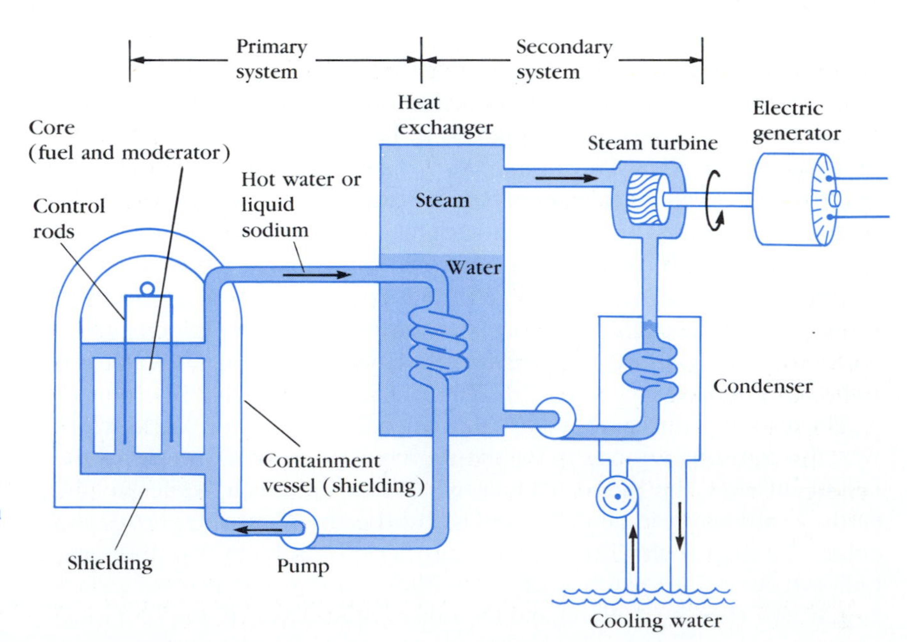

FIGURE 30-6
Nuclear reactor.
The heat generated by the fission process in the fuel rods is carried off by hot water or liquid sodium and is used to boil water to steam in the heat exchanger. The steam drives a turbine to generate electricity and is then cooled in the condenser.

close to 1.0, so the reactor is kept just barely "critical" and the energy is released slowly. This is accomplished by careful adjustment of the control rods that act to absorb neutrons. Water or other liquid (such as liquid sodium) is allowed to flow through the core. The thermal energy it absorbs is used to produce steam in the heat exchanger (Figure 30-6), so the fissionable fuel acts as the heat input for a heat engine.

Many problems are associated with nuclear power plants. There is, of course, the usual thermal pollution associated with any heat engine (See Section 6 of Chapter 15). But probably the most serious problems are associated with the radioactive fission fragments produced in the reactor and the radioactive nuclides produced by neutrons interacting with the structural parts of the reactor. The accidental release of highly radioactive fission fragments into the atmosphere poses a serious threat to health (see Section 4). Another serious problem is the disposal of the spent fuel, which contains highly radioactive fission fragments. Leakage of the radioactive wastes is possible and has already occurred. Indeed, a satisfactory method of disposal is not yet at hand. Finally, the lifetime of nuclear power plants is limited to some 30 years due to the buildup of radioactivity and the fact that the structural materials themselves are weakened by the intense internal conditions. Decommissioning of a power plant could take a number of forms. However, the cost of *any* method of decommissioning a large nuclear power plant is expected to be very great.

So-called *breeder reactors* have the same problems, but were proposed as a solution to the problem of limited supplies of fissionable uranium. A breeder reactor is one in which some of the neutrons produced in the fission of $^{235}_{92}U$ are absorbed by $^{238}_{92}U$, and $^{239}_{94}Pu$ is produced via the set of reactions shown in Figure 30-1. $^{239}_{94}Pu$ is fissionable with slow neutrons, so after separation it can be used as a fuel in a nuclear reactor. Thus a breeder reactor "breeds" new fuel† ($^{239}_{94}Pu$) from otherwise useless $^{238}_{92}U$. Since natural uranium is 99.3 percent $^{238}_{92}U$, this means that the supply of fissionable fuel could be increased by more than a factor of 100. But breeder reactors present additional problems. First, plutonium has a very long half-life of 24,000 years and is a highly toxic substance that poses considerable danger to health. Also important is the fact that plutonium thus produced can readily be used in a bomb. So the use of a breeder reactor, even more than a conventional uranium reactor, presents the danger of nuclear proliferation: even poor nations might be able to produce nuclear bombs. And the possibility of theft of fuel by terrorists who could produce a bomb further increases the probability of nuclear holocaust.

It is clear that nuclear power presents many risks. Other large-scale energy-conversion methods, such as conventional coal-burning steam

† A breeder reactor does *not* produce more fuel than it uses, a nearly endless supply, as newspaper articles often imply.

plants, also present health and environmental hazards (some of which were discussed in Chapter 15). The solution to the world's needs for energy is not only technological, but economic and political as well.

3. FUSION

The mass of every stable nucleus is less than the sum of the masses of the protons and neutrons that compose it. The mass of the helium isotope ^4_2He, for example, is less than the mass of two protons plus the mass of two neutrons. In other words, if two protons and two neutrons were to combine to form a helium nucleus, there would be a loss of mass accompanied by the release of a large amount of energy. The process of building up nuclei by bringing together individual protons and neutrons, or building larger nuclei by combining small nuclei, is called **nuclear fusion**. It is believed that all the elements in the universe were originally formed through the process of fusion. Today, fusion is continually taking place within our sun and other stars, and the intense light energy they emit is obtained from these fusion reactions.

The possibility of utilizing the energy released in fusion to make a power reactor is very attractive, but so far a successful reactor has not been achieved. The fusion reactions most likely to succeed in a reactor involve the isotopes of hydrogen, ^2_1H (deuterium) and ^3_1H (tritium), and are as follows, with the energy released given in parentheses:

$$^2_1H + ^2_1H \rightarrow ^3_1H + ^1_1H \qquad (4.0\ \text{MeV})$$

$$^2_1H + ^2_1H \rightarrow ^3_2He + n \qquad (3.3\ \text{MeV})$$

$$^2_1H + ^3_1H \rightarrow ^4_2He + n \qquad (17.6\ \text{MeV})$$

$$^3_1H + ^3_1H \rightarrow ^4_2He + 2n \qquad (11.3\ \text{MeV}).$$

The energy released in fusion reactions is greater for a given mass of fuel than in fission. Furthermore, fusion presents less of a radioactive-waste problem. For fuel, a fusion reactor could use deuterium, which is very plentiful in the water of the oceans (the natural abundance of ^2_1H is 0.015 percent, or about 1 g of deuterium per 60 liters of water).

Unfortunately, considerable difficulties still exist for making a usable fusion reactor. A major difficulty arises because all nuclei have a positive charge and thus repel each other. However, if they can be brought close enough together so that the short-range attractive nuclear force can come into play, the latter can pull the nuclei together and fusion will occur. In order for the nuclei to get close enough together they must have very high speeds. Since high speed of atoms corresponds to high temperature, very high temperatures are required for fusion to occur; hence fusion devices are often referred to as *thermonuclear devices*. The sun and other stars are very hot, many millions of degrees, so the nuclei are moving fast enough for fusion to take place and the energy released keeps the temperature high so that further fusion reactions can occur. The sun and the stars represent self-sustaining thermonuclear reactors, but on earth such high temperatures are not easily attained in a controlled manner.

Following World War II it was realized that the temperature produced within a fission (or "atomic") bomb was close to 10^8 K; this suggested that a fission bomb could be used to ignite a fusion bomb (popularly known as a thermonuclear or hydrogen bomb) to release the vast energy of fusion. The uncontrollable release of fusion energy in an H-bomb was fairly easy to obtain. But to realize usable energy from fusion at a slow and controlled rate proved to be very difficult. The high temperatures needed can now be produced by an infusion of concentrated energy, such as from a high-powered laser. The real difficulty is to contain the nuclei long enough for sufficient reactions to occur that a usable amount of energy is obtained. At the temperatures needed for fusion (about 100-million degrees, or 10^8 K) the atoms are ionized, and this collection of nuclei and electrons is referred to as a **plasma**. Ordinary materials vaporize at a few thousand degrees at best, and hence could not be used to contain a high temperature plasma. One technique is to confine a plasma with a magnetic field; unfortunately, all attempted configurations of magnetic field confinement have developed "leaks" and the charged particles leak out before sufficient fusion takes place. One promising configuration of magnetic fields is the so-called Tokamak, and confinement times are being increased. Another containment technique is to form solid pellets of fuel which are quickly heated by an intense laser or electron beam. This *inertial confinement* technique too shows promise, but controlled fusion remains elusive.

4. BIOLOGICAL USES AND DANGERS OF RADIATION; NUCLEAR MEDICINE AND DOSIMETRY

Charged particles such as α and β particles, as well as γ and X-ray photons, are referred to collectively as *radiation*. Radiation, whether from radioactive decay or produced by nuclear reactions, is a two-edged sword: it can be damaging to health; but it can also be used to treat and diagnose disease.

Radiation Damage

Charged particles, such as α and β rays and protons, can ionize the atoms and molecules of any material they pass through because of the electric force. That is, when they pass through a material they can attract or repel electrons strongly enough to remove them from the atoms of the material.

And because radiation produces ionization, it can cause considerable damage to materials, particularly to biological tissue. Neutral particles also give rise to ionization when they pass through materials. For example, X-ray and γ-ray photons can ionize atoms by knocking out electrons by means of the photoelectric and Compton effects. And if a γ ray has sufficient energy, it can undergo pair production: an electron and a positron are produced (see Section 2 of Chapter 26). The charged particles produced in all these processes can themselves go on to produce further ionization. Neutrons, on the other hand, interact with matter mainly by collisions with nuclei, with which they interact via the strong nuclear force. Often the nucleus is broken apart by such a collision, which alters the molecule of which it was a part. Radiation passing through matter can cause extensive damage. In metals and other structural materials, the strength of the material can be weakened if the radiation is very intense. This is a considerable problem for nuclear reactor power plants, and for space vehicles that must pass through areas of intense cosmic radiation.

Also very important is the radiation damage to biological organisms. The damage is due primarily to the ionization produced in cells. The knocking out of binding electrons can break a molecule apart or alter its structure so it does not perform its function. Large doses of radiation can alter so many molecules in a cell that the cell can no longer function and dies. Even small doses can have a serious effect on the DNA, the genetic material of a cell. Each alteration in the DNA affects a gene; such an alteration of the genetic material is called a mutation. Most mutations are detrimental, producing defective cells. Such defective cells may go on dividing and produce many more defective cells—the beginning of a cancerous growth—to the detriment of the organism as a whole. Thus radiation can cause cancer—the rapid production of defective cells.

FIGURE 30-7

Autoradiograph of a mature leaf of the squash plant *Cucurbita melopepo* exposed for 30 s to $^{14}CO_2$. The photosynthetic (green) tissue has become radioactive; the nonphotosynthetic tissue of the veins is free of ^{14}C and therefore does not blacken the X-ray sheet. This technique is very useful in following patterns of nutrient transport in plants.

Radiation Therapy and Nuclear Medicine

Although radiation can cause cancer, it can also be used to treat it. Rapidly growing cancer cells are more susceptible to destruction by radiation. Nonetheless, large doses are needed to kill the cancer cells, and some of the surrounding normal cells are inevitably killed as well. It is for this reason that cancer patients receiving radiation therapy often suffer side effects characteristic of radiation sickness. To minimize the destruction of normal cells, a narrow beam of γ or X rays is often used when the cancerous tumor is well localized. The beam is directed at the tumor and the source (or body) is rotated so that the beam passes through various parts of the body to keep the dose at any one place as low as possible—except at the tumor site and its immediate surroundings, where the beam passes at all times. The radiation may be from a radioactive source such as $^{60}_{27}Co$, or it may be from an X-ray machine. Protons, neutrons, electrons and pions, which are produced in particle accelerators (Chapter 31), are also being used in cancer therapy. However, in every kind of radiation treatment special precautions must be taken to avoid exposing normal tissue; some exposure may be unavoidable, but it must be kept to a minimum.

The subject of the applications of radioactivity and radiation to human beings and other organisms is known as *radiation biology*, and in medicine it is called *nuclear medicine*. When applied to the treatment of disease, as just discussed, it is called *radiation therapy*. But nuclear medicine also involves research and *diagnosis* of disease.

For example, radioactive isotopes are commonly used in biological and medical research as *tracers*. A given compound is artificially synthesized using a radioactive isotope such as ${}^{14}_{6}C$ or ${}^{3}_{1}H$. Such "tagged" molecules can then be traced as they move through an organism or as they undergo chemical reaction. The presence of these tagged molecules (or parts of them, if they undergo chemical change) can be detected by a Geiger or scintillation counter, which detects emitted radiation. The details of how food molecules are digested, and to what parts of the body they are diverted, can be traced in this way.

In a technique known as *autoradiography*, the position of the radioactive isotopes is detected on film. For example, the distribution of carbohydrates produced in the leaves of plants from absorbed CO_2 can be observed by keeping the plant in an atmosphere where the carbon atom in the CO_2 is ${}^{14}_{6}C$. After a time, a leaf is placed firmly on a photographic plate and the emitted radiation darkens the film most strongly where the isotope is most strongly concentrated (Figure 30-7). Autoradiography using labeled nucleotides (that form DNA) has revealed much about the details of DNA replication (Figure 30-8).

FIGURE 30-8
An autoradiogram of a fiber of chromosomal DNA isolated from the higher plant *Arabidopsis thaliana*. The dashed arrays of silver grains show the Y-shaped growing point of replicating DNA.

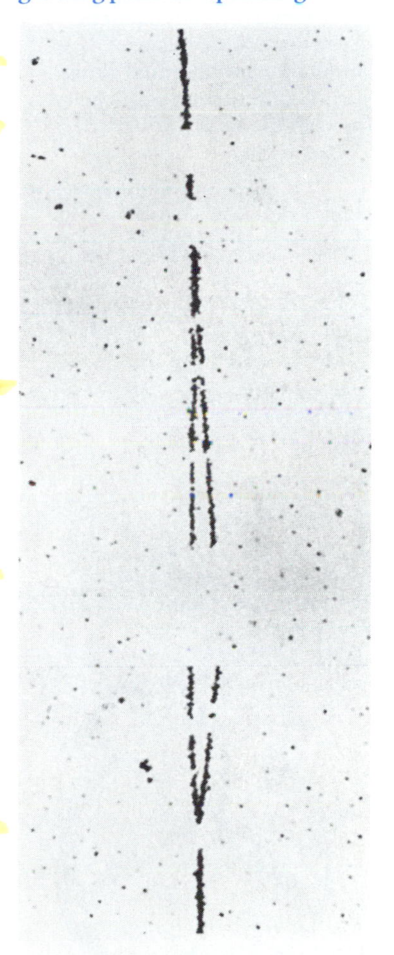

For *medical diagnosis*, the radionuclide most often used today is ${}^{99m}_{43}Tc$, a long-lived excited state of technicium-99 (the "m" in the symbol stands for "metastable" state). The great usefulness of ${}^{99m}_{43}Tc$ derives from its convenient half-life of 6 hr (short, but not too short) and the fact that it can combine with a large variety of compounds. The compound to be labeled with the radionuclide is so chosen because it concentrates in the organ or region of the anatomy to be studied. The detection of high or low densities of radioactivity may point to the overactivity or underactivity of an organ or part of an organ, or in other cases may signal the presence of a lesion or tumor.

More complex imaging techniques can actually produce images of cross sections of the human body with help from a computer. In *computerized tomography*, or *CT scanning*† ("tomography" comes from the Greek: *graph* = picture, *tomos* = slice), a narrow X-ray beam is passed through the body and the amount of the beam absorbed is measured at several

† Also called *computer-assisted tomography*, or *CAT scanning*.

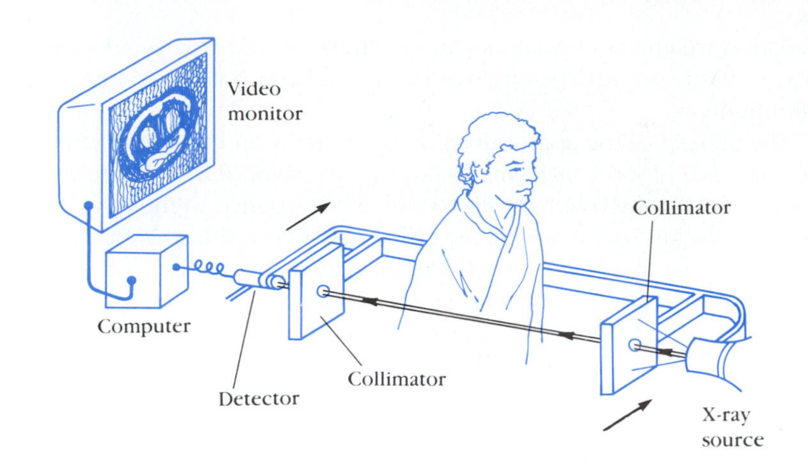

FIGURE 30-9
X-ray computer-assisted tomography (CAT Scan). The X-ray source and detector move together horizontally, the transmitted intensity being measured at a large number of points; then the source-detector assembly is rotated slightly (say 1°) and another scan is made. This is repeated for perhaps 180°. The computer reconstructs the image of the slice and it is presented on a TV monitor (cathode ray tube). More sophisticated scanners use several X-ray beams simultaneously directed along slightly different anatomical lines through the body.

positions and angles through a cross section of the body (Figure 30-9). The computer processes all this data and reconstructs an image of the cross section (Figure 30-10).

Another computerized diagnostic technique images the γ rays or β particles emitted by tracers injected into the body. For example, in *positron emission tomography* (PET) the positrons (e^+) emitted by the radioactive tracer nuclei quickly collide with ordinary electrons (e^-) of the material and annihilate them, producing two gamma rays. The two γ rays fly off in opposite directions and are detected by a ring of detectors around the patient (Figure 30-11). After detecting many such decays, a computer can reconstruct an image of that specific cross section of the body. Similar computerized diagnostic imaging techniques are being developed that use ultrasound (Chapter 17) and the magnetic properties of nuclei (nuclear magnetic resonance, or NMR, also known as magnetic resonance imaging).

From each of these imaging techniques different types of information relating to the function and structure of organs within the body can be obtained.

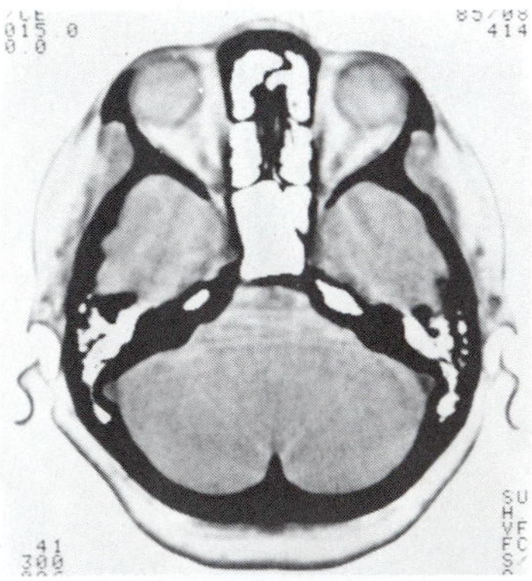

FIGURE 30-10
CAT Scan of human brain.

Dosimetry: Measurement of Radiation (optional)

Because radiation can damage biological tissue, great care must be taken to minimize exposure to normal tissue when administering radiation treatment. It is therefore important to be able to quantify the amount, or *dose*, of radiation that passes through a material. This is the subject of **dosimetry**.

There are a number of ways to measure radiation dose. The *strength* of a *source* can be specified at a given time by stating the *source activity*, or

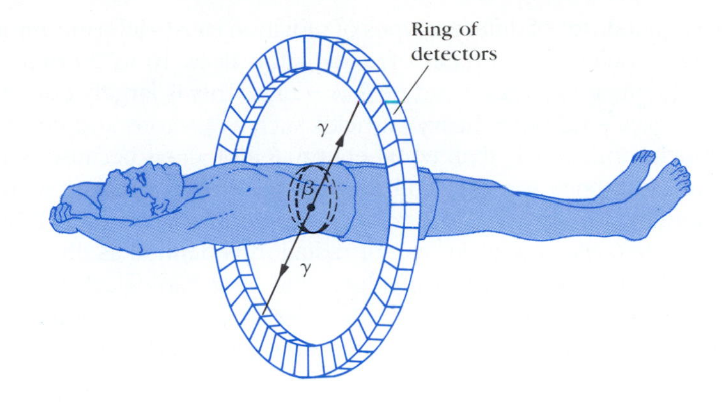

FIGURE 30-11
Positron emission tomography system showing a ring of detectors to detect the two annihilation rays ($e^+e^- \rightarrow 2\gamma$) emitted at 180° to each other.

how many disintegrations occur per second. The traditional unit is the *curie* (Ci), defined as 1 Ci = 3.70×10^{10} disintegrations per second. Thus a 2.0 μ Ci β source generates $(3.7 \times 10^{10})(2.0 \times 10^{-6}) = 7.4 \times 10^4$ β particles per second. Although the curie is still commonly used, the proper SI unit for source activity is the becquerel (Bq), defined as 1 Bq = 1 disintegration/s. Commercial suppliers of *radionuclides* (radioactive nuclides) specify the activity at a given time; since the activity decreases in time, particularly for short-lived isotopes, it is important to take this into account.

Another type of measurement, in contrast to the activity of the *source* as just discussed, is the exposure or *absorbed dose*—that is, the effect the radiation has on the absorbing material. The earliest unit of dosage was the *roentgen* (R), which was defined in terms of the amount of ionization produced by the radiation.† The roentgen has been largely superseded by another unit of absorbed dose applicable to any type of radiation, the rad: *1 rad is that amount of radiation which deposits 1.00×10^{-2} J/kg of any absorbing material*. (This is quite close to the roentgen for X and γ rays.) (The SI unit for absorbed close is the gray (Gy): 1 Gy = 1 J/kg = 100 rad, and is slowly coming into use.) The absorbed dose depends not only on the strength of a given radiation beam (number of particles per second) and the energy per particle, but also on the type of material that is absorbing the radiation. Since bone, for example, is denser than flesh and absorbs more of the radiation normally used, the same beam passing through the body deposits a greater dose in bone than in the flesh.

The gray and the rad are physical units of dose—the energy deposited per unit mass of material. They are, however, not the most meaningful units for measuring the biological damage produced by radiation. This is

† Today 1 R is defined as the amount of X or γ radiation that deposits 0.878×10^{-2} J of energy per kilogram of air.

TABLE 30-1
Quality Factor (QF) of Different Kinds of Radiation

TYPE	QF
X and γ rays	≈ 1
β (electrons)	≈ 1
Fast protons	1
Slow neutrons	≈ 3
Fast neutrons	Up to 10
α particles and heavy ions	Up to 20

because equal doses of different types of radiation cause differing amounts of damage. For example, 1 rad of α radiation does 10 to 20 times the amount of damage as does 1 rad of β or γ rays. This is largely due to the fact that α rays (and other heavy particles such as protons and neutrons) move much more slowly than equal energy β and γ rays because of their greater mass. Hence ionizing collisions occur closer together so more irreparable damage is done. The *relative biological effectiveness* (RBE) or *quality factor* (QF) of a given type of radiation is defined as the number of rads of X or γ radiation that produces the same biological damage as 1 rad of the given radiation. The QF for several types of radiation is given in Table 30-1. The numbers are approximate since they depend somewhat on the energy of the particles and on the type of damage that is used as the criterion.

The product of the dose in rads and the QF gives a unit known as the *rem* (which stands for *rad equivalent man*): effective dose (in rem) = dose (in rad) × QF. (This unit is being replaced by the SI unit for "effective dose," the sievert (Sv): effective dose (Sv) = dose (Gy) × QF.) By this definition, 1 rem of any type of radiation does approximately the same amount of biological damage. For example, 50 rem of fast neutrons does the same damage as 50 rem of γ rays. But note that 50 rem of fast neutrons is only 5 rads whereas 50 rem of γ rays is 50 rads.

We are constantly exposed to low-level radiation from natural sources: cosmic rays, natural radioactivity in rocks and soil, and naturally occurring radioactive isotopes that occur in our food such as $^{40}_{19}K$. The natural radioactive background supplies about 0.13 rem per year per person on the average. From medical X rays the average person receives about 0.07 rem per year. The United States government specifies the recommended upper limit of allowed radiation for an individual in the general populace at about 0.5 rem per year, exclusive of natural sources. However, since even low doses of radiation are believed to increase the chances of cancer or genetic defects, the prevailing attitude today is to keep the radiation dose as low as possible.

People who work around radiation—in hospitals, in power plants, in research—often are subjected to much higher doses than 0.5 rem/yr. The upper limit for such occupational exposures has been set somewhat higher, on the order of 5 rem/yr whole-body dose (presumably because such people know what they are getting into).

Large doses of radiation can cause reddening of the skin, drop in white-blood-cell count, and a large number of unpleasant symptoms such as nausea, fatigue, and loss of body hair. Such effects are sometimes referred to as radiation sickness. Large doses can also be fatal, although the time span of the dose is important. A short dose of 1000 rem is nearly always fatal. A 400-rem dose in a short period of time is fatal in 50 percent of the cases. However, the body possesses remarkable repair processes so that a 400-rem dose spread over several weeks is not usually fatal. It will nonetheless, cause considerable damage to the body.

SUMMARY

A *nuclear reaction* occurs when two nuclei collide and two or more other nuclei (or particles) are produced. In this process, as in radioactivity, *transmutation* (change) of elements occurs.

In *fission* a heavy nucleus such as uranium splits into two intermediate-sized nuclei after being struck by a neutron. ${}^{235}_{92}U$ is fissionable by slow neutrons, whereas some fissionable nuclei require fast neutrons. Much energy is released in fission because the mass of the heavy nucleus is greater than the total mass of its fission products. The fission process releases neutrons so that a *chain reaction* is possible. The *critical mass* is the minimum mass of fuel needed to sustain a chain reaction. In a *nuclear reactor* or nuclear bomb, a *moderator* is needed to slow down the released neutrons.

The *fusion* process, in which small nuclei combine to form larger ones, also releases energy. It has not yet been possible to build a fusion reactor for power generation because of the difficulty in containing the fuel long enough at the high temperature required.

Radiation can cause damage to biological tissue because it ionizes the atoms and molecules of the cell, and may cause mutations in the DNA, which in turn may lead to cancer. Radiation is also used to treat cancer. Radioactive tracers and imaging devices aid in the understanding and diagnosis of disease. It is important to quantify amounts of radiation; this is the subject of *dosimetry*. The curie (Ci) and the becquerel (Bq) are units that measure the activity or rate of decay of a source: $1\ \text{Ci} = 3.70 \times 10^{10}$ disintegrations per second whereas $1\ \text{Bq} = 1$ disintegration/s. The rad measures the amount of energy deposited per unit mass of absorbing material: 1 rad is the amount of radiation that deposits 10^{-2} J/kg of material. The rem = rad × QF where QF is the "quality factor" of a given type of radiation; 1 rem of any type of radiation does approximately the same amount of biological damage. The average dose received per person per year is about 0.20 rem.

QUESTIONS

1. Fill in the missing particles or nuclei:

$$ {}^{4}_{2}He + {}^{16}_{8}O \rightarrow {}^{19}_{10}Ne + ? $$

$$ {}^{4}_{2}He + {}^{16}_{8}O \rightarrow {}^{8}_{4}Be + ? $$

$$ {}^{2}_{1}H + {}^{2}_{1}H \rightarrow {}^{4}_{2}He + ? $$

$$ n + {}^{239}_{94}Pu \rightarrow {}^{141}_{54}Xe + {}^{97}_{40}Zr + ? $$

2. Fill in the blanks for the following reactions:

$$n + {}^{133}_{55}Cs \rightarrow ? + \gamma$$

$$n + {}^{133}_{55}Cs \rightarrow {}^{132}_{54}Xe + ?$$

$${}^{6}_{3}Li + {}^{7}_{4}Be \rightarrow {}^{4}_{2}He + ?$$

3. When ${}^{22}_{11}Na$ is bombarded by deuterons (${}^{2}_{1}H$) an α particle is emitted. What is the resulting nuclide?
4. Why are neutrons such good projectiles for producing nuclear reactions?
5. A proton strikes a ${}^{20}_{10}Ne$ nucleus, and an α particle is observed to come out. What is the residual nucleus? Write down the reaction equation.
6. Are fission fragments β^+ or β^- emitters?
7. If ${}^{235}_{92}U$ released only 1.5 neutrons per fission on the average (instead of about 2.5), would a chain reaction be possible? What would be different?
8. ${}^{235}_{92}U$ releases an average of 2.5 neutrons per fission compared to 2.7 for ${}^{239}_{94}Pu$. Pure samples of which of these two nuclei do you think would have the smaller critical mass?
9. Discuss how the course of history might have been changed if, during World War II, scientists had refused to work on developing a nuclear bomb. Do you think it would have been possible to delay the building of a bomb indefinitely?
10. Research in molecular biology is moving toward the ability to perform genetic manipulations on human beings. The moral implications of future discoveries along these lines has led to a warning that this may be the molecular biologists' "Hiroshima." Discuss.
11. What is the difference between a nuclear reactor and a nuclear bomb?
12. Why don't chain reactions occur in natural deposits of uranium ore?
13. Why must the fission process release neutrons if it is to be useful?
14. Discuss the relative merits and disadvantages, including pollution and safety, of power generation by fossil fuels, nuclear fission, and nuclear fusion.
15. What is the reason for the "secondary system" in Figure 30-6? That is, why is the water heated by the fuel in a nuclear reactor not used directly to drive the turbines?
16. The energy from nuclear fisson appears in the form of thermal energy—but the thermal energy of what?
17. Why would a porous block of uranium be more likely to explode if kept under water rather than in air?

18. Does $E = mc^2$ apply in (a) fission (b) fusion (c) nuclear reactions?
19. What breeds what in a breeder reactor?
20. Can matter be created or destroyed?
21. Light energy emitted by the sun and stars comes from the fusion process. What conditions in the interior of stars makes this possible?
22. What is the basic difference between fission and fusion?
23. How might radioactive tracers be used to locate a leak in a pipe?
24. How might radioactive tracers be used to measure (a) engine wear and (b) tire wear?
25. Why is the recommended maximum radiation dose higher for women beyond the child-bearing age than for younger women?

EXERCISES

1. When a $^{238}_{92}\text{U}$ nucleus undergoes fission, approximately 200 MeV of energy is released. What is the mass difference between the uranium nucleus and all the fission products?
2. How many fissions take place per second in a 25-MW reactor? Assume that 200 MeV is released per fission.
3. What is the energy released in the fission reaction $n + ^{235}_{92}\text{U} \rightarrow ^{141}_{56}\text{Ba} + ^{92}_{36}\text{Kr} + 3n$? (The masses of $^{141}_{56}\text{Ba}$ and $^{92}_{36}\text{Kr}$ are 140.9141 u and 91.9250 u, respectively, and that of $^{235}_{92}\text{U}$ is 235.0439.
4. Suppose that the average power consumption, day and night, of an average house is 300 W. What mass of $^{235}_{92}\text{U}$ would have to undergo fission to supply the electrical needs of such a house for a year? (Assume 200 MeV is released per fission.)
5. If an average house requires 300 W of electric power on average, how much deuterium would have to be used in a year to supply these electrical needs? Assume the second reaction of Section 3.
6. In the so-called carbon cycle that occurs in the sun, $^{4}_{2}\text{He}$ is built from four protons starting with $^{12}_{6}\text{C}$. First $^{12}_{6}\text{C}$ absorbs a proton to form nucleus X_1. X_1 decays by β^+ emission to X_2. X_2 absorbs a proton to become X_3, which itself absorbs a proton to become X_4. X_4 decays to X_5 by β^+ decay and X_5 reacts via $X_5(p, \alpha)X_6$. Determine the intermediate nuclei, write out each step in detail, and show that X_6 is again $^{12}_{6}\text{C}$, which is thus not used up in the process.
7. A 0.018-μCi sample of $^{32}_{15}\text{P}$ is injected into an animal for tracer studies. If a Geiger counter intercepts 20 percent of the emitted β particles and is 90 percent efficient in counting them, what will be the counting rate?

8. An average adult body has about 0.10 μCi of $^{40}_{19}K$, which comes from food. How many decays occur per second?

9. A dose of 500 rem of γ rays in a short period would be lethal to about half the people subjected to it. How many rads is this?

10. Fifty rads of α-particle radiation is equivalent to how many rads of X rays in terms of biological damage?

11. How many rads of slow neutrons will do as much biological damage as 50 rads of fast neutrons?

12. How much energy is deposited in the body of a 70-kg adult exposed to a 50-rad dose?

13. A 1.0-mCi source of $^{32}_{15}P$ (in $NaHPO_4$), a β^- emitter, is implanted in an organ where it is to administer 5000 rads. The half-life of $^{32}_{15}P$ is 14.3 days and 1 mCi delivers about 1 rad/min. Approximately how long should the source remain implanted?

14. In a certain town the average yearly background radiation consists of 25 mrad of X and γ rays plus 3.0 mrad of particles having a QF of 10. How many rems will a person receive per year on the average?

*15. Show that the energy released in the fusion reaction $^2_1H + ^3_1H \rightarrow ^4_2He + n$ is 17.6 MeV. The masses are ^2_1H(2.01410 u), ^3_1H(3.01605 u), ^4_2He(4.00260 u), n(1.00867 u).

*16. Calculate the energy release per gram of fuel for the reaction in Exercise 15. Compare to the energy release per gram of uranium in the fission process.

*17. In the fusion reaction

$$^2_1H + ^2_1H \rightarrow ^3_2He + n,$$

the energy release is 3.27 MeV. Determine the mass of the ^3_2He nucleus. (See Exercise 15 for masses.)

*18. What was the mass loss of $^{235}_{92}U$ actually fissioned in the first atomic bomb, whose energy was the equivalent of 20 kilotons of TNT (1 kiloton of TNT releases 5×10^{12} J)?

*19. A shielded γ-ray source yields a dose rate of 0.050 rad/hr at a distance of 1.0 m for an average-sized person. If workers are allowed a maximum dose rate of 5.0 rem/yr, how close to the source may they operate assuming a 40-hr work week? Assume that the intensity of radiation falls off as the square of the distance. (It actually falls off more rapidly than $1/r^2$ because of absorption in the air, so the preceding answer will give a better-than-permissible value.)

ELEMENTARY PARTICLES

CHAPTER

31

In this chapter we examine the exciting subject of *elementary particle* physics, which represents the human endeavor to understand the basic building blocks of all matter.

In the years following World War II, it was found that if the incoming particle in a nuclear reaction has sufficient energy, new types of particles can be produced. In order to produce high-energy particles, physicists have constructed various types of particle accelerators. Most commonly they accelerate protons or electrons, although heavy ions can also be accelerated. These high-energy accelerators have been used to probe the nucleus more deeply, to produce and study new particles, and to give us information about the basic forces and constituents of nature.

1. PARTICLE ACCELERATORS

Particle accelerators fall into two basic types: those that accelerate particles in a circular path (based on the cyclotron principle) and those that accelerate the particles along a straight line (linear accelerators).

The Cyclotron and the Synchrotron

The *cyclotron* was developed in 1930 by E. O. Lawrence (1901–1958) at the University of California at Berkeley. It uses a magnetic field to maintain the ions—usually protons—in nearly circular paths (Chapter 20). The protons move within two D-shaped cavities, as shown in Figure 31-1. Each time they pass into the gap between the "dees," a voltage is applied that accelerates them. This increases their speed and also increases the radius of curvature of their path. After many revolutions, the protons acquire high kinetic energy and reach the outer edge of the cyclotron. They then either strike a target placed inside the cyclotron or leave the cyclotron

FIGURE 31-1
Diagram of a cyclotron.
The magnetic field, applied by a large electromagnet, points out of the paper.

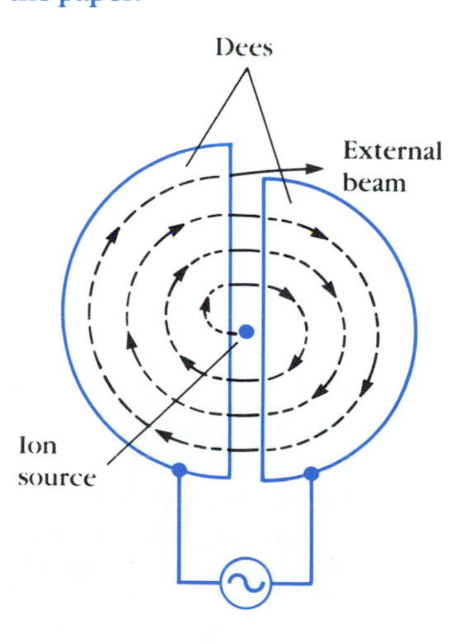

with the help of carefully placed "bending magnets", and are directed to an external target.

The voltage applied to the dees to produce the acceleration must be alternating. When the protons are moving to the right across the gap in Figure 31-1, the right dee must be negative and the left one positive. A half-cycle later the protons are moving to the left, so the left dee must be negative in order to accelerate them. The frequency of the applied voltage must be equal to that of the circulating protons; as long as the protons are not traveling at too high a speed, the required frequency remains constant as the particles are accelerated. Unfortunately, this is only true at nonrelativistic energies. For at higher speeds the mass of the ions will increase according to Einstein's formula, $m = m_0/\sqrt{1-v^2/c^2}$, where m_0 is the rest mass. Because of the increased inertia, as the particles increase in speed the frequency of the applied voltage must be reduced. To achieve large energies, complex electronics is needed to decrease the frequency as a packet of protons increases in speed and reaches larger orbits. Such a modified cyclotron is called a *synchrocyclotron*.

Another way to deal with the increase in mass accompanying speed is to increase the magnetic field B as the particles speed up. Such a device is called a *synchrotron*. The largest circular accelerators today fall into this category, and they are enormous. The Fermi National Accelerator Laboratory (Fermilab) at Batavia, Illinois has a radius of 1.0 km and that at CERN (European Center for Nuclear Research) in Geneva, Switzerland is 1.1 km in radius. They can presently accelerate protons to 500 GeV, and Fermilab is expected to soon reach 1000 GeV using superconducting magnets (it is now called the *Tevatron*, after the unit TeV: 1 TeV = 1000 GeV = 10^{12} eV). These large synchrotrons do not use enormous magnets 1 km in radius. Instead, they use a narrow ring of magnets (see Figure 31-2) which are all placed at the same radius from the center of the circle. The magnets are interrupted by gaps where high voltage accelerates the ions. After the ions are injected, they must then be maintained in motion in a circle of constant

(a)

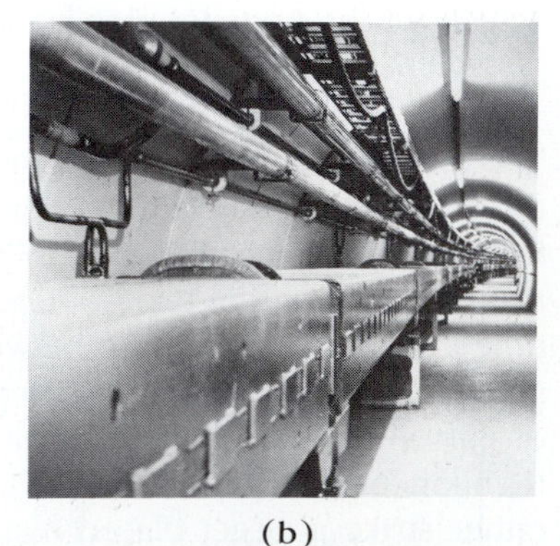

(b)

FIGURE 31-2
(a) Aerial view of Fermilab at Batavia, Illinois; the accelerator is a circular ring 2.0 km in diameter. (b) Interior photograph, showing magnets that keep ions moving in a circle.

radius. This is accomplished by giving them considerable energy initially in a much smaller accelerator and then slowly increasing the magnetic field as they speed up in the large synchrotron.

One problem with any accelerator is that accelerating electric charges radiate electromagnetic energy (see Chapter 21). Since ions or electrons are accelerated in all accelerators, we can expect considerable energy to be lost by radiation. The effect increases with speed, and is especially significant in circular machines where centripetal acceleration is present. However, it is primarily a concern in synchrotrons, and hence is called *synchrotron radiation*. Synchrotron radiation can, nevertheless, be useful. Since it consists of photons of a wide range of frequencies (typically UV and X rays), it can be used in experiments—or for treatment—where intense beams of photons are needed.

Linear Accelerator

A linear accelerator is a device in which particles are accelerated many times along a straight-line path. Figure 31-3 is a diagram of a simple "linac". The ions pass through a series of tubular conductors. The voltage applied to the tubes must be alternating so that when the ions reach a gap, the tube in front of them is negative and the one they just left is positive. This assures that they are accelerated at each gap. As the ions increase in speed, they cover a greater distance in the same amount of time. Consequently, the tubes must be longer the farther they are from the source.

Linear accelerators are of particular importance for accelerating electrons. Because of their small mass, electrons reach high speeds very quickly. (Indeed, an electron linac such as the one shown in Figure 31-3 would have tubes nearly equal in length, since the electrons would be traveling close to $c = 3.0 \times 10^8$ m/s for almost the entire distance.)

The largest electron linear accelerator is the Stanford Linear Accelerator Center (SLAC) at Stanford University. It is over 3 km (2 miles) long and can accelerate electrons to over 20 GeV.

Colliding Beams

The usual arrangement in *high-energy-physics* experiments (as this field in which high-energy accelerators are used is called) is to allow the beam of particles from an accelerator to strike a stationary target. An important way

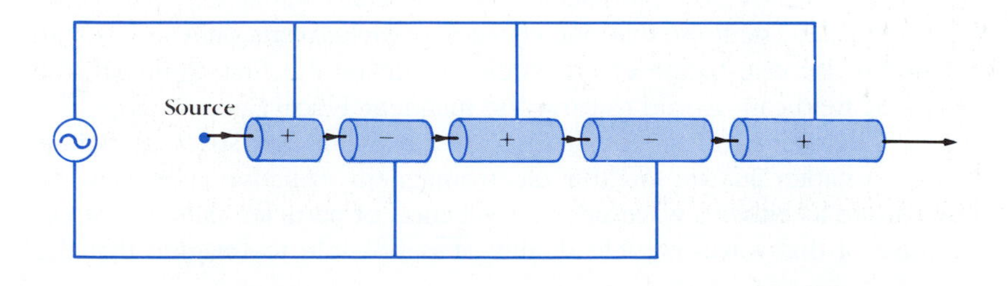

FIGURE 31-3
Simple linear accelerator.

to increase the energy of a collision is the recent development of *colliding beams*—that is, both the target particles and the projectile particles are moving. This can be accomplished with only one accelerator through the use of *storage rings*. The accelerator accelerates one type of particle (say electrons or protons) to a maximum energy and then magnets are used to steer these particles into one circular storage ring where the particles can continue to circulate for many hours. It then accelerates a second type of particle (say positrons, or it could be the same as the first type, such as protons), and these are sent to a second storage ring. The two storage rings overlap at several locations and the two beams, circulating in opposite directions, collide head-on at these intersecting points.

Storage rings for colliding-beam experiments are in use at a number of facilities around the world and continue to play an important role in recent advances in elementary particle physics.

2. PARTICLE EXCHANGE AND THE YUKAWA PARTICLE

By the mid 1930s it was recognized that all atoms can be considered to be made up of neutrons, protons, and electrons. The basic constituents of the universe were no longer considered to be atoms but rather the proton, neutron, and electron. Besides these three *elementary particles*, as they could be called, there were several others also known: the positron (a negative electron), the neutrino, and the γ particle (or photon), for a total of six elementary particles.

Looking back, things seemed fairly simple in 1935. But in the decades that followed, hundreds of other subnuclear particles were discovered. The properties and interactions of these particles, and which ones should be considered as fundamental or "elementary," became the substance of research in the field of **elementary particle physics**.

Elementary particle physics, as it exists today, can be said to have begun in 1935 when the Japanese physicist Hideki Yukawa (1907–1981) predicted the existence of a new particle that would in some way mediate the strong nuclear force. To understand Yukawa's idea, we first look at the electromagnetic force. When we first discussed electricity we saw that the electric force acts over a distance and without contact. To better perceive how a force can act over a distance, we saw that Faraday introduced the idea of a *field*. The force that one charged particle exerts on a second can be said to be due to the electric field set up by the first. Similarly, the magnetic field can be said to carry the magnetic force. Later (Chapter 21) we saw that electromagnetic fields can travel through space as waves. Finally (Chapter 26) we saw that electromagnetic radiation (light) can be considered as either a wave or as a collection of particles called photons. Because of this wave–particle duality, it is possible to imagine that the

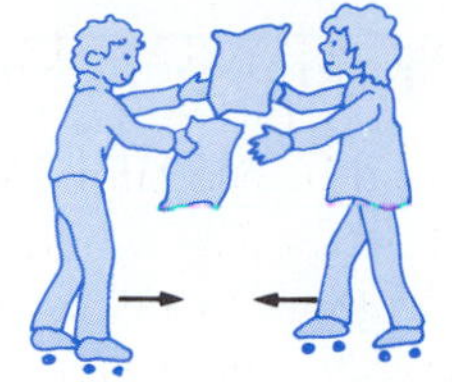

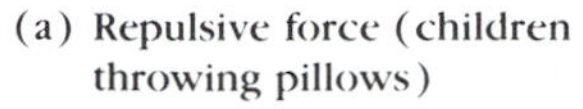

(b) Attractive force (children grabbing pillows)

FIGURE 31-4
Forces equivalent to particle exchange. (a) Repulsive force (children throwing pillows). (b) Attractive force (children grabbing pillows).

electromagnetic force between charged particles is due (1) to the EM field set up by one charge and felt by the other, or (2) to an exchange of photons or γ-particles between them. It is (2) that we concentrate on here, and an example of how an exchange of particles could give rise to a force is illustrated in the simplistic picture of Figure 31-4. In part (a) two children start throwing pillows at each other; each catch results in the child being moved backward by the impact. This is the equivalent of a repulsive force. On the other hand, if the two children exchange pillows by grabbing them out of the other's hand, they will be pulled toward each other as when an attractive force acts. For the electromagnetic force between two charged particles, it is the photon exchange between the two particles that gives rise to the force. A simple diagram describing this is shown in Figure 31-5. Such a diagram is called a *Feynman diagram*, and the theory on which it is based is called *quantum electrodynamics* (QED). The case shown is the simplest in which a single photon is exchanged. One of the charged particles emits the photon and recoils somewhat as a result; the second particle absorbs the photon. In any such collision, or *interaction*, energy and momentum are transferred from one particle to the other, and it is carried by the photon. Because the photon is absorbed by the second particle very shortly after it is emitted by the first, it is not observable and it is referred to as being a *virtual* photon, as compared to one that is free and can be detected by instruments.

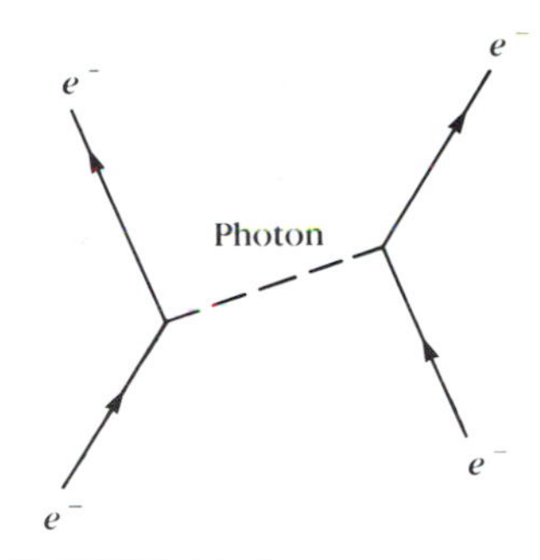

FIGURE 31-5
Feynman diagram, showing how a photon acts as carrier of electromagnetic force between two electrons.

Now to Yukawa's prediction. Drawing an analogy from photon exchange to mediate the electromagnetic force, Yukawa argued that there ought to be a particle that mediates the strong nuclear force—the force that holds nucleons together in the nucleus. Just as the photon is called the quantum of the electromagnetic field or force, so the Yukawa particle would represent the quantum of the strong nuclear force. Yukawa predicted that this new particle would have a mass intermediate between that of the electron and the proton. Hence it was called a *meson*, meaning "in the middle." Figure 31-6 is a Feynman diagram of meson exchange.

Just as photons can be observed as free particles as well as acting in an exchange, so it was expected that mesons might be observed directly. Such a meson was searched for in the cosmic radiation that enters the earth's atmosphere from the sun and other sources in the universe. It was finally

FIGURE 31-6
Meson exchange when proton and neutron interact via strong nuclear force.

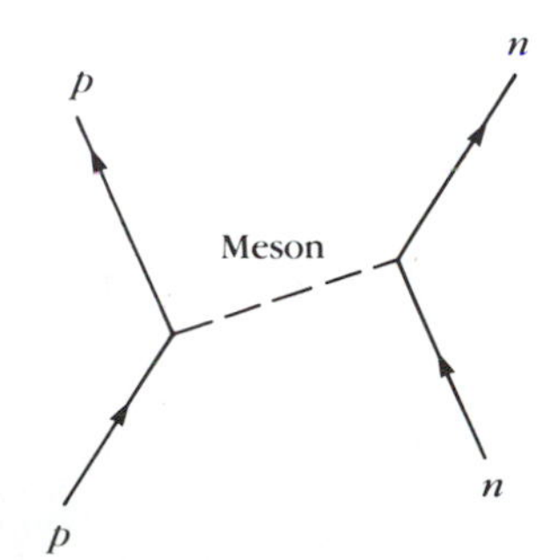

TYPE	RELATIVE STRENGTH (APPROX.)	FIELD PARTICLE
Strong nuclear	1	Mesons/gluons[‡]
Electromagnetic	10^{-2}	Photon
Weak nuclear	10^{-13}	$W^{\pm}$ and Z^0
Gravitational	10^{-40}	Graviton (?)

TABLE 31-1 The Four Forces in Nature

[‡] See Section 8.

found in 1947, and is called the π or pi meson, or simply the *pion*. It comes in three charge states: +, −, or 0. The π^+ and π^- have a mass of 139.6, MeV/c^2 and the π^0 a mass of 135.0 MeV/c^2, close to what had been predicted for it theoretically. All three interact strongly with matter. Soon after their discovery in cosmic rays, pions were produced in the laboratory using a particle accelerator. Reactions observed included

$$p + p \rightarrow p + p + \pi^0$$

$$p + p \rightarrow p + n + \pi^+.$$

The incident proton from the accelerator must have sufficient energy to produce the additional mass of the pion.

A number of other mesons were discovered in subsequent years which were also considered to mediate the strong nuclear force. (The recent theory of quantum chromodynamics, involving quarks, has replaced mesons with gluons as exchange particles for the strong force—see Section 7.)

So far we have discussed the particles that mediate the electromagnetic and strong nuclear forces. But there are four known types of force—or interaction—in nature. What about the other two: the weak nuclear force and gravity? Theorists believe that these are also mediated by particles. The particles presumed to transmit the weak force are referred to as the W^+, W^-, and Z^0. In 1983, after extensive searches, the long-awaited discovery of the W and Z particles was announced by Carlo Rubbia[†], the scientist behind the building of the proton-antiproton collider at the very high energy accelerator at CERN and leader of the large group of scientists (over 100) who were also involved in the discovery. The quantum of the gravitational force, called the *graviton*, has not yet been identified. A comparison of the four forces is given in Table 31-1, where they are listed according to their (approximate) relative strengths. Notice that although gravity may be the most obvious force in daily life (because of the huge mass of the earth), on a nuclear scale it is much the weakest of the four forces and its effect at the nuclear level can nearly always be ignored.

[†] Rubbia and Simon van der Meer were awarded the Nobel Prize in Physics in 1984 for this work.

3. ANTIPARTICLES

The positron, as we saw in Chapter 29, is basically a positive electron. That is, many of its properties are the same as for the electron, such as mass, but it has the opposite charge. The positron is said to be the **antiparticle** to the electron. After the positron was discovered in 1932, it was predicted that other particles should also have antiparticles. In 1955 the antiparticle to the proton was found, the *antiproton* ($\bar{p}$); see Figure 31-7. (The bar over the p is used to indicate antiparticle.) A large amount of energy was needed to produce this massive particle (mass = proton's mass). Its discovery (by E. Segré and O. Chamberlain) was made only after the completion of the large accelerator (the Bevatron) at the University of California at Berkeley. Soon after, the **antineutron** ($\bar{n}$) was found. Most other particles also have antiparticles. But the photon, the π^0, and a few other particles do not have distinct antiparticles—or we say that they are their own antiparticles.

Antiparticles are produced in nuclear reactions when there is sufficient energy available, and they do not live very long in the presence of matter. For example, when a positron encounters an electron, the two annihilate each other. The energy of their vanished mass, plus any kinetic energy they possess, is converted to the energy of γ rays or of other particles. Annihilation occurs for all other particle-antiparticle pairs. Far out in space

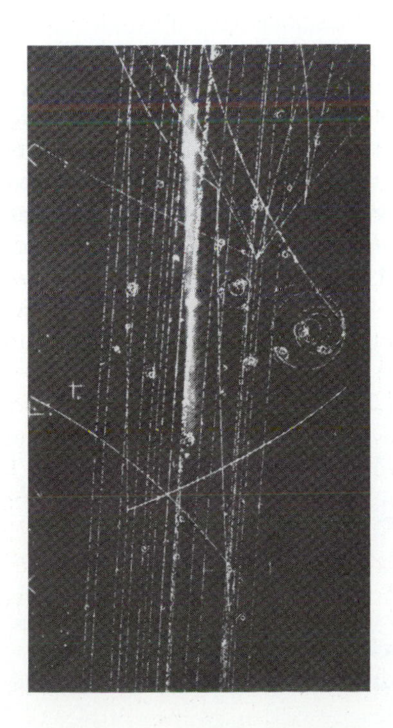

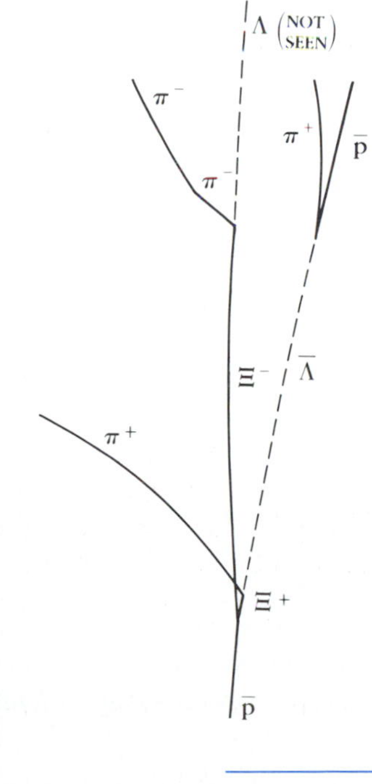

FIGURE 31-7
Liquid-hydrogen bubble-chamber photograph of an antiproton ($\bar{p}$) colliding with a proton, producing a hyperon pair ($\bar{p} + p \rightarrow \Xi^- + \Xi^+$), which subsequently decays into other particles. The key indicates the assignment of a particle to each track. Neutral-particle paths are shown by dashed lines because neutral particles produce no bubbles.

there may be **antimatter**—matter whose atoms are made up of antiparticles. Nuclei would consist of antineutrons and negatively charged antiprotons, and they would be encircled by positrons instead of electrons. These *antiatoms* would be much like ordinary atoms—they could combine chemically, emit photons, and so on. But if an antiworld made of antimatter approached our world of ordinary matter, annihilation would take place and a vast explosion would result.

4. CONSERVATION LAWS

One of the important uses of high-energy accelerators is to study the interactions of elementary particles. As a means of understanding this subnuclear world, the conservation laws are indispensable. The laws of conservation of energy, of momentum, of angular momentum, and of electric charge are found to hold precisely in all particle interactions.

A study of particle interactions has revealed a number of new conservation laws, some of which we now discuss. These new conservation laws (just like the old ones) are ordering principles: they help to explain why some reactions occur and others do not. For example, the following reaction has *never* been found to occur:

$$p + n \nrightarrow p + p + \bar{p}$$

even though charge, energy, and so on are conserved ($\bar{p}$ means an antiproton and $\nrightarrow$ means the reaction does not occur). To understand why it doesn't occur, physicists hypothesized a new conservation law, the conservation of **baryon number**. (Baryon number is the same as nucleon number, which we saw in Chapters 29 and 30 is conserved in nuclear reactions.) An important addition to this law is the proposal that whereas all nucleons have baryon number $B = +1$, all antinucleons (antiprotons, antineutrons) have $B = -1$. The reaction above does not conserve baryon number since on the left side we have $B = (+1) + (+1) = +2$ and on the right $B = (+1) + (+1) + (-1) = +1$. On the other hand, the following reaction does conserve B and *does* occur if the incoming proton has sufficient energy:

$$p + n \rightarrow p + n + \bar{p} + p$$

$$B = +1 + 1 = +1 + 1 - 1 + 1.$$

As indicated, $B = +2$ on both sides of this equation. From these and other reactions, the conservation of baryon number has been established as a basic law of physics.

Several other "number" conservation laws have been discovered, in particular the "lepton numbers" which are associated with the weak nuclear force and so-called *leptons* (including the electron and its neutrino—see the following section). Recent theoretical work suggests that the conservation laws of baryon and lepton numbers may be only approximate, rather than exact, as we shall discuss in Section 9.

5. PARTICLE CLASSIFICATION

In the decades following the discovery of the π meson in the late 1940s, a great many other subnuclear particles were discovered; today they number in the hundreds. Much theoretical and experimental work has been done to try to understand this multitude of particles. One important aid to understanding is to arrange the particles in categories according to their properties. One way of doing this is according to their interactions. Since not all particles take part in all four of the forces known in nature, this fact is used as a classification scheme. Table 31-2 lists many of their

TABLE 31-2 Elementary Particles (Stable or "long-lived")‡

CATEGORY	PARTICLE NAME	SYMBOL	ANTIPARTICLE	REST MASS (MeV/c^2)	B	S	LIFETIME (s)
Photon	Photon	γ	Self	0	0	0	Stable
Leptons	Electron	e^-	e^+	0.511	0	0	Stable
	Neutrino (e)	ν_e	$\bar{\nu}_e$	0(?)	0	0	Stable
	Muon	μ^-	μ^+	105.7	0	0	2.20×10^{-6}
	Neutrino (μ)	ν_μ	$\bar{\nu}_\mu$	0(?)	0	0	Stable
	Tau	τ^-	τ^+	1748.	0	0	$<4 \times 10^{-13}$
	Neutrino (τ)	ν_τ	$\bar{\nu}_\tau$	0(?)	0	0	Stable
Hadrons							
Mesons	Pion	π^+	π^-	139.6	0	0	2.60×10^{-8}
		π^0	Self	135.0	0	0	0.83×10^{-16}
	Kaon	K^+	K^-	493.7	0	+1	1.24×10^{-8}
		K_S^0	$\overline{K_S^0}$	497.7	0	+1	0.89×10^{-10}
		K_L^0	$\overline{K_L^0}$	497.7	0	+1	5.2×10^{-8}
Baryons	Proton	p	$\bar{p}$	938.3	+1	0	Stable‡‡
	Neutron	n	$\bar{n}$	939.6	+1	0	920
	Lambda	Λ^0	$\bar{\Lambda}^0$	1115.6	+1	−1	2.6×10^{-10}
	Sigma	Σ^+	$\bar{\Sigma}^-$	1189.4	+1	−1	0.80×10^{-10}
		Σ^0	$\bar{\Sigma}^0$	1192.5	+1	−1	6×10^{-20}
		Σ^-	$\bar{\Sigma}^+$	1197.3	+1	−1	1.5×10^{-10}
	Xi	Ξ^0	$\bar{\Xi}^0$	1315	+1	−2	2.9×10^{-10}
		Ξ^-	Ξ^+	1321	+1	−2	1.64×10^{-10}
	Omega	Ω^-	Ω^+	1672	+1	−3	0.82×10^{-10}

‡ See also Table 31-3.
‡‡ $>2 \times 10^{32}$ yr.

properties. The *photon* takes part only in the electromagnetic force, and it is in a class by itself. The **leptons** are those particles that do not interact via the strong force but do interact via the weak nuclear force (as well as the much weaker gravitational force); those that carry electric charge also interact via the electromagnetic force. The leptons include the electron, the muon (or μ, discovered in 1937 and much like an electron but 207 times heavier), and the tau (or τ, discovered in 1976 and more than 3000 times heavier than the electron), plus three types of neutrino: the electron neutrino (ν_e), the muon neutrino (ν_μ), and the tau neutrino (ν_τ). Just as in beta decay an electron is emitted along with a neutrino (an "electron neutrino"), so in other decays and interactions, the muon and tau leptons are always accompanied by their own type of neutrino. Thus we have evidence of six leptons. Each of these six has its antiparticle.

The third category of particle is the **hadron**. Hadrons are those particles that can interact via the strong nuclear force. Hence they are said to be **strongly interacting particles**. They also interact via the other forces, but the strong force predominates at short distances. The hadrons include nucleons, pions, and a large number of other particles. They are divided into two subgroups:† **baryons**, which are those particles that have baryon number $+1$ (or -1 in the case of their antiparticles), and **mesons**, which have baryon number $= 0$.

Notice that the baryons Λ, Σ, Ξ and Ω all decay to lighter-mass baryons, and eventually to a proton or neutron. All these processes conserve baryon number. Since there is no lighter particle than the proton with $B = +1$, if baryon number is strictly conserved, the proton itself cannot decay and is stable (but see Section 9).

The particles listed in Table 31-2 are those that are either stable or are rather long-lived (lifetime $\gtrsim 10^{-19}$ s). The lifetime of an unstable particle depends on which force is most active in causing the decay. When we say the strong nuclear force is stronger than the electromagnetic force, we mean that two particles will interact more quickly and more frequently if this force is acting. When a stronger force influences a decay, that decay occurs more quickly. Decays caused by the weak force have lifetimes of 10^{-10} s or longer. Particles that decay via the electromagnetic force have much shorter lifetimes, typically about 10^{-16} to 10^{-19} s. The unstable particles listed in Table 31-2 decay either via the weak or the electromagnetic interaction; those that involve a γ (photon) are electromagnetic, and the others listed decay via the weak interaction (note the lifetimes). Many particles have been found that can decay via the strong interaction. (These are not listed in Table 31-2.) Such particles decay into other strongly interacting particles (say, n, p, π but not involving γ, e, and so on)

† Originally, particles were divided according to their mass into leptons (meaning "light" particles), baryons (meaning "heavy"), and those of intermediate mass, the mesons (meaning "middle"). The newer classification according to their interactions is almost consistent with this. The muon, however, is now called a lepton (it doesn't interact strongly), although it was once called the mu meson because of its mass.

and their lifetimes are very short, typically 10^{-23} s. In fact their lifetimes are so short that they do not travel far enough to be detected before decaying. Their decay products can be detected, however, and it is from them that the existence of such short-lived particles is inferred.

6. STRANGE PARTICLES

In the early 1950s, certain of the newly found particles, namely the K, Λ, and Σ, were found to behave rather strangely in two ways. First, they were always produced in pairs. For example, the reaction

$$\pi^- + p \rightarrow K^0 + \Lambda^0$$

occured with high probability, but the reaction $\pi^- + p \rightarrow K^0 + n$ was never observed to occur; this seemed strange since no known conservation law would have been violated and there was plenty of energy available. The second feature of these *strange particles* (as they came to be called) was that although they were clearly produced via the strong interaction (that is, at a high rate), they did not decay at a rate characteristic of the strong interaction and they decayed only into strongly interacting particles (for example $K \rightarrow 2\pi$, $\Sigma^+ \rightarrow p + \pi^0$). Instead of lifetimes of 10^{-23} s as expected for strongly interacting particles, strange particles have lifetimes of 10^{-10} to 10^{-8} s, which are characteristic of the weak interaction.

To explain these observations, a new quantum number, **strangeness**, and a new conservation law, *conservation of strangeness*, were introduced. By assigning the strangeness numbers (S) indicated in Table 31-2, the production of strange particles in pairs was readily explained. Antiparticles were assigned opposite strangeness from their particles: one of each pair was assigned $S = +1$ and the other $S = -1$. For example, in the reaction $\pi^- + p \rightarrow K^0 + \Lambda^0$, the initial state has strangeness $S = 0 + 0 = 0$ and the final state has $S = +1 - 1 = 0$, so strangeness is conserved. But for $\pi^- + p \nrightarrow K^0 + n$, the initial state has $S = 0$ and the final state has $S = +1 + 0 = +1$, so strangeness would not be conserved and the reaction isn't observed.

To explain the decay of strange particles, it is assumed that strangeness is conserved in the strong interaction but is *not* conserved in the weak interaction. Thus, although the strange particles were forbidden by strangeness conservation to decay to lower mass nonstrange particles via the strong interaction, they could undergo such decay by means of the weak interaction. And because this would occur much more slowly, it would account for their longer lifetimes of 10^{-10} to 10^{-8} s.

The conservation of strangeness was the first example of a "partially conserved" quantity. In this case, the quantity strangeness is conserved by strong interactions but not by weak.

7. QUARKS

Except for the photon, observed particles fall into one of two groups: leptons or hadrons. The principal difference between these two groups is that the hadrons interact via the strong interaction whereas the leptons do not. Another important difference that physicists had to deal with in the 1960s was that there were only four known leptons (e^-, μ^-, ν_e, ν_μ; the τ and ν_τ were not yet discovered) but there are well over a hundred hadrons.

The leptons are considered to be truly elementary particles since they do not seem to break down into smaller entities, do not show any internal structure, and have no measurable size. (Attempts to determine the size of leptons have put an upper limit of about 10^{-18} m.)

The hadrons, on the other hand, are more complex. Experiments indicate they do have an internal structure. And the fact that there are so many of them suggests that they can't all be elementary. To deal with this problem, M. Gell-Mann and G. Zweig in 1963 independently proposed that none of the hadrons so far observed is elementary. Instead, they proposed that the hadrons are made up of combinations of three, more fundamental, point-like entities called **quarks**.† Quarks, then, would be considered truly elementary particles, like leptons. The three quarks were labeled *u*, *d*, *s*, and given the names *up*, *down*, and *sideways* (or, more commonly now, *strange*). They were assumed to have fractional charge: $\frac{1}{3}$ or $\frac{2}{3}$ the charge on the electron—that is, less than the previously thought smallest charge. Other properties of quarks and antiquarks are indicated in Table 31-3. All hadrons known at the time could be constructed in theory from these three types of quark. Mesons would consist of a quark–antiquark pair. For example, a π^+ meson is considered a $u\bar{d}$ pair (note that for the $u\bar{d}$ pair, $Q = \frac{2}{3}e + \frac{1}{3}e = +1e$, $B = \frac{1}{3} - \frac{1}{3} = 0$, $S = 0 + 0 = 0$, as it must for a π^+). On the other hand a $K^+ = u\bar{s}$ with $Q = +1, B = 0, S = +1$. Baryons, on the other hand, would consist of three quarks; for example a neutron is $n = ddu$ whereas an antiproton is $\bar{p} = \bar{u}\bar{u}\bar{d}$.

Soon after the quark theory was proposed, physicists began looking for these fractionally charged particles. Although there is indirect experimental evidence in favor of their existence, their direct detection remains elusive. Indeed, there are suggestions that quarks are so tightly bound together that they may not ever exist in the free state.

In 1964, several physicists proposed that there ought to be a fourth quark. Their argument was based on the expectation that there exists a deep symmetry in nature, including a connection between quarks and leptons. If there are four leptons (as was thought in the 1960s), then symmetry in nature would suggest there should also be four quarks. The fourth quark was said to be *charmed*; its charge would be $+\frac{2}{3}e$ and it

---

† Gell-Mann chose the word from the phrase "Three quarks for Muster Mark" in James Joyce's *Finnegan's Wake*.

TABLE 31-3 Properties of Quarks and Antiquarks

QUARKS

Name	Symbol	Spin	Charge	Baryon Number	Strangeness	Charm	Bottomness	Topness
Up	u	$\frac{1}{2}$	$+\frac{2}{3}e$	$\frac{1}{3}$	0	0	0	0
Down	d	$\frac{1}{2}$	$-\frac{1}{3}e$	$\frac{1}{3}$	0	0	0	0
Strange	s	$\frac{1}{2}$	$-\frac{1}{3}e$	$\frac{1}{3}$	-1	0	0	0
Charmed	c	$\frac{1}{2}$	$+\frac{2}{3}e$	$\frac{1}{3}$	0	$+1$	0	0
Bottom	b	$\frac{1}{2}$	$-\frac{1}{3}e$	$\frac{1}{3}$	0	0	$+1$	0
Top	t	$\frac{1}{2}$	$+\frac{2}{3}e$	$\frac{1}{3}$	0	0	0	$+1$

ANTIQUARKS

Name	Symbol	Spin	Charge	Baryon Number	Strangeness	Charm	Bottomness	Topness
Up	$\bar{u}$	$\frac{1}{2}$	$-\frac{2}{3}e$	$-\frac{1}{3}$	0	0	0	0
Down	$\bar{d}$	$\frac{1}{2}$	$+\frac{1}{3}e$	$-\frac{1}{3}$	0	0	0	0
Strange	$\bar{s}$	$\frac{1}{2}$	$+\frac{1}{3}e$	$-\frac{1}{3}$	$+1$	0	0	0
Charmed	$\bar{c}$	$\frac{1}{2}$	$-\frac{2}{3}e$	$-\frac{1}{3}$	0	-1	0	0
Bottom	$\bar{b}$	$\frac{1}{2}$	$+\frac{1}{3}e$	$-\frac{1}{3}$	0	0	-1	0
Top	$\bar{t}$	$\frac{1}{2}$	$-\frac{2}{3}e$	$-\frac{1}{3}$	0	0	0	-1

would have another property to distinguish it from the other three quarks. This new property, or quantum number, was called **charm** (see Table 31-3). Charm was assumed to be like strangeness: it would be conserved in strong and electromagnetic interactions but would not be conserved by the weak. The new charmed quark would have charm $C = +1$ and its antiquark $C = -1$.

The first evidence of charm came with the discovery in 1974 of the J/ψ particle, which was interpreted to be a combination of a charmed quark and its antiquark ($c\bar{c}$). The J/ψ has no net charm (the c quark has $C = +1$, $\bar{c}$ has $C = -1$). But shortly thereafter (1977) the D meson was found which does have charm ($C = +1$). More recent experiments suggest that charmed baryons also exist.

About the same time, strong evidence of a fifth lepton, the tau (τ), appeared. This lepton, like the electron and muon, presumably has a neutrino associated with it. Thus the family of leptons is at present believed to have six members. This would upset the balance between leptons and quarks, the presumed basic building blocks of matter, unless two new quarks also exist. Indeed, theoretical physicists have postulated the existence of a fifth and sixth quark. These have been named *top* and *bottom* quarks, since they resemble the "up" and "down" quarks. (Some physicists prefer the names *truth* and *beauty* for these t and b quarks.) The names apply also to the new properties (quantum numbers) that distinguish the new quarks from the old quarks. (These are included in Table 31-3.) Indeed, another new meson, Υ, has been detected, which is considered to be a $b\bar{b}$ combination. And in 1983 the B meson was observed which has "bottomness" (consisting of only one b quark plus a non-b quark). Evidence of the top (t) quark finally appeared in 1984 from Rubbia's group at CERN.

8. THE "STANDARD MODEL"

Not long after the quark theory was proposed, it was suggested that quarks have another property (or quality) called **color**. The distinction between the five or six quarks (u, d, s, c, b, t) was referred to as **flavor**. According to theory, each of the flavors of quark can have three colors, usually designated red, green, and blue. (Note that the names "color" and "flavor" have nothing to do with our senses but are purely whimsical—as are other names, like charm, in this new field.) The antiquarks are colored antired, antigreen, and antiblue. Baryons are made up of three quarks, one of each color; mesons consist of a quark–antiquark pair of a particular color and its anticolor. Thus baryons and mesons are white or colorless.

Originally the idea of quark color was proposed to preserve the Pauli exclusion principle (Chapter 28), which applies to particles having spin $\frac{1}{2}$ such as electrons and nucleons. Since quarks have spin $\frac{1}{2}$, they ought to obey the exclusion principle; yet for three particular baryons (*uuu*, *ddd*, *sss*), there was no way all three could be in different quantum states. This would seem to violate the exclusion principle. But if quarks have an additional quantum number (color), which could be different for each quark, it would serve to distinguish them and the exclusion principle would hold. Although quark color, and the resulting increase in number of quarks (threefold), was thus originally an *ad hoc* idea, it also served to bring the theory into better agreement with experiment—such as predicting the correct lifetime of the π^0 meson.

In addition, the idea of color soon became a central feature of the theory as determining the force binding quarks together in a hadron. Each quark is assumed to carry a *color charge*, analogous to electric charge, and the strong force between quarks is often referred to as the **color force**. This new theory of the strong force is called **quantum chromodynamics** (*chrome* = color in Greek), or QCD, to indicate that the force acts between color charges (and not between, say, electric charges). The strong force between hadrons† is considered to be a force between the

---

† The strong force between hadrons appears feeble, however, in comparison to the force directly between quarks within a hadron.

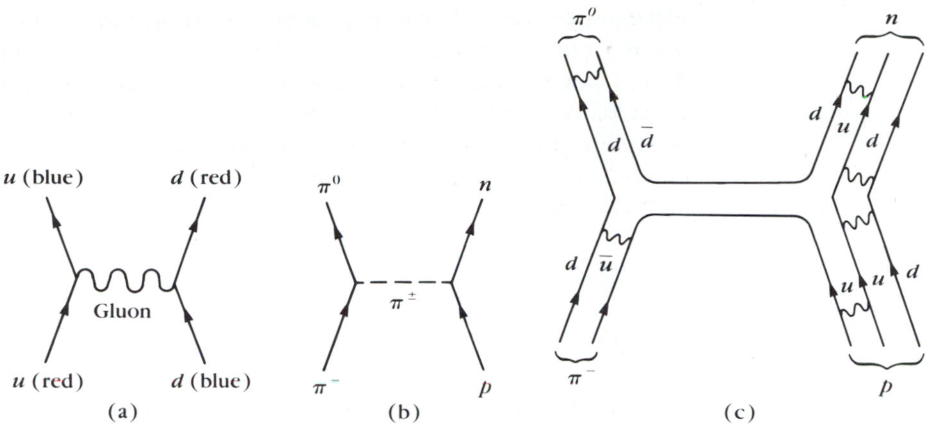

FIGURE 31-8

(a) Force between two quarks holding them together as a proton, say, is carried by a gluon which in this case involves a change in color. (b) Strong interaction $\pi^- p \rightarrow \pi^0 n$ with the exchange of a charged π meson (+ or − depending on whether it is considered moving to the left or to the right). (c) Quark representation of the same interaction $\pi^- p \rightarrow \pi^0 n$; the intermediate pion ($d\bar{u}$ or $\bar{d}u$) can be interpreted as the annihilation of a $u\bar{u}$ pair with production of a $d\bar{d}$ pair. The wavy lines between quarks represent gluon exchanges holding the hadrons together.

quarks that make them up, as suggested in Figure 31-8. The particles that transmit the force (analogous to photons for the EM force) are called **gluons** (a play on "glue"). There are eight gluons, according to the theory, all massless, and six of them have color charge.† Thus gluons have replaced mesons (Table 31-1) as the particles mediating the strong (or color) force.

The weak force, as we have seen, is thought to be mediated by the W^+, W^-, and Z^0 particles. It acts between the "weak charge" of each particle. Each elementary particle thus has electric charge, weak charge, color charge, and gravitational mass—although one or more of these could be zero. For example, all leptons have color charge of zero, so they do not interact via the strong force.

To summarize, the latest theories consider the truly elementary particles to be the photons, leptons, quarks, gluons, and $W^\pm$ and Z^0. The photons and the leptons are observed in experiments, as are the W^+, W^-, and Z^0. But so far only combinations of quarks (baryons, mesons) have been observed, and it seems likely that free quarks are unobservable. On the other hand, some physicists believe that leptons and quarks are not fundamental, but are composites of still more fundamental objects. Only the future will tell.

One important aspect of new theoretical work is the attempt to find a unified basis for the different forces in nature. This was a long-held hope of Einstein, which he was never able to fulfill. A so-called "gauge" theory that unified the weak and electromagnetic interactions was put forward in the early 1960s by Weinberg, Glashow, and Salam. In this *electroweak theory*, the weak and electromagnetic forces are seen as two different manifestations of a single, more fundamental, electroweak interaction. The

† Compare to the EM interaction where the photon has no electric charge. Because gluons have color charge, they could attract each other and form composite particles (photons cannot). Such "glueballs" are being searched for and may have been observed.

electroweak theory has had many successes, including the prediction of the $W^{\pm}$ particles, as intermediates for the weak force, having masses of 81 ± 2 GeV/c^2 in excellent agreement with the measured (1983) values of 83 ± 3 GeV/c^2. The electroweak theory and QCD for the strong interaction are referred to today as the **standard model**.

9. GRAND UNIFIED THEORIES

With the success of the unified electroweak theory, attempts have been made recently to incorporate it and QCD for the strong (color) force into a so-called **grand unified theory** (GUT). One type of such a grand unified theory of the electromagnetic, weak, and strong forces has been worked out in which there is only one class of particle—leptons and quarks belonging to the same family and able to change freely from one type to the other—and the three forces are different aspects of a single underlying force. The unity is predicted to occur, however, only on a scale of less than about 10^{-31} m. If two elementary particles (leptons or quarks) approach each other to within this *unification scale*, the apparently fundamental distinction between them would not exist at this level, and a quark could readily change to a lepton or vice versa. Baryon and lepton numbers would not be conserved. Leptons and quarks would belong to a single family. The weak, electromagnetic, and strong (color) force would blend to a force of a single strength.

How could a lepton become a quark or vice versa? The theory predicts the existence of particles that can be exchanged between a quark and a lepton that allows one to change into the other, just as the charged pion exchanged between the π^- and p in Figure 31-8(b) allows the proton to become a neutron. The mass of the new exchange particles would be about 10^{15} GeV/c^2, or 10^{15} times the proton mass. With such an incredibly large mass, there is little hope of seeing them in the laboratory. It is also this huge mass which keeps baryon and lepton numbers conserved in observed reactions, since the likelihood of producing such a massive particle, even as a virtual exchange particle, is extremely small at even the highest laboratory energies.

What happens between the unification distance of 10^{-31} m and more normal (larger) distances is referred to as *symmetry breaking*. As an analogy, consider an atom in a crystal. Deep within the atom there is great symmetry—in the innermost regions the electron cloud is spherically symmetric. Further out, this symmetry breaks down—the electron clouds are distributed preferentially along the lines joining the atoms in the crystal. In a similar way, at 10^{-31} m the force between elementary particles appears as one—it is symmetric and does not single out one type of "charge" over another. But at larger distances, that symmetry is broken and we see three distinct forces. (In the "standard model" of electroweak interactions, Section 8, the symmetry breaking between the electromagnetic and the weak interactions occurs at about 10^{-18} m.)

Since unification occurs at such tiny distances and huge energies, the theory is difficult to test experimentally. But it is not completely impossible. One possibly testable prediction is the basis for the idea, hinted at in Section 5, that the proton might decay (via, for example, $p \rightarrow \pi^0 e^+$) and violate conservation of baryon number. This could happen if two quarks approached to within 10^{-31} m of each other; but it is very unlikely at normal temperature and energy, so the decay of a proton can only be an unlikely process. In the simplest form of GUT, the theoretical estimate of the proton lifetime is $\approx 10^{31}$ yr, and this has just come within the realm of testability. Proton decays have still not been seen and experiments in 1984 put the lower limit on the proton lifetime to be 2×10^{32} yr, an order of magnitude greater than this prediction. This may seem a disappointment; on the other hand, it presents a challenge: indeed, more complex GUTs are not affected by this result.

Another interesting prediction of unified theories relates to cosmology (Chapter 33). It is thought that during the first 10^{-35} s after the theorized big bang that created the universe, the temperature was so extremely high that particles had energies corresponding to the unification scale; then baryon number would not have been conserved. This could account for the observed predominance of matter ($B > 0$) over antimatter ($B < 0$) in the universe.

This last example is fascinating, for it illustrates a deep connection between investigations at either end of the size scale: theories about the tiniest objects (elementary particles) have a strong bearing on the understanding of the universe as a whole.

Even more ambitious than grand unified theories are the attempts to incorporate gravity, and thus unify all four forces in nature into a single theory. The world of elementary particles is opening new vistas. What happens in the near future promises to be exciting.

SUMMARY

Particle accelerators are used to accelerate charged particles such as electrons and protons to very high energy. High-energy particles allowed the creation of new particles through collision (via $E = mc^2$). Linear accelerators use high voltage to accelerate particles along a line. Cyclotrons and synchrotrons use a magnetic field to keep the particles in a circular path, and they are accelerated at intervals by high voltage.

Just as the electromagnetic force can be said to be due to an exchange of photons, the strong nuclear force is thought to be carried by *mesons* that have rest mass or, according to more recent theory, by massless *gluons*.

An *antiparticle* has the same mass as a particle but opposite charge. Certain other properties may also be opposite: for example, the antiproton has *baryon number* (nucleon number) opposite to that of the proton. In all nuclear and particle reactions, the conservation laws hold: momentum, mass-energy, angular momentum, electric charge, baryon number, and lepton numbers. Certain particles have a property, called *strangeness*, which is conserved by the strong force but not by the weak force.

Particles can be classified as *leptons* and *hadrons*, plus the photon. Leptons participate in the weak and electromagnetic interactions. Hadrons participate in the strong interaction as well. The hadrons can be classified as *mesons* with baryon number zero, and *baryons* with nonzero baryon number.

All particles, except for the photon, electron, neutrinos, and (so far at least) proton, decay with measurable half-lives varying from 10^{-23} s to 10^{3} s. The half-life depends on which force is predominant in the decay. Weak decays have half-lives greater than about 10^{-10} s. Electromagnetic decays have half-lives on the order of 10^{-16} to 10^{-19} s. The shortest-lived particles, called *resonances*, decay via the strong interaction and live typically for only about 10^{-23} s.

The latest theories of elementary-particle physics postulate the existence of *quarks* as the basic building blocks of the hadrons. Initially, three quarks were proposed. Newer evidence suggests that a fourth, *charmed*, quark is needed and also a fifth and a sixth. It is expected that there are the same number of quarks as leptons, and that quarks and leptons are the truly elementary particles. Quarks are said to have *color*, and, according to *quantum chromodynamics* (QCD), the strong color force acts between their color charges and is transmitted by *gluons*. *Electroweak theory* views the weak and electromagnetic forces as two aspects of a single underlying interaction. QCD plus the electroweak theory are referred to as the *standard model*.

Grand unified theories of forces suggests that at very short distances (10^{-31} m) and very high energy, the weak, electromagnetic, and strong forces appear as a single force, and the fundamental difference between quarks and leptons disappears.

QUESTIONS

1. A proton in a synchrotron has a speed of $0.99c$. What must be done to increase its energy?
2. Give a reaction between two nucleons that could produce a $\pi-$, similar to the equation displayed in Section 2.
3. Draw a Feynman diagram for $n + p \rightarrow n + p + \pi^0$.
4. Draw a Feynman diagram for $p\bar{p}$ annihilation with the production of two pions.

5. Draw a Feynman diagram for the photoelectric effect.
6. If a large quantity of antimatter exists in the universe, we expect that it would be quite far away from our galaxy. Could it be close to us? Why?
7. Why is it that a neutron decays via the weak interaction even though the neutron and one of its decay products (proton) are strongly interacting?
8. Which of the four interactions (strong, electromagnetic, weak, gravitational) does an electron take part in? A neutrino? A proton?
9. Which of the following decays are possible? For those that are forbidden, explain which laws are violated.
   (a) $\Xi^0 \rightarrow \Sigma^+ + \pi^-$
   (b) $\Omega^- \rightarrow \Sigma^0 + \pi^-$
   (c) $\Sigma \rightarrow \Lambda + \gamma + \gamma$
10. Which of the following reactions are possible and by what interaction could they occur? For those that are forbidden, explain why.
   (a) $\pi^- p \rightarrow K^+ \Sigma^-$
   (b) $\pi^+ p \rightarrow K^+ \Sigma^+$
   (c) $\pi^- p \rightarrow \Lambda^0 K^0 \pi^0$
   (d) $\pi^+ p \rightarrow \Sigma^0 \pi^0$
   (e) $\pi^- p \rightarrow K^0 p \pi^0$
   (f) $K^- p \rightarrow \Lambda^0 \pi^0$
   (g) $K^+ n \rightarrow \Sigma^+ \pi^0 \gamma$
   (h) $K^+ \rightarrow \pi^0 \pi^0 \pi^+$
11. What are the quark combinations that can form a (a) neutron (b) antineutron (c) Λ^0 (d) Ξ^0?
12. What particles do the following quark combinations produce? (a) uud, (b) $\bar{u}\bar{u}\bar{s}$, (c) $\bar{u}s$, (d) $d\bar{u}$?
13. What is the quark combination needed to produce a D^0 meson ($Q = B = S = 0$, $C = +1$)?
14. The F^+ meson has $Q = S = C = +1$, $B = 0$. What quark combination would produce it?
15. Draw a possible quark Feynman diagram—see Figure 31-8 (c)—for the reaction $K^- p \rightarrow K^- p$.

EXERCISES

1. What is the total energy of a proton whose kinetic energy is 15 GeV?
2. The voltage across the dees of a cyclotron is 50 kV. How many revolutions do protons make to reach a kinetic energy of 10 MeV?
3. Protons are injected into the 1.0-km-radius Fermilab synchrotron with an energy of 8.0 GeV. If they are accelerated by 2.5 MV each

revolution, how far do they travel and approximately how long does it take for them to reach 400 GeV? (Take $v \approx c$.)

4. Two protons are heading toward each other with equal speeds. What minimum energy must each have if a π^0 meson is to be created in the process? (See Table 31-2.)
5. How much energy is released in the decay $\pi^+ \rightarrow \mu^+ + \nu_\mu$?
6. How much energy is released when an electron and a positron annihilate each other?
7. How much energy is released when a proton and an antiproton annihilate each other?
8. How much energy is required to produce a neutron-antineutron pair?
9. What minimum kinetic energy must a neutron and proton each have if they are traveling at the same speed toward each other, collide, and produce a K^+K^- pair in addition to themselves? (See Table 31-2.)
10. The B^- meson is presumed to be a $b\bar{u}$ quark combination. (a) Show that this is consistent for all quantum numbers. (b) What are the quark combinations for B^+, B^0, $\overline{B^0}$?
11. Draw possible Feynman diagrams using quarks—as in Figure 31-8(c)—for the reactions (a) $pn \rightarrow pn$, (b) $\bar{p}p \rightarrow \pi^+\pi^-$.

*12. What are the wavelengths of the two photons when a proton-antiproton pair at rest annihilate?

*13. Show that the so-called unification distance of 10^{-31} m in recent unified theory is equivalent to an energy of about 10^{15} GeV. Use either the uncertainty principle or de Broglie's wavelength formula, and explain how they apply.

*14. A symmetry breaking occurs in the electroweak theory at about 10^{-18} m. Show that this corresponds to an energy which is on the order of the mass of the $W^\pm$.

ASTROPHYSICS AND GENERAL RELATIVITY

CHAPTER

32

In the previous chapter we studied the tiniest bodies in the universe—elementary particles. Now we leap to the largest—stars, galaxies, and the cosmos. These two extreme realms, elementary particles and the heavens, are among the most intriguing and exciting subjects in science. And, surprising though it may seem, these distant realms affect one another in a fundamental way, as already hinted at in Chapter 31.

The study of the universe as a whole was, in fact, the first subject we studied, back in Chapter 2; so it is fitting that we end with this same subject but now as it is seen today.

Use of the techniques and ideas of physics to study the heavens is often referred to as *astrophysics*. At the base of our present theoretical understanding of the universe is Einstein's *general theory of relativity* and its theory of gravitation—for in the large-scale structure of the universe, gravity is the dominant force. This material, important in itself, serves also as the foundation for modern **cosmology**, which is the study of the universe as a whole, including, in particular, theories of the origin of the universe as well as its future; this will be the subject of the next, and last, chapter in this book. We begin our study with a close look at the heavens.

1. STARS AND GALAXIES—WHAT IS IN THE UNIVERSE

Cosmic Distances: Light-years

In the daytime, the brilliant glow of the sun lights the entire sky. But on a cloudless night, the dark sky is filled with myriads of stars. As we saw in Chapter 2, the ancients believed that the stars, except for the few that seemed to move (the planets), were fixed on a sphere beyond the last planet. The universe was neatly self-contained, and we, on earth, were at or near its center. But in the centuries following Galileo's telescopic observations of the heavens, our view of the universe has changed dramatically. We no longer place ourselves at the center; and we view the universe as vastly larger.

Indeed, the distances involved are so great that we must specify them in units much larger than meters, miles, and kilometers. Instead we use *light-time*: we specify a distance by the time it takes light to travel that distance. Since the speed of light is 3.0×10^8 m/s (three hundred million meters per second), then in one second light travels 3.0×10^8 meters. So one *light-second* is

$$1 \text{ light-second} = 3.0 \times 10^8 \text{ m}$$

or 300,000 km. In one minute light travels 60 times as far (since there are 60 s in 1 minute): $(60 \text{ s})(3.0 \times 10^8 \text{ m/s})$ or 1.8×10^{10} meters. Thus 1 *light-minute* $= 1.8 \times 10^{10}$ meters or 18,000,000 km. In one year there are 3.15×10^7 seconds $[=(365 \text{ days/yr})(24 \text{ hr/day})(3600 \text{ s/hr})]$ so one *light-year* $= (3.15 \times 10^7 \text{ s/yr})(3.0 \times 10^8 \text{ m/s}) = 9.5 \times 10^{15}$ meters (= 9,500,000,000,000 km); that is, one **light-year** (abbreviated ly) is:

$$1 \text{ ly} = 1 \text{ light-year} = 9.5 \times 10^{15} \text{ m} = 9.5 \times 10^{12} \text{ km} \simeq 10^{13} \text{ km}.$$

For specifying distances to the sun and moon, we generally use meters or kilometers; but we could specify them in terms of light. The earth-moon distance is 384,000 km, which is 1.28 light-seconds. The earth-sun distance is 1.5×10^{11} m or 150,000,000 km; this is equal to 8.3 light-minutes. To say it another way, light from the sun (traveling at its incredibly high speed) takes 8.3 minutes to reach earth.

For specifying the distances to the stars, we use light-years. Other than our Sun, the nearest star to us is about four light-years away (equal to about 40,000,000,000,000 km, or forty million million kilometers). Thus, light emitted by our nearest star takes four years just to reach us!

What We Can See

On a clear moonless night, we can see thousands of stars of varying degrees of brightness. We also can see, stretching across the heavens, that elongated cloudy stripe known as the Milky Way. As we saw in Chapter 2, it was Galileo who first observed (about 1610) that the Milky Way is comprised of countless individual stars. A century and a half later (about 1750), Thomas Wright suggested that the stars of the Milky Way formed a flat disc, a sort of thick "grindstone" of stars extending to great distances in a plane. This huge disc of stars we call the **Galaxy**. (Galaxy is Greek for "milky way.")

Our Galaxy has a diameter of almost 100,000 light-years and a thickness of about 6000 light-years. It has a bulging central nucleus and spiral arms

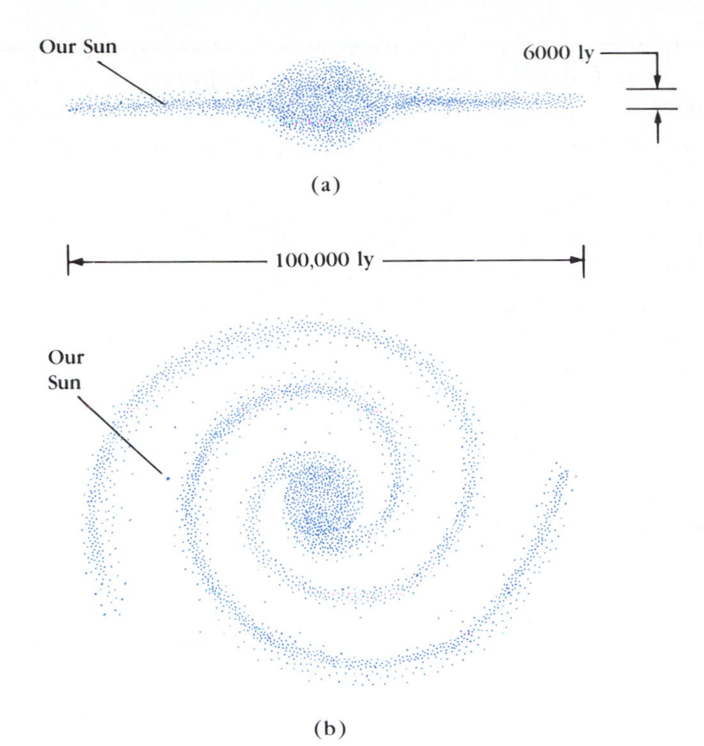

FIGURE 32-1
Our galaxy as it would appear from the outside: (a) "end view," in the plane of the disc; (b) "top view," looking down on the disc. (If only we could see it like this—from the outside!)

(Figure 32-1). Our Sun, which seems to be just another star, is located more than halfway from the center to the edge (about 30,000 ly from the center). The Galaxy contains about 10^{11} stars. Our Sun rotates about the galactic center approximately once every 200 million years; so its speed is about 250 km/s relative to the center of the Galaxy. The total mass of all the stars in our Galaxy is about 3×10^{41} kg. All of these figures are enormous numbers, almost unimaginable. We really see the necessity for the "powers of ten notation" (Appendix A).

We can see by telescope, in addition to stars both within and outside the Milky Way, many faint cloudy patches in the sky. A few of these so-called *nebulae* (nebula is Latin for "cloud") can actually be discerned with the naked eye on a clear night such as those in the constellations Andromeda and Orion. Some nebulae are *star clusters*, groups of stars that are so numerous they appear to be a cloud. Others are glowing clouds of gas. Most fascinating is the third type of nebulae: these can have a fairly regular elliptic shape and they seems to be a great distance beyond our Galaxy. Emmanuel Kant (about 1755) seem to have been the first to suggest that these nebulae might be circular discs like our Galaxy: They appear elliptical because we see them from an angle; they appear faint because they are so distant. At first it was not universally accepted that these objects were *extragalactic* (outside our Galaxy). The very large telescopes constructed in this century revealed that individual stars could be resolved

TABLE 32-1 Heavenly Distances

OBJECT	DISTANCE FROM EARTH	(IN LY)
Clouds	≈ a few km	$\approx 10^{-13}$
Moon	384,000 km	4×10^{-8}
Sun	150,000,000 km (= 8.3 light-minutes)	1.5×10^{-5}
Other planets	up to 6,000,000,000 km (Pluto)	6×10^{-4}
Nearest star	≈ 4 light-years (40,000,000,000,000 km)	4
Diameter of our Galaxy	≈ 100,000 light-years	1×10^{5}
Nearest galaxy	≈ 2,000,000 light-years	2×10^{6}
Farthest galaxy seen optically	≈ 3,000,000,000 light-years	3×10^{9}

FIGURE 32-2
See Color Plate VIII.

within these nebulae and that many of them contained spiral arms. Edwin Hubble (1889–1953), who did much of this observational work in the 1920s using the 2.5-m (100-inch) telescope[†] on Mt. Wilson near Los Angeles, California, was also able to demonstrate that these objects were indeed extragalatic because of their great distances. (The distance to the Andromeda nebulae is over 2 million light-years, a distance 20 times greater than the diameter of our Galaxy). Thus it was determined that these nebulae are galaxies similar to ours. Today the largest telescope can see about 10^{11} galaxies (100 thousand million)! See Figure 32-2, Color Plate VIII.

Galaxies tend to group in *galaxy clusters*, with only a few to many thousands of galaxies in each cluster. The galaxies nearest us are about 2 million light-years away; the farthest galaxies visible with optical telescopes are more than a thousand times farther away, about 3×10^9 ly. (A summary of comparative distances for objects beyond the earth is given in Table 32-1.)

The universe is indeed vast. So vast, according to our twentieth-century view, that we on earth are but a mere speck lost in its endless reaches. The planet Earth revolves around a star (our Sun) at a distance of approximately 8 light-minutes. Our star is embedded in a huge galaxy of 10^{11} stars, the nearest of which is about 4 light-years away. Beyond our galaxy are vast numbers of other huge galaxies, millions and hundreds of millions of light-years away. (See Table 32-1.) From this perspective, we living creatures on tiny Earth can hardly expect to feel anything but anonymity within this vast universe—a far cry from the view held but a few centuries ago that placed us at the very center of the universe.

[†] 2.5 m (= 100 inches) refers to the diameter of the curved objective mirror. The bigger the mirror, the more light it collects and the less diffraction there is—so more and fainter stars can be seen. See Chapters 23 and 24.

If this modern view of the universe seems too stark, we can perhaps take a philosophical approach: all of these observations and interpretations have been made here on earth, and the ideas created here on earth. Who is to say that we are in any way unimportant in the universe, even though we may not hold a geometrically central position?

Besides the usual stars, clusters of stars, galaxies, and clusters of galaxies, the universe contains a number of other interesting objects. Among these are strange versions of stars known as *red giants*, *white dwarfs*, *neutron stars*, *black holes* (predicted theoretically), and exploding stars called *novae* and *supernovae*. In addition there are *quasars* (quasi-stellar objects) which, if we judge their distance correctly, are galaxies thousands of times brighter than ordinary galaxies. Furthermore, there is radiation that reaches the earth but does not emanate from the bright point-like objects we call stars: it is a background radiation that seems to arrive uniformly from all directions in the universe. We discuss all these phenomena in due course.

How Distances Are Measured; The Parsec (optional)

We have talked about the various and vast distances of objects in the universe. But how do we measure these distances?

One basic technique employs simple geometry to measure the *parallax* of a star. By parallax we mean the apparent motion of a star (against the background of more distant stars) due to the earth's motion about the sun. As shown in Figure 32-3, the sighting angle of a star relative to the plane of

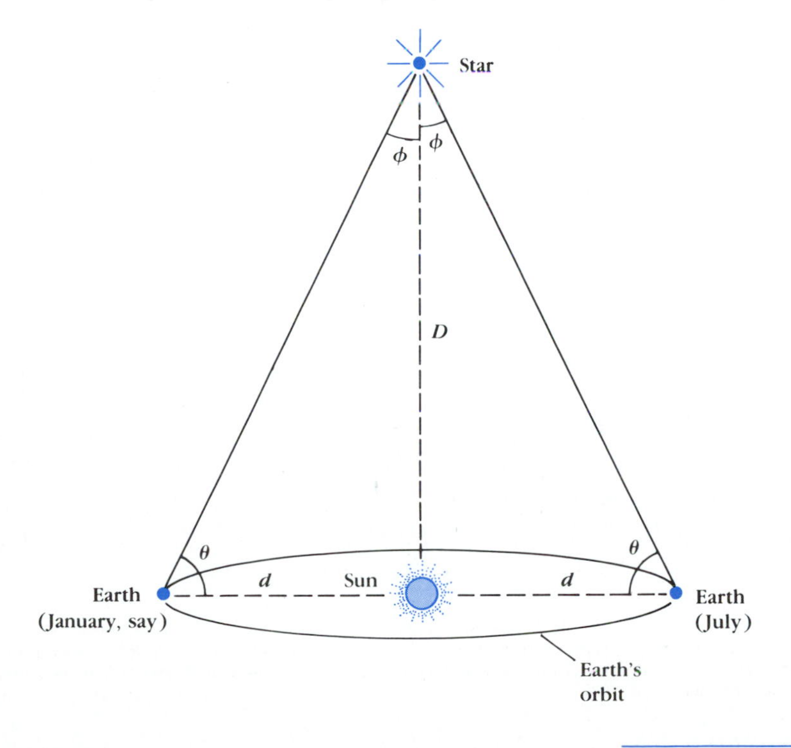

FIGURE 32-3
Distance to a star determined by *parallax*.

Earth's orbit (angle θ) is measured at different times of the year. Since we know the distance d from earth to sun, we can reconstruct the right triangles shown in the figure (that is, make a scale model on paper) and can determine[†] the distance D to the star. For example, suppose the angle θ in Figure 32-3 is measured to be 89.99994° for a particular star. Since the sum of the angles of any triangle is 180° according to Euclidean geometry, then the angle ϕ in Figure 32-3 must be 0.00006° (the third angle is 90°). We saw in Chapter 9 that we can write any small angle, in radians, as

$$\phi = \frac{d}{D}.$$

Since 1 radian = 57.3°, our angle is $\frac{0.00006}{57.3} = 0.000001$ radians. So from the equation above, the distance D is (since d = earth-sun distance = 1.5×10^8 km)

$$D = \frac{d}{\phi} = \frac{1.5 \times 10^8 \text{ km}}{10^{-6}} = 1.5 \times 10^{14} \text{ km}.$$

Because 1 light-year is about 10^{13} km, our star is about 15 ly away.

The distances of stars are often specified in terms of the parallax angle ϕ. As the preceding example shows, even the nearest stars (>4 ly) have parallax angles much less than one degree. We wrote our angle in decimal form, but it is also common to write an angle in terms of minutes ($1' = \frac{1}{60}$ of a degree) and seconds ($1'' = \frac{1}{60}$ of a minute). Thus 1 second of arc is $\frac{1}{3600}$ of a degree. Our 0.00006° angle in seconds is $(0.00006°)(3600) = 0.2''$ or $\frac{1}{5}$ of a second ($''$ stands for seconds of arc, $'$ for minutes of arc). The "parsec" unit is defined as $1/\phi$ when ϕ is given in seconds; so our star would be $1/0.2''$ or 5 parsecs away. The *parsec* unit is commonly used in astronomy; its name comes from its definition as the distance to an object that has a *par*allax angle of one *sec*ond of arc. It is easy to show that

$$1 \text{ parsec} = 3.26 \text{ light-years}.$$

In the preceding example, our star would be $(3.26 \text{ ly})/(0.2'') \approx 15$ ly away.

Parallax can be used to measure stars as far away as 100 light-years (≈30 parsecs). Beyond that distance, parallax angles are simply too small to measure. For greater distances, more subtle techniques must be employed. For example, to estimate distances to galaxies we can compare their apparent brightness. Better still, we can compare the brightest stars in galaxies or brightest galaxies in galaxy clusters. For example, it is assumed that the brightest stars in all galaxies are similar and have about

[†] This is very similar to the way the heights of mountains are determined, by "triangulation:" the top of a peak is viewed from points on earth whose altitudes are known, and the angles of view relative to the horizontal (or vertical) are carefully measured. From the measured angles and distances, the height of the mountain can be determined.

the same intrinsic brightness. Consequently, their *apparent* brightness would be a measure of how far away they are.

As we look farther and farther away, the measurement techniques are less and less reliable, so there is more and more uncertainty in the measurements of large distances.

2. THE BIRTH AND DEATH OF STARS; STELLAR EVOLUTION

The stars appear unchanging. Night after night the heavens reveal no significant variations. Indeed, on a human time scale the vast majority of stars (except novae and supernovae) change very little. Although stars seem fixed in relation to each other, many in fact do move sufficiently for the motion to be detected. This motion is almost imperceptible to us because the stars are so far away; their actual speeds, however, relative to neighboring stars can be hundreds of km/s.

Furthermore, not all stars are identical. It is clear that some are bright and some dim. The difference in brightness is due both to the amount of light a star emits and to its distance from us. A distant star will appear dimmer than a nearby star which has the same intrinsic brightness (or *absolute luminosity*, as astronomers say). Careful study of nearby stars has shown that for most stars, the absolute luminosity depends on their mass: *the bigger the star, the greater its brightness*.

We can also determine the surface temperature of stars because they emit a wide range of light-wave frequencies, just as a hot solid does. As we saw in Chapter 26, the spectrum of hotter and hotter bodies shifts from predominately lower frequencies (and longer wavelengths, such as red) to higher frequencies (and shorter wavelengths such as blue). The surface temperatures of stars typically range from about 3500 K (reddish) to perhaps 50,000 K (bluish). (Recall from Chapter 13 that K stands for kelvins, or degrees kelvin; room temperature is about 300 K, and water freezes and boils at 273 K and 373 K, respectively.)

An important astronomical discovery, made near the turn of the century, was that the color of most stars is related to their absolute luminosities, and therefore to their sizes. A useful way to present this relationship is by the so-called Hertzsprung-Russell (H-R) diagram. On the H-R diagram, one axis represents the absolute luminosity and the other represents the temperature, and each star is represented by a point on the diagram (Figure 32-4). Most stars fall along the diagonal band termed the **main sequence**. Starting at the lower right we find the coolest stars, reddish in color; they are the least luminous and therefore must be small. Farther up toward the left we find brighter and hotter stars that are yellowish-white, like our Sun. Still farther up we find even larger and brighter stars, bluish in color. Stars that fall on this diagonal band are called *main-sequence stars*. There are also stars that fall outside the main

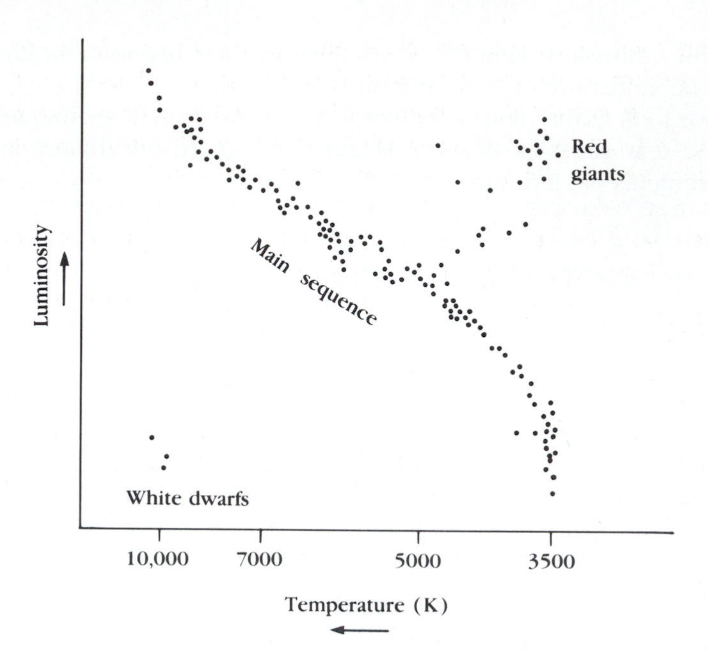

FIGURE 32-4
H-R diagram.

sequence. Above and to the right we find extremely large stars, with high luminosities but with low (reddish) color temperature; these are called *red giants*. At the lower left there are a few stars of weak luminosities (therefore small) but with high temperatures; these are the *white dwarfs*.

Why are there different types of stars? Were they all born this way, in the beginning? Or might each different type represent a different age in the life cycle of a star? Astronomers and astrophysicists today believe the latter is true. Note, however that we cannot actually follow any but the tiniest part of the life cycle of any given star since they live for ages vastly greater than ours, on the order of billions of years. Nonetheless, let us follow the process of *stellar evolution* from the birth to the death of a star, as astrophysicists have reconstructed it today.

Birth of a Star and Nucleosynthesis

Stars are born, it is believed, when gaseous clouds (mostly hydrogen) contract due to the pull of gravity. A huge gas cloud might fragment into numerous contracting masses, each mass centered in an area where the density was only slightly greater than nearby points. Once such "globules" are formed, gravity would inexorably pull each one in toward its center of mass. As the particles of a *protostar* accelerate inward, their kinetic energy (KE) increases. When the kinetic energy is sufficiently high, nuclear fusion can take place. (This corresponds to a temperature T of close to 10 million degrees: $T \approx 10^7$ K; recall from Chapter 13 that the average kinetic energy of molecules is proportional to T.) That is, as we saw in Chapter 30, the nuclei acquire sufficient kinetic energy to overcome the electric repulsion between them and to move close enough together to fuse. In such *thermonuclear* reactions, a great deal of energy is released (the same source of energy for the H-bomb). This tremendous release of energy produces a pressure sufficient to halt the gravitational contraction; and our protostar, now really a young *star*, stabilizes on the main sequence. Exactly where the star falls along the main sequence depends on its mass. To reach this point requires perhaps 30 million years; and if it is a star like our Sun, it will remain there about 10 billion years (10^{10} yr). Although most stars are billions of years old, there is evidence that stars are actually being born at this moment.

The fusion reactions occur primarily in the core of the star. For a star like our Sun, these thermonuclear reactions principally involve the conversion of hydrogen (protons) into helium, and are accompanied by the release of γ rays and neutrinos. This process of forming larger nuclei from smaller ones is called **nucleosynthesis**. The enormous luminosity of stars, including our Sun, results from the energy released during these thermonuclear reactions. This idea was first proposed in the 1930s and convincingly demonstrated by Hans Bethe (1906–   ) in 1938.

From the Main Sequence to Red Giants

As hydrogen burns (astronomers use the term "burn" to mean nuclear fusion, not the everyday "chemical" burning in oxygen), the helium that is formed is heavier and tends to accumulate in the central core. As the core of helium grows, hydrogen continues to burn in a shell around it. And as hydrogen within the core is consumed, the production of energy decreases and is no longer sufficient to prevent the huge gravitational forces from once again causing the core to contract and heat up. The hydrogen in the shell around the core then burns even more fiercely, causing the outer enevelope of the star to expand and to cool. The surface temperature, thus reduced, produces a spectrum of light shifted toward the red. By now the star has left the main sequence. It becomes redder, and as it grows in size becomes more luminous; so it moves to the right and upward in an H-R diagram, as shown in Figure 32-5. As it moves upward, it is entering the **red giant** stage. Our Sun, for example, has been on the main sequence for about $4\frac{1}{2}$ billion years. It will probably remain there another 4 or 5 billion years. (We can take comfort in that!) When our Sun leaves the main sequence, it will grow in size (as a red giant) until it occupies all the volume out to perhaps the orbit of Venus or even Earth.

As a star's envelope expands, the core is shrinking and heating up. When the temperature reaches about 100 million degrees (10^8 K), thermonuclear reactions resume. The higher temperature is required since now it is helium nuclei striking helium nuclei: since helium has two plus charges (compared to one for each H nucleus), the nuclei must hve greter KE to overcome the electric repulsion and to reach each other in order to fuse. The principal reaction is for three ${}^4_2$He nuclei to form a ${}^{12}_6$C nucleus. A ${}^4_2$He nucleus can also fuse with a ${}^{12}_6$C nucleus to form ${}^{16}_8$O. Heavier nuclei can

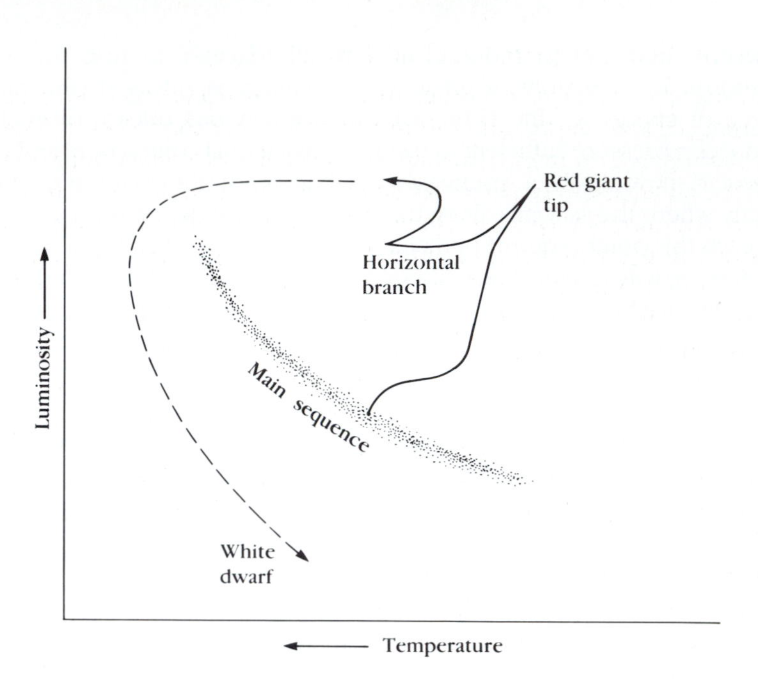

FIGURE 32-5
Evolution of a star like our Sun represented on an H-R diagram.

also be formed. But it seems that after H and He, the next most prevalant elements in the universe are oxygen and carbon. (Helium makes up about $\frac{1}{4}$ of the mass in the universe, hydrogen about $\frac{3}{4}$.)

The burning of helium begins quickly, and is referred to as the "helium flash." It causes a rapid, major change in the star, and the star moves speedily to a new position on the H-R diagram (the "horizontal branch," Figure 32-5). Once the burning of He is complete, the core cannot contract sufficiently to enter another stage of burning (of C and O) if the star's mass is less than about 0.7 solar masses. If the mass is greater than this, the core can contract and heat up further, allowing fusion reactions between C and O nuclei at around 10^9 K. Eventually nuclei as heavy as iron and nickel are formed. At this point, no further energy can be obtained from nuclear fusion. (Elements heavier than iron are believed to be made in stars by successive additions of neutrons to iron and other nuclei.)

White Dwarfs

What happens next depends on the mass of the star. If the star has a mass similar to that of the sun (or, more precisely, less than about 1.4 solar masses), no significant further contraction of the core occurs. Losing energy, the star begins to cool and typically follows the route shown in Figure 32-5, descending from the upper left downward. Here we have the birth a **white dwarf** star. It grows dimmer and dimmer, and eventually dies. Typically, white dwarfs have masses like that of our Sun, but are only about as large as the Earth.

Neutron Stars, Supernovae, and Black Holes

More massive stars (mass greater than 1.4 solar masses) follow a quite different scenario. A star with this great a mass can contract under gravity and heat up even further, reaching to extremely high temperatures. The KE of the nuclei is then so high that collisions can cause the breaking apart of iron and nickel nuclei into He nuclei, and eventually into protons and neutrons. As the core contracts further under the huge gravitational forces, electrons and protons can be squeezed together to form neutrons. This tremendous mass, now essentially an enormous nucleus made up almost exclusively of neutrons, contracts rapidly to form a **neutron star**. The contraction of the core signals a great reduction in gravtitational potential energy. Somehow this energy must be released. Indeed, it was suggested in the 1930s that core collapse may be accompanied by a catastrophic explosion whose tremendous energy could not only form virtually all elements of the periodic table but could blow away the entire outer envelope of the star, spreading its contents into interstellar space. This is probably the origin of heavy elements in our solar system.

This type of explosion is believed to be the origin of **supernovae** (there may be alternative sources as well). In a supernova explosion, a star's brightness is observed to suddenly increase billions of times in a period of just a few days and then fade away over the next few months. One such supernova was observed by Chinese astronomers in A.D. 1054. Its remains are still visible in the sky (in the Crab Nebula), in the midst of which is a *pulsar*. Pulsars are astronomical objects that emit fast pulses of radiation at regular intervals; they are now believed to be rapidly rotating neutron stars, and their discovery in 1967 has lent credence to the theory that neutron stars can be created in supernova explosions.

The core of a neutron star contracts to the point where all neutrons are as close together as they are in a nucleus. That is, the density of a neutron star is on the order of 10^{14} times that of normal solids and liquids on earth. A thimbleful of this matter would weigh millions of tons on earth! The mass of a neutron star is similar to that of our Sun, but its diameter is only 10 or 20 km.

If the mass of a neutron star is less than about 2 or 3 solar masses, its subsequent evolution is similar to that of a white dwarf. If the mass is greater than this, the gravitational force is so strong that the star contracts to an even smaller diameter and an even greater density. Gravity is so strong that light emitted from it cannot escape—it is pulled right back by the force of gravity. Since no type of radiation can escape from such a star, we cannot see it—it is black. An object may pass by it and be deflected; but if it comes too close it will be swallowed up, never to escape. This is a **black hole**.

Black holes are predicted by theory to exist. Although none has been observed for sure as yet, strong evidence published in 1985 suggests there may be a giant black hole at the very center of our Galaxy—its mass is estimated at several million times that of the sun!

3. GENERAL RELATIVITY

As we have just seen, the force of gravity plays a dominant role in the processes that occur in stars. Indeed, gravity plays a crucial role in the evolution of the universe as a whole.[†] But the force of gravity as Newton described it in his Law of Universal Gravitation, on a cosmological scale, shows some discrepancies. Einstein, in his General Theory of Relativity, developed a theory of gravity that solves these problems and that forms the basis of cosmological dynamics.

The Principle of Equivalence

In his Special Theory of Relativity, discussed earlier in Chapter 25, Einstein pointed out that there seems to be no way for an observer to determine whether his or her frame of reference is at rest or is moving at a constant velocity in a straight line. Thus, he said, the laws of physics must be the same in different reference frames that move at a uniform velocity relative to each other. (As we saw, the actual *paths* of objects are usually different in different reference frames.)

To consider only uniformly moving reference frames is somewhat restricting. What about the general case of motion, where reference frames can be *accelerating*? This is the subject of general relativity. We begin with Einstein's famous *Principle of Equivalence*:

No observer can determine by experiment if he or she is accelerating or is simply in a gravitational field.

If some observers sensed that they were accelerating (as when in a vehicle speeding around a sharp curve), they could not prove by any experiment that in fact they weren't simply experiencing the pull of a gravitational field. Conversely, we might feel we are being pulled by gravity when in fact we are undergoing an "inertial" acceleration having nothing to do with gravity. For example, pilots making a steeply banked turn in fog often have this experience, and cannot tell in which direction the earth lies.

As a thought experiment, consider a person in a freely falling elevator near the earth's surface. If our observer held out a key ring and let go of it, what would happen? Gravity would pull it downward toward the earth, but at the same rate ($g = 9.8\ \text{m/s}^2$) at which the person and elevator are falling; so the key ring would hover right next to the person's hand (Figure 32-6). The effect is exactly the same as if this reference frame were at rest and *no* forces were acting.

FIGURE 32-6
Freely falling elevator. The dropped key ring hovers next to the owner's hand.

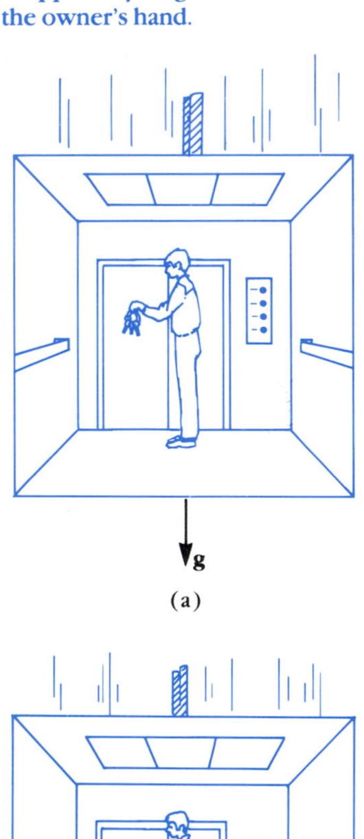

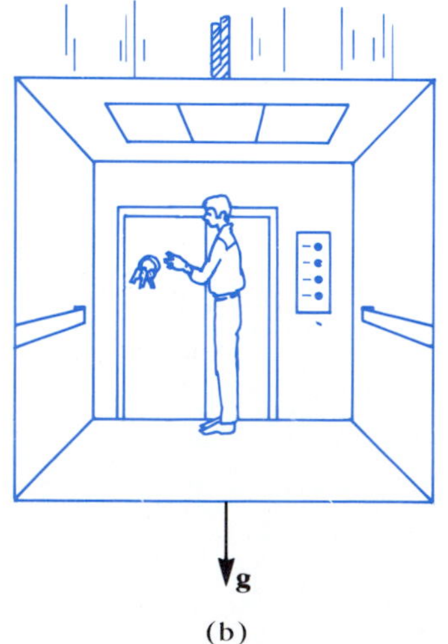

[†] The reasons gravity, and not one of the other of the four forces in nature, plays the dominant role in the universe are (1) it is long-range and (2) it is always attractive. The strong and weak nuclear forces act over short distances only, on the order of the size of a nucleus; hence they do not act over astronomical distances (although of course they act between nuclei in stars to produce nuclear reactions). The electromagnetic force, like gravity, acts over great distances; but it can be both attractive or repulsive. And since the universe does not seem to contain large areas of net electric charge, a large net force does not occur. But gravity acts between *all* masses, and there are large accumulations in the universe of only the one type of mass (not + and − as with electric charge).

On the other hand, suppose the elevator were far out in space where there is no gravitational field. If the person let go of the key ring, it would float, just as in Figure 32-6(b). If instead, the elevator were accelerating upward at an acceleration of 9.8 m/s^2, the keys, as seen by our observer, would fall to the floor with an acceleration of 9.8 m/s^2, just as if they were falling because of gravity. Indeed, according to the Principle of Equivalence, the observer could not be sure whether the key ring fell because the elevator was accelerating upward or because a gravitational field was acting downward. The two descriptions are equivalent.

Gravitational and Inertial Mass

The Principle of Equivalence is related to the concept of mass and to the idea that there are two types of mass. For any force, Newton's Second Law says that $F = ma$, where m is the mass—or more precisely, the *inertial mass*. The more inertial mass a body has, the less it is affected by a given force (the less acceleration it undergoes). You might say that inertial mass represents resistance to any type of force whatever. The second type of mass is *gravitational mass*. When one body attacts another by the gravitational force (Newton's Law of Universal Gravitation, $F \propto \frac{mm'}{d^2}$, Chapter 6), the strength of the force is proportional to the product of the gravitational masses of the two bodies. This is much like the electric force between two bodies that is proportional to the product of their electric charges. Now the property of a body termed its electric charge is certainly not related to its inertial mass; so why should we expect the gravitational mass (call it gravitational charge if you like) of a body to be related to its inertial mass? We have, up to now, assumed they were the same. Why? Because no experiment—not even high precision experiments—has been able to discern any difference between them. This, then, is another way to state the Equivalence Principle: gravitational mass is equivalent to inertial mass.

FIGURE 32-7
The bending of a light beam (exaggerated) in an elevator (b) accelerating in an upward direction. In (a) there is no acceleration.

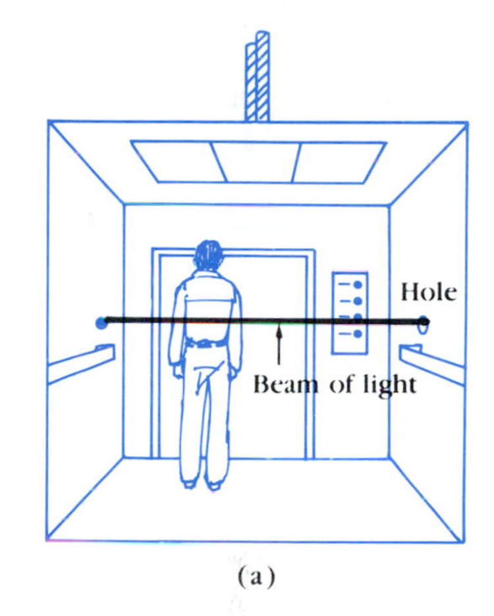

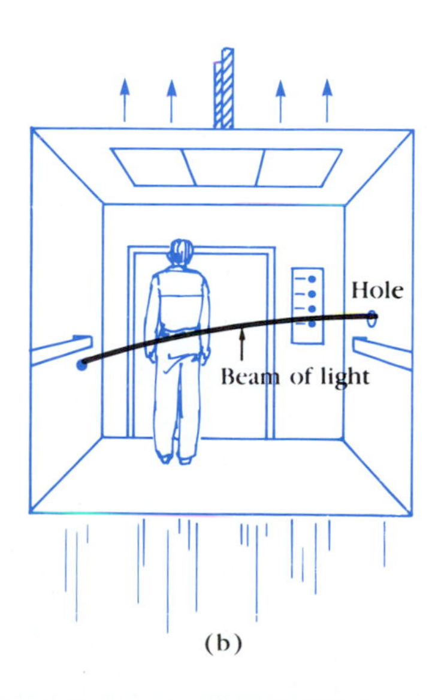

"Light Caught Bending"

Let us consider again the elevator in free space where no gravity acts. If there is a hole in the side of the elevator and a beam of light enters from outside, the beam travels straight across the elevator and makes a spot on the opposite side, if the elevator is at rest; see Figure 32-7(a). However, if the elevator is accelerating upward as in Figure 32-7(b), the light beam still travels straight across in a reference frame at rest; but in the upwardly accelerating elevator, the beam is observed to curve downward. Why? Because during the time the light travels from one side of the elevator to the other, the elevator is moving upward at ever increasing speed. Now, according to the Equivalence Principle an upwardly accelerating reference frame is equivalent to a downward gravitational field. Hence, from the thought experiment of Figure 32-7 we expect gravity to exert a force on a beam of light and to bend it out of a straight-line path! This is an

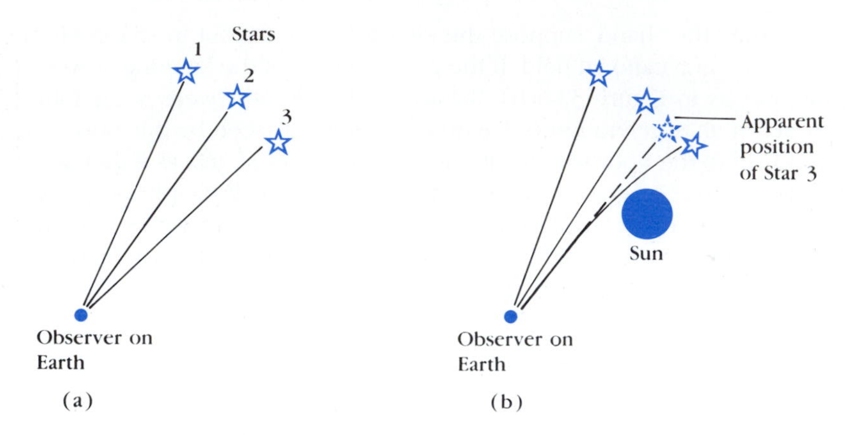

FIGURE 32-8
(a) Three stars in the sky. (b) The light from one of these same stars passes near the sun whose gravity bends the rays so the star appears higher than it actually is.

important prediction of Einstein's General Theory of Relativity. And it can be tested.

However, the amount a light beam would be deflected from a straight line path must be small even when passing a massive body. (For example, light near the earth's surface would drop only about 10^{-10} m after traveling 1 km, which is equal to the diameter of a small atom and certainly not detectable.) The most massive body near us is the sun, and it was calculated that light from a distant star would be deflected by 1.75″ of arc (tiny but detectable) as it passed near the sun (Figure 32-8). However, such a measurement could only be made during a total eclipse of the Sun, so the sun's tremendous brightness would not overwhelm the starlight passing nearby. An opportune eclipse occurred in 1919 and a group of scientists journeyed by ship to the south Atlantic to observe it. Their photos of stars around the sun revealed shifts in accordance with Einstein's prediction. One newspaper flaunted the headling "Light Caught Bending." (Newton's gravitational theory also predicts a gravitational deflection of light, considered as particles traveling at speed *c*, but the predicted deflection is less than that observed. The many observations since 1919 support Einstein's theory.)

Curved Space-Time

Einstein published his most general theory in 1916. In it, he proposed that gravity be treated as a property of space itself. As a result, space—or more precisely, four-dimensional space-time (see Section 6 of Chapter 25)—is predicted to be *curved*, particularly in the vicinity of massive bodies.

What is meant by *curved space*? To understand, let us recall that our normal method of viewing the world is via Euclidean plane geometry. In plane geometry there are many axions and theorems we take for granted, such as that the sum of the angles of any triangle is 180° (Figure 32-9). Other geometries, non-Euclidean, have also been imagined by mathemeti-

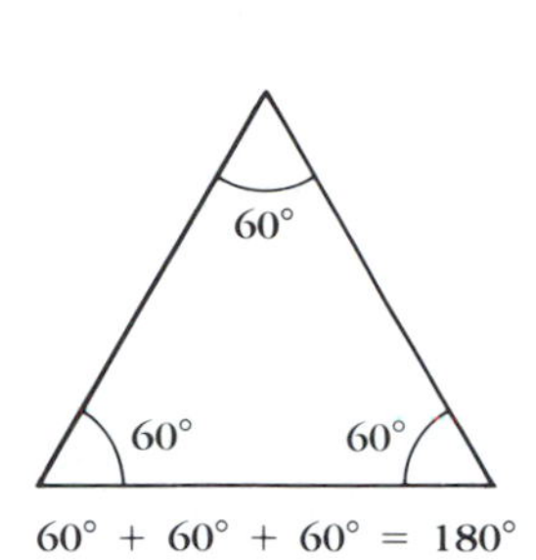

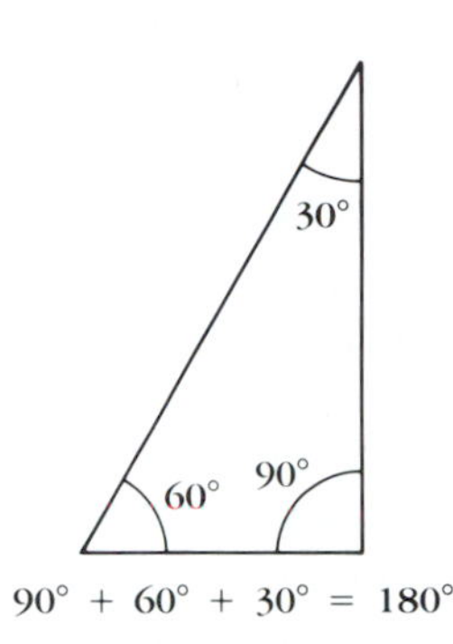

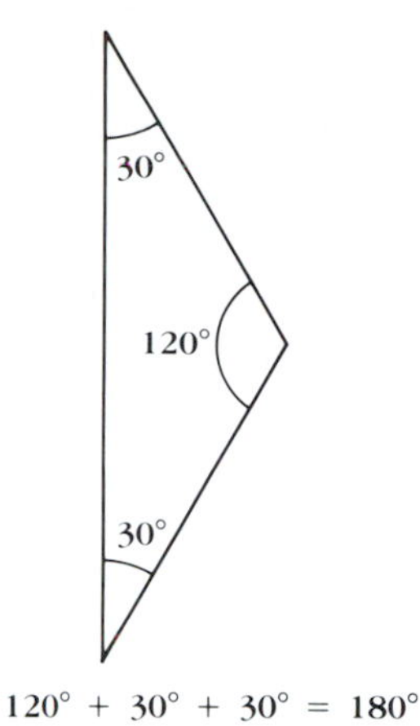

FIGURE 32-9
The sum of the angles of any (plane) triangle is 180°.

cians, which involve curved space. Now it is hard enough to imagine three-dimensional curved space, much less curved four-dimensional space-time. So let us explain the idea of curved space by using two-dimensional surfaces.

Consider, for example, the two-dimensional surface of a sphere. It is clearly curved, Figure 32-10, at least to us who view it from the outside—from our three-dimensional world. But how would hypothetical two-dimensional creatures determine whether their two-dimensional space were flat (a plane) or curved? One way would be to measure the sum of the angles of a triangle. If the surface is a plane, the sum of the angles is 180° (Figure 32-9). But if the space is curved, and a sufficiently large triangle is constructed, the sum of the angles will *not* be 180°. To construct a triangle on a curved surface, say a sphere, we must use the equivalent of a straight line: that is, the shortest distance between two points. This is called a *geodesic* and on a sphere it is an arc of a great circle (such an arc is contained in a plane passing through the center of the sphere) such as the earth's equator and the earth's longitude lines. Consider, for example, the large triangle of Figure 32-10, whose sides are two "longitude" lines passing from the "north pole" to the equator, a part of which forms the third side. The two longitude lines make 90° angles with the equator (look at a world globe to see this more clearly); if they make, say, a 90° angle with each other at the north pole, the sum of these angles is 90° + 90° + 90° + = 270°. This is clearly *not* a Euclidean space. (Note, however, that if the triangle is small in comparison to the radius of the sphere, the angles will add up to nearly 180°, and the triangle will seem flat.)

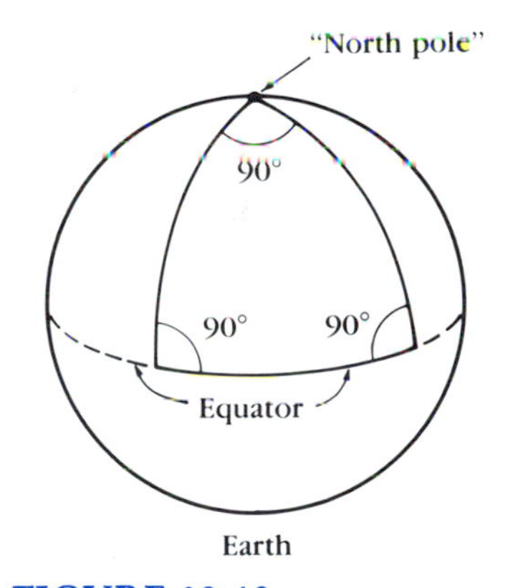

FIGURE 32-10
On a two-dimensional curved surface, the sum of the angles of a triangle may not be 180°.

Another way to test the curvature of space is to measure the radius r and circumference C of a large circle. On a plane surface, $C = 2\pi r$. But on a two-dimensional spherical surface, C is *less* than $2\pi r$, as can be seen in

Figure 32-11. Such a surface is said to have *positive curvature*. On the saddle-like surface of Figure 32-12, the circumference of a circle is greater than $2\pi r$; such a surface is said to have a *negative curvature*.

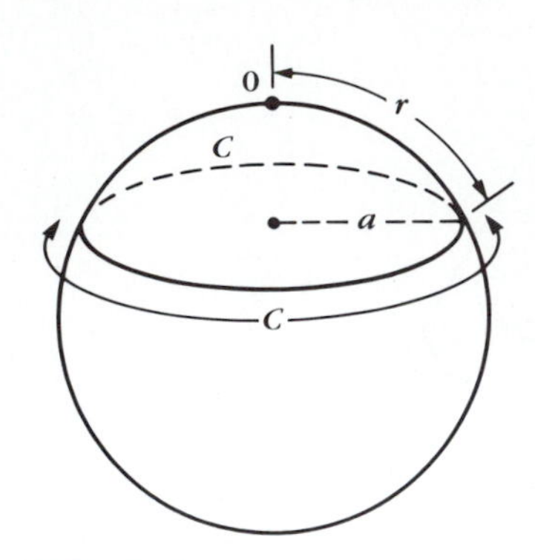

FIGURE 32-11
On a spherical surface, a circle of circumference C is drawn about point O as the center. The radius is the distance r along the surface. (Note that in our three-dimensional view we can tell that $2\pi a = C$; since $r > a$, then $2\pi r > C$.)

The Universe: Open or Closed?

Now, what about our universe? On a large scale does it have positive curvature, negative curvature, or is it flat (zero curvature)? In the nineteenth century, Carl Friedrich Gauss (1777–1855) tried to determine whether our natural three-dimensional space deviated from Euclidean space by measuring the angles of a triangle formed by three mountain peaks using light rays as sides of the triangle. He was unable to detect any deviation from 180°, nor are experiments today accurate enough to detect any deviation.

Nonetheless, the question of the curvature of space in the real world is an important one in cosmology. And the answer is still not known. If the universe has a positive curvature, then the universe is *finite*, or *closed*. This does not mean that in such a universe the stars and galaxies would extend out to a certain boundary, beyond which there is empty space. Rather, galaxies would be spread throughout the space, and the space would fold back and "close on itself." There is no boundary.[†] If a person were to start moving in a straight line in a particular direction he or she would eventually return to the starting point—albeit eons of time later. On the other hand, if the curvature is zero or negative, the universe would be *open*; it would just go on and on and never fold back on itself. An open universe would be *infinite*. Whether the universe is open or closed depends, in part, on how much total mass there is in the universe, as we will discuss in the next chapter. If the mass is great enough, it curves space into a positively curved, closed, and finite space.

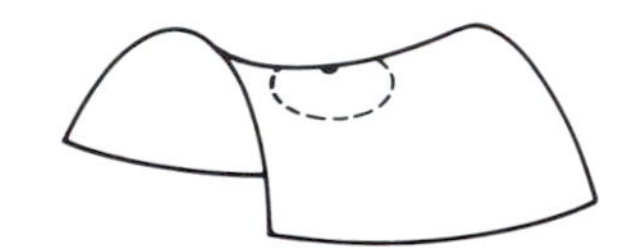

FIGURE 32-12
Example of a two-dimensional surface with negative curvature.

Gravity as Curvature of Space

According to Einstein's theory, space-time is curved, especially locally near massive bodies. To visualize this we might think of space as a thin rubber sheet; if a heavy weight is hung from it, it curves as shown in Figure 32-13. The weight corresponds to a huge mass that causes space (space itself!) to curve. Thus, in Einstein's theory* we do not speak of the "force" of gravity

---

[†] To ask "What is beyond such a closed universe?" is futile, since the space of the universe is all there is.

* Alexander Pope (1688–1744) wrote an epitaph for Newton:

"Nature and Nature's laws lay hid in night.
God said, Let Newton be! and all was light."

Sir John Squire (1884–1958), perhaps uncomfortable with Einstein's profound thoughts, added:

"It did not last; the Devil howling *Ho*
Let Einstein be, restored the status quo."

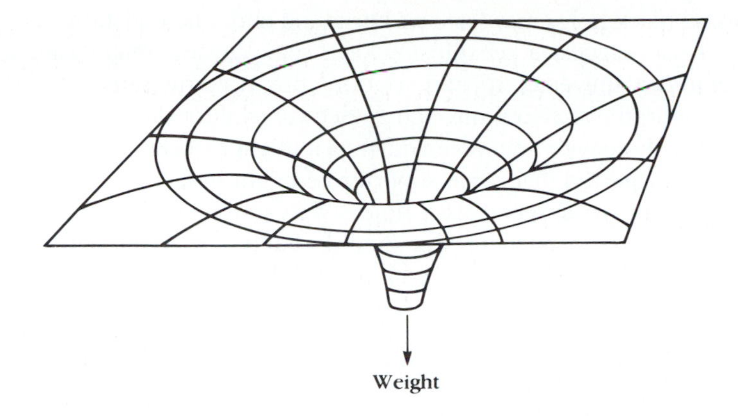

FIGURE 32-13
Rubber sheet analogy for space-time curved by matter.

acting on bodies; instead we say that bodies and light rays move along geodesics (equivalent of straight lines in plane geometry) in curved space-time. Thus, a body at rest or moving slowly near the great mass of Figure 32-13 would follow a geodesic toward that body.

Black Holes

The extreme curvature of space-time shown in Figure 32-13 could apply to a *black hole*. A black hole, as we saw in the previous section, is so dense that even light cannot escape from it. To become a black hole a body must undergo *gravitational collapse*,[†] contracting by gravitational self-attraction, to within a radius called the *Schwarzchild radius*. Matter within this radius is predicted by general relativity to fall all the way to the very central point ($r = 0$) of the body, forming an infinitely dense *singularity*. However, it is not known whether general relativity still prevails at the unimaginably dense hot conditions in a black hole. So the conclusion (of a "singularity") is open to question.

The Schwarzchild radius also represents the *event horizon* of a black hole. By event horizon we mean the surface beyond which no signals can ever reach us, and thus inform us of events that happen. As a star collapses toward a black hole, the light it emits is pulled harder and harder by gravity, but we can still see it; once the matter passes within the event horizon the emitted light cannot escape, but is pulled back in by gravity* All we can know about a black hole is its mass, its angular momentum (there could be rotating black holes), and its electric charge. No other information, no details of its structure or the kind of matter it was formed of, can be known.

[†] A black hole might also be formed by compression, say in a supernova explosion. It has also been suggested that tiny black holes may have been formed in the very early universe, just after the Big Bang (see in Chapter 33).

* According to quantum mechanics, matter and radiation could escape from a black hole by a process known as "tunneling," but the rate at which this could happen would be extremely low.

(Some black holes may have an intriguing end. The equations describing collapse suggest a reexplosion of a star outward might take place into *another* universe, related to ours but not the same. An object exploding in this way out through its Schwarzchild radius would behave like a black hole with time reversed; thus it might be called a *white hole*. It has been suggested that some supernovae might in fact be while holes.)

How might we observe black holes? We cannot see them because no light can escape from them. But they do exert a gravitational force on nearby bodies. The suspected black hole at the center of our galaxy was discovered by examining the motion of matter in its vicinity. Another technique has been to examine stars which appear to be rotating with a second sister star (a *binary* star, or system) although the sister star is invisible. If the unseen star is a black hole, it might be expected to pull off gases and other materials from its visible sister. As this matter approached the event horizon, it would be highly accelerated and should emit X rays of a characteristic type. At present there is fairly reasonable evidence of a black hole in the binary star Cygnus X-1; whether it actually is or not awaits further experiment.

SUMMARY

The night sky contains myriads of stars including those in the Milky Way, which is a "side view" of our *Galaxy* looking along the plane of the disc. Our Galaxy includes about 10^{11} stars. Beyond our Galaxy are myriads of other galaxies. Astronomical distances are measured in *light-years* ($1 \text{ ly} \approx 10^{13}$ km). The nearest star is 4 ly away; our galactic disc has a diameter of 100,000 ly; the nearest other galaxy is 2 million ly away. Distances are often specified in *parsecs*, where 1 parsec = 3.26 ly.

Stars are believed to begin life as collapsing masses of hydrogen gas (protostars). As they contract, they heat up (PE is transformed to KE). When the temperature reaches 10 million degrees, nuclear fusion begins and forms heavier elements (*nucleosynthesis*), mainly helium at first. The energy released during these reactions balances the gravitational force, and the young star stabilizes as a *main sequence* star. The tremendous luminosity of stars comes from the energy released during these thermonuclear reactions. After billions of years, as helium has collected in the core and hydrogen is used up, the core contracts and heats further; the envelope expands and cools, and the star becomes a *red giant* (larger star, redder color). The next stage of stellar evolution depends on the mass of the star. Stars of mass less than about 1.4 solar masses cool further and become *white dwarfs*, eventually fading and going out altogether. Heavier stars contract further due to their greater gravity; the density becomes equal to nuclear density, electrons are pushed into protons, and the star becomes essentially a hugh nucleus of neutrons. This is a *neuton star*, and the energy released from its final core contraction is believed to produce *supernovae*. If the star's mass is greater than 2 or 3 solar masses, it may contract even further and form a *black hole* which is so dense that no matter or light can escape from it.

In the *general theory of relativity*, the *equivalence principle* states that an observer cannot distinguish acceleration from a gravitational field. Said another way, gravitational and inertial mass are the same. The theory predicts gravitational bending of light rays to a degree consistent with experiment. Gravity is treated as a curvature in space and time, especially near massive bodies. The universe as a whole may be curved. If there is sufficient mass the curvature of the universe is positive, and the universe is *closed* and *finite*; otherwise it is *open* and *infinite*.

QUESTIONS

1. The Milky Way was once thought to be "Cloudy," but isn't any more. Explain.
2. Give an explanation for why some galaxies have arms.
3. Discuss the statement, "We are a mere speck in a vast universe."
4. If you were measuring star parallaxes from the moon instead of Earth what corrections would you have to make? What changes would occur if you were measuring parallaxes from Mars?
5. A star is in equilibrium when it radiates at its surface all the energy generated at its core. What happens when it begins to generate more energy than it radiates? Less energy? Explain.
6. Describe a red giant star. List some of its properties.
7. Select a point on the H-R diagram. Mark several directions away from this point. Now describe the change taking place in a star moving in each of these directions.
8. Does the H-R diagram reveal anything about the core of a star?
9. Why do some stars end up as white dwarfs, and others as neutron stars or black holes?
10. What is a geodesic? What is its role in general relativity?

EXERCISES

1. The circumference of Earth is about 40,000 km. How long does it take for light to travel around Earth's equator? How long is this distance in light-seconds?
2. How many light-seconds are there in a light-minute?
3. How far away is the Sun in light-seconds?

4. A star is 60 parsecs away. How long does it take for its light to reach us?
5. A star exhibits a parallax of 0.33 seconds of arc. How far away is it?
6. The parallax angle of a star is 0.00014°. How far away is the star?
7. A star is 30 parsecs away. What is its parallax angle? State (a) in seconds of arc, (b) in degrees.
8. What is the parallax angle for a star that is 4 light-years away? How many parsecs is this?
9. If one star is twice as far away from us as a second star, will the parallax angle of the first star be greater or less than that of the second star? By what factor?
10. Describe a triangle, drawn on the surface of a sphere, for which the sum of the angles is (a) 360° and (b) 180°.

COSMOLOGY

CHAPTER

33

The study of the universe as a whole is known as **cosmology**. It deals especially with the search for a theoretical framework to understand the observed universe, its origin, and its future.

The questions posed by cosmology are complex and difficult; the possible answers are often unimaginable. They are questions like "Has the universe always existed, or did it have a beginning in time?" Either alternative is extremely difficult to imagine: time going back indefinitely into the past, or an actual moment when the universe began (but, then, what was there before?). And what about the size of the universe? Is it infinite in size (hard to imagine) or is it finite in size (also hard to imagine, for if it is finite, we cannot ask what is beyond it since the universe is all there is.)

1. THE EXPANDING UNIVERSE

We discussed in Chapter 32 how individual stars evolve from their birth to their death as white dwarfs, neutron stars, and black holes. But what about the universe as a whole: is it static, or does it evolve? The evolution of stars suggests the universe as a whole evolves; let us look at the evidence.

In his theory of the universe, Newton assumed the universe was *static*: no large-scale changes occur over time. Newton recognized the difficulties in imagining a universe either as finite or as infinite. If it is finite and has a boundary, then we naturally ask "What is beyond the boundary?" Yet, why wouldn't such a region beyond also be part of the universe? Furthermore, a static finite universe would be unstable since why wouldn't gravity pull everything into one glob at the center? An infinite universe was equally hard to imagine; it too would be confronted by problems, one of which later came to be known as *Olber's paradox*.

Why the Night Sky Is Dark; Olber's Paradox

One of the most obvious astronomical observations is that *the sky at night is dark*. Why isn't it bright? Kepler had reasoned that if the universe were infinite, with stars uniformly distributed throughout, then any line of sight should fall on the surface of a star. Hence the night sky would appear bright everywhere. So he concluded the universe could not be infinite (which itself caused problems). However, it's not quite so simple: stars are not points, but have a finite size—they subtend a finite (though minute) angle. Thus the night sky could be bright even if the universe were finite.

There are possible answers to Olber's paradox, as it is called. For example, the universe might be infinite in extent but not infinitely old, so light from distant stars would not yet have reached us. It is commonly believed that this idea, together with reduced light energy from the redshift in the expansion of the universe (which we next discuss), provides a solution to Olber's paradox. Yet evidence for these two ideas—that the universe did not always exist and that it is not static—was not found until well into the twentieth century. Even Einstein, while working on solutions to his general relativity equations in 1917, assumed the universe was static.

Hubble and the Redshift

One of the most important scientific discoveries of this century proposes that distant galaxies are racing away from us, and that the further they are from us, the faster they are moving away. How astronomers reached this astonishing conclusion, and what it means for the past history of the universe as well as its future, will occupy us for the remainder of the book.

The idea that the universe is expanding was first put forth by Hubble in 1929. It was based on observations of the Doppler shift of light emitted by stars. In Chapter 17 we discussed how the frequency and wavelength of sound are altered if the source is moving toward or away from an observer. If the source moves toward us, the frequency is higher and the wavelength is shorter; if the source moves away from us, the frequency is lower and the wavelength is longer.

This *Doppler effect* also occurs for light. When a source emits light of a particular wavelength and the source is moving away from us, the wavelength appears longer to us; that is, the color of the light (if it is visible) is shifted toward the red end of the visible spectrum (see Figure 24-8), and this is called a **redshift**. If the source moves toward us, the color shifts toward the blue end of the spectrum (shorter wavelength). The amount of shift depends on the velocity of the source. For speeds not to close to the speed of light, the fractional change in wavelength is proportional to the speed of the source to or away from us (as is the case for sound; see Chapter 17). In the spectra of stars and galaxies, lines are observed that correspond to lines in the known spectra of particular atoms. What Hubble found was that the lines seen in the spectra of galaxies were generally ***redshifted***, and that the amount of shift seemed to be proportional to the distance of the galaxy from us. That is, the velocity, v, of a galaxy moving away from us is proportional to its distance, d, from us:

$$v = Hd.$$

This is known as **Hubble's law**, and the constant H is called *Hubble's constant*. Hubble's law does not work well for nearby galaxies—in fact some are actually moving toward us ("blue-shifted"); but this is believed to merely represent random motion of the galaxies. For more distant galaxies the velocity of recession (Hubble's law) is much greater than that of random motion, and so is dominant.

The value of H is not known very precisely. It is generally taken to be about

$$H \approx 50 \text{ km/s/Mpc}$$

(that is, 50 km/s per megaparsec [million parsecs] of distance). If we use light years for distance, $H \approx 15$ km/s per million light years of distance. The data, however, suggest H could be as low as 40 km/s/Mpc or as high as about 120 km/s/Mpc. This rather large uncertainty arises mainly from the uncertainty in distance measurement.

What does it mean that distant galaxies are all moving away from us, and with ever greater speed the farther they are from us? It is as if there had been a great explosion at some distant time in the past. And at first sight we seem to be in the middle of it all. But we aren't. The expansion appears the same from any other point in the universe. To understand why, let us use Figure 33-1. In part (a) we have the view from Earth (or Earth's galaxy). The velocities of surrounding galaxies are indicated by arrows, pointing away from us, and greater for galaxies more distant from us. Now what if we were on the galaxy labeled A in Figure 33-1(a)? From Earth, galaxy A appears to be moving to the right at a velocity, call it v_A, represented by the arrow pointing to the right. If we were *on* galaxy A, Earth would appear to be moving to the left at velocity v_A. To determine the velocities of other galaxies relative to A, we vectorially add the velocity vector, $-\mathbf{v}_A$ (arrow of length v_A but pointing to the left), to all the velocity arrows shown in (a). This calculation yields Figure 33-1(b). Clearly, the universe is expanding away from galaxy A as well; and the velocities of receding galaxies are also proportional to their distance from A.

Thus the expansion of the universe can be stated as follows: all galaxies are racing away from ***each other*** at an average rate of about 50 km/s per megaparsec of distance between them (or 15 km/s per million light years of separation). The ramifications of this profound discovery are enormous, and we discuss them in a moment.

FIGURE 33-1
Expansion of the universe is the same from any point in the universe.

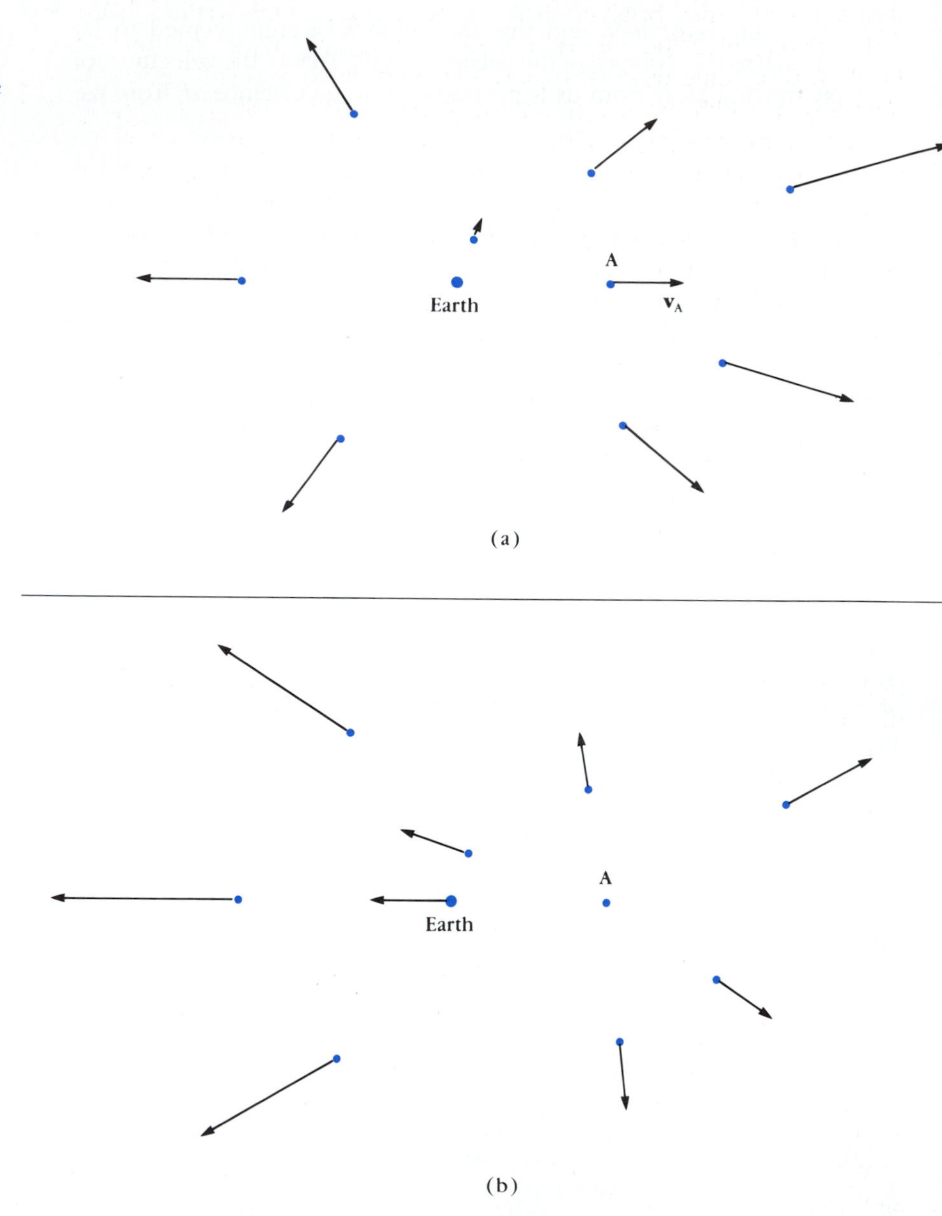

Quasars

At this point we must point out that there is a class of objects called "quasi-stellar objects" or **quasars**, that do not conform to Hubble's law. Quasars are as luminous as nearby stars but display very large redshifts. If quasars are normal participants in the general expansion of the universe according to Hubble's law, their large redshifts suggest they are very distant. If they are so far away, they must be incredibly bright—sometimes thousands of times brighter than normal galaxies. Observations using radio waves, plus the observation of a strong and rapid variation in brightness of some quasars, suggest they must be very small, much less than a thousandth the size of a normal galaxy. How such "tiny" objects could emit the energy of thousands of galaxies remains a cosmic puzzle.

A small number of astronomers maintain that some quasars, at least, are not of abnormal brightness because they are nearer than their redshifts suggest. In this case we would be trading an unresolved brightness problem for an unresolved redshift problem. An interesting piece of evidence is that the population density of quasars seems to increase with distance from us. If they are at great distances from us—as their redshifts suggest—this would merely mean they were more common in the early universe than they are now. But if instead they are much closer to us—as their brightness suggests—it would seem that *we* are in a special place in the universe, the place where quasars are least populous. Most astronomers are unwilling to accept this, for it would violate a basic assumption in cosmology, the so-called *cosmological principle*.

The Cosmological Principle

A basic assumption in cosmology is the idea that on a large scale the universe should look the same to observers at different places at the same time. In other words, the universe is both isotropic (looks the same in all directions) and homogeneous (would look the same if we were located elsewhere, say in another galaxy). This assumption is called the **cosmological principle**. On a local scale, say in our solar system or within our Galaxy, it clearly does not apply (the sky looks different in different directions). But on a large scale it appears to be valid: the average population density of galaxies and clusters of galaxies seems to be the same in different areas of the sky; and, very importantly, as we just saw, the expansion of the universe is fully consistent with the cosmological principle (see Figure 33-1). Indeed, the cosmological principle allows us to treat the universe as a single evolving entity and not as a random collection of material bodies. Another way of stating the cosmological principle is this: There is nothing special about the Earth on a cosmological scale; our large-scale observations are no different from those that might be made elsewhere in the universe.

What a change has occurred since that time, only a few hundred years ago, when we saw ourselves at the center of the universe.

Age of the Universe

The expansion of the universe, as described by Hubble's law, strongly suggests that galaxies must have been closer together in the past than they are now. Further, Hubble's law is consistent with all galaxies having been quite close together at the same time in the past. This is, in fact, the basis of the *Big Bang* theory of the origin of the universe (discussed in the next section) which pictures the beginning of the universe as a great explosion.

If we use the value of the Hubble constant, $H \approx 15$ km/s per million light years, then the time required for the galaxies to arrive at their present separations must have been one million light-years divided by 15 km/s (since $v = \frac{d}{t}, t = \frac{d}{v}$), which gives 20 billion years (20×10^9 yr). The "age" of the universe calculated in this way, which we could call the "characteristic expansion time," is probably an overestimate. The actual age should be less considering that galaxies have not been moving with fixed velocities but have been slowing down under the action of their mutual gravitational attraction. Perhaps 10 to 15 billion years would be a better estimate.

There are two other independent checks on the age of the universe. The first is determination of the age of the earth (and solar system) from radioactivity, primarily using uranium, which places the age of the solar system at about $4\frac{1}{2}$ billion years. Second, using the theory of stellar evolution, the ages of stars have been estimated to be about 10–15 billion years. These independent and unrelated determinations are consistent with a Big Bang occurring 10–15 billion years ago. The lower value determined from radioactivity is consistent since we would expect the origin of the earth (and the solidification of rocks) to have occurred somewhat after the origin of the universe as a whole.

2. THE BIG BANG; EARLY HISTORY OF THE UNIVERSE

The expansion of the universe implies that the matter of the universe was once much closer together than it is now. This is the basis for the idea that the universe began 10 to 15 billion years ago, with a huge explosion, which is affectionately referred to as the **Big Bang**.

If there was a Big Bang, it must have occurred simultaneously at all points in the universe. If the universe is *finite*, the explosion would have taken place in a tiny volume approaching a point. However, this concentrated point of extremely dense matter is not to be thought of as a mass in the midst of a much larger space around it. Rather, the initial dense point *was* the universe—the entire universe. There wouldn't have been anything else. If, on the other hand, the universe is *infinite*, then the explosion would have occurred at *all* points in the universe; for an infinite universe, even if smaller at an earlier time, would still have been infinite. In either case, when we say the universe was once smaller than it is now, we mean that the average separation between galaxies was less. Thus it is the *size of the universe itself* that has increased since the Big Bang.

What is the evidence supporting the Big Bang? First, the age of the universe as calculated from the Hubble expansion, from stellar evolution, and from radioactivity, all point to a consistent time of origin for the universe, as we saw in the last section. Another, and crucial, piece of evidence was the discovery in the 1960s of the cosmic microwave radiation background.

The 3K Cosmic Microwave Background Radiation

In 1964 Arno Penzias and Robert Wilson were experiencing difficulty with what they assumed to be background "noise" or "static" in their radio telescope (a large antenna device for detecting radio waves from the heavens). But try as they might, they could not eliminate the static. They finally became convinced that it was real and that it was coming from outside our Galaxy. They made precise measurements at a wavelength $\lambda = 7.35$ cm, which is in the microwave region of the electromagnetic spectrum. (See Figure 21-10; this radiation is called "microwave" because the wavelength, though much greater than that of visible light, is somewhat smaller than wavelengths for ordinary radio waves, which are typically meters or hundreds of meters.) The intensity of this radiation was found not to vary by day or night or time of year. Neither did it depend on direction; it came from all directions in the universe with equal intensity (within less than one part per thousand)† It could only be concluded that the radiation came from beyond our Galaxy, from the universe as a whole.

The intensity of this radiation at $\lambda = 7.35$ cm corresponded to radiation emitted by dense matter, so-called "blackbody radiation," at a temperature of 3 K (3 degrees above absolute zero). As we saw in Section 1 of Chapter

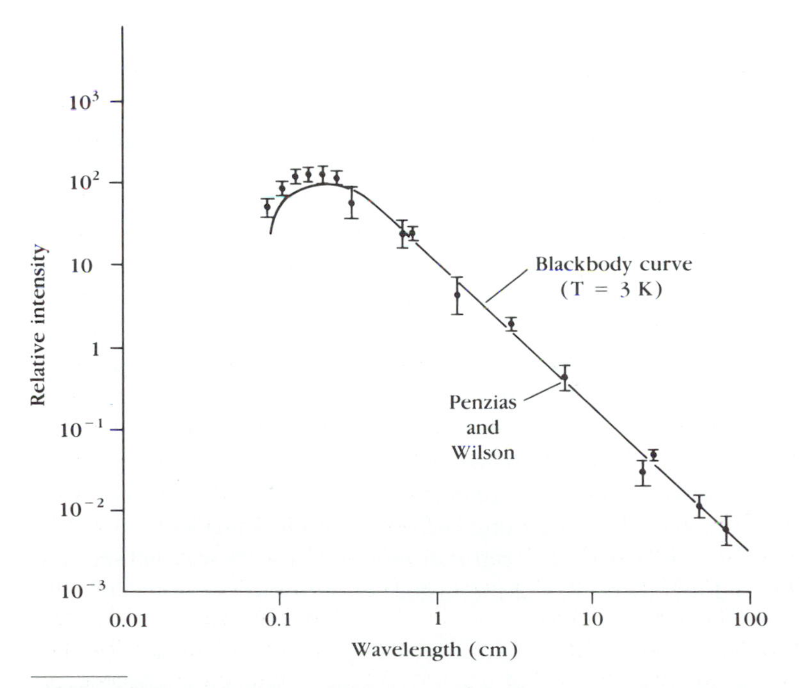

FIGURE 33-2
Spectrum of cosmic microwave background radiation, showing blackbody curve and experimental measurements including that of Penzias and Wilson. (The vertical bars represent the experimental uncertainty in each measurement.)

† The remarkable uniformity of this cosmic microwave background radiation is in accordance with the Cosmological Principle. The recently detected nonuniformity of less than 0.1% merely reflects the velocity of our galaxy (about 600 km/s) relative to the average matter and radiation of the universe.

26, dense matter emits a continuous spectrum of wavelengths, whose intensity and distribution (Figure 26-1) depend on temperature. When radiation at other wavelengths was measured, the intensities were found to fall on a blackbody curve, as shown in Figure 33-2, with a peak at around 0.1 cm corresponding to a temperature of 3 K.

The discovery of this **cosmic microwave background radiation** at a temperature of 3 K ranks as one of the two most significant cosmological discoveries of this century. (The other was Hubble's expanding universe.) It is highly significant because it provides strong evidence in support of the Big Bang, and it gives us some idea of conditions in the very early universe. In fact, in the late 1940s George Gamow and his collaborators calculated that a Big Bang origin of the universe should have generated just such a microwave background radiation.

To understand this, let us look at what a Big Bang might have been like. There must have been a tremendous release of concentrated energy. The temperature must have been extremely high, so high that there could not have been any atoms in the very early stages of the universe. Instead, the universe must have consisted solely of radiation (photons) and elementary particles. The universe would have been opaque—the photons in a sense "trapped," since as soon as they were emitted they would have been scattered or absorbed, primarily by electrons. Indeed, the microwave background radiation is strong evidence that matter and radiation were once in equilibrium at a very high temperature. As the universe expanded, the energy would have spread over an increasingly larger volume and the temperature would have dropped. Only when the temperature had reached about 3000 K could nuclei and electrons have stayed together as atoms. With the disappearance of free electrons, the radiation would have been freed—"decoupled" from matter, we say—to spread throughout the universe. As the universe expanded, so too the wavelengths of the radiation expanded, reshifting to longer wavelengths that correspond to lower temperature, until they would have reached the 3 K we observe today.

Although the total energy associated with the cosmic microwave background radiation is much larger than that from other radiation sources (such as the light emitted by stars), it is small compared to the energy, and mass (remember $E = mc^2$), associated with matter. In fact today, radiation is believed to make up less than $\frac{1}{1000}$ of the energy of the universe: today the universe is *matter-dominated*. But it was not always so. The cosmic microwave background radiation strongly suggests that early in its history the universe was *radiation-dominated*†; but, as we shall see, that period lasted less than $\frac{1}{10,000}$ of the history of the universe (thus far).

† If there had not been such intense radiation in the first few minutes of the universe, nuclear reactions might have produced a much larger percentage of heavy nuclei than we see. Instead, nearly $\frac{3}{4}$ of visible matter is hydrogen, presumably because the intense radiation blasted apart any heavy nuclei into their constituent protons and neutrons as soon as they were formed.

The Early History of the Universe: the Standard Model

It is now almost generally agreed that the evolution of the universe must have been determined in the first few moments of the Big Bang. In the last decade or two, a convincing theory of the origin and evolution of the universe has developed, now known as the **Standard Model**. Although a few cosmologists hold other views—we discuss one of these, the *Steady State Model*, later — most favor the Standard Model. Much of this theory is based on recent theoretical and experimental advances in elementary particle physics. Indeed, in the last few years cosmology and elementary particles have cross-fertilized to a surprising extent.

Let us go back now to the earliest of times—as close as possible to the Big Bang—and follow a Standard Model scenario of events as the universe expanded and cooled after the Big Bang. Initially we will be talking of extremely tiny time intervals, as well as extremely high temperatures, far beyond anything in the universe today. Figure 33-3 is a graphic representation of the events, and it may be helpful to consult it as we go along.

We begin at a time only a minuscule fraction of a second after the Big Bang, 10^{-43} s. Although this is an unimaginably short time, predictions as early as this can be made based on present theory, albeit somewhat speculatively. Earlier than this instant we can say nothing since we do not yet have a theory of quantum effects on gravity which would be needed for the incredibly high densities and temperatures then. It is imagined,

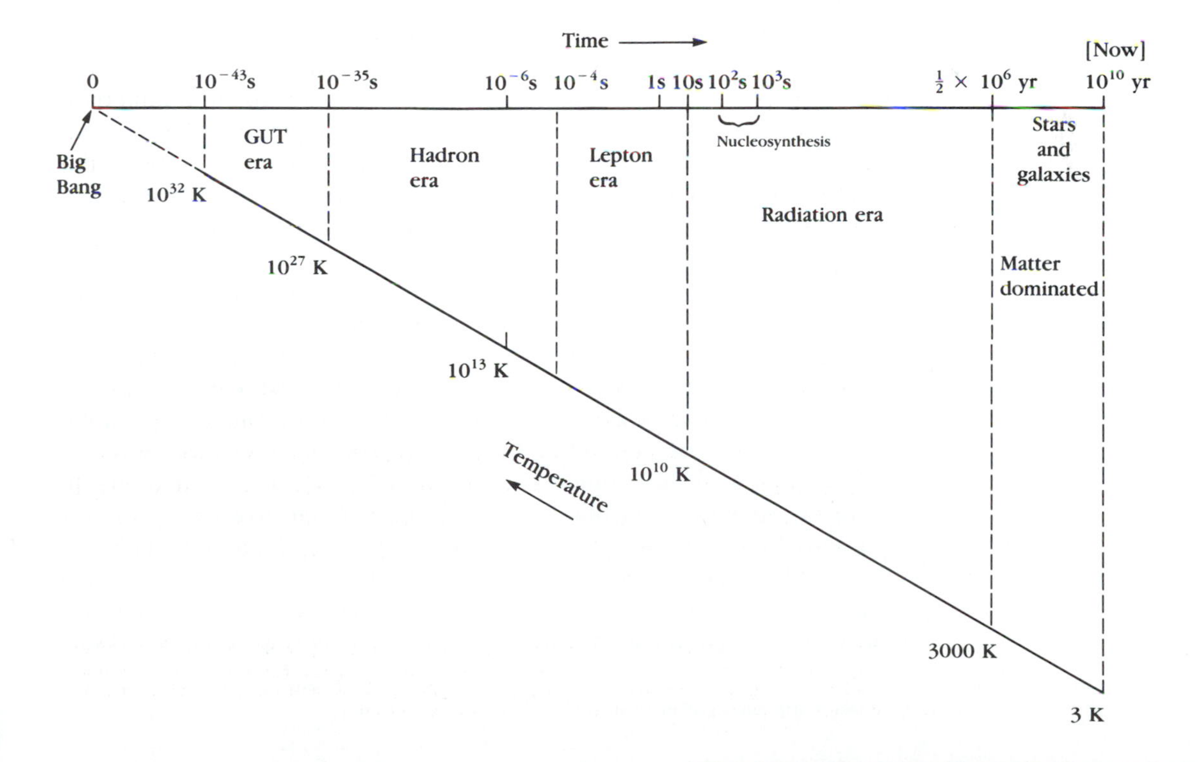

FIGURE 33-3
Graphic representation of development of the universe after the Big Bang according to the Standard Model.

however, that prior to 10^{-43} s the four forces in nature were unified—there was only one force. The temperature would have been about 10^{32} K, corresponding to particles moving about every which way with an average kinetic energy (KE) of 10^{19} GeV. (Recall from Chapter 13 that the average KE of particles is proportional to, and is a measure of, the absolute or kelvin (K) temperature[†], a fact we will use often in this discussion.) At $t = 10^{-43}$ s a kind of "phase transition" is believed to have occurred during which the gravitational force, in effect, "condensed out" as a separate force. This, and subsequent phase transitions, are somewhat analogous to the phase transitions water undergoes as it cools from a gas, condenses into a liquid, and with further cooling freezes into ice. The symmetry of the four forces was broken, but the strong, weak, and electromagnetic forces were still unified. At this point the universe entered the so-called **grand unified era** (after GUT—see Chapter 31). There was no distinction between quarks and leptons; baryon and lepton numbers were not conserved. Very shortly thereafter, as the universe expanded considerably and the temperature had dropped to about 10^{27} K, there was another phase transition during which the strong force condensed out. This probably occurred about 10^{-35} s after the Big Bang. Now the universe was filled with a soup of leptons and quarks. The leptons included electrons, muons, taus, neutrinos, and all their antiparticles. The quarks were initially free (something we have not seen in our present universe), but soon they began to "condense" into more normal particles: nucleons (protons and neutrons) and the other hadrons and their antiparticles. With this *confinement* of quarks, the universe entered the **hadron era**.

We can think of this "soup" as a grand mixture of particles and antiparticles, as well as photons—all in roughly equal numbers—colliding with one another frequently and exchanging energy.

By the time the universe had cooled to about 10^{13} K (10 trillion degrees), corresponding to an average KE of 1 GeV, the universe was only about a microsecond (10^{-6} s) old. Around this time the vast majority of hadrons disappeared. To see why, let us focus on the most familiar hadrons: nucleons and their antiparticles. When the average kinetic energy of particles was somewhat higher than 1 GeV, protons, neutrons, and their antiparticles were continually being created out of the energies of collisions ($E = mc^2$) involving photons and other particles; but just as quickly, the particles and antiparticles would annihilate (for example $p + \bar{p} \rightarrow$ photons or leptons). So the processes of creation and annihilation of nucleons were in equilibrium. The numbers of nucleons and antinucleons were high—roughly as many as there were electrons, positrons, or photons. But when the average kinetic energy dropped below about 1 GeV (1000 MeV), the minimum energy needed to create nucleons and antinucleons (more precisely the total energy needed equals the mass of a nucleon plus antinucleon, each of which is about 940 MeV—see Table

[†] 1 GeV = 1000 MeV = 10^9 eV $\approx 10^{13}$ K (10 trillion degrees kelvin).

29-1), the process of creation stopped. The process of annihilation continued, however, until there were almost no nucleons left. But not quite zero. Earlier in the universe, perhaps around 10^{-35} s after the Big Bang, a slight excess of quarks over antiquarks must have been formed.† And this resulted in a slight excess of nucleons over antinucleons. A very lucky thing for us, since we are made of these "leftover" nucleons. The excess of nucleons over antinucleons was about one part in 10 billion (1 in 10^{10}). During the hadron era there were about as many nucleons as photons; after it ended, there was thus only about one nucleon per 10^{10} photons, and this ratio has persisted to this day. Protons, neutrons, and all other heavier particles were thus tremendously reduced in number by about 10^{-6} after the Big Bang. The lightest hadrons, the pions, disappeared as the nucleons had; because they are the lightest mass hadrons (140 MeV), they were the last hadrons to go, around 10^{-4} s after the Big Bang. Lighter particles including electrons, positrons, neutrinos, photons —in roughly equal numbers—dominated, and the universe entered the **lepton era** (the era of "ligher particles").

By the time the first full second had passed (clearly the "longest" second in history!), the universe had cooled to about 10 billion degrees (10^{10} K). The average KE was about 1 MeV. This was still plenty of energy to create electrons and positrons, and balance the annihilation reactions, since their masses correspond to about 0.5 MeV. So there were about as many e^+ and e^- as there were photons. But within a few more seconds the temperature had dropped sufficiently so that e^+ and e^- could not be formed. Annihilation ($e^+ + e^- \rightarrow$ photons) continued; and, like nucleons before them, electrons and positrons all but disappeared from the universe—except, for a slight excess of electrons over positrons (eventually to join with nuclei to form atoms). Thus, about $t = 10$ s after the Big Bang, the universe entered the **radiation era**. Its major constituents were now photons and neutrinos; but the neutrinos, partaking only in the weak force, rarely interacted. So the universe, until then experiencing an energy balance between matter and radiation, became **radiation-dominated**; much more energy was contained in radiation than in matter, a situation that would last almost a million years.

Meanwhile, during the next few minutes, crucial events were taking place. Beginning about 2 or 3 minutes after the Big Bang, nuclear fusion began to occur. The temperature had dropped to a point (about 1 billion degrees, 10^9 K, corresponding to $\overline{\text{KE}} = 100$ keV) where nucleons could strike each other and be able fuse ($n + p \rightarrow$ deuterium, for example) but not so high that the newly formed nuclei could be immediately broken apart by subsequent collisions. Deuterium, helium, and very tiny amounts

† An alternative possibility is that there was perfect symmetry between quarks and antiquarks, matter and antimatter, but that the universe somehow separated into domains, some containing only matter, others only antimatter. If this were true, we would expect antiparticles from such distant domains to reach us, at least occasionally, in cosmic rays; but none has ever been detected.

of lithium were probably made. But the universe was cooling too quickly, and larger nuclei were not made. After only about a quarter of an hour, nucleosynthesis stopped, not to start again for millions of years (in stars). Thus, after the first hour or so of the universe, matter consisted mainly of bare nuclei of hydrogen (about 75%) and helium (about 25%) and electrons. But radiation (photons) continued to dominate.

[This Standard Model prediction of a 25% primordial production of helium is fully in accordance with what we observe today—the universe *does* contain about 25% He—and it is strong evidence in support of the standard Big Bang model. Furthermore, the theory says that 25% He abundance is fully consistent with there being three neutrino types, which is the number we observe so far; and it sets an upper limit of four to the maximum number of possible neutrino types. Actually the fourth could be another type of low-mass particle, a *photino* or a *gravitino*, for example. Here we have a situation where cosmology actually makes a specific prediction about fundamental physics.]

Our story is almost complete. The next important event occurred about a half-million years later. The universe had expanded to about $\frac{1}{1000}$ of its present size, and the temperature had cooled to about 3000 K. The average KE of nuclei, electrons, and photons was a few electron volts. This corresponds to ionization energies of atoms, which means that as the temperature dropped below this point electrons could orbit the bare nuclei and remain there (without being broken apart by collisions), thus forming atoms. With the birth of atoms, the photons which were continually scattering from the free electrons became much freer to spread throughout the universe. The total energy contained in radiation had been decreasing (redshifting as the universe expanded) until at this point it was about equal to the total energy contained in matter. As the photons expanded outward, they cooled further (to 3 K today, forming the cosmic microwave background radiation we detect from everywhere in the universe), and lost energy. But the mass of material particles did not decrease, so beginning at about this point the energy of the universe became increasingly concentrated in matter rather than in radiation: the universe became **matter-dominated**, as it remains today.[†]

Shortly after the birth of atoms, stars and galaxies formed—probably by self-gravitation around mass concentrations (inhomogeneities). This transpired about a million years after the Big Bang. The universe continued to evolve (see Section 2 of Chapter 32) until today, some 15,000 million years later.

This scenario is by no means "proven" in any sense. Nor does it answer all questions. But it does provide a tentative picture, for the first time, of how the universe began and evolved. Many new ideas are contained in the

[†] Although today matter contains more of the energy of the universe than does radiation, there are many more photons (perhaps 10 billion times more) than atoms, nuclei, and electrons. But each photon (at $T \approx 3$ K) has very little energy.

Standard Model, and many new speculative ideas (such as the "inflationary universe") are coming forth to help refine it and to help resolve problems.

There are, however, some questions we haven't yet treated such as: was there a stage before the Big Bang, or did time just begin with the Big Bang? And what of the future of the Universe? We'll look at these in a moment.

Steady State Model

Before looking to the future, we briefly discuss one of the alternatives to the Big Bang—the **Steady State Model**—which assumes that the universe is infinitely old and on the average looks the same now as it always has. Thus, according to the Steady-State Model, no large-scale changes have taken place in the universe as a whole, particularly no Big Bang. To maintain this view in the face of the expansion of galaxies away from each other, mass-energy conservation must be violated. That is, matter is assumed to be created continuously, keeping the density of the universe constant. The rate of mass creation required is very small—about one nucleon per second in a volume 1000 km on a side. This is much too small to be detected, and thus cannot be tested.

Although the Steady State model provided the Big Bang model with healthy competition in the 1950s, with the discovery of the cosmic microwave background radiation and other successes of the Big Bang model such as the abundance of primordial helium (neither of which the Steady State model can explain), the Steady State model has fallen into disfavor among most cosmologists.

3. AND THE FUTURE OF THE UNIVERSE?

According to the standard Big Bang model, the universe is evolving and changing. Individual stars are evolving and dying as white dwarfs, neutron stars, black holes. At the same time the universe as a whole is expanding outwardly. One important question is whether the universe will continue to expand forever. This question is connected to the curvature of space-time (Section 3 of Chapter 32) and to whether the universe is open (and infinite) or closed (and finite). These are three possibilities as shown in Figure 33-4. If curvature of the universe were *negative*, the expansion of the universe would never stop, although it would decrease due to the gravitational attraction of its parts; such a universe would be *open* and infinite. If the universe were *flat* (no curvature), it would still be open and infinite but its expansion would slowly approach a zero rate. Finally, if the universe had *positive* curvature, it would be *closed* and finite; the effect of gravity would be strong enough so that the expansion would eventually stop and the universe would begin to contract; all matter eventually would collapse back onto itself in a **big crunch**. If, in this last case, the maximum expansion of the universe corresponded to, say, an intergalactic separation twice what it is now, the maximum expansion would occur about 30 or 40

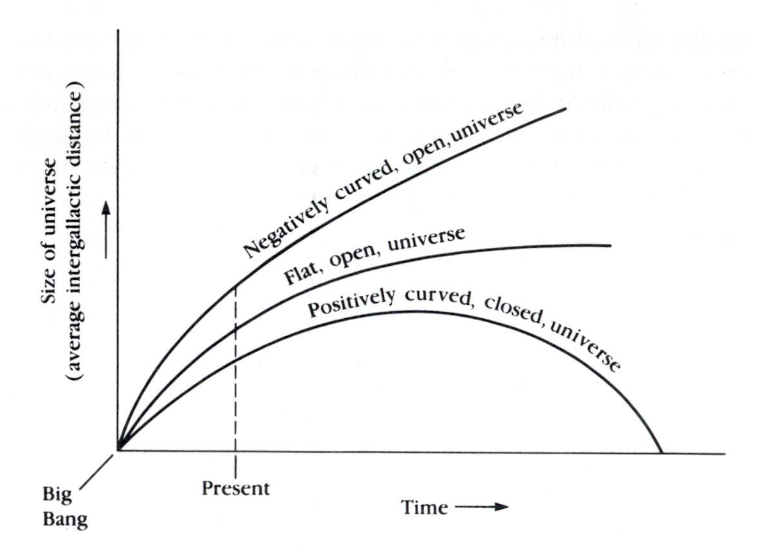

FIGURE 33-4
Three future possibilities for the universe

billion years from now. Then, as the universe began to contract, the big crunch would occur about 100 billion years after the Big Bang.

Whether we live in an open and continually expanding universe, or a closed one that eventually will contract, is a basic question in cosmology. But we don't know the answer. How might we find out? One way is to determine the average mass density in the universe. If the average mass density is above a critical value known as the *critical density*†, then gravity will prevent expansion from continuing forever, and will eventually pull the universe back into a big crunch. To say it another way, there would be sufficient mass that gravity would give space-time a positive curvature. If the actual density is equal to the critical density, the universe will be flat and open. If the actual density is less than the critical density, the universe will have negative curvature and be open, expanding forever.

Great efforts have gone into measuring the actual density of the universe. Estimates of the amount of visible matter in the universe put the actual density at only about 5% of the critical density, thereby suggesting an open universe. However, there is evidence for a significant amount of nonluminous matter* in the universe, enough to bring the density to almost exactly ρ_c. For example, observations of rotating galaxies suggest they rotate as if they had considerably more mass than we can see; and observations of the motion of galaxies within clusters also suggest they have considerably more mass than can be seen. If there is nonluminous matter, what might it be? One suggestion is that it consists of many small

† The critical density, ρ_c, has been estimated to be about $\rho_c \approx 5 \times 10^{-27}$ kg/m^3, or about 3 nucleons (or H atoms) per cubic meter.

* The needed mass is often referred to as the "missing mass."

primordial black holes made in the early stages of the universe. Another possibility is that neutrinos, once believed to be massless, may actually have rest mass. Since the universe probably contains as many neutrinos as photons (that is, a few billion times the number of nucleons, although this neutrino background has yet to be detected), neutrino masses of only a few eV would be sufficient to bring the actual density of the universe up to the critical density. Measurement of the neutrino masses within the next few years may be able to answer the question of an open or closed universe.

Another factor that, if we could measure it accurately enough, would provide the answer, is the so-called *deceleration parameter*. It is a measure of the rate at which the expansion of the universe is slowing. But to measure this rate requires looking far back in time, to the galaxies farthest away, whose light we receive now was emitted at a time closer to the beginning of the universe; at that time, the rate of expansion was much faster than today. Unfortunately, we would have to know the distance to these galaxies more precisely than is possible at present; so this method does not yield an answer to whether the universe is open or closed.

If the universe is open, how will it evolve in the future? According to the latest theories, which rely to a large extent on elementary particle theory, after about 10^{18} years, galaxies will have much of their matter knocked away and scattered throughout the universe by collisions with other stars; the remaining matter will eventually condense into massive "galactic black holes." Clusters of these will then coalesce into extremely massive "supergalactic black holes." Finally the black holes themselves will "evaporate"—the matter within them, through a slow quantum-mechanical process known as "tunneling," will "leak out." This process is so slow it would take on the order of 10^{100} years. The universe would then be a thin gas of electrons, positrons, neutrinos, and photons.

On the other hand, if the universe is closed it might well turn around and begin to contract even before all the stars have burnt out. As the universe contracts, the background radiation would increase in energy and temperature. The universe might simply retrace its steps, if it weren't for black holes. As density increases and the universe rushes toward its inevitable end in the big crunch, black holes might gobble up more and more matter until the entire universe coalesced into a single supermassive black hole—which would then be the universe.

If the universe is closed, what happens after the big crunch? We don't know, of course. What is possible, though, is a "bounce." That is, the dense fiery nucleus of the big crunch might explode again, resulting once more in an expanding universe. Thus the universe might be cyclic as shown in Figure 33-5. Such a *cyclic* or *pulsating* universe proposes a possible answer to one of our favorite "unanswerable" questions: what happened before the Big Bang? In this model there was simply a previous cycle. But left unanswered would be a number of other questions such as "When did it all begin?"

FIGURE 33-5
Cyclic model of the universe. Although the cycles shown here are the same, they could have different periods and different expansion rates.

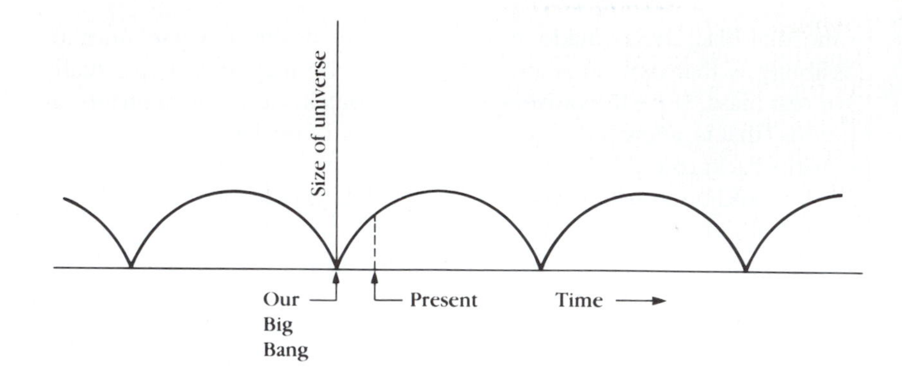

The questions raised by cosmology can seem absurd at times, they are so removed from "reality" We can always say: the sun is shining, it's going to burn on for an unimaginably long time, all is well. Nonetheless, the questions of cosmology are deep ones that fascinate the human intellect. One aspect that is especially intriguing is this: calculations on the formation and evolution of the universe have been performed that deliberately varied the values—just slightly—of certain fundamental physical constants. The result? A universe in which life could not exist. This has given rise to the so-called *Anthropic principle*, which says that if the universe were even a little different than it is, we couldn't be here. It seems the universe is exquisitely tuned, almost as if to accomodate us.

SUMMARY

Distant galaxies display a *redshift* of spectral lines, interpreted as a Doppler shift. The universe seems to be expanding, its galaxies racing away from each other at speeds proportional to the distance between them (*Hubble's law*). This expansion of the universe suggests an explosive origin, the *Big Bang*, which probably occurred 10–15 billion years ago. *Quasars* are objects with a large redshift (suggesting great distance) and high luminosity (suggesting closeness, or extraordinary and as yet unexplained energy output); their nature is disputed. The *cosmological principle* assumes the universe is homogeneous and isotropic.

Important evidence for the *Big Bang* model of the universe was the discovery of the *cosmic microwave background radiation*, which conforms to a blackbody radiation curve at a temperature of 3 K (3 degrees above absolute zero). The *Standard Model* of the Big Bang provides a possible scenario as to how the universe developed as it expanded and cooled after the Big Bang. Starting at 10^{-43} s after the Big Bang, there were a series of *phase transitions* during which previously unified forces of nature "condensed out" one by one. Until about 10^{-35} s there was no distinction between quarks and leptons. Shortly thereafter, quarks were *confined* into hadrons (the *hadron era*). About 1 μs (10^{-6} s) after the Big Bang, the majority of hadrons disappeared, introducing the *lepton era*. By the time the universe was about 10 s old, the electrons too had mostly disappeared and the universe became *dominated* by *radiation*. A couple of minutes later, nucleosynthesis began but lasted only a few minutes. It was about a half-million years before the universe was cool enough for electrons to combine with nuclei and form atoms. Also about this time, the background radiation had expanded and cooled so much that its total energy equaled the energy in matter; as the radiation cooled further, losing energy, the universe became *matter dominated* (not in numbers, but in energy). Shortly thereafter stars and galaxies formed, producing a universe not much different than it is today—10 or 15 billion years later.

If the universe is *open*, it will continue to expand indefinitely. If it is *closed*, gravity is sufficiently strong to halt expansion and the universe will eventually begin to collapse back on itself, ending in a *big crunch*. Whether the universe is open or closed depends on whether its average mass density is above or below a critical density. If the universe is closed, it may rebound from the big crunch and reexpand in a *cyclic* manner.

QUESTIONS

1. State Olber's paradox in your own words. How is it resolved?
2. If it was discovered that the redshift of spectral lines of galaxies was due to something other than expansion, how might our view of the universe change? Would there be conflicting evidence? Discuss.
3. All galaxies appear to be moving away from us. Are we therefore at the center of the universe?
4. If you were located in a galaxy nearby the boundary of our observable universe, would galaxies in the direction of the Milky Way appear to be approaching you or receding from you? Explain.
5. What is the difference between the Hubble age of the universe and the actual age? Which is greater?
6. Compare an explosion on Earth to the Big Bang. Consider such questions as: would the debris spread at a higher speed for more distant particles, as in the Big Bang? Would the debris come to rest? What type of universe would this correspond to, open or closed?
7. When the primordial nucleus exploded, thus creating the universe, into what did it expand? Discuss.
8. Explain what the 3 K cosmic microwave background radiation is. Where does it come from? Why is its temperature now so low?
9. The birth of atoms—that is, the combination of electrons with nuclei about which they orbit—occurred when the universe had cooled to

about 3000 K and is generally called *recombination*. Why is this term misleading?

10. Why were atoms unable to exist until almost a million years after the Big Bang?
11. Muons have a mass of about 100 MeV/c^2. Approximately when did most of them disappear from the universe?
12. If the universe is open, what will eventually happen to the cosmic microwave background radiation? If it is closed, what will happen to it?
13. Under what circumstances would the universe eventually collapse in on itself?

EXERCISES

1. The redshift of a galaxy indicates a velocity 2500 km/s. How far away is it?
2. If a galaxy is traveling away from us at 1% of the speed of light, how far away do you estimate it to be?
3. If a galaxy is near the "edge" of the universe, say 10 billion light-years away, what do you estimate its speed to be relative to us?
4. Make an approximate calculation for the universe using Hubble's constant. Assume (a) $H = 15$ km/s/ly and (b) $H = 30$ km/s/ly.
5. The size of the universe (the average distance between galaxies) at any one moment is believed to have been inversely proportional to the absolute temperature. Estimate the size of the universe, compared to today, at (a) $t = 10^6$ year, (b) $t = 1$ s, (c) $t = 10^{-6}$ s, (d) $t = 10^{-35}$ s.
6. Show that the critical density, $\rho_c \approx 5 \times 10^{-27}$ kg/m^3, is equivalent to about 3 hydrogen atoms per cubic meter.

A FINAL WORD

The following exchange is from congressional testimony given a few years ago by a leading physicist, Robert R. Wilson, concerning the value of building the high-energy particle accelerator research facility, Fermilab.

Senator Pastore:	Is the accelerator connected in any way with the security of our country?
Dr. Wilson:	No, sir, I do not believe so.
Senator Pastore:	It has no value in this respect?
Dr. Wilson:	It only has to do with the respect with which we regard one another, the dignity of men, our love of culture. It has to do with those things. It has nothing to do with the military, I am sorry.
Senator Pastore:	Don't be sorry for it.
Dr. Wilson:	I am not but I cannot in honesty say it has such applications; but it has to do with whether we are good painters, good sculptors, great poets, I mean all the things that we really venerate and are patriotic about in our country. In that sense, this new knowledge has everything to do with honor and country but it has nothing to do with defending our country except to help make it worth defending.

APPENDIX A
Scientific Notation

In physics we must often deal with very large or very small numbers. For example, the diameter of an atom is about 0.00000001 centimeter. On the other hand, the number of water molecules in a liter of water is about 33,000,000,000,000,000,000,000,000. And the distance from Earth to the nearest star is about 40,000,000,000,000 kilometers. Dealing with so many zeros is difficult, and it is easy to drop a zero or two when copying them. To avoid problems we use the powers of ten, commonly termed **scientific notation**.

Since ten squared is one hundred, we can write $10 \times 10 = 10^2 = 100$. Similarly, ten cubed, or ten to the third power, is one thousand: $10 \times 10 \times 10 = 10^3 = 1000$. Similarly, $50{,}000 = 5 \times 10 \times 10 \times 10 \times 10 = 5 \times 10^4$. For large numbers, it is easier to use the powers of ten, 5×10^4, instead of writing the number 50,000. The number of water molecules in a liter of water is written as 3.3×10^{25} molecules.

In general, a number is written in scientific notation as the product of a simple number times ten to some power. The exponent attached to the ten is the number of places the decimal point is moved to the right to obtain the fully written number. Thus 5.0×10^4 means moving the decimal point four places to the right to give 5.0000. (See if this works for the number 3.3×10^{25}, the number of water molecules in a liter of water.)

When the number is less than 1, say 0.001, the exponent or the power of ten is written with a negative sign; thus $0.001 = \frac{1}{10 \times 10 \times 10} = \frac{1}{10^3} = 1 \times 10^{-3}$. Similarly, $0.002 = 2 \times 10^{-3}$. A negative exponent power of ten means the decimal point is moved that number of places to the left: $2.0 \times 10^{-3} = 0.002$.; the diameter of atoms is thus about 1×10^{-8} cm.

The accompanying table is a brief summary of powers of ten used in scientific notation.

Powers of Ten

10^{12} =	1,000,000,000,000	one trillion
10^{9} =	1,000,000,000	one billion
10^{6} =	1,000,000	one million
10^{5} =	100,000	one hundred thousand
10^{4} =	10,000	ten thousand
10^{3} =	1,000	one thousand
10^{2} =	100	one hundred
10^{1} =	10	ten
10^{0} =	1	one
10^{-1} =	0.1	one-tenth
10^{-2} =	0.01	one one-hundredth
10^{-3} =	0.001	one one-thousandth
10^{-6} =	0.000001	one one-millionth

Logarithms are defined in the following way:

$$\text{if } y = A^x, \text{ then } x = \log_A y.$$

That is, the logarithm of a number y to the base A is that number which, as the exponent of A, gives back the number y. For *common logarithms*, the base is 10, so

$$\text{if } y = 10^x, \text{ then } x = \log y.$$

The subscript 10 on $\log_{10}$ is usually omitted when dealing with common logs.

Some simple rules for logarithms are as follows:

$$\log(ab) = \log a + \log b. \qquad \text{(i)}$$

This is true because if $a = 10^n$ and $b = 10^m$, then $ab = 10^{n+m}$. From the definition of logarithm, $\log a = n$, $\log b = m$, and $\log(ab) = n + m$; hence $\log(ab) = n + m = \log a + \log b$. In a similar way, we can show that

$$\log\left(\frac{a}{b}\right) = \log a - \log b \qquad \text{(ii)}$$

and

$$\log a^n = n \log a. \qquad \text{(iii)}$$

Logs were once used as a technique for simplifying certain types of calculation. Because of the advent of electronic calculators and computers, they are not oftten used any more for this purpose. However, logs do appear in certain physical equations (see, for example, Section 2 of Chapter 17), so it is helpful to know how to use them. If you do not have a calculator that calculates logs, you can easily use a *log table*, such as the small one below. The number, y is given to two digits (some tables give y to three or more digits); the first digit is in the vertical column to the left, the second digit is in the horizontal row across the top. For example, the table tells us that $\log 1.0 = 0.000$, $\log 1.1 = 0.041$, and $\log 4.1 = 0.613$; note that the table does not include the decimal point—it is understood. The table gives logs for numbers between 1.0 and 9.9; for larger or smaller numbers we use rule (i):

$$\log(ab) = \log a + \log b.$$

For example,

$$\log(380) = \log(3.8 \times 10^2) = \log(3.8) + \log(10^2).$$

From the table, $\log 3.8 = 0.580$; and from rule (iii), $\log(10^2) = 2\log(10) = 2$ since $\log(10) = 1$. [This follows from the definition of the logarithm: if $10 = 10^1$, then $1 = \log(10)$.] Thus

$$
\begin{aligned}
\log(380) &= \log(3.8) + \log(10^2) \\
&= 0.580 + 2 \\
&= 2.580.
\end{aligned}
$$

Similarly,

$$
\begin{aligned}
\log(0.081) &= \log(8.1) + \log(10^{-2}) \\
&= 0.908 - 2 = -1.092.
\end{aligned}
$$

Sometimes we need to do the reverse process: find the number y whose log is, say, 2.670. This is called "taking the antilogarithm." To do so, we separate our number 2.670 into two parts, making the separation at the decimal point:

$$
\begin{aligned}
\log y = 2.670 &= 2 + 0.670 \\
&= \log 10^2 + 0.670.
\end{aligned}
$$

We now look in the table to see what number has its log equal to 0.670; none does, so we must *interpolate*: we see that $\log 4.6 = 0.663$ and $\log 4.7 = 0.672$. So the number we want is between 4.6 and 4.7, and closer to the latter by 7/9. Approximately we can say that $\log 4.68 = 0.670$. Thus

$$
\begin{aligned}
\log y &= 2 + 0.670 \\
&= \log(10^2) + \log(4.68) = \log(4.68 \times 10^2),
\end{aligned}
$$

so $y = 4.68 \times 10^2 = 468$. If the given logarithm is negative, say, -2.180, we proceed as follows:

$$
\begin{aligned}
\log y = -2.180 &= -3 + 0.820 \\
&= \log 10^{-3} + \log 6.6 = \log 6.6 \times 10^{-3},
\end{aligned}
$$

so $y = 6.6 \times 10^{-3}$. Notice that what we did was to add to our given logarithm the next largest integer (3 in this case) so that we have an integer, plus a decimal number between 0 and 1.0 whose antilogarithm can be looked up in the table.

Short Table of Common Logarithms

y	0.0	0.1	0.2	0.3	0.4	0.5	0.6	0.7	0.8	0.9
1	000	041	079	114	146	176	204	230	255	279
2	301	322	342	362	380	398	415	431	447	462
3	477	491	505	519	531	544	556	568	580	591
4	602	613	624	633	643	653	663	672	681	690
5	699	708	716	724	732	740	748	756	763	771
6	778	785	792	799	806	813	820	826	833	839
7	845	851	857	863	869	875	881	887	892	898
8	903	908	914	919	924	929	935	940	944	949
9	954	959	964	968	973	978	982	987	991	996

APPENDIX C
Answers to Selected Questions and Odd-Numbered Exercises

Chapter 1

Questions

6. (a) inverse, (b) direct. **7.** direct proportion. **8.** direct; speed. **9.** inverse (approximate, and only for a while). **12.** π.

Chapter 2

Questions

4. Faster in winter.

Exercises

1. 1.8 yr. **3.** 162,000,000 km. **5.** 1.4 hr.

Chapter 3

Questions

3. No; rapid acceleration from rest versus high steady velocity . . . . **4.** No. Its direction may be changing. **5.** Yes. The instant an object starts moving. (See also Question 10.) **7.** No. Directions are different. **8.** Yes. Its velocity changes faster. **9.** No. Acceleration is greater on a sharp curve because the velocity changes direction more quickly. **10.** Zero; 9.8 m/s^2. **12.** They have the same acceleration.

Exercises

1. 1000. **3.** 13.7 km. **5.** 10.2 m/s; 147 s. **7.** 11.1 m/s. **9.** 15.6 hr. **11.** 108,000 km/hr; 30,100 m/s. **13.** 50 m. **15.** (a) 0.621, (b) 3.28, (c) 0.446. **17.** 1.2 m/s^2. **19.** 7.5 m/s; 9.4 m. **21.** 4.5 s. **23.** 2.6 s. **25.** 37 m/s. **27.** 8.7 m/s^2, 0.88 g.

Chapter 4

Questions

1. Yes—if no force acts to slow it down. **4.** Inertia. **5.** (a) Inertia is the name we give to this phenomenon. **6.** No (Why?). **7.** Inertia. **8.** Tendency to keep going in a straight line (inertia.) **12.** Acceleration requires more force (at constant speed, car exerts force only to overcome frictional forces). **13.** To accelerate (see Question 12). **14.** Yes; the floor, which "stretches." **15.** 200 N. **16.** You push on the ground with your feet; the ground pushes back (third law), and this force decelerates you (second law). **17.** The bag; you; downward. **19.** The rock exerts a force on your toe. **20.** The team that pushes harder against the ground.

Exercises

1. 4900 N. **3.** 5 m/s^2. **5.** 2.5 m/s^2. **7.** 6.9 m/s^2. **9.** 2.44 kg. **11.** 392 N. **13.** Slide down with an acceleration $a = 0.25\,g$. **15.** 33 N. **17.** 254 N.

Chapter 5

Questions

3. Equal to your weight; net force is zero. **4.** Zero; 50 N: 50 N. **7.** No; maybe. **9.** (a) yes, (b) no. **10.** Highest. **11.** Stay put.

Exercises

1. (a) 686 N, (b) 119 N, (c) 609 N, (d) 0. **3.** 424 N. **5.** Southwesterly. **7.** 35 N. **9.** 20 m, 7.2 m. **13.** 8.7 m. **15.** 6.4 s.

Chapter 6

Questions

2. So the force required to accelerate the car inwardly can be exerted by the road perpendicular to its surface, rather than parallel to its surface (friction). (Draw a diagram to show this.) **3.** A force is needed to keep it moving in a circular path, usually supplied by friction—which, if not great enough, means the car will tend to continue in a straight line off the curve. **4.** Force required is less. **6.** Greater around sharper curve since r is less so $a = v^2/r$ is greater. **7.** Greatest at (b), least at (a). **9.** Yes; equal to mass of apple × g in both (a) and (b). **12.** No. **14.** Less. **15.** Less; g is less. **18.** Yes. **20.** Its speed and roundness of the earth (see text, Section 3, second paragraph). **22.** Large; small. **23.** The relative change in r^2 is greater since the moon is closer.

Exercises

1. 3.1 g. **3.** 12.5 m/s^2; 25 N toward center of circle. **5.** 0.27 m/s^2, 6.8 N. **7.** 2.0×10^{20} N. **9.** 2.7×10^{-6} N. **11.** 2650 km. **13.** 4 × greater; increase. **15.** 3.8 m/s^2. **17.** 5.6 m/s. **19.** 8.9 m/s.

Chapter 7

Questions

1. Force; distance through which the force acts. **3.** The first. **5.** More work is done (force is exerted over a longer distance), so KE is greater and hence v is greater: the bullet goes straighter and farther. **6.** Yes. **7.** $Fd = \text{KE} = \frac{1}{2}mv^2$ so $d \propto v^2$ if F is more or less constant.

8. The first. **10.** Yes. It can do work on something else. **12.** KE → PE → KE → thermal. **15.** Source of energy (PE). **16.** PE; PE → KE + thermal energy. **18.** *B*; after A, at C; yes; maybe not reach C. **20.** KE; KE → thermal energy. **22.** Same height, *h*. **23.** No. **25.** Less power output required. **26.** Yes; yes.

Exercises

1. 2,250,000 J. **3.** (a) 300 J, (b) 10,800 J. **5.** 13.2 J. **7.** 4×. **9.** 9×. **11.** 386,000 J. **13.** 216 J. **15.** 5.1 m. **17.** 2.1×10^6 J. **19.** 31 m/s. **21.** 0.59 HP. **23.** The second. **25.** 375 N. **27.** 26 m/s.

Chapter 8

Questions

1. An external force is acting, so *p* not conserved. **3.** Lighter; lighter. **4.** Heavier. **6.** Yes; time is also a factor. **10.** Transferred to the Earth. **12.** Throw one (or more) coins in one direction; he'd move in the opposite direction. **13.** Fire its rockets in direction opposite to that in which it wants to go. **14.** Although total energy is conserved, KE is not, and we cannot often calculate the thermal energy involved. **16.** Conservation of *p*; throw book outward.

Exercises

1. 1.2 kg·m/s. **3.** 15 km/hr. **5.** 16 m/s. **7.** 4.5 m/s. **9.** 25,000 N; yes. **11.** 6.0×10^7 N. **13.** (a) 2.5×10^{-13} m/s, (b) 1.7×10^{-17}, (c) 0.19 J.

Chapter 9

Questions

2. Overestimates distance traveled. **3.** Longer lever arm, so greater torque for same force. **4.** Rotational inertia. **5.** (a), (c). **6.** (c). **7.** Symmetry axis; end. **9.** Its rotational inertia tends to maintain a steady speed. **10.** Its large *I* gives him time to respond and maintain balance. **12.** Solid one, since it has less rotational inertia. **13.** Speed up. **14.** Conservation of angular momentum. **15.** Lengthen it, since increase in *I* would cause decrease in ω. **16.** Forces and torques due to friction, etc. **19.** Keep c.g. over your feet (see Section 3 of Chapter 10). **22.** Your leg's mass is not uniformly distributed along its length. **24.** The c.g. is further from pivot, so torque is greater. **25.** Translation of c.g. plus rotation about c.g.

Exercises

1. (a) 0.52, (b) 1.57, (c) 6.81. **3.** 6.8 km. **5.** 209 rad/s. **7.** 0.35 m/s. **9.** 1.23 rad/s^2. **11.** (a) 2.0×10^{-7} rad/s,

(b) 7.3×10^{-5} rad/s. **13.** 400 N. **15.** 88 N·m. **17.** 1.67 m from the heavier person's end. **19.** (a) 35 N·m, (b) 25 N·m. **21.** 0.126 kg·m^2; R is so small. **23.** 12.4 kg m^2/s. **25.** 0.38 rev/s.

Chapter 10

Questions

4. No. There is no way vectors of these three lengths can sum to zero. **5.** A, unstable; B, stable; C neutral. **8.** (b); because in (a), the c.g. of the two bricks together is beyond the support, and gravity will pull them downward. **10.** The c.g. is below point of support (= stable equilibrium). **13.** To keep your c.g. over the base of support. **15.** Amount shortened is doubled. **16.** Same as 13. **17.** Wood; marble. **18.** To supply a torque, keeping the arch in equilibrium, so it doesn't collapse.

Exercises

1. 540 N. **3.** No, but it's close! **5.** (b) yes. **7.** 4.6 cm. **9.** 4 times. **11.** 67,000 N, 133,000 N. **13.** 850 N, 600 N. **15.** 9.0 m.

Chapter 11

Questions

3. Forces between them. **4.** Mercury, air, alcohol, carbon dioxide. **5.** (d). **6.** No; they might be closer together. **8.** Water. **9.** Water.

Exercises

1. Diameter of about 4×10^{-8} cm. **3.** 0.60. **5.** Platinum, 1.11 ×. **7.** 0.037 m^3. **9.** 22 kg. **11.** 100 kg. **13.** 0.89.

Chapter 12

Questions

3. The former. **5.** Pressure is the same at equal depths. **6.** No. It is density, not mass or weight, that determines whether an object sinks in a given fluid. **7.** Salt water has greater density and therefore greater buoyant force. **10.** Buoyant force in fresh water is less—boat's SG must be between 1.00 and 1.03. **12.** Because of water's buoyant force on you, you push down on the rocks with less force than your weight (and by Newton's third law, they push back on you with a force that is less than your weight). **13.** The air's buoyant force is less at higher altitudes where its density is less. The balloon stops rising when the buoyant force equals its weight. **14.** Below; lower. **16.** Pushed. **17.** Air pressure.

19. No. **20.** No atmospheric pressure to push liquid up the straw. **23.** The pressure in the rushing water between them is less than that on their outer sides (Bernoulli). **24.** Normal atmospheric pressure in still air inside is greater than reduced pressure in air rushing over the top. **26.** Speed of blood plasma greatest at center (least at edges because of drag against the walls), so pressure is least; corpuscles pushed from higher pressure areas to lower pressure areas.

Exercises

1. 1.07 kPa. **3.** 290 kPa. **5.** ≈150,000 N (≈30,000 lbs). **7.** 2,400,000 N, 20 kPa; 20 kPa. **9.** (a) 240,000 N, (b) 240,000 N. **11.** 7.4×10^7 Pa; 740 atm. **13.** 3 atm. **17.** 1009 kg/m^3. **19.** (a) 5.7 N, (b) 3.95. **21.** 4.0. **23.** 0.0025 N. **25.** 0.24 m/s. **27.** About 0.5 m/s.

Chapter 13

Questions

2. Average distance increases. **3.** The metal lid expands more than the glass, so fits less tightly. **4.** Pressure inside increases with temperature. **6.** Larger. **7.** Low. **9.** Their size is small compared to the average distance between them. **11.** Absolute temperature (K) would double (°C or °F would not double). **12.** 27%. **13.** No. $\overline{\text{KE}}$ *is* greater, but KE involves mass of molecules as well as speed.

Exercises

1. 20°C, 293 K. **3.** 3272°F. **5.** 75°. **7.** (a) 310 K, (b) 300 K, (c) 77 K. **9.** 5 × greater. **11.** Drops to $\frac{1}{3}$ of original value. **13.** 1.5 times larger. **15.** 18.0 m^3. **17.** 1.36 m^3. **19.** (a) 780 kg; (b) 65 kg leaves. **21.** 899°C. **23.** −40°C = −40°F.

Chapter 14

Questions

1. Becomes thermal energy, temperature rises. **3.** Decrease in gravitational PE transformed to thermal energy. **4.** No; not necessarily. **5.** No; the opposite (mechanical energy → thermal energy, so T rises, doesn't fall); brings cooler or less humid air to replace warmer or humid air. **8.** Water evaporating from jacket requires energy which is taken from canteen, so T drops. **10.**, **11.**, **12.** See answer to Question 8. **16.** No. **18.** (a) Convection, (b) conduction, (c) conduction, radiation; all. **19.** Convection currents. **21.** Conducts heat away faster, so foot's temperature at surface drops quickly. **22.** It is the thickness of the air layer that insulates. **23.** Thickness of air layer drastically reduced. **24.** To reach equilibrium, via conduction, with air; in the sun, absorbed radiation would cause reading to skyrocket.

25. Insulation against heat flowing *in*. **27.** Loss of energy by radiation to cool surroundings, so T drops. **30.** Dark dirt absorbs more radiation.

Exercises

1. 1.7×10^5 J. **3.** 1.67×10^6 J. **5.** 170 kg. **7.** 7800 kg. **11.** 0.75 kcal/kg·C°. **13.** 108 kg. **15.** 3100 kcal. **17.** 159 kcal. **19.** 145 W.

Chapter 15

Questions

1. Heat flows from the water vapor to the surroundings, including the glass. **3.** Yes; if heat input equals work done, U doesn't change so T doesn't change. **4.** No. **5.** Yes; no. **6.** Need a lower temperature receiver of output heat. **7.** Work must be done to achieve this—it is not spontaneous. **9.** Liquid; less orderly. **12.** No; see answer to Question 7.

Exercises

1. 115 W. **3.** 32%. **5.** 515°C. **7.** 20.8%. **9.** 687°C. **11.** 509°C. **13.** 208 m^2; possible. **15.** (a) 2.8×10^{13} J; (b) 0.39 MW. **17.** 3.3×10^{13} J.

Chapter 16

Questions

3. Yes. At the equilibrium point ($F = kx = 0$ so $a = 0$). **4.** Pointer oscillates between 0 and 20 kg, but amplitude decreases due to damping and eventually stops at 10 kg. **6.** Energy (which is conserved) is spread over greater area, plus damping. **8.** Longitudinal; transverse. **9.** Frequency; wavelength. **10.** Yes. **13.** Resonance. **14.** Sounds around you, emphasized through resonance. **18.** Natural frequencies increased. **19.** Longer wavelength, more diffraction (Figure 16-24). **20.** At nodes.

Exercises

1. 340 N/m. **3.** 31 s. **5.** 1.18 Hz. **7.** 820 N/m; 3.2 Hz. **9.** 3.3 m/s. **11.** 1.26 m. **13.** 210 m/s. **15.** 1080 m. **17.** 588 Hz, 882 Hz, 1176 Hz. **19.** 0.526 m. **21.** 170 Hz.

Chapter 17

Questions

3. Reflection. **4.** Size of mouth and throat cavity determines predominate λ (resonance); since velocity of sound wave in helium is higher, $f = v/\lambda$ is higher. **5.** Interference of waves reflected from walls, etc. **7.** 10–15 dB. **8.** 400–10,000 Hz. **9.** To make them heavier so the pitch will be lower (since v is lower). **11.** First harmonic quenched, so hear second harmonic. **13.** Length of pipe determines λ; so $f \propto v$ and f increases with temperature. **14.** Wavelength increases; frequency unchanged. **15.** At the two nodes. **17.** Different mix of frequencies. **19.** Both correspond to a strong compression of air. **20.** No; yes.

Exercises

1. 17 mm to 17 m. **3.** 340 m. **5.** 3.4 mm. **7.** 65 cm. **9.** 120 dB. **11.** 1×10^{-7} W/m^2. **13.** 30 dB. **15.** Assume 20°C: 78.0 cm; 19.5 cm. **17.** 441 Hz. **19.** (a) 107 Hz; 321 Hz, 535 Hz, 749 Hz; (b) 214 Hz; 428 Hz, 642 Hz, 856 Hz. **21.** -0.035 or 3.5% low. **23.** 65 cm. **25.** 6 Hz. **27.** 15 Hz; not by most people; 4 Hz. **29.** (a) 408 Hz, (b) 352 Hz.

Chapter 18

Questions

1. The charge on one body, charge on the other, the distance between them. **4.** The wiping process charges it. **5.** Oppositely charged parts of polar molecules attract each other. **8.** You've acquired a charge due to your shoes and carpet rubbing against each other. **15.** Some charge may transfer from ruler to paper; both then have the same charge and repel each other. **17.** Gravity, and perhaps stiffness. **18.** Toward the positive plate. **19.** Right; left. **21.** No; its charge could be different too. **22.** Higher; lower; decreases for both.

Exercises

1. 6.25×10^{14}. **3.** 1.5×10^{-8} m. **5.** (a) 1.3 N, (b) 48 N, (c) 36 N, (d) 108 N. **7.** 8.8 N, 34 N, 25 N. **9.** 5000 N/C, opp. to **F**. **11.** 813 N/C, up. **13.** 1.02×10^{-7} N/C, up. **15.** 5000 J. **17.** Twice as great. **19.** 4000 V, B. **21.** (a) 1.05×10^9 J, (b) 2,500 kg. **23.** 290 N, at 65° to horizontal.

Chapter 19

Questions

2. Charge (= current × time); 1 A·hr = (1 C/s)(3600 s) = 3600 C. **4.** Total charge they can supply, and therefore total energy. **5.** No—in fact normally doesn't have a net charge. **8.** No. **9.** Battery plays role of heart. **12.** Yes; by connecting them in series (but you need about 20). **13.** R_2; R_1. **15.** The filament breaks, so the circuit is open. **16.** 100 W. **17.** At the higher voltage, the current ($I = P/V$) for a given power P will be less, so the power lost in the wires (I^2R) will be less. **18.** A 25-A fuse will allow currents to flow that may be too large for the wiring, which may overheat and cause a fire. **21.** You can be directly in touch with the ground, and if you touch a "hot" wire on a device also connected to ground, current can flow through you.

Exercises

1. 600 C. **3.** 252,000 C. **5.** 3 V. **7.** Half. **9.** 0.6 mA. **11.** 1.1×10^{20}. **13.** 150 Ω; 600 V. **15.** 320 mA. **17.** 4, in parallel. **19.** $\frac{1}{3}$ as large; 3 × larger. **21.** 540,000 J. **23.** 1800 W. **25.** $17.52. **27.** 2.16 MJ

Chapter 20

Questions

1. Earth's magnetic field is not parallel to Earth's surface. **3.** Explanation toward end of Section 1. **4.** Either end of the unmagnetized piece will be attracted (and never repelled) to either pole of the magnets; the magnets, on the other hand, also show repulsion. **5.** Not most metals—only iron and a few others. **8.** Counterclockwise. **9.** Domains can be randomly oriented and magnetic effects cancel out. **12.** Electric and magnetic (also gravitational, but this will be small unless mass is very large). **13.** Up. **14.** No; yes. **15.** To the right. **16.** Yes; the magnetic field produced by one exerts a force on the other (and vice versa). **19.** Parallel to **B**; perpendicular to **B**. **22.** More magnetic field lines enter the earth around Norway than Egypt (where they tend to be more nearly parallel to the Earth).

Exercises

1. 0.25 N/m. **3.** 2.6 A. **5.** 10 km. **7.** 11.7 A. **9.** 4.0×10^{-15} N; 0.28 m. **11.** 0.10 T, north. **13.** 0.015 N. **15.** 0.070 T.

Chapter 21

Questions

1. With more coils, the flux is increased and the induced voltage is greater. **3.** Yes; no; yes; clockwise; counterclockwise. **4.** Yes. **7.** Yes. **9.** Voltage (and current); if dc, no changing magnetic field so no induced voltage in secondary. **10.** Measure ratio of input to output voltage; measure the resistance (≈0 for paired leads, but infinite if one lead from primary and other from secondary). **11.** No; wave in matter, such as air molecules. **12.** Yes; no. **14.** Radiowave; radiowave; microwave; microwave; IR; IR (but almost visible). **15.** Longer. **16.** AM; ≈ 500 nm (visible light). **20.** The magnet's field exerts a force on the electron beam.

Exercises

1. 0.039 V. **3.** (a) 0.063 V, (b) clockwise. **5.** 200 V. **7.** 4. **9.** 100; 1.2 V. **11.** $I_2 = \frac{1}{4} I_1$. **13.** 9.6 V, 0.64 A. **15.** 3.0×10^{10} Hz. **17.** 1.3×10^{-2} s. **19.** 200 m. **21.** Since $\lambda \approx (3 \times 10^8 \text{ m/s})/(10^8 \text{ Hz}) =$ 3 m, a $1\frac{1}{2}$ m long antenna is $\frac{1}{2}\lambda$, so just right for standing waves (see Figures 16–25 and 16–26). **23.** (a) 100 kW (i.e., most of the power would be wasted!), (b) 10 W.

Chapter 22

Questions

2. Stays the same. (But you get further away, so it looks smaller.) **5.** Reflection from planes of water at different angles. **7.** Infinite. **8.** Yes, with $f = \infty$, so $d_i = d_0$. **9.** 1. **11.** 0°. **14.** Table beneath the drop appears distorted because of refraction at the drop's surface, plus reflection from the drop's curved surface; also, all light from table beneath does not pass through the drop, so it appears darker. **15.** No. **16.** Away. **19.** Light is refracted in the air, which is moving. **20.** No, smaller.

Exercises

1. (a) 2.2×10^8 m/s, (b) 1.99×10^8 m/s. **3.** 3 m. **5.** 4.0×10^{16} m. **7.** 6.7×10^8 mph. **9.** 540 rev/s. **11.** 92 cm. **13.** 27.6 cm. **15.** 20 cm. **17.** Inside the ball, 16 cm from surface; virtual; not very. **19.** 80 cm; 75 cm. **21.** (a) 60.0 cm in front of the mirror; (b) 3× larger (4.5 cm).

Chapter 23

Questions

2. Yes; no; real images. **4.** (a) Yes, further from lens; (b) yes, larger. **6.** Yes; d_i and d_o can be interchanged and the equation retains the same form. **9.** For a closer object, the rays enter the lens at steeper angles and, since the lens bends them only so much, they exit at less steep angles and converge to a point farther from the lens. **11.** Retina. **12.** As people age, their eyes do not accommodate as well; the upper part is for distant vision, the lower for reading. **13.** Converging. **14.** Inverted.

Exercises

1. 5.26 mm. **3.** Infinity; −2.45 m (same side of lens). **5.** 0.85 mm. **7.** Converging; 12 cm; real. **9.** 6 m; no. **11.** 39 mm to 4.2 mm. **13.** 1/50 s. **15.** 53 cm. **17.** −12 cm; diverging. **19.** 3.7 D. **21.** 7.1 cm. **23.** 4.2×. **25.** $1\frac{1}{4}\times$; $3\frac{1}{8}\times$ (the magnification for the child is less because the child can see more detail without the magnifier). **27.** 67 mm. **29.** 14.1 cm.

Chapter 24

Questions

1. One model or the other more easily explains a particular phenomenon. **3.** $\frac{1}{2}$, $1\frac{1}{2}$, $2\frac{1}{2}$, etc., wavelengths. **5.** Wavelength much too long, much longer than size of molecules. **7.** Blue, because of shorter wavelength. **9.** The gap increases more quickly. **10.** Light *will* pass, although not at full intensity. **13.** Black. **14.** Reddish. **15.** Brown, chartreuse; does not correspond to a single frequency—i.e. does not appear in the natural spectrum. **17.** The wavelength. **19.** Spotlight containing no blue wavelengths. **20.** Magenta **21.** (a) blue, (b) red, (c) red. **23.** (a) green, (b) cyan, (c) magenta. **25.** No cones in the retina. **26.** Only the cones, which don't distinguish colors, detect low light levels.

Exercises

1. 6.0 mm. **3.** 613 nm. **5.** 0.097 mm. **7.** 517 nm. **9.** 228 nm. **11.** 643 nm (red-orange). **13.** Destructive.

Chapter 25

Questions

1. On the car. **3.** Either viewpoint is valid. The former allows a simpler description of the solar system. **4.** At c. **5.** No. Either view is valid. **7.** Yes. **10.** Upon return, travelers would not have aged as much as friends at home. **11.** Yes. Time dilation (Section 3). **12.** No; no; Earth observers would see changes. **14.** No; you would measure a shorter distance. **16.** Time. **17.** Yes, but the effects are extremely tiny. **19.** None would occur since $\sqrt{1 - v^2/c^2}$ would equal 1. **23.** No; mass is a form of energy. **25.** Yes. **27.** No. **29.** Mass decreases. **31.** No; yes.

Exercises

1. 6 km/hr; 14 km/hr. **3.** 800 km/hr. **5.** 0.90×10^{-8} s. **7.** 150 m. **9.** 0.94 c. **11.** (a) 36.7 yr, (b) 7.3 yr, (c) 7.16 lt-yr, (d) 0.98 c. **13.** 0.80 c. **17.** 0.87 c. **19.** $\rho = \rho_0/(1 - v^2/c^2)$.

Chapter 26

Questions

1. Bluish highest, reddish lowest, temperature. **2.** Intensity of radiation in the visible region is too small. **5.** Photons of red light do not have sufficient energy to cause the chemical changes on the film that higher frequency photons do. **6.** Higher threshold wavelength corresponds to lower work function. **8.** Photons must have greater energy to cause the chemical changes in skin that result in sunburn. **9.** Yes; increases. **13.** No; an electron is not an electromagnetic wave. **14.** and **15.** Waves: experiments such as interference and diffraction; particles: experiments such as (in 14) the photoelectric effect, (in 15) the experiments described for Figure 26-8. **17.** Proton (because *m* is larger). **18.** Electrons' path would be altered by a magnetic or electric field; an X-ray's wouldn't.

Exercises

1. 5.4×10^{-20} J, 0.34 eV. **3.** 1.5×10^{13} Hz. **5.** 4.1 eV. **7.** 270 nm. **9.** Yes; yes. **11.** 523 nm. **13.** Yes; 0.66 eV. **15.** 211 MeV; 5.9×10^{-15} m. **17.** 3.3×10^{-13} m. **19.** 1040 m/s; no. **21.** 3.6×10^{21}.

Chapter 27

Questions

1. The electric force. **2.** (a); all three, even gases if dense enough. **3.** Nearly all molecules are in the ground state. **5.** At high temperature, gases excited and emit characteristic spectral lines. **8.** Closer, because electric force exerted by nucleus is greater. **9.** Different atoms can be excited to different states. **12.** One or more electrons is not in lowest possible energy state. **13.** Yes; PE.

Exercises

1. 10.2 eV. **3.** 234 nm. **5.** 2.11 eV. **7.** 122 nm. **9.** 1.28×10^{-4} m. **11.** 1.51 eV. **13.** 4.4×10^{-40}; yes. **15.** 1.1×10^{-8}. **17.** 3.3×10^{15} Hz.

Chapter 28

Questions

1. For one thing, in the Bohr theory the electron has a definite orbit (position). **3.** Since $\Delta x \approx h/(\Delta mv)$, large m means small Δx. **4.** Its mass is so large (see Question 3). **6.** No. **8.** Not with absolute certainty. **11.** Both lack one electron of having filled outer subshell. **12.** Both have one extra electron outside a closed subshell. **14.** *n*-type. **15.** Voltage on base can stop or allow flow of current. **17.** No. **20.** No. **21.** They consist of one (or a few) wavelengths, so little mixing occurs to dim their pure color. **22.** Because the laser light is in such a narrow beam, it can put more energy per unit area on a small area of film, even at a great distance, than the street lamp which emits in all directions.

Exercises

1. 1.2×10^{-9} m. **3.** 3×10^{-30} m. **5.** 5×10^{-11} m. **7.** 60. **9.** 28. **11.** 0.0125 J; 4.0×10^{16}.

Chapter 29

Questions

2. Number of protons; number of nucleons. **3.** Same Z. **5.** (a) U, (b) N, (c) H, (d) Sr (e) Bk. **7.** They are an average over the natural abundancies for those elements. **9.** ${}^{3}_{2}$He. **10.** ${}^{64}_{29}$Cu, ${}^{64}_{30}$Zn, ${}^{64}_{28}$Ni. **11.** 144. **13.** γ. **15.** (a) ${}^{32}_{16}$S, (b) ${}^{35}_{17}$Cl, (c) ${}^{207}_{81}$Tl. **18.** Half lives short enough that nearly all have decayed since their formation. **19.** No. **20.** No.

Exercises

1. 4.8×10^{-15} m. **3.** 3727 MeV/c^2. **7.** $7\,\alpha, 4\beta^-$. **9.** $\frac{1}{4}$. **11.** 3 hr. **13.** (a) 2.3×10^{17} kg/m^3, (b) 180 m, (c) 2.6×10^{-10} m. **15.** 1.2×10^{-11} J. **17.** 4.38 MeV, 0; 0, 4.38 MeV. **19.** 3.6×10^{16}; 2.3×10^{15}.

Chapter 30

Questions

1. n; ${}^{12}_{6}$C; γ; $2n$. **3.** ${}^{20}_{10}$Ne. **5.** ${}^{17}_{9}$F; $p + {}^{20}_{10}\text{Ne} \rightarrow {}^{17}_{9}\text{F} + {}^{4}_{2}\text{He}$. **6.** γ^- **7.** Yes, but more difficult; critical mass would be much higher. **8.** ${}^{239}_{94}$Pu, since can afford to lose more neutrons through surface. **11.** Rate of release of energy is one. **13.** So can have a chain reaction (further fissions). **15.** Reduce possibility of radioactivity outside central reactor. **17.** Water could act as moderator, slowing down neutrons so more fissions could occur. **18.** Yes, all three. **19.** ${}^{238}_{92}$U breeds fissionable ${}^{239}_{94}$Pu fuel by absorption of neutrons and the reactions shown in Figure 30-1. **20.** Yes. **21.** High temperature and high density (due to gravitational force of large mass). **23.** Geiger counter would register higher counts near leak if liquid in pipe contains radioactive tracers. **25.** Lower for younger woman to reduce genetic effect on possible child.

Exercises

1. 0.2 u or 3.5×10^{-28} kg. **3.** 175 MeV. **5.** 3.6×10^{22} atoms, or 0.12 gram. **7.** 120 counts/s **9.** 500 rads. **11.** About 170 rads. **13.** About 4 days. **17.** 3.0160 u. **19.** 4.5 m.

Chapter 31

Questions

1. Do work on it and increase its mass. **2.** $p + n \rightarrow p + p + \pi^-$. **6.** We would observe explosions occuring when matter and antimatter met. **8.** em, weak, grav.; weak, grav.; all 4. **9.** (a) No, mass-energy conservation; (b) possible; (c) possible. **11.** (a) ddu, (b) $\overline{dd\bar{u}}$, (c) uds, (d) uss. **12.** (a) p, (b) $\overline{\Sigma}^-$, (c) K^-, (d) π^-. **14.** $c\bar{s}$.

Exercises

1. 16 GeV. **3.** 990,000 km; 3.3 s. **5.** 33.9 MeV. **7.** 1877 MeV. **9.** 494 MeV.

Chapter 32

Questions

1. Telescopes show it is made up of tiny dots (stars). **2.** Matter, rotating about a center, flies outward (Newton's first law). **5.** Temperature rises; falls. **6.** Relatively large, bright, and cool at the surface. **9.** Depends on their mass.

Exercises

1. 0.13 s. **3.** 500 lt-sec. **5.** 9.9 ly (3 parsecs). **7.** (a) 0.033", (b) $(9.3 \times 10^{-6})°$. **9.** Farther star has smaller parallax; $\frac{1}{2}$ as large.

Chapter 33

Questions

3. No—see discussion around Figure 33-1. **4.** Receding. **7.** Itself, since the primordial nucleus *was* the universe. **9.** *Re*-combination implies they had been combined as atoms previously, but they hadn't been. This was the first time that electrons and nuclei combined to form atoms. **10.** The temperature had to cool down; at higher temperature, the KE of particles was so high any atoms formed would have been broken apart immediately. **11.** Shortly after the pions, about 10^{-4} s after the Big Bang. **12.** Continue to cool and wavelengths increasing; may cool further and then warm again (as universe contracts) with wavelengths shortening. **13.** Positive curvature, closed and finite universe.

Exercises

1. 50 Mpc. $(1.6 \times 10^8$ ly). **3.** 150,000 km/s. **5.** (a) $\frac{1}{1000} = 10^{-3}$, (b) 10^{-10}, (c) 10^{-13}, (d) 10^{-27}.

INDEX

The abbreviation *defn* means only the definition of the term is cited; *fn* after a page reference means it is found in a footnote.

B

C

D

E

F

G

H

I

J

K

L

M

N

R

S

T

U

V

W

X

Y

Z

D 9
E 0
F 1
G 2
H 3
I 4
J 5